SMART NANODEVICES FOR POINT-OF-CARE APPLICATIONS

T0382927

SMART NANODEVICES FOR POINT-OF-CARE APPLICATIONS

Edited by
Suvardhan Kanchi
Rajasekhar Chokkareddy
Mashallah Rezakazemi

CRC Press
Taylor & Francis Group
Boca Raton London New York

CRC Press is an imprint of the
Taylor & Francis Group, an **informa** business

First edition published 2022
by CRC Press
6000 Broken Sound Parkway NW, Suite 300, Boca Raton, FL 33487-2742

and by CRC Press
4 Park Square, Milton Park, Abingdon, Oxon, OX14 4RN

CRC Press is an imprint of Taylor & Francis Group, LLC

Library of Congress Cataloging-in-Publication Data

Names: Kanchi, Suvardhan, editor. | Chokkareddy, Rajasekhar, editor. | Rezakazemi, Mashallah, editor.
Title: Smart nanodevices for point-of-care applications / edited by Suvardhan Kanchi, Rajasekhar Chokkareddy, Mashallah Rezakazemi.
Description: First edition. | Boca Raton : CRC Press, [2022] | Includes bibliographical references and index.
Identifiers: LCCN 2021059301 (print) | LCCN 2021059302 (ebook) | ISBN 9780367740245 (paperback) | ISBN 9780367744434 (hardback) | ISBN 9781003157823 (ebook)
Subjects: LCSH: Point-of-care testing--Technological innovations. | Nanomedicine. | Nanotechnology. | Biomedical engineering.
Classification: LCC RC71.7 .S63 2022 (print) | LCC RC71.7 (ebook) | DDC 616.07/5--dc23/eng/20220316
LC record available at https://lccn.loc.gov/2021059301
LC ebook record available at https://lccn.loc.gov/2021059302

ISBN: 9780367744434 (hbk)
ISBN: 9780367740245 (pbk)
ISBN: 9781003157823 (ebk)

DOI: 10.1201/9781003157823

Typeset in Times
by Deanta Global Publishing Services, Chennai, India

Contents

Preface

Nanotechnology advancements have compelled the development of novel materials with various functions. Nanomaterials have exceptional physicochemical properties due to their small size, enhanced absorption and reactivity, increased surface area, molar extinction coefficients, tunable plasmonic capabilities, quantum effects and magnetic and optical properties. However, due to their limitations (such as non-biocompatibility, poor photostability, low targeting capacity, rapid renal clearance, side effects on other organs, insufficient cellular uptake and small blood retention), it is still challenging to use nanomaterials for better therapeutics in the biomedical sector, since other types with controlled abilities, dubbed "smart" nanomaterials, must still be developed. In this context, scientists have discovered a type of nanomaterial that undergoes major reversible changes in its physical, chemical or biological properties as a result of minor environmental changes.

This book aims to give readers an overview of the most recent research on nanoscale materials that respond to various stimuli, as well as their most recent applications in the biomedical area. This book looks at how multidisciplinary science can be used to design and develop smart sensing technology using novel sensing materials and miniaturized transduction systems for novel sensing strategies. These ongoing breakthroughs in nanoscale biomedical determination gadgets described in this book will allow us to commercialize these technologies in the near future. As a result, we've highlighted various types of prognostic and diagnostic biomarkers linked to cancer, Alzheimer's disease, bacterial infections and various types of electrochemical biosensor techniques used for the early detection of potential disease biomarkers. In the present book, catering to this multidisciplinary subject, there are 25 chapters prepared by the world's leading authors. This book is designed to attract a diverse group of persons with an engineering or science background.

The book will be useful as a reference for the established research community, as well as a primer for researchers and graduate students who are just getting started in the subject of electromechanical miniaturization.

About the Editors

Suvardhan Kanchi is currently working as Associate Professor in the Department of Chemistry, Sambhram Insitute of Technology, Bengaluru, India. He obtained an MSc degree in applied chemistry from Sri Venkateswara University, India, in 2003. He received his PhD in electro-analytical chemistry from Sri Venkateswara University in 2010. He has extensive research experience in multidisciplinary fields of analytical chemistry, materials chemistry and electrochemistry. He has published 75 research articles in international journals of repute and 19 book chapters and 9 knowledge-based book editions published by renowned international publishers. He has published seven edited books with Pan Stanford, Taylor & Francis/CRC Press, Scrivener, Wiley and Elsevier. He is member of editorial boards of various journals. He has supervised two PhD and seven master's students. He has attended as well as chaired sessions at various international and national conferences. He has worked as a postdoctoral fellow, research associate and research fellow in the Department of Material Science and Engineering, Feng Chia University, Taiwan, and the Department of Chemistry, Durban University of Technology, Durban, South Africa, in the field of electrochemistry and biosensors. He has one South African patent in designing a portal sensor device to detect the hotness of chili in food materials. He is a life and senior member of several international professional organizations. His research interests include smart materials for device applications, smartphone-based sensing systems, nanodiagnostics, computational chemistry, nanoelectrochemistry, environmental chemistry and green nanotechnology.

Rajasekhar Chokkareddy is currently working as Assistant Professor in the Department of Chemistry, Aditya College of Engineering and Technology, Surampalem, Andhra Pradesh, India. He received his doctorate in chemistry on "Fabrication of Sensors for the Sensitive Electrochemical Detection of Anti-Tuberculosis Drugs" from Durban University of Technology (DUT), South Africa. He completed his MSc (organic chemistry) at Sri Venkateswara University, India. Currently, Chokkareddy is working as a Research Associate in the same department. His main research focuses on the development of electrochemical sensors/biosensors for various pharmaceutical drugs as well as thermophysical properties of ionic liquids. He has published 20 papers in various peer-reviewed international journals and 15 book chapters.

Mashallah Rezakazemi received his BSc and MSc degrees in 2009 and 2011, respectively, both in chemical engineering, from the Iran University of Science and Technology (IUST), and his PhD from the University of Tehran (UT) in 2015. In his first appointment, Rezakazemi has been serving as a Professor in the Faculty of Chemical and Materials Engineering at the Shahrood University of Technology since 2016. Rezakazemi also received his degree promotion to Associate Professor in 2019. Specifically, his research in engineered and natural environmental systems involves: (i) membrane-based processes for energy-efficient desalination, CO_2 capture, gas separation, and wastewater reuse; (ii) sustainable production of riched gas stream, water, and energy generation with the engineered membrane; (iii) environmental applications and implications of nanomaterials; and (iv) water and sanitation in developing countries.

Contributors

Annu
Department of Chemistry
Chandigarh University
Mohali, India

Sudharshan Asaithambi
Center for Data Science
New York University
New York, NY

Shaik Baji Baba
School of Chemistry and Physics
University of KwaZulu Natal
Durban, South Africa

Hari Babu Bollikolla
Department of Chemistry
Acharya Nagarjuna University
Guntur, India

P.S. Brahmanandam
Department of Physics
Shri Vishnu Engineering College for Women (Autonomous)
Vishnupur, India

Shikandar D. Bukkitgar
Department of Chemistry
KLE Institute of Technology
Hubballi, India

Rajya Lakshmi Chavakula
Department of Basic Science
Vishnu Institute of Technology
Bhimavaram, India

Basavaiah Chandu
Department of Nanotechnology
Acharya Nagarjuna University
Guntur, India

Neha Chauhan
Department of Pharmacy
Banasthali Vidyapith
Banasthali, India

Priya Chauhan
School of Studies in Environmental Chemistry
Jiwaji University
Gwalior, India

Rajasekhar Chokkareddy
Department of Humanities and Basic Sciences
Aditya College of Engineering and Technology
Surampalem, India

and

Department of Chemistry
Durban University of Technology
Durban, South Africa

Raghuraj Chouhan
Department of Chemical Sciences
Jožef Stefan Institute
Ljubljana, Slovenia

Sruchi Devi
Department of Biochemistry
Lovely Professional University
Phagwara, India

Shikha Dogra
Department of Pediatric and Preventive Dentistry
SGT University
Gurgaon, India

Manasa G.
Electrochemistry Research Group
St. Joseph's College
Bangalore, India

Mansi Gandhi
Department of Chemistry
Vellore Institute of Technology
Vellore, India

Shalini Garg
Department of Pediatric and Preventive Dentistry
SGT University
Gurgaon, India

Sourbh Suren Garg
Department of Biochemistry
Lovely Professional University
Phagwara, India

Ganesh Gollavelli
Department of Humanities and Basic Sciences
Aditya Engineering College
Surampalem, India
Jawaharlal Nehru Technological University
Kakinada, EG, AP, India

Anil Gupta
Department of Pediatric and Preventive Dentistry
SGT University
Gurgaon, India

Jeena Gupta
Department of Biochemistry
Lovely Professional University
Phagwara, India

Jayapriya J.
Department of Pharmacology
Saveetha Dental College and Hospital
Chennai, India

Smita Jain
Department of Chemical Sciences
Jožef Stefan Institute
Ljubljana, Slovenia

Sakshi Joshi
Department of Pediatric and Preventive Dentistry
SGT University
Gurgaon, India

Ashna Joy
P. Varkey Mission Hospital
Kerala, India

Vaskuri G.S. Sainaga Jyothi
Department of Pharmaceutics
National Institute of Pharmaceutical Education and Research
Hyderabad, India

K. Ganesh Kadiyala
Department of Chemistry
Shri Vishnu Engineering College for Women (Autonomous)
Vishnupur, India

Kajal Karsauliya
Pesticide Toxicology Laboratory and Regulatory Toxicology Group
and
Analytical Chemistry Laboratory and Regulatory Toxicology Group
CSIR–Indian Institute of Toxicology Research (CSIR-IITR)
Lucknow, India

Naresh Kumar Katari
School of Chemistry and Physics
University of KwaZulu Natal
Durban, South Africa

and

Department of Chemistry
Gandhi Institute of Technology and Management
Hyderabad, India

Moses Kigozi
Department of Chemistry
Busitema University
Tororo, Uganda

Raghavendra V. Kulkarni
Department of Pharmaceutics
BLDEA's SSM College of Pharmacy and Research Centre
Vijayapur, India

Malavika Kumar
Merck & Co. Inc.
Kenilworth, NJ

Malini Kumar
Sri Ramachandra Institute of Higher Education and Research
Chennai, India

Yong-Chien Ling
Department of Chemistry
National Tsing Hua University
Hsinchu, Taiwan

Tharani M.
Department Pharmacology
Saveetha Dental College and Hospital
Chennai, India

Ronald J. Mascarenhas
Electrochemistry Research Group
St. Joseph's College
Bangalore, India

Monalisa Mishra
Department of Life Science
NIT Rourkela
Rourkela, India

Nookaraju Muralasetti
Department of Humanities and Basic Sciences
Aditya College of Engineering and Technology
Surampalem, India

Puvvada Nagaprasad
Department of Applied Biology
CSIR–Indian Institute of Chemical Technology
Hyderabad, India

and

Department of Chemistry
Indrashil University
Mehsana, India

Poojith Nuthalapati
P. J. Biousys
Irving, TX

Ayan Pal
Centre for Nanotechnology
Indian Institute of Technology
Guwahati, India

G. Ramaswamy
Department of Basic Science
Vishnu Institute of Technology
Bhimavaram, India

Aishwarya Rathore
Department of Pharmacy
Banasthali Vidyapith
Banasthali, India

Dinesh Reddy
P. J. Biousys
Irving, TX

Kakarla Raghava Reddy
School of Chemical and Biomolecular Engineering
University of Sydney
Sydney, Australia

Gan G. Redhi
Department of Chemistry
Durban University of Technology
Durban, South Africa

Rajeshkumar S.
Department of Pharmacology
Saveetha Dental College and Hospital
Chennai, India

Veera Sadhu
School of Physical Sciences
Kakatiya Institute of Technology and Science
Warangal, India

Debasis Sahu
Amity Institute of Biotechnology
Amity University
Noida, India

Chiranjeevi Sainatham
P. J. Biousys
Irving, TX

Atulika Sharma
Department of Chemical Sciences
Apeejay Institute of Management and Technical Campus
Rama Mandi, India

Manu Sharma
Department of Pharmacy
Banasthali Vidyapith
Banasthali, India

Sheelu Sharma
Department of Physiotherapy
Delhi University
and
Pandit Deendayal Upadhyaya National Institute for Persons with Physical Disabilities (Divyangjan)
New Delhi, India

Swapnil Sharma
Department of Pharmacy
Banasthali Vidyapith
Banasthali, India

Nagaraj Shetti
School of Advanced Sciences
KLE Technological University
Hubballi, India

Sheelendra Pratap Singh
Pesticide Toxicology Laboratory and Regulatory Toxicology Group
and
Analytical Chemistry Laboratory and Regulatory Toxicology Group
CSIR–Indian Institute of Toxicology Research (CSIR-IITR)
Lucknow, India

Md Palashuddin Sk
Department of Chemistry
Aligarh Muslim University
Aligarh, India

Ranju Soni
Department of Pharmacy
Banasthali Vidyapith
Banasthali, India

Raghavendra Vemuri
Department of Humanities and Basic Sciences
Aditya College of Engineering and Technology
Surampalem, India

Kanika Verma
Department of Pharmacy
Banasthali Vidyapith
Banasthali, India

Neetika Verma
Department of Pediatric and Preventive Dentistry
SGT University
Gurgaon, India

Nandimalla Vishnu
Department of Chemistry
Gandhi Institute of Technology and Management
Hyderabad, India

Preeti Yendapalli
Sri Ramachandra Institute of Higher Education and Research
Chennai, India

1

Antimicrobial Applications of Nanodevices Prepared from Metallic Nanoparticles and Their Role in Controlling Infectious Diseases

Tharani M. and Rajeshkumar S.

CONTENTS

1.1 Introduction

The transformation of material science and nanotechnology has greatly advanced in creating multifunctional materials for future innovations on a nanoscale. Among various fields of study, the investigations on the combination of nano-sized particles utilizing a green science approach offer an incredible stage for creating green, conservative, recyclable and maintainable materials for future applications. Due to the unique features of metallic nanoparticles like optical density and surface plasmon resonance, they have been extensively used in targeting various diseases and also to explore more new pathways in nanotechnology.

Nanomaterials are particles having nanoscale measurement ranging from 1 to 100 nm and nanoparticles are minuscule estimated particles with upgraded catalytic reactivity, non-linear optical execution, warm conductivity and synthetic relentlessness attributable to its huge surface-region-to-volume proportion [1]. NPs have begun being considered antibiotic agents due to their antimicrobial properties. Nanoparticles have been incorporated into different mechanical, wellbeing, food, feed, space, substance and beautifying agents due to their alluring unique properties [2].

Metallic nanoparticles are novel agents that have been widely researched for their antimicrobial properties. The metal portion of the nanoparticles is assumed to be an extraordinary part in the utilization of metal nanoparticles as green catalysts because of their huge surface-region-to-volume proportion contrasted with the mass material. This empowers proficient binding, along these lines permitting reactants to bind together at metal sites, and therefore, reactions occur more effectively [3].

Various metal nanoparticles including gold, silver, zinc and copper exhibit intense antimicrobial properties. Microbial resistance seems to be a major threat worldwide and to control that, metallic nanoparticles have been planned to launch as antimicrobial agents to defeat the high spread of infectious diseases [4].

Metallic nanoparticles can be synthesized by chemical methods but it leads to various biological risks and side effects; synthesis of metallic nanoparticles by using plant extracts attracts more researchers to utilize it as an alternative method. Green synthesized metallic nanoparticles are safe, eco-friendly and have particular properties, and can be synthesized in a short period in a cost-effective manner. The green synthesized nanoparticles obtain different properties of the plant extracts and the biomolecules present in the extract act as a reducing and stabilizing agent. Existing research works confirm that biosynthesized metallic nanoparticles have been accounted for to have biocompatibility and antimicrobial activity [5–7].

DOI: 10.1201/9781003157823-1

1.2 Different Types of Metal Nanoparticles

As of late, the various sorts of metallic nanoparticles and their subsidiaries (like silver, gold, copper, nickel, platinum, titanium and zinc nanoparticles) got huge consideration for their excellent antimicrobial properties. Essentially, metal oxides and doped metal/metal composites like silver oxide, copper oxide, calcium oxide, magnesium oxide, titanium dioxide and zinc oxide have different remarkable properties, amazing potencies and showed magnificent antimicrobial capacity [8]. Moreover, the physicochemical properties and morphology of nanomaterials have been demonstrated for their antimicrobial activities. It was recognized that the nano-size metal particles convey the incredible bactericidal impact to oppose the microorganisms [9–11]. Figure 1.1 shows the different types of nanoparticles that are used for antimicrobial activity.

1.2.1 Silver Nanoparticles

The primary utilization of silver (Ag) as an antimicrobial and antibacterial specialist returns to the antiquated Greek and Roman Empires [12, 13]. Around then, the restorative and additive properties of silver were fundamentally used to shield vessels from bacterial assaults and to make water and different fluids consumable [14–18]. All around the world, it was at that point known to be an effective weapon against the development of microbe factors [19]. The antimicrobial effect of silver emerges from the association of silver particles with thiol gatherings of crucial bacterial enzymes and proteins that lead to cell demise.

Silver nanoparticles (AgNPs) are perhaps the most fundamental and entrancing nanomaterials among a few metallic nanoparticles that are associated with biomedical applications. Silver nanoparticles assume a significant part in nanoscience and nanotechnology, especially in nanomedicine. Silver nanoparticles have alluring optical, electrical and thermal properties and are being fused into products that range from photovoltaics to natural and synthetic sensors. An inexorably regular application is the utilization of silver nanoparticles for antimicrobial coatings; wound dressings, topical creams and biomedical gadgets currently contain silver nanoparticles that ceaselessly discharge a low degree of silver particles to give protection against microorganisms [18, 20].

1.2.2 Gold Nanoparticles

The utilization of gold in medication isn't restricted to gold nanoparticles, as gold mixtures were utilized sometime before the approach of nanomedicine. Robert Koch found that gold cyanide was harmful to the tuberculosis bacillus in vitro [21]. Regardless of whether it gave the idea that gold cyanide was inadequate against tuberculosis in vivo, this revelation laid the milestone in the clinical utilization of gold and examined the impacts of this valuable metal on different pathologies [22]. In this specific circumstance, Jacques Forestier confirmed that ionic gold mixtures diminish joint torment of patients experiencing rheumatoid joint inflammation [23]. From that point, gold salts treatment, otherwise called chrysotherapy, had been utilized until the 1990s, after which not so much poisonous but rather more effective medicines were created.

Gold nanoparticles have an expansive range of utilization areas including medication, food industry, water purification and organic applications [24]. Specifically, gold nanoparticles are applied to photo-thermal treatment, drug delivery, imaging, detecting, catalysis and even antimicrobials [25–27]. The real uses of gold nanoparticles are too long because of their interesting properties. Consequently, green synthesized gold nanoparticles have considerably more potential in various fields.

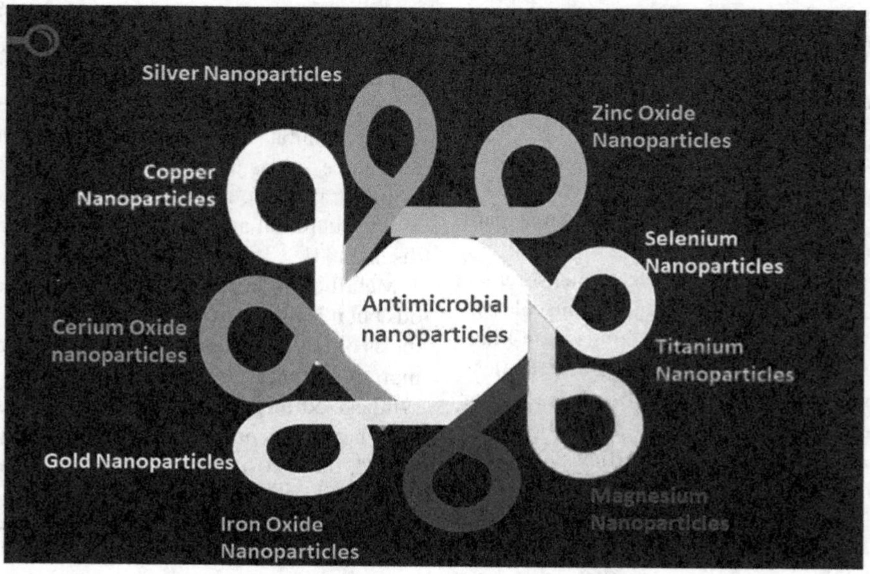

FIGURE 1.1 Different types of nanoparticles used in antimicrobial activity.

1.2.3 Zinc Nanoparticles

Quite possibly the most intriguing and promising metallic nanomaterials are zinc (Zn) and its oxide (ZnO). Zinc is a genuinely dynamic component and at the same time a powerful reducing agent; as per its reducing potential, it can undoubtedly oxidize, framing zinc oxide [28], which is useful in the formation of zinc oxide nanoparticles. Zinc assumes a significant part in humans as perhaps the most fundamental microelement [29]. It is present in all body tissues, for example in muscle and bone (85% of the entire body zinc content), in the skin (11%) and in a wide range of various tissues; it is intracellular, found primarily in the core, cytoplasm and cell layer [30]. Zinc is likewise basic for the working of metalloproteins. The zinc lacks redox movement and is viewed as generally non-poisonous, there is an expanding measure of proof that free zinc particles may cause degradation of neurons [31]. In this way, to dispose of its cytotoxic effect, combining zinc cations with bioactive ligands (for example proteins) [32] and a blend of zinc oxide nanoparticles are performed. Zinc oxide holds exceptional optical, substance detecting, semiconducting, electric conductivity and piezoelectric properties [33]. Given these properties, ZnO-NPs have acquired great interest in the biomedical field [34]. The all-encompassing use of ZnO-NPs in medication for angiogenesis, as antiplatelet agents, anti-inflammatory agents, dental materials, makeup and drug delivery has made ZnO-NPs a promising anticancer agent [35, 36].

1.2.4 Selenium Nanoparticles

Selenium is an essential trace element present in our body and is one of the fascinating compounds to incorporate with antibacterial agents [37]. Selenium plays a major role in maintaining health and growth through diet. Nanoselenium has significant attention due to being the least toxic form of selenium. Existing research has pointed out the capacity of selenium nanoparticles to show anticancer [38], antioxidant [39], antibacterial and anti-biofilm [40] properties.

1.2.5 Copper Nanoparticles

Copper is cost-effective and has gained huge attention in nanoparticles synthesis for novel catalytic antimicrobial and other optical properties. To acquire the total benefits of copper nanoparticles, an eco-friendly green combination is required. Copper is less poisonous and is a significant component of human wellbeing. Copper is a more affordable metal than others which are frequently used as nano-sized particles for improving antibiotic effectiveness. The proportion among ratio and volume is high for copper nanoparticles. The significant uses of copper nanoparticles are to exhibit as an antibacterial compound [41]. The utilization of copper nanoparticles toward wellbeing-related cycles is significant due to their antifungal and antibacterial effects [42]. The greener synthesis of different nanoparticles using various plant and green waste materials is presented in Tables 1.1–1.6.

1.3 Antimicrobial Activity of Metallic Nanoparticles

Metallic nanoparticles such as silver, gold and copper are accounted for to kill various kinds of microbes. Every one of these nanometals has its own particular properties and mechanism of action. The overall antimicrobial action of nanometals is proposed to include interruption of cell layer metabolism. Metallic nanoparticles have the capacity to enter the cell wall and disturb enzymes responsible for microbial death. Also, metal nanoparticles can produce reactive oxygen species (ROS) which leads to microbial death. Adherence of nanoparticles to the cell wall additionally diminishes bacterial replication. Significant parameters for antimicrobial nanoparticles include their shape, nanometer dimension and concentration.

1.3.1 Gold Nanoparticles

Gold nanoparticles (Au-NPs) have been generally researched because of their uniqueness particularly in biomedication [73]. In existing research works, green synthesis of gold nanoparticles was achieved using leaf extracts of both *Carica papaya* and *Catharanthus roseus*, and the antibacterial activity for biosynthesized gold nanoparticles was tested against pathogens such as *S. aureus*, *E.coli*, *Proteus vulgaris* and *Bacillus subtilus*, which resulted in exhibiting enhanced antibacterial activity [74]. *Anona muricata* leaf extract was used for gold nanoparticle synthesis and antimicrobial activity tested against some human pathogenic bacteria and fungi such as *Staphylococcus aureus*, *Clostridium sporogenes*, *Enterococcus faecalis*, *Klebsiella pneumonia*, *Aspergillus flavus*, *Penecillium cameri*, *Fusarium oxysporium* and *Candida albicans*. The *Anona muricata* mediated gold nanoparticles were reported to result in increasing order of potential antimicrobial activity as the concentration increases [50].

1.3.2 Silver Nanoparticles

Silver nanoparticles possess excellent bactericidal activity and can be incorporated into potential biomedical applications. The size, shape and capping ability determine the antimicrobial efficacy of silver nanoparticles [75]. Previous works exhibit the antimicrobial activity of silver nanoparticles against pathogenic bacteria such as *S. aureus* and *Pseudomonas aeruginosa* with a maximum zone of inhibition of 15.17 mm and 13.33 mm [76]. *Myriostachya wightiana* is a perennial salt marsh grass that is used as a reducing agent to synthesize silver nanoparticles and the antibacterial efficacy was checked against *Xanthomonas campestris* (9.2 ± 0.26 mm) and *Ralstonia solanacearum* (no zone) [77]. *Lysiloma acapulsensis* mediated silver nanoparticles reported to show higher antimicrobial activity against pathogens such as *Staphylococcus aureus* (16 ± 1.0 mm), *Pseudomonas aeruginosa* (15 ± 0.5 mm), *Escherichia coli* (18 ± 1.3 mm) and *Candida albicans* showed higher zone of inhibition [78]. Lycopene-mediated silver nanoparticles showed their maximum zone of inhibition against *S. aureus* (13 mm) and *S. mutans* (18 mm) [79].

TABLE 1.1

Green Synthesis of Silver Nanoparticles Using Various Plant Materials

S. no.	Precursor	Reducing agent	Characterization	Applications	Ref.
1.	Silver nitrate	*Azadirachta indica* leaf extract	UV spec (436–446 nm) TEM – AgNP (34 nm) DLS – AgNP (34 nm) FT-IR – 3,454 cm^{-1} is due to the N–H stretching vibration of group NH_2 and -OH the overlapping of the stretching vibration of attributed for the presence of water and *A. indica* leaf extract molecules	Antibacterial activity against *S. aureus* and *E. coli*	[43]
2.	Silver nitrate	*Lysiloma acapulcensis* extract	UV spectra – 400 nm FT-IR – presence of alkyl halides, proteins, phenolic and aromatic compounds with transmission peaks at 592, 1,631, 2,340 and 1,620 cm^{-1}, respectively TEM – 1.2–62 nm with an average size of 5 nm XRD – polycrystalline nature	Antimicrobial activity against *E. coli*, *S. aureus*, *P. aeruginosa*, *C. albicans*	[44]
3.	Silver nitrate	*Acalypha indica* leaf extract	UV spectra – 440 nm TEM, SEM, XRD – 34 nm, spherical shape FT-IR – the intense peaks showed possible compounds adsorbed on the surface of the Ag nanoparticles	Antioxidant activity by DPPH assay – IC50 value for Ag nanoparticles was 5 mg/mL Antifungal activity – *A. fumigatus* showed ZOI of 133% at 75 μL of concentration than *Aspergillus niger*, *Aspergillus flavus*	[45]
4.	Silver nitrate	Algae *Turbinaria conoides* extract	UV spectra – 420 nm SEM – spherical shape Size of about 96 nm XRD analysis FT-IR – the peak at 1,382.7 cm^{-1} indicates possible biomolecules are amines and polyphenols	Antibacterial activity against gram-positive bacteria *Bacillus subtilis* (MTCC3053) and gram-negative bacteria *Klebsiella planticola* (MTCC2277) Higher antibacterial activity in *Klebsiella planticola* than *Bacillus subtilis*	[46]
5.	Silver nitrate	*Clitoria ternatea* and *Solanum nigrum* leaf extracts	UV spectra – both AgNP showed peak between 420 and 440 nm FT-IR – absorbance bands of *Clitoria ternatea* were observed at 3,317.34 cm^{-1} assigned to O–H stretch and absorbance bands of *Solanum nigrum* were observed at 3,317.34 cm^{-1} assigned to O–H stretch SEM – both spherical shaped XRD – *Clitoria ternatea* mediated AgNP is 20 nm *Solanum nigrum* mediated AgNP is 28 nm	Both silver nanoparticles were tested against nosocomial pathogens such as *Bacillus subtilis*, *Staphylococcus aureus*, *Streptococcus pyogenes*, *Escherichia coli*, *Pseudomonas aeruginosa* and *Klebsiella aerogenes*	[47]

1.3.3 Selenium Nanoparticles

In previous studies, *Ceropagia bulbosa* Roxb. extract was used for synthesizing selenium nanoparticles and antimicrobial activity was tested against *Bacillus subtilis* and *Escherichia coli* which exhibited a clear zone of inhibition between 14 and 20 mm diameter [80]. Recently, cow urine was used as a reducing agent to synthesize selenium nanoparticles and showed higher antimicrobial activity for *Klebsiella* spp. than *S. aureus*, *E. coli*, *Serratia* and *Pseudomonas* spp. [81].

1.3.4 Zinc Nanoparticles

Catharanthus roseus leaf extract mediated zinc oxide nanoparticles were synthesized and the antibacterial activity results showed an increased zone of inhibition as concentration increased against gram-positive pathogenic bacteria such as *Staphylococcus aureus*, *Streptococcus pyogenes* and *Bacillus cereus* than gram-negative *Pseudomonas aeruginosa*, *Proteus mirabilis* and *Escherichia coli*. Zinc oxide nanoparticles were synthesized using *Cardiospermum* leaf extract as a reducing agent and tested for their antimicrobial potency against pathogenic bacteria such as *Bacillus*, *E. coli* and *S. aureus* and fungi such as *Aspergillus flavus*, *Aspergillus fumigatus* and *Aspergillus niger*. Gram-positive organisms resulted in exhibiting higher sensitivity to biosynthesized zinc nanoparticles than gram-negative bacteria. Among the other two fungi, *Aspergillus fumigatus* showed a high zone of inhibition proving its sensitivity to zinc nanoparticles [82].

1.3.5 Copper Nanoparticles

Since copper nanoparticles are potent antimicrobial agents, the antimicrobial efficiency is increased by using *Citrus medica* Linn. (Idilimbu) juice and the green synthesized copper nanoparticles were tested against *E. coli* which showed high sensitivity followed by *Klebsiella pneumonia*, *Pseudomonas aeruginosa*, *P. acne* and *S. typhi* [83]. The antifungal efficacy

TABLE 1.2
Green Synthesis of Gold Nanoparticles Using Various Green Waste Materials

S. no.	Precursor	Reducing agent	Characterization	Applications	Ref.
1.	Hydrogen tetra chloroaurate (III) hydrate ($HAuCl_4.3H_2O$)	*Garcinia mangostona* fruit peels extract	UV spectra – 540–550 nm XRD – peaks at 2θ = 38.48°, 44.85°, 66.05° and 78.00° corresponding to (111), (200), (220) and (311) planes confirm AuNP synthesis TEM – size of about 32.96 ± 5.25 nm with a spherical shape FT-IR – the peaks obtained correspond to phenols, flavonoids, benzophenones and anthocyanins	–	[48]
2.	Hydrogen tetra chloroaurate (III) hydrate ($HAuCl_4.3H_2O$)	*Cryptolepis buchanani Roem.* extract	UV spectra – 530 nm TEM – spherical shape Size – 11 nm Zeta potential – 30.28 mV EDS analysis XRD – the diffraction peaks at 38.02°, 44.9°, 65.26° and 78.3° corresponding to the (111), (200), (220) and (311) planes indicate the crystalline nature of AuNPs FT-IR – indicated the presence of phenol compounds responsible for reduction and stabilization of AuNP	Antibacterial activity against *Staphylococcus aureus*, methicillin-resistant *Staphylococcus aureus* and *Acinetobacter baumannii* Catalytic activity of green synthesized AuNP using Methylene blue dye	[49]
3.	Hydrogen tetra chloroaurate (III) hydrate ($HAuCl_4.3H_2O$)	*Anona muricata* leaf extract	UV spectra TEM – spherical monodispersed structure with average size of about 25.5 nm FT-IR – reveal band at 3,271.14, 2,111.91 and 1,637.82 cm^{-1} corresponding to –N–H, –C=C and –C–N functional groups	Antibacterial activity against *Staphyloccus aureus > Entrococcus faecalis > Klebsiella pneumonia > Clostridium sporogenes* Antifungal activity against *Aspergillus flaws > Candida albican > Fusarium oxysperium > Penicillium camemeri*	[50]
4.	Hydrogen tetra chloroaurate (III) hydrate ($HAuCl_4.3H_2O$)	*Platycodon grandiflorum* extract (balloon flower plant)	UV spectra – 545 nm SEM and TEM – spherical shape, 15 nm EDAX analysis X-ray diffraction and X-ray photoelectron spectroscopy – confirms the face central cubic crystalline nature and elemental composition	Antibacterial activity against *Escherichia coli* (16 mm) and *Bacillus subtilis* (11 mm)	[51]
5.	Hydrogen tetra chloroaurate (III) hydrate ($HAuCl_4.3H_2O$)	Pomegranate peel extract	UV spectra – 530 nm TEM – spherical shape with 16 nm EDX spectra confirm the presence of gold level of about 59.90% by weight	Antioxidant activity and anticancer activity against Hep G-2 liver malignant cell line	[52]

of copper nanoparticles mediated by *Celastrus paniculatus* leaf extract was checked against plant pathogenic fungi *Fusarium oxysporium* which showed maximum mycelial inhibition around 76.92 ± 1.52 mm [84].

Figure 1.2 shows the mechanism of action of nanoparticles against different pathogenic bacteria.

1.4 Metallic Nanoparticles and Their Role in Controlling Infectious Pathogens

Infectious diseases caused by microbes such as fungi, bacteria, viruses and parasites are responsible for generating health crises every year which leads to high mortality rates worldwide. More than emerging infectious diseases, re-emerging infectious diseases lead to a high death rate due to their resistance to antibiotics, and treatment, in this case, is very difficult to enact or control. Majorly, infections get transmitted when pathogens enter the body through natural openings leading to the multiplication of microorganisms at the site of entry which results in tissue damage [85–87].

Existing research works reported that silver nanoparticles enter the microbial cell wall and cause structural damage and result in cell death. Some other researchers reported that silver nanoparticles deal with the production of free radicals which exhibit the destruction of cells [19, 88, 89]. Silver nanoparticles have been accounted to be powerful for the treatment of bacterial diseases like tuberculosis, gonorrhea, chlamydia, syphilis and infections that occur in the urinary tract. Tuberculosis is a disease that affects the lungs and it is caused by the bacterium *Mycobacterium tuberculosis*. The treatment of tuberculosis is disturbed by drug resistance which occurs due to the usage of antibiotics throughout an extensive period. Metal-based nanoparticles have been reported to control antibiotic resistance. In previous studies, silver and zinc oxide nanoparticles are combined at specific ratios which

TABLE 1.3
Green Synthesis of Zinc Nanoparticles Using Various Green Waste Materials

S. no.	Precursor	Reducing agent	Characterization	Applications	Ref.
1.	Zinc nitrate	Orange fruit peel extract	TEM – small spherical particles with size ranges from 10 to 20 nm XRD – confirms hexagonal wurtzite structure FT-IR – vibration bands at around 450cm^{-1}, which indicates the stretching vibration of Zn–O bonding	Bactericidal activity toward *E. coli* and *S. aureus* – showed strong antibacterial activity toward *Escherichia coli* and *Staphylococcus aureus* without UV illumination at an NP concentration of 0.025 mg mL^{-1}	[53]
2.	Zinc acetate dihydrate	*Lippia adoensis* leaf extract	TGA analysis – ZnO nanoparticles are thermally stable after 400°C FT-IR analysis XRD – confirms hexagonal wurtzite structure SEM – spherical shape ZnONP EDS analysis – zinc (59.43%) and oxygen (40.57%) TEM – 19.78 nm SAED analysis – confirms polycrystalline nature of the synthesized ZnO-NPs	Antibacterial activity against clinical and standard strains of *Escherichia coli*, *Klebsiella pneumonia*, *Staphylococcus aureus* and *Enterococcus faecalis* with a maximum inhibition zone of 14 mm and 12 mm	[54]
3.	Zinc nitrate hexahydrate	Aqueous extract of *Kalanchoe blossfeldiana*	UV spectra – 430 nm SEM and TEM – the shapes were found to be nanometer scale and in rods, hexagonal or other various shapes XRD – hexagonal wurtzite phase of ZnO-NPs EDX analysis – 0.5 and 9.6 keV, which confirms the elemental distribution of the ZnO-NPs DLS analysis – ZnO-NPs (94 nm)	Antimicrobial activity *S. aureus* (13 mm) (gram-positive bacteria); *E. coli* (11 mm) and *P. aeruginosa* (15 mm) (gram-negative bacteria); and *Fusarium solani* (45 mm), *Alternaria alternate* (28 mm) and *Helmenthosporium sp. Fungi* (22.5 mm)	[55]
4.	Zinc acetate dihydrate	*Peganum harmala* plant extract	XRD – hexagonal wurtzite structure SEM – spherical shape with 39.94 nm EDX analysis – confirms the presence of 29.39% oxygen element and 70.61% zinc element	Antibacterial activity against *S. aureus* (17 mm), *E. coli* (10 mm), *L. monocytogenes* (13 mm) and *S. typhi* (10 mm)	[56]
5.	Zinc chloride	Garlic skin extract	UV spec – 370 nm TEM – rod- and hexagon-shaped nanoparticles having an average size of 7.77 nm EDX analysis – confirms the formation of highly pure ZnONP XRD – hexagonal wurtzite phase with crystallite size of 12.61 nm FT-IR analysis – the FT-IR peak at 441 cm^{-1} corresponds to ZnO bonding which confirms the presence of ZnO particles	–	[57]

resulted in exhibiting potential antibacterial activity against *Mycobacterium tuberculosis* [90].

Nanoparticles of comparative sizes are skilled to bind with COVID-19 infections and disrupt their structure with a mix of infrared light treatment. That primary disruption would then end the capacity of the infection to endure and recreate in the body. Advancement in the field of nanotechnology has decorated the need of using metallic nanoparticles for the detection and treatment of illnesses. Nanoparticles can harm these microorganisms before they can cause any damage to the body, as they clutch various objects and surfaces. Gold nanoparticles and carbon quantum spots (CQDs) are outstanding decisions for communicating with the infections and forestalling their entrance into cells due to their high explicit surface area and chances of being functionalized with an expansive scope of functional groups. These carbon quantum dots having a normal measurement of 10 nm and their great dissolvability in water can be the ideal candidates to overcome coronavirus, because they can without much of a stretch enter the cell through endocytosis and connect with the virus's protein, accordingly forestalling viral genome replication [91]. Other nanomaterials have likewise been found to have viable antiviral impacts. For example, a Chinese research team has created novel gold nanorods as a class of peptide-based inhibitors, which cautiously focus on COVID's S protein and retard its movement [92].

Green synthesized zinc oxide nanoparticles were effective in controlling infectious diseases such as tuberculosis. *Limonia acidissima* leaf extract had been used to synthesize zinc oxide nanoparticles and found it was effective in inhibiting the growth of *Mycobacterium tuberculosis* at 12.5 μg/mL concentration [93]. Existing research works have reported that zinc oxide nanoparticles are capable of destructing microbial biofilms which are highly resistant to antibiotics and host defense mechanisms [94].

Green synthesized selenium nanoparticles have been recognized to treat cancer which is reported to be responsible for 9.6 million deaths in 2018 worldwide [95]. Green synthesis of

TABLE 1.4

Green Synthesis of Selenium Nanoparticles Using Various Green Waste Materials

S. no.	Precursor	Reducing agent	Characterization	Applications	Ref.
1.	Sodium selenite	*Emblica officinalis* fruit extract	XRD – amorphous in nature Zeta potential – found to be negatively charged (–24.4 mV) SEM – spherical in shape TEM – 15–40 nm	Antibacterial activity – (*Escherichia coli* MTCC 41, *Listeria monocytogenes* MTCC 657, *Staphylococcus aureus* MTCC 96, and *Enterococcus faecalis* MTCC 439) and fungi (*Aspergillus brasiliensis* MTCC 1344, *A. flavus* MTCC 1883, *A. oryzae* MTCC 634, *A. ochraceus* MTCC 10276, *Fusarium anthophilum* MTCC 10129, and *Rhizopus stolonifer* MTCC 4886) EC_{50}15.67 ± 1.41 and 18.84 ± 1.02 µg/mL for DPPH and ABTS assays	[58]
2.	Sodium selenite	Black tea extract	UV spectra – 380 nm	Antibacterial activity against *Streptococcus mutans*, *Staphylococcus aureus*, *Enterococcus feacalis* Antixidant activity by DPPH method – increased concentration leads to increased scavenging activity	[59]
3.	Sodium selenite	*Allium sativa* extract	UV spectra – 400 nm TEM – polydispersed nanoparticles with size ranging from 8 to 52 nm SAED analysis – confirms the crystalline nature of the selenium nanoparticles FT-IR	Antimicrobial activity against *Escherichia coli* (29 mm) and *Salmonella typhi* (27 mm) at 100 µl concentration and lesser at 25 µl concentration *Escherichia coli* (11 mm) and *Salmonella typhi* (13 mm)	[60]
4.	Sodium selenite	Parsley leaves extract	UV spectra – 270 nm FT-IR analysis AFM and DLS – spherical shape, with diameter around 400 nm Zeta potential – (–14.2 mV)	–	[61]
5.	Sodium selenite	Aloe vera leaf extract	FT-IR – peak at 1,635.52 cm^{-1}, indicates the stabilization of synthesized selenium nanoparticles UV spectra – 323 nm Zeta potential value – (–18 mV) TEM – spherical shape with 50 nm	Antibacterial activity against *S. aureus* (12 mm) and *E. coli* (10 mm) Antifungal activity against *P. digitatum* and *C. coccodes*	[62]

TABLE 1.5

Green Synthesis of Copper Nanoparticles Using Various Plant Materials

S. no.	Precursor	Reducing agent	Characterization	Applications	Ref.
1.	Copper sulfate	*Cissus vitiginea* leaves extract	UV spectra – 340 nm SEM – monodispersed spherical nature with size of 20 nm EDX – confirms the presence of copper elements with carbon and oxygen XRD – high crystalline in nature TEM – spherical shape with size ranging from 10 to 20 nm FT-IR analysis and AFM analysis	Antibacterial activity against urinary tract pathogens such as *E. coli* (22.2mm), *Enterococcus* sp. (20.3mm), *Proteus* sp. (16.33mm) and *Klebsiella* sp. (18.5mm)	[63]
2.	Copper sulfate	Fresh aloe vera gel	UV spectra – around 250–350 nm FT-IR analysis – the peak at 517 cm^{-1} indicated the formation of CuO nanostructures	Anti-cariogenic activity against *S. aureus*, *S. mutans*, *Pseudomonas* sp., *E. faecalis*	[64]
3.	Cupric nitrate trihydrate	*Caesalpinia bonducella* seed extract	UV spectra –250 nm FE-SEM – exhibits a rice-grain-shaped morphology EDX spectra reveals the weight percentage of copper and oxygen to be 73.15 and 22.17% FT-IR – the narrow peaks at 457, 526, 600 and 784 cm^{-1} confirms the formation of pure CuO Nps XRD – CuONP are monoclinic and crystalline in nature	Antibacterial activity against *Staphylococcus aureus* and *Aeromonas*. *Aeromonas* showed higher zone of inhibition than *S. aureus*	[65]
4.	$CuSO_4.5H_2O$	*Ocimum sanctum* leaf extract	XRD – diffraction peaks of 22.3°, 25.9°, 28.3° and 44.8°, which correspond to the characteristic face centered cubic (FCC) of copper lines indexed at (111), (200), (210) and (222), respectively FT-IR – confirms the presence of terpenoids, alcohols, ketones, aldehydes and carboxylic acid	–	[66]
5.	Cupric nitrate trihydrate	*Hagenia abyssinica* (Brace) JF. Gmel. leaf extract	UV spectra – 403 nm FT-IR – confirms the presence of polyphenols, tannins and glycosides in the leaf extract XRD – crystalline in nature SEM and TEM – spherical, hexagonal, triangular, cylindrical and irregularly shaped copper nanoparticles Size – 34.76 nm	Antibacterial activity against *E. coli* (12.7 nm), *Pseudomonas aeruginosa* (12.7 nm), *Staphylococcus aureus* (14.7 nm) and *Bacillus subtilis* (14.2 nm)	[67]

TABLE 1.6

Green Synthesis of Other Metallic Nanoparticles Using Various Plant Materials

S. no.	Precursor	Reducing agent	Characterization	Applications	Ref.
1.	Ferric chloride hexahydrate	*Lagenaria siceraria* leaves extract	UV spectra – 658 nm EDX analysis – 48% iron, 31% oxygen SEM – cubic shaped with size of about 30–100 nm FT-IR – confirms the presence of –OH and –COOH as functional groups present on the nanoparticles	Antibacterial activity against *S. aureus* (8 mm), *E. coli* (10 mm)	[68]
2.	Titanium dioxide	*Cassia fistula* leaves extract	UV spectra – 350 nm Thermogravimetric analysis Atomic force microscopic analysis XRD – synthesized titanium NPs were face-centered cubic (FCC) and crystalline in nature SEM – titanium nanoparticles are agglomerated and nearly spherical in shape	Antibacterial activity against *Escherichia coli* and *Staphylococcus aureus*	[69]
3.	Zirconium oxynitrate	*Sargassum wightii* seaweed extract	XRD analysis – 4.8 nm with tetragonal structure UV spectra – 330 nm FT-IR – the peak at 1,653 cm^{-1} indicates carboxylate ions which are responsible for stabilization of zirconia nanoparticles TEM – monodispersed spherical shape with particle size of about 5 nm	Antibacterial activity against *Bacillus subtilis, Escherichia coli, Salmonella typhi*	[70]
4.	Magnesium nitrate	*Rhizophora lamarckii* leaves extract	UV spectra – 301 nm TEM – polydispersed nanoparticles with size 20 nm and 50 nm XRD – nanocrystalline in nature FT-IR – the functional groups of the mangrove, amine and alkane are found to act as reductants and stabilizers	Antibacterial activity against gram-positive bacteria (*S. aureus, S. pneumoniae*) and gram-negative bacteria (*E. coli, S. typhi*)	[71]
5.	Cerium nitrate	*Origanum majorana* leaf extract	TEM – spherical shape with size of about 20 nm FE-SEM – pseudo-spherical shape XRD – 3,427, 1,592, 1,383 and 1,131 cm^{-1} corresponds to the O–H, aliphatic C–H and C = O stretching vibration of flavonoid and phenolic groups, which confirms the presence of aromatic groups along with the CeO-NP	Antioxidant activity by DPPH assay and ABTS assay Anticancer activity against MDA-MB-231 cancer cells compared to HUVEC normal cells	[72]

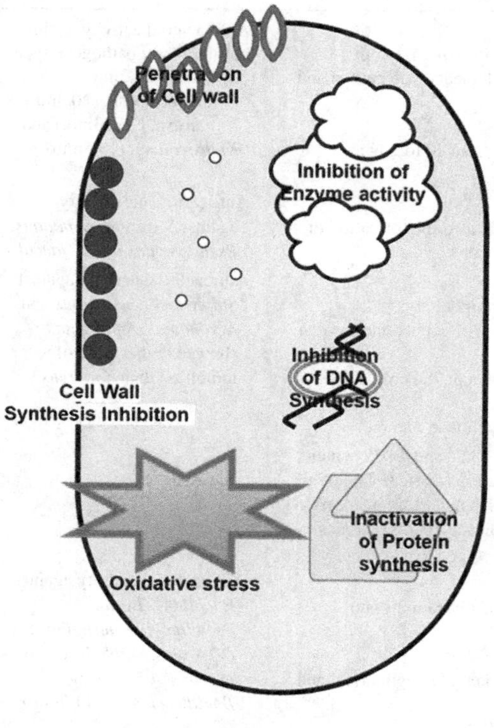

FIGURE 1.2 Mechanism of nanoparticles against bacterial pathogens.

SeNPs by utilizing the leaf concentrate of *Withania somnifera* is known to diminish the viability of adeno-carcinomic human alveolar basal epithelial cancer cells (AJ 549). It was clarified that the SeNPs initiate instability of chromosomes and mitotic arrest in the human alveolar basal malignant growth cells [96]. Another examination showed that the breast cancer cells were seriously affected through *Carica papaya* latex interceded SeNPs due to apoptotic changes, for example, the development of the apoptotic body in the breast malignancy cells, chromatic condensation and discontinuity. Remarkable adjustments in nuclear morphology and nuclear condensation occur by plant-based SeNPs which causes cytotoxicity in the cancer cells [97].

Other metallic nanoparticles such as zirconium nanoparticles were also reported to reduce inflammation caused by the pathogenic H5N1 influenza virus. Pre-treatment with zirconia nanoparticles has been proven to activate mature dendritic cells and expression of cytokines along with antiviral response and innate immunity in H5N1 influenza virus-infected mice [98]. In previous research works, the combination of silver and titanium oxide nanoparticles synthesized through *Euphorbia prostrata* extract showed a shift from apoptosis to G0/G1 arrest accomplished by necrotic cell death in *Leishmania donovani*, a parasitic infection [99]. Recently, *Cissus vitiginea* extract mediated copper nanoparticles exhibited effectiveness in killing or inhibiting the growth of urinary tract infection-causing

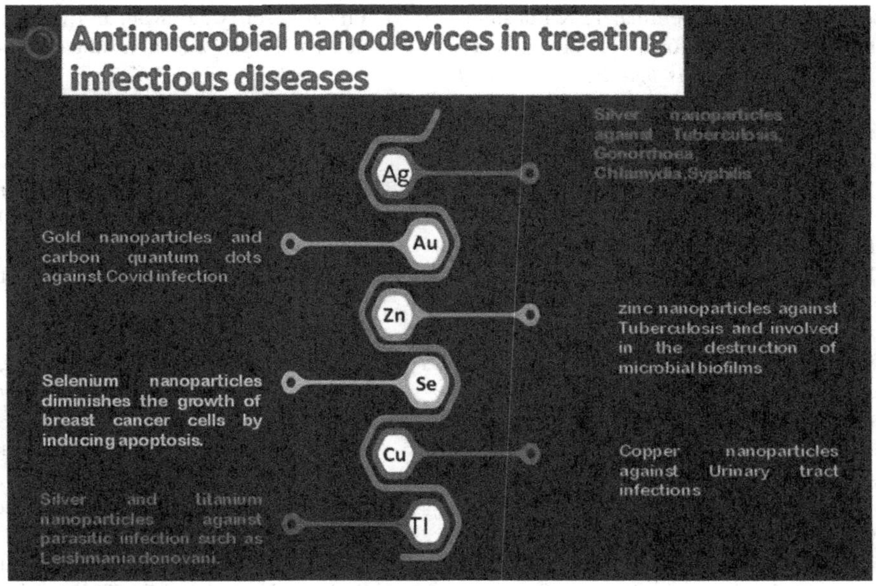

FIGURE 1.3 Different nanodevices treating infectious diseases.

pathogens such as *E. coli*, *Enterococcus* spp., *Proteus* spp. and *Klebsiella* spp. Figure 1.3 lists the nanoparticles used in infectious diseases.

1.5 Conclusion

This book chapter elucidates the importance of using metallic nanoparticles as antimicrobial nanodevices in the biomedical field. It also emphasizes the importance of plant-mediated synthesis of metallic nanoparticles and also its indispensable role in fighting against infectious pathogens. Synthesis of metallic nanoparticles is important and highly demanded due to their unique electrical, optical, magnetic and chemical properties. Even though current examinations have shown that metallic nanoparticles can give huge benefits, the uncertainty in creating them for a specific target with high explicitness is as yet challenging. Therefore, future research works should also focus on addressing the safety and biocompatibility of metallic nanoparticles, in particular long-term toxicities.

REFERENCES

1. Khan, S. T., Musarrat, J., & Al-Khedhairy, A. A. (2016). Countering drug resistance, infectious diseases, and sepsis using metal and metal oxides nanoparticles: Current status. *Colloids and Surfaces B: Biointerfaces*, *146*, 70–83.
2. Sastry, M., Ahmad, A., Khan, M. I., & Kumar, R. (2003). Biosynthesis of metal nanoparticles using fungi and actinomycete. *Current Science*, 162–170.
3. Hong, X., Liu, J., Zheng, B., Huang, X., Zhang, X., Tan, C.,... & Zhang, H. (2014). A universal method for preparation of noble metal nanoparticle-decorated transition metal dichalcogenide nanobelts. *Advanced Materials*, *26*(36), 6250–6254.
4. Rajeshkumar, S., & Bharath, L. V. (2017). Mechanism of plant-mediated synthesis of silver nanoparticles: A review on biomolecules involved, characterisation and antibacterial activity. *Chemico-Biological Interactions*, *273*, 219–227.
5. Amini, S. M. (2019). Preparation of antimicrobial metallic nanoparticles with bioactive compounds. *Materials Science and Engineering: C*, 109809.
6. Zhang, D., Ma, X., Gu, Y., Huang, H., & Zhang, G. (2020). Green synthesis of metallic nanoparticles and their potential applications to treat cancer. *Frontiers in Chemistry*, *8*.
7. Salem, S. S., & Fouda, A. (2021). Green synthesis of metallic nanoparticles and their prospective biotechnological applications: An overview. *Biological Trace Element Research*, *199*, 344–370.
8. Dizaj, S. M., Lotfipour, F., Barzegar-Jalali, M., Zarrintan, M. H., & Adibkia, K. (2014). Antimicrobial activity of the metals and metal oxide nanoparticles. *Materials Science and Engineering C*, *44*, 278–284.
9. Mohammadi, G., Nokhodchi, A., Barzegar-Jalali, M., Lotfipour, F., Adibkia, K., Ehyaei, N., et al. (2011). Physicochemical and anti-bacterial performance characterization of clarithromycin nanoparticles as colloidal drug delivery system. *Colloids and Surfaces B: Biointerfaces*, *88*, 39–44.
10. Fellahi, O., Sarma, R. K., Das, M. R., Saikia, R., Marcon, L., Coffinier, Y., et al. (2013). The antimicrobial effect of silicon nanowires decorated with silver and copper nanoparticles. *Nanotechnology*, *24*, 495101–495112.
11. Besinis, A., De Peralta, T., & Handy, R. D. (2014). The antibacterial effects of silver, titanium dioxide and silica dioxide nanoparticles compared to the dental disinfectant chlorhexidine on Streptococcus mutans using a suite of bioassays. *Nanotoxicology*, *8*, 1–16.
12. Sivarajan, M., Rajeshkumar, S., & Ezhilarasan, D. (2021). Role of silver nanoparticles in antioxidant property. *Annals of the Romanian Society for Cell Biology*, 693–703.

13. Rajeshkumar, S., Tharani, M., Sivaperumal, P., & Lakshmi, T. (2020). Synthesis of antimicrobial silver nanoparticles by using flower of *Calotropis figantea*. *Journal of Complementary Medicine Research*, *11*(5), 8–16.
14. Alexander, J. W. (2009). History of the medical use of silver. *Surgical Infections*, *10*, 289–292.
15. Khaydarov, R. R., Khaydarov, R. A., Estrin, Y., Evgrafova, S., Scheper, T., Endres, C., & Cho, S. Y. (2009). Silver nanoparticles. In *Nanomaterials: Risks and Benefits* (pp. 287–297). Dordrecht: Springer.
16. Castellano, J. J., Shafii, S. M., Ko, F., Donate, G., Wright, T. E., Mannari, R. J., … & Robson, M. C. (2007). Comparative evaluation of silver-containing antimicrobial dressings and drugs. *International Wound Journal*, *4*, 114–122.
17. Marambio-Jones, C., & Hoek, E. M. V. (2010). A review of the antibacterial effects of silver nanomaterials and potential implications for human health and the environment. *Journal of Nanoparticle Research*, *12*, 1531–1551.
18. Quang Huy, T., van Quy, N., & Anh-Tuan, L. (2013). Silver nanoparticles: Synthesis, properties, toxicology, applications and perspectives. *Advances in Natural Sciences: Nanoscience and Nanotechnology*, *4*, 033001.
19. Prabhu, S., & Poulose, E. K. (2012). Silver nanoparticles: Mechanism of antimicrobial action, synthesis, medical applications, and toxicity effects. *International Nano Letters*, *2*, 32.
20. Rai, M., Yadav, A., & Gade, A. (2009). Silver nanoparticles as a new generation of antimicrobials. *Biotechnology Advances*, *27*, 76–83.
21. Gibier, P. (1890). Dr. Joch's discovery. *North American Review*, *151*, 726–731.
22. Benedek, T. G. (2004). The history of gold therapy for tuberculosis. *Journal of History of Medicine and Allied Sciences*, *59*, 50–89.
23. Forestier, J. (1932). The treatment of rheumatoid arthritis with gold salts injections. *Lancet*, *219*, 441–444.
24. Yeh, Y. C., Creran, B., Rotello, V. M. (2012). Gold nanoparticles: Preparation, properties, and applications in bionanotechnology. *Nanoscale*, *4*(6), 1871–1880.
25. Madkour, L. H. (2018). Applications of gold nanoparticles in medicine and therapy. *Pharmacy and Pharmacology International Journal*, *6*(3), 157–174.
26. Nadeem, M., Abbasi, B. H., Younas, M., Ahmad, W., & Khan, T. (2017). A review of the green syntheses and anti-microbial applications of gold nanoparticles. *Green Chemistry Letters and Reviews*, *10*(4), 216–227.
27. Król, A., Pomastowski, P., Rafińska, K., Railean-Plugaru, V., & Buszewski, B. (2017). Zinc oxide nanoparticles: Synthesis, antiseptic activity and toxicity mechanism. *Advances in Colloid and Interface Science*, *249*, 37–52.
28. Agarwal, H., Venkat Kumar, S., & Rajeshkumar, S. (2017). A review on green synthesis of zinc oxide nanoparticles: An eco-friendly approach. *Resource-Efficient Technologies*, *3*(4), 406–413.
29. Maret, W. (2011). Metals on the move: zinc ions in cellular regulation and in the coordination dynamics of zinc proteins. *Biometals*, *24*, 411–418.
30. Tapiero, H., & Tew, K. D. (2003). Trace elements in human physiology and pathology: Zinc and metallothioneins. *Biomedicine and Pharmacotherapy*, *57*, 399–411.
31. Frederickson, C. J., Koh, J. Y., & Bush, A. I. (2005). The neurobiology of zinc in health and disease. *Nature Reviews Neuroscience*, *6*, 449–62.
32. Pomastowski, P., Sprynskyy, M., & Buszewski, B. (2014). The study of zinc ions binding to casein. *Colloids and Surfaces B: Biointerfaces*, *120*, 21–7.
33. Fan, Z., & Lu, J. G. (2005). Zinc oxide nanostructures: synthesis and properties. *Journal of Nanoscience Nanotechnology*, *5*, 1561–73.
34. Liu, D. et al. (2008). Surface functionalization of ZnO nanotetrapods with photoactive and electroactive organic monolayers. *Langmuir*, *24*(9), 5052–5059.
35. Bisht, G., & Rayamajhi, S. (2016). ZnO nanoparticles: A promising anticancer agent. *Nanobiomedicine (Rij)*, *3*, 9–20.
36. Sonia, S., Ruckmani, K., & Sivakumar, M. (2017). Antimicrobial and antioxidant potentials of biosynthesized colloidal zinc oxide nanoparticles for a fortified cold cream formulation: A potent nanocosmeceutical application. *Materials Science and Engineering: C*, *79*, 581–589.
37. Skalickova, S., Milosavljevic, V., Cihalova, K., Horky, P., Richtera, L., & Adam, V. (2017). Selenium nanoparticles as a nutritional supplement. *Nutrition*, *33*, 83–90.
38. Yu, B., Zhang, Y., Zheng, W., Fan, C., & Chen, T. (2012). Positive surface charge enhances selective cellular uptake and anticancer efficacy of selenium nanoparticles. *Inorganic Chemistry*, *51*(16), 8956–8963.
39. Forootanfar, H., Adeli-Sardou, M., Nikkhoo, M., Mehrabani, M., Amir-Heidari, B., Shahverdi, A. R., & Shakibaie, M. (2014). Antioxidant and cytotoxic effect of biologically synthesized selenium nanoparticles in comparison to selenium dioxide. *Journal of Trace Elements in Medicine and Biology*, *28*(1), 75–79.
40. Shakibaie, M., Forootanfar, H., Golkari, Y., Mohammadi-Khorsand, T., & Shakibaie, M. R. (2015). Anti-biofilm activity of biogenic selenium nanoparticles and selenium dioxide against clinical isolates of *Staphylococcus aureus*, *Pseudomonas aeruginosa*, and *Proteus mirabilis*. *Journal of Trace Elements in Medicine and Biology*, *29*, 235–241.
41. Chandran, T., Arivarasu, L., & Rajeshkumar, S. (2020). Green synthesis, anti inflammatory and antioxidant property of *Cardiospermum halicacabum* mediated copper nanoparticle. *Plant Cell Biotechnology and Molecular Biology*, 17–21.
42. Kishore, S. O. G., Priya, A. J., Narayanan, L., Kumar, S. R., & Devi, G. (2020). Controlling of oral pathogens using turmeric and tulsi herbal formulation mediated copper nanoparticles. *Plant Cell Biotechnology and Molecular Biology*, 33–37.
43. Ahmed, S., Saifullah, A. M., Swami, B. L., & Ikram, S. (2016). Green synthesis of silver nanoparticles using *Azadirachta indica* aqueous leaf extract. *Journal of Radiation Research and Applied Sciences*, *9*(1), 1–7.
44. Garibo, D., Borbón-Nuñez, H. A., de León, J. N. D., Mendoza, E. G., Estrada, I., Toledano-Magaña, Y., … & Susarrey-Arce, A. (2020). Green synthesis of silver nanoparticles using *Lysiloma acapulcensis* exhibit high-antimicrobial activity. *Scientific Reports*, *10*(1), 1–11.
45. Menon, S., Agarwal, H., Kumar, S. R., & Kumar, S. V. (2017). Green synthesis of silver nanoparticles using medicinal plant *Acalypha indica* leaf extracts and its application as an antioxidant and antimicrobial agent against foodborne pathogens. *International Journal of Applied Pharmacy*, *9*(5), 42–50.

46. GnanaJobitha, G., Annadurai, G., & Kannan, C. (2012). Green synthesis of silver nanoparticle using *Elettaria cardamomom* and assessment of its antimicrobial activity. *International Journal of Pharmaceutical Sciences and Research (IJPSR), 3*(3), 323–330.
47. Krithiga, N., Rajalakshmi, A., & Jayachitra, A. (2015). Green synthesis of silver nanoparticles using leaf extracts of *Clitoria ternatea* and *Solanum nigrum* and study of its antibacterial effect against common nosocomial pathogens. *Journal of Nanoscience, 2015.*
48. Xin Lee, K., Shameli, K., Miyake, M., Kuwano, N., Bt Ahmad Khairudin, N. B., Bt Mohamad, S. E., & Yew, Y. P. (2016). Green synthesis of gold nanoparticles using aqueous extract of garcinia mangostana fruit peels. *Journal of Nanomaterials, 2016*, 8489094, 7 pages.
49. Wongyai, K., Wintachai, P., Maungchang, R., & Rattanakit, P. (2020). Exploration of the antimicrobial and catalytic properties of gold nanoparticles greenly synthesized by *Cryptolepis buchanani* Roem. and Schult extract. *Journal of Nanomaterials, 2020.*
50. Folorunso, A., Akintelu, S., Oyebamiji, A. K., Ajayi, S., Abiola, B., Abdusalam, I., & Morakinyo, A. (2019). Biosynthesis, characterization and antimicrobial activity of gold nanoparticles from leaf extracts of *Annona muricata. Journal of Nanostructure in Chemistry*, 9(2), 111–117.
51. Anbu, P., Gopinath, S. C., & Jayanthi, S. (2020). Synthesis of gold nanoparticles using *Platycodon grandiflorum* extract and its antipathogenic activity under optimal conditions. *Nanomaterials and Nanotechnology, 10*, 184798 0420961697.
52. Rajeshkumar, S., Lakshmi, T., Tharani, M., & Sivaperumal, P. (2020). Green synthesis of gold nanoparticles using pomegranate peel extract and its antioxidant and anticancer activity against liver cancer cell line. *Alinteri Journal of Agricultural Sciences, 35*(2), 164–169.
53. Thi, T. U. D., Nguyen, T. T., Thi, Y. D., Thi, K. H. T., Phan, B. T., & Pham, K. N. (2020). Green synthesis of ZnO nanoparticles using orange fruit peel extract for antibacterial activities. *RSC Advances, 10*(40), 23899–23907.
54. Demissie, M. G., Sabir, F. K., Edossa, G. D., & Gonfa, B. A. (2020). Synthesis of zinc oxide nanoparticles using leaf extract of *Lippia adoensis* (koseret) and evaluation of its antibacterial activity. *Journal of Chemistry, 2020.*
55. Aldalbahi, A., Alterary, S., Ali Abdullrahman Almoghim, R., Awad, M. A., Aldosari, N. S., Fahad Alghannam, S., … & Abdulrahman Alrashed, R. (2020). Greener synthesis of zinc oxide nanoparticles: Characterization and multifaceted applications. *Molecules, 25*(18), 4198.
56. Mehar, S., Khoso, S., Qin, W., Anam, I., Iqbal, A., & Iqbal, K. (2019). Green synthesis of zinc oxide nanoparticles from *Peganum harmala*, and its biological potential against bacteria. *Frontiers in Nanoscience and Nanotechnology, 6*, 1–5.
57. Modi, S., & Fulekar, M. H. (2020). Green synthesis of zinc oxide nanoparticles using garlic skin extract and its characterization. *Journal of Nanostructures, 10*(1), 20–27.
58. Gunti, L., Dass, R. S., & Kalagatur, N. K. (2019). Phytofabrication of selenium nanoparticles from *Emblica officinalis* fruit extract and exploring its biopotential applications: Antioxidant, antimicrobial, and biocompatibility. *Frontiers in Microbiology, 10*, 931.
59. Rajeshkumar, S., Tharani, M., & Sivaperumal, P. (2020). Green synthesis of selenium nanoparticles using black tea (*Camellia sinensis*) and its antioxidant and antimicrobial activity. *Journal of Complementary Medicine Research, 11*(5), 75–82.
60. Jay, V., & Shafkat, R. (2018). Synthesis of selenium nanoparticles using *Allium sativum* extract and analysis of their antimicrobial property against gram positive bacteria. *Pharma Innovation, 7*(9), 262–266.
61. Fritea, L., Laslo, V., Cavalu, S., Costea, T., & Vicas, S. I. (2017). Green biosynthesis of selenium nanoparticles using parsley (*Petroselinum crispum*) leaves extract. *Studia Universitatis "Vasile Goldis" Arad. Seria Stiintele Vietii (Life Sciences Series), 27*(3), 203–208.
62. Fardsadegh, B., & Jafarizadeh-Malmiri, H. (2019). Aloe vera leaf extract mediated green synthesis of selenium nanoparticles and assessment of their in vitro antimicrobial activity against spoilage fungi and pathogenic bacteria strains. *Green Processing and Synthesis, 8*(1), 399–407.
63. Wu, S., Rajeshkumar, S., Madasamy, M., & Mahendran, V. (2020). Green synthesis of copper nanoparticles using Cissus vitiginea and its antioxidant and antibacterial activity against urinary tract infection pathogens. *Artificial Cells, Nanomedicine, and Biotechnology, 48*(1), 1153–1158.
64. Rajeshkumar, S., Tharani, M., Jeevitha, M., & Santhoshkumar, J. (2019). Anticariogenic activity of fresh aloe vera gel mediated copper oxide nanoparticles. *Indian Journal of Public Health Research & Development, 10*(11).
65. Sukumar, S., Rudrasenan, A., & Padmanabhan Nambiar, D. (2020). Green-synthesized rice-shaped copper oxide nanoparticles using *Caesalpinia bonducella* seed extract and their applications. *ACS Omega, 5*(2), 1040–1051.
66. Nirmala, M., Kavitha, B., Nasreen, S., Kiruthika, S., Divya, B., & Manochitra, E. (2020). Green synthesis of copper nanoparticles using *Ocimum sanctum* L. (tulsi) and *Piper nigrum* L. (pepper seed) for pollution free environment. *World Scientific News, 150*, 105–117.
67. Murthy, H. C., Desalegn, T., Kassa, M., Abebe, B., & Assefa, T. (2020). Synthesis of green copper nanoparticles using medicinal plant *Hagenia abyssinica* (Brace) JF. Gmel. Leaf extract: Antimicrobial properties. *Journal of Nanomaterials, 2020.*
68. Kanagasubbulakshmi, S., & Kadirvelu, K. (2017). Green synthesis of iron oxide nanoparticles using *Lagenaria siceraria* and evaluation of its antimicrobial activity. *Defence Life Science Journal, 2*(4), 422–427.
69. Swathi, N., Sandhiya, D., Rajeshkumar, S., & Lakshmi, T. (2019). Green synthesis of titanium dioxide nanoparticles using *Cassia fistula* and its antibacterial activity. *International Journal of Research in Pharmaceutical Sciences, 10*(2), 856–860.
70. Kumaresan, M., Anand, K. V., Govindaraju, K., Tamilselvan, S., & Kumar, V. G. (2018). Seaweed *Sargassum wightii* mediated preparation of zirconia (ZrO2) nanoparticles and their antibacterial activity against gram positive and gram negative bacteria. *Microbial Pathogenesis, 124*, 311–315.
71. Prasanth, R., Dinesh Kumar, S., Jayalakshmi, A., Singaravelu, G., Govindaraju, K., & Ganesh Kumar, V. (2019). Green synthesis of magnesium oxide nanoparticles and their antibacterial activity. *Indian Journal of Geo Marine Sciences, 48*(8).

72. Aseyd Nezhad, S., Es-haghi, A., & Tabrizi, M. H. (2020). Green synthesis of cerium oxide nanoparticle using *Origanum majorana* L. leaf extract, its characterization and biological activities. *Applied Organometallic Chemistry, 34*(2), e5314.
73. Paciotti, G. F., Myer, L., Weinreich, D., Goia, D., Pavel, N., McLaughlin, R. E., & Tamarkin, L. (2004). Colloidal gold: A novel nanoparticle vector for tumor directed drug delivery. *Drug Delivery, 11*(3), 169–183.
74. Muthukumar, T., Sambandam, B., Aravinthan, A., Sastry, T. P., & Kim, J. H. (2016). Green synthesis of gold nanoparticles and their enhanced synergistic antitumor activity using HepG2 and MCF7 cells and its antibacterial effects. *Process Biochemistry, 51*(3), 384–391.
75. Asha, S., Thirunavukkarasu, P., & Rajeshkumar, S. (2017). Green synthesis of silver nanoparticles using *Mirabilis jalapa* aqueous extract and their antibacterial activity against respective microorganisms. *Research Journal of Pharmacy and Technology, 10*(3), 811–817.
76. Bhuyar, P., Rahim, M. H. A., Sundararaju, S., Ramaraj, R., Maniam, G. P., & Govindan, N. (2020). Synthesis of silver nanoparticles using marine macroalgae *Padina* sp. and its antibacterial activity towards pathogenic bacteria. *Beni-Suef University Journal of Basic and Applied Sciences, 9*(1), 1–15.
77. Vadlapudi, V., & Amanchy, R. (2017). Phytofabrication of silver nanoparticles using *Myriostachya wightiana* as a novel bioresource, and evaluation of their biological activities. *Brazilian Archives of Biology and Technology, 60.*
78. Murthykumar, K., & Malaiappan, S. (2020). Antioxidant and antibacterial effect of lycopene mediated silver nanoparticle against *Staphylococcus aureus* and *Streptococcus mutans*: An in vitro study. *Plant Cell Biotechnology and Molecular Biology*, 90–98.
79. Cittrarasu, V., Kaliannan, D., Dharman, K., Maluventhen, V., Easwaran, M., Liu, W. C., … & Arumugam, M. (2021). Green synthesis of selenium nanoparticles mediated from *Ceropegia bulbosa* Roxb extract and its cytotoxicity, antimicrobial, mosquitocidal and photocatalytic activities. *Scientific Reports, 11*(1), 1–15.
80. Menon, S., Agarwal, H., Rajeshkumar, S., Rosy, P. J., & Shanmugam, V. K. (2020). Investigating the antimicrobial activities of the biosynthesized selenium nanoparticles and its statistical analysis. *Bionanoscience, 10*(1), 122–135.
81. Gupta, M., Tomar, R. S., Kaushik, S., Mishra, R. K., & Sharma, D. (2018). Effective antimicrobial activity of green ZnO nano particles of *Catharanthus roseus. Frontiers in Microbiology, 9*, 2030.
82. Rajeshkumar, S., Sivaperumal, P., Tharani, M., & Lakshmi, T. (2020). Green synthesis of zinc oxide nanoparticles by cardiospermum. *Journal of Complementary Medicine Research, 11*(5).
83. Shende, S., Ingle, A. P., Gade, A., & Rai, M. (2015). Green synthesis of copper nanoparticles by *Citrus medica* Linn. (Idilimbu) juice and its antimicrobial activity. *World Journal of Microbiology and Biotechnology, 31*(6), 865–873.
84. Mali, S. C., Dhaka, A., Githala, C. K., & Trivedi, R. (2020). Green synthesis of copper nanoparticles using *Celastrus paniculatus* Willd. leaf extract and their photocatalytic and antifungal properties. *Biotechnology Reports, 27*, e00518.
85. Morens, D. M., & Fauci, A. S. (2013). Emerging infectious diseases: Threats to human health and global stability. *PLoS Pathogens, 9*(7), e1003467.
86. Fauci, A. S., & Morens, D. M. (2012). The perpetual challenge of infectious diseases. *New England Journal of Medicine, 366*(5), 454–461.
87. Parrish, C. R., Holmes, E. C., Morens, D. M., Park, E. C., Burke, D. S., Calisher, C. H., … & Daszak, P. (2008). Cross-species virus transmission and the emergence of new epidemic diseases. *Microbiology and Molecular Biology Reviews, 72*(3), 457–470.
88. Sondi, I., & Salopek-Sondi, B. (2004). Silver nanopartiklar som antimikrobiellt medel: En fallstudie på E. coli som odel för gramnegativa bakterier. *Journal of Colloid and Interface Science, 275*(1), 177–182.
89. Kim, J. S., Kuk, E., Yu, K. N., Kim, J. H., Park, S. J., Lee, H. J., … & Cho, M. H. (2007). Antimicrobial effects of silver nanoparticles. *Nanomedicine: Nanotechnology, Biology, and Medicine, 3*, 95–101.
90. Jafari, A. R., Mosavi, T., Mosavari, N., Majid, A., Movahedzade, F., Tebyaniyan, M., … & Arastoo, S. (2016). Mixed metal oxide nanoparticles inhibit growth of Mycobacterium tuberculosis into THP-1 cells. *International Journal of Mycobacteriology, 5*, S181–S183.
91. Vijayakumar, S., & Ganesan, S. (2012). Gold nanoparticles as an HIV entry inhibitor. *Current HIV research, 10*(8), 643–646.
92. Ahmed, S. R., Kim, J., Suzuki, T., Lee, J., & Park, E. Y. (2016). Detection of influenza virus using peroxidase-mimic of gold nanoparticles. *Biotechnology and Bioengineering, 113*(10), 2298–2303.
93. Patil, B. N., & Taranath, T. C. (2016). Limonia acidissima L. leaf mediated synthesis of zinc oxide nanoparticles: A potent tool against Mycobacterium tuberculosis. *International Journal of Mycobacteriology, 5*(2), 197–204.
94. Bhattacharyya, P., Agarwal, B., Goswami, M., Maiti, D., Baruah, S., & Tribedi, P. (2018). Zinc oxide nanoparticle inhibits the biofilm formation of *Streptococcus pneumonia. Antonie Van Leeuwenhoek, 111*(1), 89–99.
95. Javed, B. (2020). Synergistic effects of physicochemical parameters on bio-fabrication of mint silver nanoparticles: Structural evaluation and action against HCT116 colon cancer cells. *International Journal of Nanomedicine, 15*, 3621.
96. Alagesan, V., & Venugopal, S. (2019). Green synthesis of selenium nanoparticle using leaves extract of *Withania somnifera* and its biological applications and photocatalytic activities. *Bionanoscience, 9*(1), 105–116.
97. Rajasekar, S., & Kuppusamy, S. (2020). Eco-friendly formulation of selenium nanoparticles and its functional characterization against breast cancer and normal cells. *Journal of Cluster Science*, 1–9.
98. Huo, C., Xiao, J., Xiao, K., Zou, S., Wang, M., Qi, P., … & Hu, Y. (2020). Pre-treatment with zirconia nanoparticles reduces inflammation induced by the pathogenic H5N1 influenza virus. *International Journal of Nanomedicine, 15*, 661.
99. Zahir, A. A., Chauhan, I. S., Bagavan, A., Kamaraj, C., Elango, G., Shankar, J., … & Singh, N. (2015). Green synthesis of silver and titanium dioxide nanoparticles using *Euphorbia prostrata* extract shows shift from apoptosis to G0/G1 arrest followed by necrotic cell death in *Leishmania donovani. Antimicrobial Agents and Chemotherapy, 59*(8), 4782–4799.

2

Asthma Epidemiology, Etiology, Pathophysiology and Management in the Current Scenario

Manu Sharma, Aishwarya Rathore, Sheelu Sharma, Kakarla Raghava Reddy, Veera Sadhu and Raghavendra V. Kulkarni

CONTENTS

2.1 Introduction

Asthma is one of the most common chronic lower respiratory diseases associated with airway hyper-responsiveness which results in recurrent episodes of wheezing, chest tightness, persistent cold and cough, breathlessness, etc. It is a major healthcare problem observed in both children and adults. The disease is non-communicable and greatly affects the quality of life of the patient. Worldwide 339 million people suffer from asthma and the high mortality rate is observed in the elderly and patients belonging to the low-/middle-income group [1]. Elderly patients have a high mortality rate because of poor diagnosis of the disease. Many elderly citizens confuse asthma symptoms with the signs of aging and end up with a severe asthmatic condition. Symptoms of asthma aggravate over time if they remain untreated for a long period of time. It is predicted by many healthcare professionals and researchers that the disease is likely to shoot up to 439 million patients by the end of 2025 [2].

The Global Initiative for Asthma (GINA) classified asthma into four major types based on the severity of the disease, i.e. mild intermittent, mild persistent, moderate persistent and severe persistent. Mild intermittent is the least severe whereas severe persistent asthma is the most severe form of asthma (Table 2.1) [3, 4]. Despite various advances and modernization in healthcare treatments and facilities, there is still no permanent cure for the disease. Current treatments can provide only symptomatic treatment with a decrease in the progression of the disease. However, there is still a long way to go to improve current treatments, diagnostic tools and patient awareness.

2.2 Epidemiology of Asthma

The disease has shown comparatively lower prevalence (2–4%) in Asian countries like India and a high prevalence rate (15–20%) in developed countries like Canada, the United Kingdom, Australia, etc. [5]. However, the mortality rates are higher for the Asian countries like India compared to the developed countries. Worldwide nearly 11.6% of children aged between six and seven years suffer from asthma. The disease is more prevalent in boys than girls [3]. In the United States alone, there were 12.9% (25 million) total reported cases of asthma in both children and adults. The ethnicity of individuals also plays an important role in the prevalence rate of the disease. In the United States, it was observed that the disease is more prevalent in Puerto Rican Hispanic (13.3%) and black (8.7%) than white individuals (7.6%) [3, 5].

DOI: 10.1201/9781003157823-2

TABLE 2.1
Types of Asthma

Type of asthma	Symptom recurrence
Mild intermittent	Twice a week
Mild persistent	More than twice a week but not more than once a day
Moderate persistent	Once a day and more than once in night
Severe persistent	Throughout the day and frequently at night

Asthma prevalence and severity also differ based on gender. After puberty, women are twice susceptible to asthma as men of the same age. Asthma also depends on the BMI of an individual and the risk and susceptibility of the disease increase with an increase in obesity. Moreover, it was also observed that 41% of asthmatic patients had a history of smoking in the past [6].

2.3 Etiology and Risk Factors of Asthma

The main fundamental cause of asthma is unknown but it is believed that the highest risk factor for developing asthma is a combination of genetic predisposition and environmental exposure to certain pollutants and allergens.

It has been widely observed that asthma runs in families and recently a genome-wide association study has identified the ORMDL3 gene, which is significantly associated with asthma [7]. Some children develop asthma due to prenatal maternal smoking which drastically affects the lungs of the newborn and leads to breathing difficulties. Cigarette smoking is also a main cause of asthma in adults and nearly 41% of asthmatics in the United States had a history of smoking [8, 9]. Asthma is also one of the occupational hazards in places where people are exposed to high fumes and dust of cadmium, silica, flour, etc. Allergens like food, dust, pollen and pet dander provoke allergic asthma [10]. Air pollution is another crucial reason for asthma and an increase in air pollutants due to industrialization has eventually increased the rate of asthma worldwide.

2.4 Pathophysiology of Asthma

Asthma is chronic recurrent bronchospasm that occurs on the encounter of certain triggering agents. The pathophysiology of asthma is quite complex and has many overlapping pathophysiologies which eventually lead to airway obstruction, airway hyper-responsiveness, bronchospasm and inflammatory reactions. As illustrated in Figure 2.1, asthma is caused by the recruitment and activation of immune cells like mast cells, eosinophils, T-lymphocytes, neutrophils, dendritic cells, etc. that cause inflammatory and cellular infiltration in the airways causing asthma. Asthma is associated with specifically Th2 (T helper cell type-2) immune response. The TH2 releases IL-4, IL-5 and IL-13 which in turn activates the B-cells to mature into IgE-producing plasma cells. The IgE cells bind to mast cells and on exposure to a certain allergen, the IgE cell provokes the mast cell to degranulate and release chemical mediators like histamine, cysteinyl leukotriene, etc. These released chemical mediators remodel the airway with deposition of extracellular protein, increased goblet cell which is responsible for mucus production, and smooth muscle hypertrophy which causes bronchospasm. Frequent exposure to the allergen decreases lung function and hence frequent assessment and

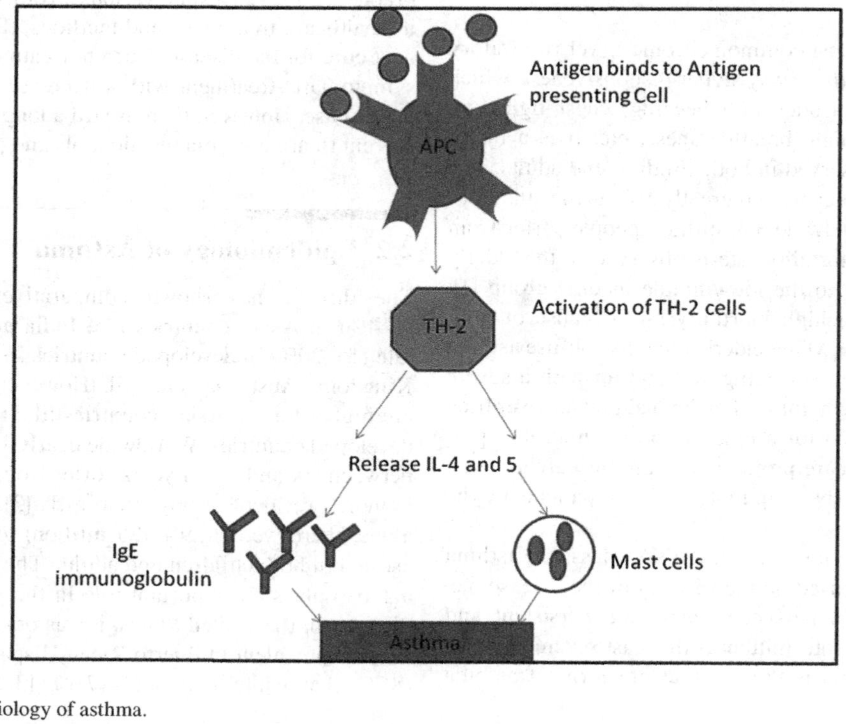

FIGURE 2.1 Pathophysiology of asthma.

avoiding exposure to the allergen are important for asthmatic patients [11].

Evidence also suggests that asthma can be inherited through genes. The mutated gene which increases the production of T helper 2 cells and production of IgE cells are majorly responsible for asthma. Another major reason identified for asthma is smoking. The toxic fumes of tobacco can themselves act as an allergen and activate the cascade of immune response. Tobacco smoking can also damage the epithelial lining and cilia of the lungs aggravating the asthmatic condition [12].

2.4.1 Diagnosis of Asthma

Despite being one of the common pulmonary diseases, asthma remains undiagnosed in many patients due to a lack of knowledge in patients and inadequate diagnosis technique. Many people confuse the symptoms of asthma with the signs of weakness and aging and delay the diagnosis step. There are different diagnostic tests available for asthma (Figure 2.2) but spirometry is the most reliable diagnostic test. It measures the total amount of air exhaled per second and the speed of exhalation. The limitation of spirometry is its inability to test children. Another diagnostic test is the peak flow test. The peak flow test measures how fast a person breathes out before and after administration of a bronchodilator. Generally, asthmatic patients have a lower peak flow value which improves with the use of a bronchodilator. Both these tests are grouped under the lung function test [13, 14].

Apart from the lung function test, there are many additional tests to diagnose asthma and these tests include fractional exhaled nitric oxide test, methacholine challenge, X-ray imaging, allergy testing, provocative test for exercise, and cold-induced asthma and sputum test. The fractional exhaled nitric oxide test is a quick, non-invasive test that measures the amount of nitric oxide in exhaled breath. The level of nitric oxide is higher in patients suffering from asthma. In methacholine challenge and allergy testing, the patient is exposed to methacholine which is a direct stimulant of airway smooth muscle and a triggering agent [15, 16]. Asthmatic patients would react to methacholine and face mild constriction of airways [14]. Similar to the methacholine test, an allergy test can also be performed by exposing the patient to certain allergens followed by a skin test and blood test. The provocative test for exercise and cold-induced asthma involves testing of exercise and cold bronchoconstriction. The test involves measuring the airway obstruction before and after a patient is subject to rigorous exercise or several breaths of cold air. Apart from this, a simple blood and sputum test can also be used as a confirmatory test. For the blood sample test, a small peg-like device is attached to the finger of a patient and the device measures the amount of oxygen present in the blood. In the sputum test, the patient's sputum or mucus is tested for the presence of eosinophils with the help of eosin dye.

Chest X-ray imaging and high-resolution computerized tomography can also be used to image lungs and identify any abnormality or presence of pulmonary disease.

2.4.2 Current Treatments Available for Asthma

2.4.2.1 First-Line Allopathic Treatments

Bronchodilators, steroids, mucolytic agents and oxygen therapy are first-line treatments for asthma. They are prescribed based on the severity of the disease, age, presence of other co-morbid diseases, etc. These therapies provide symptomatic treatments and can only relieve a patient from future exacerbations.

Bronchodilators are the primary treatment medicines for asthma. They facilitate dilation of the airway by relaxing the bronchial muscle and help the patient to breathe without difficulty. There are two forms of bronchodilator: short-acting bronchodilator and long-acting bronchodilator. Short-acting bronchodilators are also called rescue medicines as they immediately relieve the patient from bronchoconstriction. They work within 10–15 min and their effect lasts for 4–6 h whereas long-acting bronchodilators do not provide immediate

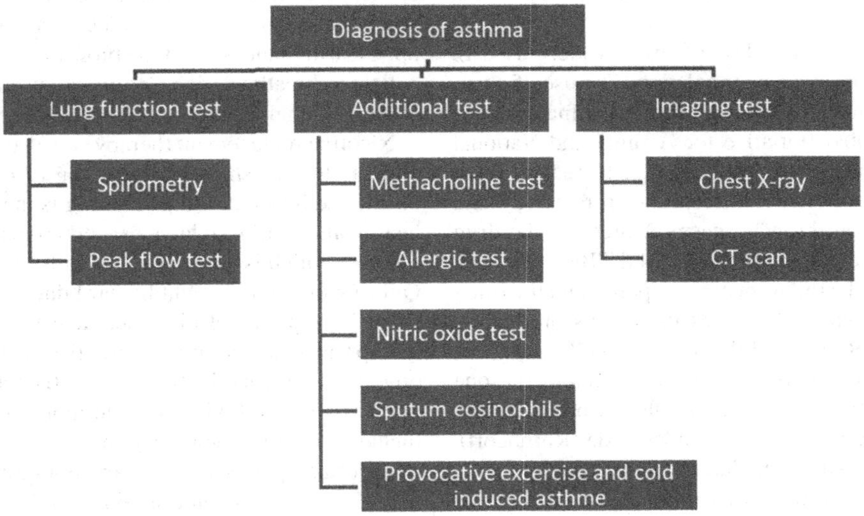

FIGURE 2.2 Diagnosis tests available for asthma.

relief but have a longer duration of action. Generally, bronchodilators can be classified into beta-adrenergic bronchodilators, anticholinergic bronchodilators and xanthine derivatives. Beta-adrenergic bronchodilators stimulate the β-2 adrenergic receptor which relaxes the bronchial smooth muscle by decreasing intracellular calcium in the lung, increasing potassium membrane conductance, activating myosin light-chain phosphatase and inactivating myosin light-chain kinase. Beta-adrenergic bronchodilators are highly prescribed medication and show good pharmacological activity in both children and adults. Available beta-adrenergic bronchodilators include albuterol, pirbuterol, levalbuterol, salmeterol and formoterol. Nowadays there are many combinations of short-acting and long-acting bronchodilators available in the market to increase patient compliance. In combination therapy, a short-acting bronchodilator provides immediate effect while a long-acting bronchodilator provides the effect for a longer duration of time. Examples of some marketed products are Adavir (combination of salmeterol and fluticasone) and Symbicort (combination of formoterol and budesonide). Side effects of beta-2 agonists include nervousness, hyperactivity, tachycardia and insomnia [17–19].

Another class of bronchodilators available in the market is anticholinergic bronchodilators. These bronchodilators block the action of acetylcholine neurotransmitters on the M3 receptor and result in bronchodilation. Marketed anticholinergics include Atrovent HFA (ipratropium bromide) and Spiriva (tiotropium bromide). Atrovent HFA should be administered four times a day whereas Spiriva is administered only once a day. They cannot be used as a rescue medication and their side effects include dry throat, constipation, urinary retention, dizziness, etc. [20].

Xanthine derivatives are the agents which resemble caffeine, theobromine and methylxanthines. It relieves from bronchoconstriction and also suppresses the exaggerated response of bronchial tubes to the triggering agents. They block the phosphodiesterase enzyme and cause bronchodilation. Theophylline and aminophylline are widely used xanthine derivatives. Common side effects associated with xanthine derivatives can be headache, nausea, palpitation, flushing, etc. [21].

Corticosteroids are another class of drugs widely used in asthma management orally or by inhalation. The use of corticosteroids in the treatment of acute persistent asthma has been recommended by Global Initiative for Asthma and National Asthma Education and Prevention Program. It treats asthma by enhancing the beta-adrenergic response in the body and relieves the patient from bronchospasm. Moreover, the drug also inhibits the release of leukotriene LTC4, LTB4 and LTD4, which in turn inhibits the inflammatory response and the function of secretagogue which decreases the mucus production and downregulation of eosinophils and mast cell activation. Prednisolone is an oral corticosteroid whereas beclomethasone (Qvar, Qnasl), mometasone (Asmanex), fluticasone (Arnuity Ellipta), flunisolide (Aerospan HFA), budesonide (Rhinocort), ciclesonide (Omnaris, Alvesco, Zetonna) and triamcinolone are widely used as inhalation corticosteroids. Corticosteroids are used for only a short period of time as long-term usage of the drug results in severe lethal side effects which include osteoporosis, increase in blood sugar, adrenal suppression, weight gain, psychosis, hypertension, etc. [22, 23].

Leukotriene-modifying drugs are also used in the treatment of asthma. They inhibit the synthesis of leukotriene or inhibit the binding of leukotriene to leukotriene receptors. Leukotrienes are lipid mediator which is responsible for bronchoconstriction, inhibiting its action can relieve the patient from bronchospasm. Montelukast (Singulair), zileuton (Zyflo) and zafirlukast (Accolate) are the common leukotriene-modifying drugs available on the market. In certain cases, these drugs can be associated with side effects like hallucination, depression and suicidal tendencies [24].

In cases of extreme asthma with recurrent exacerbations, the patient might need oxygen therapy. Oxygen therapy helps to overcome hypoxia in an asthmatic patient. The patient inhales saturated oxygen (90%) with the help of a nebulizer or venture mixture.

2.4.2.2 Additional Allopathic and Surgical Therapies

Apart from the first-line treatment mentioned above, the patient can also be administered additional medications to relieve and prevent further exacerbation. The additional therapies include surgical methods and medications like mucolytic agents, antibacterial agents and nicotine replacement therapy.

Surgical procedures are available for permanent cure of asthma but due to very poor success rates, they are rarely performed. Bronchial thermoplasty is one of the main surgical methods to treat asthma. It is an endoscopic procedure in which a small tube called a bronchoscope is placed down the bronchial tubes through the mouth. The tube is heated gently to shrink the bronchial smooth muscle and decrease its ability to contract aggressively on encountering an allergen in future [25–27].

Sometimes during severe persistent asthma, bronchodilators and corticosteroids may not be enough to provide relief to patients and in such cases, the patient might be administered with mucolytic agents as an additional therapy to relieve bronchospasm in asthmatic patients. Mucolytic agents break down the mucus and make it less viscous and easier to cough up. Clearing mucus from the bronchial tubes can also help the patient to breathe easily. Commonly used mucolytic drugs are N-Acetylcysteine, carbocysteine, erdosteine, etc. [28, 29].

Nicotine replacement therapy can also be used by asthmatic patients to quit smoking. Smoking can trigger an asthmatic attack and hence quitting smoking is the best way to improve health and avoid sudden exacerbations. Cigarettes contain nicotine which is addictive and leads to physical dependence. Quitting cigarettes might be hard due to the unpleasant withdrawal symptoms of nicotine. Hence a nicotine replacement therapy provides a good alternative to nicotine without any other harmful chemicals present in the tobacco. Currently, the market is flooded with nicotine replacement therapies which include nicotine chewing gums, lozenges, nasal spray and transdermal patches. They are non-prescription drugs and help the patient to quit smoking. Table 2.2 summarizes all the first-line and additional allopathic treatments available for asthma.

TABLE 2.2

First-Line and Additional Allopathic Treatments for Asthma

Class of the drug	Example	Mechanism of action	Side effects	Ref.
Short-acting β2 agonist	Albuterol (AccuNeb, Proventil HFA, ProAir HFA, Ventolin HFA) and levalbuterol (Xopenex, Xopenex HFA)	Stimulates the beta-2 adrenergic receptor and relax bronchial muscle	Nervousness, headache, tremors, tachycardia and insomnia	[22]
Long-acting β2 agonist	Arformoterol (Brovana), Formoterol (Perforomist, Oxeze, Foradil)			[22]
Muscarinic antagonists	Ipratropium bromide (Atrovent), Aclidinium (Tudorza)	Blocks the action of acetylcholine on M3 receptor and result in bronchodilation	Dry throat, headache, constipation, urinary retention	[20]
Xanthine derivatives	Theophylline and aminophylline	Block the phosphodiesterase enzyme and cause bronchodilation	Flushing, restlessness, insomnia, tachycardia	[21]
Corticosteroids	Beclomethasone (Qvar, Qnasl), mometasone (Asmanex), fluticasone (Arnuity Ellipta), flunisolide (Aerospan HFA), budesonide (Rhinocort), ciclesonide (Omnaris, Alvesco, Zetonna)	Enhance beta-adrenergic response, inhibits the release of Leuctrines LTC4, LTB4, and LTD4, inhibits inflammatory cascade, inhibit secretagogue and decreases mucus production, downregulates eosinophils and mast cell activation	Osteoporosis, glaucoma, increase in blood sugar, psychotic problems	[22]
Leukotrine modifiers	Montelukast (Singulair), zileuton (Zyflo) and zafirlukast (Accolate)	Inhibit leukotriene synthesis or binding leukotriene to leukotriene receptor	Hallucination, depression and suicidal tendencies	[24]
Mucolytic agents	N-Acetylcysteine, carbocysteine, erdosteine	Breaks down the mucus and make it less viscous and easier to cough out	Ulcers, throat irritation, frequent dosing, high first-pass metabolism	[29]
Nicotine replacement therapy	Nicorette, Nicorette Plus, Nicoderm	Nicotine alternative, helps in smoking cessation	Requires frequent dosing, irritation in mouth and bad taste especially after nicotine lozenges	[30]
Oxygen therapy	InogenOne G4 Portable Oxygen Concentrator	Provides direct oxygen in extreme cases	Explosion risk, expensive	[31]

2.5 Aromatherapy

Essential oils have a wide range of health benefits. They can help the bronchial muscle to relax in asthmatic patients. Although aromatherapy is only a side-line treatment for comforting the patients and cannot replace the first-line treatments, it can relieve stress-induced asthma and help to calm down the patient. The essential oils are found more beneficial when other aspects like diet and regular medication are given consideration. Aromatherapy is for external use and should not be ingested. It should be also avoided by patients who are allergic to essential oils. Common essential oils used to relax bronchial muscles and relieve chest tightness are described in Table 2.3.

2.6 Ayurvedic Treatments

According to Ayurveda, asthma occurs due to an imbalance in Kapha dosha. Ayurvedic treatments are widely accepted alternative treatment options over allopathic medicines. The main reason behind this is the inexpensiveness and safety of the medicine. Unlike allopathic medicines, Ayurvedic medicines are produced from herbs, vitamins, minerals and protein and hence have fewer side effects. The demand for Ayurvedic medicines has increased several folds throughout the world and more researchers are working with naturally based ingredients to treat chronic diseases like asthma. Table 2.4 summarizes the available Ayurvedic medicines for the treatment of asthma.

2.7 Yogas and Aasnas

Bronchial asthma is called "Thamaka Shvasa" in Ayurveda. According to Ayurveda, Tamaka Shvasa is caused by the aggravation of Kapha dosha which leads to obstruction in airflow.

As per Ayurvedic concepts, bronchial asthma is a Vatakaphaja disease and it begins in the stomach and progresses to the bronchial tubes and lungs. Hence for the treatment of asthma, the excess Kapha should be sent back to the stomach and then eliminated. Different yogas and asanas can help eliminate the excess of Kapha and maintain the balance between the doshas. It also improves lung function and decreases the number of flare-ups during the day as well as the night.

Pranayam is a breathing exercise and it controls the motion of inhalation and exhalation of energy. Practicing pranayam like Kapalabhati, AnulomaViloma (three-part breathing) and Ujayyi (breathing) improves respiratory efficiency and removes dead space volume of lungs. It also increases blood circulation and increases immunity. Bhujangasana, Savasana, Shalabhasana, Paschimotasana are useful asanas for Pranavaha Srotas as it accelerates the blood circulation of the lungs and thus increases the vital capacity of the lung [55].

TABLE 2.3

Essential Oils and Their Benefits in Asthma

S. no.	Essential oil	Activity	MOA in asthma	Ref.
1	Lavender	Relaxing and sedative properties	Relieves from airway inflammation and relaxes the bronchial muscle	[32]
2	Peppermint	Decongestant, antihistaminic	Relieves from airway inflammation, decrease mucus formation and prevent histamine release	[33]
3	Tea tree	Expectorant	Helps to cough out the mucus and makes airway clean	[34]
4	Eucalyptus	Mucolytic	Breaks down mucus and helps to cough up the mucus easily	[35, 36]
5	Chamomile	Antispasmodic	Relaxes constricted bronchial muscle	[37]
6	Clove	Anti-inflammatory, antispasmodic	Relaxes bronchial muscle and inhibits activation of inflammatory cascade	[36]
7	*Pistacia integerrima*	Antispasmodic and antihistaminic activity	Prevents histamine release and relaxes bronchial muscle	[38]
8	Rosemary	Relax smooth muscles	Relaxes constricted bronchial muscle	[39]
9	Camphor	Anti-inflammatory and mucolytic	Inhibits the activation of inflammatory cascade and relaxes bronchial muscle, clears mucus from the airway	[36]

According to Ayurveda, the patient with bronchial spasms should also perform Shodhana (biopurification). It should be performed in accordance with the patient's strength; strong patients may undergo Vamana, Virechana and Niruha basti as well as Nasya therapy in appropriate seasons. These types of therapies alleviate Kapha and may dry up excess mucous secretions in the Pranavaha Srotasa (respiratory system) [56, 57].

2.8 Diet

According to GINA (Global Initiative for Asthma), diet plays a very vital role in improving the immunity of the patient and preventing future flare-ups. A good nutritious healthy diet improves the overall health of the patient and helps the patient to not become obese or underweight. Food can also play a vital role as an allergen in many asthmatic patients and these patients should avoid the particular allergen to maintain good health. According to a survey by nutritionists in the United States, patients who consumed a high amount of vitamins C, D and E, beta-carotene, flavonoids, magnesium, selenium, nuts, fruits, vegetables and omega-3 fatty acids had lower rates of flare-ups compared to those who consumed fatty junk food rich in trans fat and omega-6-fatty acid [58, 59].

The patient should routinely visit a nutritionist and get a personalized diet chart to improve the asthmatic condition. Generally, food like green leafy vegetables, fruits, nuts, honey, beans and pulses, egg, milk, carrots, fish, broccoli and other sources of vitamin A, D, E, magnesium, etc. should be included in patients' diets after reviewing the patients' allergen profiles. The patients are also advised to avoid food like caffeine, spices, alcohol, etc. to avoid the development and exaggeration of gastroesophageal reflux disorder (GERD) [60, 61].

2.9 Recently Approved Monoclonal Antibodies for Asthma Treatment

A greater advancement in medical equipment and a better understanding of the pathophysiology of the disease has resulted in the development of many novel treatment techniques for better treatment of asthma.

An anti-IgE monoclonal antibody-like Omalizumab was recently approved by USFDA as an add-on therapy for severe persistent allergic asthmatic patients. Generally, patients with elevated serum IgE levels above six years of age are prescribed Omalizumab along with corticosteroids or other bronchodilators. Omalizumab has a good safety profile and is usually well tolerated by patients above six years of age [62].

Anti-interleukin-5 antibodies are other emerging treatments for asthma treatments. Eosinophils play a major role in asthma pathogenesis causing airway remodeling and airway inflammation leading to breathlessness. Anti-interleukin-5 monoclonal antibodies like reslizumab and benralizumab inhibit the action of interleukin-5 resulting in a decrease in the intensity of the flare-ups. It has been approved as an add-on therapy for patients above 12 years of age [63].

Similarly, other monoclonal antibodies like anti-IL-13 antibodies, anti-IL-3 antibodies and anti-IL-4 antibodies have also depicted positive results in treating asthma exacerbations. These monoclonal antibodies are highly promising and can be given as supplemental treatments.

2.10 Recent Research and Novel Treatments

Since allopathic synthetic drugs like long-acting β2 agonists, anticholinergics, phosphodiesterase (PDE) inhibitors, corticosteroids, etc. have many drawbacks like high cost, frequent dosing and higher side effects, physicians and patients have directed their interest towards natural and plant-based products. These products are cost-effective and have a broad range of pharmacological benefits with fewer side effects. These products are highly favored in the treatment of chronic disorders as they have fewer side effects even when administered for a longer period of time. Due to the increase in demand for plant-based products and their ever-increasing global market, many scientists have tried to study the pharmacological activity of the chemical constituent extracted from the natural source and tried to incorporate them into a better drug delivery system for better results. In the last few decades, a shift from

TABLE 2.4

Ayurvedic and Nutraceutical Therapies for Asthma Treatment

Name of the product	Manufacturing company	Main ingredients	Mechanism of action	Dose	Ref.
Vasadi syrup	Ayush Herbs Pharmaceuticals	Haridra, vasa, dhanyaka, bharangi, shunthi, guduchi, kantakari, pipppali	Antioxidant, mucolytic, expectorant bronchodilator	15ml thrice a day after meals	[40]
Shwasaghna dhuma	N/A	Kantakari, ajowan, dhatura, khurasani ajowan, kalmishora, haridra	Mucolytic, expectorant bronchodilator, antioxidant, immunity booster	2 puffs/1 g twice daily after meal	[40]
Curcumin capsule	Planet Ayurveda	*Curcuma longa*	Immune boosting, antioxidant, manages cough and wheezing	2 capsules twice daily after meal	[41]
Tulsi capsule	Planet Ayurveda	*Ocimum tenuiflorum*	Antioxidant	1 capsule twice daily after meal	[42]
Praanrakshak churna	Planet Ayurveda	*Albizia lebbeck*, *Adhatoda vasica*, *Tylphora asthmatica*, *Cinnamomum zeylanica*, *Clerodandrum serratum*, *Solanum xanthocarpum*, *Glycyrrhiza glabra*	Antioxidant, improves exchange of oxygen throughout the blood, improves lung function	1 teaspoon (tsp) twice daily after meal with water or honey	[43]
Khaas Har churna	Planet Ayurveda	*Bambusa arudinacea*, crystallized sugar lumps, *Piper longum*, *Eletteria cardamomum*, *Cinnamomum zeylanica*	Anti-inflammatory, antioxidant, antimicrobial, immune booster, improve lung function	1 tsp twice daily after meal with water or honey	[44]
Vasaka capsule	Planet Ayurveda	*Adhatoda vasica*	Expectorant, mucolytic properties, clear respiratory channels, antioxidant	2 capsules twice daily after meal	[45]
Sitopaladi churna	Dabur/Zandu ayurveda	Sugar crystals, pepper, vankshalochana, cardamom, cinnamon	Improves lung function, boost immunity, anti-inflammatory	1–2 g/day after meal	[46]
Pushkarmool (*Inula racemosa*)	Heilen Biopharm	Pushkarmool	Expectorant, antibacterial, mucolytic, antioxidant	2 grams/day	[47]
Abhrak bhasma	Patanjali Ayurveda	Abhrakpatra, harad, amla, baheda, dhaan, vatankur, kadlimool, ark mool vatjata/vatmool	Immunity booster, improves lung function, expectorant, anti-inflammatory	1–2 g/day	[48]
Tankan bhasma	Patanjali Ayurveda	Suhaga	Expectorant, anti-inflammatory	500 mg/day	[46]
Shringa bhasma	Patanjali Ayurveda	Deer horns	Expectorant, mucolytic	125 mg twice a day	[46]
Divya Lavangadi Vati tablet	Patanjali Ayurveda	Clove	Demulcent, anti-inflammatory and expectorant	1–2 tablets thrice daily with lukewarm water	[49, 50]
Divya Swasari Pravahi	Patanjali Ayurveda	Black pepper, *Saccharum officinarium*, *Cassia fistula*, *Cordia dichotoma*, *Eclipta alba*, *Zingiber officinale*	Bronchodilator, immunity booster, antioxidant, anti-inflammatory, antimicrobial, mucolytic	5 to 10 ml/day twice or thrice a day	[46]
Dashamoola taila	BANYAN Botanicals	Bilva, agnimantha, shyonaka, patala, gambhari, bruhati, kantakari, gokshura, shalaparni and prishniparni	Calms Vata dosha and has bronchodilator activity	2–5 g/day with warm water	[51]
Mahalakshmi vilas ras	Zandu Ayurveda	Swarna bhasma, raupya bhasma, abhrak bhasma, tamra bhasma, bang bhasma, loha bhasma, mandur bhasma, kanta loha bhasma, naga bhasma, mukta bhasma	Improves lung function, mucolytic, bronchodilator, antioxidant	125–250 mg/day	[52]
Swasa kasa chintamani	Dabur	Purified mercury, calx of copper and iron, gold bhasma, pearl bhasma, purified sulphur, mica bhasma, iron bhasma, juice extract of: yellow berried nightshade, goat milk juice of licorice, betel leaf juice	Balances Vata and Kapha	125–250 mg/day	[53]
Panax ginseng	Nutrazee	Ginseng	Antioxidants, bronchodilator	100 mg–5 g/day	[54]

TABLE 2.5

Natural Potential Chemical Ingredients for Asthma Management

Chemical constituent	Source	Dose	Pulmonary disease treated	MOA	Ref.
Quercetin	Red wine, onions, green tea, apples, berries, etc.	10 mg/kg	Asthma, emphysema, bronchial asthma	Antioxidant, anti-inflammatory, inhibits MMP production and improves lung function	[64, 65]
Rutin	Buckwheat, apricot, citrus fruits	100 mg/kg	Asthma, emphysema, chronic bronchitis	Antioxidant, anti-inflammatory, inhibits MMP production, regulates NF-κB signaling and improves lung function	[66, 67]
Vitamin E	Almonds, peanuts and hazelnuts, sunflower oil, soybean oil, etc.	400 IU daily	Asthma, emphysema, chronic bronchitis	Antioxidant	[68, 69]
Delphinidin-3-glucoside	*Eugenia brasiliensis* fruits, blackcurrant, blueberry, huckleberry	100 mg/kg	Asthma	Antioxidant, anti-inflammatory	[70]
Curcumin	Turmeric	100 mg/kg	Allergic asthma, bronchial asthma	Antioxidant, anti-inflammatory, immunomodulator	[71]
N-acetylcysteine	Beans, lentils, spinach, bananas, salmon, tuna, chicken, yogurt	600–1,200 mg per day per patient	Prevent exacerbations in bronchial asthma	Antioxidant and anti-inflammatory	[72]
Vitamin C	Citrus fruits, spinach, broccoli, etc.	250 mg/day per patient	Supplemental therapy to prevent exacerbations in bronchial asthma	Antioxidant	[73]
Vitamin D	Fatty fish, like tuna, mackerel and salmon; cheese, egg yolk, etc.	2,000 IU daily	Supplemental therapy to prevent exacerbations in COPD	Improves lung function, anti-inflammatory, inhibits MMP production	[74]
Piperine	Black pepper	50 mg/kg	Bronchitis and asthma	Inhibition of PDE enzyme and Ca2+ influx hence has bronchodilator effect	[75, 76]
Bromelain	Pineapple	6 mg/kg	Supplemental therapy for asthma	Anti-inflammatory	[77, 78]
Coniferin	*Pinellia ternata*	100 mg/ml	Supplemental therapy for asthma and bronchitis	Decreases mucus secretion, anti-inflammatory	[79]
Cinnamaldehyde	Cinnamon	50 mg/kg	Supplemental therapy for cough and bronchial asthma	Antioxidant, anti-microbial and anti-inflammatory	[80]
Eugenol	Clove, nutmeg, cinnamon, basil, bay leaf, etc.	40–80 mg/kg	Supplemental therapy to improve lung condition, rhinitis, occupational asthma	Anti-inflammatory, antioxidant properties and improves lung function	[81, 82]
Catechin	Acacia catechu	100 mg/kg	Allergic asthma	Inhibits histidine decarboxylase enzyme and thereby histamine synthesis	[83, 84]
Ferulic acid	Rice, wheat, oats, pineapple, coffee beans, etc.	100 mg/kg	Allergic airway inflammation	Anti-inflammatory, antioxidant	[85]
Gallic acid	Tea leaves, black radish, onion, etc.	100 mg/kg	Bronchitis, allergic airway inflammation	Improves proinflammatory cell infiltration, inactivates ILC2s and suppresses release of IL-5 and IL-13 via the IL-33/MyD88/NF-κB signaling pathway	[86]
Epigallocatechin gallate	Green tea, nuts, dark chocolate, red wine, etc.	10–20 mg/kg	Emphysema, mucus hypersecretion in smokers, asthma, lung injury	Suppresses the production of MIP-2 and TNF-α and thereby is anti-inflammatory	[87]
Gingerol	Ginger	100 mg/kg	Allergic rhinitis, bronchial asthma	Potential immunosuppressive agent and calms down cytokine-mediated immune responses	[88, 89]
Coenzyme Q10	Fatty fish, liver, kidney, spinach, chicken, broccoli, etc.	320 mg/day	Supplemental therapy for exacerbation of bronchial asthma	Antioxidant, strengthens muscle	[90]
Triptolide	*Tripterygium wilfordii*	40 μg/kg	Asthma, cough	Inhibits airway goblet cell hyperplasia and Muc5ac expression, anti-inflammatory	[91, 92]
Caffeic acid	Phenethyl ester honeybee hives, propolis, etc.	5 mg/kg	Supplemental therapy for asthma	Anti-inflammatory, antioxidant	[93–95]

synthetic research to research based on natural chemical constituents has been observed. Different natural chemical constituents have been studied for their efficacy in the treatment of different pulmonary diseases and areas of research interest for scientific groups (Table 2.5).

2.11 Conclusion

Asthma is a respiratory disease associated with high mortality. Asthma can't be cured but an early diagnosis can help in better management of the disease. Beta-adrenergic bronchodilators and corticosteroids are the primary treatment medicines for asthma. For patients with severe exacerbations, combination therapy is used. Add-on medicines like xanthine derivatives, leukotriene-modifying drugs, etc. are also potential treatment options. Contrary to this, the current treatment methods are expensive, need frequent dosing, have varied side effects including tremors, dry mouth, anxiety, etc. This has drifted the attention of both physicians and patients towards the plant-based nutraceutical product. Nutraceutical products have a broad range of pharmacological benefits with lesser side effects even when administered for a longer period of time. The future of asthma treatment looks into the use of personalized, easy-to-use naturally based medicine with novel drug delivery systems with effective site-specific delivery.

REFERENCES

1. Vos, T., Abajobir, A. A., Abate, K. H., Abbafati, C., Abbas, K. M., Abd-Allah, F., … & Criqui, M. H. (2017). Global, regional, and national incidence, prevalence, and years lived with disability for 328 diseases and injuries for 195 countries, 1990–2016: A systematic analysis for the global burden of disease study 2016. *The Lancet*, *390*(10100), 1211–1259.
2. To, T., Cruz, A. A., Viegi, G., McGihon, R., Khaltaev, N., Yorgancioglu, A., … & Bousquet, J. (2018). A strategy for measuring health outcomes and evaluating impacts of interventions on asthma and COPD: Common chronic respiratory diseases in Global Alliance against Chronic Respiratory Diseases (GARD) countries. *Journal of Thoracic Disease*, *10*(8), 5170.
3. Koebnick, C., Fischer, H., Daley, M. F., Ferrara, A., Horberg, M. A., Waitzfelder, B., … & Gould, M. K. (2016). Interacting effects of obesity, race, ethnicity and sex on the incidence and control of adult-onset asthma. *Allergy, Asthma & Clinical Immunology*, *12*(1), 1–16.
4. Jeong, A., Imboden, M., Hansen, S., Zemp, E., Bridevaux, P. O., Lovison, G., … & Probst-Hensch, N. (2017). Heterogeneity of obesity-asthma association disentangled by latent class analysis, the SAPALDIA cohort. *Respiratory Medicine*, *125*, 25–32.
5. Geier, D. A., Kern, J. K., & Geier, M. R. (2019). Demographic and neonatal risk factors for childhood asthma in the USA. *The Journal of Maternal-Fetal & Neonatal Medicine*, *32*(5), 833–837.
6. To, T., Stanojevic, S., Moores, G., Gershon, A. S., Bateman, E. D., Cruz, A. A., & Boulet, L. P. (2012). Global asthma prevalence in adults: Findings from the cross-sectional world health survey. *BMC Public Health*, *12*(1), 1–8.
7. Schedel, M., Michel, S., Gaertner, V. D., Toncheva, A. A., Depner, M., Binia, A., … & Kabesch, M. (2015). Polymorphisms related to ORMDL3 are associated with asthma susceptibility, alterations in transcriptional regulation of ORMDL3, and changes in TH2 cytokine levels. *Journal of Allergy and Clinical Immunology*, *136*(4), 893–903.
8. Jaakkola, J. J., Hernberg, S., Lajunen, T. K., Sripaijboonkij, P., Malmberg, L. P., & Jaakkola, M. S. (2019). Smoking and lung function among adults with newly onset asthma. *BMJ Open Respiratory Research*, *6*(1), e000377.
9. Hsu, J., Qin, X., Beavers, S. F., & Mirabelli, M. C. (2016). Asthma-related school absenteeism, morbidity, and modifiable factors. *American Journal of Preventive Medicine*, *51*(1), 23–32.
10. Virchow, J. C., Backer, V., Kuna, P., Prieto, L., Nolte, H., Villesen, H. H., … & de Blay, F. (2016). Efficacy of a house dust mite sublingual allergen immunotherapy tablet in adults with allergic asthma: A randomized clinical trial. *JAMA*, *315*(16), 1715–1725.
11. Sibilano, R., Gaudenzio, N., DeGorter, M. K., Reber, L. L., Hernandez, J. D., Starkl, P. M., … & Galli, S. J. (2016). A TNFRSF14-FcεRI-mast cell pathway contributes to development of multiple features of asthma pathology in mice. *Nature Communications*, *7*(1), 1–15.
12. Coleman, S. L., & Shaw, O. M. (2017). Progress in the understanding of the pathology of allergic asthma and the potential of fruit proanthocyanidins as modulators of airway inflammation. *Food & Function*, *8*(12), 4315–4324.
13. Thorat, Y. T., Salvi, S. S., & Kodgule, R. R. (2017). Peak flow meter with a questionnaire and mini-spirometer to help detect asthma and COPD in real-life clinical practice: A cross-sectional study. *NPJ Primary Care Respiratory Medicine*, *27*(1), 1–7.
14. Tamura, K., Endo, Y., Masuda, T., Takahashi, S., Tanaka, Y., Watanabe, H., … & Shirai, T. (2018). Fewer bronchodilator responses with forced oscillation technique following methacholine challenge test predict asthma exacerbations. In *B65. Asthma: Pathophysiology and Clinical Trials* (pp. A3926–A3926). American Thoracic Society.
15. Heaney, L. G., Busby, J., Bradding, P., Chaudhuri, R., Mansur, A. H., Niven, R., … & Costello, R. W. (2019). Remotely monitored therapy and nitric oxide suppression identifies nonadherence in severe asthma. *American Journal of Respiratory and Critical Care Medicine*, *199*(4), 454–464.
16. Borrill, Z., Clough, D., Truman, N., Morris, J., Langley, S., & Singh, D. (2006). A comparison of exhaled nitric oxide measurements performed using three different analysers. *Respiratory Medicine*, *100*(8), 1392–1396.
17. Mohamed-Hussein, A. A., Sayed, S. S., Eldien, H. M. S., Assar, A. M., & Yehia, F. E. (2018). Beta 2 adrenergic receptor genetic polymorphisms in bronchial asthma: Relationship to disease risk, severity, and treatment response. *Lung*, *196*(6), 673–680.
18. Bandaru, S., Tarigopula, P., Akka, J., Marri, V. K., Kattamuri, R. K., Nayarisseri, A., … & Sagurthi, S. R. (2016). Association of Beta 2 adrenergic receptor (Thr164Ile) polymorphism with Salbutamol refractoriness in severe asthmatics from Indian population. *Gene*, *592*(1), 15–22.

19. Hussein, M. H., Sobhy, K. E., Sabry, I. M., El Serafi, A. T., & Toraih, E. A. (2017). Beta2-adrenergic receptor gene haplotypes and bronchodilator response in Egyptian patients with chronic obstructive pulmonary disease. *Advances in Medical Sciences*, *62*(1), 193–201.
20. Santhosh, V. G., Gaude, G. S., & Hattiholi, J. (2017). Clinical effectiveness of anticholinergic tiotropium bromide as an add-on therapy in patients with severe bronchial asthma: A randomized controlled trial. *Indian Journal of Health Sciences and Biomedical Research (KLEU)*, *10*(1), 44.
21. Calzetta, L., Hanania, N. A., Dini, F. L., Goldstein, M. F., Fairweather, W. R., Howard, W. W., & Cazzola, M. (2018). Impact of doxofylline compared to theophylline in asthma: A pooled analysis of functional and clinical outcomes from two multicentre, double-blind, randomized studies (DOROTHEO 1 and DOROTHEO 2). *Pulmonary Pharmacology & Therapeutics*, *53*, 20–26.
22. Stanford, R. H., Averell, C., Parker, E. D., Blauer-Peterson, C., Reinsch, T. K., & Buikema, A. R. (2019). Assessment of adherence and asthma medication ratio for a once-daily and twice-daily inhaled Corticosteroid/Long-Acting β-Agonist for asthma. *The Journal of Allergy and Clinical Immunology: In Practice*, *7*(5), 1488–1496.
23. Bleecker, E. R., FitzGerald, J. M., Chanez, P., Papi, A., Weinstein, S. F., Barker, P., ... & SIROCCO Study Investigators. (2016). Efficacy and safety of benralizumab for patients with severe asthma uncontrolled with high-dosage inhaled corticosteroids and long-acting β2-agonists (SIROCCO): A randomised, multicentre, placebo-controlled phase 3 trial. *The Lancet*, *388*(10056), 2115–2127.
24. Dahlén, B., Kumlin, M., Ihre, E., Zetterström, O., & Dahlén, S. E. (1997). Inhibition of allergen-induced airway obstruction and leukotriene generation in atopic asthmatic subjects by the leukotriene biosynthesis inhibitor BAYx 1005. *Thorax*, *52*(4), 342–347.
25. Thomson, N. C., Rubin, A. S., Niven, R. M., Corris, P. A., Siersted, H. C., Olivenstein, R., ... & Cox, G. (2011). Long-term (5 year) safety of bronchial thermoplasty: Asthma Intervention Research (AIR) trial. *BMC Pulmonary Medicine*, *11*(1), 1–9.
26. Pavord, I. D., Cox, G., Thomson, N. C., Rubin, A. S., Corris, P. A., Niven, R. M., ... & RISA Trial Study Group*. (2007). Safety and efficacy of bronchial thermoplasty in symptomatic, severe asthma. *American Journal of Respiratory and Critical Care Medicine*, *176*(12), 1185–1191.
27. Castro, M., Rubin, A. S., Laviolette, M., Fiterman, J., De Andrade Lima, M., Shah, P. L., ... & Cox, G. (2010). Effectiveness and safety of bronchial thermoplasty in the treatment of severe asthma: A multicenter, randomized, double-blind, sham-controlled clinical trial. *American Journal of Respiratory and Critical Care Medicine*, *181*(2), 116–124.
28. Laforest, L., Van Ganse, E., Devouassoux, G., El Hasnaoui, A., Osman, L. M., Bauguil, G., & Chamba, G. (2008). Dispensing of antibiotics, antitussives and mucolytics to asthma patients: A pharmacy-based observational survey. *Respiratory Medicine*, *102*(1), 57–63.
29. Snoek, A. P., & Brierley, J. (2015). Mucolytics for intubated asthmatic children: A national survey of United Kingdom paediatric intensive care consultants. *Critical Care Research and Practice*, *2015*. Article ID 396107.
30. Winickoff, J. P., Buckley, V. J., Palfrey, J. S., Perrin, J. M., & Rigotti, N. A. (2003). Intervention with parental smokers in an outpatient pediatric clinic using counseling and nicotine replacement. *Pediatrics*, *112*(5), 1127–1133.
31. Perrin, K., Wijesinghe, M., Healy, B., Wadsworth, K., Bowditch, R., Bibby, S., ... & Beasley, R. (2011). Randomised controlled trial of high concentration versus titrated oxygen therapy in severe exacerbations of asthma. *Thorax*, *66*(11), 937–941.
32. Ueno-Iio, T., Shibakura, M., Yokota, K., Aoe, M., Hyoda, T., Shinohata, R., ... & Kataoka, M. (2014). Lavender essential oil inhalation suppresses allergic airway inflammation and mucous cell hyperplasia in a murine model of asthma. *Life Sciences*, *108*(2), 109–115.
33. Meamarbashi, A. (2014). Instant effects of peppermint essential oil on the physiological parameters and exercise performance. *Avicenna Journal of Phytomedicine*, *4*(1), 72–78.
34. Hart, P. H., Brand, C., Carson, C. F., Riley, T. V., Prager, R. H., & Finlay-Jones, J. J. (2000). Terpinen-4-ol, the main component of the essential oil of *Melaleuca alternifolia* (tea tree oil), suppresses inflammatory mediator production by activated human monocytes. *Inflammation Research*, *49*(11), 619–626.
35. Juergens, U. R., Dethlefsen, U., Steinkamp, G., Gillissen, A., Repges, R., & Vetter, H. (2003). Anti-inflammatory activity of 1.8-cineol (eucalyptol) in bronchial asthma: A double-blind placebo-controlled trial. *Respiratory Medicine*, *97*(3), 250–256.
36. Burrow, A., Eccles, R., & Jones, A. S. (1983). The effects of camphor, eucalyptus and menthol vapour on nasal resistance to airflow and nasal sensation. *Acta Oto-laryngologica*, *96*(1–2), 157–161.
37. Li, Q., Zhao, S., Lu, J., Kang, X., Zhang, G., Zhao, F., ... & Aisa, H. A. (2020). Quantitative proteomics analysis of the treatment of asthma rats with total flavonoid extract from chamomile. *Biotechnology Letters*, *42*(6), 1–12.
38. Shirole, R. L., Shirole, N. L., Kshatriya, A. A., Kulkarni, R., & Saraf, M. N. (2014). Investigation into the mechanism of action of essential oil of *Pistacia integerrima* for its anti-asthmatic activity. *Journal of Ethnopharmacology*, *153*(3), 541–551.
39. Erkan, N., Ayranci, G., & Ayranci, E. (2008). Antioxidant activities of rosemary (*Rosmarinus Officinalis L.*) extract, blackseed (*Nigella sativa L.*) essential oil, carnosic acid, rosmarinic acid and sesamol. *Food Chemistry*, *110*(1), 76–82.
40. Sharma, P. K., Johri, S., & Mehra, B. L. (2010). Efficacy of vasadi syrup and shwasaghna dhuma in the patients of COPD (Shwasa Roga). *Ayu*, *31*(1), 48.
41. Sharafkhaneh, A., Lee, J. J., Liu, D., Katz, R., Caraway, N., Acosta, C., ... & Kurie, J. M. (2013). A pilot double-blind, randomized, placebo-controlled trial of curcumin/bioperine for lung cancer chemoprevention in patients with chronic obstructive pulmonary disease. *Advances in Lung Cancer*, *2*, 62–69.
42. Vinaya, M., Kamdod, M. A., Swamy, M., & Swamy, M. (2017). Bronchodilator activity of *Ocimum sanctum Linn.* (tulsi) in mild and moderate asthmatic patients in comparison with salbutamol: A single-blind cross-over study. *International Journal of Basic & Clinical Pharmacology*, *6*(3), 511.

43. Kumar, S., Bansal, P., Gupta, V., Sannd, R., & Rao, M. (2010). The clinical effect of *Albizia lebbeck* stem bark decoction on bronchial asthma. *International Journal of Pharmaceutical Sciences and Drug Research*, *2*(1), 48–50.
44. Muniappan, M., & Sundararaj, T. (2003). Antiinflammatory and antiulcer activities of *Bambusa arundinacea*. *Journal of Ethnopharmacology*, *88*(2–3), 161–167.
45. Bangar, M. S., Balsaraf, C. D., Dhamdhere, S. K., Patel, S. G., Jadhav, S. L., & Gaikwad, D. D. (2019). Formulation and evaluation of Vasaka granules for asthma. *Journal of Drug Delivery and Therapeutics*, *9*(2-s), 296–299.
46. Kaur, G., Gupta, V., & Bansal, P. (2017). Innate antioxidant activity of some traditional formulations. *Journal of Advanced Pharmaceutical Technology & Research*, *8*(1), 39.
47. Khurana, P., Singh, A., & Saroch, V. (2015). Role of Pushkarmool in the asthma management: A conceptual study. *Journal of Traditional & Natural Medicines*, *1*(1), 10–12.
48. Subedi, R. P., Vartak, R. R., & Kale, P. G. (2017). Study of general properties of Abhrak Bhasma: A nanomedicine. *International Journal of Pharmaceutical Sciences Review and Research*, *44*(2), 238–242.
49. Saini, A., Sharma, S., & Chhibber, S. (2009). Induction of resistance to respiratory tract infection with *Klebsiella pneumoniae* in mice fed on a diet supplemented with tulsi (*Ocimum sanctum*) and clove (*Syzgium aromaticum*) oils. *Journal of Microbiology, Immunology, and Infection*, *42*(2), 107–113.
50. Weismayer, C., & Pezenka, I. (2017). Identifying emerging research fields: A longitudinal latent semantic keyword analysis. *Scientometrics*, *113*(3), 1757–1785.
51. Mehta, C. L., Shirode, P., Kumar, A., & Kumar, S. A conceptual Ayurvedic study on dashmoola taila and yava kshara. *World Journal of Pharmaceutical and Medical Research*, *5*(9), 201–212.
52. Singh, M. (2014). A study to assess the effectiveness of stress management protocol on the perceived level of stress among the professional and non-professional students in selected colleges of Rajasthan. *International Journal of Health Sciences and Research*, *4*(3), 174–181.
53. Kumar, Y., Singh, B. M., & Gupta, P. (2014). Clinical and metabolic markers based study of Swas Kasa Chintamani Rasa (an Ayurvedic herbometallic preparation) in childhood bronchial asthma (Tamak Swas). *International Journal of Green Pharmacy (IJGP)*, *8*(1).
54. Kim, D. Y., & Yang, W. M. (2011). Panax ginseng ameliorates airway inflammation in an ovalbumin-sensitized mouse allergic asthma model. *Journal of Ethnopharmacology*, *136*(1), 230–235.
55. Turan, G. B., & Tan, M. (2020). The effect of yoga on respiratory functions, symptom control and life quality of asthma patients: A randomized controlled study. *Complementary Therapies in Clinical Practice*, *38*, 101070.
56. Prem, V., Sahoo, R. C., & Adhikari, P. (2013). Comparison of the effects of Buteyko and pranayama breathing techniques on quality of life in patients with asthma: A randomized controlled trial. *Clinical Rehabilitation*, *27*(2), 133–141.
57. Yüce, G. E., & Taşcı, S. (2020). Effect of pranayama breathing technique on asthma control, pulmonary function, and quality of life: A single-blind, randomized, controlled trial. *Complementary Therapies in Clinical Practice*, *38*, 101081.
58. Ma, J., Strub, P., Lv, N., Xiao, L., Camargo, C. A., Buist, A. S., ... & Rosas, L. G. (2016). Pilot randomised trial of a healthy eating behavioural intervention in uncontrolled asthma. *European Respiratory Journal*, *47*(1), 122–132.
59. Özbey, Ü., Balaban, S., Sözener, Z. Ç., Uçar, A., Mungan, D., & Mısırlıgil, Z. (2020). The effects of diet-induced weight loss on asthma control and quality of life in obese adults with asthma: A randomized controlled trial. *Journal of Asthma*, *57*(6), 618–626.
60. Stoodley, I., Garg, M., Scott, H., Macdonald-Wicks, L., Berthon, B., & Wood, L. (2020). Higher omega-3 index is associated with better asthma control and lower medication dose: A cross-sectional study. *Nutrients*, *12*(1), 74.
61. Gold, D. R., Litonjua, A. A., Carey, V. J., Manson, J. E., Buring, J. E., Lee, I. M., ... & Luttmann-Gibson, H. (2016). Lung VITAL: Rationale, design, and baseline characteristics of an ancillary study evaluating the effects of vitamin D and/or marine omega-3 fatty acid supplements on acute exacerbations of chronic respiratory disease, asthma control, pneumonia and lung function in adults. *Contemporary Clinical Trials*, *47*, 185–195.
62. Maltby, S., Gibson, P. G., Powell, H., & McDonald, V. M. (2017). Omalizumab treatment response in a population with severe allergic asthma and overlapping COPD. *Chest*, *151*(1), 78–89.
63. FitzGerald, J. M., Bleecker, E. R., Nair, P., Korn, S., Ohta, K., Lommatzsch, M., ... & CALIMA Study Investigators. (2016). Benralizumab, an anti-interleukin-5 receptor α monoclonal antibody, as add-on treatment for patients with severe, uncontrolled, eosinophilic asthma (CALIMA): A randomised, double-blind, placebo-controlled phase 3 trial. *The Lancet*, *388*(10056), 2128–2141.
64. Ganesan, S., Faris, A. N., Comstock, A. T., Chattoraj, S. S., Chattoraj, A., Burgess, J. R., ... & Sajjan, U. S. (2010). Quercetin prevents progression of disease in elastase/LPS-exposed mice by negatively regulating MMP expression. *Respiratory Research*, *11*(1), 1–15.
65. Yang, T., Luo, F., Shen, Y., An, J., Li, X., Liu, X., ... & Wen, F. (2012). Quercetin attenuates airway inflammation and mucus production induced by cigarette smoke in rats. *International Immunopharmacology*, *13*(1), 73–81.
66. Lv, H. Y., Chen, J., & Wang, T. (2017). Rutin has anti-asthmatic effects in an ovalbumin-induced asthmatic mouse model. *Tropical Journal of Pharmaceutical Research*, *16*(6), 1337–1347.
67. Jung, C. H., Lee, J. Y., Cho, C. H., & Kim, C. J. (2007). Anti-asthmatic action of quercetin and rutin in conscious guinea-pigs challenged with aerosolized ovalbumin. *Archives of Pharmacal Research*, *30*(12), 1599–1607.
68. Wu, T.-C., Huang, Y.-C., Hsu, S.-Y., Wang, Y.-C., & Yeh, S.-L. (2007). Vitamin E and vitamin C supplementation in patients with chronic obstructive pulmonary disease. *International Journal for Vitamin and Nutrition Research*, *77*(4), 272–279.
69. Peh, H. Y., Tan, W. D., Chan, T. K., Pow, C. W., Foster, P. S., & Wong, W. F. (2017). Vitamin E isoform γ-tocotrienol protects against emphysema in cigarette smoke-induced COPD. *Free Radical Biology and Medicine*, *110*, 332–344.

70. Flores, G., Dastmalchi, K., Paulino, S., Whalen, K., Dabo, A. J., Reynertson, K. A., … & Kennelly, E. J. (2012). Anthocyanins from Eugenia brasiliensis edible fruits as potential therapeutics for COPD treatment. *Food Chemistry, 134*(3), 1256–1262.
71. Zhang, M., Xie, Y., Yan, R., Shan, H., Tang, J., Cai, Y., … & Li, Y. (2016). Curcumin ameliorates alveolar epithelial injury in a rat model of chronic obstructive pulmonary disease. *Life sciences, 164*, 1–8.
72. Cazzola, M., Calzetta, L., Facciolo, F., Rogliani, P., & Matera, M. G. (2017). Pharmacological investigation on the anti-oxidant and anti-inflammatory activity of N-acetylcysteine in an ex vivo model of COPD exacerbation. *Respiratory Research, 18*(1), 1–10.
73. Koike, K., Ishigami, A., Sato, Y., Hirai, T., Yuan, Y., Kobayashi, E., … & Seyama, K. (2014). Vitamin C prevents cigarette smoke-induced pulmonary emphysema in mice and provides pulmonary restoration. *American Journal of Respiratory Cell and Molecular Biology, 50*(2), 347–357.
74. Rafiq, R., Prins, H. J., Boersma, W. G., Daniels, J. M., den Heijer, M., Lips, P., & de Jongh, R. T. (2017). Effects of daily vitamin D supplementation on respiratory muscle strength and physical performance in vitamin D-deficient COPD patients: A pilot trial. *International Journal of Chronic Obstructive Pulmonary Disease, 12*, 2583.
75. Panahi, Y., Ghanei, M., Hajhashemi, A., & Sahebkar, A. (2016). Effects of curcuminoids-piperine combination on systemic oxidative stress, clinical symptoms and quality of life in subjects with chronic pulmonary complications due to sulfur mustard: A randomized controlled trial. *Journal of Dietary Supplements, 13*(1), 93–105.
76. Chauhan, P. S., Jaiswal, A., & Singh, R. (2018). Combination therapy with curcumin alone plus piperine ameliorates ovalbumin-induced chronic asthma in mice. *Inflammation, 41*(5), 1922–1933.
77. Secor Jr, E. R., Shah, S. J., Guernsey, L. A., & Schramm, C. M. (2012). Bromelain limits airway inflammation in an ovalbumin-induced murine model of established asthma. *Alternative Therapies in Health and Medicine, 18*(5), 9.
78. Secor, E. R., Carson, W. F., Singh, A., Pensa, M., Guernsey, L. A., Schramm, C. M., & Thrall, R. S. (2008). Oral bromelain attenuates inflammation in an ovalbumin-induced murine model of asthma. *Evidence-based Complementary and Alternative Medicine, 5*(1), 61–69.
79. Lee, J. H., Sun, Y. N., Kim, Y. H., Lee, S. K., & Kim, H. P. (2016). Inhibition of lung inflammation by acanthopanax divaricatus var. albeofructus and its constituents. *Biomolecules & Therapeutics, 24*(1), 67.
80. Molania, T., Moghadamnia, A. A., Pouramir, M., Aghel, S., Moslemi, D., Ghassemi, L., & Motallebnejad, M. (2012). The effect of Cinnamaldehyde on mucositis and salivary antioxidant capacity in gamma-irradiated rats (a preliminary study). *DARU Journal of Pharmaceutical Sciences, 20*(1), 1–5.
81. Pan, C., & Dong, Z. (2015). Antiasthmatic effects of eugenol in a mouse model of allergic asthma by regulation of vitamin D3 upregulated protein 1/NF-κB pathway. *Inflammation, 38*(4), 1385–1393.
82. Zin, W. A., Silva, A. G., Magalhães, C. B., Carvalho, G. M., Riva, D. R., Lima, C. C., … & Faffe, D. S. (2012). Eugenol attenuates pulmonary damage induced by diesel exhaust particles. *Journal of Applied Physiology, 112*(5), 911–917.
83. Patel, S., & Patel, V. (2019). Inhibitory effects of catechin isolated from Acacia catechu on ovalbumin induced allergic asthma model: Role of histidine decarboxylase. *Nutrition & Food Science, 49*(1), 18–31.
84. Siebert, D. A., Paganelli, C. J., Queiroz, G. S., & Alberton, M. D. (2021). Anti-inflammatory activity of the epicuticular wax and its isolated compounds catechin and gallocatechin from Eugenia brasiliensis Lam.(Myrtaceae) leaves. *Natural Product Research, 35*(22), 4720–4723.
85. Ko, J. W., Shin, N. R., Park, S. H., Lee, I. C., Ryu, J. M., Cho, Y. K., … & Shin, I. S. (2017). Ssanghwa-Tang, a traditional herbal formula, suppresses cigarette smoke-induced airway inflammation via inhibition of MMP-9 and Erk signaling. *Molecular & Cellular Toxicology, 13*(3), 295–304.
86. Singla, E., & Naura, A. S. (2020). Protective effects of Gallic acid against COPD linked inflammation and emphysema. *The FASEB Journal, 34*(S1), 1–1.
87. Bae, H. B., Li, M., Kim, J. P., Kim, S. J., Jeong, C. W., Lee, H. G., … & Kwak, S. H. (2010). The effect of epigallocatechin gallate on lipopolysaccharide-induced acute lung injury in a murine model. *Inflammation, 33*(2), 82–91.
88. Kawamoto, Y., Ueno, Y., Nakahashi, E., Obayashi, M., Sugihara, K., Qiao, S., … & Takeda, K. (2016). Prevention of allergic rhinitis by ginger and the molecular basis of immunosuppression by 6-gingerol through T cell inactivation. *The Journal of Nutritional Biochemistry, 27*, 112–122.
89. Yamprasert, R., Chanvimalueng, W., Mukkasombut, N., & Itharat, A. (2020). Ginger extract versus Loratadine in the treatment of allergic rhinitis: A randomized controlled trial. *BMC Complementary Medicine and Therapies, 20*(1), 1–11.
90. De Benedetto, F., Pastorelli, R., Ferrario, M., de Blasio, F., Marinari, S., Brunelli, L., … & Celli, B. R. (2018). Supplementation with Qter® and creatine improves functional performance in COPD patients on long term oxygen therapy. *Respiratory Medicine, 142*, 86–93.
91. Chen, M., Lv, Z., & Jiang, S. (2011). The effects of triptolide on airway remodelling and transforming growth factor-β1/Smad signalling pathway in ovalbumin-sensitized mice. *Immunology, 132*(3), 376–384.
92. Chen, M., Lv, Z., Zhang, W., Huang, L., Lin, X., Shi, J., … & Jiang, S. (2015). Triptolide suppresses airway goblet cell hyperplasia and Muc5ac expression via NF-κB in a murine model of asthma. *Molecular Immunology, 64*(1), 99–105.
93. Ma, Y., Zhang, J. X., Liu, Y. N., Ge, A., Gu, H., Zha, W. J., … & Huang, M. (2016). Caffeic acid phenethyl ester alleviates asthma by regulating the airway microenvironment via the ROS-responsive MAPK/Akt pathway. *Free Radical Biology and Medicine, 101*, 163–175.
94. Jung, W. K., Lee, D. Y., Choi, Y. H., Yea, S. S., Choi, I., Park, S. G., … & Choi, I. W. (2008). Caffeic acid phenethyl ester attenuates allergic airway inflammation and hyper-responsiveness in murine model of ovalbumin-induced asthma. *Life Sciences, 82*(13–14), 797–805.
95. Saluja, B., Li, H., Desai, U. R., Voelkel, N. F., & Sakagami, M. (2014). Sulfated caffeic acid dehydropolymer attenuates elastase and cigarette smoke extract–induced emphysema in rats: Sustained activity and a need of pulmonary delivery. *Lung, 192*(4), 481–492.

3

Recent Trends in Evaluating the Mechanistic Aspects of Alzheimer's Disease and Its Diagnosis with Smart Devices

Poojith Nuthalapati, Preeti Yendapalli, Malini Kumar, Ashna Joy and Chiranjeevi Sainatham

CONTENTS

3.1 Introduction

Alzheimer's disease (AD) is a progressive neurological disorder that affects behavioral and cognitive skills such as linguistic, thinking ability, understanding, and memory; there is no early diagnostic approach that could prevent long-term neurological consequences [1, 2]. As a result of neurodegeneration, the process of brain damage begins many years before symptoms such as dementia occur [3]. According to physicians, dementia and decline in cognitive functions are the early clinical signs and symptoms of Alzheimer's disease [4]. Hence, dementia and mild cognitive impairment (MCI) have been considered markers to diagnose AD.

Overall, pre-clinical AD research and diagnosis are very important as they may help achieve early-stage treatment. Patients who present with dementia or without dementia are diagnosed by the occurrence of amyloid plaques in the neocortical regions of the brain. Patients below the age of 70 show these plaques to about 50%, while amyloid plaques may not appear until the age of 85 in a few cases.

There occur some neurofibril tangles and senile plaques that are responsible for the onset of brain damage [5, 6]. The 42-amino-acid isoform of β-amyloid (Aβ42) is a major component of senile plaques. Increased amounts of beta-amyloid induce amyloid aggregation, which causes neuronal damage, and tangles have a stronger link to AD than plaques. These are the key histological characteristics of AD [7]. These factors cause inflammation in the neural and axonal areas of brain cells, resulting in receptor loss at synaptic locations and nerve impulse transmission failure [8, 9], as illustrated in Figures 3.1–3.2. As a result of this, patients will develop dementia, cognitive impairment, and the inability to make decisions. The accumulation of plaques in the brain, as well as neuronal loss, aggravates the illness and, in rare circumstances, results in death [10, 11].

DOI: 10.1201/9781003157823-3

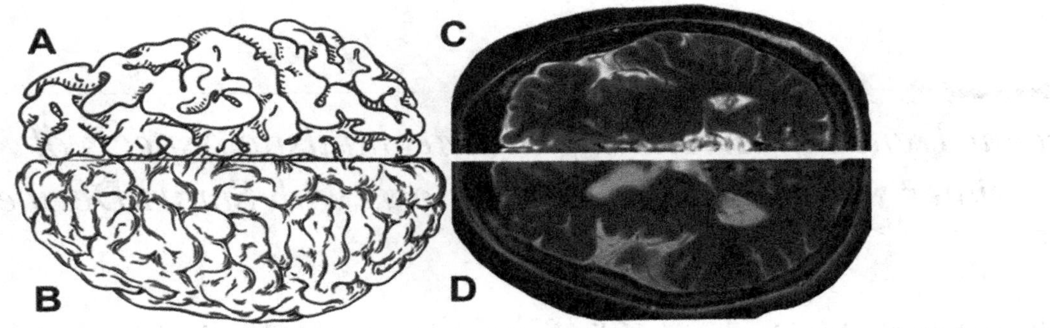

FIGURE 3.1 Comparison of the brains of normal and Alzheimer's patients. (A) Brain of Alzheimer's patient, (B) brain of normal individual, (C) MRI image of Alzheimer's brain, (D) MRI image of normal individual.

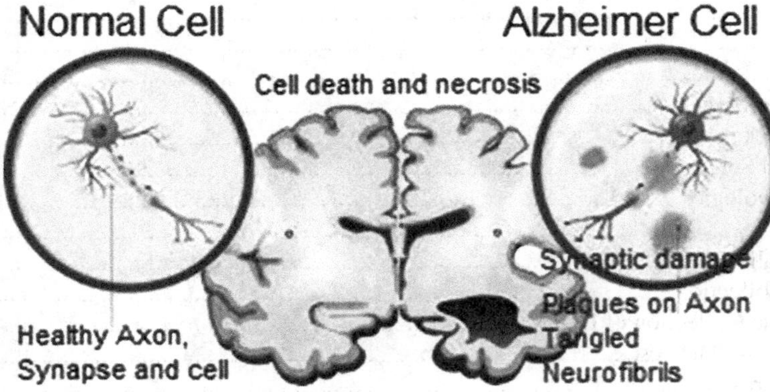

FIGURE 3.2 Various events leading to Alzheimer's disease.

3.2 Epidemiology

Alzheimer's disease is the world's most common disease nowadays. The global incidence of dementia is estimated to be as high as 24 million people, with the number expected to quadruple by 2050. In 2017, it was estimated that 46 million people worldwide had dementia caused by AD. In 2019, the number of patients was estimated to reach over 50 million. According to predictions, the AD population would increase to above 75 million in 2030 and 135 million in 2050 [12]. The disease affects both men and women in the elderly population [13]. According to studies, 58% of those who suffer from the disease are from the lower and middle economic groups. The disease is quite common among the elderly, and nations with a high elderly population suffer further more [14]. According to the Alzheimer's Association, there are 5.5 million Alzheimer's patients in the United States alone. The numbers clearly demonstrate that there is a rising market for Alzheimer's medicines, as well as an increase in the number of patients who will require more reliable and quick diagnostic procedures in the coming years. Novel neuroimaging approaches and biomarkers, according to research, might significantly improve the pre-clinical diagnosis of Alzheimer's disease [15].

3.3 Biomarkers

A diagnostic method that detects the pathology of the disease and dementia in the early stages of AD is in high demand. Biomarkers are the most effective diagnostic tools for determining the severity of AD. There is a direct correlation between the concentration of biomarkers and the extent of the disease pathogenesis [16, 17]. Hence, biomarker estimates can be used to monitor or detect disease progression. The body's response to a specific therapy may also be monitored by sensitively estimating the biomarker molecules. Therefore, the current focus has shifted to the development and validation of biomarkers for early detection of the disease. Though biomarkers for AD are being developed, there are a few drawbacks that prevent biomarkers from being used as an efficient way of disease estimation. Such drawbacks include overlapping diseases that can cause dementia, as well as differences in biomarkers that respond to a various range of neurological symptoms and a lack of understanding of the disease's pathological process [18, 19]. Apart from these limitations, biomarkers are the most reliable way of predicting Alzheimer's disease. They are divided into the following categories:

i) Biochemical biomarkers which employ the estimation of cerebrospinal fluid and blood-based analysis: various methods, like mass spectrometry and ELISA, can be used to quantify Aβ42 present in the CSF. Reduced levels of Aβ42 have been found in the CSF of individuals with mild cognitive impairment (MCI) and in the early stages of Alzheimer's disease (AD). Blood biomarkers for Aβ pathology have been challenging to develop. When cerebral β-amyloidosis is evaluated immunochemically, the association between Aβ proteins and cerebral

TABLE 3.1

Qualities of an Ideal Biomarker According to the Literature

1	Should be able to mark the basic character of the pathogenesis of the disease
2	Should validate the neuropathology of confirmed cases of AD
3	Must be sensitive to detect AD and to differentiate other kinds of dementia
4	Should be able to detect even the starting stage/early stages of AD
5	The biomarkers are to be remarkable, reproducible, and inexpensive in routine practice

β-amyloidosis is absent or weak. The production of plasma Aβ by platelets and other extracerebral tissues is believed to impact its concentration. P-tau in the CSF is currently regarded to be the most specific biomarker for Alzheimer's disease. Except for herpes encephalitis and superficial CNS siderosis, no other condition causes this biomarker to rise in a predictable pattern [20].

ii) Neuroanatomical biomarkers which include the estimation of brain activity using computed tomography and MRI scans.

iii) Metabolic biomarkers include the estimation using positron emission tomography and single-photon emission computerized tomography scans: as demonstrated by autopsy and *in vivo* amyloid positron emission tomography (PET) imaging investigations in patients, the decrease in Aβ42 reflects sequestration of Aβ42 in senile plaques in the brain. Pittsburgh Compound-B, an ^{11}C-labeled modified derivative of the amyloid-binding histology dye thioflavin-T, was the first chemical probe for amyloid PET (PiB, also known as ^{11}C-PiB). The short half-life of ^{11}C, on the other hand, makes it difficult to employ ^{11}C-PiB outside of specialized research centers with an on-site cyclotron and radiochemistry expertise. As a result, ^{18}F-labeled probes with a half-life of 110 minutes have been developed. Although they show greater levels of nonspecific binding to white matter than 11C-PiB, they have shown a good correlation with amyloid plaque burden at autopsy [20].

iv) Genetic biomarkers involve the estimation of amyloid precursor protein (APP), presenilin gene 1 (PSEN1), presenilin gene 2 (PSEN2), apolipoprotein E4 allele (APOE4), etc.

v) Neuropsychological biomarkers involve performing various tests like episodic memory tests, attention estimation tests, executive function tests, cognitive tests, etc.

3.3.1 Nonspecific Biomarkers

The nonspecific biomarkers are utilized to predict or estimate the prognosis of AD that includes:

i) Estimating the CSF cell counts.
ii) CSF to serum albumin ratios.
iii) Intrathecal immunoglobulin synthesis, etc.

These biomarkers aren't specific to abnormalities caused by AD, but they can help rule out other disorders with comparable symptoms. In general, vascular diseases leading to dementia and inflammations induce elevation in the serum albumin ratio [17, 21].

3.3.2 Specific Biomarkers

In addition to nonspecific biomarkers, specific biomarkers have been established to diagnose the disease, as shown in Table 3.1 and Figure 3.3. They are as follows:

- Phosphorylated-tau shows a remarkable increase when there is a presence of neurofibril tangles. These tangles are specific to Alzheimer's disease [22, 23].
- When there is any type of brain injury or damage, total-tau, like phosphorylated-tau, exhibits a substantial increase in levels. Even though Alzheimer's disease causes neural damage, other diseases can also cause it. As a result, this biomarker is only relevant to neuronal mechanisms and cannot be used to estimate AD [20, 21].
- Aβ1-42 is the estimate of the extent of amyloid plaques. The more plaque formation, the lesser is the amount of Aβ1-42 in the brain [24, 25].
- F2 isoprostanes are specific biomarkers that may be used to assess mitochondrial dysfunction, and their levels rise as the dysfunction worsens.
- The synthesis of amyloid plaques in the brain is measured using truncated amyloid β isomers. The levels of TAIs increase with the formation of plaques in the brain [26, 27].
- CSF-BACE1 behaves similarly to TAIs where their levels also increase in the brain with the formation and development of amyloid plaques [28, 29].
- Synaptic dysfunction elevates the levels of biomarkers of synaptic degeneration in the brain.

3.4 Digital Biomarkers and Sensors

With an increase in the demography of the elderly population, especially those affected by Alzheimer's disease, both developed and developing nations are adversely affected [30]. There has been a strong link between age and the prevalence of the disease. Around 19 million healthcare professionals serve in the primary assistance jobs of the affected to help them get through their daily lives. To avoid the direct involvement of healthcare workers in managing AD patients, the use of technology is helpful in most cases. Digital gadgets, smart acting sensors, artificial intelligence apps, and other technologies are assisting the affected individuals in

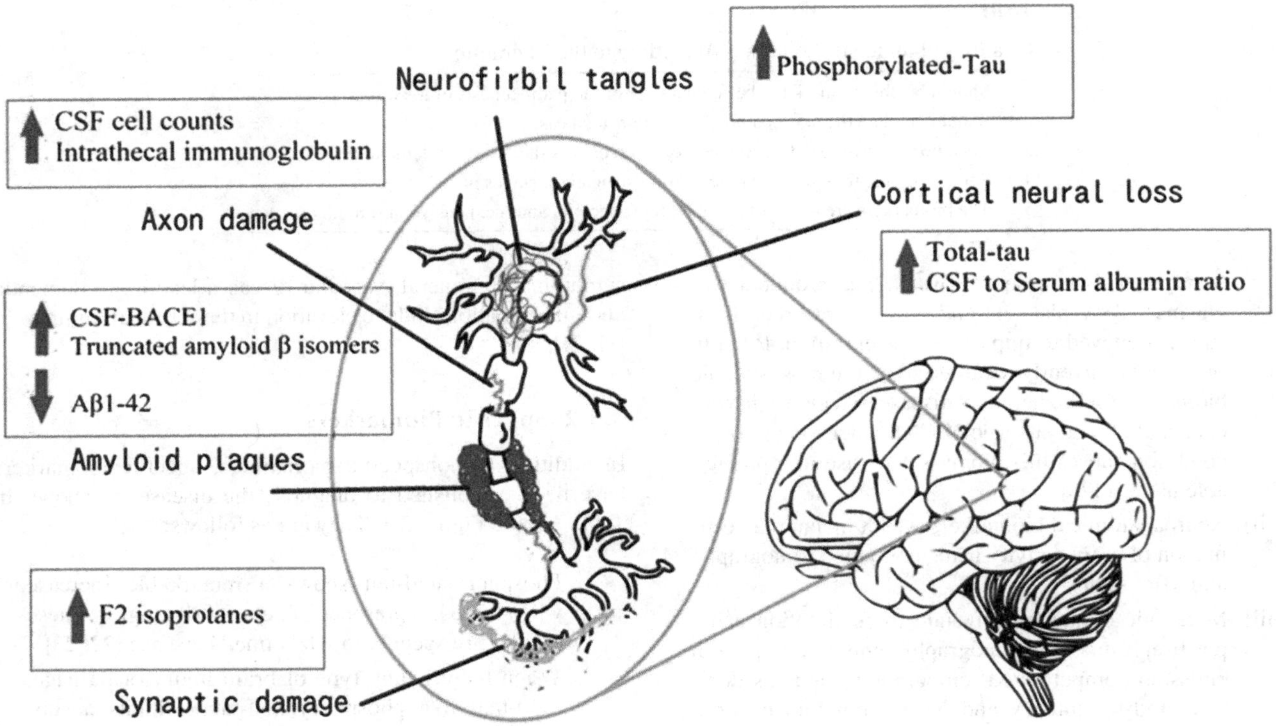

FIGURE 3.3 Biomarker levels indicative of Alzheimer's disease.

their homes. The development of strategies to link and synchronize such devices into a single technological solution might significantly reduce the day-to-day issues that AD patients experience, as well as enhance their health, monitoring convenience, and safety.

Behavioral abnormalities are the first symptoms of dementia caused by AD and/or other neurological disorders. Patients experience memory loss, learning difficulties, and poor decision-making abilities as the disease progresses. Patients with Alzheimer's disease show reduced mental abilities and cognitive function, making communication difficult. The consequences of the disease may vary as it advances, depending on the patient's lifestyle and prior health state. With the worsening of their condition, some individuals may lose their memories completely or even behave erratically.

Patients benefit greatly from simple technologies such as automated messaging, recordings, IVR, and read-out-loud. There are gadgets that read out messages and provide medication time reminders; pillbox technology helps patients remember to take their medications on a regular basis. There are also sophisticated systems with automatic pill dispensers to assist patients with medicine administration. Cameras, sensors, and scanners are used to monitor the patient's movement patterns and keep track of the patient's movements [31].

3.5 Recent Marketed Technologies

There have been a number of recent technological advancements [32] such as:

Smart homes – Such homes employ activity sensors in the patient's home, as well as video cameras and audio equipment, to monitor patient movement and properly guide them around the house. These are beneficial in assisting the patient with daily tasks [31]. Smart home monitoring technology may be able to detect cognitive changes as they happen, resulting in a more cost-effective and prompter diagnosis. More research on this aspect might contribute to technological development which will help in the early detection of symptoms associated with AD [33].

Home controllers – These are devices that control the house ambiance and interact directly with the patient to ensure the patient's comfort while in the house.

Wearables – Sleep monitors, object motion sensors, walking sensors, and other wearable implants are among these devices. They are used to identify any irregularities in the patient's everyday activities and to keep track of the patient's health and location information. It also allows doctors to keep a constant monitor on the patient's condition when he/she visits the clinic independently [31]. Sensors used in wearable devices allow live and near-to-high-accuracy data collection in large populations, as well as real-time behavioral monitoring to measure cognitive abilities [34].

Robotic assistance – Robot is a type of technology developed to help patients with their everyday tasks. It also helps in the treatment of speech and communication problems by allowing patients to interact directly with them through gestures. However, the main disadvantage of robotic systems is their high cost of installation and maintenance in individual houses.

Smart clinical decision support systems (SCDSS) – These cameras are based on sensing devices that are useful in recording and monitoring the patient's medical records, medication management, and intervention. It acquires the data and delivers the information to physicians and clinical investigators about the patient's condition and their health indices.

Mobility aids and assistive robots – These devices are beneficial in improving social communication and interaction. They also assist the patient's mobility and route tracking by using GPS and assisting with navigation.

Brain games and telecare – This technology promotes the development of brain cells by stimulating through games that involve more brain activity and improves the patient's cognitive function.

Assistive technologies and information and communication technologies (ICT) – They assist the patient in promoting physical health and enhancing the emotional status of the patient.

Wearable health trackers – They are employed in the use of the sphygmomanometer and oximeter to measure the blood pressure (BP) and blood oxygen levels, respectively, and are also used in body mass index (BMI) calculators. They share the information with the caretaker and directly sync the information.

Passive positioning alarm – This technology is used in alarm systems to improve the patient's faith in the device's effectiveness and enhance the patient's ability to operate it.

3.6 Data Collection

Apart from the devices that assist AD patients, there are also devices that track the data and share the information with the authorized caretaker or physician. Mobiles and other wearable devices that are used by the patients extract various types of data to detect the neurocognitive decline in a timely and cost-effective manner.

Advantages of data collecting and monitoring devices include:

i) Data can be shared instantly whenever it is needed, and information may be synchronized across all devices with the help of internet connectivity.
ii) Easy to maintain and facilitate data sharing between patients and physicians in a cost-effective manner.

3.6.1 Active Data Collection

The data is collected when the patient takes specific tests and the results are shared with the caregiver. This type of data is only collected as a result of a test or activity designed by physicians or other healthcare professionals. For example, a digital assessment of cognitive tests is performed to collect the data about a patient's cognitive decline and mental health status.

When specific words are prompted for the patient, other devices such as voice probes track the tremors in the voice and evaluate the strength of the vocal cords. Overall, data collection devices actively force the patient to perform some activity, through which test data are collected. These tests are designed based on the symptoms and correlations established with the progression of the disease.

3.6.2 Passive Data Collection

The various ranges of data are collected without the patient's knowledge. Physical activities or tasks designed may or may not be included in this data collection method. Using a smartwatch, for example, enables the collection of heart rate data and its changes during any physical activity. Wearable devices, such as smart rings, monitor the data regarding daily activity and other signs of the disease.

The physician's ability to extract data in the active mode gets more challenging as a patient's cognitive functions and responses worsen as the disease progresses. Passive data collection is sometimes more convenient and advantageous than active data collection since it does not require the patient's cooperation or condition to estimate the state of the disease. The usage of passive devices that collect data from patients for treatment and clinical studies has increased. They have the advantages of being reliable and simple to use, as well as being affordable and easy to handle, resulting in reduced load for patients and higher chances of accurately estimating disease in normal conditions.

3.6.3 Concerns for the Collection of Data

Between the devices, sensors, and physicians, a huge amount of data is exchanged. Platforms such as e-medical records, online assessment tools, and internet tech capture the data such as browsing history and typing speed, which is important in assessing disease progression and response to therapy. In addition to the benefits, there are potential risks and ethical challenges associated with the use and collection of data. It increases the clinical trial's planning and documentation stages. With these new technologies, ethical consents, privacy and data sharing, as well as terms and conditions, play a significant role. Despite the fact that data is protected, privacy is still a major concern. In general, Alzheimer's patients are willing to share data via devices in an active manner due to the spontaneity and convenience of sharing, but they refuse to permit data collection in a passive manner [35]. However, there are certain restrictions, such as the fact that Alzheimer's patients are concerned about their data privacy, which may limit the sharing of their personal data.

There are always consents that are informed and signed, and ethical privacy documentation that requires a patient's consent [36]. They provide a significant benefit to physicians who manage patient data. However, these consent signs and documentation parts have major flaws, such as:

i) Routine processes of documentation are time- and energy-consuming.
ii) Limited possibility of getting consent from patients who are suffering from cognitive impairment.
iii) Large amounts of textual documentation work and a narrow margin to make choices.
iv) Common parts of the informed consent make the documentation part unreliable and uncommon for all the patients.
v) It is difficult to manage misunderstandings about privacy and confidentiality.

3.6.4 Condition-Specific Metrics

There are various metrics that are measured and detected using different types of sensors and devices [37], as presented in Tables 3.2–3.3 and Figure 3.4.

3.6.4.1 Cameras

Facial features may be visualized with the use of cameras. When reading, saccades can be used to estimate pupillary movement, deformation, and twitching in the facial muscles. The pupillary reflex and eye movements have been studied in neurological disease research for decades. It was discovered that AD patients had considerably reduced pupil constriction velocity and acceleration in an active test established earlier, in which the light flashes while the subject's eyes are recorded using a high-speed camera. Camera sensors might aid in the collection of these passive data from patients. Longitudinal blink rate variations may be detected by the face camera when interacting in front of a tablet or phone screen and can be used in the digital biomarker arsenal for early disease diagnosis.

3.6.4.2 Accelerometer/Gyrometer

These are used to monitor the movement patterns of AD patients by measuring gait parameters, energy metrics, and tremors in body parts. Physical activity including steps taken and intensity of the patient's movement can also be measured and recorded. It also differentiates between the types of activity performed [38].

3.6.4.3 Global Positioning System (GPS)

These are used to assess driving ability and location patterns. Analyzing these metrics provides info about the patient's navigational ability, reaction time, and spatial patterns. It will also help AD patients to reach their destination and enhance their driving ability over time.

TABLE 3.2

Variations in Biomarkers Used in Estimating the Level of Alzheimer's Disease

Type of biomarker	Change in levels of biomarkers	Stage of Alzheimer's	Application
Phosphorylated-tau	Increase	Second stage	Infers neurofibril tangles
Total-tau	Increase	Second stage (nonspecific)	Nerve damage
Aβ1-42	Decrease	Advanced	Amyloid plaques
F2 isoprotanes	Increase	Second stage	Mitochondrial dysfunction
Truncated amyloid β isomers	Increase	Second stage	Plaques
CSF-BACE1	Increase	Starting stage	Amyloid plaques
Intrathecal immunoglobulin	Increase	Starting stage (nonspecific)	Neural inflammation and damage
CSF to serum albumin ratios	Increase	Starting-advanced	Nerve cell apoptosis
CSF cell counts	Abnormalities	Any stage	Dementia

TABLE 3.3

Sensors and Their Applications in Alzheimer's Disease

Sensor	Tracking	Application
Camera	Video and photo	• To instruct patients via calls • To know the activity • To see the motion of the patient • To identify facial abnormalities
GPS	Location and motion	• To track the location of demented patients • To study the motor and driving skills • To study the movement habits
Accelerometer	Speed and pace	• To study the motion of the patient and his gait
Gyrometer	Orientation	• To track the orientation of the patient • To study the body balance
Microphone	Voice	• To study the voice tremor • To give assistance to patients based on their verbal communication
ECG	Heart rate	• To track the patient's heart rate and beat
Speaker	Audibility	• To test patient audibility • To give instructions • To test the cognitive function
Thermometer	Temperature	• To check on the body function • BMR • To check immunity condition
EMG	Muscle movement	• To test the mind and muscle coordination • To check the muscle power and activity

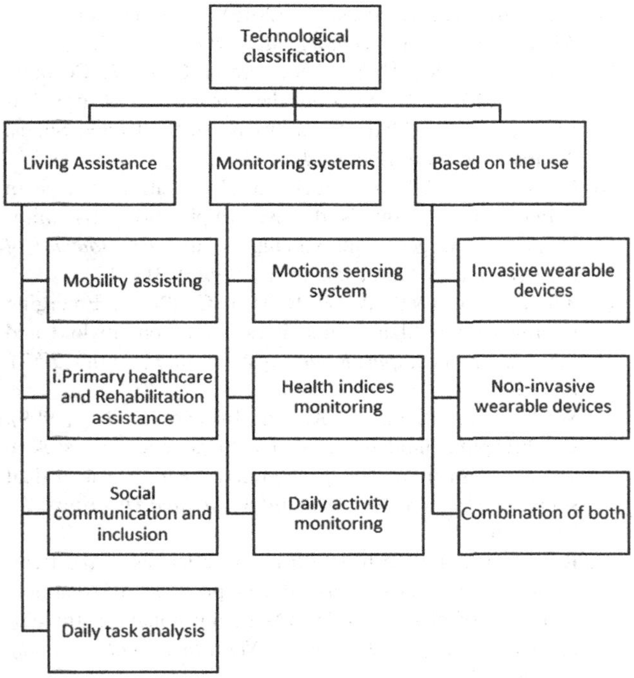

FIGURE 3.4 Classification of technology in AD.

3.6.4.4 Microphones

These are the instruments that are used to analyze the vocal cord health and tremors in the voice. They also estimate the response and cognitive function of the patient.

3.6.4.5 Electrocardiogram (ECG)

It is used to measure the heart rate and pulse pattern. Sleep patterns, electrical activity in the heart, and other functional disabilities are measured using these devices.

3.6.4.6 Thermometers

Usual thermometers measure the body temperature of the patients that corresponds to the hormonal activity, infections, and inflammation in the body of the patient. Whereas infrared (IR) thermometers are used to measure the temperature without any physical contact with the patient.

3.6.4.7 Electromyogram (EMG)

EMG is useful in estimating body movement and muscular involvement. It also helps to estimate functional tremor and disease-induced tremor. Seizure episodes and unusual muscle movements can also be measured which can help determine the extent of disease progression in the patient (Figure 3.5).

FIGURE 3.5 Connectivity of sensors and supporting devices using the Internet of Things (IOT). (a) Smart home devices, (b) voice assistant, (c) assistive robots, (d) satellite connection, (e) smart belts, (f) smartwatch and Fitbit, (g) positioning system, (h) tablet to provide active data. (1) Smartwatch with physician to sync data from patient via the internet, (2) tablet to monitor active data.

3.7 Future Prospects and Conclusion

Alzheimer's is a multi-faced disease requiring multidimensional digital sensing technologies. The focus of research should be on developing new devices and technologies that detect biomarkers that are more sensitive and rapid, with a small margin of error. The sensors are also designed in such a manner that they can extract a large amount of data at once without compromising the patient's privacy. There are sensors that can collect data from a single system or organ. This reflects the future scope of designing systems, which can collect data and analyze readings from multiple data sources with the use of a single device. Therefore, there is a need to design predictors that can sense, record, and also predict the data without even recording or collecting actual metrics from the patients. This saves a lot of time and energy in terms of estimating disease progression and effectively treating disease based on the prediction-based diagnosis. The statistical predictions that are being generated now aren't reliable, thus necessitating technological interventions in this aspect. During the early stages of Alzheimer's disease, patients often give out a very weak signal to the sensors, making diagnosis and evaluation difficult. New technologies must be developed to detect the weakest signal from biomarkers or to amplify the available signals detected by the devices. Overall, there is a need to establish a correlation between the data collected and the progression of the disease. Ground evidence must be established toward the relationship and accuracy of the data collected and the progression of the disease which is estimated accurately using conventional techniques. This enhances the reliability of the sensory devices and provides a scope for the development and effective application of those technologies.

REFERENCES

1. Jack, C. R., Therneau, T. M., Weigand, S. D., Wiste, H. J., Knopman, D. S., Vemuri, P., … & Petersen, R. C. (2019). Prevalence of biologically vs clinically defined Alzheimer spectrum entities using the National Institute on Aging–Alzheimer's Association research framework. *JAMA Neurology*, *76*(10), 1174–1183.
2. Barnes, L. L., Leurgans, S., Aggarwal, N. T., Shah, R. C., Arvanitakis, Z., James, B. D., … & Schneider, J. A. (2015). Mixed pathology is more likely in black than white decedents with Alzheimer dementia. *Neurology*, *85*(6), 528–534.
3. Santana, I., Farinha, F., Freitas, S., Rodrigues, V., & Carvalho, Á. (2015). The epidemiology of dementia and Alzheimer disease in Portugal: Estimations of prevalence and treatment-costs. *Acta Medica Portuguesa*, *28*(2), 182–188.
4. Albert, M. S., Moss, M. B., Tanzi, R., & Jones, K. (2001). Preclinical prediction of AD using neuropsychological tests. *Journal of the International Neuropsychological Society*, *7*(5), 631–639.
5. Alzheimer, A., Stelzmann, R. A., Schnitzlein, H. N., & Murtagh, F. R. (1995). An English translation of Alzheimer's 1907 paper, "Uber eine eigenartige Erkankung der Hirnrinde". *Clinical Anatomy*, *8*(6), 429–431.
6. Kontsekova, E., Zilka, N., Kovacech, B., Novak, P., & Novak, M. (2014). First-in-man tau vaccine targeting structural determinants essential for pathological tau–tau interaction reduces tau oligomerisation and neurofibrillary degeneration in an Alzheimer's disease model. *Alzheimer's Research & Therapy*, *6*(4), 1–12.
7. Howell, A. (2021). *Ethics and End-of-Life Care for Alzheimer's Disease Patients: A Phenomenological Study of Clinical Healthcare Managers' Viewpoints* (Doctoral dissertation, Colorado Technical University).
8. Marsh, J., & Alifragis, P. (2018). Synaptic dysfunction in Alzheimer's disease: The effects of amyloid beta on synaptic vesicle dynamics as a novel target for therapeutic intervention. *Neural Regeneration Research*, *13*(4), 616.
9. Zhang, B., Carroll, J., Trojanowski, J. Q., Yao, Y., Iba, M., Potuzak, J. S., … & Brunden, K. R. (2012). The microtubule-stabilizing agent, epothilone D, reduces axonal dysfunction, neurotoxicity, cognitive deficits, and Alzheimer-like pathology in an interventional study with aged tau transgenic mice. *Journal of Neuroscience*, *32*(11), 3601–3611.
10. Serrano-Pozo, A., Qian, J., Monsell, S. E., Blacker, D., Gómez-Isla, T., Betensky, R. A., … & Hyman, B. T. (2014). Mild to moderate Alzheimer dementia with insufficient neuropathological changes. *Annals of Neurology*, *75*(4), 597–601.
11. James, B. D., Wilson, R. S., Boyle, P. A., Trojanowski, J. Q., Bennett, D. A., & Schneider, J. A. (2016). TDP-43 stage, mixed pathologies, and clinical Alzheimer's-type dementia. *Brain*, *139*(11), 2983–2993.
12. Chokkareddy, R., Thondavada, N., Kabane, B., & Redhi, G. G. (2020). Nanotechnology-based devices in the treatment for Alzheimer's disease. In *Nanomaterials in Diagnostic Tools and Devices* (pp. 241–256). Elsevier.
13. Chêne, G., Beiser, A., Au, R., Preis, S. R., Wolf, P. A., Dufouil, C., & Seshadri, S. (2015). Gender and incidence of dementia in the Framingham Heart Study from mid-adult life. *Alzheimer's & Dementia*, *11*(3), 310–320.
14. Guerreiro, R., & Bras, J. (2015). The age factor in Alzheimer's disease. *Genome Medicine*, *7*(1), 1–3.
15. Brookmeyer, R., Abdalla, N., Kawas, C. H., & Corrada, M. M. (2018). Forecasting the prevalence of preclinical and clinical Alzheimer's disease in the United States. *Alzheimer's & Dementia*, *14*(2), 121–129.
16. Reddy, P. H. (2006). Mitochondrial oxidative damage in aging and Alzheimer's disease: Implications for mitochondrially targeted antioxidant therapeutics. *Journal of Biomedicine and Biotechnology*, *2006*(3), 31372.
17. Khan, A., Corbett, A., & Ballard, C. (2017). Emerging treatments for Alzheimer's disease for non-amyloid and non-tau targets. *Expert Review of Neurotherapeutics*, *17*(7), 683–695.
18. Wallin, A., Blennow, K., & Rosengren, L. (1999). Cerebrospinal fluid markers of pathogenetic processes in vascular dementia, with special reference to the subcortical subtype. *Alzheimer Disease and Associated Disorders*, *13*, S102–S105.
19. Kuo, Y. C., & Rajesh, R. (2019). Challenges in the treatment of Alzheimer's disease: Recent progress and treatment strategies of pharmaceuticals targeting notable pathological factors. *Expert Review of Neurotherapeutics*, *19*(7), 623–652.
20. Lashley, T., Schott, J. M., Weston, P., Murray, C. E., Wellington, H., Keshavan, A., … & Zetterberg, H. (2018). Molecular biomarkers of Alzheimer's disease: Progress and prospects. *Disease Models & Mechanisms*, 11(5):031781.
21. Moretti, R., & Caruso, P. (2020). Small vessel disease-related dementia: An invalid neurovascular coupling?. *International Journal of Molecular Sciences*, *21*(3), 1095.
22. Sämgård, K., Zetterberg, H., Blennow, K., Hansson, O., Minthon, L., & Londos, E. (2010). Cerebrospinal fluid total tau as a marker of Alzheimer's disease intensity. *International Journal of Geriatric Psychiatry: A Journal of the Psychiatry of Late Life and Allied Sciences*, *25*(4), 403–410.
23. Singh, A. K., Mishra, G., Maurya, A., Awasthi, R., Kumari, K., Thakur, A., … & Singh, S. K. (2019). Role of TREM2 in Alzheimer's disease and its consequences on β-amyloid, tau and neurofibrillary tangles. *Current Alzheimer Research*, *16*(13), 1216–1229.
24. Sekiya, M., Wang, M., Fujisaki, N., Sakakibara, Y., Quan, X., Ehrlich, M. E., … & Iijima, K. M. (2018). Integrated biology approach reveals molecular and pathological interactions among Alzheimer's Aβ42, Tau, TREM2, and TYROBP in drosophila models. *Genome Medicine*, *10*(1), 1–20.
25. Tapiola, T., Alafuzoff, I., Herukka, S. K., Parkkinen, L., Hartikainen, P., Soininen, H., & Pirttilä, T. (2009). Cerebrospinal fluid β-amyloid 42 and tau proteins as biomarkers of Alzheimer-type pathologic changes in the brain. *Archives of Neurology*, *66*(3), 382–389.
26. Hansson, O., Zetterberg, H., Buchhave, P., Andreasson, U., Londos, E., Minthon, L., & Blennow, K. (2007). Prediction of Alzheimer's disease using the CSF Aβ42/Aβ40 ratio in patients with mild cognitive impairment. *Dementia and Geriatric Cognitive Disorders*, *23*(5), 316–320.
27. Cable, J., Holtzman, D. M., Hyman, B. T., Tansey, M. G., Colonna, M., Kellis, M., … & Tanzi, R. E. (2020). Alternatives to amyloid for Alzheimer's disease therapies: A symposium report. *Annals Reports*, *1475*(1), 3–14.

28. Zetterberg, H., Andreasson, U., Hansson, O., Wu, G., Sankaranarayanan, S., Andersson, M. E., & Blennow, K. (2008). Elevated cerebrospinal fluid BACE1 activity in incipient Alzheimer's disease. *Archives of Neurology*, *65*(8), 1102–1107.
29. Forlenza, O. V., Radanovic, M., Talib, L. L., Aprahamian, I., Diniz, B. S., Zetterberg, H., & Gattaz, W. F. (2015). Cerebrospinal fluid biomarkers in Alzheimer's disease: Diagnostic accuracy and prediction of dementia. *Alzheimer's & Dementia: Diagnosis, Assessment & Disease Monitoring*, *1*(4), 455–463.
30. Maresova, P., Klimova, B., Novotny, M., & Kuca, K. (2016). Alzheimer's and Parkinson's diseases: Expected economic impact on Europe: A call for a uniform European strategy. *Journal of Alzheimer's Disease*, *54*(3), 1123–1133.
31. Soukup, O., Jun, D., Zdarova-Karasova, J., Patocka, J., Musilek, K., Korabecny, J., … & Kuca, K. (2013). A resurrection of 7-MEOTA: A comparison with tacrine. *Current Alzheimer Research*, *10*(8), 893–906.
32. Lazarou, I., Stavropoulos, T. G., Mpaltadoros, L., Nikolopoulos, S., Koumanakos, G., Tsolaki, M., & Kompatsiaris, I. Y. (2021). Human factors and requirements of people with cognitive impairment, their caregivers, and healthcare professionals for mhealth apps including reminders, games, and geolocation tracking: A survey-questionnaire study. *Journal of Alzheimer's Disease Reports*,*5*(1), 497–513.
33. Alberdi, A., Weakley, A., Schmitter-Edgecombe, M., Cook, D. J., Aztiria, A., Basarab, A., & Barrenechea, M. (2018). Smart home-based prediction of multidomain symptoms related to Alzheimer's disease. *IEEE Journal of Biomedical and Health Informatics*, *22*(6), 1720–1731.
34. Sun, S., Folarin, A. A., Ranjan, Y., Rashid, Z., Conde, P., Stewart, C., … & RADAR-CNS Consortium. (2020). Using smartphones and wearable devices to monitor behavioral changes during COVID-19. *Journal of Medical Internet Research*, *22*(9), e19992.
35. Albrecht, U. V., von Jan, U., Jungnickel, T., & Pramann, O. (2013). App-synopsis-standard reporting for medical apps. *Studies in Health Technology and Informatics*, *192*, 1154–1154.
36. Gold, M., Amatniek, J., Carrillo, M. C., Cedarbaum, J. M., Hendrix, J. A., Miller, B. B., … & Czaja, S. J. (2018). Digital technologies as biomarkers, clinical outcomes assessment, and recruitment tools in Alzheimer's disease clinical trials. *Alzheimer's & Dementia: Translational Research & Clinical Interventions*, *4*, 234–242.
37. Kourtis, L. C., Regele, O. B., Wright, J. M., & Jones, G. B. (2019). Digital biomarkers for Alzheimer's disease: The mobile/wearable devices opportunity. *NPJ Digital Medicine*, *2*(1), 1–9.
38. Koohsari, M. J., Nakaya, T., McCormack, G. R., Shibata, A., Ishii, K., Yasunaga, A., & Oka, K. (2019). Cognitive function of elderly persons in Japanese neighborhoods: The role of street layout. *American Journal of Alzheimer's Disease & Other Dementias*, *34*(6), 381–389.

4

Eco-Friendly Synthesis of Metal Nanoparticles for Smart Nanodevices in the Treatment of Diseases

Jayapriya J. and Rajeshkumar S.

CONTENTS

4.1 Introduction

Atoms are the building blocks of matter; everything on earth is made up of atoms. Nanoparticles range between 1 and 100 nm [1]. Nanoparticles occur widely in nature and are studied in different fields of sciences such as biology, physics, chemistry and geology [2]. Nanotechnology and nanomedicine has been the most emerging branch of science from the last decade to date. Nanotechnology has proved its part in many scientific areas especially in nanobiomedicines [3]. Nanoparticles can be trigged in different ways (physical, chemical and biological) [4]. Though there are three different ways that have been established, a few drawbacks were noticed regarding environmental concerns. In both physical and chemical ways, the toxic substances used affected the globe badly [5]. Further, the eco-friendly approach was needed to overcome the harmful situation [6]. It is otherwise called "green nanotechnology". Developing this eco-friendly method has created a successful impact around the world [7]. Also, green nanotechnology is environmentally friendly, non-toxic, simple, cost-effective and deals with an important branch of sciences [8]. It is a one-step procedure with enhanced stability and eliminates high temperature, pressure and pH. But physical, chemical and biological ways definitely follow one of these two approaches: top-down or bottom-up approach [9]. Noble metal nanoparticles such as gold, silver, platinum and copper are still a new topic of research since the 2000s [10]. One of the major reasons for its rapid development is its physicochemical properties [11]. In this chapter, we are discussing eco-friendly metal nanoparticles and their therapeutic applications. The first part of this chapter talks about different metal nanoparticle biosynthesised from plants, microbes and algae. The second part of the chapter focuses on the different metal nanoparticles and their biomedical applications.

4.2 Nanotechnology

In coexistence with biology, nanotechnology opened up a novel dimension in nanobiotechnology. It deals with the particles of size ranging from 1 to 100 nm. It also involves both prokaryotic and eukaryotic organisms such as plants [12], bacteria [13], fungi [14], viruses [15], algae [14], yeasts, cyanobacteria [15, 16] and actinomycetes [17]. The evolution of metal nanoparticles along with biological material by an eco-friendly approach has huge significance in the research field [18, 19]. This technology holds great promise for medical and non-medical applications.

4.2.1 Nanoparticles

Several hundreds of atoms form a nanoparticle; it appears in different morphologies such as crystalline, amorphous, spherical, round, hexagonal, rod, needle, etc. [20]. Depending upon its size and shape it shows various potential and useful properties. It is widely applied in food safety, medicine, health care, electronics, transportation, agriculture, cosmetics and other fields [21]. Generally, two strategies are used for the synthesis of NPs: top-down and bottom-up. In the top-down method, materials are broken down into smaller ones. The bottom-up

DOI: 10.1201/9781003157823-4

TABLE 4.1
Plant-Mediated Synthesis of Nanoparticles

Metal/metal oxide	Plant/part synthesized	Specifications/characterization	Applications	Citation
Ag	*Ananas comosus*/leaves	UV-vis, FTIR, XRD, TEM, surface plasmon bands (SPB)	Antibacterial activity	[116]
Ag	*Artemesia squamosa*/leaves	UV-vis, FTIR, TEM, zeta potential, EDX	Antibiotic and antimicrobial activity	[117]
Ag	*Artemesia nilagirica*/leaves	SEM, EDX	Anti-agglomeration	[118]
Ag	*Banana*/peel	UV-vis, FTIR, XRD, TEM, DLS, EDX	Antimicrobial activity	[119]
Ag	Beet/root	UV-vis, XRD, TEM, SAED	Antibacterial	[120]
Ag	*Boerhaavia diusa*/whole plant	UV-vis, FTIR, SEM, EDAX, XRD, TEM	Antibacterial activity	[121]
Ag	*Coriandrum sativum*/seed	UV-vis, SEM, PDS, XRD, TEM	Antimicrobial activity	[122]
Ag	*Emblica officinalis*/fruit	UV-vis, FTIR, SEM, XRD, AFM	Antibacterial activity	[123]
Ag	*Illicium verum*/seed	UV-vis, FTIR, SAED, XRD, TEM, EDS, HRTEM, Raman	Molecular sensors and nanophotonic devices	[124]
Ag	*Lansium domesticum*/fruit peel	UV-vis, TEM, SAED, zeta potential	Antimicrobial activity, pharmaceutical, biotechnological and biomedical applications	[125]
Ag	*Latana camara*/leaves	UV-vis, FTIR, SAED, XRD, TEM, XPS, AFM	Antibacterial activity	[126]
Ag	*Neem*/gum	UV-vis, FTIR, XRD, TEM, AFM	Therapeutic applications	[127]
Ag	*Tribulus terrestris*/fruit	UV-vis, FT-IR, TEM, XRD, AFM	Antimicrobial activity	[128]
Au	*Abelmoschus esculentus*	UV-vis, FT-IR, XRD, AFM, FESEM, EDX	Antifungal activity	[129]
Au	*Curcuma pseudomontana*	UV-vis, FT-IR, SEM, HRTEM	Anticancer activity	[130]
Au	*Terminalia arjuna*/leaves	UV-vis, FT-IR, XRD, AFM, TEM	Mitotic cell division	[131]
Au	*Hovenia dulcis*/fruit	UV-vis, FT-IR, TEM, XRD, SAED, EDX	Antibacterial and antioxidant	[132]
Au	*Lansium domesticum*/fruit peel	UV-vis, FT-IR, TEM, SAED	Antimicrobial activity	[133]
Au	Edible mushroom/whole plant	UV-vis, FT-IR, TEM, SAED, XRD, photoluminescence	Therapeutic applications	[134]
Au	*Phoenix dactylifera*/leaves	UV-vis, FT-IR, TEM	Catalytic activity	[135]
Au	*Pistacia integerrima*/gall	UV-vis, FT-IR, SEM	Biological activity	[136]
Cu	*Euphorbia esula* L/leaves	UV-vis, FT-IR, XRD, TEM	Catalytic activity	[137]
Cu	*Ginkgo biloba* L/leaves	UV-vis, FT-IR, TEM, EDS	Catalytic activity	[138]
Cu	*Ocimum sanctum*/leaves	UV-vis, FT-IR, DLS, SEM, EDS	Antibacterial activity	[139]
FeO	*Lawsonia inermis*/leaves	UV-vis, FT-IR, TEM	Antimicrobial activity	[140]
FeO	*Ruellia tuberosa*/leaves	UV-vis, FT-IR, TEM, FESEM, EDX, DLS, DSC	Antimicrobial activity	[141]
Pd	*Syzygium aromaticum*/leaves	UV-vis, XRD, SEM, TEM	Anticancer activity	[142]
Pd	*Solanum nigurum*/leaves	UV-vis, FT-IR, XRD, SEM, TEM	Antimicrobial and nanobiomedicine applications	[143]
Pd	Andean blackberry/leaves	UV-vis, TEM, DLS, XRD	Catalytic activity	[144]
Pd	Piper betle/leaves	UV-vis, FT-IR, TEM, DLS, EDX, XRD, HRTEM, SAED, XPS	Antifungal activity	[145]
Pd	*Delonix regia*/leaves	UV-vis, FT-IR, DLS, XRD, TEM, EDS	Catalytic activity	[146]

method involves the assembly of atoms into molecular structures [22]. Biological synthesis comes under the bottom-up approach. All biological methods have the advantages of being more stable and eco-friendly [23], as shown in Tables 4.1–4.3 and Figure 4.1.

4.2.1.1 Metal Nanoparticles

In this 21st century, the hotspot area of "green synthesis" research is metal nanoparticles. It is the ideal and important route in nanobiomedicine and also in other fields of research and development [18]. A metal nanoparticle is nano-sized with various shapes, sizes and dimensions ranging from 1 to 100 nm [24]. In 1857, Faraday first explored the presence of metal nanoparticles [25]. To date, researchers and scientists have been continuously passionate about this topic. These metal nanoparticles can be synthesized with various chemical reactions to give rise to drugs of interest, antibodies and ligands [26]. It is also widely applied in various ranges of biotechnology [27], targeted drug delivery [28], vehicles for gene and drug delivery [29] and is very important in diagnostic imaging [30].

4.2.1.2 Different Types of Metal and Metal Oxide Nanoparticles

4.2.1.2.1 Silver

Silver is a precious metal. It is an attractive, inorganic, white, lustrous transition metal [31]. AgNP plays a vital role in

TABLE 4.2
Flower-Mediated Nanoparticle Synthesis

Metal nanoparticle	Flower name	Specifications	Application	Citation
Ag	*Lablab purpureus*	UV-vis, FT-IR, SEM, XRD, TEM	Antibacterial activity	[147]
Ag	*Cassia angustifolia*	UV-vis, FT-IR, DLS, XRD, TEM, EDX	Antioxidant activity and antibacterial activity	[148]
Ag	*Caesalpinia pulcherrima*	UV-vis, FT-IR, DLS, XRD, TEM	Antifungal activity	[149]
Au	*Tussilago farfara*	UV-vis, FT-IR, DLS, XRD, TEM, SEM	Antibacterial and cytotoxity activity	[150]
Au	*Plumeria alba* Linn.	UV-vis, FT-IR, DLS, XRD, TEM, EDS	Antibacterial activity	[151]
Cu	*Mimusops elengi*	UV-vis, FT-IR, DLS, TEM, SEM	Antibactrial activity; antifungal activity; antioxidant activity; thrombolytic activity; anti-larval activity; cytotoxicity activity	[152]
Fe	*Piliostigma thonningii*	UV-vis, FT-IR, DLS, XRD, SEM	Antibacterial	[153]
Zn	*Nyctanthes arbor-tristis*	UV-vis, FT-IR, DLS, XRD, TEM	Antifungal activity	[154]
Zn	*Syzygium aromaticum*	UV-vis, FT-IR, DLS, XRD, TEM, SEM	Antifungal activity	[155]
Cd	*Tagetes* sp.	UV-vis, FT-IR, SEM	Larvicidal activity	[156]
Ti	*Calotropis gigantean*	UV-vis, FT-IR, DLS, XRD, TEM, EDX, SEM	Acaricidal activity	[157]

TABLE 4.3
Microbially Mediated Nanoparticle Synthesis

Metal nanoparticle	Bacteria/fungi/yeast/algae	Specification	Applications	Citation
Ag	*Aspergillus niger*	UV-vis, FT-IR, XRD, TEM, EDX	Antibacterial activity	[158]
Ag	*Aspergillus* sp.	UV-vis, FT-IR, EDX, XRD, TEM	Antifungal activity	[159]
Ag	*Cochliobolus lunatus*	UV-vis, FT-IR, XRD, TEM	Mosquito larvicidal	[160]
Ag	*Penicillium chrsogenum*	UV-vis, FT-IR, XRD, TEM, DLS, SEM	Antifungal activity	[161]
Lead	*Aspergillus* sp.	UV-vis, FT-IR, XRD	Antimicrobial	[162]
Ag	*Saccharomyces cerevisiae*	UV-vis, FT-IR, SEM, TEM	Antimicrobial activity	[163]
Au	*Yarrowia lipolytica* NCYC 789	UV-vis, FT-IR, XRD, SEM-EDS	Antibacterial and antibiofilm activity	[163]
Ag	*Bacillus licheniformis*	UV-vis, FT-IR, TEM, EDXS	Antimicrobial and antiviral activity	[164]
Ag	*Deinococcus radiodurans*	UV-vis, FT-IR, XRD, TEM, SEM, EDXS	Antibacterial, antibiofouling, anticancer activity	[165]
Ag	*Nocardiopsis* sp. MBRC-1	UV-vis, FT-IR, XRD, FE-SEM, EDX	Antimicrobial and anticancer activity	[166]
Ag	*Serratia nematodiphila*	UV-vis, FT-IR, XRD, TEM	Antibacterial activity	[167]
Ag	*Spirulina*	UV-vis, FT-IR, EDX, SEM	Antibacterial activity	[168]
Ag	*Chrorella pyrenoidosa*	UV-vis, FT-IR, XRD, TEM, EDX	Antibacterial activity and photocatalytic	[169]
Au	*Streptomyces clavuligerus*	UV-vis, FT-IR, XRD, TEM	Anticancer activity	[170]
Au	*Streptomyces fulvissimus*	UV-vis, FT-IR, XRD, TEM, EDX	Antibacterial activity	[171]

commercial applications. Biosynthesized AgNPs are more acceptable for medical applications such as antimicrobial [32], anti-inflammatory [33], anti-diabetic agents [34], anticancer therapy [35], and are also applied in the promotion of wound repair and bone healing [36], as the vaccine adjuvant and as biosensors [37].

4.2.1.2.2 Gold

Gold is a chemical element with the symbol Au. It is a precious metal used mostly for jewelry, coinage and other artwork [38]. Among the successful development of metal nanoparticles, AuNP explored various nanotechnology-related biomedical applications, and also has non-toxic [39], unique optical, physicochemical and biological properties [40]. Applications include tumor destruction [41], environmental, health care, gene delivery, optics, food industry and space industry [42]. In the 19th century, it has been used as an "anxiolytic" therapy for nervous disorders [43]. And recently, many reports have stated that it can cross the BBB, may interact with DNA and produce genotoxic effects [44].

4.2.1.2.3 Copper

The word copper came from the Latin word "*cuprum*". This is why the chemical symbol for copper is Cu [45]. It has outstanding properties including good electrical conductivity, good thermal conductivity and corrosive resistance [46]. Also, it is a tough, ductile, non-magnetic, attractive, recyclable and easily joined catalytic metal [47]. CuNPs have a high antibacterial

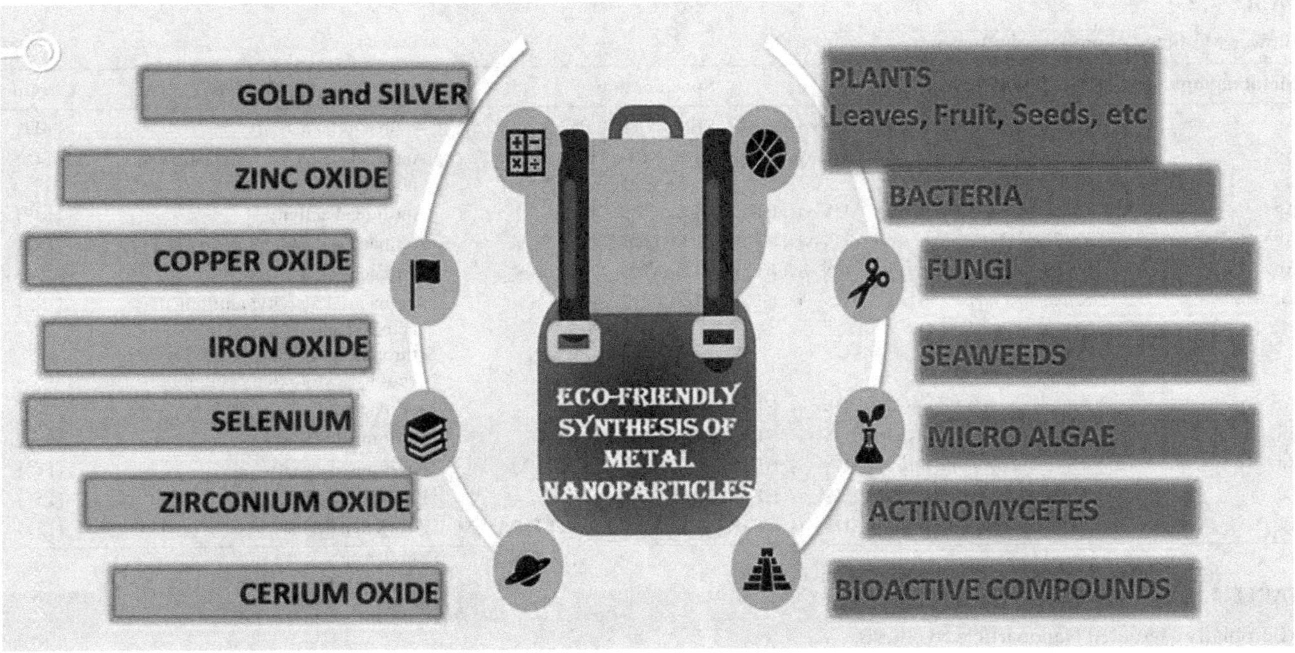

FIGURE 4.1 Eco-friendly synthesis of metal and metal oxide nanoparticles.

strength when compared to other metals [48]. Copper oxide NP has been reported in both in vivo and in vitro studies such as in cancer and neurodegenerative diseases; also it is a potential diagnostic and therapeutic metal NP [49].

4.2.1.2.4 Iron

Iron is also an important metal; it has been applied and utilized in different ways such as environmental [50], textiles [51], health care [52], food industry [53], electronics, renewable energy [54], etc. The body of an adult human contains about 4 g of iron, mostly in hemoglobin and myoglobin [55]. The role of iron in cancer defense can be described as a "double-edged sword" because of its pervasive presence in non-pathological processes [56]. It has a great attraction in biomedical applications due to its non-toxic role in biological systems [57]. The iron oxide nanoparticles have both magnetic behavior and semiconductor properties which lead to multifunctional biomedical applications [58]. The iron oxide nanoparticles used in biomedical fields such as antibacterial, antifungal [59] and anticancer [60] were reviewed. Medical applications and biotechnological advances, including magnetic resonance imaging [61], cell separation and detection [62], tissue repair, magnetic hyperthermia and drug delivery [63], have strongly benefited from employing iron oxide nanoparticles (IONPs) due to their remarkable properties, such as superparamagnetism, size and possibility of receiving a biocompatible coating [64].

4.2.1.2.5 Zinc Oxide

It is used as an additive in numerous materials and products including cosmetics [65], food supplements [66], rubbers, plastics [67], ceramics, glass, cement, lubricants, paints [68], ointments, adhesives [69], pigments, food [70], batteries, fire retardants and first-aid tapes. ZnO NPs with comparatively inexpensive and relatively less toxic properties exhibit excellent biomedical applications, such as anticancer [71], drug delivery [72], antibacterial [73], anti-diabetic, anti-inflammation [74], wound healing, bioimaging, antioxidant and optic properties [75].

4.2.1.2.6 Platinum

Platinum is also a chemical element with the symbol Pt [76]. Platinum was used in biomedical applications [77] as well as being utilized in vehicle emissions control devices, jewelry [78], and petroleum refining [79] and for electrical applications such as hard disk drives [80]. The PtNPs are highly remarkable owing to their intrinsic physicochemical and biological properties making them an effective candidate for catalytic and biomedical applications [81, 82] like targeted drug delivery systems [83], anticancer conjugates, anticancer theranostic agents [84] and anti-Alzheimer's disease compounds [85].

4.2.1.2.7 Palladium

Palladium is a chemical element with many commercial benefits. It has chemical applications, also used in electronics [86], medicine [87], dentistry [88], watch-making [89], blood sugar test strips, aircraft spark plugs, surgical instruments, hydrogen purification, ground treatment and jewelry [89]. Palladium is a key component of fuel cells, which react with hydrogen with oxygen to produce electricity, heat and water [90]. Palladium nanoparticles (PdNPs) have intrinsic features, such as brilliant catalytic, electronic, physical, mechanical and optical properties, as well as diversity in shape and size [91, 92]. The initial research proved that PdNPs have impressive potential for the development of novel photo-thermal agents and photoacoustic agents [93, 94], antimicrobial/antitumor agents [87], gene/drug carriers, prodrug activators and biosensors [95]. In the biomedical field, palladium has primarily been used as a component of alloys for dental prostheses [96]. However, recent research has shown the utility of palladium alloys for devices such as vascular stents that do not distort magnetic resonance images [97]. Therefore, technologies based on alternative metals are now being evaluated for their potential in medical applications.

4.2.1.2.8 Titanium

Titanium and titanium alloys exhibit a unique combination of strength and biocompatibility, which enables their use in medical applications and accounts for their extensive use as implant materials in the last 50 years [98–100]. It has many medical uses, including surgical implements and implants, such as hip balls and sockets (joint replacement and dental implants) [100, 101]. Modern advancements in additive manufacturing techniques have increased potential for titanium use in orthopedic implant applications [101, 102]. Recent advances in the biomedical applications of TiO_2 include photodynamic therapy for cancer treatment [103], drug delivery systems, cell imaging, biosensors for biological assay and genetic engineering [103–105]. The characterizations and applications of TiO_2 nanoparticles, as well as nanocomposites and nanosystems, have been prepared by different modifications to improve the function of TiO_2. Research has supported the significance of nano topography (TiO_2 nanotube diameter) in cell adhesion and cell growth, and suggests that the mechanics of focal adhesion formation are similar among different cell types [106, 107].

4.2.1.2.9 Selenium

Selenium (Se) is an essential trace element. It is mostly available in nuts, cereals and mushrooms. Nanotechnology has established the benefits of selenium in many research articles [108–110]. It is the necessary supplementation for cellular function in animals and humans [111]. In the year 2015, the US Food and Drug Administration (FDA) published the minimum and maximum levels of selenium in infant formula [112]. The content of the human body should range between 13 and 20 mg. Selenium has interacted with iodine and vitamin E [113]. It has been suggested that selenium supplements may prevent different kinds of cancers [114]. Accordingly, it has played a role in nano-therapeutics, diagnostic and medical devices. It has anticancerous, antimicrobial and anti-inflammatory properties [114, 115].

4.3 Biomedical Applications

Despite there being an improvement in technologies in and around medicine and the medical field, people still die due to diseases. The main cause is the negative side of misconsumption of medicines. Therefore, in the case of cancer, targeted and personalized NP-based drugs can be introduced [172, 173], owing to the fact that currently more attention has been received on the topic of NP-based drugs for targeting cancer cells [174, 175]. Likewise, for other diseases too NP-based drugs have proved their part in an eco-friendly method using metal NP. Inorganic nanomaterials, including metal oxides and metal (zinc oxide, iron oxide, titanium dioxide, gold, silver and nickel particles), are some of the promising materials applicable in medicine, such as in cell imaging, biosensing, gene or drug delivery and cancer therapy [176, 177]. Here are some applications highlighted under each property emphasizing green synthesis metal NP. Figure 4.2 shows the different nanoparticles used in infectious diseases.

4.3.1 Antitumor and Anticancer Activity

Many studies established that tumor and cancer cell growth is controlled by eco-friendly synthesized NPs. Green synthesis

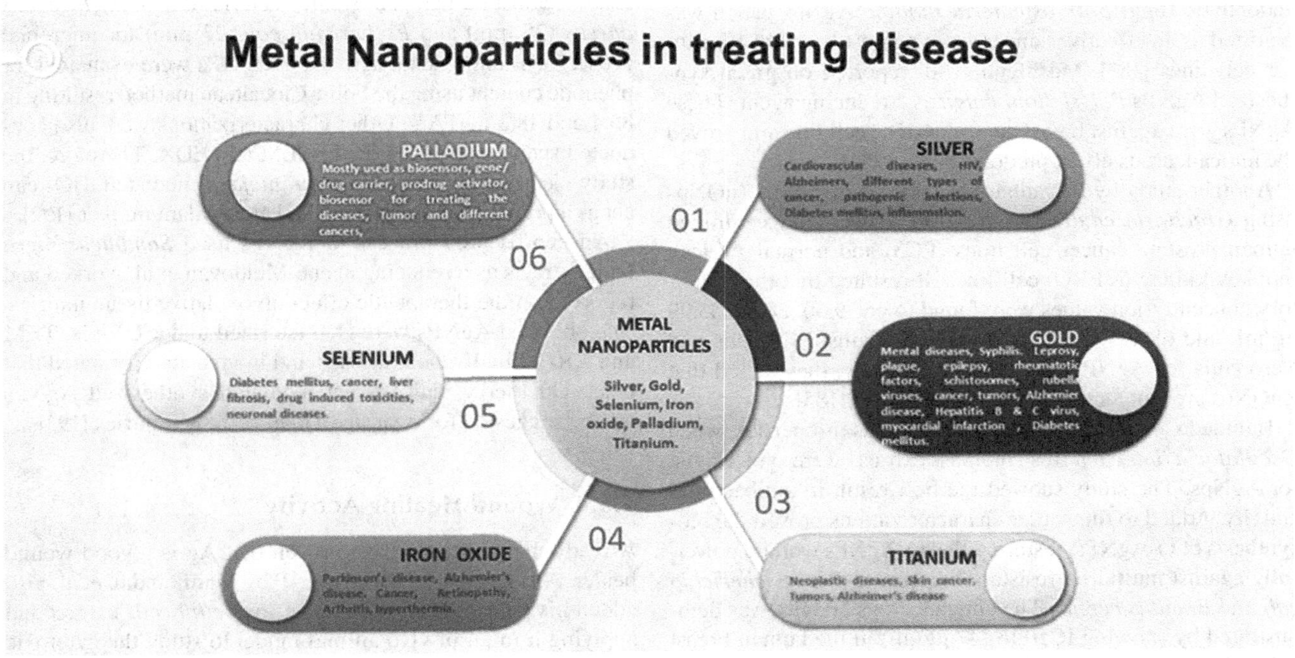

FIGURE 4.2 Nanoparticles used against infectious diseases.

NPs seem to be a novel tool for prognostic biomarkers for cancer diagnosis and drug delivery in tumor cells.

Sharma et al. reported that tungsten nanoparticles mediated aqueous extracts of *Moringa oleifera*. Here, they mentioned various bioactive compounds which were revealed through UV, SEM, TEM, FT-IR and XRD. In the antimicrobial activity, it was potent against *Bacillus subtilis* (18 mm) at 80 μg/ml. For fungus the activity was maximum and noted against *Fusarium oxysporium* (20 mm). In the prothrombin (PT) and activated partial thromboplastin time (APTT) assay, the cytotoxic effect was studied on MCF-7 (breast cancer cell line) and in 3T3 (fibroblast cell line); both showed the result as 200 μg/ml in MCF-7 and 500 μg/ml in 3T3 cell lines. Hence it proved that this method is non-toxic and can be applied as a biomedical application [178].

Rajeshkumar et al. suggested eco-friendly ZnO NPs using *Mangifera indica* extract. In the study, they proved ZnO NPs are a potential and therapeutic marker for treating cancer-causing agents. Briefly revealed, in the MTT assay, ZnO NP was compared with a standard drug (cyclophosphamide). Finally, an MTT assay reported that ZnO NP acted as an anticancer agent against lung (A549) cancer cell lines [179].

Bilal et al. reported on the *Euphorbia helioscopia* methanolic leaf extract using AgNps loaded with chitosan alginate. The study revealed the impact of its anticancerous activity against human epithelial cells (HeLa) and L929 cells [180]. In the in vitro research by Singh et al. *Carissa carandas* leaves were used as a reducing agent which was biofabricated with AgNPs. Here, the research proved that biosynthesized AgNPs were tested against cancer cells like hepatic cell lines (HUH-7) and renal cell lines (HEK-293) [181].

Reporting on *Rheum emodi* roots, Sharma et al. worked on the cytotoxicity of TiO2 NPs. Further, the study revealed that this biosynthesized nano drug can be used to treat liver cancer cells (HepG2) [182]. Akther et al. reported on novel endophytic fungi *Botryosphaeria rhodina* AgNps which was reported as an effective cancer agent against lung (A549) cancer cell lines [183]. Moshfegh et al. reported on green synthesized AgNPs *Polysiphonia alga* as a reducing agent. These AgNPs were against breast cancer MCF-7 cell line and proved the anticancerous effect predominantly [184].

Another study by Priyadharshini et al. worked on ZnONps using *Gracilaria edulis* as a reducing agent. They examined human prostate cancer cell lines (PC3) and normal African monkey kidney (VERO) cell lines. It resulted that the inhibitory concentration values were found to be 39.60, 28.55, 53.99 μg/mL and 68.49, 88.05, 71.98 μg/mL against PC3 cells and Vero cells for AgNPs and ZnONPs. Finally, they stated that ZnONps are a potential anticancerous agent [185].

Hamauda et al. revealed anticancerous properties when *Oscillatoria limnetica* fresh biomass extracts were synthesized for AgNps. The study showed the best result in antibacterial activity. Added to this, other characterizations proved the biosynthesis of O-AgNPs, resulting that O-AgNPs worked powerfully against multidrug-resistant bacteria such as *Escherichia coli* and *Bacillus cereus*. The anticancerous activity was demonstrated by showing IC50 (6.147 μg/ml) in the human breast (MCF-7) cell line and IC50 (5.369 μg/ml) in the human colon cancer (HCT-116) cell line. Hence, proved that O-AgNPs are a notable marker for cancer [186].

4.3.2 Anti-Inflammatory Activity

Kameswari et al. biosynthesized *Acalypha indica*-mediated selenium nanoparticles. The SeNP showed the potential anti-inflammatory property assessed by albumin denaturation assay and α, α-diphenyl-β-picrylhydrazyl free radical scavenging assay [187].

Kumar et al. also reported on the anti-inflammatory property. AgNPs were the first reported green synthesized from the *Holoptelea integrifolia* aqueous extract. Remarkably, the study revealed the binding constant as $2.60 \pm 0.05 \times 10^{-4}$ in the anti-inflammatory activity [188].

In an in vivo study by Zhu et al., SeNPs were coated with *Ulva lactuca* polysaccharide and treated to mice with acute colitis. By increasing the antioxidant activity the selenoproteins reduced the inflammatory responses. They resulted that ULP-SeNPs showed potential antioxidant activity also inhibited inhibit PG-E2 which is an activator of pro-inflammatory cytokines like IL-6 and TNF-α [189].

By using the albumin denaturation assay technique, Rajesh Kumar et al. worked on the anti-inflammatory activity of AgNPs mediated from cumin oil. The result stated that there is an increase in surface plasmon absorption band concentration. It showed a good sign for anti-inflammatory activity [190].

4.3.3 Antimicrobial and Antioxidant Activity

CuNPs were synthesized by Rajesh Kumar et al. using *Persea americana* of seed extract for the purpose of antimicrobial and antioxidant study. The study found that CuNP has a remarkable range against fungal and bacterial pathogens. DPPH activity was noted for antioxidant properties [191].

The research article by Santhosh Kumar et al. worked on aqueous leaf extract of *Psidium guajava* mediated biosynthesis of titanium dioxide nanoparticles (TiO_2 NPs). The extract was tested against human pathogenic bacteria like *Staphylococcus aureus* (25 mm) and *Escherichia coli* (23 mm) for microbial activity. For antioxidant activity, TiO_2 NPs were evaluated for phenolic content using the Folin-Ciocalteau method resulting in 85.4 and 18.3 mgTA/g. Other characterizations were also conducted such as XRD, FTIR, FESEM and EDX. Therefore, the study clearly stated that *Psidium guajava* mediated TiO_2 can act as a promising antibacterial and antioxidant marker [192].

An eco-friendly process of AgNPs used *Sambucus nigra* fruit extracts as a reducing agent. Moldovan et al. worked and reported on the therapeutic effect on oxidative tissue injuries. The obtained AgNPs were characterized under UV-vis, TEM and XRD. Finally, both in vitro and in vivo activity stated that black elderberry fruit extract mediated synthesized AgNPs can be suggested for treating oxidative tissue injuries [193].

4.3.4 Wound Healing Activity

Already many articles have proven that Ag is a good wound healer. An effective study on AgNP by Manikandan et al. also added his point in using *Caulerpa scalpelliformis* extract and applying it in an in vivo animal model to study the cytotoxic effect; rats that were exposed to 40 μg/ml of AgNPs found 50% cell death against MCF-7 cells. In the in vitro study, it showed mRNA level expression of Bax/ Bcl-2 and protein levels of caspase-3 and -9. It also mentioned that *C. scalpelliformis* extract has many biomedical applications such as anti-proliferative, pro-apoptotic and antitumor effects. Hence, it has a potential role in healing chronic and diabetic cutaneous wounds [194].

Another article published by Nei et al. worked with *Impatiens balsamina* extract. For this study, they relied on thermal and excision wound models to prove the wound healing ability of AuNPs. 20 mg of AuNp concentration for eight days significantly proved that AuNP can be applied as a wound dresser after surgery [195].

Shao et al. revealed that ZnO is a promising wound healing agent. In this eco-friendly synthesis method, ZnO was biosynthesized in *Barleria gibsoni* leaf extract. Characterizations such as UV-vis, FT-IR, XRD and TEM showed the results for ZnO NP. Antibacterial activity was tested against bacterial pathogens and proved nano-ZnO gel as an absolute effect in burn infections. It also added that formulated nano-ZnO gel exhibited predominantly in rat models [196].

4.4 Conclusion

In recent years, nanobiotechnology has emerged as an interdisciplinary field of science and played a major role in the production of nanoparticles that are non-toxic, cost-effective and eco-friendly. This can be produced on a large scale in various fields such as renewable energy, therapeutics, pharmaceuticals and commercial products. Using plants, microbes and algae that have the ability to produce metal nanoparticles such as Ag, Au, Pd, Ti, Cu, ZnO, Se, etc., this chapter provides information on the overall eco-friendly synthesis of metal nanoparticle nanodevices in treating diseases. Considering the usefulness of all the metal and metal oxides mediated from plants and other microbes significantly proved their role in each disease. MNPs and MONPs synthesized from plant extracts are dependable, efficacious, foreseeable, scalable, reproducible and safe to be applied through many fields, so they must find ingenious solutions to all the affronts they will meet to open new horizons. Still, there are many varieties to explore its biomedical properties. In the future we can also expect greater observation in the research and development under this topic.

REFERENCES

1. Sudha, P. N., Sangeetha, K., Vijayalakshmi, K., & Barhoum, A. (2018). Nanomaterials history, classification, unique properties, production and market. In *Emerging Applications of Nanoparticles and Architecture Nanostructures* (pp. 341–384). Eds: Abdel Salam Hamdy Makhlouf, Ahmed Barhoum Elsevier.
2. Hochella, M. F., Mogk, D. W., Ranville, J., Allen, I. C., Luther, G. W., Marr, L. C., ... & Yang, Y. (2019). Natural, incidental, and engineered nanomaterials and their impacts on the Earth system. *Science*, *363*(6434).
3. Zottel, A., Videtič Paska, A., & Jovčevska, I. (2019). Nanotechnology meets oncology: Nanomaterials in brain cancer research, diagnosis and therapy. *Materials*, *12*(10), 1588.
4. Liao, J., Jia, Y., Wu, Y., Shi, K., Yang, D., Li, P., & Qian, Z. (2020). Physical-, chemical-, and biological-responsive nanomedicine for cancer therapy. *Wiley Interdisciplinary Reviews: Nanomedicine and Nanobiotechnology*, *12*(1), e1581.
5. Rajeshkumar, S., & Naik, P. (2018). Synthesis and biomedical applications of cerium oxide nanoparticles–A review. *Biotechnology Reports*, *17*, 1–5.
6. Shanmuganathan, R., Karuppusamy, I., Saravanan, M., Muthukumar, H., Ponnuchamy, K., Ramkumar, V. S., & Pugazhendhi, A. (2019). Synthesis of silver nanoparticles and their biomedical applications-a comprehensive review. *Current Pharmaceutical Design*, *25*(24), 2650–2660.
7. Singh, A., Gautam, P. K., Verma, A., Singh, V., Shivapriya, P. M., Shivalkar, S., ... & Samanta, S. K. (2020). Green synthesis of metallic nanoparticles as effective alternatives to treat antibiotics resistant bacterial infections: A review. *Biotechnology Reports*, *25*, e00427.
8. Ahmed, S., Ahmad, M., Swami, B. L., & Ikram, S. (2016). A review on plants extract mediated synthesis of silver nanoparticles for antimicrobial applications: A green expertise. *Journal of Advanced Research*, *7*(1), 17–28.
9. Tahir, M. B., Kiran, H., & Iqbal, T. (2019). The detoxification of heavy metals from aqueous environment using nano-photocatalysis approach: A review. *Environmental Science and Pollution Research*, *26*(11), 10515–10528.
10. Tauran, Y., Brioude, A., Coleman, A. W., Rhimi, M., & Kim, B. (2013). Molecular recognition by gold, silver and copper nanoparticles. *World Journal of Biological Chemistry*, *4*(3), 35.
11. Doria, G., Conde, J., Veigas, B., Giestas, L., Almeida, C., Assunção, M., ... & Baptista, P. V. (2012). Noble metal nanoparticles for biosensing applications. *Sensors*, *12*(2), 1657–1687.
12. Ocsoy, I., Tasdemir, D., Mazicioglu, S., & Tan, W. (2018). Nanotechnology in plants. In *Plant Genetics and Molecular Biology* (pp. 263–275). Eds: Rajeev K. Varshney, Manish K. Pandey, Annapurna Chitikineni, Cham: Springer.
13. Singh, R., Smitha, M. S., & Singh, S. P. (2014). The role of nanotechnology in combating multi-drug resistant bacteria. *Journal of Nanoscience and Nanotechnology*, *14*(7), 4745–4756.
14. Boroumand Moghaddam, A., Namvar, F., Moniri, M., Azizi, S., & Mohamad, R. (2015). Nanoparticles biosynthesized by fungi and yeast: A review of their preparation, properties, and medical applications. *Molecules*, *20*(9), 16540–16565.
15. Azarudeen, R. M. S. T., Govindarajan, M., Amsath, A., Muthukumaran, U., & Benelli, G. (2017). Single-step biofabrication of silver nanocrystals using *Naregamia alata*: A cost effective and eco-friendly control tool in the fight against malaria, Zika virus and St. Louis encephalitis mosquito vectors. *Journal of Cluster Science*, *28*(1), 179–203.
16. Kalimuthu, K., Cha, B. S., Kim, S., & Park, K. S. (2020). Eco-friendly synthesis and biomedical applications of gold nanoparticles: A review. *Microchemical Journal*, *152*, 104296.
17. Hassan, S. E. D., Salem, S. S., Fouda, A., Awad, M. A., El-Gamal, M. S., & Abdo, A. M. (2018). New approach for antimicrobial activity and bio-control of various pathogens by biosynthesized copper nanoparticles using endophytic actinomycetes. *Journal of Radiation Research and Applied Sciences*, *11*(3), 262–270.
18. Thakkar, K. N., Mhatre, S. S., & Parikh, R. Y. (2010). Biological synthesis of metallic nanoparticles. *Nanomedicine: Nanotechnology, Biology and Medicine*, *6*(2), 257–262.

19. Ingale, A. G., & Chaudhari, A. N. (2013). Biogenic synthesis of nanoparticles and potential applications: An eco-friendly approach. *Journal of Nanomedicine and Nanotechnology, 4*(165), 1–7.
20. Masoomi, M. Y., & Morsali, A. (2013). Morphological study and potential applications of nano metal–organic coordination polymers. *RSC Advances, 3*(42), 19191–19218.
21. Kulkarni, N., & Muddapur, U. (2014). Biosynthesis of metal nanoparticles: A review. *Journal of Nanotechnology, 2014*, Article ID 510246.
22. Fu, X., Cai, J., Zhang, X., Li, W. D., Ge, H., & Hu, Y. (2018). Top-down fabrication of shape-controlled, monodisperse nanoparticles for biomedical applications. *Advanced Drug Delivery Reviews, 132*, 169–187.
23. Jamkhande, P. G., Ghule, N. W., Bamer, A. H., & Kalaskar, M. G. (2019). Metal nanoparticles synthesis: An overview on methods of preparation, advantages and disadvantages, and applications. *Journal of Drug Delivery Science and Technology, 53*, 101174.
24. Guo, Z., Chen, Y., Wang, Y., Jiang, H., & Wang, X. (2020). Advances and challenges in metallic nanomaterial synthesis and antibacterial applications. *Journal of Materials Chemistry B, 8*(22), 4764–4777.
25. Rana, A., Yadav, K., & Jagadevan, S. (2020). A comprehensive review on green synthesis of nature-inspired metal nanoparticles: Mechanism, application and toxicity. *Journal of Cleaner Production, 272*, 122880.
26. Schaming, D., & Remita, H. (2015). Nanotechnology: From the ancient time to nowadays. *Foundations of Chemistry, 17*(3), 187–205.
27. Madkour, L. H. (2018). Ecofriendly green biosynthesized of metallic nanoparticles: Bio-reduction mechanism, characterization and pharmaceutical applications in biotechnology industry. *Global Drugs and Therapeutics, 3*(1), 1-11.
28. Rajeshkumar, S. (2016). Anticancer activity of eco-friendly gold nanoparticles against lung and liver cancer cells. *Journal of Genetic Engineering and Biotechnology, 14*(1), 195–202.
29. Panda, M. K., Panda, S. K., Singh, Y. D., Jit, B. P., Behara, R. K., & Dhal, N. K. (2020). Role of nanoparticles and nanomaterials in drug delivery: An overview. *Advances in Pharmaceutical Biotechnology*, 247–265, Eds: Patra J., Shukla A., Das G, Springer, Singapore.
30. Patil, M. P., & Kim, G. D. (2017). Eco-friendly approach for nanoparticles synthesis and mechanism behind antibacterial activity of silver and anticancer activity of gold nanoparticles. *Applied Microbiology and Biotechnology, 101*(1), 79–92.
31. Li, Z., Capretto, D. A., & He, C. (2010). Silver-mediated oxidation reactions: Recent advances and new prospects. *ChemInform, 41*(27), 1-6
32. Abdelghany, T. M., Al-Rajhi, A. M., Al Abboud, M. A., Alawlaqi, M. M., Magdah, A. G., Helmy, E. A., & Mabrouk, A. S. (2018). Recent advances in green synthesis of silver nanoparticles and their applications: About future directions. A review. *BioNanoScience, 8*(1), 5–16.
33. Kedi, P. B. E., Meva, F. E. A., Kotsedi, L., Nguemfo, E. L., Zangueu, C. B., Ntoumba, A. A., ... & Maaza, M. (2018). Eco-friendly synthesis, characterization, in vitro and in vivo anti-inflammatory activity of silver nanoparticle-mediated *Selaginella myosurus* aqueous extract. *International Journal of Nanomedicine, 13*, 8537.
34. Govindappa, M., Hemashekhar, B., Arthikala, M. K., Rai, V. R., & Ramachandra, Y. L. (2018). Characterization, antibacterial, antioxidant, antidiabetic, anti-inflammatory and antityrosinase activity of green synthesized silver nanoparticles using *Calophyllum tomentosum* leaves extract. *Results in Physics, 9*, 400–408.
35. Yesilot, S., & Aydin, C. (2019). Silver nanoparticles: A new hope in cancer therapy? *Eastern Journal of Medicine, 24*(1), 111–116.
36. Ovais, M., Ahmad, I., Khalil, A. T., Mukherjee, S., Javed, R., Ayaz, M., ... & Shinwari, Z. K. (2018). Wound healing applications of biogenic colloidal silver and gold nanoparticles: Recent trends and future prospects. *Applied Microbiology and Biotechnology, 102*(10), 4305–4318.
37. Sundar, S., & Kumar Prajapati, V. (2012). Drug targeting to infectious diseases by nanoparticles surface functionalized with special biomolecules. *Current Medicinal Chemistry, 19*(19), 3196–3202.
38. Tjoa-Bonatz, M. L., & Lockhoff, N. (2019). JAVA: ARTS AND REPRESENTATIONS. Art historical and archaeometric analyses of ancient jewellery (7–16th C.): The prillwitz collection of Javanese gold. *Archipel. Études interdisciplinaires sur le monde insulindien, 97*, 19–68.
39. Fadeel, B., & Garcia-Bennett, A. E. (2010). Better safe than sorry: Understanding the toxicological properties of inorganic nanoparticles manufactured for biomedical applications. *Advanced Drug Delivery Reviews, 62*(3), 362–374.
40. Elahi, N., Kamali, M., & Baghersad, M. H. (2018). Recent biomedical applications of gold nanoparticles: A review. *Talanta, 184*, 537–556.
41. Huang, X., Jain, P. K., El-Sayed, I. H., & El-Sayed, M. A. (2007). Gold nanoparticles: Interesting optical properties and recent applications in cancer diagnostics and therapy. *Nanomedicine, 2*(5). 681-93.
42. Santhoshkumar, J., Rajeshkumar, S., & Kumar, S. V. (2017). Phyto-assisted synthesis, characterization and applications of gold nanoparticles–A review. *Biochemistry and Biophysics Reports, 11*, 46–57.
43. Cryan, J. F., & Sweeney, F. F. (2011). The age of anxiety: Role of animal models of anxiolytic action in drug discovery. *British Journal of Pharmacology, 164*(4), 1129–1161.
44. Cabuzu, D., Cirja, A., Puiu, R., & Mihai Grumezescu, A. (2015). Biomedical applications of gold nanoparticles. *Current Topics in Medicinal Chemistry, 15*(16), 1605–1613.
45. Lipshutz, B. H., & Yamamoto, Y. (2008). Introduction: Coinage metals in organic synthesis. *Chemical Reviews, 108*(8), 2793–2795.
46. Qu, X. H., Zhang, L., Mao, W. U., & Ren, S. B. (2011). Review of metal matrix composites with high thermal conductivity for thermal management applications. *Progress in Natural Science: Materials International, 21*(3), 189–197.
47. Wang, Z., Zhao, S., Pang, H., Zhang, W., Zhang, S., & Li, J. (2019). Developing eco-friendly high-strength soy adhesives with improved ductility through multiphase core–shell hyperbranched polysiloxane. *ACS Sustainable Chemistry & Engineering, 7*(8), 7784–7794.
48. Santhoshkumar, J., Agarwal, H., Menon, S., Rajeshkumar, S., & Kumar, S. V. (2019). A biological synthesis of copper nanoparticles and its potential applications. In *Green Synthesis, Characterization and Applications of Nanoparticles* (pp. 199–221). Eds: Ashutosh Shukla, Siavash Iravani Elsevier.

49. Sulaiman, G. M., Tawfeeq, A. T., & Jaaffer, M. D. (2018). Biogenic synthesis of copper oxide nanoparticles using olea europaea leaf extract and evaluation of their toxicity activities: An in vivo and in vitro study. *Biotechnology Progress*, *34*(1), 218–230.
50. Tarafdar, J. C., & Raliya, R. (2013). Rapid, low-cost, and ecofriendly approach for iron nanoparticle synthesis using *Aspergillus oryzae* TFR9. *Journal of Nanoparticles*, *2013*, Article ID 141274.
51. Fouda, A., Hassan, S. E. D., Saied, E., & Azab, M. S. (2021). An eco-friendly approach to textile and tannery wastewater treatment using maghemite nanoparticles (γ-Fe2O3-NPs) fabricated by *Penicillium expansum* strain (Kw). *Journal of Environmental Chemical Engineering*, *9*(1), 104693.
52. Zou, Y., Wang, X., Khan, A., Wang, P., Liu, Y., Alsaedi, A., ... & Wang, X. (2016). Environmental remediation and application of nanoscale zero-valent iron and its composites for the removal of heavy metal ions: A review. *Environmental Science & Technology*, *50*(14), 7290–7304.
53. Rashidi, L., & Khosravi-Darani, K. (2011). The applications of nanotechnology in food industry. *Critical Reviews in Food Science and Nutrition*, *51*(8), 723–730.
54. Zheng, X. F., Liu, C. X., Yan, Y. Y., & Wang, Q. (2014). A review of thermoelectrics research–Recent developments and potentials for sustainable and renewable energy applications. *Renewable and Sustainable Energy Reviews*, *32*, 486–503.
55. Abbaspour, N., Hurrell, R., & Kelishadi, R. (2014). Review on iron and its importance for human health. *Journal of Research in Medical Sciences: The Official Journal of Isfahan University of Medical Sciences*, *19*(2), 164.
56. Kievit, F. M., & Zhang, M. (2011). Surface engineering of iron oxide nanoparticles for targeted cancer therapy. *Accounts of Chemical Research*, *44*(10), 853–862.
57. Sangaiya, P., & Jayaprakash, R. (2018). A review on iron oxide nanoparticles and their biomedical applications. *Journal of Superconductivity and Novel Magnetism*, *31*(11), 3397–3413.
58. Ling, D., Lee, N., & Hyeon, T. (2015). Chemical synthesis and assembly of uniformly sized iron oxide nanoparticles for medical applications. *Accounts of Chemical Research*, *48*(5), 1276–1285.
59. Arias, L. S., Pessan, J. P., Vieira, A. P. M., Lima, T. M. T. D., Delbem, A. C. B., & Monteiro, D. R. (2018). Iron oxide nanoparticles for biomedical applications: A perspective on synthesis, drugs, antimicrobial activity, and toxicity. *Antibiotics*, *7*(2), 46.
60. Moaca, E. A., Coricovac, E. D., Soica, C. M., Pinzaru, I. A., Pacurariu, C. S., & Dehelean, C. A. (2018). Preclinical aspects on magnetic iron oxide nanoparticles and their interventions as anticancer agents: Enucleation, apoptosis and other mechanism. *Iron Ores and Iron Oxide Materials*, *229*. Eds: Volodymyr Shatokha, IntechOpen.
61. Zhao, Z., Zhou, Z., Bao, J., Wang, Z., Hu, J., Chi, X., ... & Gao, J. (2013). Octapod iron oxide nanoparticles as high-performance T2 contrast agents for magnetic resonance imaging. *Nature Communications*, *4*(1), 1–7.
62. Arias, L. S., Pessan, J. P., Vieira, A. P. M., Lima, T. M. T. D., Delbem, A. C. B., & Monteiro, D. R. (2018). Iron oxide nanoparticles for biomedical applications: A perspective on synthesis, drugs, antimicrobial activity, and toxicity. *Antibiotics*, *7*(2), 46.
63. Vallabani, N. S., & Singh, S. (2018). Recent advances and future prospects of iron oxide nanoparticles in biomedicine and diagnostics. *3 Biotech*, *8*(6), 1–23.
64. Singh, N., Jenkins, G. J., Asadi, R., & Doak, S. H. (2010). Potential toxicity of superparamagnetic iron oxide nanoparticles (SPION). *Nano Reviews*, *1*(1), 5358.
65. Lu, P. J., Huang, S. C., Chen, Y. P., Chiueh, L. C., & Shih, D. Y. C. (2015). Analysis of titanium dioxide and zinc oxide nanoparticles in cosmetics. *Journal of Food and Drug Analysis*, *23*(3), 587–594.
66. Wegmüller, R., Tay, F., Zeder, C., Brnić, M., & Hurrell, R. F. (2014). Zinc absorption by young adults from supplemental zinc citrate is comparable with that from zinc gluconate and higher than from zinc oxide. *The Journal of Nutrition*, *144*(2), 132–136.
67. Begum, P. S., Joseph, R., & Yusuff, K. M. (2008). Preparation of nano zinc oxide, its characterization and use in natural rubber. *Progress in Rubber Plastics and Recycling Technology*, *24*(2), 141–152.
68. Tiwari, N., Pandit, R., Gaikwad, S., Gade, A., & Rai, M. (2017). Biosynthesis of zinc oxide nanoparticles by petals extract of *Rosa indica* L., its formulation as nail paint and evaluation of antifungal activity against fungi causing onychomycosis. *IET Nanobiotechnology*, *11*(2), 205–211.
69. Reddy, A. K., Kambalyal, P. B., Patil, S. R., Vankhre, M., Khan, M. Y., & Kumar, T. R. (2016). Comparative evaluation and influence on shear bond strength of incorporating silver, zinc oxide, and titanium dioxide nanoparticles in orthodontic adhesive. *Journal of Orthodontic Science*, *5*(4), 127.
70. Swain, P. S., Rao, S. B., Rajendran, D., Dominic, G., & Selvaraju, S. (2016). Nano zinc, an alternative to conventional zinc as animal feed supplement: A review. *Animal Nutrition*, *2*(3), 134–141.
71. Mishra, P. K., Mishra, H., Ekielski, A., Talegaonkar, S., & Vaidya, B. (2017). Zinc oxide nanoparticles: A promising nanomaterial for biomedical applications. *Drug Discovery Today*, *22*(12), 1825–1834.
72. Fahimmunisha, B. A., Ishwarya, R., AlSalhi, M. S., Devanesan, S., Govindarajan, M., & Vaseeharan, B. (2020). Green fabrication, characterization and antibacterial potential of zinc oxide nanoparticles using Aloe socotrina leaf extract: A novel drug delivery approach. *Journal of Drug Delivery Science and Technology*, *55*, 101465.
73. Lingaraju, K., Naika, H. R., Nagabhushana, H., & Nagaraju, G. (2019). *Euphorbia heterophylla* (L.) mediated fabrication of ZnO NPs: Characterization and evaluation of antibacterial and anticancer properties. *Biocatalysis and Agricultural Biotechnology*, *18*, 100894.
74. Khatami, M., Varma, R. S., Zafarnia, N., Yaghoobi, H., Sarani, M., & Kumar, V. G. (2018). Applications of green synthesized Ag, ZnO and Ag/ZnO nanoparticles for making clinical antimicrobial wound-healing bandages. *Sustainable Chemistry and Pharmacy*, *10*, 9–15.
75. Kalpana, V. N., & Devi Rajeswari, V. (2018). A review on green synthesis, biomedical applications, and toxicity studies of ZnO NPs. *Bioinorganic Chemistry and Applications*, *2018*,. Article ID 3569758.
76. Naseer, A., Ali, A., Ali, S., Mahmood, A., Kusuma, H. S., Nazir, A., ... & Iqbal, M. (2020). Biogenic and eco-benign synthesis of platinum nanoparticles (Pt NPs) using plants

aqueous extracts and biological derivatives: Environmental, biological and catalytic applications. *Journal of Materials Research and Technology*, *9*(4), 9093–9107.

77. Pasha, S. S., Fageria, L., Climent, C., Rath, N. P., Alemany, P., Chowdhury, R., … & Laskar, I. R. (2018). Evaluation of novel platinum (II) based AIE compound-encapsulated mesoporous silica nanoparticles for cancer theranostic application. *Dalton Transactions*, *47*(13), 4613–4624.
78. Zito, D., Carlotto, A., Loggi, A., Sbornicchia, P., Maggian, D., & Progold, S. A. (2014). Optimization of SLM technology main parameters in the production of gold and platinum jewelry. In *The Santa Fe Symposium on Jewelry Manufacturing Technology 2014* (pp. 439–470). Met-Chem Research. based AIE compound-encapsulated mesoporous silica nanoparticles for cancer theranostic application. *Dalton Transactions*, *47*(13), 4613–24.
79. Motaghed, M., Mousavi, S. M., Rastegar, S. O., & Shojaosadati, S. A. (2014). *Platinum and rhenium* extraction from a spent refinery catalyst using *Bacillus megaterium* as a cyanogenic bacterium: Statistical modeling and process optimization. *Bioresource Technology*, *171*, 401–409.
80. Guo, C. Y., Wan, C. H., He, W. Q., Zhao, M. K., Yan, Z. R., Xing, Y. W., … & Han, X. F. (2020). A nonlocal spin Hall magnetoresistance in a platinum layer deposited on a magnon junction. *Nature Electronics*, *3*(6), 304–308.
81. Puja, P., & Kumar, P. (2019). A perspective on biogenic synthesis of platinum nanoparticles and their biomedical applications. *Spectrochimica Acta Part A: Molecular and Biomolecular Spectroscopy*, *211*, 94–99.
82. Phan, T. T. V., Huynh, T. C., Manivasagan, P., Mondal, S., & Oh, J. (2020). An up-to-date review on biomedical applications of palladium nanoparticles. *Nanomaterials*, *10*(1), 66.
83. Xiao, H., Yan, L., Dempsey, E. M., Song, W., Qi, R., Li, W., … & Chen, X. (2018). Recent progress in polymer-based platinum drug delivery systems. *Progress in Polymer Science*, *87*, 70–106.
84. Kumar, R., Shin, W. S., Sunwoo, K., Kim, W. Y., Koo, S., Bhuniya, S., & Kim, J. S. (2015). Small conjugate-based theranostic agents: An encouraging approach for cancer therapy. *Chemical Society Reviews*, *44*(19), 6670–6683.
85. Barnham, K. J., Kenche, V. B., Ciccotosto, G. D., Smith, D. P., Tew, D. J., Liu, X., … & Cappai, R. (2008). Platinum-based inhibitors of amyloid-β as therapeutic agents for Alzheimer's disease. *Proceedings of the National Academy of Sciences*, *105*(19), 6813–6818.
86. Li, C., Hu, Q., Chen, Q., Yu, W., Xu, J., & Xu, Z. X. (2021). Tetrapropyl-substituted palladium phthalocyanine used as an efficient hole transport material in perovskite solar cells. *Organic Electronics*, *88*, 106018.
87. Lazarević, T., Rilak, A., & Bugarčić, Ž. D. (2017). Platinum, palladium, gold and ruthenium complexes as anticancer agents: Current clinical uses, cytotoxicity studies and future perspectives. *European Journal of Medicinal Chemistry*, *142*, 8–31.
88. Wataha, J. C., & Shor, K. (2010). Palladium alloys for biomedical devices. *Expert Review of Medical Devices*, *7*(4), 489–501.
89. Houghton, O. S., & Greer, A. L. (2021). A conflict of fineness and stability: Platinum-and palladium-based bulk metallic glasses for jewellery. *Johnson Matthey Technology Review, 65(4),* 506-518.
90. Staffell, I., Scamman, D., Abad, A. V., Balcombe, P., Dodds, P. E., Ekins, P., … & Ward, K. R. (2019). The role of hydrogen and fuel cells in the global energy system. *Energy & Environmental Science*, *12*(2), 463–491.
91. Phan, T. T. V., Huynh, T. C., Manivasagan, P., Mondal, S., & Oh, J. (2020). An up-to-date review on biomedical applications of palladium nanoparticles. *Nanomaterials*, *10*(1), 66.
92. Maduraiveeran, G., Sasidharan, M., & Ganesan, V. (2018). Electrochemical sensor and biosensor platforms based on advanced nanomaterials for biological and biomedical applications. *Biosensors and Bioelectronics*, *103*, 113–129.
93. Phan, T. T. V., Nguyen, Q. V., & Huynh, T. C. (2020). Simple, green, and low-temperature method for preparation of palladium nanoparticles with controllable sizes and their characterizations. *Journal of Nanoparticle Research*, *22*(3), 1–11.
94. Nguyen, H. T., Soe, Z. C., Yang, K. Y., Dai Phung, C., Nguyen, L. T. T., Jeong, J. H., … & Kim, J. O. (2019). Transferrin-conjugated pH-sensitive platform for effective delivery of porous palladium nanoparticles and paclitaxel in cancer treatment. *Colloids and Surfaces B: Biointerfaces*, *176*, 265–275.
95. Xu, M., & Li, N. (2020). Metal-based nanocontainers for drug delivery in tumor therapy. In *Smart Nanocontainers* (pp. 195–215). Eds:Phuong Nguyen-Tri, Trong-On Do and Tuan Anh Nguyen Elsevier.
96. Wataha, J. C., & Shor, K. (2010). Palladium alloys for biomedical devices. *Expert Review of Medical Devices*, *7*(4), 489–501.
97. Balakrishnan, P., Sreekala, M. S., & Thomas, S. (Eds.). (2018). *Fundamental Biomaterials: Metals*. Woodhead Publishing.
98. Kulkarni, M., Mazare, A., Gongadze, E., Perutkova, Š., Kralj-Iglič, V., Milošev, I., … & Mozetič, M. (2015). Titanium nanostructures for biomedical applications. *Nanotechnology*, *26*(6), 062002.
99. Kulkarni, M., Mazare, A., Schmuki, P., & Iglič, A. (2014). Biomaterial surface modification of titanium and titanium alloys for medical applications. *Nanomedicine*, *111*, 111.
100. Liu, X., Chen, S., Tsoi, J. K., & Matinlinna, J. P. (2017). Binary titanium alloys as dental implant materials—A review. *Regenerative Biomaterials*, *4*(5), 315–323.
101. Lee, H. H., Lee, S., Park, J. K., & Yang, M. (2018). Friction and wear characteristics of surface-modified titanium alloy for metal-on-metal hip joint bearing. *International Journal of Precision Engineering and Manufacturing*, *19*(6), 917–924.
102. Dharme, M. R., Kuthe, A. M., & Dahake, S. W. (2013). Comparison of fatigue analysis of hip joint implant for stainless steel, cobalt chrome alloys and titanium alloys. *Trends in Biomaterials & Artificial Organs*, *27*(2), 58-61.
103. Nejad, S. M., Takahashi, H., Hosseini, H., Watanabe, A., Endo, H., Narihira, K., … & Tachibana, K. (2016). Acute effects of sono-activated photocatalytic titanium dioxide nanoparticles on oral squamous cell carcinoma. *Ultrasonics Sonochemistry*, *32*, 95–101.
104. Ghosh, S., & Das, A. P. (2015). Modified titanium oxide (TiO2) nanocomposites and its array of applications: A review. *Toxicological & Environmental Chemistry*, *97*(5), 491–514.

105. Jafari, S., Mahyad, B., Hashemzadeh, H., Janfaza, S., Gholikhani, T., & Tayebi, L. (2020). Biomedical applications of TiO2 nanostructures: Recent advances. *International Journal of Nanomedicine*, *15*, 3447.
106. Minagar, S., Wang, J., Berndt, C. C., Ivanova, E. P., & Wen, C. (2013). Cell response of anodized nanotubes on titanium and titanium alloys. *Journal of Biomedical Materials Research Part A*, *101*(9), 2726–2739.
107. Voltrova, B., Hybasek, V., Blahnova, V., Sepitka, J., Lukasova, V., Vocetkova, K., ... & Filova, E. (2019). Different diameters of titanium dioxide nanotubes modulate Saos-2 osteoblast-like cell adhesion and osteogenic differentiation and nanomechanical properties of the surface. *RSC Advances*, *9*(20), 11341–11355.
108. Aliasgharpour, M., & Rahnamaye Farzami, M. (2013). Trace elements in human nutrition: A review. *International Journal of Medical Investigation*, *2*(3), 1-7.
109. Shahri, S. H., Hajinezhad, M. R., & Jamshidian, A. (2021). Comparison of the histopathological effects of selenium nanoparticles and cerium oxide nanoparticles in cadmium-intoxicated rabbits. *Journal of Environmental Treatment Techniques*, *9*(2), 528–533.
110. El-Ramady, H., Abdalla, N., Taha, H. S., Alshaal, T., El-Henawy, A., Salah, E. D. F., ... & Schnug, E. (2016). Selenium and nano-selenium in plant nutrition. *Environmental Chemistry Letters*, *14*(1), 123–147.
111. Benstoem, C., Goetzenich, A., Kraemer, S., Borosch, S., Manzanares, W., Hardy, G., & Stoppe, C. (2015). Selenium and its supplementation in cardiovascular disease—What do we know? *Nutrients*, *7*(5), 3094–3118.
112. Lönnerdal, B., Vargas-Fernández, E., & Whitacre, M. (2017). Selenium fortification of infant formulas: Does selenium form matter? *Food & Function*, *8*(11), 3856–3868.
113. Hess, S. Y. (2010). The impact of common micronutrient deficiencies on iodine and thyroid metabolism: The evidence from human studies. *Best Practice & Research Clinical Endocrinology & Metabolism*, *24*(1), 117–132.
114. Vinceti, M., Filippini, T., Del Giovane, C., Dennert, G., Zwahlen, M., Brinkman, M., ... & Crespi, C. M. (2018). Selenium for preventing cancer. *Cochrane Database of Systematic Reviews*, *2014*(3), 1-11.
115. Ramya, S., Shanmugasundaram, T., & Balagurunathan, R. (2015). Biomedical potential of actinobacterially synthesized selenium nanoparticles with special reference to anti-biofilm, anti-oxidant, wound healing, cytotoxic and anti-viral activities. *Journal of Trace Elements in Medicine and Biology*, *32*, 30–39.
116. Emeka, E. E., Ojiefoh, O. C., Aleruchi, C., Hassan, L. A., Christiana, O. M., Rebecca, M., ... & Temitope, A. E. (2014). Evaluation of antibacterial activities of silver nanoparticles green-synthesized using pineapple leaf (*Ananas comosus*). *Micron*, *57*, 1–5.
117. Vijayakumar, M., Priya, K., Nancy, F. T., Noorlidah, A., & Ahmed, A. B. A. (2013). Biosynthesis, characterisation and anti-bacterial effect of plant-mediated silver nanoparticles using *Artemisia nilagirica*. *Industrial Crops and Products*, *41*, 235–240.
118. Ahmad, N., Sharma, S., Alam, M. K., Singh, V. N., Shamsi, S. F., Mehta, B. R., & Fatma, A. (2010). Rapid synthesis of silver nanoparticles using dried medicinal plant of basil. *Colloids and Surfaces B: Biointerfaces*, *81*(1), 81–86.
119. Ibrahim, H. M. (2015). Green synthesis and characterization of silver nanoparticles using banana peel extract and their antimicrobial activity against representative microorganisms. *Journal of Radiation Research and Applied Sciences*, *8*(3), 265–275.
120. Bindhu, M. R., & Umadevi, M. (2015). Antibacterial and catalytic activities of green synthesized silver nanoparticles. *Spectrochimica Acta Part A: Molecular and Biomolecular Spectroscopy*, *135*, 373–378.
121. Kumar, P. V., Pammi, S. V. N., Kollu, P., Satyanarayana, K. V. V., & Shameem, U. (2014). Green synthesis and characterization of silver nanoparticles using *Boerhaavia diffusa* plant extract and their antibacterial activity. *Industrial Crops and Products*, *52*, 562–566.
122. Nazeruddin, G. M., Prasad, N. R., Prasad, S. R., Shaikh, Y. I., Waghmare, S. R., & Adhyapak, P. (2014). Coriandrum sativum seed extract assisted in situ green synthesis of silver nanoparticle and its anti-microbial activity. *Industrial Crops and Products*, *60*, 212–216.
123. Ramesh, P. S., Kokila, T., & Geetha, D. (2015). Plant mediated green synthesis and antibacterial activity of silver nanoparticles using *Emblica officinalis* fruit extract. *Spectrochimica Acta Part A: Molecular and Biomolecular Spectroscopy*, *142*, 339–343.
124. Luna, C., Chávez, V. H. G., Barriga-Castro, E. D., Núñez, N. O., & Mendoza-Reséndez, R. (2015). Biosynthesis of silver fine particles and particles decorated with nanoparticles using the extract of *Illicium verum* (star anise) seeds. *Spectrochimica Acta Part A: Molecular and Biomolecular Spectroscopy*, *141*, 43–50.
125. Shankar, S., Jaiswal, L., Aparna, R. S. L., Prasad, R. G. S. V., Kumar, G. P., & Manohara, C. M. (2015). Wound healing potential of green synthesized silver nanoparticles prepared from *Lansium domesticum* fruit peel extract. *Materials Express*, *5*(2), 159–164.
126. Ajitha, B., Reddy, Y. A. K., & Reddy, P. S. (2015). Green synthesis and characterization of silver nanoparticles using *Lantana camara* leaf extract. *Materials Science and Engineering: C*, *49*, 373–381.
127. Philip, D. (2009). Biosynthesis of Au, Ag and Au–Ag nanoparticles using edible mushroom extract. *Spectrochimica Acta Part A: Molecular and Biomolecular Spectroscopy*, *73*(2), 374–381.
128. Gopinath, V., MubarakAli, D., Priyadarshini, S., Priyadharsshini, N. M., Thajuddin, N., & Velusamy, P. (2012). Biosynthesis of silver nanoparticles from *Tribulus terrestris* and its antimicrobial activity: A novel biological approach. *Colloids and Surfaces B: Biointerfaces*, *96*, 69–74.
129. Jayaseelan, C., Ramkumar, R., Rahuman, A. A., & Perumal, P. (2013). Green synthesis of gold nanoparticles using seed aqueous extract of *Abelmoschus esculentus* and its antifungal activity. *Industrial Crops and Products*, *45*, 423–429.
130. Muniyappan, N., & Nagarajan, N. S. (2014). Green synthesis of gold nanoparticles using *Curcuma pseudomontana* essential oil, its biological activity and cytotoxicity against human ductal breast carcinoma cells T47D. *Journal of Environmental Chemical Engineering*, *2*(4), 2037–2044.
131. Gopinath, K., Venkatesh, K. S., Ilangovan, R., Sankaranarayanan, K., & Arumugam, A. (2013). Green synthesis of gold nanoparticles from leaf extract of *Terminalia*

arjuna, for the enhanced mitotic cell division and pollen germination activity. *Industrial Crops and Products*, *50*, 737–742.
132. Basavegowda, N., Idhayadhulla, A., & Lee, Y. R. (2014). Phyto-synthesis of gold nanoparticles using fruit extract of *Hovenia dulcis* and their biological activities. *Industrial Crops and Products*, *52*, 745–751.
133. Shankar, S., Jaiswal, L., Aparna, R. S. L., & Prasad, R. G. S. V. (2014). Synthesis, characterization, in vitro biocompatibility, and antimicrobial activity of gold, silver and gold silver alloy nanoparticles prepared from *Lansium domesticum* fruit peel extract. *Materials Letters*, *137*, 75–78.
134. Philip, D. (2009). Biosynthesis of Au, Ag and Au–Ag nanoparticles using edible mushroom extract. *Spectrochimica Acta Part A: Molecular and Biomolecular Spectroscopy*, *73*(2), 374–381.
135. Zayed, M. F., & Eisa, W. H. (2014). *Phoenix dactylifera* L. leaf extract phytosynthesized gold nanoparticles; controlled synthesis and catalytic activity. *Spectrochimica Acta Part A: Molecular and Biomolecular Spectroscopy*, *121*, 238–244.
136. Islam, N. U., Jalil, K., Shahid, M., Muhammad, N., & Rauf, A. (2019). *Pistacia integerrima* gall extract mediated green synthesis of gold nanoparticles and their biological activities. *Arabian Journal of Chemistry*, *12*(8), 2310–2319.
137. Nasrollahzadeh, M., Sajadi, S. M., & Khalaj, M. (2014). Green synthesis of copper nanoparticles using aqueous extract of the leaves of *Euphorbia esula* L. and their catalytic activity for ligand-free Ullmann-coupling reaction and reduction of 4-nitrophenol. *RSC Advances*, *4*(88), 47313–47318.
138. Nasrollahzadeh, M., & Sajadi, S. M. (2015). Green synthesis of copper nanoparticles using *Ginkgo biloba* L. leaf extract and their catalytic activity for the Huisgen [3+ 2] cycloaddition of azides and alkynes at room temperature. *Journal of Colloid and Interface Science*, *457*, 141–147.
139. Patel, B. H., Channiwala, M. Z., Chaudhari, S. B., & Mandot, A. A. (2016). Biosynthesis of copper nanoparticles; its characterization and efficacy against human pathogenic bacterium. *Journal of Environmental Chemical Engineering*, *4*(2), 2163–2169.
140. Chauhan, S., & Upadhyay, L. S. B. (2019). Biosynthesis of iron oxide nanoparticles using plant derivatives of *Lawsonia inermis* (Henna) and its surface modification for biomedical application. *Nanotechnology for Environmental Engineering*, *4*(1), 8.
141. Vasantharaj, S., Sathiyavimal, S., Senthilkumar, P., LewisOscar, F., & Pugazhendhi, A. (2019). Biosynthesis of iron oxide nanoparticles using leaf extract of *Ruellia tuberosa*: Antimicrobial properties and their applications in photocatalytic degradation. *Journal of Photochemistry and Photobiology B: Biology*, *192*, 74–82.
142. Shanthi, K., Sreevani, V., Vimala, K., & Kannan, S. (2017). Cytotoxic effect of palladium nanoparticles synthesized from *Syzygium aromaticum* aqueous extracts and induction of apoptosis in cervical carcinoma. *Proceedings of the National Academy of Sciences, India Section B: Biological Sciences*, *87*(4), 1101–1112.
143. Vijilvani, C., Bindhu, M. R., Frincy, F. C., AlSalhi, M. S., Sabitha, S., Saravanakumar, K., … & Atif, M. (2020). Antimicrobial and catalytic activities of biosynthesized gold, silver and palladium nanoparticles from *Solanum nigurum* leaves. *Journal of Photochemistry and Photobiology B: Biology*, *202*, 111713.
144. Kumar, B., Smita, K., Cumbal, L., & Debut, A. (2015). Ultrasound agitated phytofabrication of palladium nanoparticles using Andean blackberry leaf and its photocatalytic activity. *Journal of Saudi Chemical Society*, *19*(5), 574–580.
145. Mallikarjuna, K., Sushma, N. J., Reddy, B. S., Narasimha, G., & Raju, B. D. P. (2013). Palladium nanoparticles: Single-step plant-mediated green chemical procedure using Piper betle leaves broth and their anti-fungal studies. *International Journal of Chemical and Analytical Science*, *4*(1), 14–18.
146. Dauthal, P., & Mukhopadhyay, M. (2013). Biosynthesis of palladium nanoparticles using *Delonix regia* leaf extract and its catalytic activity for nitro-aromatics hydrogenation. *Industrial & Engineering Chemistry Research*, *52*(51), 18131–18139.
147. Muruganantham, N., Govindharaju, R., Anitha, P., & Anusuya, V. (2018). Synthesis and characterization of silver nanoparticles using *Lablab purpureus* flowers (Purple colour) and its anti-microbial activities. *International Journal of Biological Sciences*, *5*, 1–7.
148. Bharathi, D., & Bhuvaneshwari, V. (2019). Evaluation of the cytotoxic and antioxidant activity of phyto-synthesized silver nanoparticles using *Cassia angustifolia* flowers. *BioNanoScience*, *9*(1), 155–163.
149. Moteriya, P., & Chanda, S. (2017). Synthesis and characterization of silver nanoparticles using *Caesalpinia pulcherrima* flower extract and assessment of their in vitro antimicrobial, antioxidant, cytotoxic, and genotoxic activities. *Artificial cells, Nanomedicine, and Biotechnology*, *45*(8), 1556–1567.
150. Lee, Y. J., Song, K., Cha, S. H., Cho, S., Kim, Y. S., & Park, Y. (2019). Sesquiterpenoids from *Tussilago farfara* flower bud extract for the eco-friendly synthesis of silver and gold nanoparticles possessing antibacterial and anticancer activities. *Nanomaterials*, *9*(6), 819.
151. Nagaraj, B., Malakar, B., Divya, T. K., Krishnamurthy, N. B., Liny, P., & Dinesh, R. (2012). Environmental benign synthesis of gold nanoparticles from the flower extracts of Plumeria alba Linn, (Frangipani) and evaluation of their biological activities. *International Journal of Drug Development & Research*, *4*(1), 144-150.
152. Sarah, S. L. R., & Iyer, P. R. (2019). Green synthesis of copper nanoparticles from the flowers of Mimusops elengi. *International Journal of Recent Scientific Research*, *10*, 32956–32963.
153. Nwamezie, O. U. I. F. (2018). Green synthesis of iron nanoparticles using flower extract of *Piliostigma thonningii* and their antibacterial activity evaluation. *Chemistry International*, *4*(1), 60–66.
154. Jamdagni, P., Khatri, P., & Rana, J. S. (2018). Green synthesis of zinc oxide nanoparticles using flower extract of *Nyctanthes arbor-tristis* and their antifungal activity. *Journal of King Saud University-Science*, *30*(2), 168–175.
155. Lakshmeesha, T. R., Kalagatur, N. K., Mudili, V., Mohan, C. D., Rangappa, S., Prasad, B. D., … & Niranjana, S. R. (2019). Biofabrication of zinc oxide nanoparticles with Syzygium aromaticum flower buds extract and finding its

novel application in controlling the growth and mycotoxins of *Fusarium graminearum*. *Frontiers in Microbiology, 10*, 1244.
156. Hajra, A., Dutta, S., & Mondal, N. K. (2016). Mosquito larvicidal activity of cadmium nanoparticles synthesized from petal extracts of marigold (*Tagetes* sp.) and rose (*Rosa* sp.) flower. *Journal of Parasitic Diseases, 40*(4), 1519–1527.
157. Marimuthu, S., Rahuman, A. A., Jayaseelan, C., Kirthi, A. V., Santhoshkumar, T., Velayutham, K., ... & Rao, K. V. B. (2013). Acaricidal activity of synthesized titanium dioxide nanoparticles using *Calotropis gigantea* against *Rhipicephalus microplus* and *Haemaphysalis bispinosa*. *Asian Pacific Journal of Tropical Medicine, 6*(9), 682–688.
158. Vanaja, M., Rajeshkumar, S., Paulkumar, K., Gnanajobitha, G., Chitra, K., Malarkodi, C., & Annadurai, G. (2015). Fungal assisted intracellular and enzyme based synthesis of silver nanoparticles and its bactericidal efficiency. *International Research Journal of Pharmaceutical and Biosciences, 2*(3), 8–19.
159. Prabavathy, D., Niveditha, R., & Vaishnavie, R. (2015). Antimicrobial activity of silver nanoparticles synthesized by endophytic *Aspergillus* sp. isolated from *Justicia beddomei*. *Journal of Chemical and Pharmaceutical Research, 7*(3), 784–788.
160. Salunkhe, R. B., Patil, S. V., Salunke, B. K., Patil, C. D., & Sonawane, A. M. (2011). Studies on silver accumulation and nanoparticle synthesis by *Cochliobolus lunatus*. *Applied Biochemistry and Biotechnology, 165*(1), 221–234.
161. Pereira, L., Dias, N., Carvalho, J., Fernandes, S., Santos, C., & Lima, N. (2014). Synthesis, characterization and antifungal activity of chemically and fungal-produced silver nanoparticles against *Trichophyton rubrum*. *Journal of Applied Microbiology, 117*(6), 1601–1613.
162. Pavani, T., Rao, K. V., Chakra, C. S., & Prabhu, Y. T. (2015). Microbial synthesis of ZnO nanoparticles by yeast: *Sacchromyces cerevisiae*. *Journal of NanoScience, NanoEngineering & Applications, 5*(2), 1–5.
163. Siddiqi, K. S., Rashid, M., Rahman, A., Husen, A., & Rehman, S. (2018). Biogenic fabrication and characterization of silver nanoparticles using aqueous-ethanolic extract of lichen (*Usnea longissima*) and their antimicrobial activity. *Biomaterials Research, 22*(1), 1–9.
164. Elbeshehy, E. K., Elazzazy, A. M., & Aggelis, G. (2015). Silver nanoparticles synthesis mediated by new isolates of *Bacillus* spp., nanoparticle characterization and their activity against Bean Yellow Mosaic Virus and human pathogens. *Frontiers in Microbiology, 6*, 453.
165. Kulkarni, R. R., Shaiwale, N. S., Deobagkar, D. N., & Deobagkar, D. D. (2015). Synthesis and extracellular accumulation of silver nanoparticles by employing radiation-resistant *Deinococcus radiodurans*, their characterization, and determination of bioactivity. *International Journal of Nanomedicine, 10*, 963.
166. Manivasagan, P., Venkatesan, J., Senthilkumar, K., Sivakumar, K., & Kim, S. K. (2013). Biosynthesis, antimicrobial and cytotoxic effect of silver nanoparticles using a novel *Nocardiopsis* sp. MBRC-1. *BioMed Research International, 2013*.
167. Malarkodi, C., Rajeshkumar, S., Paulkumar, K., Vanaja, M., Jobitha, G. D. G., & Annadurai, G. (2013). Bactericidal activity of bio mediated silver nanoparticles synthesized by *Serratia nematodiphila*. *Drug Invention Today, 5*(2), 119–125.
168. Sudha, S. S., Rajamanickam, K., & Rengaramanujam, J. (2013). Microalgae mediated synthesis of silver nanoparticles and their antibacterial activity against pathogenic bacteria. *Indian Journal of Experimental Biology, 51*(5), 393–9.
169. Aziz, N., Faraz, M., Pandey, R., Shakir, M., Fatma, T., Varma, A., ... & Prasad, R. (2015). Facile algae-derived route to biogenic silver nanoparticles: Synthesis, antibacterial, and photocatalytic properties. *Langmuir, 31*(42), 11605–11612.
170. Kumar, S. A., Peter, Y. A., & Nadeau, J. L. (2008). Facile biosynthesis, separation and conjugation of gold nanoparticles to doxorubicin. *Nanotechnology, 19*(49), 495101.
171. Soltani, N. M., Shahidi, B. G., & Khaleghi, N. (2015). Biosynthesis of gold nanoparticles using *Streptomyces fulvissimus* isolate. *Nanomedicine Journal, 2*(2), 153–159.
172. Ryu, J. H., Lee, S., Son, S., Kim, S. H., Leary, J. F., Choi, K., & Kwon, I. C. (2014). Theranostic nanoparticles for future personalized medicine. *Journal of Controlled Release, 190*, 477–484.
173. Yallapu, M. M., Gupta, B. K., Jaggi, M., & Chauhan, S. C. (2010). Fabrication of curcumin encapsulated PLGA nanoparticles for improved therapeutic effects in metastatic cancer cells. *Journal of Colloid and Interface Science, 351*(1), 19–29.
174. Yao, S., Li, L., Su, X. T., Wang, K., Lu, Z. J., Yuan, C. Z., ... & Song, K. (2018). Development and evaluation of novel tumor-targeting paclitaxel-loaded nano-carriers for ovarian cancer treatment: In vitro and in vivo. *Journal of Experimental & Clinical Cancer Research, 37*(1), 1–13.
175. Aghamiri, S., Jafarpour, A., Malekshahi, Z. V., Mahmoudi Gomari, M., & Negahdari, B. (2019). Targeting siRNA in colorectal cancer therapy: Nanotechnology comes into view. *Journal of Cellular Physiology, 234*(9), 14818–14827.
176. Yaqoob, A. A., Ahmad, H., Parveen, T., Ahmad, A., Oves, M., Ismail, I. M., ... & Mohamad Ibrahim. (2020). Recent advances in metal decorated nanomaterials and their various biological applications: A review. *Frontiers in Chemistry, 8*, 341.
177. Salem, S. S., & Fouda, A. (2021). Green synthesis of metallic nanoparticles and their prospective biotechnological applications: An overview. *Biological Trace Element Research, 199*(1), 344–370.
178. Sharma, A. K., Swami, A. K., Jangir, D., Saran, M., Upadhyay, T. K., Prajapat, R. K., ... & Mathur, M. (2020). An eco-friendly green synthesis of tungsten nanoparticles from *Moringa oleifera* Lam. and their pharmacological studies. *GMJ, 31*, 719–725.
179. Rajeshkumar, S., Kumar, S. V., Ramaiah, A., Agarwal, H., Lakshmi, T., & Roopan, S. M. (2018). Biosynthesis of zinc oxide nanoparticles using *Mangifera indica* leaves and evaluation of their antioxidant and cytotoxic properties in lung cancer (A549) cells. *Enzyme and Microbial Technology, 117*, 91–95.
180. Bilal, M., Rasheed, T., Iqbal, H. M., Li, C., Hu, H., & Zhang, X. (2017). Development of silver nanoparticles loaded chitosan-alginate constructs with biomedical potentialities. *International Journal of Biological Macromolecules, 105*, 393–400.
181. Barabadi, H., Vahidi, H., Rashedi, M., Mahjoub, M. A., Nanda, A., & Saravanan, M. (2020). Recent advances in biological mediated cancer research using silver nanoparticles as a promising strategy for hepatic cancer therapeutics: A systematic review. *Nanomedicine Journal, 7*(4), 251–262.

182. Sharma, D., Parveen, K., Oza, A., & Ledwani, L. (2018). Synthesis of anthraquinone-capped TiO 2 nanoparticles using *R. emodi* roots: Preparation, characterization and cytotoxic potential. *Rendiconti Lincei. Scienze Fisiche e Naturali*, *29*(3), 649–658.
183. Akther, T., Mathipi, V., Kumar, N. S., Davoodbasha, M., & Srinivasan, H. (2019). Fungal-mediated synthesis of pharmaceutically active silver nanoparticles and anticancer property against A549 cells through apoptosis. *Environmental Science and Pollution Research*, *26*(13), 13649–13657.
184. Moshfegh, A., Jalali, A., Salehzadeh, A., & Jozani, A. S. (2019). Biological synthesis of silver nanoparticles by cell-free extract of Polysiphonia algae and their anticancer activity against breast cancer MCF-7 cell lines. *Micro & Nano Letters*, *14*(5), 581–584.
185. Priyadharshini, R. I., Prasannaraj, G., Geetha, N., & Venkatachalam, P. (2014). Microwave-mediated extracellular synthesis of metallic silver and zinc oxide nanoparticles using macro-algae (*Gracilaria edulis*) extracts and its anticancer activity against human PC3 cell lines. *Applied Biochemistry and Biotechnology*, *174*(8), 2777–2790.
186. Hamouda, R. A., Hussein, M. H., Abo-Elmagd, R. A., & Bawazir, S. S. (2019). Synthesis and biological characterization of silver nanoparticles derived from the cyanobacterium *Oscillatoria limnetica*. *Scientific Reports*, *9*(1), 1–17.
187. Kameswari, S., Narayanan, A. L., & Rajeshkumar, S. (2020). Free radical scavenging and anti-inflammatory potential of *Acalypha indica* mediated selenium nanoparticles. *Drug Invention Today*, *13*(2), 348-351.
188. Kumar, V., Singh, S., Srivastava, B., Bhadouria, R., & Singh, R. (2019). Green synthesis of silver nanoparticles using leaf extract of *Holoptelea integrifolia* and preliminary investigation of its antioxidant, anti-inflammatory, antidiabetic and antibacterial activities. *Journal of Environmental Chemical Engineering*, *7*(3), 103094.
189. Zhu, C., Zhang, S., Song, C., Zhang, Y., Ling, Q., Hoffmann, P. R., … & Huang, Z. (2017). Selenium nanoparticles decorated with Ulva lactuca polysaccharide potentially attenuate colitis by inhibiting NF-κB mediated hyper inflammation. *Journal of Nanobiotechnology*, *15*(1), 1–15.
190. Keerthiga, N., Anitha, R., Rajeshkumar, S., & Lakshmi, T. (2019). Antioxidant activity of cumin oil mediated silver nanoparticles. *Pharmacognosy Journal*, *11*(4), 787-789.
191. Morais, E. D. S. (2019). Nanopartículas metálicas biossintetizadas por fungos endofíticos isolados de amêndoas de *Bertholletia excelsa* ducke. *Trabalhos nas áreas de fronteira da química,* 8, 98-111.
192. Santhoshkumar, T., Rahuman, A. A., Jayaseelan, C., Rajakumar, G., Marimuthu, S., Kirthi, A. V., … & Kim, S. K. (2014). Green synthesis of titanium dioxide nanoparticles using *Psidium guajava* extract and its antibacterial and antioxidant properties. *Asian Pacific Journal of Tropical Medicine*, *7*(12), 968–976.
193. Moldovan, B., David, L., Achim, M., Clichici, S., & Filip, G. A. (2016). A green approach to phytomediated synthesis of silver nanoparticles using *Sambucus nigra* L. fruits extract and their antioxidant activity. *Journal of Molecular Liquids*, *221*, 271–278.
194. Manikandan, R., Anjali, R., Beulaja, M., Prabhu, N. M., Koodalingam, A., Saiprasad, G., … & Arumugam, M. (2019). Synthesis, characterization, anti-proliferative and wound healing activities of silver nanoparticles synthesized from *Caulerpa scalpelliformis*. *Process Biochemistry*, *79*, 135–141.
195. Nie, S., Wei, R., Zhou, H., Zhang, L., Chen, Z., & Hou, L. (2020). Eco-friendly synthesis of AuNPs for cutaneous wound-healing applications in nursing care after surgery. *Green Processing and Synthesis*, *9*(1), 366–374.
196. Shao, F., Yang, A., Yu, D. M., Wang, J., Gong, X., & Tian, H. X. (2018). Bio-synthesis of *Barleria gibsoni* leaf extract mediated zinc oxide nanoparticles and their formulation gel for wound therapy in nursing care of infants and children. *Journal of Photochemistry and Photobiology B: Biology*, *189*, 267–273.

5

Raman SERS Nanodevices: The Next-Generation Multiplex Tools for Cancer Diagnostics

Basavaiah Chandu, Puvvada Nagaprasad and Hari Babu Bollikolla

CONTENTS

5.1 Introduction

Cancer is one of the most dangerous threats to modern civilization, causing a large number of deaths even though advanced drugs and phototherapy techniques are available. Nowadays, the number of cancer cases and deaths have increased manifold due to the inefficiencies of these therapeutics. So an important means of cancer survival is early diagnosis because of the limited therapeutic possibilities existing for advanced or metastatic cancer stages. Patients need to be given a prognosis after treatment to avoid the re-occurrence of cancer. Therefore, the diagnosis of a minor amount of cancer cells is another challenge to prevent the risk [1]. Many of the present diagnosis techniques, particularly for cancer, are very complicated, time-consuming, and very costly with minimal flexibility and specificity. Driven by next-generation healthcare and point-of-care diagnosis demands, simple, rapid, sensitive, multiple detections with high specificity and cost-effective techniques for disease identification need to be developed [2].

Recently, biomedical devices based on surface-engineered biocompatible nanoparticles have provided great improvements in this regard. Nanomaterials that possess tunable properties associated with the changes in size, shape, and composition offer the development of therapeutic nanoparticles (or nanodevices) for advanced multifunctional diagnosis methods. These nanodevices conjugated with specific ligands, antibodies, drugs, and analytes are useful in binding to specific cells, thereby facilitating identification, imaging, targeted drug delivery, and monitoring with high affinity and specificity [3]. Current research on cancer diagnosis is mostly fluorescence and sensor detection centric. Among them, surface-enhanced Raman scattering (SERS) associated with metal nanoparticles offers high sensitivity, non-invasive, no photobleaching and toxicity, more simplicity, easy processing, low cost, and recognizes multiple targets in a single scan. The SERS technique has proven to diagnose traces of tumor metabolism and to distinguish the subtle changes between normal and cancer tissues. The SERS technique can be applied in the diagnosis of blood, salvia, and urine samples and is best used for the screening of various stages of cancer, size of the tumors, SERS imaging, and the prognosis of patients after treatment. SERS has demonstrated excellent flexibility, multiple and ultrasensitive detections of cancer biomarkers in body fluids [3, 4].

5.1.1 Raman Spectroscopy

Raman spectroscopy was discovered by the great Indian physicist, Sir C. V. Raman, in 1958. The vibrational and rotational changes of a molecule are detected in Raman spectroscopy and the information obtained is proficient in probing the chemical composition of the molecules. The narrow peak patterning in Raman spectroscopy is considered the chemical fingerprints

DOI: 10.1201/9781003157823-5

for various biomolecules, so the Raman shifts often provide fingerprint information for the analytes [5]. Further, Raman spectroscopy is very simple, very sensitive, and is capable of molecular-level detection; a single molecule can also be detected. Therefore, Raman spectroscopy is extremely advantageous at molecular-level identification and quantification of molecules; thus it requires only a micro-sized sample for diagnosis.

5.1.2 SERS Technology

The Raman peak intensities can be amplified manifold in magnitude simply by adsorbing a Raman active molecule on a metallic surface. This great enhancement of Raman signal intensity is SERS as SERS activity. Several biomolecules, dyes, food colorants, diseased body fluids, etc. give characteristic enhancement in their Raman signals in the presence of metal nanoparticles. Raman SERS was first observed for pyridine molecules adsorbed on the roughened silver electrode, and the intensity of Raman signals of pyridine was amplified manifold [6]. The amplification of Raman signal intensities is ascribed to the electromagnetic field and chemical enhancements [7, 8]. The SERS enhancement is heterogeneous throughout the surface which ranges from one million folds (10^6) to one billion (10^9) folds [9]. When a surface (hot spot) of a material with high field concentration adsorbs a molecule, then under resonance conditions even the Raman signal for a single molecule can also be observed, which is now famous as molecular-level detection using Raman SERS behavior [10]. This enables the detection of even trace amounts of biomolecules like cancer cells that leads to early detection or prognosis of cancer patients for micro-level detection. SERS signals are almost instantaneous making it a reliable technique for the detection of trace amounts of biomolecules. The use of Raman spectroscopy in diagnosis is made possible with the SERS discovery as shown in Figure 5.1.

The excitation energy of Raman scattering instantaneously relaxes at a faster rate thereby reducing the photodegradation of SERS molecules, so the SERS nanodevices are more stable [11]. The Raman signals have very narrow widths compared to fluorescence peaks; thus the spectrum will be very clear with no overlapping [12]. Further, multiple Raman active molecules incorporated into a single SERS label with non-superimposable SERS signatures in the same spectral range facilitate the multiplex detection in a single scan. Raman spectroscopy better suits bioimaging than infrared spectroscopy. These advantages make the SERS technique the best choice of advanced diagnosis technique.

5.1.3 SERS Enhancement Mechanism

SERS phenomenon is a combination of light-metal and light-molecule interactions. The two mechanisms that contribute to the Raman signal strength enhancement are chemical enhancement (short-range) and electromagnetic enhancement (long-range). Raman active molecules adsorbed at the hot spots will couple electronically with the metal surface and enhance its Raman signal strength when illuminated with a Raman laser. This type of signal enhancement is called chemical enhancement (CE) mechanism and it is believed to be similar to the resonance Raman scattering [8]. Generally, molecules that are directly attached to the metal surface show the CE and thus the CE is a very short-ranged effect, observed within a few angstroms length from the metal surface. In addition, the CE factors are weak, typically in the range of 100 to 1,000 [13]. The following four processes contribute to the overall CE [3]:

i) Adsorbed molecules on the metal surface enhance the molecular polarizability.
ii) Charge transfer resonance of the metal-molecule complex.
iii) Enhancement due to resonance scattering.
iv) Enhancement due to plasmon resonance.

The impinged Raman laser on the metal surface drives the delocalized conduction electrons of metals into oscillating free-electron pools. When the intrinsic frequency of the electron pools matches with the incident laser frequency,

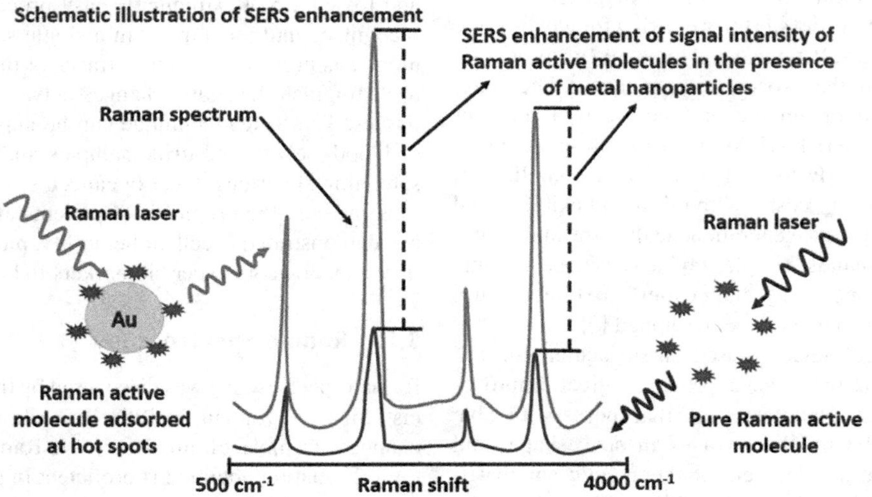

FIGURE 5.1 Schematic representation of SERS enhancement in Raman spectra.

then surface plasmon resonance (SPR) exists. The SPR of the metal nanostructure is highly localized within a specific space which is known as localized SPR (LSPR) [4]. Metals with a strong LSPR effect are termed plasmonic metals and Ag, Au, and Cu with strong SPR in visible to near-IR ranges are some of the best examples. This LSPR effect is called electromagnetic enhancement. Usually, electromagnetic enhancement is the major contributor to SERS with a giant enhancement range from 10^{10}–10^{15} [7]. The chemical and electromagnetic enhancements coexist and together contribute to the overall SERS enhancement, as depicted in Figure 5.2.

5.2 Design and Fabrication of SERS Labels

SERS labels inherently carry Raman signals and they can also perform identification of selective molecules with specific binding functional groups [14]. SERS labels are commonly employed in spectroscopic identification of various disease biomarkers and molecular imaging of tumor cells [15]. SERS labels are advantageous compared to fluorescent probes; SERS labels offer more sensitivity, greater multiplexing capability, and quantitative detection. The SERS labels mainly contain a plasmonic metal nanomaterial as the core, Raman reporter molecules adsorbed on the metal surface, an outer protective shell that encases the Raman reporter, metalcore to enhance the stability of the overall SERS label, and an additional biocompatible layer [16]. The biocompatible layer surface is further functionalized with specific receptors, analytes, biomarkers, and antibodies for specific binding. When the SERS label is illuminated with a suitable laser, the plasmonic field generated from the core amplifies the Raman signal of the Raman reporter and thus produces the SERS signal, characteristic of the Raman molecules, as represented in Figure 5.3.

5.2.1 Choice of Metal

Among plasmonic materials, colloidal solutions of gold and silver nanoparticles are frequently used in the identification of Raman SERS signals for biosensing and bioimaging applications. This is mainly due to the strong amplification of Raman signals at vertices or sharp edges where the electromagnetic fields are confined. Though the plasmonic activity of silver is high in the visible region, its activity is lost in the NIR region. Furthermore, the silver surface is more susceptible to oxidation and is toxic to mammalian cells [17]. On the other hand, gold nanoparticles are nontoxic, soluble in water, have ease of functionalization, are chemically inherent, and even large doses can be administered to humans safely during diagnosis and therapy [18]. Therefore, gold nanoparticles are the preferred choice as the core in a SERS label.

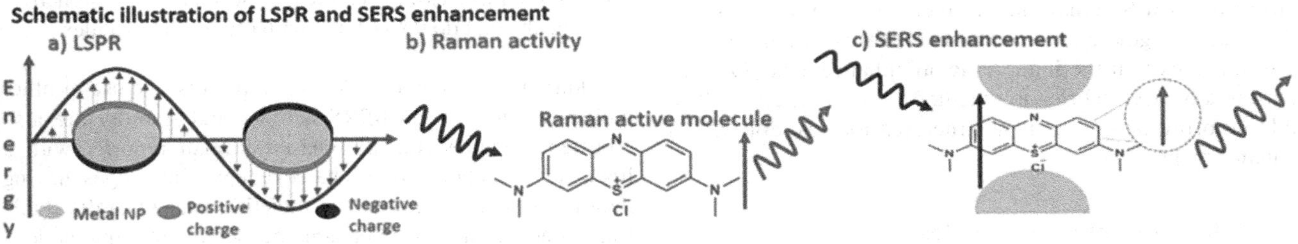

FIGURE 5.2 Schematic representation of (a) LSPR effect, (b) Raman activity, (c) SERS electromagnetic enhancement.

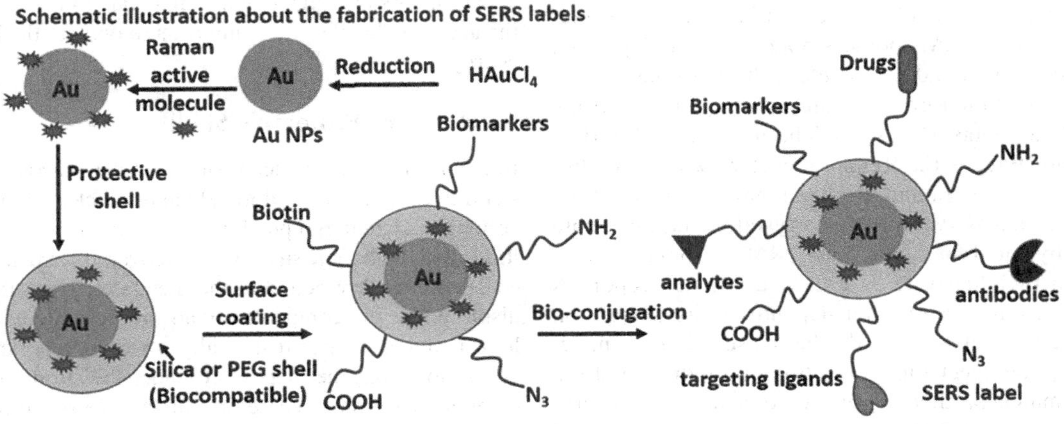

FIGURE 5.3 The design of a biofunctionalized SERS label containing metalcore, Raman reporters, protective shell, surface engineered with various analytes.

Colloidal solutions of gold and silver are generally obtained by reduction of chloroauric acid and silver nitrate with citrate solution which also acts as a stabilizing agent. Then the addition of salt leads to the aggregation of particles; however, these NPs lack uniformity, reproducibility, and the distribution of hot spots are heterogeneous with varying SERS intensities. Monodisperse nanoparticles deposited on a substrate in a controlled fashion using polymer surfaces, drop-casting colloidal solutions or Langmuir-Blodgett films, and template synthesis via lithography are the other routine methods used to fabricate uniform SERS metals with tailored hot spots and a high level of reproducibility.

5.2.2 Hot Spots

The entire surface of a nanomaterial will not show the SERS effect. Only some areas of the surface with high field concentration show this SERS effect. These spots are called hot spots and the distribution of hot spots on a single nanomaterial is trivial. The sharp edges, sharp tips, and vertices present in the anisotropic nanostructure (rods, flowers, cubes, triangles, stars, etc.) are quite attractive for SERS applications because of their high Raman signal enhancement and thus possess good hot spots [7]. However, the presence of hot spots at crevices, junctions, and gaps in nanoparticle aggregation or when they are near each other is most common. Several factors such as the size of the gaps, neighboring particle size, surrounding dielectric properties, particle orientation influenced polarization of light, etc. will contribute to the SERS signal enhancement from these hot spots. For example, gold and silver nanospheres with a 40–60 nm size range exhibit the highest signal enhancement [19]. The current advanced fabrication methods have allowed the flexible tailoring of several nanostructures for desirable SERS enhancements.

5.2.3 Raman Active Molecules

The desirable Raman active molecules (RAMs)/Raman reporters are the molecules that exhibit inherent Raman spectra and also have high affinity toward metal surfaces, large Raman cross-sections that are non-superimposable with the other biomolecules (characteristic Raman signatures) [20]. Often, RAMs possess polarizable conjugated π electronic structures and also contain thiol, amine, or isothiocyanate functional groups which are capable of strong affinity for the metasurfaces [7]. The most crucial part of a SERS nanodevice is the Raman reporter owing to its fingerprint Raman spectra and the signal enhancement. Many organic dyes such as cyanine dyes, crystal violet, malachite green, methylene blue, Nile blue, R6G, aromatic compounds, and their derivates are used as Raman reporters [3]. The selection of a reporter depends on its adsorption range and the wavelength of the laser used. For example, if the absorption spectrum of the Raman reporter and the laser line matches, then a surface-enhanced resonance Raman scattering (SERRS) enhancement of 10–100-fold is also observed [21]. The Raman reporters are anchored to the metal core through chemical or physical adsorption. But under robust biological environments, these molecules may desorb and the biological molecules may get adsorbed on the core which destroys the SERS label. This situation will drastically reduce the SERS signal and/or the signal is completely lost. To prevent the destabilization of SERS labels, Raman reporters are commonly adsorbed on metal surfaces either with strong covalent bonds through sulfur and nitrogen atoms of Raman reporters or strong physical adsorption using molecules capable of large surface binding affinity [3]. For instance, diethylthiatricarbocyanine (DTTC) dye possessing positively charged sulfur moiety can electrostatically anchor to negatively charged citrate capped gold core [22]. On the other hand, water-soluble Raman reporters which will undergo desorption from the core under robust biological conditions are also not suitable for biological applications. So, low water-soluble molecules such as 4-nitrophenol, malachite green, etc. are preferred for in vitro and in vivo SERS devices [7].

The important characteristics for good Raman reporter molecules are:

i) They should exhibit strong SERS signals.
ii) They must possess a strong affinity with the metal surface. Chemisorption is a preferred mode of linking to physisorption.
iii) The reporters must be stable to high-intensity laser and high local fields.
iv) They must exhibit characteristic and well-assigned peaks that do not overlap with other biomolecules.

In addition to the common Raman reporters, the use of other molecules with strong SERS activity and stability are also reported – triphenylmethine-based Raman reporter with a lipoic acid anchor group, chalcogenopyrylium dyes having multiple selenium and sulfur moieties, etc. Overall, stable Raman active molecules capable of electrostatic interactions and covalent linkage with the metal surface are preferred in the fabrication of Raman labels. The high electric fields at the hotspots may lead to the destruction of the Raman active molecules that result in the loss of SERS activity. Therefore, to avoid the loss of signal enhancement due to the breakdown of the active molecules, it is important to operate the laser at low energy.

5.2.4 Outer Protective Shell

In colloidal solutions, electrostatic repulsions contribute to the stability of ligand-coated metal nanoparticles. However, these colloids may undergo precipitation or aggregation in the body fluids with high ionic strength. Moreover, if Raman active molecules are weakly bound to the metal core, then they may be displaced by the competitively adsorbing proteins in the biological fluids or electrostatically attached reporters undergo desorption from the metal core under the strong ionic nature of biological fluids. Therefore, under robust biological environments, the Raman active molecules may undergo desorption which destroys the SERS label. This situation can be avoided by having a protective outer layer that stabilizes the SERS label. Further, functionalization of this outer shell with antibodies for biocompatibility or specific functional groups for targeted identification of the biomolecules is necessary. Diverse molecules that include antibodies, biological derivatives, polymers, polyethylene glycol, silica, and graphene derivatives are attractive as stable protective layer materials to enhance the biocompatibility and stability of SERS labels [7]. The additional attractions associated with these protective shell materials are their biocompatibility, low toxicity, and surface engineering with known reactions to host a diversity of molecules. These modifications will enable the SERS labels to apply in the broad category of applications in sensors, diagnosis, and multiplex identification of biomolecules. Moreover, the stabilized SERS labels can be advantageous for reproducibility and long-term use under robust biological environments.

SERS labels without protection shells are called native SERS labels and protective shells are called sandwich SERS labels. The biosensing applications of native SERS labels are initially reported by the Porter group and the Mirkin group [23, 24]. The Porter group used different thiol derivatives as the Rama reporters which bind onto the gold surface through a more facile covalent linkage of Au-S. This SERS label is further functionalized with antibodies for targeting ability and demonstrated the multiplex detection capability of ELISA-like immunoassay. The use of dyes and oligonucleotides as Raman reporters is exemplified by the Mirkin group for highly selective multiplex detection of oligonucleotides. Though these native SERS labels showed excellent biosensing and bioimaging capability, they suffered from poor stability and reproducibility. To overcome these problems, a protective layer encapsulating the native SERS label is introduced in sandwich SERS labels. Sandwich SERS labels are advantageous with high stability, flexibility for further functionalization. The first sandwich SERS label with a glass protective shell was reported by the Natan group [25] and then by the Nie group [11]. The Nie group also reported the use of polyethylene glycol (PEG) as the protective shell [26]. Layer-by-layer coating of polyelectrolyte is also reported as a protective shell in addition to proteins, carbohydrates, lipids, and peptides [7].

5.2.5 Bioconjugation

Further functionalization with several targeted ligands, biomarkers, antibodies, drugs, various analytes, and surface functional groups on the surface of the protective shell is vital for the specific application of the Raman labels. Selective binding of SERS labels to disease-specific biomolecules in the body fluids is possible; the biomarkers like antibodies, proteins, peptides, and aptamers detect the specific disease molecules and the drugs attached on the surface will be delivered to the targeted cells. Various analytes attached to the SERS label are useful in vivo in monitoring the stages of the disease and the efficacies of the therapeutics. Thus, bioconjugation as illustrated in Figure 5.4 plays an important role in SERS technology.

5.3 SERS Labels in Cancer Diagnosis

SERS spectra of cancer and normal cells are different because of the higher metabolic rates, transformations in metabolic pathways, and variable levels of surface receptor regulations [27]. So the diagnosis, profiling, and quantification of cancer are possible with specific Raman signatures exhibited in SERS spectra. The rapidness and simple operating processes associated with the SERS technique are advantageous in cancer diagnosis compared to mass spectrometry, fluorescence, PCR, flow cytometry (FCM), and other methods. SERS labels with excellent selectivity, specificity, sensitivity, efficiency, reliability, reproducibility, low cost, and multiplex detection capability seem to be the next-generation diagnostic tool for better diagnosis, prognosis, and imaging of cancer cells. Some of the important SERS applications toward cancer are presented here. There are high chances for the reversion of cancer cells even after the patient has undergone advanced treatment techniques like chemotherapy or bone marrow transplantation. The small quantities of malignant cells that remain in the body relapse the disease again thus reduce the long-term survival rate of the cancer patients. This in vivo presence of small amounts of the malignant cells after the treatment is called "minimal residual disease" (MRD) [28]. Moreover, early detection of cancer depends on the identification of very small amounts of the disease cells. So molecular level detection with "limit of detection" (LOD) values ranging from 1 cell/mL to 1,000 cells/mL are highly appreciated in early diagnosis. Therefore, accurate and sensitive techniques are in high demand for MRD and early detection which could be a great benefit in the prognosis of cancer patients thereby enhances their long-term survival rate.

5.3.1 Cancer Screening

Several disease-specific biomarkers such as proteins, nucleic acids, metabolites, circulating tumor cells, and many small molecules will be produced by the disease cells in the body which are then shed into the body fluids. So disease screening, monitoring, and efficacy of therapeutic treatment can be evaluated by quantification of these biomarkers. For example, SERS technology is used to identify several biomarkers in the biological fluids like saliva, blood, and urine to provide fingerprint information about the cancer cells, a useful technology for cancer screening. Further, several biomarkers successfully classified the normal, liver cancer, and nasopharyngeal cancer cells with 95% accuracy and showed 90% diagnostic accuracy when these cell lines are mixed with silver nanoparticles. CTC phenotypic changes of melanoma cancer cells with a handheld Raman instrument during the treatment are observed by screening several biomarkers coated gold nanoparticles through the SERS technique. This could lead to simple, efficient, and affordable techniques to monitor the patients for drug resistance.

Oral squamous cell carcinoma (OSCC) is a widespread head and neck squamous cell carcinoma. Xue et al. studied OSCC

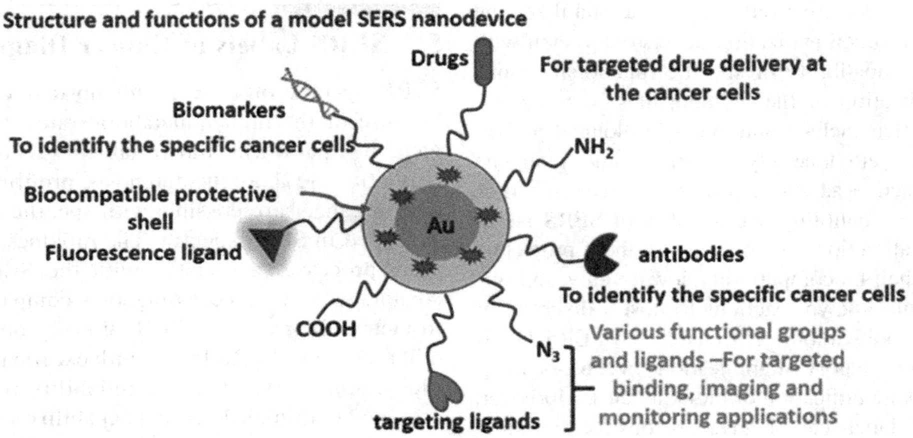

FIGURE 5.4 A model depiction of SERS nanodevice with multiplex capabilities.

disease blood fluids mixed with gold nanoparticles using the Raman SERS technique [29]. About 150 blood samples are collected from patients and after centrifugation, the blood serum sample is collected and mixed with gold nanoparticles. The mixture is agitated for uniform mixing, then a drop of it is used for acquiring Raman data. The healthy blood serum sample showed no enhancement in its signal intensity when mixed with gold nanoparticles whereas the diseased blood serum sample mixed with gold nanoparticles showed profound signal enhancement in their Raman peaks. In these SERS spectra, the Raman signatures of biomolecules in the serum are utilized to differentiate the different cancer cells and stages. Using the principal component and linear discriminant analyses (PCA, LDA) techniques, the collected blood samples were categorized into four groups as T1, T2, T3, and T4 based on the tumor size as shown in Figure 5.5.

The quantity of mucin protein MUC4 is used as a biomarker in the diagnosis and prognosis of pancreatic cancer. A SERS label fabricated with Au NPs, 4-nitrobenzenethiol as Raman active molecule, and 8G7 antibody could effectively identify the MUC4 biomarker with a 33 ng/mL of LOD [30]. Multiple biomarker detection with high selectivity is desired in diagnosis to conserve the time, patient specimen. Three different breast cancer biomarkers are diagnosed in a single scan using a gold nanostar-based SERS label with 0.99 U/mL, 0.13 U/mL, 0.05 ng/mL LOD values [31]. The low quantities of genetic and epigenetic biomarkers need challengeable and highly sensitive diagnostic tools. Silver nano rice-based SERS label is reported for the detection of Hepatitis B virus DNA detection [32]. Danish et al. demonstrated the in vivo and in vitro multiplex detection capability of Raman SERS fabricated with Au NPs and three different Raman reporters – Cy5, MGITC, and Rh6B – that are encapsulated with PEG outer shell [33]. The PEGlated protective shell is further functionalized with characteristic antibodies for specific detection of the three cancer biomarkers EGFR, CD44, and TGFβRII in the breast cancer model as shown in Figure 5.6.

Wang et al. [51] achieved MRD diagnosis by simultaneous detection of two surface markers CD19 and CD20 on the B cell hematological malignancies employing the SERS technique. Raji cells CD19 and CD20 along with K562 control cells with varying concentrations from 5 to 5,000 cells/mL are used for model detection. Blood samples from patients with B lymphoid leukemia and B lymphoma are obtained under the supervision of senior physicians. The SERS labels are prepared by adding the two Raman receptors MBA and DNTB to Ag NPs followed by subsequent coating with PVP, SiO_2, and decoration with anti-CD19 and anti-CD20 antibodies. These individual and mixture (for simultaneous analysis) of SERS probes, Ag NPs@SiO_2@antibodies are then mixed with cell lines and real blood samples in PBS and drop-coated onto a glass slide five times. Then the SERS signals are measured to assess the expression levels of CD19 and CD20. The CD19 (at 1,078 cm^{-1}) and CD20 (at 1,330 cm^{-1}) SERS spectral detections detected for even 5 cells/mL and the SERS signal intensity increased with an increase in concentration up to 5,000 cells/mL, as shown in Figure 5.7. Further, the simultaneous detection with the mixture of two SERS probes exhibited good accuracy without any cross-reactivity between the two cell lines. Further, the experiments on 13 patients showed a linear correlation with the results obtained by the FCM technique. It proves that SERS technology could be a potential cancer diagnosis technique with very low LOD values, as tabulated in Table 5.1.

Ruan et al. [52] have reported the detection of CTC using triangular Ag nanoprisms (AgNPR) and superparamagnetic iron oxide nanoparticles (SPION)-based SERS probes with very low LOD values. CTC detection is very challenging owing to their low availability in blood and their identification could be beneficial in the diagnosis of early tumors [2]. So, the SERS technique which is very sensitive in molecular level detection is applied for the diagnosis of CTCs. A SERS probe based on the mixture of AgNPR-MBA-reductive BSA-folic acid (AgNPR@MBA@rBSA@FA) and SPION@rBSA@FA is employed to capture, enrich, and detect HeLa CTCs in rabbit blood samples with 1 cell/mL, the lowest LOD value. This supersensitive SERS probe is also employed in the diagnosis of ovarian cancer, kidney cancer, breast cancer, and lung cancer, as shown in Figures 5.8–5.9.

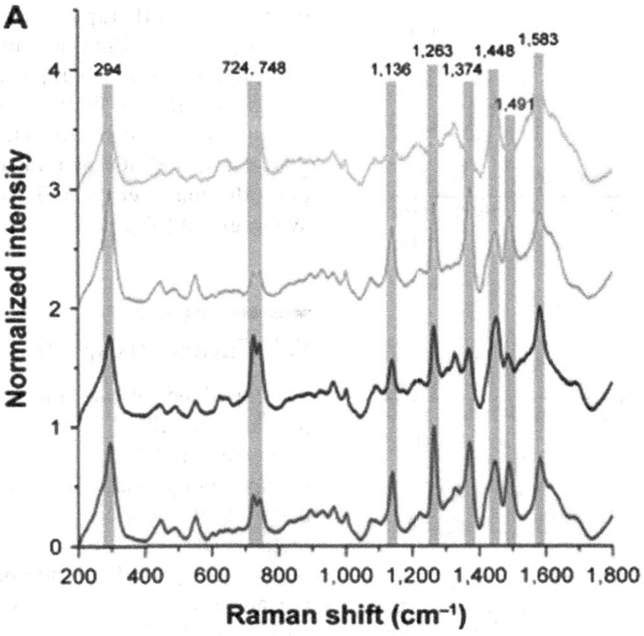

FIGURE 5.5 The SERS spectra are grouped into T1, T2, T3, and T4 groups (top to bottom). Here based on the intensity of peaks toward the right end, the top two groups are identified as early stages and the bottom two groups are identified as advanced stages of cancer. Reproduced from [29]. Copyright 2018, Xue et al., Dove Medical Press Limited.

5.3.2 Imaging Technique Based on SERS Detection

The SERS technique with unique fingerprint Raman spectra of RAMs has garnered much attention in the recent literature for swift biosensing and imaging of cancer disease. The surface engineering of SERS labels with various molecules possessing biocompatible and specific binding capable surface functional groups incorporates good biocompatibility and selective binding with the target tumor cells. Thus, surface-engineered SERS labels with exceptional biocompatibility, superior specificity, and sensitivity are then introduced into the targeting systems. These SERS labels with targeted cell identification functionality are known to accumulate on the malignant tissues owing to their enhanced permeability and retention (EPR) effect [53]. When the SERS labels are administered into the body fluids, using their specific binding nature, will bind to the targeted tumors. Then the tumors are imaged with suitable lasers to exactly identify the location, stage, and size. The first SERS imaging technique is demonstrated by the Schlucker group [54].

Tumor imaging using Raman labels fabricated with a gold core and three different Raman active molecules – methylene blue, 4,4-azobis(pyridine), and 4,4-dipyridyl – is demonstrated by Lim et al. [55]. The Raman label is further modified with mPEG thiol, RGD peptide, MLS peptide, and NLS peptide for selectively targeting cytoplasm and mitochondria. The incorporated Raman reporters exhibited strong and uniform Raman signals when illuminated with low power laser and brief exposure times. The high-resolution Raman images enabled the rapid monitoring of cell morphology changes and also clearly depicted the distribution of Raman labels on the cytoplasm, mitochondria, and nucleus without damaging them. Yang et al. also reported cancer cell imaging using a low-cost Raman label engineered by gold core, covalently linked Raman dye, mPEG-SH, and Au-Ag outer shell [56]. The Raman label demonstrated strong, uniform, reproducible SERS signals for multiplex detection and imaging of cancer cells. Further M. Li et al. demonstrated the development of a low cytotoxicity Raman imaging label employed for imaging the live prostate cancer cell by selective recognition of prostate-specific membrane antigen (PSMA) [57]. The SERS label is conjugated with urea-based small molecule inhibitors to create the SERS imaging agent capable of PSMA specific recognition.

The Jeong group established a fluorescence-Raman endomicroscopic system for multiple target detection and imaging [58], the Ren group demonstrated an Au@organosilica Raman label for live-cell imaging [59], the Zhang group developed a Fe_2O_3@Aubased Raman label for tumor imaging [60], and the Chourpa and Cui groups developed fluorescence and Raman dual-loaded magnetic labels capable of multiplex detection, separation, and imaging [61, 62]. A significant improvement of Raman SERS imaging technology is the multiplex SERS imaging, collecting images of multiple cancer cells in a single experiment. Multiplex imaging is demonstrated by Gambhir et al. [63] using 10 Raman signatures encoded into SERS labels. Five prominent SERS labels with characteristic spectral features are simultaneously injected into a mouse and its liver is imaged after 24 h. The multiplex SERS images clearly showed the accumulation of these five SERS labels. This report proved the cost-effective multiplex detection and imaging capacity of Raman SERS labels.

5.3.3 Multifunctional Applications of SERS Labels

In addition to imaging and sensing techniques, the inherent properties of the metal NPs present in SERS labels

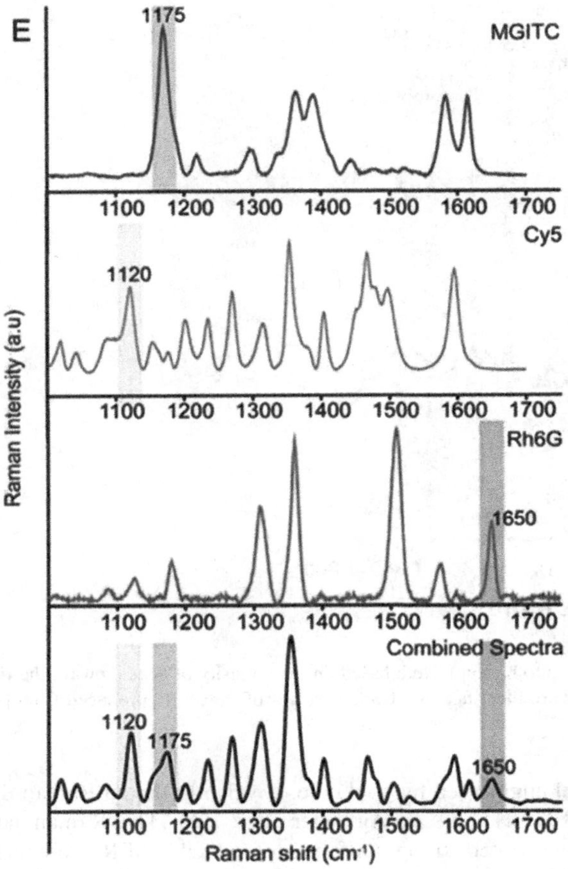

FIGURE 5.6 The multiplex detection using a characteristic Raman signature of three Raman reporters MGITC, Cy5, Rh6B. These molecules showed fingerprint peaks at 1,120, 1,175, and 1,650 cm^{-1} in their individual as well as combined Raman SERS spectra. Reproduced from [33]. Copyright 2018, Springer Nature.

enabled photo-thermal therapy, photodynamic therapy, magnetic separation, drug delivery, and monitoring of the tumor cells, as shown in Figure 5.10. Hence SERS labels with diagnosis, treatment, and monitoring capabilities will emerge as the "all in one" multifunctional devices for cancer patients [4]. Drezek and Sayed's groups have studied photo-thermal therapy research on cancer cells using SERS labels [64, 65]. Simultaneous multimodal tumor detection and photodynamic therapy using Au nanorod-based SERS labels containing fluorescent labels and photosensitizers are demonstrated by the He group [66]. A double-walled gold nanocage/SiO_2 nanorattle-based SERS label capable of photothermal therapy and drug delivery is fabricated by the Wang group [67].

5.4 Future Prospects

As explained above, the rich spectroscopic information provided by SERS can be applied in efficient diagnosis, prognosis, and imaging of cancer cells with high sensitivity, specificity, and multiplexing capacity. As stated, the distribution of hot spots and their density is very small on the metal surface, the information from the remaining surface is zero though a huge number of Raman active molecules are adsorbed on the surface. Also, the uniform substrates with high SERS activity offer uniform SERS intensities and high sensitivity of the analytes [6]. Hence more efforts are needed to fabricate uniform substrates with a high density of hotspots. Perhaps the greatest problem of human in vivo use of SERS labels is their biocompatibility and toxicity. Numerous modifications with biomolecules are available for the biocompatibility of SERS but the toxicity needs to be investigated. Gold is more nontoxic than Ag and large quantities of Au can be administered to humans for treatment and imaging the cancer tumors but more efforts are required to fabricate more biocompatible and nontoxic SERS labels [68]. Other pivotal factors in the fabrication of simple and cost-effective SERS-based nanodevices for cancer diagnosis are the specificity, sensitivity, and multiplexing capability of SERS labels. Though many biocompatible and nontoxic SERS labels are reported, their clinical trials in humans still need to be addressed. SERS labels with specific binding capacity can exhibit multi-model applications in tumor imaging, targeted drug delivery, and monitoring the therapeutic responses, so future research must be concentrated in this area which is expected to be the possible treatment technique for cancer in the future.

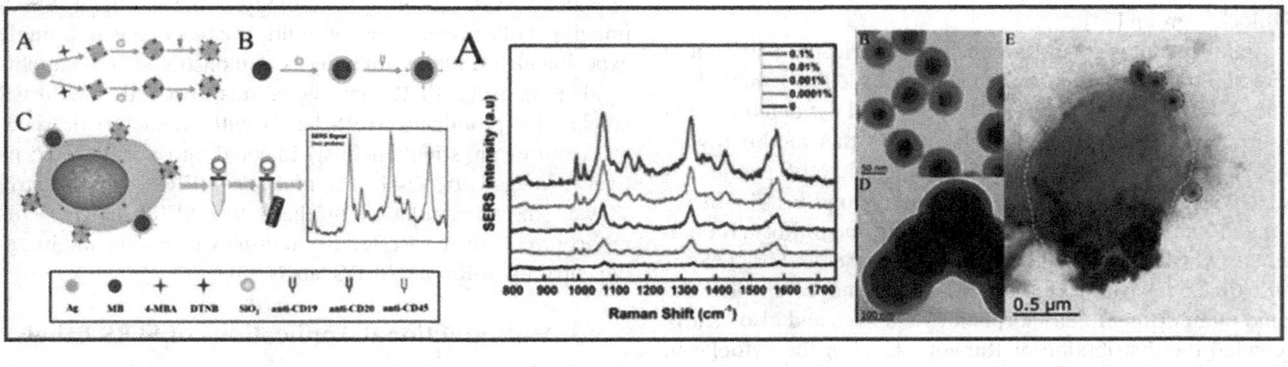

FIGURE 5.7 The schematic showing the synthesis of SERS probes, their SERS signals, and the TEM images of the SERS probes. Reproduced with permission from [51].

TABLE 5.1
Various Reports on SERS Detection of Cancer Cells

S. no.	Type of cancer cells	SERS substrate	Diagnosis type	LOD/accuracy	Ref.
1	Prostate cancer	Ag NPs	Screening	94%	[34]
2	Ovarian cancer	Ag NPs	Screening	96%	[35]
3	Breast cancer Cancer antigen 15-3 Cancer antigen 27-29 Cancer ambryonic antigen	Au nanostars	Multiplex imaging	0.99 U/mL 0.13 U/mL 0.05 ng/mL	[31]
4	Hepatitis B virus	Ag nanorice and Au triangles	Screening	50 aM	[32]
5	Prostate cancer	Au NPs	Multiplex detection	1 ng/mL to 10 μg/mL	[36]
6	Colorectal cancer	Au NPs	Screening	–	[37]
7	Lung cancer	Au NPs	Screening	–	[38]
8	Mouse IgG antibodies	Au nanostars	Screening	10 fg/mL	[39]
9	Prostate cancer	Ag NPs	Screening	–	[40]
10	Oral cancer	Ag NPs coated on TiO_2 NPs	Screening, classification	97%	[41]
11	Pancreatic cancer	Au NPs	Early detection	87%	[42]
12	Nasopharyngeal cancer	Ag NPs	Screening	83%	[43]
13	Liver cancer AFP protein CEA protein FER protein	Au NPs	Multiplex detection	0.15 pg/mL 20 pg/mL 4 pg/mL	[44]
14	Prostate cancer	Au nanosperes	Screening	10^{-15} M	[45]
15	Nasopharyngeal cancer	Ag NPs coated on Si NPs	Screening and isolation	10^{-13} M	[46]
16	Lung cancer	Ag films	Screening	2×10^{-16} M	[47]
17	Breast cancer	Gold nanopillar	Screening	10^{-7} M	[48]
18	Breast cancer	Au NPs	Screening	–	[49]
19	Breast cancer Prostate cancer Colorectal cancer	Au shell magnetic beads	Multiplex detection	32 exosomes/μL 73 exosomes/μL 203 exosomes/μL	[50]
20	Bladder cancer	Au NPs	Screening	–	[1]

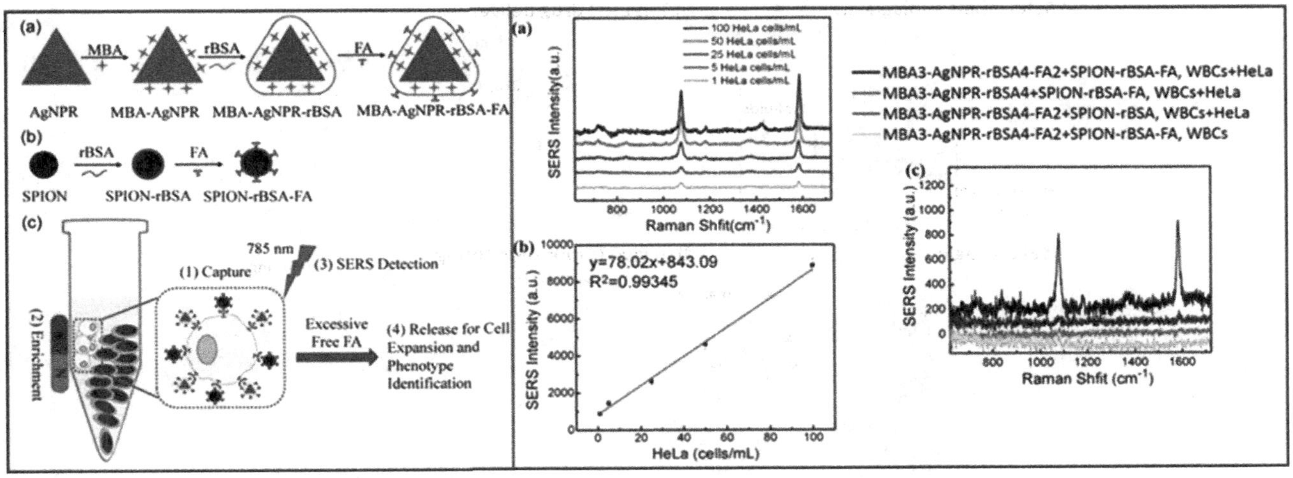

FIGURE 5.8 The schematic showing the fabrication of different SERS nanodevices (left box), SERS sensitivity, and specificity spectra (right box). The SERS signal showed selectivity for rabbit blood with HeLa cells. Reproduced with permission from [52]. Copyright 2018, American Chemical Society.

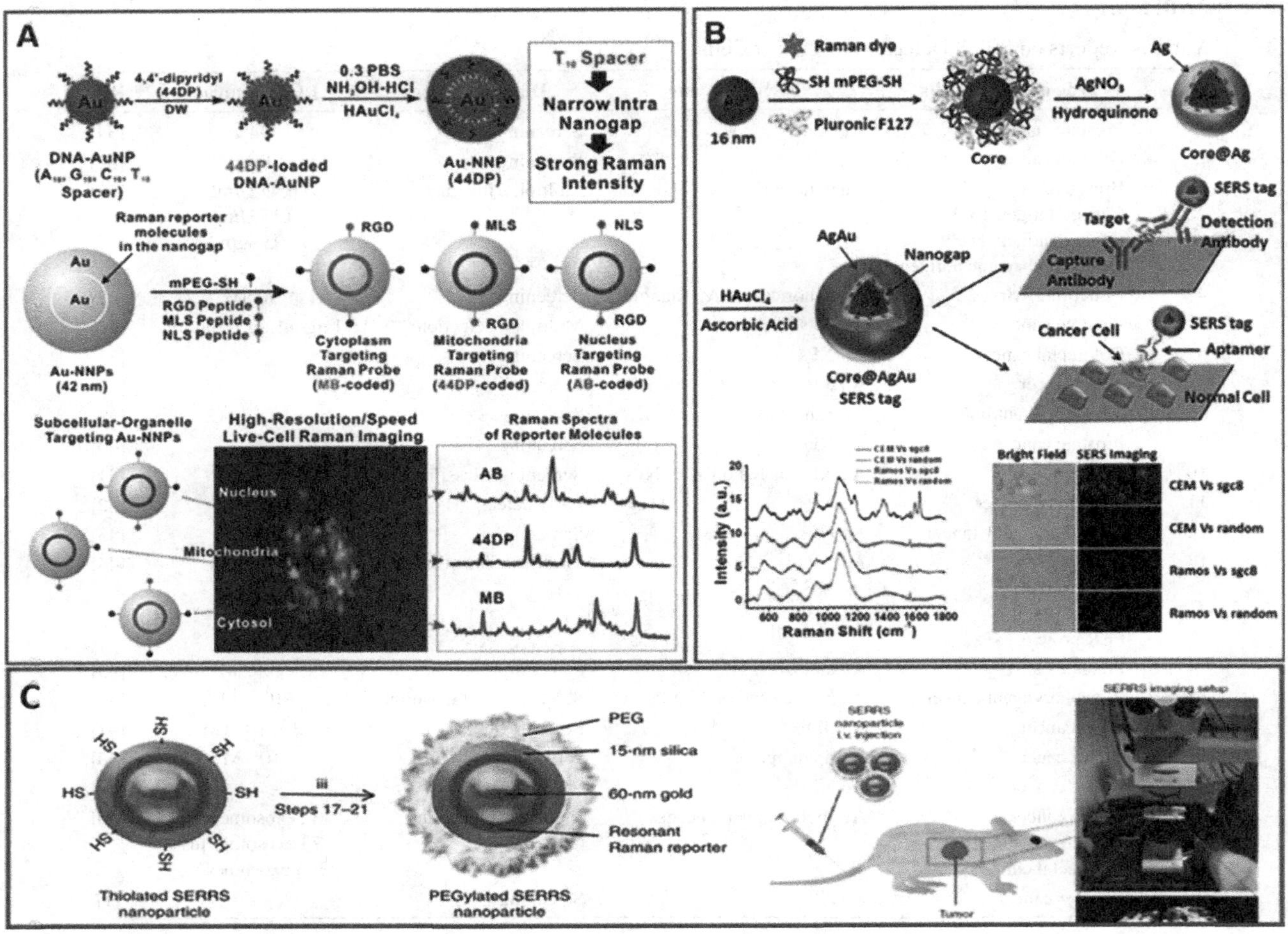

FIGURE 5.9 Cancer cell imaging techniques using targeted Raman SERS labels. Reprinted with permission from [7]. Copyright 2018, Elsevier.

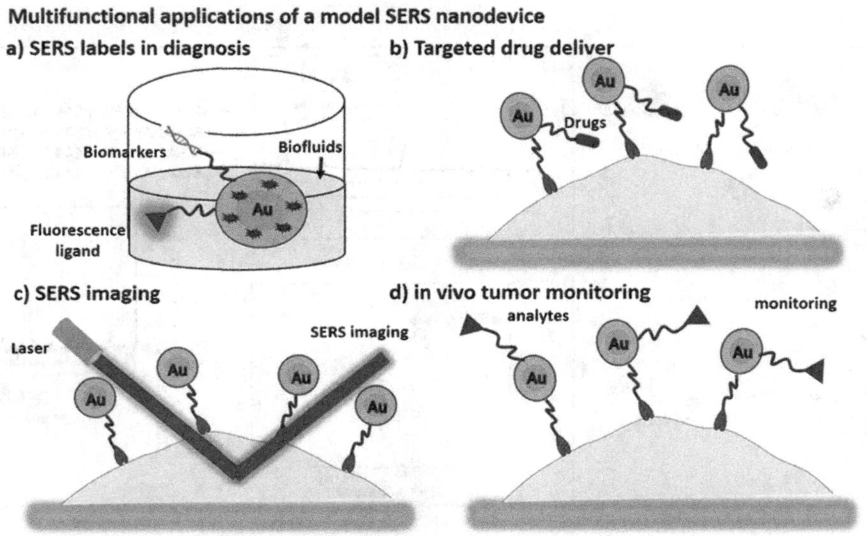

FIGURE 5.10 Schematic illustration of various applications of SERS labels.

REFERENCES

1. Davis, R. M., Kiss, B., Trivedi, D. R., Metzner, T. J., Liao, J. C., & Gambhir, S. S. (2018). Surface enhanced Raman scattering nanoparticles for multiplexed imaging of bladder cancer tissue permeability and molecular phenotype. *ACS Nano, 12*(10), 9669–9679.
2. Shen, Z., Wu, A., & Chen, X. (2017). Current detection technologies for circulating tumor cells. *Chemical Society Reviews, 46*(8), 2038–2056.
3. Lane, L. A., Qian, X., & Nie, S. (2015). SERS nanoparticles in medicine: From label-free detection to spectroscopic tagging. *Chemical Reviews, 115*(19), 10489–10529.
4. Zong, C., Xu, M., Xu, L. J., Wei, T., Ma, X., Zheng, X. S., Hu, R., & Ren, B. (2018). Surface enhanced Raman spectroscopy for bioanalysis: Reliability and challenges. *Chemical Reviews, 118*(10), 4946–4980.
5. Wang, H., Zhang, S., Wan, L., Sun, H., Tan, J., & Su, Q. (2018). Screening and staging for non-small cell lung cancer by serum laser Raman spectroscopy. *Spectrochimica Acta Part A: Molecular and Biomolecular Spectroscopy, 201*, 34–38.
6. Fleischmann, M., Hendra, P. J., & McQuillan, A. J. (1974). Raman spectra of pyridine adsorbed at a silver electrode. *Chemical Physics Letters, 26*, 163–166.
7. Shan, B., Pu, Y., Chen, Y., Liao, M., & Li, M. (2018). Novel SERS labels: Rational design, functional integration and biomedical applications. *Coordination Chemistry Reviews, 371*, 11–37.
8. Albrecht, M. G., & Creighton, J. A. (1977). Anomalously intense Raman spectra of pyridine at a silver electrode. *Journal of the American Chemical Society, 99*(15), 5215–5217.
9. Gersten, J., & Nitzan, A. (1980). Electromagnetic theory of enhanced Raman scattering by molecules adsorbed on rough surfaces. *The Journal of Chemical Physics, 73*(7), 3023–3037.
10. Nie, S., & Emory, S. R. (1997). Probing single molecules and single nanoparticles by surface enhanced Raman scattering. *Science (New York, N.Y.), 275*(5303), 1102–1106.
11. Doering, W. E., & Nie, S. (2003). Spectroscopic tags using dye embedded nanoparticles and surface enhanced Raman scattering. *Analytical Chemistry, 75*(22), 6171–6176.
12. McCreery, R. L. (2005). *Raman Spectroscopy for Chemical Analysis*. John Wiley & Sons.
13. Morton, S. M., & Jensen, L. (2009). Understanding the molecule-surface chemical coupling in SERS. *Journal of the American Chemical Society, 131*(11), 4090–4098.
14. Li, M., Cushing, S. K., & Wu, N. (2014). Plasmon-enhanced optical sensors: A review. *Analyst, 140*(2), 386–406.
15. Doering, W. E., Piotti, M. E., Natan, M. J., & Freeman, R. G. (2007). SERS as a foundation for nanoscale, optically detected biological labels. *Advanced Materials, 19*(20), 3100–3108.
16. Li, M., Cushing, S. K., Zhang, J., Lankford, J., Aguilar, Z. P., Ma, D., & Wu, N. (2012). Shape-dependent surface-enhanced Raman scattering in gold–Raman-probe–silica sandwiched nanoparticles for biocompatible applications. *Nanotechnology, 23*(11), 115501.
17. Ahamed, M., Karns, M., Goodson, M., Rowe, J., Hussain, S. M., Schlager, J. J., & Hong, Y. (2008). DNA damage response to different surface chemistry of silver nanoparticles in mammalian cells. *Toxicology and Applied Pharmacology, 233*(3), 404–410.
18. Cormode, D. P., Roessl, E., Thran, A., Skajaa, T., Gordon, R. E., Schlomka, J. P., Fuster, V., Fisher, E. A., Mulder, W. J. M., Proksa, R., & Fayad, Z. A. (2010). Atherosclerotic plaque composition: Analysis with multicolor CT and targeted gold nanoparticles. *Radiology, 256*(3), 774–782.
19. Samanta, A., Jana, S., Das, R. K., & Chang, Y. T. (2014). Wavelength and shape dependent SERS study to develop ultrasensitive nanotags for imaging of cancer cells. *RSC Advances, 4*(24), 12415–12421.
20. Wang, Z., Zong, S., Wu, L., Zhu, D., & Cui, Y. (2017). SERS activated platforms for immunoassay: Probes, encoding methods, and applications. *Chemical Reviews, 117*(12), 7910–7963.
21. McNay, G., Eustace, D., Smith, W. E., Faulds, K., & Graham, D. (2011). Surface enhanced Raman scattering (SERS) and surface enhanced resonance Raman scattering (SERRS): A review of applications. *Applied Spectroscopy, 65*(8), 825–837.
22. Jiang, L., Qian, J., Cai, F., & He, S. (2011). Raman reporter-coated gold nanorods and their applications in multimodal optical imaging of cancer cells. *Analytical and Bioanalytical Chemistry, 400*(9), 2793–2800.
23. Ni, J., Lipert, R. J., Dawson, G. B., & Porter, M. D. (1999). Immunoassay readout method using extrinsic Raman labels adsorbed on immunogold colloids. *Analytical Chemistry, 71*(21), 4903–4908.
24. Cao, Y. C., Jin, R., & Mirkin, C. A. (2002). Nanoparticles with Raman spectroscopic fingerprints for DNA and RNA detection. *Science, 297*(5586), 1536–1540.
25. Mulvaney, S. P., Musick, M. D., Keating, C. D., & Natan, M. J. (2003). Glass-coated, analyte-tagged nanoparticles: A new tagging system based on detection with surface enhanced Raman scattering. *Langmuir, 19*(11), 4784–4790.
26. Qian, X., Peng, X.-H., Ansari, D. O., Yin-Goen, Q., Chen, G. Z., Shin, D. M., Yang, L., Young, A. N., Wang, M. D., & Nie, S. (2008). In vivo tumor targeting and spectroscopic detection with surface enhanced Raman nanoparticle tags. *Nature Biotechnology, 26*(1), 83–90.
27. Han, G., Liu, R., Han, M.-Y., Jiang, C., Wang, J., Du, S., Liu, B., & Zhang, Z. (2014). Label-free surface-enhanced Raman scattering imaging to monitor the metabolism of antitumor drug 6-mercaptopurine in living cells. *Analytical Chemistry, 86*(23), 11503–11507.
28. Thompson, P. A., & Wierda, W. G. (2016). Eliminating minimal residual disease as a therapeutic end point: Working toward cure for patients with CLL. *Blood, 127*(3), 279–286.
29. Xue, L., Yan, B., Li, Y., Tan, Y., Luo, X., & Wang, M. (2018). Surface-enhanced Raman spectroscopy of blood serum based on gold nanoparticles for tumor stages detection and histologic grades classification of oral squamous cell carcinoma. *International Journal of Nanomedicine, 13*, 4977–4986.

30. Wang, G., Lipert, R. J., Jain, M., Kaur, S., Chakraboty, S., Torres, M. P., Batra, S. K., Brand, R. E., & Porter, M. D. (2011). Detection of the potential pancreatic cancer marker MUC4 in serum using surface enhanced Raman scattering. *Analytical Chemistry*, *83*(7), 2554–2561.
31. Li, M., Woong Kang, J., Sukumar, S., Rao Dasari, R., & Barman, I. (2015). Multiplexed detection of serological cancer markers with plasmon enhanced Raman spectro-immunoassay. *Chemical Science*, *6*(7), 3906–3914.
32. Li, M., Cushing, S. K., Liang, H., Suri, S., Ma, D., & Wu, N. (2013). Plasmonic nanorice antenna on triangle nanoarray for surface enhanced Raman scattering detection of hepatitis B virus DNA. *Analytical Chemistry*, *85*(4), 2072–2078.
33. Dinish, U. S., Balasundaram, G., Chang, Y. T., & Olivo, M. (2014). Actively targeted in vivo multiplex detection of intrinsic cancer biomarkers using biocompatible SERS nanotags. *Scientific Reports*, *4*(1), 4075.
34. Stefancu, A., Moisoiu, V., Couti, R., Andras, I., Rahota, R., Crisan, D., Pavel, I. E., Socaciu, C., Leopold, N., & Crisan, N. (2018). Combining SERS analysis of serum with PSA levels for improving the detection of prostate cancer. *Nanomedicine*, *13*(19), 2455–2467.
35. Liz-Marzán, L. M., Murphy, C. J., & Wang, J. (2014). Nanoplasmonics. *Chemical Society Reviews*, *43*(11), 3820–3822.
36. Cheng, Z., Choi, N., Wang, R., Lee, S., Moon, K. C., Yoon, S.-Y., Chen, L., & Choo, J. (2017). Simultaneous detection of dual prostate specific antigens using surface enhanced Raman scattering based immunoassay for accurate diagnosis of prostate cancer. *ACS Nano*, *11*(5), 4926–4933.
37. Hong, Y., Li, Y., Huang, L., He, W., Wang, S., Wang, C., Zhou, G., Chen, Y., Zhou, X., Huang, Y., Huang, W., Gong, T., & Zhou, Z. (2020). Label-free diagnosis for colorectal cancer through coffee ring-assisted surface enhanced Raman spectroscopy on blood serum. *Journal of Biophotonics*, *13*(4), e201960176.
38. Qiao, X., Su, B., Liu, C., Song, Q., Luo, D., Mo, G., & Wang, T. (2018). Selective surface enhanced Raman scattering for quantitative detection of lung cancer biomarkers in superparticle@MOF structure. *Advanced Materials*, *30*(5), 1702275.
39. Pei, Y., Wang, Z., Zong, S., & Cui, Y. (2013). Highly sensitive SERS-based immunoassay with simultaneous utilization of self-assembled substrates of gold nanostars and aggregates of gold nanostars. *Journal of Materials Chemistry B*, *1*(32), 3992–3998.
40. Koo, K. M., Wang, J., Richards, R. S., Farrell, A., Yaxley, J. W., Samaratunga, H., Teloken, P. E., Roberts, M. J., Coughlin, G. D., Lavin, M. F., Mainwaring, P. N., Wang, Y., Gardiner, R. A., & Trau, M. (2018). Design and clinical verification of surface enhanced Raman spectroscopy diagnostic technology for individual cancer risk prediction. *ACS Nano*, *12*(8), 8362–8371.
41. Chundayil Madathil, G., Iyer, S., Thankappan, K., Gowd, G. S., Nair, S., & Koyakutty, M. (2019). A novel surface enhanced Raman catheter for rapid detection, classification, and grading of oral cancer. *Advanced Healthcare Materials*, *8*(13), e1801557.
42. Carmicheal, J., Hayashi, C., Huang, X., Liu, L., Lu, Y., Krasnoslobodtsev, A., Lushnikov, A., Kshirsagar, P. G., Patel, A., Jain, M., Lyubchenko, Y. L., Lu, Y., Batra, S. K., & Kaur, S. (2019). Label-free characterization of exosome via surface enhanced Raman spectroscopy for the early detection of pancreatic cancer. *Nanomedicine: Nanotechnology, Biology and Medicine*, *16*, 88–96.
43. Lin, D., Wu, Q., Qiu, S., Chen, G., Feng, S., Chen, R., & Zeng, H. (2019). Label-free liquid biopsy based on blood circulating DNA detection using SERS-based nanotechnology for nasopharyngeal cancer screening. *Nanomedicine: Nanotechnology, Biology and Medicine*, *22*, 102100.
44. Bai, X. R., Wang, L. H., Ren, J. Q., Bai, X. W., Zeng, L. W., Shen, A. G., & Hu, J. M. (2019). Accurate clinical diagnosis of liver cancer based on simultaneous detection of ternary specific antigens by magnetic induced mixing surface enhanced Raman scattering emissions. *Analytical Chemistry*, *91*(4), 2955–2963.
45. Li, J., Koo, K. M., Wang, Y., & Trau, M. (2019). Native microRNA targets trigger self-assembly of nanozyme-patterned hollowed nanocuboids with optimal interparticle gaps for plasmonic activated cancer detection. *Small*, *15*(50), 1904689.
46. Lee, S. W., Chen, Y. W., Kuan, E. C., & Lan, M. Y. (2019). Dual-function nanostructured platform for isolation of nasopharyngeal carcinoma circulating tumor cells and EBV DNA detection. *Biosensors and Bioelectronics*, *142*, 111509.
47. Zhang, J., Dong, Y., Zhu, W., Xie, D., Zhao, Y., Yang, D., & Li, M. (2019). Ultrasensitive detection of circulating tumor DNA of lung cancer via an enzymatically amplified SERS based frequency shift assay. *ACS Applied Materials & Interfaces*, *11*(20), 18145–18152.
48. Lee, J. U., Kim, W. H., Lee, H. S., Park, K. H., & Sim, S. J. (2019). Quantitative and specific detection of exosomal mirnas for accurate diagnosis of breast cancer using a surface enhanced Raman scattering sensor based on plasmonic head-flocked gold nanopillars. *Small*, *15*(17), 1804968.
49. Li, M., Wu, J., Ma, M., Feng, Z., Mi, Z., Rong, P., & Liu, D. (2019). Alkyne- and nitrile-anchored gold nanoparticles for multiplex SERS imaging of biomarkers in cancer cells and tissues. *Nanotheranostics*, *3*(1), 113–119.
50. Wang, Z., Zong, S., Wang, Y., Li, N., Li, L., Lu, J., Wang, Z., Chen, B., & Cui, Y. (2018). Screening and multiple detection of cancer exosomes using an SERS-based method. *Nanoscale*, *10*(19), 9053–9062.
51. Wang, Y., Zong, S., Li, N., Wang, Z., Chen, B., & Cui, Y. (2019). SERS-based dynamic monitoring of minimal residual disease markers with high sensitivity for clinical applications. *Nanoscale*, *11*(5), 2460–2467.
52. Ruan, H., Wu, X., Yang, C., Li, Z., Xia, Y., Xue, T., Shen, Z., & Wu, A. (2018). A Supersensitive CTC analysis system based on triangular silver nanoprisms and spion with function of capture, enrichment, detection, and release. *ACS Biomaterials Science & Engineering*, *4*(3), 1073–1082.
53. Wilhelm, S., Tavares, A. J., Dai, Q., Ohta, S., Audet, J., Dvorak, H. F., & Chan, W. C. W. (2016). Analysis of nanoparticle delivery to tumours. *Nature Reviews Materials*, *1*(5), 1–12.
54. Schlücker, S., Küstner, B., Punge, A., Bonfig, R., Marx, A., & Ströbel, P. (2006). Immuno-Raman microspectroscopy: In situ detection of antigens in tissue specimens by surface-enhanced Raman scattering. *Journal of Raman Spectroscopy*, *37*(7), 719–721.

55. Kang, J. W., So, P. T. C., Dasari, R. R., & Lim, D. K. (2015). High resolution live cell raman imaging using subcellular organelle-targeting SERS-sensitive gold nanoparticles with highly narrow intra-nanogap. *Nano Letters*, *15*(3), 1766–1772.
56. Li, J., Zhu, Z., Zhu, B., Ma, Y., Lin, B., Liu, R., Song, Y., Lin, H., Tu, S., & Yang, C. (2016). Surface-enhanced Raman scattering active plasmonic nanoparticles with ultrasmall interior nanogap for multiplex quantitative detection and cancer cell imaging. *Analytical Chemistry*, *88*(15), 7828–7836.
57. Li, M., Ray Banerjee, S., Zheng, C., M. Pomper, & Barman, I. (2016). Ultrahigh affinity Raman probe for targeted live cell imaging of prostate cancer. *Chemical Science*, *7*(11), 6779–6785.
58. Jeong, S., Kim, Y., Kang, H., Kim, G., Cha, M. G., Chang, H., Jung, K. O., Kim, Y. H., Jun, B. H., Hwang, D. W., Lee, Y. S., Youn, H., Lee, Y. S., Kang, K. W., Lee, D. S., & Jeong, D. H. (2015). Fluorescence-Raman dual modal endoscopic system for multiplexed molecular diagnostics. *Scientific Reports*, *5*(1), 1–9.
59. Cui, Y., Zheng, X. S., Ren, B., Wang, R., Zhang, J., Xia, N. S., & Tian, Z. Q. (2011). Au@organosilica multifunctional nanoparticles for the multimodal imaging. *Chemical Science*, *2*(8), 1463–1469.
60. Huang, J., Guo, M., Ke, H., Zong, C., Ren, B., Liu, G., Shen, H., Ma, Y., Wang, X., Zhang, H., Deng, Z., Chen, H., & Zhang, Z. (2015). Rational design and synthesis of γFe2O3@Au magnetic gold nanoflowers for efficient cancer theranostics. *Advanced Materials*, *27*(34), 5049–5056.
61. Wang, Z., Zong, S., Chen, H., Wang, C., Xu, S., & Cui, Y. (2014). SERS-fluorescence joint spectral encoded magnetic nanoprobes for multiplex cancer cell separation. *Advanced Healthcare Materials*, *3*(11), 1889–1897.
62. Carrouée, A., Allard-Vannier, E., Même, S., Szeremeta, F., Beloeil, J. C., & Chourpa, I. (2015). Sensitive trimodal magnetic resonance imaging surface enhanced resonance Raman scattering-fluorescence detection of cancer cells with stable magneto-plasmonic nanoprobes. *Analytical Chemistry*, *87*(22), 11233–11241.
63. Zavaleta, C. L., Smith, B. R., Walton, I., Doering, W., Davis, G., Shojaei, B., Natan, M. J., & Gambhir, S. S. (2009). Multiplexed imaging of surface enhanced Raman scattering nanotags in living mice using noninvasive Raman spectroscopy. *Proceedings of the National Academy of Sciences*, *106*(32), 13511–13516.
64. Loo, C., Lin, A., Hirsch, L., Lee, M. H., Barton, J., Halas, N., West, J., & Drezek, R. (2004). Nanoshell-enabled photonics-based imaging and therapy of cancer. *Technology in Cancer Research & Treatment*, *3*(1), 33–40.
65. Huang, X., Neretina, S., & Ma, E. S. (2009). Gold nanorods: From synthesis and properties to biological and biomedical applications. *Advanced Materials*, *21*(48), 4880–4910.
66. Zhang, Y., Qian, J., Wang, D., Wang, Y., & He, S. (2013). Multifunctional gold nanorods with ultrahigh stability and tunability for in vivo fluorescence imaging, SERS detection, and photodynamic therapy. *Angewandte Chemie International Edition*, *52*(4), 1148–1151.
67. Hu, F., Zhang, Y., Chen, G., Li, C., & Wang, Q. (2015). Double-walled Au nanocage/SiO_2 nanorattles: Integrating SERS imaging, drug delivery and photothermal therapy. *Small*, *11*(8), 985–993.
68. Hi Grieve, S., Puvvada, N., Phinyomark, A., Russell, K., Murugesan, A., Zed, E., Hassan, A., Legare, J. F., P.C. Kienesberger, Pulinilkunnil, T., & Reiman, T. (2021). Nanoparticle surface-enhanced Raman spectroscopy as a noninvasive, label-free tool to monitor hematological malignancy. *Nanomedicine*, *16*(24), 2175–2188.

6

Smartphone-Based Nanodevices for Point-of-Care Diagnostics

Ayan Pal and Md Palashuddin Sk

CONTENTS

6.1 Introduction

Biosensors can be defined as analytical tools consisting of sensing probes for biological sample detections [1]. Over the past few decades, numerous biologically derived substances like enzymes, antigen–antibodies, and nucleic acid have been integrated with optical, electrochemical, or mechanical detectors to devise biosensors showing high selectivity and sensitivity [2–5]. Indeed, such devices are powerful detecting tools and have been successfully utilized in various fields such as clinical diagnostics, environmental monitoring. However, current research on the biological sensing tools mostly focuses on the development of point-of-care (POC), miniaturized, and easy-to-use detection platforms which are cost-effective as well. Simply, by amalgamating the concepts of micro-electro-mechanical systems (MEMS) and nanotechnology, the miniaturization of biosensors into micro- and nano-scale is possible nowadays. Nanomaterials such as nanoparticles (NPs), nanorods (NRs), nanowires, nanotubes, and various nanocomposites have been extensively used to fabricate chemosensors and biosensors [6–9]. Their ease of synthesis, excellent photophysical properties, high surface-to-volume ratios, conductivity, mechanical strength, and electrochemical activity have made corresponding devices easy to fabricate, precise, and efficient.

On the other hand, smartphone-based diagnostic techniques are gaining wide interest as compared to traditional laboratory-based testing equipment due to economic considerations, portability, and availability in resource-limited areas [10]. Research in recent years has demonstrated that a smartphone in combination with a biosensor has the potential to offer high data accuracy and susceptivity for healthcare diagnostics. As smartphone technologies are becoming more and more powerful, only a few accessories are needed to design a phone-based bioanalytical platform such as illumination sources, high-resolution cameras, signal detectors, and/or data processing units. With the help of advanced complementary metal-oxide-semiconductor (CMOS) image sensor (CIS), central processing unit (CPU) and graphics processing unit (GPU), and light-emitting diode (LED) sources, smartphones have enabled us to capture high-resolution images, produce laboratory standard reports, detect and analyze the changes, indispensable for field-based rapid testing [11, 12].

Thus, the unification of nanotechnology with smartphone technology has become a very active area of research to fabricate flexible optoelectronic devices, wearable sensing devices, antimicrobial nanofilms, and coating for medical testing. This chapter focuses on recent developments in smartphone-based diagnostic techniques and the fabrication of miniaturized devices for real-time applications.

6.2 Smartphone-Based Optical Sensors

6.2.1 Colorimetric Biosensors and Nanodevices

Probably, smartphone-based colorimetric diagnostic tools are the most primitive optical systems where smartphones have been used as detectors. In this process, the camera attached to the smartphone is directly used to record colorimetric changes. A colorimetric POC testing device is associated with a smartphone comprising of a complementary CMOS image

DOI: 10.1201/9781003157823-6

sensor, optical grating, and image processing unit. External equipment such as LED illumination and external lenses is also required to control proper light conditions along with a battery supply [13]. It is important to note that proper white balance and color correction are crucial parameters for achieving the color accuracy of the captured images. That is why for colorimetric detection and quantification, maintaining proper illumination conditions is essential. During the process, the device measures the color change of the analyte solution or paper strip depending upon sample concentration. Next, color differences are analyzed by using pre-installed smartphone software. Generally, colorimetric analysis of the solution or paper strip is done by using Beer–Lambert's law. Such a technique is believed to have strong potential applications in POC and real-time monitoring for customized diagnosis. This is the general principle to perform colorimetry analysis by using smartphone technology. In this section, we shall discuss the utilization of colorimetric nanodevices for diagnosis.

Precise and rapid detection methods of proteins are of great importance in medical science. In this regard, the devices that we use have to be low-cost, fast, miniaturized, and portable. Utilizations of colorimetric technology in developing theranostic tools are quite effective as they are sensitive, cost-effective, and are devoid of sophisticated instrumentations. On the other hand, the uses of nanomaterials have made such devices lower-cost, portable, and highly sensitive. Citrate-stabilized gold nanoparticles (AuNPs; size ~13 nm) mediated by varying concentrations of NaCl salt have been designed as an effective differential receptor for multiple protein discrimination. Due to the differences in ionic strength, AuNPs exhibit different aggregation behavior leading to diverse colors. Such systems can effectively discriminate 12 different types of proteins at 200 nM concentration in human urine samples with 100% accuracy. Obvious color changes due to different proteins can be viewed by the naked eye whereas the smartphone can easily extract RGB intensity changes for quantitative colorimetric analysis [14]. A smartphone-controlled theranostic device has been prepared for photodynamic therapy (PDT) using a silver nanorod (Ag NR; size ~ 3.71 ± 0.58 nm) as the photosensitizer. Ag NRs show a strong plasmon absorption band at 632 nm and therefore, they are photo-irradiated at this wavelength using an LED source. Upon irradiation with a 632 nm LED source singlet, O_2 generation has been identified which further results in a significant reduction in cell viability in cancerous HeLa cells from 90% to 63% due to cell death. Such cytotoxic effect on HeLa cells becomes more significant when treated in presence of chemotherapeutic drug doxorubicin (Dox). The image of the device used for PDT is presented in Figure 6.1. Here, the smartphone-operated theranostic device consists of an illumination unit and an optical measurement unit. In the illumination chamber samples retained in 96-well plates are irradiated with a switchable LED array (1.4 mW/cm^2). Simultaneously, the phototransistor array set in the measurement compartment records the absorbance of 96 samples in a single scan. The device is operated wirelessly through a smartphone app providing digital output reading of the measured values. The designed system can operate as a POC platform for both therapy and diagnosis and can function in complex environments with minimal human intervention [15].

Tuberculosis (TB) continues to be one of the most infectious diseases according to the World Health Organization (WHO). Recent advances in TB molecular diagnostics correspond to nanotechnology-based systems that reduce the laboratory diagnostics time from weeks to days. Au-nanoprobes have been used to prepare a paper-based platform, capable of *Mycobacterium tuberculosis* (*M. tuberculosis*) complex (MTBC) detection at point-of-need. In this work, wax printed microplate paper platform has been used to develop an Au-nanoprobe assay. In presence of salt, the SPR of the Au-nanoprobe disappears due to aggregation. However, in the presence of a complementary DNA target sequence of *M. tuberculosis*, strong red color further appears due to hybridization between the Au-nanoprobe and the DNA target sequence. Indeed, such hybridization protects Au-nanoprobes from salt-induced aggregation. Here, the detection method completely relies on the visible color changes of the nanoprobe system following aggregation. Distinct color changes are then analyzed with a smartphone device at a minimum MTBC sample DNA concentration of 10 μg mL^{-1} [16]. The process of Kaposi's sarcoma (KS) disease diagnosis has been simplified by creating a smartphone-based POC system capable of showing actionable diagnostic output. Here, Au NPs have been utilized for the colorimetric detection and quantification of pathogen DNA. For precise detection, a disposable microfluidic chip, smartphone application software, and other working accessories have also been developed in this work [17].

Over the past decade, the amplitude of Ebola virus disease (EVD) outbreaks has been vastly increased. Recent studies based on human survivors have confirmed that recovery from EVD is related to the generation of cell-mediated and humoral immune responses. Recently, a smartphone-based POC device has been proposed for quick detection of IgG antibodies against Ebola (Figure 6.2A–B). The testing device includes a lateral flow test strip and a smartphone. When an Ebola virus antibody is dropped over the strip, a red-purple line appears. Then, a smartphone application determines the relative strength of the test line and shows the complete result within 15 min. This platform provides quick and portable testing, data storage, and holds great potential for field diagnosis, vaccination, and therapeutic evaluation [18].

Vitamin B12 is compulsory for the generation of red blood cells, neural myelination, brain development, and DNA formation in the human body. Poor vitamin B12 level is associated with anemia, cognitive impairments, paresthesia, muscle weakness, and behavioral changes. To curb the limitations to the diagnosis of such micronutrient deficiencies concepts of nanotechnology have been applied. A "NutriPhone" mobile platform has been developed for POC assessment of vitamin B12. The testing system consists of a smartphone accessory with a smartphone application and test strip. In this process, test strips are prepared by using the monoclonal anti-vitamin B12 IgG conjugated with Au NPs (anti-B12-Au NP). During the testing, at first, blood samples are directly dropped onto the test strip's inlet. During this period, sample B12 interacts with anti-B12-Au NP. After 4 min, the user applies chase buffer droplets following which subtle change in the colorimetric signal can be observed in the subsequent 6 min. In this process, a silver enhancement solution is used for signal amplification.

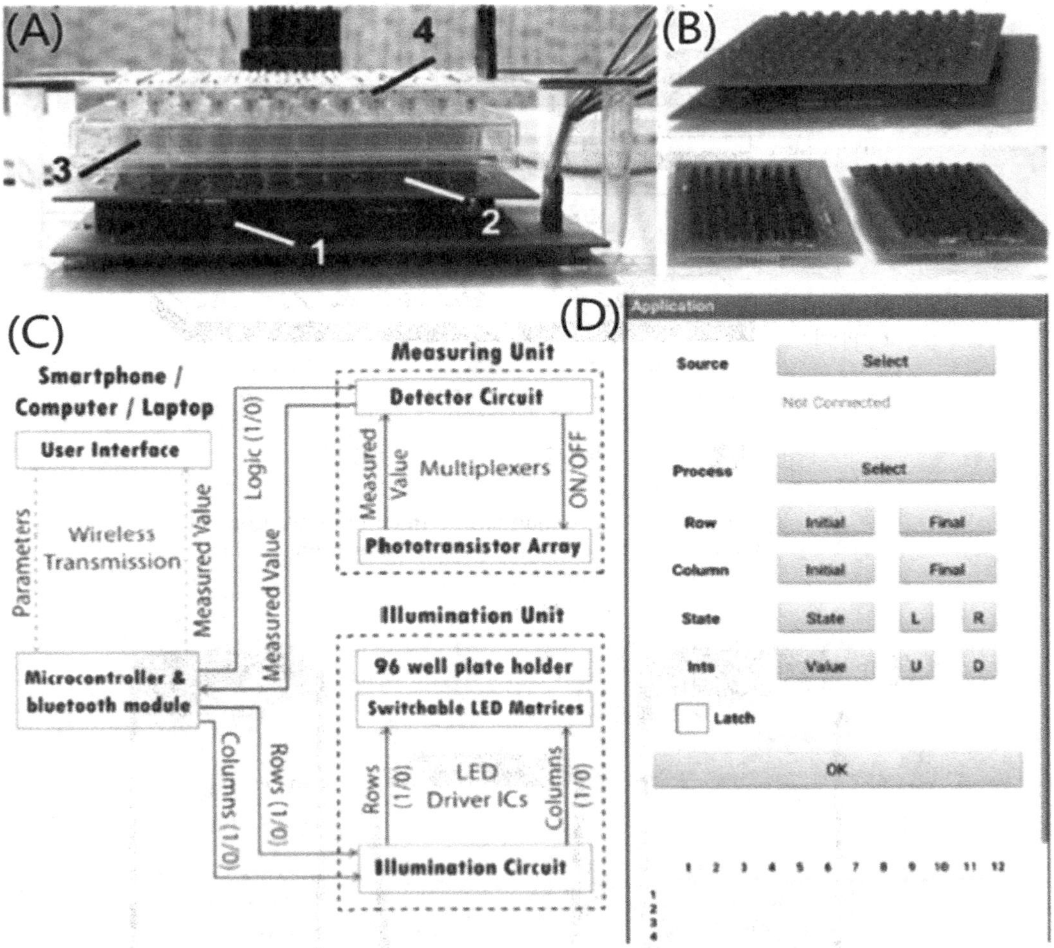

FIGURE 6.1 Smartphone-based theranostic device. (A–B) Image of the device and switchable LED arrays. (C–D) Units and interconnections in the device and the smartphone application interface. Reprinted with permission from [15].

Then, the test strip is introduced to the NutriPhone accessory for measurement. An inbuilt smartphone application captures and processes the image to calculate vitamin B12 concentration in real samples with respect to the calibration curve. It is important to note that the complete process takes less than 15 min for detection in sub-nmol/L ranges [19]. Similarly, an Au NP (size ~40 nm) conjugation kit has been developed for the smartphone-based measurement of serum ferritin levels from a drop of blood sample [20]. Recently, a POC testing device has been developed for hematocrit level detection in human blood samples by using the gray-scale-valuation (GSV) method. With the help of an image processing program, the device can successfully determine hematocrit levels varying from 10% to 65%. By applying the concept of microfluidic effect, a quick and sensitive colorimetric POC testing device has been developed here, showing promises of convenient and accurate measurement of hematocrit of human blood [21].

6.2.2 Fluorescence-Based Nanodevices

Fluorometric analysis provides a platform for high sensitivity, strong selectivity, and very simple operation technique for diagnosis. For such applications, nowadays, smartphones are quite efficient in showing diagnostically relevant results following the attachment with appropriate equipment. Typically, a fluorescence-based diagnostic device consists of an excitation source (mostly LED sources), cutoff filters along with external lenses, sample chamber, and finally, a smartphone camera with a pre-installed software application. Background interferences due to autofluorescence or scattering can be removed by tilted illumination at a higher angle. During image acquisition, the time of exposure, camera ISO, and f-stop number are maintained strictly to acquire an appropriate relationship between pixel brightness and sample concentration. The measurement process can either be solution-based or paper-based where highly fluorescent nanoscale particles act as the sensing probe. A schematic diagram for the fluorescence-based testing process is represented in Figure 6.3.

Semiconductor quantum dots (Qdots) are one of the promising nanomaterials and have been used as luminescent nanoprobes for smartphone-based POC diagnostics. These inorganic nanocrystals can absorb light energy, become excited, and emit photons during relaxation to ground state resulting in fluorescence. Their unique properties include tunable and stable fluorescence, good surface-to-volume ratio, broad absorbance, and inorganic interface which can be easily

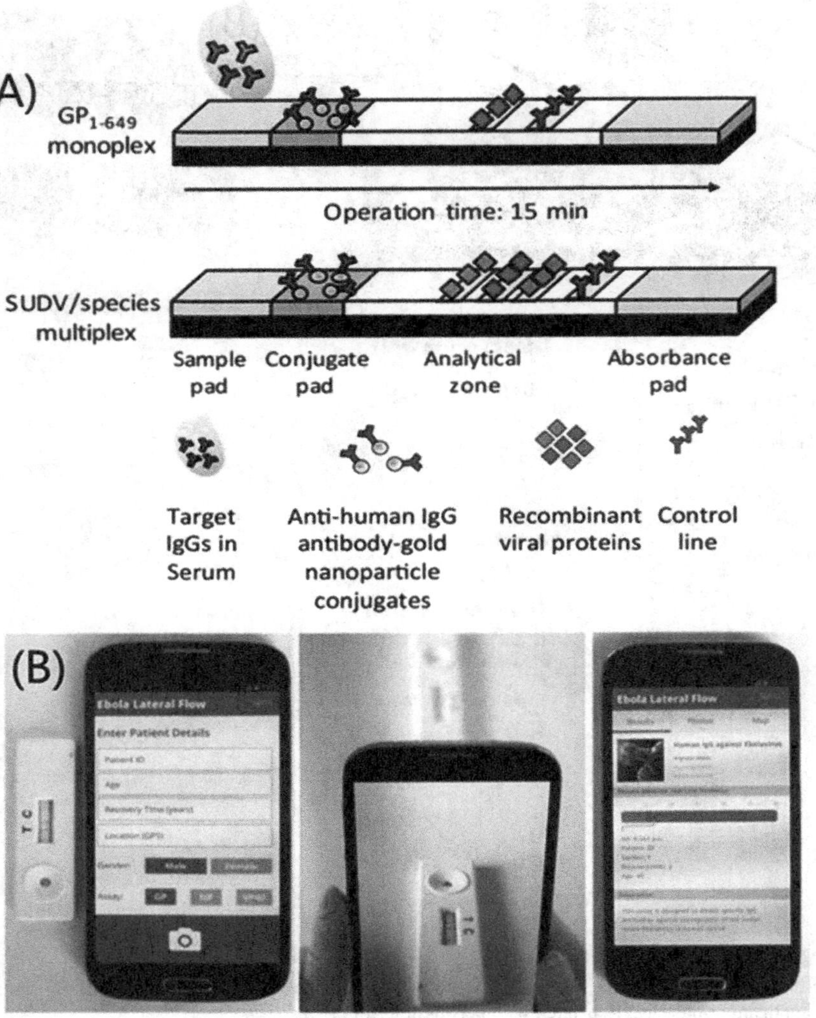

FIGURE 6.2 (A) Schematic illustration of lateral flow test strip used for the detection of Ebola virus IgG. (B) The smartphone application interface that is used to record patient details and analyze results. Reprinted with permission from [18].

conjugated with various biomolecules [10]. For example, the fluorescence intensity changes of Qdots can be utilized for the real-time monitoring of proteolytic activity. Qdots are excited with a portable UV lamp (λ_{ex} = 365 nm) and fluorescence images are acquired in a smartphone camera. In such a technique, during the image acquisition, appropriate cutoff filters are required in front of the camera lens to block stray lights. Indeed, such a smartphone-based diagnostic system has removed the instrumental impediments for POC testing and also demonstrates the use of Qdots for routine bioanalysis [22]. A field-portable design has been employed to perform Qdot-based enzyme-linked immunosorbent assay (ELISA) to detect *Escherichia coli* (*E. coli*). In this technique, glass capillaries functionalized with anti-*E. coli* O157:H7 antibodies have been used as solid substrates. During testing, sample particles are flushed through capillaries at ~50 μL min^{-1} flow rate (for 20 min) followed by labeling with streptavidin-conjugated Qdots. The fluorescence emission of Qdot attached to *E. coli* is then quantified by using a cost-effective and portable microscope system attached to the cell phone (Figure 6.4). As shown in Figure 6.4, approximately 6 to 10 capillary tubes are inserted into the sample chamber and are microscopically imaged. Qdots inside the capillary tubes are excited with UV LEDs and emitted fluorescence is imaged by the phone camera. The device is associated with an additional lens of 15 mm focal length and a long pass glass filter to minimize UV scattered light and produce a dark-field background. Finally, the captured images are processed in ImageJ software for the quantification of *E. coli*. In such testing devices Qdots show bright and stable fluorescence and, on the other hand, a high surface-to-volume ratio of glass capillaries increases the bacteria capture efficiency, thus improving the device sensitivity [23].

Förster resonance energy transfer (FRET) is a distance-dependent nonradiative energy transfer process between a donor-acceptor pair. A paper-based platform has been developed by utilizing FRET-directed ratiometric changes in Qdot emission for monitoring the nucleic acid hybridization [24].

Carbon quantum dots (CQdots) are bright and stable optical tools for fluorescence-based diagnostic applications. CQdots are biocompatible, highly fluorescent, and less toxic

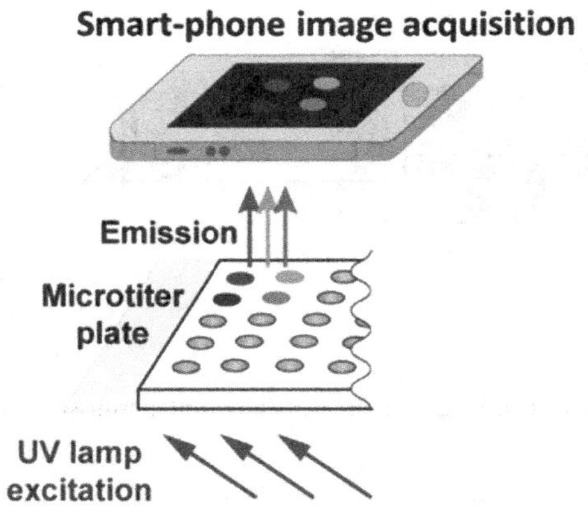

FIGURE 6.3 Schematic representation of fluorescence-based detection process using the smartphone camera. Reprinted with permission from [22].

as compared to semiconductor Qdots. Their bright emission originates from various surface functional groups (generating surface states), polymeric core, and sometimes from organic fluorophores attached to surfaces [25]. Recently, their tunable and bright emission characteristics have been successfully utilized for smartphone-based detections as well. For example, a disposable fluorescent nanopaper-based assay kit has been developed for the diagnosis of jaundice. Fluorescent CQdots are embedded in a bacterial cellulose substrate the luminescence of which are quenched in presence of bilirubin. Following irradiation with a 470 nm light source, unconjugated bilirubin is converted to colorless oxidation products leading to fluorescence recovery of the paper strip. Recovery of the emission can be monitored by an integrated smartphone camera. By taking advantage of the nontoxicity and excellent physicochemical properties of CQdots, the assay kit provides a portable sensing platform for the diagnosis of jaundice at POC and routine clinical laboratories [26]. Similarly, ultrabright polymer dots have been used for real-time glucose monitoring via smartphone [27].

Metal nanoclusters (NCs) are also used for developing fluorometric smartphone-based POC testing devices as they feature tunable emission and high biocompatibility. For example, phenylboronic acid templated Au nanoclusters, termed as PB-Au NCs, have been used as fluorescent probes for smartphone-assisted mucin detection [28]. In another work, a direct read-out portable device has been developed for on-site detection of glutathione levels in human serum. The device consists of a smartphone, a dark chamber, a filter holder, a UV lamp chamber, and a sample slot to place the paper strip. The sensing probe solution is prepared by combining CQ dots and Au NCs and is fabricated over the paper strip. In the presence of different concentrations of analyte, the paper strip shows ratiometric color changes. A smartphone camera captures corresponding photos and uses a pre-installed color recognizer application to show RGB values for quantitative analysis [29]. NC-based diagnosis by using smartphone technology can also be seen for detection of septicemia in infants [30], determination of trypsin [31], and detection of microRNA levels in cancer cells [32].

6.2.3 Smartphone-Based Imaging in Nanodevices

Smartphone-based microscopic diagnostics are quite popular for POC treatment. There are two different types of microscopic devices, which are commonly known as brightfield microscopic imaging devices and fluorescent microscopic imaging devices. During the brightfield microscopic imaging, samples are illuminated with white LEDs from below. The light transmits through the sample and visible contrasts are generated due to the attenuation of light in dense areas. On the other hand, during fluorescent microscopic imaging, LEDs excite the objects and signal-to-noise ratios are quantified from the emitted light [11]. In order to develop a smartphone-based microscope, different optical components should be integrated with the smartphone camera. By simply integrating external lenses and light sources imaging can be done in a smartphone. In this technique, the smartphone camera functions as a charge-coupled device (CCD)/CMOS camera, and the external lenses are analogous to the objective lens in a conventional microscope [11, 12]. Magnification of the camera can be expressed as $M = f_1/f_2$, where f_1 is the focal length of the inbuilt camera lens and f_2 is the external lens focal length. Such lens modules are cost-effective, readily available, and produce standard image quality. Power-efficient LEDs can be applied as the light source for imaging. Microscopic images

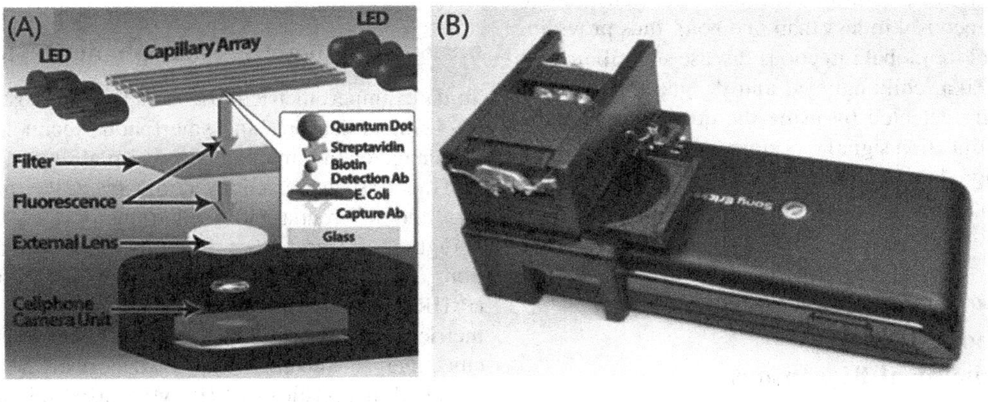

FIGURE 6.4 (A–B) Schematic representation and a photograph of optical attachment used for *E. coli* detection and quantification on a cell phone device using the Qdot-based assay in glass capillary tubes. Reprinted with permission from [23].

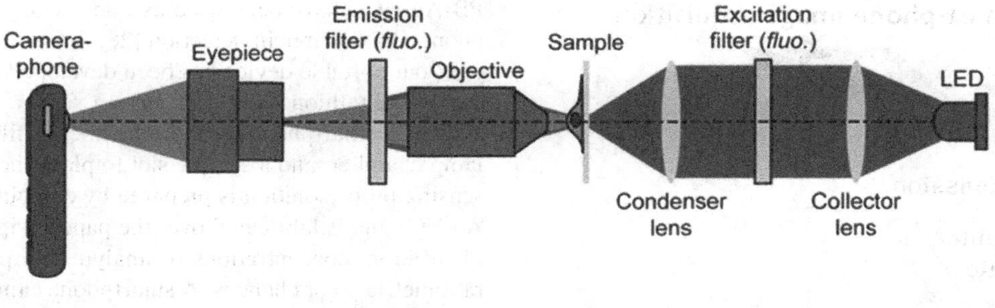

FIGURE 6.5 Schematic illustration of microscopy optical layout for the phone-based fluorescence imaging. Reprinted with permission from [33].

are analyzed in real-time by using the advanced computational system of the phone providing details of object color, brightness, and size. Recently, efforts have been made on developing such smartphone integrated portable devices for producing microscopic images in POC diagnostics [10, 13]. A schematic diagram for the microscopy-based imaging device has been presented in Figure 6.5. In the following section, we shall discuss a few smartphone-based imaging techniques for real-time diagnosis of diseases.

A microfluidic chip has been developed for smartphone-based detection of CD4 glycoprotein under brightfield. Here, poly (methyl methacrylate) (PMMA) sheets, double-sided adhesive (DSA) sheets, and silanized glass slides have been utilized for microfluidic chip fabrication. The device consists of several optical accessories and smartphone applications for imaging and estimation of the quantity of cells. The system is inexpensive and can be easily used in resource-limited areas where the entire detection process takes less than 10 s. Such a device can also recognize other living cells [34]. Similarly, researchers have developed a device for the fluorescence-based screening of cells for the identification of their types and activities. Red blood cells and cancer cells stained with calcein have been imaged under bright and dark fields, respectively [35]. Bar code devices enable clinicians to efficiently manage the spread of diseases by a proper diagnosis of pathogens. A chip-based wireless testing device has been developed by combining Qdot barcode technology with smartphones for the diagnosis of HIV or hepatitis B in infected patients (Figure 6.6). Here, Qdots and Alexa Fluor 647 dye have been used as emission probes for detecting biomarker targets [36]. By using this process, several infectious pathogens can be detected simultaneously in less than one hour, thus providing an important tool for global infectious disease surveillance. In another study, Zika, chikungunya, and dengue viruses have been successfully detected by using the quenching of unincorporated amplification signal reporters (QUASR) technique. Fluorescence signals are analyzed by software applications based on chromaticity. This platform can be used to analyze blood, urine, and saliva samples [37].

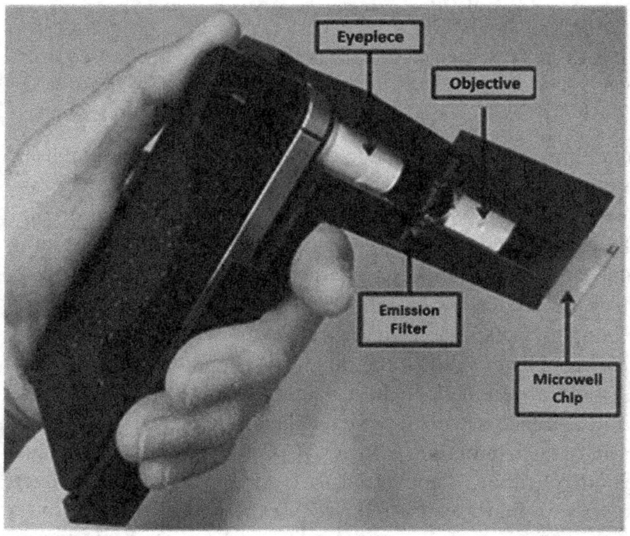

FIGURE 6.6 Digital photograph of a microscopic imaging tool using smartphone technology. Reprinted with permission from [36].

6.3 Smartphone-Based Electrochemical Biosensors

Electrochemistry has a wide range of applications for quantitative analysis of important biomolecules such as nucleic acids, proteins, and metabolites. The results obtained by this technique are reliable and devices are portable, low-cost, and easy to handle. Besides, the utilization of smartphone technology with several excellent capacities has made electrochemical analysis simpler with hand-held devices. Based on the analysis technique, smartphone-based electrochemical biosensors can be classified into three categories, namely amperometric, potentiometric, and impedimetric.

6.3.1 Amperometric Smartphone Devices

To date, amperometric biosensors are the most reported electrochemical sensor using smartphone technology. The main advantage of this method is the availability of various detection modes such as cyclic voltammetry (CV), chronoamperometry, and differential pulse voltammetry (DPV).

Figure 6.7 represents a prototype device used for the detection of *Plasmodium falciparum* histidine-rich protein 2 (*Pf*HRP2), a prominent biomarker for malaria. Here, amperometric measurements are done by using an integrated circuit, plugged into the micro-USB port of the mobile phone. A poly(dimethylsiloxane) (PDMS)-based microfluidic chip is used that offers a disposable platform during sample analysis. In Figure 6.7, the arrow points to the microfluidic chip.

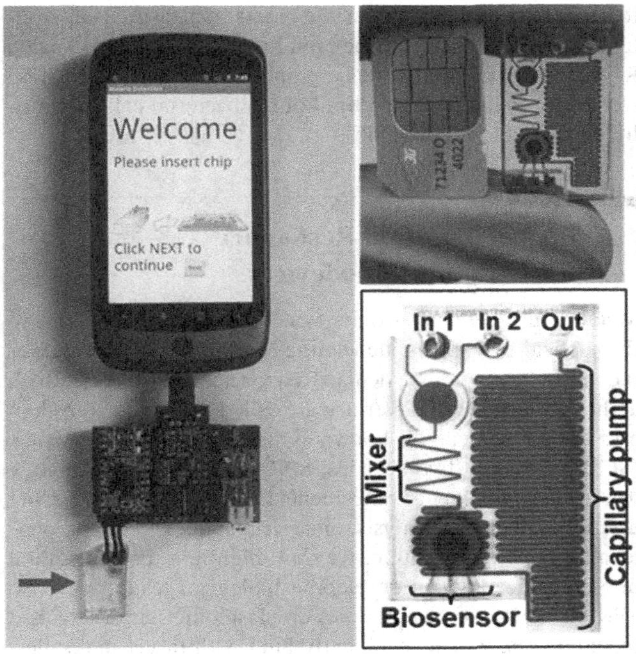

FIGURE 6.7 A microfluidic chip-based amperometric device for the detection of *Plasmodium falciparum* histidine-rich protein 2 (*Pf*HRP2). Reprinted with permission from [38].

As shown, the size of the as-prepared chip is similar to the size of a subscriber identity module (SIM) card. Inlets are designed in the chip for loading the sample. Once the sample is inserted and the measurement process is selected, the user needs to adjust a few parameters such as the applied voltage and measurement time. Next, this information is transferred to a microcontroller via a USB connection. Once the connection is set up between the microcontroller and phone, the system measures current signals from the chip and transfers the acquired data to the microcontroller. The detection technique requires only 15 min to complete the diagnosis process with a measured limit of detection of 16 ng/mL [38].

Recently, a device has been developed for label-free precise counting of white blood cells (WBC) by using the CV technique. The sensor uses a gold deposited three-electrode system, patterned over polyvinylidene fluoride (PVDF) membrane and ferricyanide/ferrocyanide as the electrolyte. During quantification, WBCs are separated from the blood sample and are trapped within the membrane electrode. Here, the sensing mechanism is based on the diffusion blockage by trapped cells in the membrane electrode and corresponding voltammetry signal changes with the WBC concentrations. A portable potentiostat is used to generate voltammetry signals. This smartphone-based wireless technique allows rapid quantification of WBC within 1 min by taking only 10 μL of the sample [39]. On the other hand, utilization of amperometric sensors for cancer biomarker detection has been done recently. A biosensing platform consisting of the rGO/Au composite-tailored electrode and an Android phone has been constructed to detect microRNA-21 (miR-21). During measurement, rGO/Au composite-tailored electrode is connected with a circuit board and DPV results are transferred to the smartphone via Bluetooth. As the concentration of analyte increases, DPV peak current reduces, leading to sensitive detection of miR-21 in saliva samples [40]. An inexpensive sensing platform illustrates the utilization of amperometric methods in the universal mobile detector in resource-limited areas. The device can detect (i) glucose level in blood samples, (ii) sodium level in urine, (iii) heavy metals, and (iv) malarial antigens. For testing, the device is connected with commercially available electrodes loaded with samples, and changes are recorded by using an electrochemical detector. Notably, the mobile phone device uses chronoamperometry, potentiometry, and electrochemical ELISA for glucose level measurement, sodium level analysis, and malarial antigen detection respectively. Overall, the techniques used here are inexpensive and simple which can be used as an efficient platform to detect multiple diseases [41].

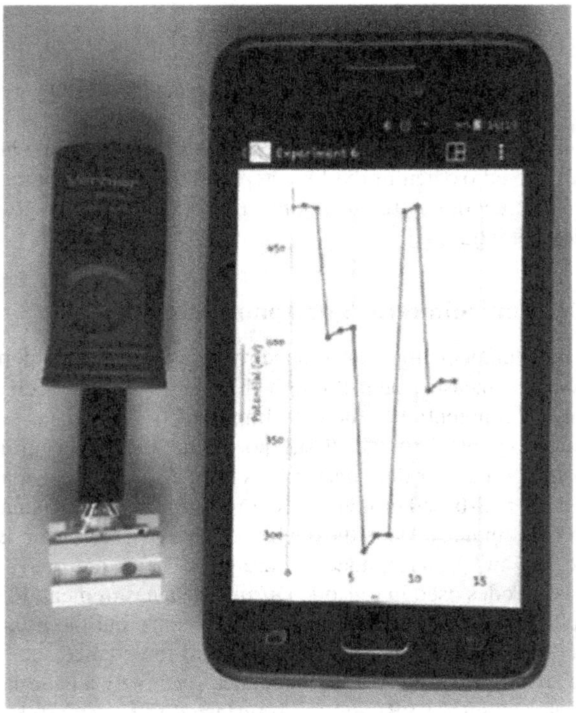

FIGURE 6.8 Digital photograph of the device consisting of the sensor (in the lower section) and the wireless potentiometer (upper section). Data plot obtained in the wirelessly connected mobile during sample analysis has also been presented. Reprinted with permission from [42].

6.3.2 Potentiometric Smartphone Devices

Potentiometric sensing is an application where cumulative electrical charges produce electrical potential differences in the dielectric layer. On-field biological detection can be done successfully by using such potentiometric devices. Glucose levels can be monitored in biological fluid by using a paper-based platform where the device follows a potentiometric technique for detection (Figure 6.8). A working electrode is prepared by using a Nafion coated platinized filter paper which is able to trap glucose oxidase. A polyvinyl butyral casted conductive paper is used as the reference electrode. Here, the sensing is done based on the following equations.

$$H_2O_2 \rightarrow 2H^+ + O_2 + 2e^-$$

$$H_2O + O_2 + \text{Glucose} \rightarrow \text{Gluconolactone} + H_2O_2$$

As mentioned, oxidation of glucose takes place with the help of dissolved oxygen in the first step. Subsequently, as a result of H_2O_2 generation, the platinum surface identifies changes in redox potential [42].

6.3.3 Impedimetric Smartphone Devices

Electrochemical impedance spectroscopy (EIS) corresponds to current measurement through the electrochemical cell by applying a potential to the cell. Impedimetric biosensor on a smartphone platform was first reported in 2013 for histamine sensing [43]. After that, researchers have developed a fast and sensitive EIS-based protein detection platform by using smartphone technology. Here, the device consists of a biosensor, EIS detector, and a smartphone. Figure 6.9A shows the image of the electrodes used in sensors. During the measurement, these electrodes are modified with bio-components and the protein reactions are detected by using electrical impedance signals. The signals are recorded as impedance spectra by a hand-held detector (Figure 6.9B) and are transferred to the smartphone via Bluetooth. After receiving the data signals, the smartphone displays the electrochemical measurements in the form of a Nyquist plot (Figure 6.9C). The system can also monitor the enzymatic activities of thrombin. By using impedance technique, this device can detect protein and thrombin up to 1.78 µg/mL and 2.97 ng/mL concentrations, respectively [44].

The smartphone-based impedimetric technique has also been utilized for in-field quantification of bacteria species and pathogens. Here, the system includes a microfluidic sensor, impedance converter chip, a microcontroller, and a cell phone. Interdigitated microelectrodes have been designed on a piece of silicon chip giving a good signal-to-noise ratio and low resistance. During measurement, an impedance analyzer chip and a microcontroller perform the EIS analysis. An Android application enables data recording and visualization of results in the smartphone. In this process, *E. coli* concentration in a real sample solution is calculated based on measured electron-transfer resistance (R_{et}) of the sample solution and a pre-loaded R_{et} vs. concentration reference graph. The device shows good sensitivity as the detection limit of ten bacterial cells per milliliter has been achieved [45].

6.4 Surface Plasmon Resonance (SPR)-Based Nanodevices

Metallic surface/nanoparticles, which are employed to improve the optical responses depending upon the interaction of incoming light with particular frequencies and free electrons at the metal-dielectric interface are called plasmonic transducers. The phenomenon is known as surface plasmon resonance (SPR). In the last two decades, SPR-based sensor technology has made significant advancements in various fields including electronic devices, catalysis, biological applications, and photocatalysis. In this section, we shall highlight the application of SPR technology in developing biological testing devices. SPR technology typically relies on plasmonic sensor-attached biomolecules, which specifically bind with target molecules. When the target analyte binds with the plasmonic sensor-attached biomolecule refractive index of the local environment changes enabling detection of the analyte molecule. Most of the SPR-based biosensors consist of gold (Au) as the plasmonic material because of prominent plasmon resonance in the visible region, non-toxicity, excellent resistance to oxidation, and ability to be easily fabricated into nanostructures. A plasmonic biosensing platform has been developed consisting of ionic gold embedded inside agarose gel scaffolding. Once the as-prepared probe reacts with ascorbic acid, ionic gold is reduced into plasmonic nanoparticles. Thus, using smartphone technology, ascorbic acid concentration in eye fluid can be quantified [46]. Researchers have utilized plasmon-enhanced fluorescence microscopy for smartphone-based detection of DNA origami structures. In this device, a silver-coated glass slide connected with a spacer is used to place the samples. Samples are excited with a laser-diode source from the backside giving rise to surface plasmon polaritons. The plasmon-enhanced

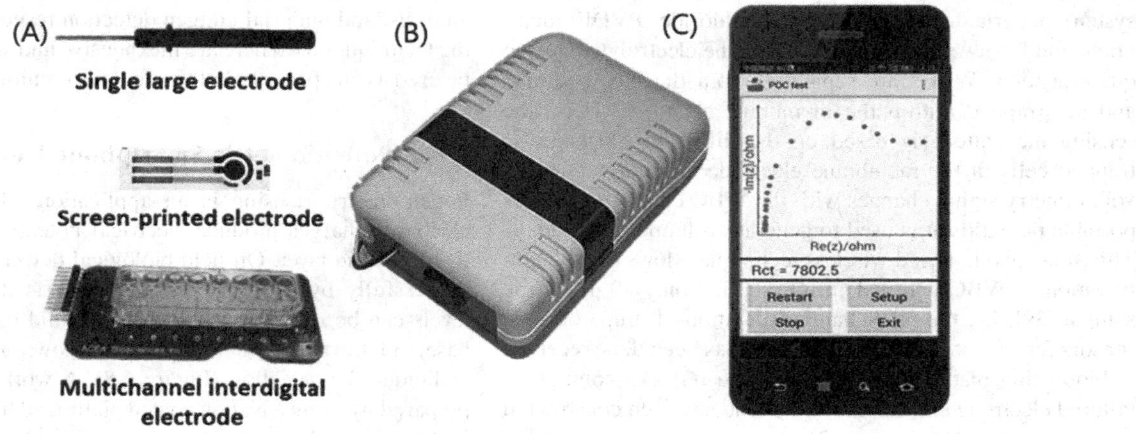

FIGURE 6.9 (A) Three different types of electrodes used in the impedimetric protein sensor. (B) Image of the hand-held detector. (C) The smartphone screen demonstrating the impedance measurement. Reprinted with permission from [44].

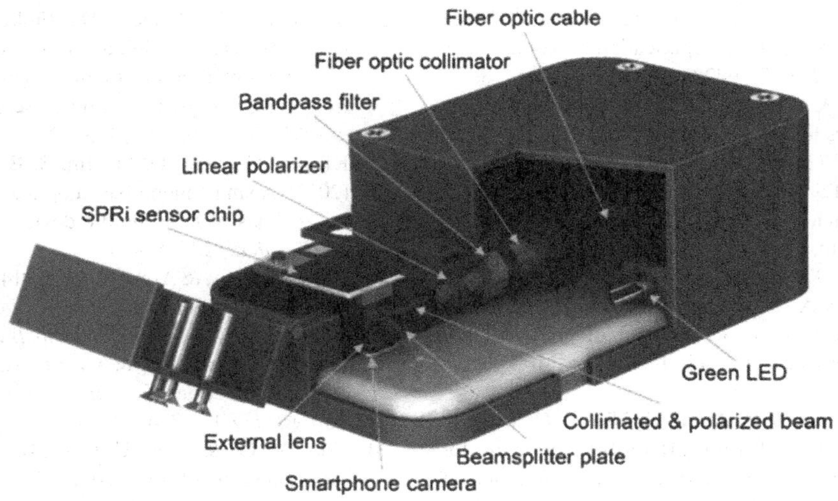

FIGURE 6.10 Schematic illustration of SPR imaging platform attached with a smartphone. Reprinted with permission from [50].

fluorescence can be optimized by tuning film thickness, excitation angle, spacer distance, and polarization. Such techniques enhance the detection sensitivity of fluorescence microscopy [47]. β2 microglobulin is an important biomarker for cancers and kidney diseases. Recently, an SPR instrument has been attached to a smartphone for POC diagnosis of β2 microglobulin [48]. The device consists of a lab-on-a-chip system, an optical coupler, and a smartphone for detection. On the other hand, researchers have utilized a nanoplasmonic imaging platform for smartphone-based detection of biochemicals. The device works based on the changes in refractive index and absorbance enhancement. By using this portable device, high protein concentration in urine can be detected. In this device, the sensing mechanism is based on refractive index changes by using SPR and localized SPR (LSPR) phenomena leading to higher sensitivity [49]. Figure 6.10 represents an SPR-based POC testing device for nanomolar level detection of IgG antibodies. Here, a low-cost SPR sensor chip has been developed by using an optical storage disk. The device is associated with an LED source, collimator, filter, polarizer, and a lens for imaging which can be attached to a smartphone. Silver/gold (Ag/Au) bilayer structure has been coated over Blu-ray disc to operate plasmon resonance imaging. The system uses an Android application to measure intensity changes and interpret the acquired data [50].

6.5 Conclusion

As discussed in this chapter, smartphone technology provides a very simple, even efficient, and miniaturized platform for clinical diagnosis. Furthermore, the uses of nanomaterials in the sensor devices have made the testing processes cost-effective, error-free, and quicker than usual. In other words, smartphone devices with nanosensors have unfolded a new paradigm in the field of POC disease detection. Thus, advances in field-based diagnosis in the twenty-first century are the amalgamation of unprecedented physicochemical properties of nanoprobes, concepts of microfluidics, and the utilization of phone cameras with high-resolution imaging capability. Besides, the latest technology holds great potential for geographical tagging, data storage, and sharing, essential for effective disease management.

Despite the ample progress made, several challenges still exist. For example, present analytical biosensors are equipped with hand-made accessories which do not have unified standards. Sensitivity of the instrument, data reproducibility, and accuracy should be guaranteed strictly so that biosensors can be brought into service for clinical or real-life applications. The diagnostic setup has to be more compact, even not too complicated. Moreover, the integration of multimodal sensors can help in achieving more analytical capabilities than the present sensors.

Overall, due to the extensive penetration of smartphones in our daily lives and the induction of advanced technologies, smartphone-based biosensors are strongly believed to revolutionize diagnostic and healthcare systems in the near future.

Acknowledgment

AP acknowledges the Ministry of Electronics and Information Technology, Government of India (MEITY Grant No. 5[9]/2012 – NANO). MPS is thankful to the Department of Chemistry (DRS-II [SAP], DST [FIST & PURSE] funded), AMU, Aligarh, for research and funding support.

REFERENCES

1. Soleymani, L., & Li, F. (2017). Mechanistic challenges and advantages of biosensor miniaturization into the nanoscale. *ACS Sensors*, *2*(4), 458–467.
2. Qureshi, A., Gurbuz, Y., & Niazi, J. H. (2012). Biosensors for cardiac biomarkers detection: A review. *Sensors and Actuators B: Chemical*, *171*, 62–76.
3. Saha, K., Agasti, S. S., Kim, C., Li, X., & Rotello, V. M. (2012). Gold nanoparticles in chemical and biological sensing. *Chemical Reviews*, *112*(5), 2739–2779.

4. Liu, Q., Wu, C., Cai, H., Hu, N., Zhou, J., & Wang, P. (2014). Cell-based biosensors and their application in biomedicine. *Chemical Reviews*, *114*(12), 6423–6461.
5. Tokel, O., Inci, F., & Demirci, U. (2014). Advances in plasmonic technologies for point of care applications. *Chemical Reviews*, *114*(11), 5728–5752.
6. Farka, Z., Jurik, T., Kovář, D., Trnkova, L., & Skládal, P. (2017). Nanoparticle-based immunochemical biosensors and assays: Recent advances and challenges. *Chemical Reviews*, *117*(15), 9973–10042.
7. Nusz, G. J., Curry, A. C., Marinakos, S. M., Wax, A., & Chilkoti, A. (2009). Rational selection of gold nanorod geometry for label-free plasmonic biosensors. *ACS Nano*, *3*(4), 795–806.
8. Verardo, D., Lindberg, F. W., Anttu, N., Niman, C. S., Lard, M., Dabkowska, A. P., … & Linke, H. (2018). Nanowires for biosensing: Lightguiding of fluorescence as a function of diameter and wavelength. *Nano Letters*, *18*(8), 4796–4802.
9. Yang, N., Chen, X., Ren, T., Zhang, P., & Yang, D. (2015). Carbon nanotube based biosensors. *Sensors and Actuators B: Chemical*, *207*, 690–715.
10. Liu, J., Geng, Z., Fan, Z., Liu, J., & Chen, H. (2019). Point-of-care testing based on smartphone: The current state-of-the-art (2017–2018). *Biosensors and Bioelectronics*, *132*, 17–37.
11. Huang, X., Xu, D., Chen, J., Liu, J., Li, Y., Song, J., Ma, X., & Guo, J. (2018). Smartphone-based analytical biosensors. *Analyst*, *143*(22), 5339–5351.
12. Hernández-Neuta, I., Neumann, F., Brightmeyer, J., Ba Tis, T., Madaboosi, N., Wei, Q., Ozcan, A., & Nilsson, M. (2019). Smartphone-based clinical diagnostics: Towards democratization of evidence-based health care. *Journal of Internal Medicine*, *285*(1), 19–39.
13. Zhang, D., & Liu, Q. (2016). Biosensors and bioelectronics on smartphone for portable biochemical detection. *Biosensors and Bioelectronics*, *75*, 273–284.
14. Wang, F., Lu, Y., Yang, J., Chen, Y., Jing, W., He, L., & Liu, Y. (2017). A smartphone readable colorimetric sensing platform for rapid multiple protein detection. *Analyst*, *142*(17), 3177–3182.
15. Sailapu, S. K., Dutta, D., Simon, A. T., Ghosh, S. S., & Chattopadhyay, A. (2019). Smartphone controlled interactive portable device for theranostics in vitro. *Biosensors and Bioelectronics*, *146*, 111745.
16. Veigas, B., Jacob, J. M., Costa, M. N., Santos, D. S., Viveiros, M., Inácio, J., … & Baptista, P. V. (2012). Gold on paper–paper platform for Au-nanoprobe TB detection. *Lab on a Chip*, *12*(22), 4802–4808.
17. Mancuso, M., Cesarman, E., & Erickson, D. (2014). Detection of Kaposi's sarcoma associated herpesvirus nucleic acids using a smartphone accessory. *Lab on a Chip*, *14*(19), 3809–3816.
18. Brangel, P., Sobarzo, A., Parolo, C., Miller, B. S., Howes, P. D., Gelkop, S., … & Stevens, M. M. (2018). A serological point-of-care test for the detection of IgG antibodies against Ebola virus in human survivors. *ACS Nano*, *12*(1), 63–73.
19. Lee, S., O'Dell, D., Hohenstein, J., Colt, S., Mehta, S., & Erickson, D. (2016). NutriPhone: A mobile platform for low-cost point-of-care quantification of vitamin B 12 concentrations. *Scientific Reports*, *6*(1), 1–8.
20. Srinivasan, B., O'Dell, D., Finkelstein, J. L., Lee, S., Erickson, D., & Mehta, S. (2018). ironPhone: Mobile device-coupled point-of-care diagnostics for assessment of iron status by quantification of serum ferritin. *Biosensors and Bioelectronics*, *99*, 115–121.
21. Kim, S. C., Jalal, U. M., Im, S. B., Ko, S., & Shim, J. S. (2017). A smartphone-based optical platform for colorimetric analysis of microfluidic device. *Sensors and Actuators B: Chemical*, *239*, 52–59.
22. Petryayeva, E., & Algar, W. R. (2014). Multiplexed homogeneous assays of proteolytic activity using a smartphone and quantum dots. *Analytical Chemistry*, *86*(6), 3195–3202.
23. Zhu, H., Sikora, U., & Ozcan, A. (2012). Quantum dot enabled detection of *Escherichia coli* using a cell-phone. *Analyst*, *137*(11), 2541–2544.
24. Noor, M. O., & Krull, U. J. (2014). Camera-based ratiometric fluorescence transduction of nucleic acid hybridization with reagentless signal amplification on a paper-based platform using immobilized quantum dots as donors. *Analytical Chemistry*, *86*(20), 10331–10339.
25. Pal, A., Sk, M. P., & Chattopadhyay, A. (2020). Recent advances in crystalline carbon dots for superior application potential. *Materials Advances*, *1*(4), 525–553.
26. Tabatabaee, R. S., Golmohammadi, H., & Ahmadi, S. H. (2019). Easy diagnosis of jaundice: A smartphone-based nanosensor bioplatform using photoluminescent bacterial nanopaper for point-of-care diagnosis of hyperbilirubinemia. *ACS Sensors*, *4*(4), 1063–1071.
27. Sun, K., Yang, Y., Zhou, H., Yin, S., Qin, W., Yu, J., … & Wu, C. (2018). Ultrabright polymer-dot transducer enabled wireless glucose monitoring via a smartphone. *ACS Nano*, *12*(6), 5176–5184.
28. Dutta, D., Sailapu, S. K., Chattopadhyay, A., & Ghosh, S. S. (2018). Phenylboronic acid templated gold nanoclusters for mucin detection using a smartphone-based device and targeted cancer cell theranostics. *ACS Applied Materials & Interfaces*, *10*(4), 3210–3218.
29. Chu, S., Wang, H., Du, Y., Yang, F., Yang, L., & Jiang, C. (2020). Portable smartphone platform integrated with a nanoprobe-based fluorescent paper strip: Visual monitoring of glutathione in human serum for health prognosis. *ACS Sustainable Chemistry & Engineering*, *8*(22), 8175–8183.
30. Sheini, A. (2021). A point-of-care testing sensor based on fluorescent nanoclusters for rapid detection of septicemia in children. *Sensors and Actuators B: Chemical*, *328*, 129029.
31. Li, H., Yang, M., Kong, D., Jin, R., Zhao, X., Liu, F., Yan, X., Lin, Y., & Lu, G. (2019). Sensitive fluorescence sensor for point-of-care detection of trypsin using glutathione-stabilized gold nanoclusters. *Sensors and Actuators B: Chemical*, *282*, 366–372.
32. Li, Y., Tang, D., Zhu, L., Cai, J., Chu, C., Wang, J., Xia, M., Cao, Z., & Zhu, H. (2019). Label-free detection of miRNA cancer markers based on terminal deoxynucleotidyl transferase-induced copper nanoclusters. *Analytical Biochemistry*, *585*, 113346.
33. Breslauer, D. N., Maamari, R. N., Switz, N. A., Lam, W. A., & Fletcher, D. A. (2009). Mobile phone based clinical microscopy for global health applications. *PloS One*, *4*(7), e6320.

34. Kanakasabapathy, M. K., Pandya, H. J., Draz, M. S., Chug, M. K., Sadasivam, M., Kumar, S., … & Shafiee, H. (2017). Rapid, label-free CD4 testing using a smartphone compatible device. *Lab on a Chip*, *17*(17), 2910–2919.
35. Knowlton, S., Joshi, A., Syrrist, P., Coskun, A. F., & Tasoglu, S. (2017). 3D-printed smartphone-based point of care tool for fluorescence-and magnetophoresis-based cytometry. *Lab on a Chip*, *17*(16), 2839–2851.
36. Ming, K., Kim, J., Biondi, M. J., Syed, A., Chen, K., Lam, A., Ostrowski, M., Rebbapragada, A., Feld, J. J., & Chan, W. C. (2015). Integrated quantum dot barcode smartphone optical device for wireless multiplexed diagnosis of infected patients. *ACS Nano*, *9*(3), 3060–3074.
37. Priye, A., Bird, S. W., Light, Y. K., Ball, C. S., Negrete, O. A., & Meagher, R. J. (2017). A smartphone-based diagnostic platform for rapid detection of Zika, chikungunya, and dengue viruses. *Scientific Reports*, *7*(1), 1–11.
38. Lillehoj, P. B., Huang, M. C., Truong, N., & Ho, C. M. (2013). Rapid electrochemical detection on a mobile phone. *Lab on a Chip*, *13*(15), 2950–2955.
39. Wang, X., Lin, G., Cui, G., Zhou, X., & Liu, G. L. (2017). White blood cell counting on smartphone paper electrochemical sensor. *Biosensors and Bioelectronics*, *90*, 549–557.
40. Low, S. S., Pan, Y., Ji, D., Li, Y., Lu, Y., He, Y., Chen, Q., & Liu, Q. (2020). Smartphone-based portable electrochemical biosensing system for detection of circulating microRNA-21 in saliva as a proof-of-concept. *Sensors and Actuators B: Chemical*, *308*, 127718.
41. Nemiroski, A., Christodouleas, D. C., Hennek, J. W., Kumar, A. A., Maxwell, E. J., Fernández-Abedul, M. T., & Whitesides, G. M. (2014). Universal mobile electrochemical detector designed for use in resource-limited applications. *Proceedings of the National Academy of Sciences*, *111*(33), 11984–11989.
42. Cánovas, R., Parrilla, M., Blondeau, P., & Andrade, F. J. (2017). A novel wireless paper-based potentiometric platform for monitoring glucose in blood. *Lab on a Chip*, *17*(14), 2500–2507.
43. Broeders, J., Croux, D., Peeters, M., Beyens, T., Duchateau, S., Cleij, T. J., Wagner, P., Thoelen, R., & De Ceuninck, W. (2013). Mobile application for impedance-based biomimetic sensor readout. *IEEE Sensors Journal*, *13*(7), 2659–2665.
44. Zhang, D., Lu, Y., Zhang, Q., Liu, L., Li, S., Yao, Y., Jiang, J., Liu, G. L., & Liu, Q. (2016). Protein detecting with smartphone-controlled electrochemical impedance spectroscopy for point-of-care applications. *Sensors and Actuators B: Chemical*, *222*, 994–1002.
45. Jiang, J., Wang, X., Chao, R., Ren, Y., Hu, C., Xu, Z., & Liu, G. L. (2014). Smartphone-based portable bacteria pre-concentrating microfluidic sensor and impedance sensing system. *Sensors and Actuators B: Chemical*, *193*, 653–659.
46. Misra, S. K., Dighe, K., Schwartz-Duval, A. S., Shang, Z., Labriola, L. T., & Pan, D. (2018). In situ plasmonic generation in functional ionic-gold-nanogel scaffold for rapid quantitative bio-sensing. *Biosensors and Bioelectronics*, *120*, 77–84.
47. Wei, Q., Acuna, G., Kim, S., Vietz, C., Tseng, D., Chae, J., … & Ozcan, A. (2017). Plasmonics enhanced smartphone fluorescence microscopy. *Scientific Reports*, *7*(1), 1–10.
48. Preechaburana, P., Gonzalez, M. C., Suska, A., & Filippini, D. (2012). Surface plasmon resonance chemical sensing on cell phones. *Angewandte Chemie International Edition*, *51*(46), 11585–11588.
49. Wang, X., Chang, T. W., Lin, G., Gartia, M. R., & Liu, G. L. (2017). Self-referenced smartphone-based nanoplasmonic imaging platform for colorimetric biochemical sensing. *Analytical Chemistry*, *89*(1), 611–615.
50. Guner, H., Ozgur, E., Kokturk, G., Celik, M., Esen, E., Topal, A. E., Ayas, S., Uludag, Y., Elbuken, C., & Dana, A. (2017). A smartphone-based surface plasmon resonance imaging (SPRi) platform for on-site biodetection. *Sensors and Actuators B: Chemical*, *239*, 571–577.

7

Current and Future Prospects in the Treatment of Chronic Obstructive Pulmonary Disorders

Manu Sharma, Aishwarya Rathore, Sheelu Sharma, Kakarla Raghava Reddy and Veera Sadhu

CONTENTS

7.1 Introduction

Lifestyle-associated changes like increased stress, smoking, air pollution, etc. have elevated the probability of occurrence of lung diseases. Lung diseases often lead to difficulty in breathing. If it remains untreated, lung disease may worsen and progress to lung cancer and ultimately death due to poor functioning of the lungs. According to the World Health Organization (WHO), the top five lung diseases are chronic obstructive pulmonary disorders (COPD), asthma, acute lower respiratory tract infection, tuberculosis, and lung cancer [1]. These lung diseases can be classified based on the part they affect (Figure 7.1).

According to the American Lung Association, COPD is the third highest reason for death in the United States and, currently, 15.3 million people are diagnosed with COPD [2]. COPD develops slowly and worsens with time, so diagnosis at the right time can be useful in tackling the disease. Unfortunately, patients and even healthcare professionals misdiagnose the disease to be tuberculosis, asthma, etc., making the recovery of patients even more difficult. If COPD remains undiagnosed, the patient may have difficulty breathing and over some time would be prevented from doing normal routine activities like walking, cooking, running, etc.

DOI: 10.1201/9781003157823-7

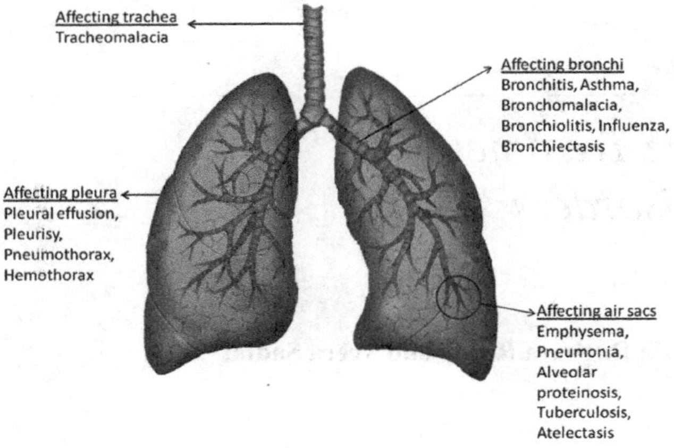

FIGURE 7.1 Common lung diseases impairing the livelihood of an individual globally.

7.2 Respiratory System

Lungs are a pair of spongy air-filled organs responsible for gaseous exchange between the blood and inhaled air. The right lung has three lobes and is larger than the left lung which has two lobes. Hence, the right lung holds more air volume. The respiratory tract comprises conducting regions (trachea, bronchi, and bronchioles) and respiratory (peripheral) regions (respiratory bronchioles and alveolar regions) (Figure 7.2). The upper respiratory tract comprises the nose, throat, pharynx, and larynx whereas the trachea, bronchi, bronchioles, and alveolar regions constitute lower respiratory tract. Airways can be simply described by a symmetric model according to which each airway divides into two equivalent branches. The trachea branches into two main bronchi. The right bronchus is wider and leaves the trachea at a smaller angle than the left bronchus and thus is more likely to receive inhaled material. The bronchi branch into terminal bronchioles which further divide to produce respiratory bronchioles. The bronchioles connected to alveolar ducts end in a tiny air sac called alveoli. Alveoli are elastic air sacs that resemble a small balloon where the exchange of gases with blood occurs [3]. Approximately $2-6 \times 10^8$ alveoli are present in an adult male. The conducting

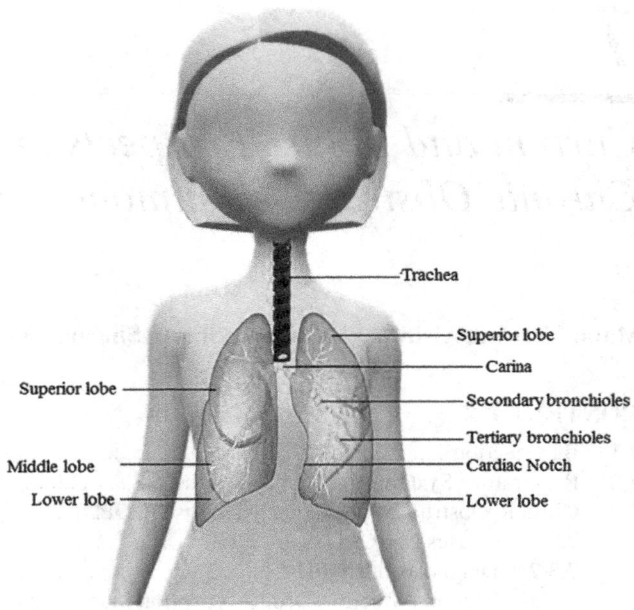

FIGURE 7.2 Lungs and their anatomy.

airways are lined by ciliated epithelium cells which facilitate sweeping upward of insoluble particles trapped in mucus deposited on airways by beating cilia and swallowing [4].

7.3 Chronic Obstructive Pulmonary Disease (COPD)

COPD is a progressive inflammatory lung disease characterized by chronic bronchitis, airway thickening, and emphysema. Chronic bronchitis is associated with bronchial damage (Figure 7.3). The irritation and swelling of bronchial tubes with loss of cilia movements lead to bronchitis associated with intense coughing and shortness of breath. Frequent coughing makes the tubes more irritated, secretes more mucus, and worsens breathing [5]. While during emphysema, walls inside the alveoli disappear, leading to the collapse of small sacs to form a larger sac (Figure 7.3). Larger sacs exhibit poor oxygen absorption

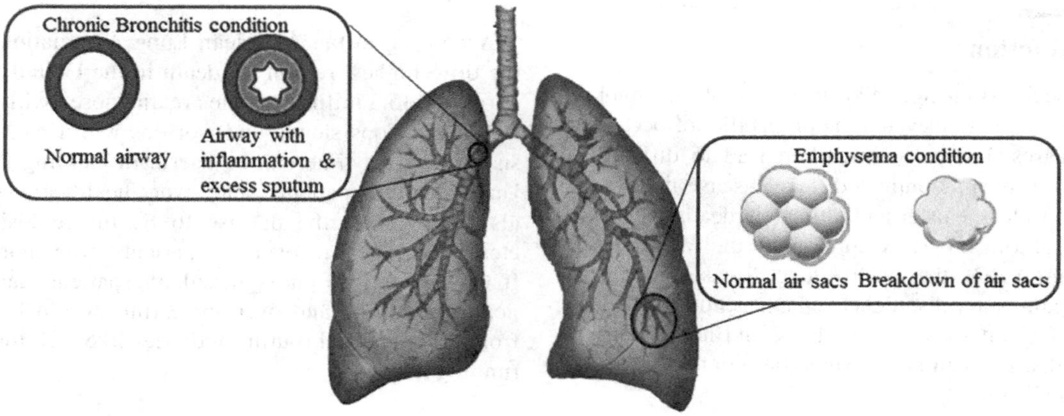

FIGURE 7.3 Chronic bronchitis and emphysema conditions.

compared to smaller sacs. Moreover, due to alveoli damage, lungs get stretched out and lose their elasticity which eventually makes the airway flabby and traps air in the lungs leading to difficulty in breathing. Thus, COPD is majorly characterized by symptoms like shortness of breath, wheezing, frequent coughing (with and without mucus), persistent chesty cough with phlegm, frequent cold, and respiratory infection. Under worsened patient conditions, lips or fingernail turn blue or gray due to insufficiency of oxygen level in the blood, fast heartbeat, reduction of mental alertness, and fatigue [6]. The pathogenesis of COPD is not clear. However, literature reports high inflammatory cell infiltration in the central airway, the predominance of $CD8^+$ cells and macrophage in bronchial mucosa, tumor necrosis factor-α, and perforins which eventually activate –Fas ligand apoptotic pathway which would damage lungs and cause emphysema. The circulating neutrophils count also increases which potentiates the secretion of cathepsin G, neutrophil elastase, proteinase-3, matrix metalloproteinase 8 (MMP-8), and MMP-9. The elevated level of proteases promotes the destruction of alveoli and stimulates mucus secretion and inflammation [7].

7.3.1 Causes of COPD

The probability of occurrence of COPD is mainly associated with the degree of damage or inflammation of the lungs and the airways. Inhalation of toxic substances activates inflammatory immune response, particularly if the normal repair mechanisms are overwhelmed or defective. Inflammatory changes in lung tissues are associated with mucus hypersecretion, vascular changes, airway narrowing and fibrosis, and other physiological abnormalities (Figure 7.4).

Smoking is the first and foremost cause of COPD due to oxidative stress generated by harmful chemicals in cigarettes. COPD is also one of the occupational hazards in places where people are exposed to high fumes and dust of cadmium, silica, coal, grain and flour dust, isocyanates, etc. Air pollution is another crucial reason for COPD which has increased several-fold with the advent of modernization. Apart from these reasons, individuals with alpha-1-antitrypsin deficiency have a higher chance of having COPD [8].

7.3.2 Diagnosis of COPD

Healthcare officials often suggest the patient undergo a certain test to ensure the disease with which the patient is suffering (Figure 7.5). The first and foremost test prescribed by the healthcare professional is the lung function test which is also called a spirometry test. This test indicates how well the lung functions. The patient breathes out into the mouthpiece of the machine after inhaling the bronchodilator. The spirometer would measure the air volume one breathes out per second and also the total air amount one breathes out. Apart from spirometry other lung function tests include measurement of lung volumes, diffusing capacity, and pulse oximetry [9]. Similar to this is a peak flow test that measures how fast a person breathes out. Blood oxygen tests can determine the level of oxygen in the blood sample. In this test, a peg-like device is attached to the finger of the patient and it measures the oxygen level in the blood. A simple blood test can also be done to determine if the patient has polycythemia, alpha-1-antitrypsin deficiency, etc. [10]. A phlegm (sputum) sample can also be tested to check for signs of a chest infection.

Chest X-ray imaging is used to determine the lung problems causing symptoms resembling COPD. Similarly, the CT scan also helps in detecting emphysema and whether surgery would benefit the patient or not.

7.3.3 Factors Affecting Drug Absorption in the Respiratory System

Different factors that may affect drug absorption through the respiratory system can be broadly divided into physiological factors, physicochemical factors, and pharmaceutical factors.

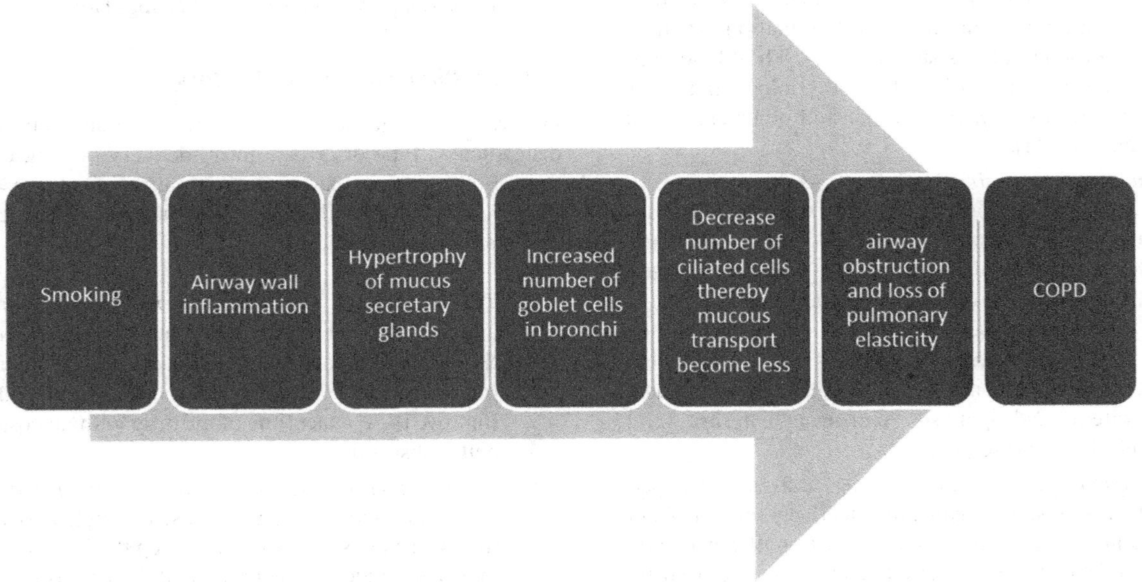

FIGURE 7.4 Correlation between toxic substance/smoking and COPD.

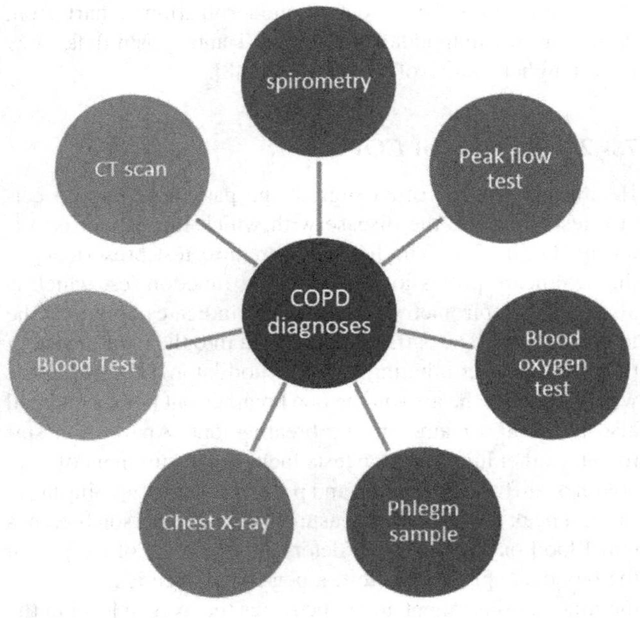

FIGURE 7.5 Different techniques to diagnose COPD.

7.3.3.1 Physiological Factors

Physiological factors are the factors that are related to the physiological characteristics of the body. These factors include:

i) *Blood flow*: The respiratory system is richly supplied with blood vessels making it an optimum site for drug absorption. The absorption of the drug is directly proportional to the blood supply in a particular organ. Since the majority of the drug is absorbed via the passive diffusion method, a higher blood supply is required to maintain the concentration gradient. The blood supply to the lungs is affected by many physiological changes like change in temperature, humidity, inflammation, trauma, and presence of vasoactive drugs. The blood supply can also be affected by physiological factors like fear, anxiety, frustration, and other emotional behaviors [11].
ii) *Enzymatic activity*: Different enzymes present in the lungs' mucosa can affect the pharmacodynamic and pharmacokinetic profile of the drug. The drug administered via the pulmonary route can undergo significant metabolism when it crosses the epithelial barrier due to the presence of different metabolic enzymes like cytochrome P450 dependent monooxygenase, oxidoreductase, lactate dehydrogenase, hydrolase, esterase, and acid phosphatase [12].
iii) *Mucociliary clearance*: Drug particles trapped into the mucosal layer are effectively cleared from the lungs due to the combined action of mucus and cilia. This defense system becomes a barrier to drug absorption. The normal mucociliary transit time in humans has been reported to be 12–15 min [13].

7.3.3.2 Physicochemical Factors

The physicochemical nature of the drug also affects the rate and extent of absorption of the drug. Molecular weight, size, pH, partition coefficient, drug solubility, and dissolution rate are the main physicochemical properties governing the therapeutic efficacy of the drug.

i) *Drug molecular weight and size*: Different scientists have worked on the effect of particle size on the absorption of the drug administered. Different literature data have concluded that absorption of the smaller particle (molecular weight approx. 100 Daltons) through the pulmonary route was 80% higher and absorption declined rapidly as the molecular weight increased [14]. The rule fits best for the drug molecules which are absorbed by paracellular diffusion. Thus, only molecules with smaller molecular weight and size than the channels can diffuse through the barrier and get absorbed.
ii) *Effect of pH and partition coefficient*: Molecular size is an important factor for the absorption of hydrophilic particles since they are absorbed by paracellular diffusion. On the other side lipophilic particle is absorbed via transcellular diffusion and hence partition coefficient is the major affecting factor for its absorption [15]. The lipophilic drugs can easily penetrate the lipophilic membrane and therefore unionized lipophilic drugs have better absorption properties as they can be absorbed easily by a passive diffusion process.
iii) *Drug solubility and dissolution rate*: Solubility and dissolution rate are important parameters for drug absorption. The solid particles must dissolve before they cross the mucosal layer otherwise they would undergo rapid clearance. Moreover, other factors like powder size and powder morphology are some other factors that affect the surface area of contact in the lungs [16].

7.3.3.3 Pharmaceutical Factors

Different pharmaceutical factors affect the absorption of the drug such as type of dosage form, delivery system, the concentration of the preparation, the volume of administration, viscosity of the formulation, pH of the final pharmaceutical preparation, etc.

i) *Type of dosage form and delivery system*: Solution and sprays are absorbed better and have lesser irritation when compared to the powder sprays. Recently, many hydrogel systems have been developed to improve the contact time of the drug with mucosa for better absorption.
ii) *Drug concentration and volume of administration*: Lungs can withstand a large dose but highly concentrated products can irritate the epithelium mucosa increasing mucosal secretion and thereby loss of the drug. The ideal dose-volume range for pulmonary administration is up to 2 ml [16].

iii) *pH of the formulation*: pH and pKa are two important factors for drug absorption. The right pH preparation can avoid irritation and thereby avoid drainage of the drug due to increased mucosal secretion. The pH of the plural cavity is 7.64; therefore, preparations should have a pH between 4.5 and 6.5 to avoid irritation and have maximum absorption.

7.3.4 Treatment Available for COPD

Presently no permanent cure or treatment for COPD is available. The accessible treatment can only either slow the progression of the disease or control the symptoms of the disease. However, correct diagnosis and treatment can be helpful in the management of the disease. Out of the treatments available in the market, inhaler and nebulizer dosage forms are the mainstream treatment method meant to be administered via the pulmonary route. They are more effective than any other type of medication. Different types of devices like metered-dose inhalers, dry powder inhalers, soft mist inhalers, and nebulizers are commonly used for the delivery of a variety of bronchodilator medicines in COPD treatment.

7.4 Devices Used for Drug Delivery

7.4.1 Metered-Dose Inhalers (MDIs)

An MDI is a hand-held device useful to deliver potent inhaled medicines accurately in a reproducible amount to the lungs in aerosol form. An MDI is a pressurized canister of medicine attached to a spacer and mouthpiece to deliver the drug to the lungs. The ability of metered-dose inhalers to administer the different types of COPD medicines like steroids, bronchodilators, and combination medicines and their reproductivity of dose makes these inhalers highly popular (Table 7.1). However, patients need to maintain coordination between the delivery of aerosolized medicine and breathing. Deep and slow breathing improves the efficacy of the medicine. Faster breathing leads the medicine to reach only the upper part of the respiratory tract. Another disadvantage is that the patient may need to use a spacer to coordinate the breathing and inhalation of the medicine [17].

7.4.2 Dry Powder Inhalers

A dry powder inhaler (DPI) facilitates the delivery of medicines in dry powder form to the lungs when one breathes in through the device (Figure 7.6). A DPI, instead of using a propellant to deliver medicine into the lungs, utilizes an inward breath for delivery of the medicines. They are available in single-dose and multiple-dose devices. Multiple-dose devices contain up to 200 doses. They are easy to use, spacers are not needed, and coordination between breathing and pressing the device is not required [18]. However, patients have to breathe in deeper while using a DPI compared to an MDI. Furthermore, the reproducibility of the dose is also difficult to achieve each time with a DPI. Humidity and other environmental factors also govern the efficacy of DPIs.

7.4.3 Soft Mist Inhalers

A soft mist inhaler (SMI) is a device used to create a cloud of medicine that one can inhale without the use of a propellant. The mist contains finer particles than MDIs and DPIs facilitating faster drug absorption through the lungs. Thus, it lowers the dose of medicine. Mist is formed slowly. Thus, the device doesn't require any coordination between breathing and mist formation, and hence no spacer is required [19]. SMIs are used to administer different short-acting bronchodilators and steroids (Figure 7.7). Some new inhalers containing a combination of a long-acting β-2 agonist and anti-muscarinic drugs to treat COPD symptoms are also available like the salmeterol and fluticasone (Advair) inhaler and the formoterol and budesonide (Symbicort) inhaler.

7.4.4 Nebulizers

Nebulizers were introduced in 1930 to make drug administration easy for patients with respiratory disorders. Nebulizers deliver a relatively larger amount of drug solutions and suspensions. They are preferred for drugs that cannot be conveniently manufactured as metered-dose or dry powder inhalers, or when the therapeutic dose is too large. Different categories of nebulizers used by COPD patients are jet, ultrasonic, and vibrating mesh nebulizers. Depending on the type of medication being administered, nebulization usually takes 25 min or less to inhale the drug. The device should be thoroughly cleaned with soap and water after use [20, 21]. Nebulizers still stand as the preferred option for drug delivery to patients with severe COPD, bed-ridden and intellectually disabled patients, as well as children, since it allows inhalation of a drug at normal tidal breathing through a facemask or mouthpiece. However, nebulizers are not easy to carry, and hence metered-dose and dry powder inhalers came into the market (Table 7.1).

7.5 Supplementary Therapies

Apart from the aerosolized inhalable dosage form of medicines, the patients can also be treated with drugs administered via oral route medication like theophylline tablets, carbocisteine mucolytic tablets, antibiotic tablets, etc. Other treatment measures might also include smoking cessation medication, anxiolytics, opioids, aromatherapy, etc. (Table 7.2) [22, 23].

During extreme COPD conditions when bronchodilators are unable to calm down the symptoms, oxygen therapy or a non-invasive ventilation technique is used. A patient can directly inhale the pure form of oxygen through a portable oxygen cylinder during oxygen therapy. While during non-invasive ventilation, the patients are admitted to the hospital and nebulized with a pure form of oxygen [24].

7.6 Surgical Therapies

Surgical procedures are also available for permanent cure of COPD but due to very poor success rates, they are rarely

TABLE 7.1
Inhalation Therapies for COPD Treatment

S. no.	Drug	Marketed preparation	Dose	Route of administration	Advantage	Disadvantage	Contraindicated: Drug	Contraindicated: Patient	Mechanism of action	Ref.
1	Albuterol (also known as salbutamol)	Proventil HFA ProAir Airomir Ventolin HFA	90–180 mcg (1–2 puffs) every 4–6 h and do not exceed 12 puffs/day	Inhalation Dry powder inhalation Inhalation Inhalation	Rapid relief; HFA used instead of CFC propellant; treat exercise resulted in asthma	Beta receptor down-regulation, misuse, unpleasant taste	Anti-hypertensive, anti-anginal drugs, steroids, and diuretics	Diabetic patient	Short-acting β2 agonist acts as a bronchodilator	[27]
2	Levalbuterol	Xopenex	0.31–0.63 mg thrice a day by nebulization; in severe cases 1.25 mg thrice a day	Inhalation using metered-dose inhaler or nebulizer	Quick relief	Chills, abnormal ECG, diarrhea, lymphadenopathy, myalgia, anxiety	Sympathomimetic, epinephrine, β-blockers, non-potassium-sparing diuretics, MAO inhibitors, tricyclic antidepressants	Children below six years, CVS disorder, diabetes	Short-acting β2 agonist acts as a bronchodilator	[28, 29]
3	Albuterol and ipratropium	Combivent DuoNeb	Two inhalations four times/day 3 ml vial four times/day via nebulization	Inhalation using Respimat Inhalation using Ellipta inhaler	Effective in a patient with high exacerbation, the low dose of individual drug administered	Paradoxical bronchospasm, hives, difficulty urinating	Atropine and its derivative	Hypersensitivity	Combination of short-acting muscarinic antagonist and short-acting βagonist	[30, 31]
4	Umeclidinium	Incruse	One inhalation (62.5 mcg) once daily	Inhalation using Ellipta inhaler	Once-daily; cost-effective; no dosage adjustment required	No instant action; no relief from acute symptoms	Anti-cholinergic drugs	Hypersensitivity, glaucoma, lactose intolerance	Long-acting muscarinic antagonists which act as a bronchodilator	[32]
5	Glycopyrrolate	Seebri	15.6 mcg of powder content in one capsule inhaled twice daily	Inhalation using Respimat inhaler	No dose adjustment required, potent than ipratropium bromide and tiotropium bromide	No instant action; dry mouth, sore throat, irregular heartbeat; paradoxical bronchospasm	Antidepressant anticholinergics, sedatives, pramlintide, revefenacin	Benzyl alcohol hypersensitivity, glaucoma, myasthenia gravis, neuropathy	Long-acting muscarinic antagonists which act as bronchodilator	[33, 34]
6	Tiotropium	Spiriva	One time daily, two puffs (2.5 mcg/puff)	Inhalation using Respimat or Handihaler inhaler	Once a day administration, high patient compliance	No instant relief; dry mouth, dizziness, painful urination, paradoxical bronchospasm	Umeclidinium bromide, glucagon, revefenacin	Alcoholics; breastfeeding mother, UTI disorders, glaucoma	Long-acting muscarinic antagonists which act as bronchodilator	[35]
7	Aclidinium	Tudorza	400 mcg twice daily (approx. 12 h apart)	Inhalation using press air inhaler	Low systemic exposure; prolonged effect; cost-effective; well tolerated; no dosage adjustment required	Not for instant relief; paradoxical bronchospasm; painful urination, dizziness	Anticholinergics	Glaucoma, urinary problem; below 18 yrs age	Long-acting muscarinic antagonists which act as bronchodilator	[36, 37]

(Continued)

TABLE 7.1 (CONTINUED)

Inhalation Therapies for COPD Treatment

S. no.	Drug	Marketed preparation	Dose	Route of administration	Advantage	Disadvantage	Contraindicated		Mechanism of action	Ref.
							Drug	Patient		
8	Ipratropium bromide	Atrovent	Two puffs four times a day, not more than 12 puffs a day	Inhalation	Eco-friendly; rapid onset of action i.e., 15–30 min	Short acting; not an instant relief medicine; dizziness, nausea, stomach upset, dry mouth, constipation	Anticholinergics; tricyclic antidepressants, anti-Parkinson's drugs, and quinidine	Narrow-angle glaucoma, urinary problem; below 18 years age	Muscarinic antagonists which act as bronchodilator	[38, 39]
9	Indacaterol	Arcapta	75 mcg inhaled orally each day; not to exceed once daily	Inhalation using Neohaler inhaler	Once-daily administration, quick relief within 5 min; no dose adjustment required	Tremor, paradoxical bronchospasm, headache, b.p. rise, cough	Adrenergic drugs, xanthine derivatives, steroids, diuretics, MAO inhibitor, tricyclic antidepressants, β-blockers	Children, diabetes, CVS disorders, hyper-thyroidism; seizures	Long-acting β2 agonists which acts as bronchodilator	[40]
		Onbrez	150–300 mg capsule once daily using inhaler	Inhalation						
10	Arformoterol	Brovana	15 mcg twice daily by nebulization; dose > 30 mcg is not recommended	Nebulization	Long acting; no dose adjustment required	No instant relief; nervousness, dizziness, tremor, headache, nausea, dry mouth, dose adjustment required	β agonists, steroids, adrenergic drugs, xanthine derivatives, diuretics, antidepressants	Hypersensitivity, diabetes, thyrotoxicosis, CVS disorders, seizures, asthma	Long-acting beta2 agonists which act as bronchodilator	[41, 42]
11	Formoterol	Perforomist	20 mcg vial twice daily	Nebulization	Long acting; used in maintenance therapy, exercise-induced bronchospasm	Tremor, headache, dizziness, nervousness, dry mouth, stomach upset, hoarseness	Theophylline, β- blockers	Children below 12 years	Long-acting β-2 agonists which act as bronchodilator	[43]
		Oxeze	6–12 mcg every 12 h	Inhalation						
		Foradil	12–24 mcg twice a day; dose not more than 48 mcg	Inhalation						
12	Salmeterol	Serevent	One inhalation (50 mcg) twice daily; not more than twice daily	Inhalation (metered-dose inhalers)	Long-term maintenance treatment, exercise-induced bronchospasm	No instant relief, hoarseness, throat irritation, rapid heartbeat, cough, dry mouth/throat, upset stomach	Other long-acting inhaled β-agonists, antibiotics, antidepressant	CVS disorder, diabetes, liver and thyroid problem	Long-acting β-2 agonists which act as bronchodilator	[44]
13	Olodaterol	Stiverdi	5 mcg (two actuations)/day, not to exceed two inhalations a day	Inhalation using Respimat inhaler	Long acting, no dose adjustment required	Nervousness, tremor, and trouble sleeping may occur	Long-acting β-agonists, adrenergic, steroids, diuretics, antidepressants	Glaucoma, diabetes, CVS disorder, thyrotoxicosis	Long-acting β-2 agonists which act as bronchodilator	[45]
14	Umeclidinium and vilanterol	Anoro	One actuation/day (62.5 mcg of Umeclidinium and 25 mcg of Vilanterol)	Inhalation using Ellipta inhaler	Long-acting	No instant relief, lack of flexibility in dosing, paradoxical bronchospasm	Ketoconazole, CYP3A4 inhibitors, MAO inhibitors, tricyclic antidepressants	CVS disorder, narrow-angle glaucoma, urinary problem	Long-acting muscarinic antagonists and β-2 agonists	[46]

(Continued)

TABLE 7.1 (CONTINUED)
Inhalation Therapies for COPD Treatment

S. no.	Drug	Marketed preparation	Dose	Route of administration	Advantage	Disadvantage	Contraindicated: Drug	Contraindicated: Patient	Mechanism of action	Ref.
15	Olodaterol and tiotropium	Stiolto	Two inhalations once daily; not to exceed two inhalations per day	Inhalation using Respimat inhaler	Long-acting maintenance treatment; hand held, pocket-sized	Lack of flexibility in dosing; convulsions, decreased urine, dry mouth, irregular heartbeat, loss of appetite, muscle pain	Adrenergic drugs, xanthine derivatives, steroids, diuretics, antidepressants, anticholinergics, β blockers	Urinary disorders, CVS disorder, glaucoma	Long-acting muscarinic antagonists and long-acting β 2 agonists	[47]
16	Indacaterol and glycopyrrolate	Utibron	One capsule twice daily using neohaler	Inhalation using Neohaler inhaler	Long acting; soft whirring sound ensures proper drug administration	No instant effect, lack of flexibility in dosing, worsen glaucoma, renal impairment, hypokalemia	Antihistamines, β-blocker, MAO inhibitor, ipratropium, tiotropium,	Glaucoma, renal impairment, hypokalemia	Long-acting muscarinic antagonists and long-acting β 2 agonists	[48]
17	Glyco-pyrrolate and formoterol	Bevespi	Two inhalations twice daily. Each inhalation contains 9 mcg glycopyrrolate and 4.8 mcg formoterol fumarate	Inhalation using Aerosphere inhaler	Long acting	Lack of flexibility in dosing; no instant relief; coughing, tremor, irregular heartbeat, painful urination, increased thirst/urination, muscle cramp	Anticholinergic drugs or LABA drugs (such as salmeterol, vilanterol), βblocker, diuretics	Children below 12 years	Long-acting muscarinic antagonists and long-acting beta2 agonists which acts as bronchodilator	[49]
18	Fluticasone and salmeterol	Advair	One inhalation twice daily	Inhalation	Long-acting; improves lung function; lesser S/E than corticosteroid alone, hand-breath coordination not necessary, suitable for above three years	Tremor, fast heartbeat, thrush, sore throat, reduced adrenal; not for instant relief, expensive	Amiodarone, diuretics, HIV medicines, antidepressants, antifungal medications, or beta-blockers	Seizures, diabetes, liver disorder hyperthyroidism, CVS disorder, milk protein allergies	Combination of bronchodilator and corticosteroid	[50]
19	Vilanterol/ fluticasone furoate	Breo Ellipta	One inhalation per day	Inhalation	Once daily; improves lung function; eco-friendly as no propellant added	Fungal infection; tremor, fast heartbeat, thrush, sore throat, weak immunity; not for instant relief	Tuberculosis, herpes simplex infection, glaucoma, osteoporosis	Milk protein allergy patient and children below 18 years	Combination of bronchodilator and corticosteroid	[51, 52]
20	Budesonide and formoterol fumarate	Symbicort	Two inhalation twice daily (12 h apart)	Inhalation	Works within 15 min; spacers can be used; first-line treatment	Tremor, thrush, sore throat, hand-breath coordination req.	Diabetes, thyrotoxicosis, CVS disorder, seizure	Children below 12 years	Combination of bronchodilator and corticosteroid	[53, 54]
21	Supplemental oxygen	InogenOne G4 Portable Oxygen Concentra-tor	oxygen treatment using a Venturi mask at inflow rate 4 L/min with an initial target saturation of 88–92%	Venturi mask or nasal cannula attached to portable O_2 cylinder	Relive severe breathlessness; beneficial during sleep, exercise, etc., comfortable	Explosion risk, respiratory depression in few; expensive	None	Neonates and pediatric patient with nasal obstruction	Oxygen is a treatment for hypoxaemia	[55, 56]

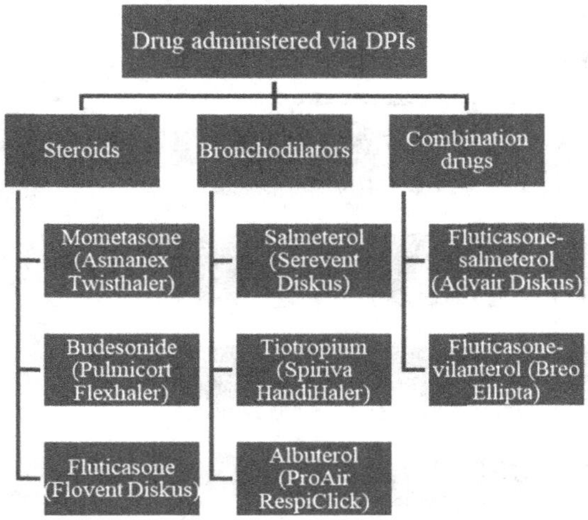

FIGURE 7.6 Drugs administered via DPIs.

performed. The three main types of surgery performed to treat COPD are bullectomy, lung volume reduction surgery, and lung transplant. The removal of air pockets from one of the lungs to make breathing more comfortable is known as bullectomy [25]. On the other hand, removal of the badly damaged area of the lung to facilitate the better working of healthier parts is achieved by lung volume reduction surgery, while lung transplant is an operation to remove and replace a damaged lung with a healthy lung from a donor [26].

7.7 Other Therapies

Lifestyle changes and certain home remedies can also help to relieve the symptoms of COPD when followed along with regular medication (Figure 7.8). Herbal and natural medicines have been increasingly popular in the 21st century due to higher safety index, cost-effectiveness, and synergistic actions when combined with daily exercise.

7.7.1 Exercise

COPD can be managed by regular breathing exercises which strengthen and improve the endurance of respiratory muscles and make breathing easy for the patients. Different breath controlling exercises that must be practiced by COPD patients include pursed-lip breathing exercises (inhaling through the nose and exhaling through tightly pursed lips), diaphragm breathing (expanding the belly while breathing in and deflating while breathing out), pranayama (a controlled breathing yoga technique), etc. [69, 70]. Apart from these exercises, singing, swimming, etc. are also a type of breathing exercise. COPD patients can also join the pulmonary rehabilitation centers running specialized programs of exercise training and education. A typical program includes physical exercise training tailored according to the need of the patients which include walking, cycling, strength exercise, etc., dietary advice, and counseling. Certain Chinese martial arts practices like Tai Chi can also help relieve COPD symptoms. Tai Chi emphasizes the use of "mind" to control breathing and circular body motions to facilitate the flow of internal energy. Tai Chi is beneficial for reducing dyspnea and improving exercise capacity and physiological and psychosocial well-being among people with COPD.

7.7.2 Diet

A healthy diet should be paid attention to as diet improves immunity and helps avoid any flare-ups in COPD. Patients with COPD should strictly maintain their diet and prevent obesity or underweight conditions since both conditions affect the patient's breathing rate. However, sometimes the sensitivity of COPD patients to specific foods leads to allergic reactions and exaggerates the breathing difficulty. Thus, such a particular food should be avoided. Use of Katu, Lavana, Ushna, Snigdha, Laghu, Ahara, vegetables including gourd, bottle gourd, spinach, methi, garlic, ginger, karvellaka, patola, shigru, and pulses like Mudaga daal and Kulatha should be encouraged for consumption in the patients of COPD [71]. Additional supplements like vitamin D, coenzyme Q10, and creatine are useful in reducing inflammation of airways and symptoms of COPD. In addition, vitamin D supplements can improve the body's ability to clear the bacteria. Apart from these antioxidants, vitamins and omega-3 fatty acid can also be given. Both these supplements can improve lung conditions by reducing oxidative stress in COPD patients.

7.7.3 Avoiding Pollution

Airway obstruction can also be prevented by avoiding smoke and air pollution. Avoiding pollution and smoke can also prevent further damage to the lungs [72]. COPD symptoms can flare up even with small air pollutants. Certain components like paints, cleaning products, pesticides, dust, etc. inside the home can make it difficult for the patient to breathe. So, a patient can improve breathing by limiting contact with such irritants. Washing of bed linens, curtains, etc. regularly can also remove dust mites and potential air pollutants. Cleaning the airway can also help relieve COPD patients from breathlessness due to airway obstruction by mucus. Mucus can be cleared by controlled coughing, gargling, drinking plenty of water, using a humidifier, etc.

7.8 Aromatherapy

Nowadays even aromatherapy is used for the treatment of COPD. It involves the use of many essential oils which help the patient to calm down and improve breathing. Essential oils like eucalyptus oil and Myrtol standardized essential oil can help patients with COPD (Figure 7.8). Eucalyptus oil contains the natural ingredient eucalyptol which has antioxidant, anti-inflammatory, and mucolytic properties. It also opens up the airway in the lungs and prevents flare-ups of COPD.

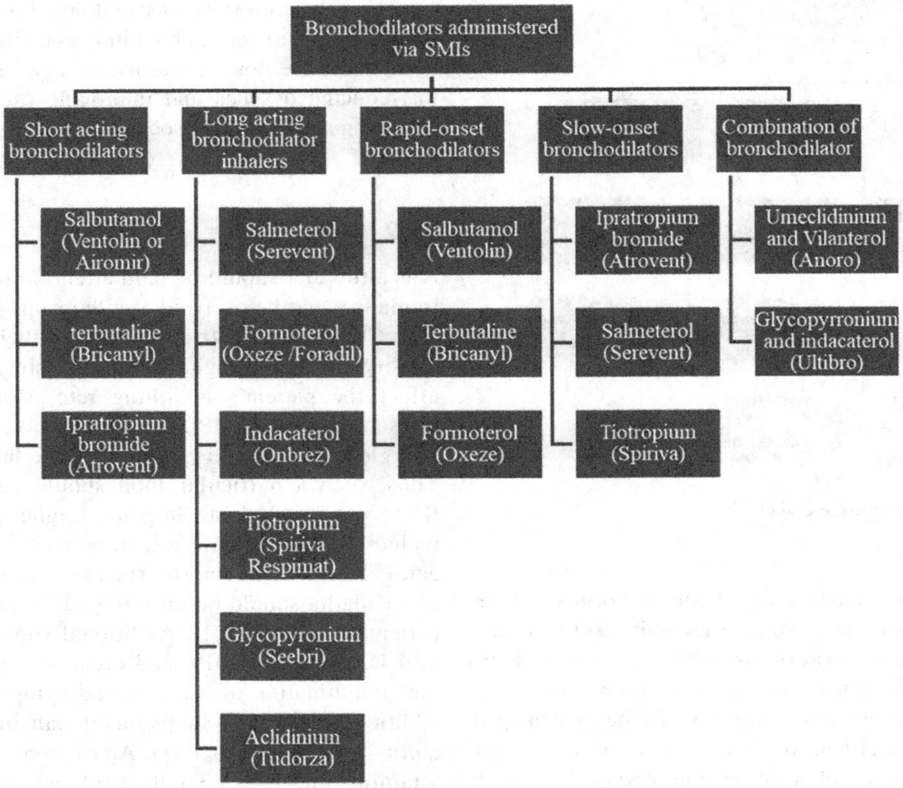

FIGURE 7.7 SMI of the different bronchodilators.

7.9 Homeopathy Treatment for COPD

Homeopathy medicines, similar to allopathy medicines, can only prevent further damage and associated disease complications. They also claim to reduce the dependence on a bronchodilator and steroidal drugs along with the treatment of distressing symptoms without the fear of side effects. The selection of remedy is based upon the theory of individualization and symptoms similarity by using a holistic approach [73]. The aim of homeopathy is not only to treat COPD symptoms but to address its underlying cause and individual susceptibility. Commonly used homeopathic remedies in COPD are depicted in Table 7.3.

7.10 Novel Approaches to Treat COPD

Advancements in technology and research have provided many new treatment methods for COPD (Figure 7.9). Recently approved novel techniques to treat COPD include:

i) *Once-daily treatment*: Olodaterol has been found to be a good drug for once-daily COPD treatment after two 48-week sessions of Phase III studies. Apart from olodaterol, glycopyrronium is another recently approved once-daily maintenance therapy [49]. Another recently US FDA-approved once-daily nebulized bronchodilator for COPD treatment by Mylan NV and Theravance Biopharma, Inc. is YUPELRI, containing revefenacin [84, 85].

ii) Kinase inhibitors are gaining attention as recent treatment techniques. Since Toll receptor ligands (e.g., lipopolysaccharide) and cytokines activate kinase pathways such as p38 MAPKs, PI3K, Janus kinase (JAK)/signal transducer and activator of transcription and Rho kinase, resulting in downstream activation of transcription factors such as NFkB and increasing pro-inflammatory mediators. Many oral p38 inhibitors like losmapimod and PH797804 have undergone a phase II clinical trial for COPD treatment, but the result found was variable. Similarly, another method for treating COPD is by inhibiting JAK as cytokine and inflammatory mediators signal act via JAK/STAT pathway. Tofacitinib, a JAK inhibitor, has shown promising potential in inflammatory diseases like ulcerative colitis and rheumatoid arthritis although response in COPD has not been investigated yet [86, 87].

iii) *Vitamin D*: Vitamin D supplement improves lung function in patients with COPD by strengthening muscles. Hence, vitamin D deficiency may be a major risk factor for COPD [88].

iv) *Matrix metalloproteinase (MMP) inhibitors*: The rise of MMPs facilitates the degradation of most components of the extracellular matrix leading to emphysema and alveolar destruction. Thus, targeting the MMPs using broad-spectrum MMP inhibitors RS113456 or PKF242-484 or AZD1236 may be an alternative therapy for COPD [89, 90].

TABLE 7.2

Different Supplemental Allopathic Treatments Are Available for COPD Treatment

S. no.	Drug	Marketed preparation	Dose	Advantage	Disadvantage	Contraindicated		Mechanism of action	Ref.
						Drug	Patient		
1	Prednisone	Deltasone	30 to 50 mg once daily orally for five to 14 days	Easy administration; less costly, long and rapid-acting	Increased infection risk, psychological effects, weak bones decrease sex drive	NSAIDS, anticoagulant medicine	Tuberculosis, herpes simplex infection, diabetes, glaucoma	Decrease airway swelling	[57, 58]
2	Trimethoprim and sulfamethazole	Bactrim	One double-strength tablet orally every 12 hrs for 10–14 days	Easy administration; no inhalation device needed, less costly, treat amoxicillin resistant bacteria; don't stain teeth	Incomplete treatment leads to resistance; not a first-line treatment; muscle weakness, kidney damage, dizziness	Sulfamethoxa-zole, diuretics, methotrexate, cycloporin, ACE inhibitors	Pregnant, lactating mother, thrombocytopenia, folate deficiency	Broad-spectrum antibiotics	[59]
3	Roflumilast	Daliresp/Daxas	500 mg once daily orally	Inexpensive, improve lung function, well-tolerated than cilomilast	GI harms, headache, weight loss, high risk of psychiatric events; dose adjustment required, first-pass metabolism	CYP inducers, psychiatric drugs	Insomnia/anxiety/ de-pression	Phosphodiesterase-4 (PDE-4) inhibitor (anti-inflammatory)	[60]
4	Cilomilast	Ariflo	15 mg twice daily orally	Ten times more selective for PDE4D; improve lung function	Diarrhea, abdominal pain, dose-limiting gastrointestinal toxicity, first-pass metabolism	Riociguat	GI problems, ulcers	Phosphodiesterase-4 (PDE-4) inhibitor so act as an anti-inflammatory	[61]
5	Aminophylline hydrate (theophylline derivative)	Phyllocontin	10–20 μg/ml intravenous injection	Bronchodilator and anti-inflammatory agent; instant action	Narrow therapeutic index, large inter individual variability; regular monitoring of plasma drug conc. required; dose adjustment required	Fluoroquino-lones, macrolides, lithium, ephedrine, rifampicin, and anticonvulsants	Children below six months, porphyria	Non-selective inhibition of cyclic neucleotide phosphodies-terases and competitive antagonism of adenosine receptors (bronchodilator)	[62]
6	Ambroxol HCl	Acorex	30 mg to 120 mg taken in two to three divided doses orally	Easy administration; no inhalation device needed; less costly, different dosage form available	Not a first-line treatment, the onset of action – 30 mins, occasional mild gastrointestinal side effects; first-pass metabolism	Antitussive like codeine	Pregnant women, gastric ulcer	Mucolytic agents	[63]
7	Erdosteine	Erdozet	300 mg twice daily orally	Easy administration; no inhalation device needed, less costly, dual-action mucolytic agent, and antioxidant	Gastrointestinal side effects; undergo first-pass metabolism; dose adjustment required	Alcohol	Pregnant and breastfeeding mothers, gastric ulcer	Mucolytic agents	[64, 65]

(*Continued*)

TABLE 7.2 (CONTINUED)

Different Supplemental Allopathic Treatments Are Available for COPD Treatment

						Contraindicated			
S. no.	**Drug**	**Marketed preparation**	**Dose**	**Advantage**	**Disadvantage**	**Drug**	**Patient**	**Mechanism of action**	**Ref.**
8	Carbocisteine	Mucodyne	Two 375 mg capsules/15 ml syrup thrice a day orally	Easy administration; inexpensive, rapid absorption; different dosage forms available	Not a first-line treatment for COPD; undergo first-pass metabolism	Chloramphenicol, acetohexamide, disulfiram, glimepiride	Pregnant and breastfeeding mothers, gastric ulcer	Mucolytic agents	[66]
9	Guaifenesin	Mucinex	Dose – 1–2 tablets (600–1,200 mg) every 12 hrs. Max – four tablets/day	Easy administration; inexpensive, high patient compliance	Not a first-line treatment; first-pass metabolism; GI disorders, rashes, dizziness, headache	Alcohol, phenylephrine, methotrimeprazine, mirtazapine, nefazodone	Kidney stones, CVS disorder, diabetes, glaucoma	Mucolytic agents	[67]
10	Nicotine	Nicorette, Nicorette Plus	One pack cigarette/ day – 4 mg pieces, chew hourly first two weeks and then decrease the number	Easy administration; inexpensive; safe; available in different dose and flavors; instant relief from craving	Frequent administration, not a first-line treatment; mouth or throat irritation, bad after taste, aggravate the dental problem, need proper chewing to extract nicotine	Imipramine, oxazepam, propranolol, prazosin, insulin, theophylline	Hiatal hernia, CVS disorder GI disorder, hypersensitivity with nicotine, non-smokers	Nicotine replacement therapy to cease smoking	[68]
		NicoDerm, Nicotrol, and Habitrol	> 10 cigarettes/day and 21 mg patch daily for six weeks with gradual dose reduction	Steady release for 18–24 h, no frequent dosing					

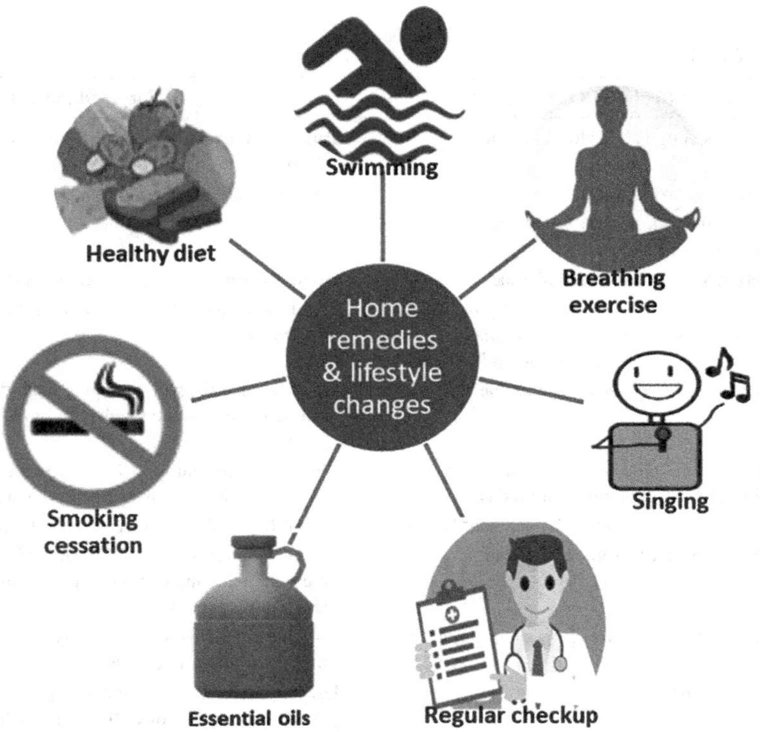

FIGURE 7.8 Lifestyle changes and natural remedies to treat COPD symptoms.

v) *New combination therapy*: QVA149, a dual bronchodilator combination of the long-acting β_2 adrenoceptor agonist indacaterol and long-acting muscarinic receptor antagonist glycopyrronium, has been approved as once-daily maintenance therapy for COPD patients. Furthermore, a once-daily fixed-dose combination of olodaterol and tiotropium showed bronchodilatory, anti-inflammatory, and anti-proliferative activity. GSK961081, a muscarinic agonist-β_2 agonist a combination of tiotropium and salmeterol is a well-tolerated bronchodilator [91, 92].

vi) *Combination of corticosteroid and bronchodilator*: A newer combination of fluticasone furoate and vilanterol has been sanctioned by the FDA for the remedy of COPD [93].

vii) *Anticholinergic drugs as an add-on therapy for COPD*: A combination therapy for 16 weeks consisting of tiotropium, salmeterol, and fluticasone propionate reduced the airway thickening and improved lung function test compared to treatment with tiotropium and salmeterol or salmeterol and fluticasone propionate [94].

viii) Tanreqing (Chinese medicine) had better clinical effectiveness in combination with routine treatment for COPD. This combination can remarkably ameliorate oxygen partial pressure, carbon dioxide pressure, and lung function of patients with COPD. Moreover, no serious adverse reactions are reported [95, 96].

ix) *Tumor necrosis factor α (TNF-α) inhibitors*: Humanized TNF antibodies to target the increased TNFα levels during COPD have been under clinical trial [97].

x) N-acetylcysteine (NAC) has an antioxidant effect and reduces the number of exacerbations. Other new antioxidants used include α-phenyl N-tert-butyl nitrone [97].

xi) *Prostanoid inhibitors*: Isoprostanes, potent constrictors of human airways, are directly formed from arachidonic acid in the absence of cyclooxygenase due to excessive oxidative stress during COPD. Therefore, seratrodast and Bay u3405, thromboxane receptor antagonists, might be beneficial in COPD [98].

xii) *Protease inhibitors*:

a. Neutrophil elastase (NE) inhibitors: ICI 200355 and nonpeptide inhibitors, such as ONO-5046 of high potency, have been developed as NE inhibitors [99].

b. Cathepsin inhibitors: Cathepsin G and proteinase 3, having elastolytic activity, must necessarily be inhibited together with NE. Suramin, a hexasulfonated naphthylurea, is a potent inhibitor of

TABLE 7.3
Homeopathic Treatments for COPD

Homeopathy remedy	Effect on COPD patient	Targeted patient	Ref.
Rumex	Decreases cough and calms breathlessness	Patients who have dry teasing cough initially, followed by a stringy cough	[74]
Calcarea carb	Help relieve from breathlessness	Patient with severe breathlessness	[75]
Sulfur	Helps to treat the respiratory disorder and strengthens immunity	Weak patients with dyspnea	[76]
Arsenic alb	Treats anxiety derived breathlessness	Good for patient who has anxiety issue along with COPD	[77]
Aspidosperma Q	Tonic for lungs, removes temporary obstruction to the oxidation of blood by stimulating respiratory centers, relieves breathlessness	Good for a patient having weakness and emphysema at the same time	[77]
Antimonium Tart 30	Lung tonic and relieves breathlessness	Good for an aged patient having emphysema and having mucus deposit in the lungs with rapid, short, difficult breathing	[78]
Bryonia Alba 30	Relieves from quick, difficult respiration with pain in the chest and rust colored sputum	Meant for a patient who desires to take long breath. Good when there is dry, barking cough, worse at night	[78]
Coca 30	Treats hoarseness or loss of voice and dyspnea	Useful for aged sportsmen and alcoholics	[78]
Naphthaline 30	Treats dyspnea and sighing respiration	Useful for emphysema of the aged with asthma and when there are long and continued paroxysms of coughing and tenacious expectoration	[74–78]
Senega 30	Treats dyspnea and strengthens chest muscle	Useful for patients having sharp contractile pain in chest muscle or resistant cough	[74–78]
Lobelia Q	Treats dyspnea	Useful for patients having cough along with vomiting or even when asthma or COPD is preceded by prickling all over, hyperventilation, panting, fear of suffocation, fear of death, constricted chest, etc.	[79]
Antimonium Ars 30	Decrease mucus secretion and thereby relieves breathlessness	Patient with excess mucus secretion and emphysema and the condition worsen on eating or lying down	[80]
Chininium Ars 30	Relieves from dyspnea and exhaustion	Good for an aged patient who faces exhaustion along with breathlessness	[81]
Strychninum 30	Relieves dyspnea, persistent cough, and sharp contractive pains in the muscles of the chest	Good for a patient with emphysema, asthma, breathing problem	[78–81]
Curare 6	Treats short breath, dry cough, very distressing dyspnea	Patients who have threatened cessation of respiration on falling asleep	[82]
Kali Bichromicum	Relieves from frequent cough	Good for a patient having yellow mucus along with breathlessness	[83]

cathepsin G, proteinase 3, and NE (35). Novel and specific cathepsin inhibitors are under development [99, 100].

c. Serpins (serum protease inhibitor): Elafin, an elastase-specific inhibitor, can be used as a bronchodilator [101].

d. Secretory leukoprotease inhibitor [101].

xiii) *New anti-inflammatory treatments*:

a. Phosphodiesterase inhibitor: Rolipram, cilomilast and roflumilast (PDE4 inhibitors), and RPL554 (a dual PDE3/PDE4 inhibitor) are more targeted approaches to provide relief from COPD via anti-inflammatory activity. Roflumilast has been sanctioned for the treatment of COPD by both the European Medicines Agency and the US Food and Drug Administration [102–104].

b. Adhesion molecule blockers: Monoclonal antibodies to CD18, ICAM-1, and E-selectin inhibit neutrophil-mediated inflammation in COPD [105].

c. Interleukin 10 (IL-10): It is a cytokine with a wide spectrum of anti-inflammatory actions with remarkable tolerability over several weeks. Drugs like theophylline and PDE4 inhibitors also increase the secretion of IL-10 [106].

d. p38 MAP kinase inhibitors: Nonpeptide inhibitors of p38 MAP kinase-like SB 203580, SB 220025, and RWJ 67657 have a broad range of anti-inflammatory effects [107].

e. Neutrophil inhibitors: Colchicine potently inhibits neutrophil activation, enzyme release, and chemotaxis by disrupting cytoskeletal microtubule structure [108].
f. Fudosteine: It has been used as a mucolytic and antioxidant [109].
g. Erdosteine: Erdosteine is a mucoactive thiol antioxidant. Its oral dose of 300 mg twice daily for eight months produced a remarkable improvement in the quality of health and a reduction in exacerbations compared to placebo [110].
h. Spin traps and INOS inhibitors: Spin traps are chemical agents that can put an end to free radicals. Most of them have a nitrone nucleus or nitroxide nucleus or are their derivatives. However, earlier spin traps had very short half-lives and produce toxic hydroxyl radicals on decay. The newer moieties having electron-withdrawing groups around the core pyrroline ring-like STANZ and NXY-059 have overcome the drawbacks [111].
i. Redox sensors like enzymatic thioredoxin and redox effector factor-1 (Ref-1) can be used as adjuvant therapy for COPD by acting as antioxidants [111].
j. Enzymatic antioxidants: Cellular reactive oxygen species can be successfully negated by antioxidant enzymes, such as superoxide dismutase, catalase, and glutathione peroxidase [111].

xiv) *Lipid peroxidation and protein carbonylation inhibitors [112]*:
a. Edaravone (MC-186) is a potent free-radical and carbonyl scavenger and inhibitor of lipid peroxidation with antioxidant and anti-inflammatory activity.
b. Lazaroids: The protective effects of lazaroids are mainly contributed by the inhibition of lipid peroxidation, prevention of free radical formation, and release of TNF-α by alveolar macrophages. Thus, lazaroids can be a part of an effective therapeutic strategy to treat COPD.

xv) *Muco-regulators*:
a. Tachykinin antagonists: CP-99,994 and SR 140333 are potent tachykinins inhibitors leading to suppression of mucus secretion in humans [113, 114].
b. Mediator and enzyme inhibitors: A cyclooxygenase inhibitor like indomethacin reduces hypersecretion of mucus in patients with COPD [115].
c. Mucolytic agent: In vitro studies proved the mucolytic behavior of N-acetylcysteine, methylcysteine, and carbocisteine is effective in reducing the viscosity of mucus in vitro. Another study has shown that DNase also lessens sputum viscosity especially when it is infected [116].
d. Macrolide antibiotics: Erythromycin suppresses goblet cells involved in mucin secretion.

xvi) Alveolar repair: Preclinical studies have proven the potential of retinoic acid and hepatocyte growth factors in the growth of alveoli in fatal lung conditions and may be future drugs for COPD treatment.

The novel treatments for COPD treatments are summarized in Figure 7.9.

7.11 Future Prospects for COPD

Every year cases of COPD are increasing at a faster rate. Thus, COPD has been announced as a major global epidemic disease. It is predicted that COPD becomes the third most common reason for death and chronic disability by 2020. Despite the enormous global impact of COPD, no therapies are available to stop disease progression. However, discoveries by researchers and pharmaceutical industries are in a state of progress to understand cellular and molecular mechanisms of COPD and in the development of novel targeted therapies (Table 7.4).

Various preclinical and clinical phase trials result have provided great hope and promise to treat pulmonary disorders with great ease with the application of nanotechnology by enhancing the pharmacologic and therapeutic potency of the drugs. Nanoparticles are one of the most promised nanotherapeutics for COPD treatment with many unique ways and capabilities to target lung tissues. Therefore, a huge amount of investment is done in finding new and better nanotherapeutics for COPD. The success and emergence of new nanotherapeutic for COPD treatment depends on their ability to overcome the pulmonary barriers to reach the target site of action (Figure 7.10). Hypersecretion of mucus as well as inflammatory cells and pH of lung changes during COPD act as a barrier for the entry of nanoparticles. Subepithelial fibrosis can also act as a mechanical barrier for the nanoparticles. The activated immune cells like macrophages can engulf the nanoparticle and render them inactive.

Recently, potential multifunctional polymeric vesicles loaded with prednisolone and theophylline showed good results in COPD management. Furthermore, the successful fabrication of dimethyl fumarate dry powder for inhalation using advanced particle engineering design technology for targeted delivery to lower airways and treat COPD is another example [117].

Metallic nanoparticles are also gaining attention as a potential carrier for COPD drugs. They can also be used for imaging and thereby diagnosing COPD. Antibody-conjugated superparamagnetic iron oxide (SPIO) nanoparticles can be used in

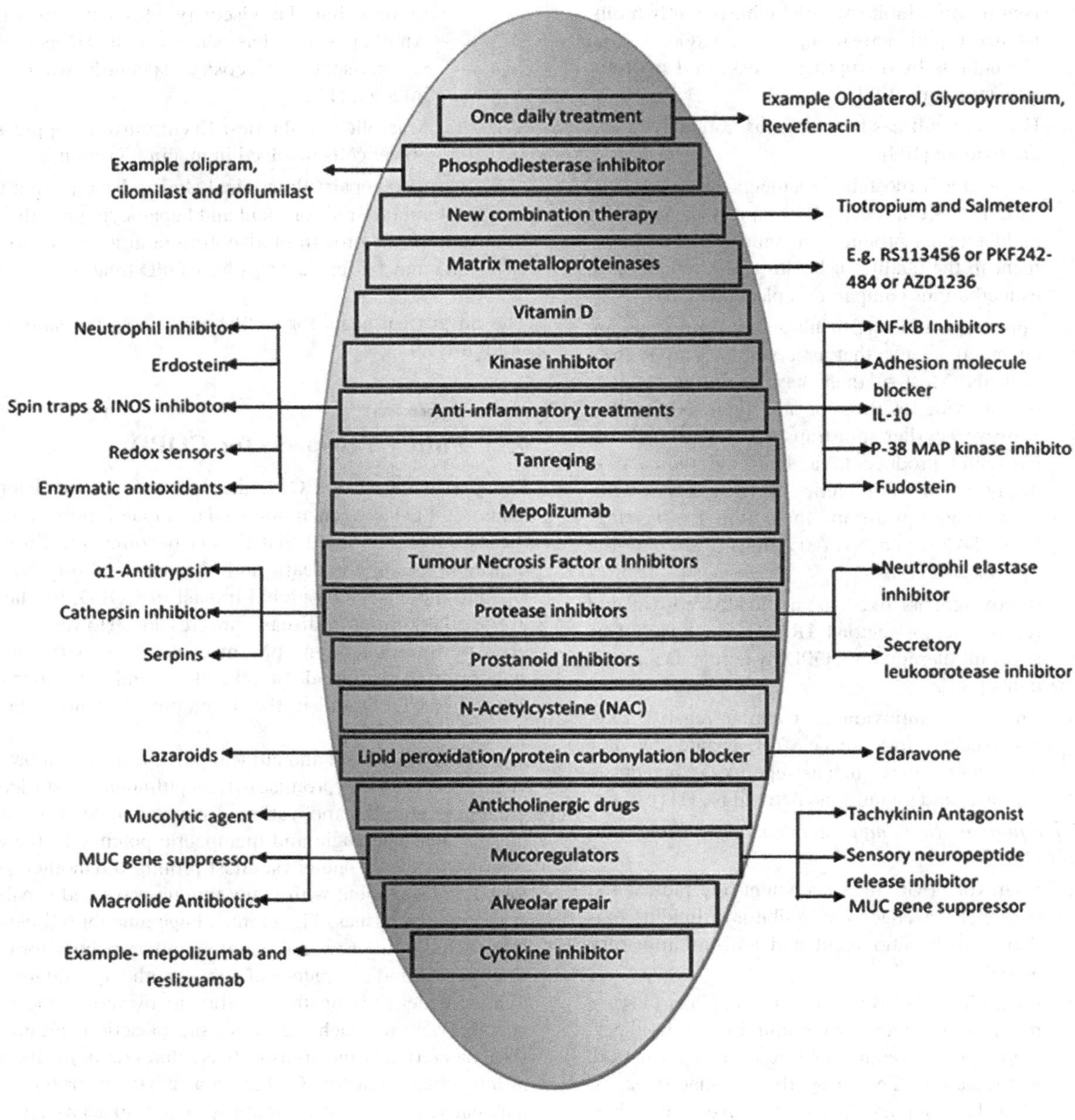

FIGURE 7.9 Novel treatments for COPD.

non-invasive MRI of macrophage subpopulation to diagnose early pulmonary inflammatory diseases [118]. Surface individualization with targeting molecules bestows selective delivery of nanoparticles for contrast molecular imaging and targeted therapy.

7.12 Conclusions

Globally COPD will continue to be a significant public health problem for a few more foreseeable years. The disease majorly affects the world and is associated with both economic and social burdens. The inefficiency to diagnose the disease and confusing symptoms with other diseases delay its treatment and worsen the health of patients. No permanent cure is available for the disease to date. Currently available treatments can only stop the further progression of the disease but can't reverse the already caused damage. Therefore, it is necessary to divert the research toward finding the correct pathogenesis of the disease and thereafter finding more effective ways to diagnose and treat the disease efficiently and permanently. The research should also focus on novel drug delivery systems to target the drug at the right site of action. Future research can also be drifted toward naturally based products because of their broad pharmacological benefits and lesser side effects.

TABLE 7.4

Patented Molecules Effective for COPD Treatment

Patent number	Title of the patent	Publication year	Inventors	Activity	Dose	Safety	Availability in the market	Ref.
WO2015181262A1 WIPO (PCT)	Fluticasone furoate in the treatment of COPD	2015	Neil Christopher, BARNES Steven, John PASCOE (GSK)	Improves lung function and used in a patient with eosinophil count of > 150 cells/μ I	Two actuations in one nostril once daily and total dose shouldn't exceed 110 mcg/day	When given once daily via dry powder inhaler, the drug was safe to use	ALISADE 27.5 micrograms/spray	[119]
WO2011067212A1 WIPO (PCT)	Combinations of a muscarinic receptor antagonist and a beta-2 adrenoreceptor agonist	2011	Darrell Baker, Mark Bruce, Glenn Crater, Brian Noga, Marian Thomas, Patrick Wire(GSK)	Vilanterol triphenylacetate and umeclidinium bromide can help the patient breathe properly because of bronchodilator activity	Once daily; each inhalation contains 62.5 mcg of umeclidinium bromide and 25 mcg of vilanterol triphenyl acetate	Favorable safety profile	ANORO ELLIPTA	[120]
USRE44874E1	Phenethanolamine derivatives for the treatment of respiratory diseases	2014	Philip Charles Box, Diane Mary Coe, Brian Edgar Looker, Panayiotis Alexandrou Procopiou (Glaxo group)	Vilanterol trifenatate is the selected phenethanolamine. It improves lung function and relieves from exacerbation due to its bronchodilator activity	0.01 mg to 1 mg per day	They are safe for human use	Many combination marketed products contain Vilanterol trifenatate, e.g., Anoro, Breo Ellipta, Trelegy	[121]
US5096916A	Treatment of chronic obstructive pulmonary disease (COPD) by inhalation of an imidazoline	1992	AEGIS TECHNOLOGY Inc	Tolazoline has vasodilator and an alpha-adrenergic blocking potential which helps in the treatment of COPD	Initial dose of about 4 to 12 mg of an imidazoline compound later 5 to 100 mg as required	Not so safe for human use	No product in the market	[122]
WO2007042811A1	Pharmaceutical compositions for the treatment of COPD	2007	Geoffrey Guy, Philip Robson	Cannabidiol (CBD) and delta-9-tetrahydrocannabinol (THC)	CBD – ≤ 120 mg THC – ≤ 130 mg	Well tolerated and good safety profile	Sativex	[123]
JP4819699B2 Japan	Anticholinergics and glucocorticoids of the combination for the long-term treatment of asthma and COPD	2011	Jurgen Knota, Peter Gode, Joachimszelenui, Istofanmouth, Joachim	Anticholinergics inhibit bronchoconstriction by relaxing the smooth muscles and cause considerable bronchodilation effects e.g budesonide. Inhaled glucocorticoid like fluticasone has potent anti-inflammatory action and together they show an additive effect	160 mcg and 4.5 mcg of budesonide and formoterol fumarate respectively. Dose – two inhalations twice daily	Good safety profile	Symbicort	[124]
US20120052122A1 United States	Treatment of chronic obstructive pulmonary disease with phosphodiesterase-4 inhibitor	2012	AstraZeneca AB	Relieves from COPD exacerbation due to its anti-inflammatory action	500 micrograms per day of roflumilast	Good safety profile	DALIRESP	[125]

(Continued)

TABLE 7.4 (CONTINUED)

Patented Molecules Effective for COPD Treatment

Patent number	Title of the patent	Publication year	Inventors	Activity	Dose	Safety	Availability in the market	Ref.
US10098837B2 United States	Combination therapy for COPD	2018	Mario Scuri, Pierfrancesco Coli, Giuseppe DELMONTE	The combination of glycopyrronium bromide, formoterol and beclometasone dipropionate are useful for the prevention or treatment of moderate/severe chronic obstructive pulmonary disease	Beclometasone dipropionate 100 μg/actuation, formoterol fumarate dihydrate 6 μg/actuation and glycopyrronium bromide 12.5 μg/actuation b.i.d.	Good safety profile	Riarify by Chiesi Farmaceutici S.p.A.	[126]
WO2015122031A1 WIPO (PCT)	An optically active form of mepenzolate, and ameliorating agent for COPD which contains same as an active ingredient	2015	Mizushima Toru	Optically active Mepenzolate bromide relieves from COPD exacerbation because of its bronchodilation and anti-inflammatory activity	3.8 μg/kg	Good safety profile	Torankoron	[127]
WO2015173153A1 WIPO (PCT)	Combinations of tiotropium bromide, formoterol and budesonide for the treatment of COPD	2015	Michiel ULLMANN	This combination gives long-term treatment of COPD and can be used as a rescue medicine for acute exacerbation of COPD	1–50 μg tiotropium bromide, 1–20 μg formoterol, 50–500 μg budesonide per actuation	Good safety profile	Under clinical trials	[128]
WO2017108917A1 WIPO (PCT)	Pharmaceutical compositions for use in the treatment of COPD	2017	Gonzalo De, Miquel Serra	Combination of long-acting muscarinic antagonist and long-acting β-2 agonist has better action on COPD symptoms and reduce the number of exacerbations	340 mcg aclidinium bromide and 12 mcg formoterol fumarate	Good safety profile	Duaklir Genuair®	[129]
US20130125882A1 United States	Method and composition for treating asthma and COPD	2013	Jonathan Matz	A combination of beclomethasone disproportionate with albuterol sulfate is a good bronchodilator for COPD treatment. The two product combines to show an additive effect	Albuterol – 200 mg and 400 mg Beclomethasone	Good safety profile	Aerocort	[130]
EP1212089B1 European Patent Office	Synergistic combination of roflumilast and salmeterol	2006	Rolf Beume, Daniela Bundschuh, Armin Hatzelmann, Christian Schudt, Christian Weimar, Ulrich Kilian	Combined administration of PDE inhibitors and β2 adrenoceptor agonists can relieve from COPD exacerbation because of anti-inflammatory and bronchodilator property	Salmeterol – 3 μmol/kg and roflumilast – 0.3 μmol/kg	Good safety profile	No product in market yet	[131]
EP3193835B1 European Patent Office	Liquid inhalation formulation comprising rpl554	2018	Peter Lionel Spargo, Edward James French, Phillip A Haywood	Ensifentrine (RPL554), a dual phosphodiesterase 3/4 inhibitor exhibiting both bronchodilator and anti-inflammatory activity	0.01 mg/mL to 40 mg/mL	Presently in phase 2 trials in the UK as of June 4, 2019	To be released soon by Verona Pharma	[132]

(Continued)

TABLE 7.4 (CONTINUED)

Patented Molecules Effective for COPD Treatment

Patent number	Title of the patent	Publication year	Inventors	Activity	Dose	Safety	Availability in the market	Ref.
BRPI0618804A2 Brazil	Treatment of asthma and COPD using triple combination therapy	2011	Stephen Paul Collingwood, Barbara Haeberlin	Maintenance treatment in adult patients with moderate to severe COPD in a combination of an inhaled corticosteroid	92 mg fluticasone furoate, 65 mg umeclidinium bromide, and 22 mg vilanterol (as trifenatate)	Safe for use	Trelegy Ellipta	[133]
US20180291108A1 United States	Methods for treating COPD using Benralizumab	2018	Rene van der Merwe, Christine Ward, Ubaldo Martin, Lorin Roskos, Bing Wang	Benralizumab is a humanized, afucosylated monoclonal antibody (mAb) specifically binds to the alpha chain of human interleukin-5 receptor α expressed on eosinophils. It induces apoptosis of these cells via antibody-dependent cell cytotoxicity	30 mg of dose and dose per day should not exceed 100 mg	Safe for use	Fasenra	[134]
US20110124589A1 United States	Inhaled fosfomycin/tobramycin for the treatment of COPD	2011	Elizabeth Peters Bhatt, David Lloyd MacLeod, Matthew Thomas McKevitt, Terry Glenn Newcomb	Fosfomycin/tobramycin combination prevents any further lung infection and thereby prevent any further exacerbation of COPD	10 to 160 mg of fosfomycin and from 2.5 to 40 mg of tobramycin	Safe to use	FTI by Gilead Sciences	[135]
EP2265258A2 European Patent Office	Inhalation composition containing aclidinium for treatment of asthma and COPD	2015	Gonzalo De Miquel Serra, Rosa Lamarca Casado	Long-acting muscarinic antagonists which act as bronchodilator	400 mcg twice daily	Safe to use	Tudorza	[136]

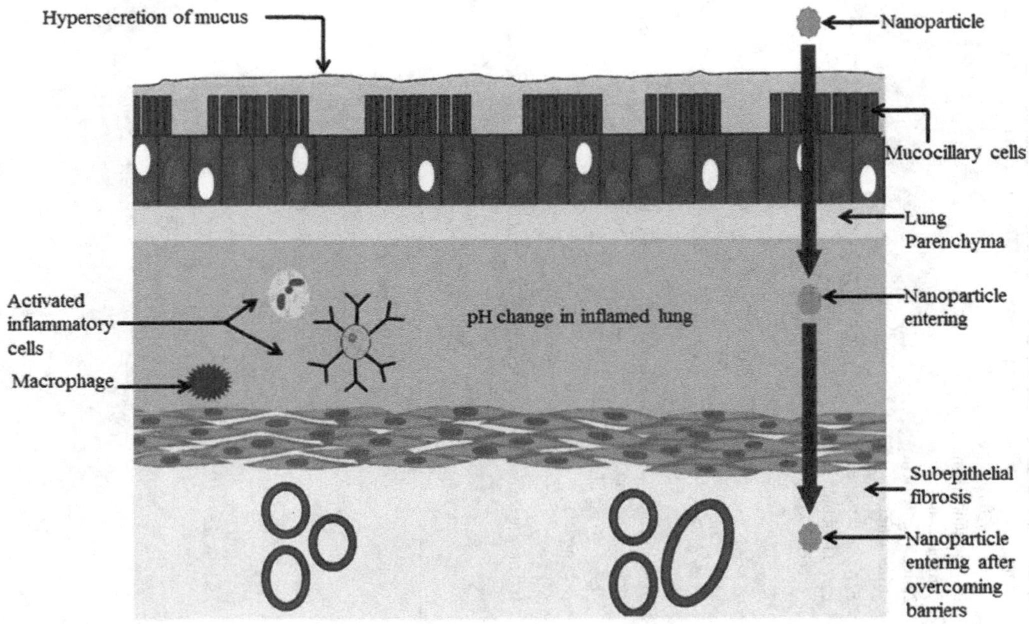

FIGURE 7.10 Pulmonary barriers faced by drug and drug carriers to reach a target site of action.

REFERENCES

1. Maghsoudloo, M., Jamalkandi, S. A., Najafi, A., & Masoudi-Nejad, A. (2020). Identification of biomarkers in common chronic lung diseases by co-expression networks and drug-target interactions analysis. *Molecular Medicine*, *26*(1), 1–19.
2. Ford, E. S., Murphy, L. B., Khavjou, O., Giles, W. H., Holt, J. B., & Croft, J. B. (2015). Total and state-specific medical and absenteeism costs of COPD among adults aged 18 years in the United States for 2010 and projections through 2020. *Chest*, *147*(1), 31–45.
3. Tu, J., Inthavong, K., & Ahmadi, G. (2013). The human respiratory system. In *Computational Fluid and Particle Dynamics in the Human Respiratory System* (pp. 19–44). Eds: Jiyuan Tu Kiao Inthavong Goodarz Ahmadi Springer, Dordrecht.
4. Benam, K. H., Novak, R., Nawroth, J., Hirano-Kobayashi, M., Ferrante, T. C., Choe, Y., … & Ingber, D. E. (2016). Matched-comparative modeling of normal and diseased human airway responses using a microengineered breathing lung chip. *Cell Systems*, *3*(5), 456–466.
5. Semenzato, U., Biondini, D., Bonato, M., Tinè, M., Bazzan, E., Oliani, K. L., … & Turato, G. (2019). Chronic bronchitis affects mortality in smokers with and without COPD. *European Respiratory Journal*, *54*, PA3901
6. Kato, K., Tsutsui, M., Noguchi, S., Naito, K., Ogoshi, T., Wang, K. Y., … & Yatera, K. (2019). Protective role of nitric oxide synthases in pulmonary emphysema in mice. *European Respiratory Journal*, *54*, PA4105
7. Parameswaran, G. I., Wrona, C. T., Murphy, T. F., & Sethi, S. (2009). Moraxella catarrhalis acquisition, airway inflammation and protease-antiprotease balance in chronic obstructive pulmonary disease. *BMC Infectious Diseases*, *9*(1), 1–10.
8. Hiller, A. M., Piitulainen, E., Jehpsson, L., & Tanash, H. (2019). Decline in FEV1 and hospitalized exacerbations in individuals with severe alpha-1 antitrypsin deficiency. *International Journal of Chronic Obstructive Pulmonary Disease*, *14*, 1075.
9. Vestbo, J., Hurd, S. S., Agustí, A. G., Jones, P. W., Vogelmeier, C., Anzueto, A., … & Rodriguez-Roisin, R. (2013). Global strategy for the diagnosis, management, and prevention of chronic obstructive pulmonary disease: GOLD executive summary. *American Journal of Respiratory and Critical Care Medicine*, *187*(4), 347–365.
10. Parr, D. G., Stoel, B. C., Stolk, J., & Stockley, R. A. (2004). Pattern of emphysema distribution in α1-antitrypsin deficiency influences lung function impairment. *American Journal of Respiratory and Critical Care Medicine*, *170*(11), 1172–1178.
11. Labiris, N. R., & Dolovich, M. B. (2003). Pulmonary drug delivery. Part I: Physiological factors affecting therapeutic effectiveness of aerosolized medications. *British Journal of Clinical Pharmacology*, *56*(6), 588–599.
12. Morita, T., Yamamoto, A., Takakura, Y., Hashida, M., & Sezaki, H. (1994). Improvement of the pulmonary absorption of (Asu 1, 7)-eel calcitonin by various protease inhibitors in rats. *Pharmaceutical Research*, *11*(6), 909–913.
13. Furubayashi, T., Inoue, D., Nishiyama, N., Tanaka, A., Yutani, R., Kimura, S., … & Sakane, T. (2020). Comparison of various cell lines and three-dimensional mucociliary tissue model systems to estimate drug permeability using an in vitro transport study to predict nasal drug absorption in rats. *Pharmaceutics*, *12*(1), 79.
14. Bruinsmann, F. A., Pigana, S., Aguirre, T., Dadalt Souto, G., Garrastazu Pereira, G., Bianchera, A., … & Sonvico, F. (2019). Chitosan-coated nanoparticles: Effect of chitosan molecular weight on nasal transmucosal delivery. *Pharmaceutics*, *11*(2), 86–88.

15. Chen, Y., Cheng, G., Hu, R., Chen, S., Lu, W., Gao, S., … & Fang, W. (2019). A nasal temperature and pH dual-responsive in situ gel delivery system based on microemulsion of huperzine A: Formulation, evaluation, and in vivo pharmacokinetic study. *AAPS PharmSciTech*, *20*(7), 1–12.
16. Zhang, H., Huang, X., Sun, Y., Lu, G., Wang, K., Wang, Z., … & Gao, Y. (2015). Improvement of pulmonary absorption of poorly absorbable macromolecules by hydroxypropyl-β-cyclodextrin grafted polyethylenimine (HP-β-CD-PEI) in rats. *International Journal of Pharmaceutics*, *489*(1–2), 294–303.
17. Maltais, F., Ferguson, G. T., Feldman, G. J., Deslee, G., Bourdin, A., Fjällbrant, H., … & Martin, U. J. (2019). A randomized, double-blind, double-dummy study of glycopyrrolate/formoterol fumarate metered dose inhaler relative to umeclidinium/vilanterol dry powder inhaler in COPD. *Advances in Therapy*, *36*(9), 2434–2449.
18. Dal Negro, R. W., Turco, P., & Povero, M. (2019). Patients' usability of seven most used dry-powder inhalers in COPD. *Multidisciplinary Respiratory Medicine*, *14*(1), 1–9.
19. Iwanaga, T., Tohda, Y., Nakamura, S., & Suga, Y. (2019). The Respimat® soft mist inhaler: Implications of drug delivery characteristics for patients. *Clinical Drug Investigation*, *39*(11), 1021–1030.
20. Menon, M., Naik, I., Rajawat, G. S., Nagarsenker, M., & Krishnaprasad, K. (2019). Nebulized glycopyrronium and formoterol, budesonide aerosol aerodynamic assessment with vibrating mesh and compressor air nebulizer: Anderson cascade impactor study. *Journal of Drug Delivery and Therapeutics*, *9*(6), 79–82.
21. Bodet-Contentin, L., Guillon, A., Boulain, T., Frat, J. P., Garot, D., Le Pennec, D., … & CRICS-TRIGGERSEP network. (2019). Salbutamol nebulization during noninvasive ventilation in exacerbated chronic obstructive pulmonary disease patients: A randomized controlled trial. *Journal of Aerosol Medicine and Pulmonary Drug Delivery*, *32*(3), 149–155.
22. Singh, S., Verma, S. K., Kumar, S., Ahmad, M. K., Nischal, A., Singh, S. K., & Dixit, R. K. (2017). Evaluation of oxidative stress and antioxidant status in chronic obstructive pulmonary disease. *Scandinavian Journal of Immunology*, *85*(2), 130–137.
23. Balkissoon, R. (2017). Journal club: COPD and oxygen therapy. *Chronic Obstructive Pulmonary Diseases: Journal of the COPD Foundation*, *4*(1), 71–75.
24. McKenna Jr, R. J., Brenner, M., Gelb, A. F., Mullin, M., Singh, N., Peters, H., … & Schein, M. J. (1996). A randomized prospective trial of stapled lung reduction versus laser bullectomy for diffuse emphysema. *The Journal of Thoracic and Cardiovascular Surgery*, *111*(2), 317–322.
25. Caviezel, C., Aruldas, C., Franzen, D., Ulrich, S., Inci, I., Schneiter, D., … & Opitz, I. (2018). Lung volume reduction surgery in selected patients with emphysema and pulmonary hypertension. *European Journal of Cardio-Thoracic Surgery*, *54*(3), 565–571.
26. Calverley, P. M., Anderson, J. A., Celli, B., Ferguson, G. T., Jenkins, C., Jones, P. W., … & Vestbo, J. (2007). Salmeterol and fluticasone propionate and survival in chronic obstructive pulmonary disease. *New England Journal of Medicine*, *356*(8), 775–789.
27. Berger, W. E., Milgrom, H., Skoner, D. P., Tripp, K., Parsey, M. V., Baumgartner, R. A., & Xopenex Pediatric Asthma Group. (2006). Evaluation of levalbuterol metered dose inhaler in pediatric patients with asthma: A double-blind, randomized, placebo-and active-controlled trial. *Current Medical Research and Opinion*, *22*(6), 1217–1226.
28. Virk, M., Hotz, J., Khemani, R., Newth, C., & Ross, P. (2017). Change in oxygen consumption following inhalation of albuterol in comparison with levalbuterol in healthy adult volunteers. *Lung*, *195*(2), 233–239.
29. Ferguson, G. T., Ghafouri, M., Dai, L., & Dunn, L. J. (2013). COPD patient satisfaction with ipratropium bromide/albuterol delivered via Respimat: A randomized, controlled study. *International Journal of Chronic Obstructive Pulmonary Disease*, *8*, 139.
30. Zhang, R., Hu, J., Deng, L., Li, S., Chen, X., Liu, F., … & Tan, W. (2020). Aerosol characteristics and physico-chemical compatibility of Combivent® (containing salbutamol and ipratropium bromide) mixed with three other inhalants: Budesonide, beclomethasone or N-acetylcysteine. *Pharmaceutics*, *12*(1), 78.
31. Maltais, F., Bjermer, L., Kerwin, E. M., Jones, P. W., Watkins, M. L., Tombs, L., … & Vogelmeier, C. F. (2019). Efficacy of umeclidinium/vilanterol versus umeclidinium and salmeterol monotherapies in symptomatic patients with COPD not receiving inhaled corticosteroids: The E_{MAX} randomised trial. *Respiratory Research*, *20*(1), 1–15.
32. Mahler, D. A., Kerwin, E., Murray, L., & Dembek, C. (2019). The impact of twice-daily indacaterol/glycopyrrolate on the components of health-related quality of life and dyspnea in patients with moderate-to-severe chronic obstructive pulmonary disease. *Chronic Obstructive Pulmonary Diseases: Journal of the COPD Foundation*, *6*(4), 308.
33. Leaker, B. R., Singh, D., Nicholson, G. C., Hezelova, B., Goodin, T., Ozol-Godfrey, A., & Barnes, P. J. (2019). Evaluation of systemic absorption and bronchodilator effect of glycopyrronium bromide delivered by nebulizer or a dry powder inhaler in subjects with chronic obstructive pulmonary disease. *Respiratory Research*, *20*(1), 1–12.
34. Muro, S., Yoshisue, H., Kostikas, K., Olsson, P., Gupta, P., & Wedzicha, J. A. (2020). Indacaterol/glycopyrronium versus tiotropium or glycopyrronium in long-acting bronchodilator-naïve COPD patients: A pooled analysis. *Respirology*, *25*(4), 393–400.
35. Gelb, A. F., Tashkin, D. P., Make, B. J., Zhong, X., Gil, E. G., Caracta, C., & LAS-MD-35 study investigators. (2013). Long-term safety and efficacy of twice-daily aclidinium bromide in patients with COPD. *Respiratory Medicine*, *107*(12), 1957–1965.
36. Kamei, T., Nakamura, H., Nanki, N., Minakata, Y., Matsunaga, K., & Mori, Y. (2019). Clinical benefit of two-times-per-day aclidinium bromide compared with once-a-day tiotropium bromide hydrate in COPD: A multicentre, open-label, randomised study. *BMJ Open*, *9*(7), e024114.
37. Schilero, G. J., Hobson, J. C., Singh, K., Spungen, A. M., Bauman, W. A., & Radulovic, M. (2018). Bronchodilator effects of ipratropium bromide and albuterol sulfate among subjects with tetraplegia. *The Journal of Spinal Cord Medicine*, *41*(1), 42–47.

38. Maesen, F. P. V., Smeets, J. J., Bernsen, R., & Cornelissen, P. J. G. (1986). Ipratropium bromide (Atrovent®) as inhalation powder: A double-blind study of comparison with ipratropium as a pressure aerosol in patients with reversible airways obstruction. *Allergy*, *41*(1), 37–42.
39. Salomon, J., Stolz, D., Domenighetti, G., Frey, J. G., Turk, A. J., Azzola, A., … & Brutsche, M. (2017). Indacaterol and glycopyrronium versus indacaterol on body plethysmography measurements in COPD—A randomised controlled study. *Respiratory Research*, *18*(1), 1–7.
40. Donohue, J. F., Ganapathy, V., Bollu, V., Stensland, M. D., & Nelson, L. M. (2017). Health status of patients with moderate to severe COPD after treatment with nebulized arformoterol tartrate or placebo for 1 year. *Clinical Therapeutics*, *39*(1), 66–74.
41. Gilmer, T. P., Celli, B. R., Xu, Z., Cho-Reyes, S., Dembek, C., & Navaie, M. (2019). Predictors of nebulized arformoterol treatment: A retrospective analysis of medicare beneficiaries with chronic obstructive pulmonary disease. *COPD: Journal of Chronic Obstructive Pulmonary Disease*, *16*(2), 140–151.
42. Hanania, N. A., Sethi, S., Koltun, A., Ward, J. K., Spanton, J., & Ng, D. (2019). Long-term safety and efficacy of formoterol fumarate inhalation solution in patients with moderate-to-severe COPD. *International Journal of Chronic Obstructive Pulmonary Disease*, *14*, 117–127.
43. Amore, E., Manca, M. L., Ferraro, M., Valenti, D., La Parola, V., Di Vincenzo, S., … & Pace, E. (2019). Salmeterol xinafoate (SX) loaded into mucoadhesive solid lipid microparticles for COPD treatment. *International Journal of Pharmaceutics*, *562*, 351–358.
44. Koch, A., Watz, H., Maleki-Yazdi, M. R., Bothner, U., Tetzlaff, K., Voß, F., & McGarvey, L. (2017). Comprehensive assessment of the safety of olodaterol 5 μg in the Respimat® device for maintenance treatment of COPD: Comparison with the long-acting β 2-agonist formoterol. *NPJ Primary Care Respiratory Medicine*, *27*(1), 1–6.
45. Feldman, G. J., Sousa, A. R., Lipson, D. A., Tombs, L., Barnes, N., Riley, J. H., … & Navarrete, B. A. (2017). Comparative efficacy of once-daily umeclidinium/vilanterol and tiotropium/olodaterol therapy in symptomatic chronic obstructive pulmonary disease: A randomized study. *Advances in Therapy*, *34*(11), 2518–2533.
46. Calverley, P. M., Anzueto, A. R., Carter, K., Grönke, L., Hallmann, C., Jenkins, C., … & Rabe, K. F. (2018). Tiotropium and olodaterol in the prevention of chronic obstructive pulmonary disease exacerbations (DYNAGITO): A double-blind, randomised, parallel-group, active-controlled trial. *The Lancet Respiratory Medicine*, *6*(5), 337–344.
47. Rajagopalan, K., Bloudek, L., Marvel, J., Dembek, C., & Kavati, A. (2018). Cost-effectiveness of twice-daily indacaterol/glycopyrrolate inhalation powder for the treatment of moderate to severe COPD in the US. *International Journal of Chronic Obstructive Pulmonary Disease*, *13*, 3867.
48. Reisner, C., Pearle, J., Kerwin, E. M., St Rose, E., & Darken, P. (2018). Efficacy and safety of four doses of glycopyrrolate/formoterol fumarate delivered via a metered dose inhaler compared with the monocomponents in patients with moderate-to-severe COPD. *International Journal of Chronic Obstructive Pulmonary Disease*, *13*, 1965–1977.
49. Anzueto, A. R., Sense, W., Yates, J., Brown, C., & Knobil, K. (2004). Efficacy of Advair Diskus 250/50 (fluticasone propionate/salmeterol) in patients previously naive to COPD maintenance therapy. *Chest*, *126*(4), 808S.
50. Grant, A. C., Walker, R., Hamilton, M., & Garrill, K. (2015). The ELLIPTA® dry powder inhaler: Design, functionality, in vitro dosing performance and critical task compliance by patients and caregivers. *Journal of Aerosol Medicine and Pulmonary Drug Delivery*, *28*(6), 474–485.
51. Dobler, C. C. (2018). The IMPACT of triple versus dual single-inhaler therapy on exacerbations of COPD. *Breathe*, *14*(4), 333–335.
52. Rosenhall, L., Elvstrand, A., Tilling, B., Vinge, I., Jemsby, P., Stahl, E., … & Bergqvist, P. B. F. (2003). One-year safety and efficacy of budesonide/formoterol in a single inhaler (Symbicort® Turbuhaler®) for the treatment of asthma. *Respiratory Medicine*, *97*(6), 702–708.
53. Maeda, K., Yamaguchi, M., Nagase, H., Yasuno, N., Itagaki, F., & Watanabe, M. (2018). Utility and effectiveness of Symbicort® Turbuhaler® (oral inhalation containing budesonide and formoterol) in a patient with severe asthma after permanent tracheostomy. *Journal of Pharmaceutical Health Care and Sciences*, *4*(1), 1–5.
54. Brown, H. V., & Ziment, I. (1978). Evaluation of an oxygen concentrator in patients with COPD. *Respiratory Therapy*, *8*(5), 55–57.
55. Martin, D. C. (2019). Contemporary portable oxygen concentrators and diverse breathing behaviours--a bench comparison. *BMC Pulmonary Medicine*, *19*(1), 1–11.
56. Abroug, F., Ouanes-Besbes, L., Fkih-Hassen, M., Ouanes, I., Ayed, S., Dachraoui, F., … & ElAtrous, S. (2014). Prednisone in COPD exacerbation requiring ventilatory support: An open-label randomised evaluation. *European Respiratory Journal*, *43*(3), 717–724.
57. Takiguchi, H., Chen, V., Obeidat, M. E., Hollander, Z., FitzGerald, J. M., McManus, B. M., … & Sin, D. D. (2019). Effect of short-term oral prednisone therapy on blood gene expression: A randomised controlled clinical trial. *Respiratory Research*, *20*(1), 1–10.
58. Ronaldson, S. J., Raghunath, A., Torgerson, D. J., & Van Staa, T. (2017). Cost-effectiveness of antibiotics for COPD management: Observational analysis using CPRD data. *ERJ Open Research*, *3*(2), 1-8.
59. Reddy, A. T., Lakshmi, S. P., Banno, A., & Reddy, R. C. (2020). Glucocorticoid receptor α mediates roflumilast's ability to restore dexamethasone sensitivity in COPD. *International Journal of Chronic Obstructive Pulmonary Disease*, *15*, 125.
60. Liu, F. C., Yu, H. P., Lin, C. Y., Elzoghby, A. O., Hwang, T. L., & Fang, J. Y. (2018). Use of cilomilast-loaded phosphatiosomes to suppress neutrophilic inflammation for attenuating acute lung injury: The effect of nanovesicular surface charge. *Journal of Nanobiotechnology*, *16*(1), 1–14.
61. Seidenfeld, J. J., Jones, W. N., Moss, R. E., & Tremper, J. (1984). Intravenous aminophylline in the treatment of acute bronchospastic exacerbations of chronic obstructive pulmonary disease. *Annals of Emergency Medicine*, *13*(4), 248–252.
62. Puścińska, E., Radwan, L., & Zieliński, J. (1994). Effect of intravenous ambroxol hydrochloride on lung function and exercise capacity in patients with severe chronic obstructive pulmonary disease. *Pneumonologia i Alergologia Polska*, *62*(5–6), 246–249.

63. Dal Negro, R. W., & Visconti, M. (2016). Erdosteine reduces the exercise-induced oxidative stress in patients with severe COPD: Results of a placebo-controlled trial. *Pulmonary Pharmacology & Therapeutics*, *41*, 48–51.
64. Calverley, P. M., Page, C., Dal Negro, R. W., Fontana, G., Cazzola, M., Cicero, A. F., ... & Wedzicha, J. A. (2019). Effect of erdosteine on COPD exacerbations in COPD patients with moderate airflow limitation. *International Journal of Chronic Obstructive Pulmonary Disease*, *14*, 2733.
65. Hooper, C., & Calvert, J. (2008). The role for S-carboxymethylcysteine (carbocisteine) in the management of chronic obstructive pulmonary disease. *International Journal of Chronic Obstructive Pulmonary Disease*, *3*(4), 659.
66. Storms, W. W., & Miller, J. E. (2018). Daily use of guaifenesin (Mucinex) in a patient with chronic bronchitis and pathologic mucus hypersecretion: A case report. *Respiratory Medicine Case Reports*, *23*, 156–157.
67. Tønnesen, P., Mikkelsen, K., & Bremann, L. (2006). Nurse-conducted smoking cessation in patients with COPD using nicotine sublingual tablets and behavioral support. *Chest*, *130*(2), 334–342.
68. Pezzuto, A., & Carico, E. (2020). Effectiveness of smoking cessation in smokers with COPD and nocturnal oxygen desaturation: Functional analysis. *The Clinical Respiratory Journal*, *14*(1), 29–34.
69. Alison, J. A., McKeough, Z. J., Leung, R. W., Holland, A. E., Hill, K., Morris, N. R., ... & McDonald, C. F. (2019). Oxygen compared to air during exercise training in COPD with exercise-induced desaturation. *European Respiratory Journal*, *53*(5), 1802429.
70. Tabak, C., Smit, H. A., Heederik, D., Ocke, M. C., & Kromhout, D. (2001). Diet and chronic obstructive pulmonary disease: Independent beneficial effects of fruits, whole grains, and alcohol (the MORGEN study). *Clinical & Experimental Allergy*, *31*(5), 747–755.
71. van de Bool, C., Rutten, E. P., van Helvoort, A., Franssen, F. M., Wouters, E. F., & Schols, A. M. (2017). A randomized clinical trial investigating the efficacy of targeted nutrition as adjunct to exercise training in COPD. *Journal of Cachexia, Sarcopenia and Muscle*, *8*(5), 748–758.
72. Beghi, G. M., & Morselli-Labate, A. M. (2016). Does homeopathic medicine have a preventive effect on respiratory tract infections? A real life observational study. *Multidisciplinary Respiratory Medicine*, *11*(1), 1–10.
73. Sahreen, S., Khan, M. R., & Khan, R. A. (2014). Comprehensive assessment of phenolics and antiradical potential of Rumex hastatus D. Don. roots. *BMC Complementary and Alternative Medicine*, *14*(1), 1–11.
74. Zanasi, A., Mazzolini, M., Tursi, F., Morselli-Labate, A. M., Paccapelo, A., & Lecchi, M. (2014). Homeopathic medicine for acute cough in upper respiratory tract infections and acute bronchitis: A randomized, double-blind, placebo-controlled trial. *Pulmonary Pharmacology & Therapeutics*, *27*(1), 102–108.
75. Diez, S. C., Casas, A. V., Rivero, J. L. G., Caro, J. C. L., Portal, F. O., & Saez, G. D. (2019). Impact of a homeopathic medication on upper respiratory tract infections in COPD patients: Results of an observational, prospective study (EPOXILO). *Respiratory Medicine*, *146*, 96–105.
76. Huang, W. L. (2018). The treatment of asthma based on traditional Chinese medicine and homeopathy. *Journal of Pediatrics and Infants*, *1*(1), 24–30.
77. Kadam, N., Nadgauda, S., & Jadhav, A. B. (2020). Effectiveness of Senega 30C in chronic smokers in cases of COPD according to grade 2 global initiative for chronic obstructive lung disease (COLD). *Cough*, *8*, 26–66.
78. Brown, D. P., Rogers, D. T., Pomerleau, F., Siripurapu, K. B., Kulshrestha, M., Gerhardt, G. A., & Littleton, J. M. (2016). Novel multifunctional pharmacology of lobinaline, the major alkaloid from *Lobelia cardinalis*. *Fitoterapia*, *111*, 109–123.
79. Gupta, J., Rao, M. P., Raju, K., Prasad, R. V. R., Arya, J. S., Mondal, B. K., ... & Roja, V. (2019). Management of early years of simple and mucopurulent chronic bronchitis with pre-defined homeopathic medicines-a prospective observational study with 2-years follow-up. *International Journal of High Dilution Research*, *18*(3-4), 47-62.
80. Shafei, H. F., AbdelDayem, S. M., & Mohamed, N. H. (2012). Individualized homeopathy in a group of Egyptian asthmatic children. *Homeopathy*, *101*(04), 224–230.
81. Sa'ed, H. Z., Al-Jabi, S. W., Sweileh, W. M., Awang, R., & Waring, W. S. (2015). Bibliometric profile of the global scientific research on methanol poisoning (1902–2012). *Journal of Occupational Medicine and Toxicology*, *10*(1), 1–8.
82. Frass, M., Dielacher, C., Linkesch, M., Endler, C., Muchitsch, I., Schuster, E., & Kaye, A. (2005). Influence of potassium dichromate on tracheal secretions in critically ill patients. *Chest*, *127*(3), 936–941.
83. Van Noord, J. A., Smeets, J. J., Drenth, B. M., Rascher, J., Pivovarova, A., Hamilton, A. L., & Cornelissen, P. J. G. (2011). 24-hour bronchodilation following a single dose of the novel β2-agonist olodaterol in COPD. *Pulmonary Pharmacology & Therapeutics*, *24*(6), 666–672.
84. Ferguson, G. T., Feldman, G. J., Hofbauer, P., Hamilton, A., Allen, L., Korducki, L., & Sachs, P. (2014). Efficacy and safety of olodaterol once daily delivered via Respimat® in patients with GOLD 2–4 COPD: Results from two replicate 48-week studies. *International Journal of Chronic Obstructive Pulmonary Disease*, *9*, 629.
85. Onions, S. T., Ito, K., Charron, C. E., Brown, R. J., Colucci, M., Frickel, F., ... & Williams, J. G. (2016). Discovery of narrow spectrum kinase inhibitors: New therapeutic agents for the treatment of COPD and steroid-resistant asthma. *Journal of Medicinal Chemistry*, *59*(5), 1727–1746.
86. Ratcliffe, M. J., & Dougall, I. G. (2012). Comparison of the anti-inflammatory effects of Cilomilast, Budesonide and a p38 Mitogen activated protein kinase inhibitor in COPD lung tissue macrophages. *BMC Pharmacology and Toxicology*, *13*(1), 1–7.
87. Said, A. F., & Abd-Elnaeem, E. A. (2015). Vitamin D and chronic obstructive pulmonary disease. *Egyptian Journal of Chest Diseases and Tuberculosis*, *64*(1), 67–73.
88. Vernooy, J. H., Lindeman, J. H., Jacobs, J. A., Hanemaaijer, R., & Wouters, E. F. (2004). Increased activity of matrix metalloproteinase-8 and matrix metalloproteinase-9 in induced sputum from patients with COPD. *Chest*, *126*(6), 1802–1810.
89. Brajer, B., Batura-Gabryel, H., Nowicka, A., Kuznar-Kaminska, B., & Szczepanik, A. (2008). Concentration of matrix metalloproteinase-9 in serum of patients with

chronic obstructive pulmonary disease and a degree of airway obstruction and disease progression. *Journal of Physiology and Pharmacology, 59*(Suppl 6), 145–152.
90. Beeh, K. M., Korn, S., Beier, J., Jadayel, D., Henley, M., D'Andrea, P., & Banerji, D. (2014). Effect of QVA149 on lung volumes and exercise tolerance in COPD patients: The BRIGHT study. *Respiratory Medicine, 108*(4), 584–592.
91. Donohue, J. F., van Noord, J. A., Bateman, E. D., Langley, S. J., Lee, A., Witek Jr, T. J., … & Towse, L. (2002). A 6-month, placebo-controlled study comparing lung function and health status changes in COPD patients treated with tiotropium or salmeterol. *Chest, 122*(1), 47–55.
92. Dransfield, M. T., Bourbeau, J., Jones, P. W., Hanania, N. A., Mahler, D. A., Vestbo, J., … & Calverley, P. M. (2013). Once-daily inhaled fluticasone furoate and vilanterol versus vilanterol only for prevention of exacerbations of COPD: Two replicate double-blind, parallel-group, randomised controlled trials. *The Lancet Respiratory Medicine, 1*(3), 210–223.
93. Hoshino, M., & Ohtawa, J. (2013). Effects of tiotropium and salmeterol/fluticasone propionate on airway wall thickness in chronic obstructive pulmonary disease. *Respiration, 86*(4), 280–287.
94. Liu, S., Shergis, J., Chen, X., Yu, X., Guo, X., Zhang, A. L., … & Xue, C. C. (2014). Chinese herbal medicine (weijing decoction) combined with pharmacotherapy for the treatment of acute exacerbations of chronic obstructive pulmonary disease. *Evidence-Based Complementary and Alternative Medicine, 2014*, Article ID 257012.
95. Yang, L., Li, J., Li, Y., Tian, Y., Li, S., Jiang, S., … & Song, X. (2015). Identification of metabolites and metabolic pathways related to treatment with Bufei Yishen formula in a rat COPD model using HPLC Q-TOF/MS. *Evidence-Based Complementary and Alternative Medicine, 2015,* Article ID 956750.
96. Aaron, S. D., Vandemheen, K. L., Maltais, F., Field, S. K., Sin, D. D., Bourbeau, J., … & Mallick, R. (2013). TNFα antagonists for acute exacerbations of COPD: A randomised double-blind controlled trial. *Thorax, 68*(2), 142–148.
97. Coleman, R. A., & Sheldrick, R. L. G. (1989). Prostanoid-induced contraction of human bronchial smooth muscle is mediated by TP-receptors. *British Journal of Pharmacology, 96*(3), 688–692.
98. Kuna, P., Jenkins, M., O'Brien, C. D., & Fahy, W. A. (2012). AZD9668, a neutrophil elastase inhibitor, plus ongoing budesonide/formoterol in patients with COPD. *Respiratory Medicine, 106*(4), 531–539.
99. de Garavilla, L., Greco, M. N., Sukumar, N., Chen, Z. W., Pineda, A. O., Mathews, F. S., … & Maryanoff, B. E. (2005). A novel, potent dual inhibitor of the leukocyte proteases cathepsin G and chymase: Molecular mechanisms and anti-inflammatory activity in vivo. *Journal of Biological Chemistry, 280*(18), 18001–18007.
100. Suzuki, T., Wang, W., Lin, J. T., Shirato, K., Mitsuhashi, H., & Inoue, H. (1996). Aerosolized human neutrophil elastase induces airway constriction and hyperresponsiveness with protection by intravenous pretreatment with half-length secretory leukoprotease inhibitor. *American Journal of Respiratory and Critical Care Medicine, 153*(4), 1405–1411.
101. Demizu, S., Asaka, N., Kawahara, H., & Sasaki, E. (2019). TAS-203, an oral phosphodiesterase 4 inhibitor, exerts anti-inflammatory activities in a rat airway inflammation model. *European Journal of Pharmacology, 849*, 22–29.
102. Rennard, S. I., Calverley, P. M., Goehring, U. M., Bredenbröker, D., & Martinez, F. J. (2011). Reduction of exacerbations by the PDE4 inhibitor roflumilast-the importance of defining different subsets of patients with COPD. *Respiratory Research, 12*(1), 1–10.
103. Singh, D., Beeh, K. M., Colgan, B., Kornmann, O., Leaker, B., Watz, H., … & Nandeuil, M. A. (2019). Effect of the inhaled PDE4 inhibitor CHF6001 on biomarkers of inflammation in COPD. *Respiratory Research, 20*(1), 1–12.
104. Riise, G. C., Larsson, S., Lofdahl, C. G., & Andersson, B. A. (1994). Circulating cell adhesion molecules in bronchial lavage and serum in COPD patients with chronic bronchitis. *European Respiratory Journal, 7*(9), 1673–1677.
105. Rémy, G., Grandjean, T., Kervoaze, G., Pichavant, M., Chamaillard, M., & Gosset, P. (2013). Implication of interleukin-10 in the development of COPD induced by cigarette smoke exposure in mice. *European Respiratory Journal, 42*(1), 611–614.
106. Lomas, D. A., Lipson, D. A., Miller, B. E., Willits, L., Keene, O., Barnacle, H., … & Tal-Singer, R. (2012). An oral inhibitor of p38 MAP kinase reduces plasma fibrinogen in patients with chronic obstructive pulmonary disease. *The Journal of Clinical Pharmacology, 52*(3), 416–424.
107. Luisetti, M., Sturani, C., Sella, D., Madonini, E., Galavotti, V., Bruno, G., … & Grassi, C. (1996). MR889, a neutrophil elastase inhibitor, in patients with chronic obstructive pulmonary disease: A double-blind, randomized, placebo-controlled clinical trial. *European Respiratory Journal, 9*(7), 1482–1486.
108. Esposito, A., Valentino, M. R., Bruzzese, D., Bocchino, M., Ponticiello, A., Stanziola, A., & Sanduzzi, A. (2016). Effect of CArbocisteine in Prevention of exaceRbation of chronic obstructive pulmonary disease (CAPRI study): An observational study. *Pulmonary Pharmacology & Therapeutics, 37*, 85–88.
109. Rogliani, P., Matera, M. G., Page, C., Puxeddu, E., Cazzola, M., & Calzetta, L. (2019). Efficacy and safety profile of mucolytic/antioxidant agents in chronic obstructive pulmonary disease: A comparative analysis across erdosteine, carbocysteine, and N-acetylcysteine. *Respiratory Research, 20*(1), 1–11.
110. Gouzi, F., Maury, J., Héraud, N., Molinari, N., Bertet, H., Ayoub, B., … & Hayot, M. (2019). Additional effects of nutritional antioxidant supplementation on peripheral muscle during pulmonary rehabilitation in COPD patients: A randomized controlled trial. *Oxidative Medicine and Cellular Longevity, 2019,* Article ID 5496346.
111. Wise, R. A., Holbrook, J. T., Criner, G., Sethi, S., Rayapudi, S., Sudini, K. R., … & Broccoli Sprout Extract Trial Research Group. (2016). Lack of effect of oral sulforaphane administration on Nrf2 expression in COPD: A randomized, double-blind, placebo controlled trial. *PloS One, 11*(11), e0163716.

112. De Swert, K. O., Bracke, K. R., Demoor, T., Brusselle, G. G., & Joos, G. F. (2009). Role of the tachykinin NK 1 receptor in a murine model of cigarette smoke-induced pulmonary inflammation. *Respiratory Research, 10*(1), 1–12.
113. Bolser, D. C., DeGennaro, F. C., O'Reilly, S., McLeod, R. L., & Hey, J. A. (1997). Central antitussive activity of the NK1 and NK2 tachykinin receptor antagonists, CP-99,994 and SR 48968, in the guinea-pig and cat. *British Journal of Pharmacology, 121*(2), 165–170.
114. Belvisi, M. G., Saunders, M. A., Haddad, E. B., Hirst, S. J., Yacoub, M. H., Barnes, P. J., & Mitchell, J. A. (1997). Induction of cyclo-oxygenase-2 by cytokines in human cultured airway smooth muscle cells: Novel inflammatory role of this cell type. *British Journal of Pharmacology, 120*(5), 910–916.
115. Malerba, M., Ponticiello, A., Radaeli, A., Bensi, G., & Grassi, V. (2004). Effect of twelve-months therapy with oral ambroxol in preventing exacerbations in patients with COPD. Double-blind, randomized, multicenter, placebo-controlled study (the AMETHIST Trial). *Pulmonary Pharmacology & Therapeutics, 17*(1), 27–34.
116. Muralidharan, P., Hayes, D., Black, S. M., & Mansour, H. M. (2016). Microparticulate/nanoparticulate powders of a novel Nrf2 activator and an aerosol performance enhancer for pulmonary delivery targeting the lung Nrf2/Keap-1 pathway. *Molecular Systems Design & Engineering, 1*(1), 48–65.
117. Al Faraj, A., Shaik, A. S., Afzal, S., Al Sayed, B., & Halwani, R. (2014). MR imaging and targeting of a specific alveolar macrophage subpopulation in LPS-induced COPD animal model using antibody-conjugated magnetic nanoparticles. *International Journal of Nanomedicine, 9*, 1491.
118. Vestbo, J., Leather, D., Diar Bakerly, N., New, J., Gibson, J. M., McCorkindale, S., ... & Woodcock, A. (2016). Effectiveness of fluticasone furoate–vilanterol for COPD in clinical practice. *The New England Journal of Medicine, 375*, 1253–1260.
119. Baker, D., Bruce, M., Crater, G., Noga, B., Thomas, M., & Wire, P. (2012). *U.S. Patent Application No. 13/510,962* AbbVie Biotechnology Ltd, The United States of America.
120. Yang, J., Guan, J., Pan, L., Jiang, K., Cheng, M., & Li, F. (2008). Enantioseparation and impurity determination of the enantiomers of novel phenylethanolamine derivatives by high performance liquid chromatography on amylose stationary phase. *Analytica Chimica Acta, 610*(2), 263–267.
121. Skupin, A. H. (1992). *U.S. Patent No. 5,096,916*. Washington, DC: U.S. Patent and Trademark Office. US5096916A.
122. Brandsma, C. A., Van den Berge, M., Hackett, T. L., Brusselle, G., & Timens, W. (2020). Recent advances in chronic obstructive pulmonary disease pathogenesis: From disease mechanisms to precision medicine. *The Journal of Pathology, 250*(5), 624–635.
123. Novelli, F., Malagrinò, L., Dente, F. L., & Paggiaro, P. (2012). Efficacy of anticholinergic drugs in asthma. *Expert Review of Respiratory Medicine, 6*(3), 309–319.
124. Bredenbroeker, D. (2012). *U.S. Patent Application No. 13/216,936*.
125. Papi, A., Vestbo, J., Fabbri, L., Corradi, M., Prunier, H., Cohuet, G., ... & Singh, D. (2018). Extrafine inhaled triple therapy versus dual bronchodilator therapy in chronic obstructive pulmonary disease (TRIBUTE): A double-blind, parallel group, randomised controlled trial. *The Lancet, 391*(10125), 1076–1084.
126. Mizushima, T. (2016). *U.S. Patent Application No. 15/117,461*.
127. Ullmann, M. (2017). *U.S. Patent Application No. 15/310,133*.
128. Slinde, F., Grönberg, A., Engström, C. P., Rossander-Hulthén, L., & Larsson, S. (2005). Body composition by bioelectrical impedance predicts mortality in chronic obstructive pulmonary disease patients. *Respiratory Medicine, 99*(8), 1004–1009.
129. Matz, J. (2013). *U.S. Patent Application No. 13/298,543*.
130. Suzuki, É. Y., Simon, A., da Silva, A. L., Amaro, M. I., de Almeida, G. S., Agra, L. C., ... & de Sousa, V. P. (2020). Effects of a novel roflumilast and formoterol fumarate dry powder inhaler formulation in experimental allergic asthma. *International Journal of Pharmaceutics, 588*, 119771.
131. Spargo, P. L., French, E. J., & Haywood, P. A. (2018). *U.S. Patent No. 9,956,171*. Washington, DC: U.S. Patent and Trademark Office.
132. Collingwood, S. P., & Haeberlin, B. (2008). *U.S. Patent Application No. 12/093,663*.
133. Maselli, D. J., Keyt, H., & Restrepo, M. I. (2017). Inhaled antibiotic therapy in chronic respiratory diseases. *International Journal of Molecular Sciences, 18*(5), 1062.
134. Bhatt, E. P., MacLeod, D. L., McKevitt, M. T., & Newcomb, T. G. (2011). *U.S. Patent Application No. 12/943,778*.
135. Chokkareddy, R., Thondavada, N., Kabane, B., & Redhi, G. G. (2021). A novel ionic liquid based electrochemical sensor for detection of pyrazinamide. *Journal of the Iranian Chemical Society, 18*(3), 621–629.
136. Casado, R.L., & de Miquel Serra, G. (2008). European Patent Application No.EP2265258A2.

8

Screening and Pharmacological Management of Neuropathic Pain

Manu Sharma, Ranju Soni, Kakarla Raghava Reddy, Veera Sadhu and Raghavendra V. Kulkarni

CONTENTS

8.1 Introduction

Pain is an obnoxious sensory and emotional experience that significantly affects the general and psychological health of individuals along with their social and economic wellbeing [1]. Pain may arise due to some external stimulus, like thermal, mechanical, or chemical, approaching harmful intensity or damage to system conducting signals, i.e. neurons [2]. It is a complex sensory response that often occurs due to extreme or injurious stimuli [3]. Pain also serves as a warning to indicate damage or impending danger to nerves. It can be categorized into two types, namely neuropathic and nociceptive [4]. Numerous diseases like multiple sclerosis, spinal cord injury, stroke, and central lesion are responsible for neuronal injury

DOI: 10.1201/9781003157823-8

or ailments in peripheral or central nervous systems and are the main cause of neuropathic pain [5, 6]. On the other hand, detrimental stimuli like any chemical, mechanical, or thermal stimuli at sensory endings of tissue result in nociceptive pain. Pain affects millions of people around the world. It may be associated with various diseases like diabetes, heart disease, cancer, stroke, stress, burn, etc., or any external harmful stimulus [7]. Thus, it affects the wellbeing, performance, and socioeconomic status of the sufferer and creates a huge burden on the health system and state economies [8].

8.2 Pathophysiology of Pain

Induction of pain is a notification to an individual for the protection of tissue from further injury. Generally acute or chronic pain is stimulated from a harmful or noxious stimulus in normal tissue or nerve [9]. The harmful stimuli such as mechanical (squeezing or painful pressure of the tissue), thermal (cold, heat), and chemical (chili flakes) activate sensory receptors (Ad and C fibers) in peripheral nerve or tissue. The sensory endings of tissue are also known as free nerve endings as they are not equipped with corpuscular end organs. These sensory endings have sensor molecules that transduce external stimuli (mechanical, thermal, and chemical) into adequate sensor action potential conveyed from the axon to the dorsal horn of the spinal cord to give action reflexes [10] (Figure 8.1).

8.3 Types of Pain

Commonly, pain is categorized as psychogenic, nociceptive, and neuropathic pain. Nociceptive and neuropathic pain are associated with tissue or nerve injury, respectively [11] (Figure 8.2), whereas psychogenic pain is a disorder of psychological factors.

8.3.1 Psychogenic Pain

Psychogenic pain, or psychalgia, is a chronic pain condition in people suffering from a mental disorder or illness such as depression or anxiety even though the physical cause is not known. Its diagnosis is possible only when all other causes of pain are ruled out. It is often difficult to treat. Its treatment requires consecutive consultation with a physician along with a mental health specialist [12]. The most common types of psychogenic pain are headache, muscle pains, back pain, and stomach pains.

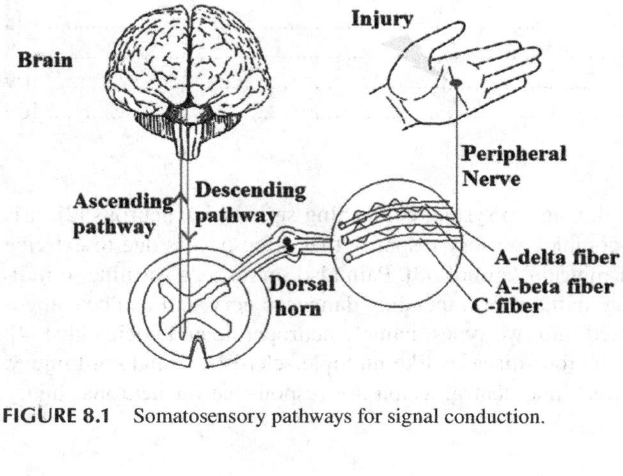

FIGURE 8.1 Somatosensory pathways for signal conduction.

8.3.2 Nociceptive Pain

This pain is elicited through activation of nociceptive nerve fibers or endings by physical, chemical, or mechanical triggers or by tissue damage. It arises as a result of the lesion(s) in the skin, soft tissue, muscle, or bone. Nociceptive pain is closely confined spasmodic or persistent pain that may be aggravated by movement. It is generally expressed as aching, squeezing, and cramping [13].

Examples of nociceptive pain include:

i) *Intense heat*, such as when a hand touches a hot stovetop surface.
ii) *Sharp mechanical stimulation*, such as a razor blade nicking the skin during shaving.
iii) *Pressure*, which is experienced when force is applied, such as a pinch to the back of the arm.
iv) *Chemical stimulation*, such as when salt comes into contact with an open cut or wound.

Nociceptive pain covers most leg, arm, and back pain. It is further categorized as somatic, radicular, and visceral pain. Somatic pain happens when any of the pain receptors in tissues, such as muscles, bone, or skin, are activated. Headaches and cuts are both considered somatic pain. This type of pain is usually localized but often stimulated by movement [1] whereas radicular pain radiates to the lower extremity along the sciatic nerve down the back to the leg. Compression, inflammation, and/or injury to the spinal nerve root are common causes of radicular pain. It is often associated with numbness, muscle weakness, and loss of reflexes. Injuries or inflammation of internal organs such as involuntary muscles in the heart lead to visceral pain. Normally, this pain is characterized by aching with an unclear location.

8.3.3 Neuropathic Pain

Neuropathic pain is also known as neuralgia which emerges due to damage in one or many nerves affecting the somatosensory nervous system either centrally or peripherally. Central neuropathic pain usually occurs due to spinal cord injury, multiple sclerosis, stroke, etc., whereas peripheral neuropathy is generally associated secondary to multiple etiologies like diabetes, infections, chemotherapy-associated, herpes zoster infection, complex regional pain syndrome, amyloid neuropathy, and many more. The nature of neuropathic pain may be intermittent or constant and spontaneous or provoked. It is usually sensed by the sufferer as itching, tingling, and shooting like an electric shock, burning, and pricking sensation [14]. It can persist for extended periods of time without improvement or apparent utility for the body. It has been characterized

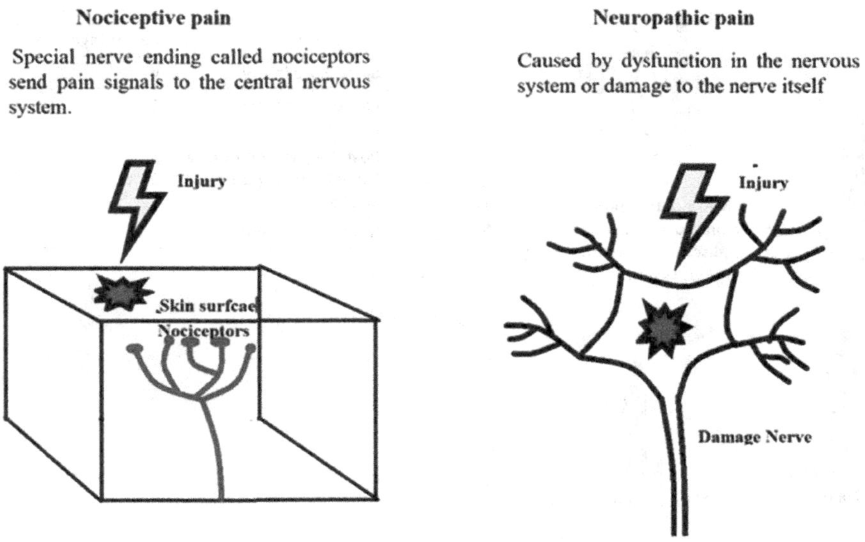

FIGURE 8.2 Diagrammatic representation of difference in nociceptive and neuropathic pain.

by different signs and symptoms like allodynia, hypoalgesia, hyperalgesia, paresthesia, and dysesthesia [15, 16].

i) Allodynia is a kind of neuropathic pain experienced with a touch or stimulus which generally does not induce pain. For example, slightly touching on the face and brushing hairs may provoke pain in individuals with trigeminal neuralgia, or bedclothes pressure can induce pain in patients with diabetic neuropathy.
ii) Hypoalgesia is a condition of reduced pain or numbness from a painful stimulus.
iii) Hyperalgesia is a dreadful condition of pain from a stimulus or touch which causes slight discomfort.
iv) Paraesthesia is a neuropathic pain condition where the patient usually has uncomfortable or distressing feelings of pricking needles or electric shock although nothing is in contact.
v) Dysesthesia is an extemporaneous or induced unusual abnormal disagreeable sensation.

8.4 Causes of Neuropathic Pain Conditions

Neuropathic pain may occur due to one or multiple etiologies (Figure 8.3). Diabetes, chemotherapeutic drugs, trigeminal neuralgia, HIV, and herpes infections are the most common conditions resulting in peripheral neuropathic pain while stroke, spinal cord injury, and trigeminal neuralgia are usually responsible causes of central neuropathic pain. Although physiology may differ, patients have similar indications and expressions of neuropathic pain [17].

8.4.1 Diabetes

Diabetes is amalgamated with a variety of peripheral nerve complications. The incidence of peripheral neuropathy is nearly 50% of patients suffering from diabetes. Distal symmetric polyneuropathy and autonomic neuropathy are the usual conditions affecting diabetic patients. Common symptoms like numbness, tingling, poor balance, and pain expressed as burning and electric shock sensations are often reported by patients. However, the precise mechanism of happening of neuropathic pain is unclear. It has been postulated in the literature that metabolic dysfunction leads to oxidative and inflammatory stress in cells responsible for ultimate damage to nerve cells. The quality of life of the sufferer is usually diminished due to depletion of sensation, movement interruption, and foot ulceration [18].

8.4.2 HIV Infection

HIV infection associated with neurological disorders is another endemic sensory peripheral neuropathy as a consequence of nerve injury by HIV. Individuals treated for infection suffer symmetrically with distal polyneuropathy similar to untreated individuals with HIV infection. This might be contributed by treatment-induced mitochondrial dysfunction of nerve cells. The probability of happening of HIV-sensory neuropathy depends on the vulnerability to neurotoxic chemical entities, progressing age, malnourishment, race, genetic factors, and metabolic dysfunction [19, 20].

8.4.3 Chemotherapy-Induced

With the evolution of newer cancer therapies, the survival rate of cancer patients has increased but long-term treatments are usually assisted with peripheral neuropathy. The intensity of peripheral neuropathy is often dose-dependent and progressive with chemotherapeutic agents like platinum drugs, vinca alkaloids, bortezomib, and a taxane. Disruption of microtubule mediated axonal transport, degradation of axons, direct degradation of dorsal root ganglion and disrupted mitochondrial

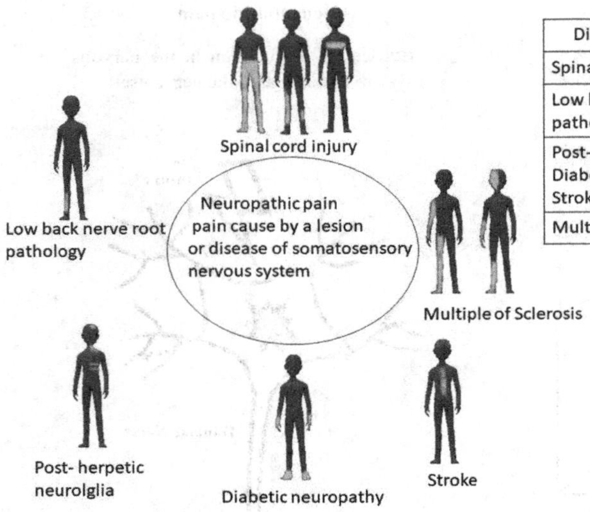

Disease	Percentage of patients
Spinal cord injury	67%
Low back nerve root pathology	10-17%
Post- herpetic neurolglia Diabetic neuropathy Stroke	8% 26% 8%
Multiple of Sclerosis	28%

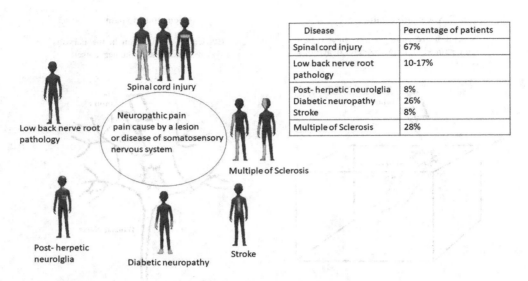

FIGURE 8.3 Multiple etiologies affecting different areas during neuropathic pain along with its incidence rate.

dysfunction of nerves are the major cause of peripheral neuropathy during cancer chemotherapy. Symptoms of peripheral neuropathy may worsen with continued cancer therapy and need disruption. The discontinuation of cancer therapy may show improvement in neuropathy although peripheral neuropathy may persist or progress even after discontinuation of cisplatin and oxaliplatin [21].

8.4.4 Herpes Infection

Herpes zoster is the causative organism infecting or reactivating inside the body to cause postherpetic neuralgia. The virus remains latent in dorsal root ganglions and gets activated as the immune system is disrupted or compromised with exposure to any mechanical or chemical stimuli or disease. The virus sensitizes nerves to become hyperexcited and facilitates sensory nerve death. Pain is typically disseminated unilaterally along spinal dermatomes or the ophthalmic branch of the trigeminal nerve [22].

8.4.5 Damage or Injury to Trigeminal Nerve

Injury or lesion to the fifth cranial nerve, termed the trigeminal nerve, causes chronic, extraordinary, periodic, unexpected burning or trauma on the face and is termed trigeminal neuralgia. The episodes of such pain occur in swift sessions and facial pain persists from a few seconds to a couple of minutes. Sometimes compression of the cranial nerve by vasculature or tumors and demyelination of nerves are responsible for symptoms experienced by patients. Such painful assaults are normally provoked by non-painful stimuli and assisted by reflexive spasms or contractions of facial muscles [23, 24].

8.4.6 Spinal Cord Injury

Spinal cord injury (SCI) leads to varying degrees of motor or sensory deficits and paralysis. SCI may lead to lifelong loss of function, autonomic disturbances, and reduced quality of life, as well as increased morbidity and mortality. Pain is common in patients with SCI. The pain may be of the nociceptive or neuropathic type or a combination of the two. Neuropathic pain following SCI is caused by damage to or dysfunction of the nervous system, while nociceptive pain is caused by damage to nonneural tissue either musculoskeletal due to bone, joint, muscle trauma or inflammation, mechanical instability, or muscle spasm [25].

8.4.7 Central Post-Stroke Pain

Central post-stroke pain (CPSP) has been referred to as thalamic pain. It is characterized by an intense spontaneous or evoked pain localized in the affected extremities and can affect the entire side of the body with an aching and burning quality. The sensory disturbance is a major component of CPSP, including abnormal temperature sensation, dysesthesia, and hypersensitivity to cutaneous stimuli. The pain appears to be alleviated with relaxation and worsened with emotional and physical stress. Commonly prescribed oral medications for use in post-stroke pain include antidepressants and anticonvulsants, while opioids are not felt to be effective.

8.5 Current Screening Tools for Neuropathic Pain

8.5.1 Leeds Assessment of Neuropathic Symptoms and Signs (LANSS)

LANSS pain scale is a validated report tool to differentiate neuropathic pain from non-neuropathic pain based on examination of the description of sensitivity and its deficit. It is based on an evaluation of five groups of symptoms at painful sites like allodynia, paroxysmal pain, dysesthesia, autonomic changes, and burning sensation. Allodynia and alteration in pain threshold at the site of needle pricking are commonly considered symptoms due to their ease of physical evaluation. Questionnaires used in this study are binary and assessed on a scale with a score ranging from zero to 24. A score equal to 12 or more identifies patients suffering from neuropathic pain. LANSS was able to identify 80% of neuropathic cases with good sensitivity and specificity compared to clinical diagnosis. However, the use of LANSS has been criticized due to the use of sharp needles and the need for training to apply it in some clinical situations and research.

8.5.2 Douleur Neuropathique Four Questions (DN4)

Similar to LANSS, Douleur Neuropathique 4 questions (DN4) is a contrivance focusing on the identification of neuropathic pain patients. The ease of its use makes it more accessible to healthcare professionals as well as patients. It is made up of seven items related to symptoms and three items related to physical evaluation. The scoring of every item is 1 if the answer is positive and 0 for negative answers. A score ranging from 0 to 10 is observed with a cutoff point at 4, i.e. scores equal to or above 4 suggest neuropathic pain. Due to the discriminative property of the first seven items, these could be used in some types of clinical trials; however, this still needs to be validated.

8.5.3 ID-Pain

ID-Pain is a self-report questionnaire to evaluate patients with pain. It relates patients' pain characteristics with scores varying from –1 to 5. The endorsed cut-off score for neuropathic pain is above 3. This scale has sensitivity and specificity of approximately 70 to 80% and may be a useful tool to diagnose neuropathic pain. ID-Pain comprises six items out of which five items are sensory descriptor items and one relates to the location of pain in joints to recognize nociceptive pain. It also does not necessitate clinical examination to interpret the outcome.

8.5.4 Neuropathic Pain Scale (NPS)

The Neuropathic Pain Scale (NPS) was generated to evaluate different pain qualities associated with neuropathic pain. The NPS includes 10 items, out of which two determine pain dimensions (intensity and discomfort) and eight evaluate NP quality (stabbing, burning, freezing, boring, tender, itching, deep pain, and superficial pain). Items are assessed by a numeric scale from 0 to 10. Zero is assigned for no pain sensation (not hot) whereas 10 for very hot, or worst imaginable sensation to describe "too hot" pain.

8.5.5 Pain Quality Assessment Scale (PQAS)

This is a self-report tool derived from the NPS. The PQAS was developed to evaluate the quality of neuropathic pain not evaluated by the NPS scale. The PQAS is able to evaluate qualities or domains affected by pain management. It has 20 descriptors to evaluate two global aspects (intensity and discomfort), two spatial aspects (superficial and deep), and 16 quality domains: jumping (pricking, drilling), burning (on fire), dull, cold (freezing), tender (such as open sore), as a wound itching (such as mosquito bite), tugging, numbness, shock (lightning, spark), tingling, cramping (crushing, pressing), radiating, throbbing, hurting (such as toothache), and heavy (pressure). PQAS has also an item to evaluate pain temporal pattern (intermittent with no pain in other moments, minimum pain whole time with exacerbation periods, and constant pain which does not change a lot from one moment to the other). Each item is evaluated by the verbal numeric scale where 0 means no pain or no painful sensation and 10 represents the worst imaginable pain sensation.

8.6 Management of Neuropathic Pain

Neuropathic pain management requires a focused approach to recognize the cause and alleviate the symptoms of neuropathic pain. Comorbidities like anxiety, depression, stroke, diabetes, herpes infection, etc. are severe and persistent in nature and associated with injury to the nervous system [26]. It is also necessary to identify suitable secondary treatments of these diseases to improve the quality of life. The primary aim of pain management is to make pain tolerable. Various pharmacological interventions along with neuromodulation, nerve block, psychological and physical therapies are commonly used in neuropathic pain management.

8.6.1 Pharmacological Interventions

Currently available therapeutic options for neuropathic pain management include adjuvant (anticonvulsants, antidepressants, and local anesthetics), opioids, and non-steroidal anti-inflammatory drugs [25, 27]. Medicaments available are classified clinically into four classes: first-line (certain antidepressants and anticonvulsants), second-line (serotonin, noradrenaline reuptake inhibitors, and topical lidocaine), third-line (tramadol and opioids), and fourth-line analgesics (cannabinoids, methadone, and anticonvulsants like lamotrigine, topiramate, and valproic acid) [28] (Table 8.1).

8.6.2 Antidepressants

Tricyclic antidepressant (TCA) drugs have shown sufficient therapeutic potential in relieving pain due to diabetic, postherpetic, and trigeminal neuralgia. Their ability to induce analgesia depends upon blockage of noradrenaline and serotonin reuptake and blockage of N-methyl-D-aspartate agonists and sodium channels [29]. Drugs like amitriptyline, venlafaxine, duloxetine, and imipramine show simultaneous inhibition of the reuptake of noradrenaline and serotonin whereas desipramine and nortriptyline are selective inhibitors of noradrenaline uptake. Continuous use of TCA is usually associated with dry mouth, constipation, nausea, vomiting, and dizziness but is of low clinical significance at doses used in neuropathic pain management [30].

TABLE 8.1
Pharmacotherapy of Neuropathic Pain

S. no.	Drug	Class	Brand name	Dose	Side effect
First-line analgesics					
Antidepressants					
1.	Amitritytline	Tricyclic antidepressants	Vantatrip	100 mg QD	Abdominal or stomach pain, black or tarry stools, blood in urine or stools, blurred vision
2.	Desipramine	Tricyclic antidepressants	Norpramin	100–200 mg per day	Abdominal or stomach pain, black or tarry stools, blood in urine or stools, blurred vision, anxiety
3.	Imipramine	Tricyclic antidepressants	Tofranil	200 mg per day	Allergy reaction, hives, difficult breathing
Anticonvulsants					
4.	Carbamazepine	Carboxamids	Tegretol	200 mg BID	Nausea, vomiting, dizziness, drowsiness, constipation, dry mouth
5.	Clonazepam	Benzodiazepine	Klonopin	1 mg per day	Ear congestion, cough, difficulty in breathing, body aches
6.	Gabapentin	Anti-epileptic agent	Neurontin	300–600 mg TID	Dizziness, drowsiness, fatigue, fever, sedated state
7.	Pregabalin	Anti-epileptic agent	Lyrica	50 mg TID	Loosening of the skin, cough, dizziness, difficulty in swallowing, hives, itching
8.	Topiramate	Anti-epileptic agent	Topamex	200–400 mg BID	Confusion, dizziness, drowsiness, memory problems, menstrual changes
9.	Oxcarbazepine	Carboxamide	Oxtellar XR	300 mg TID	Change in walking balance, change in vision, mental depression, dizziness
10.	Lamotrigine	Anti-epileptic agent	Lamictal	25 mg per day	Abnormal gait, constipation, drowsiness, insomnia, vomiting
Second-line analgesics					
Serotonin-norepinephrine reuptake inhibitor					
11.	Venlafaxine	Selective serotonin-norepinephrine reuptake inhibitor	Effexor XR	25 mg TID	High fever, convulsion, irritability, menstrual change, nervousness
12.	Duloxetine	Serotonin-norepinephrine reuptake inhibitor	Cymbalta	60 mg per day	Nausea, dry mouth, constipation, loss of appetite
13.	Bupropion	Selective serotonin reuptake inhibitor	Aplenzin	100 mg TID	Constipation, stomach pain, dizziness, trembling, decrease in appetite
14.	Lidocain	Local anesthesia	UAD caine	50–100 mg per day	Unusually warm skin, itching skin, small red spot on skin
Third-line analgesics					
15.	Tramadol	Opioid	ConZip	50–100 mg QID	Bloating, blood pressure, blurred vision, difficulty in urination
16.	Morphine	Opioid	Arymo ER	10–30 mg QID	Upset stomach, blurred vision, change to ability to see color, confusion, chest pain, dizziness
17.	Oxycodone	Opioid	Oxaydo	5–15 mg QID	Confusion, difficulty in breathing, twitching, fever, stomach pain
18.	Hydromorphone	Opioid	Dilaudid	2–4 mg QID	Agitation, bloody and black stool and urination, change in behavior, convulsion, dry mouth
19.	Fentanyl	Opioid	Actiq	20–50 mg per day	Confusion, difficulty in breathing, twitching, confusion, stomach pain
20.	Codeine	Narcotic analgesic		15–60 mg QID	Bloating, blurred vision, confusion, constipation, dizziness
Fourth-line analgesics					
21.	Methadone	Opioid	Dolophine	2.5–10 mg per day	Bleeding gum, black stool, blood in urine or stool, confusion, dry mouth
22.	Cannabinoid	Herbal drug	Cannabinoid	300 mg per day	Dry mouth, low blood pressure, drowsiness

8.6.3 Anticonvulsants

Anticonvulsants manifest analgesic activity to relieve neuropathic pain by reducing neuronal excitability. Drugs belonging to this class exhibit different mechanisms to reduce neuronal excitability. For example, gabapentin modulates neuronal calcium channels, and carbamazepine and lamotrigine act on sodium channels, while topiramate acts on both [31, 32].

Gabapentin and pregabalin, structural analogs of gamma-aminobutyric acid, are widely utilized in chronic neuropathic pain management. They exhibit great affinity to bind with the α2-δ subunit of voltage-dependent calcium channels to reduce the action potential of nerves. The analgesic potential of gabapentin and pregabalin is sufficient to relieve pain during postherpetic neuralgia, diabetic neuropathy, spinal cord injury, and trigeminal neuralgia. It is well tolerated but common side effects during the initiation phase of therapy are mild to moderate dizziness and somnolence, ataxia, and confusion. It is usually prescribed orally at a dose of 300 mg daily. A dose of gabapentin can be increased until adequate analgesia is achieved or side effects are experienced. However, pregabalin's effective dose ranges from 150 to 600 mg/day orally [33].

Topiramate another anticonvulsant drug work via multiple pathways, i.e. by modulating sodium and calcium channels, amplification of action of inhibitory neurotransmitter GABA, blocking excitatory amino acid glutamate, and inhibiting carbonic anhydrase. Topiramate has been found to decrease allodynia in preclinical models of neuropathic pain.

Carbamazepine and oxcarbazepine, a ketoanalogue of carbamazepine, impede Na^+ channels while their metabolite also suppresses K^+ channels. A randomized controlled clinical trial of oxcarbazepine revealed the analgesic potential in trigeminal neuralgia while another showed equipotency in analgesic property with amitriptyline in cancer-related neuropathic pain. Another neuronal sodium channel blocker lamotrigine exhibits a use-dependent analgesic effect in painful diabetic neuropathy, trigeminal neuralgia, and complex regional pain syndrome [34].

8.6.4 Opioids

During intense or persistent pain, neurons in the dorsal horn liberate endogenous opioid peptides (beta-endorphin, enkephalins, dynorphins) to diminish the perception of pain. These endogenous opioids modulate pain-related signals by inhibiting the transmission of signals at synapses. Opioid receptors (μ, κ, and δ) are present in different areas of the brain, brain stem, spinal cord, and peripheral nervous system. Opioids exhibit mood-elevating properties because of specific affinity towards μ receptors in the brain whereas they enhance the performance of cells in the brain stem involved in inhibition of descending pain. Opioids also inhibit the transmission of nociceptive pain signals at the spinal cord and peripheral nervous system. Agonistic binding of opioids to all receptors elicits the closure of Ca^{+2} channels which diminishes neurotransmitter release and inhibits postsynaptic neurons [35–37].

Methadone is widely utilized in neuropathic pain management. It is an opioid drug having agonistic activity at both μ and δ opioid receptors, N-methyl-D-aspartate (NMDA) antagonist activity, and impedes reuptake of monoamines. The therapeutic potential of reducing pain by modulating and inhibiting transmission of nerve signals along with low cost has increased the interest of physicians in using methadone for the treatment of neuropathic or chronic pain. Side effects associated with methadone therapy are similar to those of opioids like dizziness, sleepiness, vomiting, and sweating while opioid abuse and respiratory depression are serious side effects [38].

Tramadol is a safe and potent opioid derivative analgesic effective in moderate to severe chronic pain conditions during osteoarthritis, low back pain, fibromyalgia, and diabetic neuropathy. It acts dually by binding to μ opioid receptors and inhibiting the reuptake of noradrenaline and serotonin. Tramadol exhibited a strong ability to reduce peripheral pains compared to central pain due to its weak agonistic activity at μ-opioid receptors. Common side effects of tramadol therapy are nausea, drowsiness, dizziness, headache, dry mouth, pruritis, diarrhea, and constipation. Serious risks include respiratory depression, blurring of vision, urine retention [39].

Dextromethorphan hydrobromide, an NMDA receptor antagonist, has shown good therapeutic efficacy in relieving peripheral pain. It acts by downregulating NMDA receptors by inhibiting glutamate release and repeated firing of peripheral afferent fibers responsible for nerve injury [40].

8.6.5 Muscle Relaxants

Muscle relaxants are widely used to reduce the symptoms of muscle spasms, pain, and hyperreflexia. Drugs having spasmolytic and neuromuscular blocking activity are categorized under muscle relaxants. Neuromuscular blockers do not exhibit any central nervous system activity in contrast to spasmolytic. They work by blocking the transmission of the signal at the neuromuscular endplate. They are often used during surgical procedures and in intensive care and emergency medicine to cause temporary paralysis [41].

Meprobamate, an anxiolytic drug, exhibits affinity to bind GABA receptors to pepper the neuronal transmission in the spinal cord. This symptomatically lessens pain sensation and produces sedation. Similarly, methocarbamol, a central muscle relaxant, is used to treat skeletal muscle spasms caused by musculoskeletal disorders, tetanus, and injury [42]. The GABA-derivative drug, baclofen, is mainly effective in managing muscle spasticity associated with spinal cord injury. Metaxalone is extensively used to relax muscle during strains, sprains, and other musculoskeletal conditions, although its mechanism of action is unclear [43]. Dantrolene is a hydantoin derivative. It produces muscle relaxation by dissociating excitation-contraction coupling possibly by restricting secretion of Ca^{+2} from sarcoplasmic reticulum by binding to ryanodine receptorl. It has been found suitable for the treatment of fulminant hypermetabolism of skeletal muscles [44].

8.6.6 Non-Steroidal Anti-Inflammatory Drugs

Conventional NSAIDs like aspirin and ibuprofen have limited utility in neuropathic pain management. However, the use of other NSAIDs is unpromising in long-term use due to their renal toxicities. Randomized controlled clinical trials have reported adverse cardiovascular events with prolonged use of COX-2 inhibitors. Thus, COX-2 inhibitors cannot be endorsed for the treatment of patients with neuropathic pain syndromes [45].

8.6.7 Corticosteroids

Corticosteroids are also used as adjuvant therapy for metastatic bone pain, neuropathic pain, and visceral pain. They reduce inflammation and vascular permeability by inhibiting prostaglandin synthesis and ultimately pain [46].

8.6.8 Topical Analgesics

Controlled trials have revealed the competence of local anesthetics like topical NSAIDs, capsaicin, and doxepin in relieving peripheral neuropathic pain if an area of pain is comparatively small or confined. The topical combination of amitriptyline and ketamine has shown a suitable analgesic effect in neuropathic pain. Topical NSAIDs have also attained success in the treatment of chronic pain of arthritis and rheumatism associated with cutaneous side effects. Studies have reported that topical lidocaine (5%) facilitates a remarkable reduction in pain during postherpetic neuralgia. The safety index of topical lidocaine was comparatively high over topical NSAIDs. Capsaicin creams are also effective in alleviating neuropathic pain and need to be applied carefully due to their high irritating potential [47].

8.6.9 Newer Pharmacological Interventions

Several studies have emphasized the neuropathic pain-relieving potential of acetyl-L-carnitine, α-lipoic-acid, cannabis products, botulinum toxin, and angiotensin II type 2 receptor antagonists. Notably, proofs relative to the efficacy of most of these emerging treatments for neuropathic pain are insufficient. Thus, further new studies need to be performed to prove their efficacy and tolerability, in addition to providing newer therapeutic options to manage pain refractory to current drugs.

Voltage-gated sodium channels (VGSC) are a necessary component of nerve impulse conduction. The utilization of VGSC blockers can augment pain relief during neuropathic pain conditions. Since α subunits of VGSC encode genes potentially relevant for neuropathic pain are namely NaV1.3, NaV1.7, NaV1.8, and NaV1.9. NaV1.3. However, researchers have considered NaV1.3 as a suitable target for pain therapeutics due to its upregulated overexpression in dorsal horn sensory neurons after injury in the nervous system. NaV1.7 is present at excessive levels in growth cones of small-diameter neurons whereas the NaV1.8 channel is particularly expressed in primary sensory neurons to conduct most of the inward current during an action potential. The knockdown of NaV1.8 channel expression in rats reversed injury-induced hyperalgesia in a neuropathic pain model and decreased bladder hyperactivity in a visceral pain model. The peripheral sodium channel NaV1.9 exclusively resides in the dorsal root ganglia and emphasizes neurotrophin (BDNF)-evoked depolarization and excitation. Thus, the development of specific VGSC blockers can be an alternative treatment for pain relief.

8.6.10 Combination Pharmacotherapy

Drugs belonging to different classes are utilized in the treatment of chronic pain (Table 8.1). Treatment options available have their limitations due to incomplete efficacy and dose-limiting side effects. Thus, combination therapies having synergistic activity can provide superior analgesia and lesser side effects by targeting multiple sites of pain modulation than monotherapy. Several combination therapies have been evaluated for their efficacy and tolerability in neuropathic pain, fibromyalgia, arthritis, and other disorders. Few studies proved the superiority of combination therapies over monotherapy, while others had not shown any improvement.

8.7 Neuromodulation Techniques

Neuromodulation is a normal physiological process regulating the functioning of diverse populations of neurons. Neurotransmitters like dopamine, serotonin, acetylcholine, histamine, and nor-epinephrine are neuromodulators. They exhibit a modulatory effect on target areas such as decorrelation of spiking, an increase of firing rate, sharpening of spatial tuning curves, and maintenance of increased spiking during working memory. A neuromodulator that is not re-absorbed by pre-synaptic neurons or metabolized can spend a remarkable duration in cerebrospinal fluid and modulate the activity of other neurons like serotonin ad acetylcholine.

8.7.1 Nerve Block Therapy

The nerve block technique involves a procedure designed to interfere with neural conduction to diminish or hamper pain. Both afferents as well as efferent conduction may be interrupted. Local anesthetics are the most commonly injected substance for nerve blocks and the effect is prolonged on the addition of a corticosteroid. Nerve block techniques have been classified into four categories, i.e. diagnostic, prognostic, prophylactic, and therapeutic nerve block. Diagnostic nerve blocks can define more clearly the anatomical etiology of the pain and help in distinguishing peripheral and central pain syndromes. Prognostic nerve blocks are performed to help to predict response to a procedure that may have a greater duration of action than a nerve block with a local anesthetic. On the other hand, prophylactic nerve blocks or preemptive analgesia are techniques employed to prevent the development of significant pain following surgery or trauma. Therapeutic nerve blocks may be used to diminish pain and promote functional rehabilitation in combination with physiotherapy during acute or chronic pain syndromes.

8.7.2 Psychological Therapies

Recurrent pain may bestow dysfunctional cognition and behavior which deteriorates physical functioning and increases psychiatric distress and episodes of pain. Therefore, psychotherapy along with medical interventions is essentially required to improve physical, emotional, social, and occupational functioning. Psychological interventions have been classified into four categories, i.e. operant behavior, cognitive behavior, mindfulness-based, and acceptance and commitment therapy. Nowadays psychological therapies consider diverse domains like physical functioning, type of pain medicaments used, mood, perception patterns, and quality of life, although an alteration in pain intensity may be secondary. Thus, psychological therapies are commonly used as complementary treatments along with therapeutic medicaments to improve quality of life.

8.7.3 Physical Therapy

Physical therapy and exercise can diminish chronic pain during a variety of chronic pain conditions like neuropathic pain, osteoarthritis, rheumatoid arthritis, fibromyalgia, and chronic headache. These therapies include a number of different types of pain management methods including massage, manual therapy, cold laser therapy, microcurrent stimulation, movement therapy, and exercise. Thus, an integration of physical therapies with fundamental pain management approaches improves quality of life and prevents future pathology and physiologic changes which often result in significant pain syndromes.

REFERENCES

1. Loeser, J. D., & Treede, R. D. (2008). The Kyoto protocol of IASP basic pain terminology. *Pain, 137*(3), 473–477.
2. Van Hecke, O., Austin, S. K., Khan, R. A., Smith, B. H., & Torrance, N. (2014). Neuropathic pain in the general population: A systematic review of epidemiological studies. *PAIN, 155*(4), 654–662.
3. Gilron, I., Baron, R., & Jensen, T. (2015, April). Neuropathic pain: Principles of diagnosis and treatment. In *Mayo Clinic Proceedings* (Vol. 90, No. 4, pp. 532–545). Elsevier.
4. Woolf, C. J., Bennett, G. J., Doherty, M., Dubner, R., Kidd, B., Koltzenburg, M., … & Torebjork, E. (1998). Towards a mechanism-based classification of pain. *Pain, 2*, 11–15.
5. Serpell, M. G., Makin, A., & Harvey, A. (1998). Acute pain physiology and pharmacological targets: The present and future. *Acute Pain, 1*(3), 31–47.
6. Bruce, J., Krukowski, Z. H., Al-Khairy, G., Russell, E. M., & Park, K. G. M. (2001). Systematic review of the definition and measurement of anastomotic leak after gastrointestinal surgery. *Journal of British Surgery, 88*(9), 1157–1168.
7. Deng, Y., Luo, L., Hu, Y., Fang, K., & Liu, J. (2015). Clinical practice guidelines for the management of neuropathic pain: A systematic review. *BMC Anesthesiology, 16*(1), 1–10.
8. Drenth, J. P., te Morsche, R. H., Guillet, G., Taieb, A., Kirby, R. L., & Jansen, J. B. (2005). SCN9A mutations define primary erythermalgia as a neuropathic disorder of voltage gated sodium channels. *Journal of Investigative Dermatology, 124*(6), 1333–1338.
9. Schaible, H. G., & Richter, F. (2004). Pathophysiology of pain. *Langenbeck's Archives of Surgery, 389*(4), 237–243.
10. Tsuda, M., Inoue, K., & Salter, M. W. (2005). Neuropathic pain and spinal microglia: A big problem from molecules in 'small'glia. *Trends in Neurosciences, 28*(2), 101–107.
11. Kendall, N. A. (1999). Psychosocial approaches to the prevention of chronic pain: The low back paradigm. *Best Practice & Research Clinical Rheumatology, 13*(3), 545–554.
12. Kanchiku, T., Suzuki, H., Imajo, Y., Yoshida, Y., Nishida, N., & Taguchi, T. (2017). Psychogenic low-back pain and hysterical paralysis in adolescence. *Clinical Spine Surgery, 30*(8), E1122–E1125.
13. Yan, Y. Y., Li, C. Y., Zhou, L., Ao, L. Y., Fang, W. R., & Li, Y. M. (2017). Research progress of mechanisms and drug therapy for neuropathic pain. *Life Sciences, 190*, 68–77.
14. Chokkareddy, R., Bhajanthri, N. K., Kabane, B., & Redhi, G. G. (2018). Bio-sensing performance of magnetite nanocomposite for biomedical applications. *Nanomaterials: Biomedical, Environmental, and Engineering Applications,1*, 165-196.
15. Cr, C., & Gavrin, J. (1999). Suffering: The contributions of persistent pain. *Lancet, 353*, 2233–2237.
16. Rolke, R., Baron, R., Maier, C. A., Tölle, T. R., Treede, R. D., Beyer, A., … & Wasserka, B. (2006). Quantitative sensory testing in the German Research Network on Neuropathic Pain (DFNS): Standardized protocol and reference values. *Pain, 123*(3), 231–243.
17. Marks, R. (1992). Peripheral articular mechanisms in pain production in osteoarthritis. *Australian Journal of Physiotherapy, 38*(4), 289–298.
18. Schreiber, A. K., Nones, C. F., & Reis, R. C. (2015). Diabetic neuropathic pain: Physiopathology and treatment. *World Journal of Diabetes, 6*(3), 432–444.
19. Hesdorffer, D. C. (2016). Comorbidity between neurological illness and psychiatric disorders. *CNS Spectrums, 21*(3), 230–238.
20. Chokkareddy, R., Thondavada, N., Thakur, S., & Kanchi, S. (2019). Cholesterol-based enzymatic and nonenzymatic sensors. In *Advanced Biosensors for Health Care Applications* (pp. 315–339). Eds: Shakeel Ahmed, Elsevier.
21. Sisignano, M., Baron, R., Scholich, K., & Geisslinger, G. (2014). Mechanism-based treatment for chemotherapy-induced peripheral neuropathic pain. *Nature Reviews Neurology, 10*(12), 694–707.
22. Steckbeck, J. D., Deslouches, B., & Montelaro, R. C. (2014). Antimicrobial peptides: New drugs for bad bugs?. *Expert Opinion on Biological Therapy, 14*(1), 11–14.
23. Cecelja, M., & Chowienczyk, P. (2016). Molecular mechanisms of arterial stiffening. *Pulse, 4*(1), 43–48.
24. Nizam, S. A., & Ziccardi, V. B. (2015). Trigeminal nerve injuries: Avoidance and management of iatrogenic injury. *Oral and Maxillofacial Surgery Clinics, 27*(3), 411–424.
25. Ali, A., Arif, A. W., Bhan, C., Kumar, D., Malik, M. B., Sayyed, Z., … & Ahmad, M. Q. (2018). Managing chronic pain in the elderly: An overview of the recent therapeutic advancements. *Cureus, 10*(9), 1-9.
26. Nishikawa, N., & Nomoto, M. (2017). Management of neuropathic pain. *Journal of General and Family Medicine, 18*(2), 56–60.
27. Attal, N. (2019). Pharmacological treatments of neuropathic pain: The latest recommendations. *Revue Neurologique, 175*(1–2), 46–50.
28. Finnerup, N. B., Attal, N., Haroutounian, S., McNicol, E., Baron, R., Dworkin, R. H., … & Wallace, M. (2015). Pharmacotherapy for neuropathic pain in adults: A systematic review and meta-analysis. *The Lancet Neurology, 14*(2), 162–173.

29. Obata, H. (2017). Analgesic mechanisms of antidepressants for neuropathic pain. *International Journal of Molecular Sciences, 18*(11), 2483.
30. Kremer, M., Salvat, E., Muller, A., Yalcin, I., & Barrot, M. (2016). Antidepressants and gabapentinoids in neuropathic pain: Mechanistic insights. *Neuroscience, 338*, 183–206.
31. Billinger, S. A., Arena, R., & Bernhardt, J. (2014). A statement for healthcare professionals from the American Heart Association/American Stroke Association. *Stroke, 45*, 2532–2553.
32. Moore, R. A., Wiffen, P. J., Derry, S., & Lunn, M. P. (2015). Zonisamide for neuropathic pain in adults. *Cochrane Database of Systematic Reviews, 1*, 10
33. Wiffen, P. J., Derry, S., Moore, R. A., Aldington, D., Cole, P., Rice, A. S., ... & Kalso, E. A. (2013). Antiepileptic drugs for neuropathic pain and fibromyalgia-an overview of Cochrane reviews. *Cochrane Database of Systematic Reviews, 11*, 10567–10574.
34. Chogtu, B., Bairy, K. L., Smitha, D., Dhar, S., & Himabindu, P. (2011). Comparison of the efficacy of carbamazepine, gabapentin and lamotrigine for neuropathic pain in rats. *Indian Journal of Pharmacology, 43*(5), 596–598.
35. Eisenberg, E., McNicol, E. D., & Carr, D. B. (2006). Opioids for neuropathic pain. *Cochrane Database of Systematic Reviews, 3*, 6144–6149.
36. Wiffen, P. J., Derry, S., Moore, R. A., Stannard, C., Aldington, D., Cole, P., & Knaggs, R. (2015). Buprenorphine for neuropathic pain in adults. *Cochrane Database of Systematic Reviews, 9*, 11669–11672.
37. Ratnasabapathy, Y., Chi-Lun Lee, A., Feigin, V., & Anderson, C. (2003). Blood pressure lowering interventions for preventing dementia in patients with cerebrovascular disease (Protocol). *Cochrane Database of Systematic Reviews*, (2 Art. No.: CD004130.).
38. McNicol, E. D., Ferguson, M. C., & Schumann, R. (2017). Methadone for neuropathic pain in adults. *Cochrane Database of Systematic Reviews, 5*, 12499–12509.
39. Duehmke, R. M., Derry, S., Wiffen, P. J., Bell, R. F., Aldington, D., & Moore, R. A. (2017). Tramadol for neuropathic pain in adults. *Cochrane Database of Systematic Reviews, 15*(6), 3726–3732.
40. Yang, P. P., Yeh, G. C., Huang, E. Y. K., Law, P. Y., Loh, H. H., & Tao, P. L. (2015). Effects of dextromethorphan and oxycodone on treatment of neuropathic pain in mice. *Journal of Biomedical Science, 22*(1), 1–13.
41. Uhl, R. L., Roberts, T. T., Papaliodis, D. N., Mulligan, M. T., & Dubin, A. H. (2014). Management of chronic musculoskeletal pain. *JAAOS-Journal of the American Academy of Orthopaedic Surgeons, 22*(2), 101–110.
42. Cohen, S. P., & Hooten, W. M. (2017). Advances in the diagnosis and management of neck pain. *BMJ, 14*(358), 3221–3227.
43. Yu, H., Zhang, P., Chen, Y. R., Wang, Y. J., Lin, X. Y., Li, X. Y., & Chen, G. (2019). Temporal changes of spinal transcriptomic profiles in mice with spinal nerve ligation. *Frontiers in Neuroscience, 13*, 13–17.
44. Bathen, M. E., & Linder, J. (2017). Spin Seebeck effect and thermoelectric phenomena in superconducting hybrids with magnetic textures or spin-orbit coupling. *Scientific Reports, 7*(1), 1–13.
45. Borah, J. C., Mujtaba, S., Karakikes, I., Zeng, L., Muller, M., Patel, J., ... & Zhou, M. M. (2011). A small molecule binding to the coactivator CREB-binding protein blocks apoptosis in cardiomyocytes. *Chemistry & Biology, 18*(4), 531–541.
46. Han, Y., Zhang, J., Chen, N., He, L., Zhou, M., & Zhu, C. (2013). Corticosteroids for preventing postherpetic neuralgia. *Cochrane Database of Systematic Reviews, 28*(3), 5582–5589.
47. Knezevic, N. N., Tverdohleb, T., Nikibin, F., Knezevic, I., & Candido, K. D. (2017). Management of chronic neuropathic pain with single and compounded topical analgesics. *Pain Management, 7*(6), 537–558.

9

Clinical Use of Innovative Nanomaterials in Dentistry

Shikha Dogra, Anil Gupta, Shalini Garg, Sakshi Joshi and Neetika Verma

CONTENTS

9.1 Introduction

The word "nano", which is derived from the Greek word *nannos* meaning "dwarf", is a prefix that refers to one-billionth of physical size. Nanomaterials have distinct advantages as they exhibit unique physicochemical properties when compared to their bulk counterparts due to their nanoscale sizes and high surface-area-to-volume ratio. This includes increased reactivity, greater solubility, biomimetic features, and the ability to be functionalized with other materials such as drugs, bioactive molecules, and photosynthesizers [1]. Nanotechnology gives us the ability to arrange atoms as we desire and subsequently to achieve effective, complete control of the structure of matter. Therefore, the application of nanotechnology in dentistry enhanced the quality of oral health services by using nanomaterials and biotechnologies, including tissue engineering and nanorobots.

Applications of nanotechnology in dentistry are vast including diagnostics, preventive, dental materials, endodontics, periodontics, regenerative dentistry, etc.

9.2 Nanomaterial for Caries Arresting Agents

Nanoparticles can penetrate into biofilms and interact electrostatically with bacterial cell walls, which leads to cell membrane damage, increased cell permeability through the generation of reactive oxygen species which interfere with cellular functions destroying proteins, DNA damage, and ultimately cell death [1]. Silver is known for its antimicrobial action against oral bacteria. The high surface-to-volume ratio of silver nanoparticles results in increased effectiveness against cariogenic bacteria. Ginjupalli et al. (2018) showed that silver nanoparticles of size 80–100 nm have superior antimicrobial properties [2].

A metal ion-based topical fluoride preparation named silver diamine fluoride (SDF) has been drawing increased attention contemporarily, due to its efficacy in arresting the progression of dental caries. The major drawback of SDF is the staining of carious tissue to dark black due to the oxidation process of ionic silver contained in its formulation, along with ulceration and reversible staining of oral tissues which are painful. An

DOI: 10.1201/9781003157823-9

advancement of nanomaterials in dentistry has led to the modification of conventional SDF which can be used as an upcoming approach for caries arresting procedures. Nanosilver fluoride can be synthesized using chitosan, silver nitrate, and NaF in the laboratory or it can be synthesized using ready-made nanoparticles. Nanosilver (AgNPs) is one of the potent antimicrobial agents and its efficacy against cariogenic bacteria such as *S. mutans* is established in vitro. In an in vitro comparison of antimicrobial efficacy and cytotoxicity between SDF and a nanosilver-based fluoride varnish preparation carried out by Targino et al. (2014) [3], the antimicrobial efficacy of the nanosilver-based fluoride varnish preparation was better than SDF; however, the cytotoxicity of the nanosilver-based fluoride varnish preparation was less when compared to silver diamine fluoride. The nanosilver-based fluoride has advantages over SDF in the context of patient acceptance as it does not cause immediate dark staining of dentinal tissue – the reason being nanosilver does not form oxides – there is no metallic taste, no painful ulceration if comes in contact with oral mucosa, and it is relatively economical [4].

FIGURE 9.1 Caries arresting agent (SDF – silver diamine fluoride).

9.3 Innovative Nanomaterials for Restoration of Dental Caries

9.3.1 Bioactive Nanocomposites for Root Caries

In patients with periodontitis gingival recession often leads to exposure of root surfaces to the oral environment which consequently results in the increasing risk of root caries. Root caries can be treated with a Class V restoration, whose margins are often subgingival [5].

Microbial dental biofilms are the principal aetiological factor of periodontitis. The current resin-based restorations for treating root caries usually have no antibacterial function, and instead may even accumulate more biofilms/plaques in vivo, which could aggravate the progress of periodontitis.[5] Therefore, nano-enhanced restorative materials for root caries are developed to overcome these limitations of conventional materials.

Wang et al. (2016) developed a bioactive nanocomposite for Class V restorations to inhibit periodontitis-related pathogens [6]. This nanocomposite contained a combination of antimicrobial monomers (dimethylaminohexadecyl methacrylate) and amorphous calcium phosphate nanoparticles. This nanocomposite showed a strong inhibiting effect against all six species of periodontitis-related pathogens, i.e., *Porphyromonas gingivalis*, *Prevotella intermedia*, *Prevotella nigrescens*, *Aggregatibacter actinomycetemcomitans*, *Fusobacterium nucleatum*, and *Enterococcus faecalis*. This composite may have potential in Class V restorations to restore root caries and combat periodontitis.

9.3.2 Nano-Modified GIC

Glass ionomer cement (GIC) was developed in the 1960s by Alan Wilson and his team of co-workers as a replacement for dental silicate cement. GIC has been a popular restorative material due to its aesthetic properties, self-adhesive capability, antibacterial properties, and good biocompatibility. However, low mechanical properties and sensitivity to moisture have been a major hurdle for the widespread clinical applicability of this restorative material [7].

To improve the properties of conventional GIC, nanosilver (nano-Ag) particles were synthesized and added to conventional GIC (cGIC) by Paiva et al. (2018). They reported that the higher concentration of nano-Ag particles (0.50% wt) improved handling characteristics of the modified cement and increased the compressive strength (CS) by 32% of nano-Ag-added GIC along with significant inhibition of microbial growth [8].

Hydroxyapatite (HA) has remarkable biological properties; it has a similar composition and crystal structure to the natural apatite found in human dental hard tissues. It has been reported that nano-HA crystals favor the remineralization of enamel. Nano-HA has also been linked to the enhanced mechanical properties exhibited by nano-HA-added GIC. It was suggested that this increase in mechanical properties of nano-HA-added GIC was due to the ionic interaction between the polyacrylic acid and the apatite crystals. It was suggested that decreasing the particle size of HA from micrometer to nanometer scale increases their surface area remarkably. This could lead to the infiltration of the nano-crystals into dentine as well as enamel which may enhance bonding of GIC to the tooth at the tooth–ionomer interface [7], as shown in Figure 9.1.

9.4 Nanomaterials in Minimal Invasive Dentistry for Management of Non-Pitted White Spot Lesions

White spot lesions on tooth surface enamel are early signs of demineralization. They can be related to various etiologies, for instance, incipient carious lesions, initial tooth trauma, developmental defect of enamel (dental fluorosis, molar inicial hypomineralization), etc. Appearance of these white spot lesions to naked eyes is an optical phenomenon of difference in the refractive index of the lesion area and the adjacent normal tooth enamel. Noncavitated lesions as well as caries extending up to the dentinoenamel junction can be arrested. In context of the surface topography, they are covered with hypermineralized surface enamel below which there are subsurface microporosities. Therefore, on further acid insult by a fall in the pH of the saliva, if these lesions are left untreated, they can lead to breakage of surface enamel layer resulting in the formation of cavitated lesions exposing the underlying dentin. Therefore, these lesions need to be addressed by suitable treatment therapy at an early stage. Various conventional options have been put forward in clinical practice to halt the progression of these lesions.

9.4.1 Nanomaterials for Enamel Remineralization

In an incipient stage, prior to cavitation, when there is loss of minerals from the dental hard tissues, demineralization takes place. However, the repair of a lesion can occur when the calcium and phosphate gradients are reversed and they diffuse inwards rather than outwards; this process is termed remineralization. Products containing calcium, phosphate, and fluoride in their bioavailable forms have claimed to enhance remineralization [8].

i) *Nano-hydroxyapatite* – With the recent advances in nanotechnology, the size of particles has decreased to usually 0.1 to 100 nm and some modifications in their shape, yielding highly bioactive calcium and phosphate compounds have higher potential for penetration into the porosities of the demineralized area with a potential of remineralization [9]. The physical and chemical properties of the nano-hydroxyapatite (NHAP) are similar to those of natural tooth hydroxyapatite. NHAP can stably release Ca^{2+} and $PO4^{3-}$ to promote remineralization. NHAP enhanced the shear bond strength to remineralized enamel in the etched enamel model. NHAP can significantly enhance tooth remineralization and increase microhardness in a dose-dependence manner. In vivo and in vitro, toothpaste containing NHAP has proven to be a valuable prevention measure against dental caries in primary dentition [10], as depicted in Figure 9.2.
ii) *Nano calcium fluoride* – Compared with traditional fluoride, NcaF materials can keep fluorine release at a better level for a long time. In a rat caries model, NcaF was observed to substantially decrease caries by scanning electron micrographs. Thus, NcaF materials are promising to inhibit enamel demineralization, white spot lesions, and caries [10].
iii) *Nano amorphous calcium phosphate* – Compared with amorphous calcium phosphate, NACP can release higher levels of Ca^{2+} and $PO4^{3-}$ with lower filler. Lee et al. found that NACP can release more ions at low pH, with the acid invasion neutralization, increasing the pH value from 4 to 6.5 to resist dental caries [11]. When NACP combined with antibacterial components is added to the adhesive, the new binder has dual functions of both antibacterial and remineralization [10].

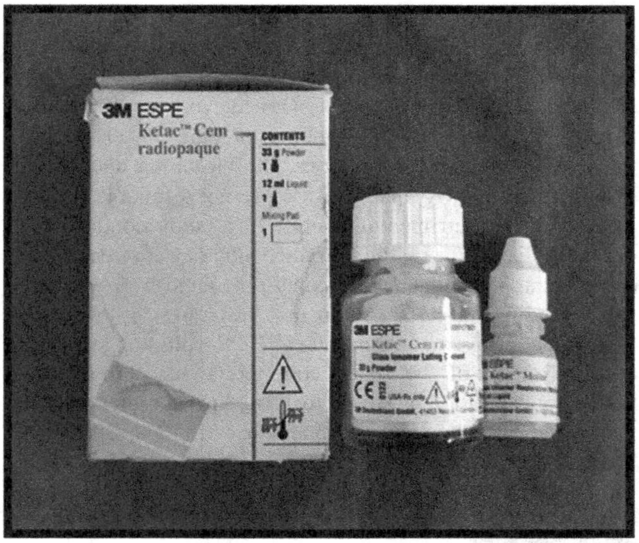

FIGURE 9.2 Nano-modified GIC (Ketac molar) [26].

9.4.2 Resin Infiltration Technique with Nano Enhancement

Resin infiltrants have been effectively applied in dentistry to manage non-cavitated carious lesions. Infiltrants are light-curable resins that occlude the porous lesions, sealing inside the lesion body and its surface. They have a higher flow because they lack inorganic filler [12].

In a study conducted by Cuppini et al. [12], a novel resin infiltrant with microcapsules loaded with an ionic liquid (MC-IL) was developed and analyzed the physical properties and cytotoxicity of the dental resin. ILs, mainly those based on the imidazolium cation, are organic salts with low melting points and 3D structural organization due to the imidazolium cation interaction with a weakly coordinating anion. The MC-IL showed a mean particle size of 1.64 ($\pm$ 0.08) μm, shriveled aspect, and a de-agglomeration profile suggestive of nanoparticles' presence in the synthesized powder. The antibacterial role of IL cation is via the interaction with the negatively charged species of bacterial membrane and wall, acting via direct contact with the microorganism [12].

The addition of MC-IL increased the contact angle and decreased the SFE of the resin infiltrant. This feature is interesting since the higher hydrophobicity and lower wettability of polymers may favor hydrolytic stability in the oral environment. MC-IL incorporation changed the resin infiltrants' surface properties without affecting the polymer's tensile strength and cytotoxicity [12].

9.4.3 Nano-Incorporated Tooth Bleaching Agents

Tooth whitening is accomplished by the physical removal of stains or chemical reactions to whiten the tooth structure. Commonly used dental bleaching agents include hydrogen peroxide, carbamide peroxide, etc. It has been observed that these agents increase the porosity of enamel and decrease its microhardness [13]. Bleaching agents can generate patients' discomfort and dental hard tissue damages, not achieving an efficient and long-lasting treatment with optimum whitening effect [14].

To overcome these limitations, the bleaching agents containing nano-hydroxyapatite can represent a reliable solution to avoid these detrimental effects. In a study conducted by Giulia Orilisi et al. [14], human third molars were treated with commercial bleaching agents, containing nano-hydroxyapatite (nHA) and 6% (at-home treatment), 12%, and 18% (in-office treatments) of hydrogen peroxide (HP). Results suggested that the application of the tested commercial bleaching agents, with a concentration of HP up to 12%, does not alter the morphological and chemical composition of the enamel surface and maintains its crystallinity [14].

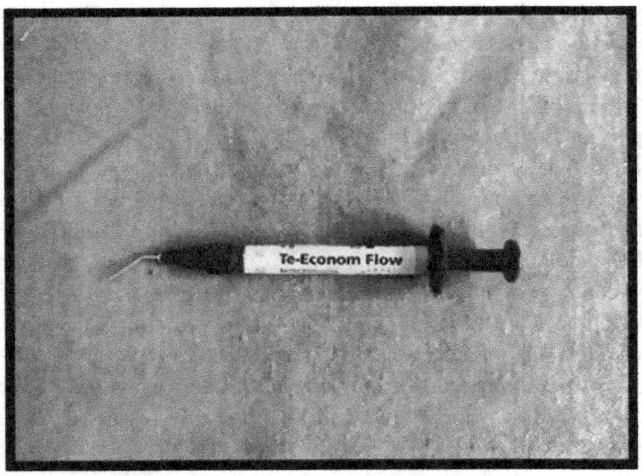

FIGURE 9.3 Flowable composite resin (Ivoclar Te-Econom Flow).

9.5 Nanomaterials for Esthetic Intervention

9.5.1 Pitted Enamel Defects

Pitted enamel defects in the anterior teeth region are a major esthetic concern for patients. Various treatment options are available, like composite resin veneers, full coverage restorations, etc. Composite veneers present considerable advantages, such as ultra-minimally invasive properties and excellent esthetic appearances. The success seems to depend on a combination of sound adhesive principles, fracture resistance, and adequate design of the restoration [15].

Raorane et al. [16] in a study presented a green and facile method for the synthesis of TiO_2 nanohybrid particles that can be successfully used as fillers in an experimental light-curing resin matrix for enhancing its dental properties [16]. The experimental light-curing nanocomposites with 5 wt% nanohybrid surface-modified filler particles with BisGMA (60 wt%), TEGDMA (20 wt%), and UDMA (20 wt%) resin composition provided increased physical strength and durability with higher compressive stress 195.56 MPa and flexural stress 83.30 MPa. Furthermore, the dental property, such as polymerization shrinkage (PS) obtained from the volumetric method was decreased up to 3.4% by the addition of nanohybrid fillers. In addition to this, the biocompatible and antimicrobial nature of TiO_2 and its aesthetic properties such as tooth-like color make TiO_2 favorable to use as fillers.

9.5.2 Fragment Reattachment

Coronal fractures of the anterior teeth are a common form of dental trauma that mainly affects the maxillary incisors because of their position in the arch. One of the options for managing coronal tooth fractures, especially when there is minimal or no violation of the biological width, and the fractured fragment is retained, is the reattachment of the dental fragment. Reattachment of a fragment to the fractured tooth can provide good and long-lasting esthetics because the tooth's original anatomic form, color, and surface texture are maintained. As to the materials used for bonding, different studies use different types of adhesive systems (multimode, total-etch, or self-etch) and different intermediate materials (conventional composite resin, flowable composite resin [Figure 9.3], resin cement, or glass ionomer cement). Zhao et al. in an in vitro study used flowable nanocomposite for reattachment of fractured crowns [17].

9.5.3 Esthetic Buildup of Fractured Anterior Teeth

The evolution of adhesive dentistry and patients' esthetic demands have made resin composite the most used restorative material in current dental practice. The layering of different shades allows the conventional resin composites to mimic the natural tooth.

Recently, a novel supra-nano filled resin composite, claiming to rely mainly on structural color technology for shade matching, is gaining attention from clinicians. In a study conducted by Chen et al., the supra-nano filled resin composite showed a better shade matching ability with all the shades of resin composite frames compared to clustered-nano filled resin composite and microhybrid filled resin composite[18].

9.6 Nano-Modified Caries Vaccine

Vaccines are immuno-biological substances designed to produce specific protection against a given disease. They stimulate the production of a protective antibody and other immune mechanisms. Vaccines are prepared from live, inactivated, or killed organisms, modified organisms extracted cellular fractions, toxoids, or a combination thereof. *Mutans streptococci* are the primary etiological agents of dental caries [19]. Protection has been attributed to salivary IgA antibodies which can inhibit sucrose-independent or sucrose-dependent mechanisms of streptococcal accumulation on tooth surfaces according to the choice of vaccine antigen [20].

DNA vaccines were found to be an effective, safe, stable, and inexpensive immunogenic strategy in inducing both humoral and cellular immune responses. DNA vaccines, however, have poor immunogenicity in large animals and human beings. Nanotechnology has been employed to tailor delivery vehicles (e.g., anionic liposomes in chitosan/DNA nanoparticle complex were used as a delivery vehicle) to enhance the immunogenicity of anti-caries DNA vaccine [21].

Due to the lack of effective targeting, the effectiveness of the mucosal administration of "naked" DNA is limited. In order to improve the function of the "naked" DNA vaccine, a plasmid was loaded into chitosan nanoparticles. The chitosan-DNA nanoparticles are suitable for mucosal anti-caries DNA vaccination, which could induce the oral specific immune responses with biocompatible and non-cytotoxic. In further research, human clinical trials are needed to implement regular anti-caries strategies [10].

9.7 Nano-Enhanced Orthodontic Materials

9.7.1 Nano-Coated Orthodontic Archwires

Orthodontic treatment duration can be shortened by reducing the friction between wire and bracket, thus increasing the desired tooth movement. In recent years, nanoparticles have been used as a component of dry lubricants. These solid-phase materials are capable of reducing the friction between two sliding surfaces without the need for a liquid medium. One of the many examples are inorganic fullerene-like tungsten sulfide nanoparticles (IF-WS2) that are used as self-lubricating coatings for orthodontic stainless steel wires and nano-coated nitinol [22].

9.7.2 Silver Nanoparticle-Coated Orthodontic Appliances

Silver nanoparticles (AgNPs) have been shown to be materials with excellent antimicrobial properties in a wide variety of microorganisms. In the orthodontic field, studies have incorporated AgNPs (17 nm) into orthodontic elastomeric modules, orthodontic brackets and wires, and others against a wide variety of bacterial species, concluding that these orthodontic appliances with AgNPs could potentially combat the dental biofilm decreasing the incidence of dental enamel demineralization during and after the orthodontic treatments [23, 24].

9.8 Dental Nanorobots

Dental nanorobots use specific motility mechanisms to penetrate human tissue with navigational precision, acquire energy, and sense and manipulate their surroundings in real time. An onboard nanocomputer that executes preprogrammed instructions in response to local sensor stimuli could be utilized to control the nanorobot functions. Also, the dentist could issue strategic orders directly to the nanorobots in vivo via acoustic signals [25].

9.8.1 Nano Anesthesia

When nanotechnology or nanorobots are used to induce anesthesia, the gingiva of the patient is instilled with a colloidal suspension containing millions of active, analgesic, micron-sized dental robots that respond to input supplied by the dentist. Nanorobots in contact with the surface of the crown or mucosa can reach the pulp via the gingival sulcus, lamina propria, or dentinal tubules. Once in the pulp, they shut down all sensations by establishing control over nerve-impulse traffic in any tooth that requires treatment. After completion of treatment, they restore this sensation, thereby providing the patient

TABLE 9.1

Summarization of Conventional versus Nano Modification of Dental Materials Being Used in Daily Practices

Material	Conventional form	Nano modification
Caries arresting agent	SDF	Nanosilver fluoride
Restorative materials	Composite	Bioactive nanocomposite
	GIC	Nano-Ag-added GIC Nano-HA-added GIC Nano Ketac [27]
Enamel remineralizing agents	Calcium Phosphate Fluoride Hydroxyapatite	Nano-hydroxyapatite [28] Nano calcium fluoride Nano amorphous calcium phosphate
Resin infiltration	Resin infiltrants	Resin infiltrant with microcapsules loaded with an ionic liquid (MC-IL)
Bleaching agents	Hydrogen peroxide Carbamide peroxide	Nano-hydroxyapatite (nHA) added bleaching agents
Pitted enamel defects	Conventional composite veneer	TiO_2 nanohybrid surface-modified filler particles containing resins
Tooth fragment reattachment	Conventional composite resin, flowable composite resin, resin cement, or glass ionomer cement	Flowable nano-modified composite resin
Esthetic crown buildup	Conventional composite resin	Supra-nano filled resin composite
Caries vaccine	DNA vaccine	Chitosan-DNA nanoparticles vaccine
Orthodontic materials	Stainless steel archwires NiTi archwires	Tungsten sulfide nanoparticles coated archwires Nano-coated nitinol [29] Silver nanoparticle coated orthodontic appliances
Dental anesthesia	Local anesthetic injection	Nanorobots
Dentrifices		Dentifrobots

with anxiety-free and needleless comfort. The anesthesia is fast-acting and reversible, with no side effects or complications associated with its use [25].

9.8.2 Nanorobotic Dentrifices (Dentifrobots)

Nanorobotic dentifrices, when delivered either by mouthwash or toothpaste, can cover all subgingival surfaces, thereby metabolizing trapped organic matter into harmless and odorless vapors. Properly configured dentifrobots can identify and destroy pathogenic bacteria that exist in the plaque and elsewhere. These invisibly small dentifrobots are purely mechanical devices that safely deactivate themselves when swallowed [25].

9.9 Conclusion

There is an increasing need for awareness about the futuristic and advanced forms of treatment provided today. The future utilization of the advantages of nanotechnology will facilitate improvements in oral health. Advanced restorative materials, new diagnostic and therapeutic techniques, and pharmacologic approaches will improve dental care. The conclusion of all dental materials mentioned in this chapter showing conventional versus nano modification of dental materials being used in daily practices is presented in Table 9.1.

REFERENCES

1. Wong, J., Zou, T., Lee, A. H. C., & Zhang, C. (2021). The potential translational applications of nanoparticles in endodontics. *International Journal of Nanomedicine*, *16*, 2087–2106.
2. Ginjupalli, K., Shaw, T., Tellapragada, C., Alla, R., Gupta, L., & Perampalli, N. U. (2018). Does the size matter? Evaluation of effect of incorporation of silver nanoparticles of varying particle size on the antimicrobial activity and properties of irreversible hydrocolloid impression material. *Dental Materials*, *34*(7), e158–e165.
3. Targino, A. G. R., Flores, M. A. P., dos Santos Junior, V. E., Bezerra, F. D. G. B., de Luna Freire, H., Galembeck, A., & Rosenblatt, A. (2014). An innovative approach to treating dental decay in children. A new anti-caries agent. *Journal of Materials Science: Materials in Medicine*, *25*(8), 2041–2047.
4. Tirupathi, S., Nirmala, S. V. S. G., Rajasekhar, S., & Nuvvula, S. (2019). Comparative cariostatic efficacy of a novel nano-silver fluoride varnish with 38% silver diamine fluoride varnish a double-blind randomized clinical trial. *Journal of Clinical and Experimental Dentistry*, *11*(2), e105–112.
5. Wang, L., Melo, M. A., Weir, M. D., Xie, X., Reynolds, M. A., & Xu, H. H. (2016). Novel bioactive nanocomposite for class-V restorations to inhibit periodontitis-related pathogens. *Dental Materials*, *32*(12), e351–e361.
6. Chokkareddy, R., Bhajanthri, N. K., Kabane, B., & Redhi, G. G. (2018). Bio-sensing performance of magnetite nanocomposite for biomedical applications. *Nanomaterials: Biomedical, Environmental, and Engineering Applications*, *165*, 1-60
7. Moheet, I. A., Luddin, N., Ab Rahman, I., Kannan, T. P., Abd Ghani, N. R. N., & Masudi, S. M. (2019). Modifications of glass ionomer cement powder by addition of recently fabricated nano-fillers and their effect on the properties: A review. *European Journal of Dentistry*, *13*(03), 470–477.
8. Paiva, L., Fidalgo, T. K. S., da Costa, L. P., Maia, L. C., Balan, L., Anselme, K., ... & Thiré, R. M. S. M. (2018). Antibacterial properties and compressive strength of new one-step preparation silver nanoparticles in glass ionomer cements (nanoAg-GIC). *Journal of Dentistry*, *69*, 102–109.
9. Thimmaiah, C., Shetty, P., Shetty, S. B., Natarajan, S., & Thomas, N. A. (2019). Comparative analysis of the remineralization potential of CPP–ACP with fluoride, tri-calcium phosphate and nano hydroxyapatite using SEM/EDX–An in vitro study. *Journal of Clinical and Experimental Dentistry*, *11*(12), e1120–1126.
10. Chen, H., Gu, L., Liao, B., Zhou, X., Cheng, L., & Ren, B. (2020). Advances of anti-caries nanomaterials. *Molecules*, *25*(21), 5047.
11. Lee, J. H., Seo, S. J., & Kim, H. W. (2016). Bioactive glass-based nanocomposites for personalized dental tissue regeneration. *Dental Materials Journal*, *35*(5), 710–720.
12. Cuppini, M., Garcia, I. M., de Souza, V. S., Zatta, K. C., Visioli, F., Leitune, V. C. B., ... & Collares, F. M. (2021). Ionic liquid-loaded microcapsules doped into dental resin infiltrants. *Bioactive Materials*, *6*(9), 2667–2675.
13. Vargas-Koudriavtsev, T., Fonseca-Jiménez, P., Barrantes-Delgado, P., Ruiz-Delgado, B., Conejo-Barboza, G., & Herrera-Sancho, Ó. A. (2021). Effects of bleaching gels on dental enamel crystallography. *Oral Health & Preventive Dentistry*, *19*(1), 7–14.
14. Orilisi, G., Tosco, V., Monterubbianesi, R., Notarstefano, V., Özcan, M., Putignano, A., & Orsini, G. (2021). ATR-FTIR, EDS and SEM evaluations of enamel structure after treatment with hydrogen peroxide bleaching agents loaded with nano-hydroxyapatite particles. *PeerJ*, *9*, e10606.
15. Duque, F. L., & Ardila, C. M. (2011). Oral myiasis caused by the screwworm *Cochliomyia hominivorax* treated with subcutaneous ivermectin and creolin: Report of six cases after trauma. *Dental Traumatology*, *27*(5), 404–407.
16. Raorane, D. V., Chaughule, R. S., Pednekar, S. R., & Lokur, A. (2019). Experimental synthesis of size-controlled TiO2 nanofillers and their possible use as composites in restorative dentistry. *The Saudi Dental Journal*, *31*(2), 194–203.
17. Xiao, H. U. A. N. G., Hui, C. H. E. N., & Jun, W. A. N. G. (2017). Effect of flowablenano-composite on the bonding strength of tooth reattachment on fractured crowns. *Shanghai Journal of Stomatology*, *26*(4), 404–408.
18. Chen, F., Toida, Y., Islam, R., Alam, A., Chowdhury, A. F. M. A., Yamauti, M., & Sano, H. (2021). Evaluation of shade matching of a novel supra-nano filled esthetic resin composite employing structural color using simplified simulated clinical cavities. *Journal of Esthetic and Restorative Dentistry*, *33*(6), 874–883.
19. Nguyen, S. V., Icatlo Jr, F. C., Nakano, T., Isogai, E., Hirose, K., Mizugai, H., ... & Chiba, I. (2011). Anti-cell-associated glucosyltransferase immunoglobulin Y suppression of salivary mutans streptococci in healthy young adults. *The Journal of the American Dental Association*, *142*(8), 943–949.

20. Sujith, R., Naik, S., & Rajanikanth, P. (2014). Caries vaccine–A review. *Indian Journal of Mednodent and Allied Sciences*, *2*(2), 198–203.
21. Abou Neel, E. A., Bozec, L., Perez, R. A., Kim, H. W., & Knowles, J. C. (2015). Nanotechnology in dentistry: Prevention, diagnosis, and therapy. *International Journal of Nanomedicine*, *10*, 6371.
22. Redlich, M., Katz, A., Rapoport, L., Wagner, H. D., Feldman, Y., & Tenne, R. (2008). Improved orthodontic stainless steel wires coated with inorganic fullerene-like nanoparticles of WS2 impregnated in electroless nickel–phosphorous film. *Dental Materials*, *24*(12), 1640–1646.
23. Hernández-Gómora, A. E., Lara-Carrillo, E., Robles-Navarro, J. B., Scougall-Vilchis, R. J., Hernández-López, S., Medina-Solís, C. E., & Morales-Luckie, R. A. (2017). Biosynthesis of silver nanoparticles on orthodontic elastomeric modules: Evaluation of mechanical and antibacterial properties. *Molecules*, *22*(9), 1407.
24. Mhaske, A. R., Shetty, P. C., Bhat, N. S., Ramachandra, C. S., Laxmikanth, S. M., Nagarahalli, K., & Tekale, P. D. (2015). Antiadherent and antibacterial properties of stainless steel and NiTi orthodontic wires coated with silver against *Lactobacillus acidophilus*—an in vitro study. *Progress in Orthodontics*, *16*(1), 1–6.
25. Shetty, N. J., Swati, P., & David, K. (2013). Nanorobots: Future in dentistry. *The Saudi Dental Journal*, *25*(2), 49–52.
26. Ketac™ Nano Light-Curing Glass Ionomer Restorative [Internet]. (2021). [Cited 8 October 2021]. Available from: https://www.3m.com/3M/en_US/p/d/espe_ketac_nano/
27. Singh, A., Shetty, B., Mahesh, C. M., Reddy, V. P., Chandrashekar, B. S., & Mahendra, S. (2017). Evaluation of efficiency of two nanohydroxyapatite remineralizing agents with a hydroxyapatite and a conventional dentifrice: A comparative In vitro study. *Journal of Indian Orthodontic Society*, *51*(2), 92–102.
28. Estelite Universal Flow | Tokuyama Dental | Estelite Composite [Internet]. Tokuyama Dental America Inc. (2021). [Cited 8 October 2021]. Available from: https://www.tokuyama-us.com/estelite-universal-flow-dental-composite/
29. Nano Coated Nitinol | Round | Right Form [Internet]. Ortho Arch. (2021). [Cited 8 October 2021]. Available from: https://orthoarchshop.com/store/round-se-nitinol-right-form-nano-coated-details.html

10

Graphene-Based Electrochemicals and Biosensors for Multifaceted Applications in Healthcare

G. Manasa, Nagaraj Shetti, Ronald J. Mascarenhas and Kakarla Raghava Reddy

CONTENTS

10.1 Introduction to Electrochemical Sensors and Their Significance in Healthcare

Electrochemistry has a crucial role in many fields of science and technology that are intimately related to the sustainable future of humanity [1]. Biological and chemical processes are crucial for their numerous functional applications. Thereupon, subtle detection of such processes is significant in diagnostics [2]. Of the available methodologies, electrochemical (EC) sensing is promising as it converts a chemical or a biological response into an informative electrical signal. Sensors have possessed a potential role since their discovery, which is creditable to some of their key advantages such as the tailored architecture of electrode surfaces, ease of miniaturization, simple assay procedure and measurement, rapid response time, high sensitivity, and selectivity. The utilization of electrochemistry provides sensitivity, accuracy, speed, low cost, versatility, practicality, and portable sensor systems. According to the International Union of Pure and Applied Chemistry (IUPAC), a chemical sensor is defined as "a device that transforms chemical information, ranging from the concentration of a specific sample component to total composition analysis, into an analytically useful signal" [3]. An EC sensor belongs to the category of chemical sensors, designed by coupling the recognition or receptor part of the device to an EC transducer to convert the EC reaction into a measurable electrical signal.

Since specific biomarker concentrations can be used to define disease type or the body's response to treatment, EC sensors provide limitless opportunities for monitoring biomarkers of medical interest. For example, (i) the development of highly sensitive sensors is imperative in our daily lives; an example for that is the outbreak of coronavirus-19 (COVID-19) disease globally which has been designated as a global health emergency by the World Health Organization (WHO) [4]; (ii) EC urine analysis is sufficient to diagnose a patient with high sodium concentrations [5], prostate cancer [6], cholesterol, and/or diabetes [7]. In general, determining more than one biomarker is required for an accurate and precise diagnosis, increasing the number of clinical tests. As a result of these challenges, EC sensor-based strategies have gained traction because they enable reliable simultaneous analysis of multiple biomarkers relevant in medical diagnostics. Wearable multi-sensing electrodes are among those for simultaneous monitoring of blood pressure, heart rate, pulse rate, and temperature of the body, which could reflect the condition of a patient for the diagnosis of various diseases [8].

EC sensors based on graphene nanocomposites mainly include: (i) EC sensors based on electrocatalysis of graphene materials for investigating organic molecules such as exogenous antioxidants/polyphenols [9, 10] and drugs [11–15]; (ii) enzyme biosensors; (iii) DNA and protein sensors; (iv) EC immunosensors [4]. A single chapter cannot present the diverse work being done on EC detection utilizing graphene. Therefore, we have tried to account for part of it to highlight the significance of graphene-based materials in developing EC sensors. The chapter thus leverages the attributes of EC techniques toward the development of graphene-based smart nanoelectrodes for healthcare.

This chapter specifically focuses on recent research advancements in the last two years (from May 2019 to May 2021) in voltammetric sensing of important biomolecule dopamine (DOPA), as well as virus diagnosis using graphene-based EC and biosensors. In this context, the working principle of graphene-based EC electrodes, as well as the application of various graphene-based materials for EC detection, will be introduced and discussed in detail.

DOI: 10.1201/9781003157823-10

10.2 Graphene: An Efficient Electrode Modifier for EC Sensing

EC sensor research is a growing field that is gaining traction in both nanoscience and nanotechnology. As smart analytical devices capable of detecting and quantifying a variety of important analytes, EC sensors have been developed for medical applications such as amino acids [16–18] and neurotransmitters [19, 20]. On the other hand, a pharmaceutical sensor is an essential branch of analytical chemistry [21]. It is crucial in the quality control of drugs and has major implications on public health [22–24]. Advances in nanomaterial engineering and synthesis are driving the blooming development of EC sensors [25–28]. This has resulted in nanostructured materials with distinct properties that can be used in the design of EC sensors to meet specific requirements for a specific application.

The working electrode is the most important component of an EC sensor, and the most common working electrodes from various carbon-based electrodes are carbon paste electrode (CPE), glassy carbon electrode (GCE), screen-printed electrode (SPE), pencil graphite electrode (PGE), and solid metallic electrodes such as a gold electrode. Oxidation and reduction of electroactive species is the principle behind EC detection. This process often causes the fouling of the working electrode surface, which is a major concern, as the reaction products block the active surface area of the electrode and delay significant electrode processes. In order to circumvent the fouling phenomena, electrode modification has been widely used [29, 30].

Electrode surface modification is a process wherein a bare working electrode is covered with a chemical (chemical sensor) or biological (biosensor) component termed as a modifier. The two important characteristics of a modifier are [29, 30]: first, it should possess a large specific surface area, and second, it should enhance the electron transfer rates, and augment the currents, thereby decreasing the overpotential. A carbon-based material is considered an ideal electrode material for such type of application due to its chemical inertness over a wide anodic potential window, low residual current, easy availability, and reduced cost. When reviewing the essential features of an electrode material for widespread application in electrochemistry, graphene (and its derivatives) has gained considerable attention by researchers due to its exotic properties, which can be read in detail elsewhere [7, 31–33]. Graphene renders a high density of electroactive sites over a large area that can accommodate active species and facilitate favorable electron transfer, making it an idealistic transduction element and supporting substrate for EC sensing [9]. Graphene-oriented sensors contribute to high sensitivity that emerges due to two primary reasons [9, 11–15, 34]: (i) the two-dimensional structure of graphene provides total exposure of all its carbon atoms to the target analyte, thus providing a potential increase in sensitivity; (ii) it's inherently a material with low noise attributed to the quality of its crystal lattice. Moreover, graphene is a good conductor of electrical charge that favors heterogeneous electron transfer, which refers to the transfer of electrons between graphene and the target molecule in solution – a requisite for oxidation or reduction. This heterogeneous electron transfer occurs at the plane defects or the graphene edges rather than the basal plane. These advantages have motivated researchers to develop graphene-based EC sensors and biosensors, but graphene alone cannot meet all the requirements of EC determination. This is due to certain limitations, such as the difficulty in dispersion, curling, or graphene lamellae and agglomeration stacking. Therefore, by modifying the surface chemistry of graphene through functionalization or defect engineering, one can tailor the specific response to a target analyte with extremely high sensitivity [35].

Researchers have discovered three potential forms of graphene, namely, graphene oxide (GO), reduced graphene oxide (rGO), and functionalized graphene. Depending on the functional groups present or the extent of reduction, each of these forms has different EC characteristics. The ability of graphene to integrate with other nanomaterials to form composites with desired EC properties is key to its successful application [36]. Thus, graphene is generally functionalized or combined with some organic or inorganic functional materials to form a nanocomposite. Functionalized graphene-nanomaterials not only have a synergistic effect between conductivity, catalytic activity, and biocompatibility to accelerate signal transduction, but they also amplify biorecognition events. On the other hand, graphene nanocomposites have a great performance in immobilizing biomolecules and maintaining their bioactivities [37, 38]. The advantages of implementing graphene-based EC sensor substrate have been widely reported for sensitive detection of diverse analytes such as several biomolecules, drug analysis, and miscellaneous inorganic, organic compounds in favor of human healthcare [8, 11–15, 26, 39–42].

10.3 Functionalized Graphene as an EC Sensor

As mentioned earlier, to take full advantage of graphene's superior properties, it is necessary to functionalize pristine graphene to form multifunctional hybrid material. Many methods by which graphene, GO, rGO, and functionalized graphene are prepared can be referred from several reports in the literature [43–46]. The two major methods of functionalized graphene nanocomposite are: (a) covalent functionalization – it is based on the grafting of molecules onto the sp^2 carbon atoms of the pi-conjugated skeleton; (b) noncovalent functionalization – it is obtained by adsorption of polycyclic aromatic compounds or surfactants through pi-pi and hydrophobic interactions on the carbon framework. The inferred advantages of graphene functionalization were [44–46]: (i) possibility to effectively tune its electrical properties and enhance its electrocatalytic performances; (ii) the EC active sites produced by functional groups, triggers the anchoring of other functional moieties, i.e., adsorption or activation of analytes; (iii) accelerate the charge transfer between electrode and analyte. As a consequence of these merits, enhanced EC sensing performance can be achieved.

10.4 Classification of Electrochemical and Biosensors Based on Transduction

EC sensor is an analytical device comprising a recognition component, EC transducer, amplifier, signal processor, and display unit [47, 48]. The recognition component, in the case of an EC sensor, is graphene, and its composites. While in an EC biosensor, it is any biomolecule-incorporated graphene. The transducer serves as an interface between the recognition element and the amplifier, acting as a detector. The EC transducer's primary function is to convert the signal generated by the interaction between the recognition element and the analyte into a quantifiable physical output. Eventually, the amplifier amplifies the transducer signal via a signal processor which is then displayed in a user-friendly display unit. The transduction material choice depends on the chemical approach to develop the sensing layer on the transducer surface.

Depending on the EC property to be determined by the detector unit, the EC sensor or biosensor is categorized as potentiometric, conductometric, voltammetric, amperometric, and electrochemical impedance spectroscopy (EIS) sensors. The description of the EC techniques is given in Table 10.1, as this knowledge is fundamental for developing and optimizing EC sensors [21, 49–51]. An EC sensor's performance is usually experimentally evaluated based on its sensitivity, selectivity, limit of detection (LOD), dynamic range, response time, reproducibility, and operational and storage stability. Figure 10.1 summarizes the key parameters for a typical EC sensor.

It is important to note the factors that affect the sensitivity of EC sensors, which are [51]: (i) choice of the recognition element – nanomaterials or biomolecules; (ii) surface modification technique – physical or chemical modification of the electrode substrate; (iii) EC transduction mechanism – the surface architectures that link the sensing component with the biological samples. The three-electrode system is represented in Figure 10.2.

TABLE 10.1

Classification of EC and Biosensors Based on Transduction

Electrochemical method	Quantity controlled	Quantity measured
Two-electrode system: This is the simplest approach to studying current or voltage characteristics.		
Potentiometry: Potentiometry involves measuring the potential (E) of a cell at equilibrium to detect the concentration of an analyte [49] directly. Basic types of potentiometric sensors are ion-selective electrodes (ISE) and field-effect transistors (FETs).	I = 0	E
Conductometry: Conductometry measures the change in electrical conductance of the analyte (as a result of a chemical reaction), which can be measured between a pair of electrodes [78].	–	Conductivity (K)
Three-electrode system: A three-electrode-system arrangement is the most common EC cell setup used in electrochemistry and generally consists of a working electrode, counter/auxiliary electrode, and reference electrode, as represented in Figure 10.2. This three-electrode system has a distinct experimental advantage over the two-electrode system because it measures only one-half of the cell, i.e., the working electrode's potential is measured independent of changes at the counter electrode.		
Voltammetry: The collective term "voltammetry" encompasses all methods based on the evaluation of current-potential curves. A cyclic voltammetric technique records the electrical current (I) as the electrode potential (E) varies cyclically with time (t) between two potential limits at a constant scan rate. The peak current's height is proportional to the concentration of the electroactive species. The advantages of this technique are: (i) easy to construct and simple to operate, (ii) rapid and robust, (iii) measures the low concentration of analytes in minute sample volumes, (iv) highly sensitive and simultaneous determination of multiple analytes is possible, (v) economical and portable device. Over the last decade, cyclic voltammetry (CV), linear sweep voltammetry (LSV) [17], square wave voltammetry (SWV), differential pulse voltammetry (DPV), and so forth have been acknowledged as well-known methods to determine trace analyte concentrations.	E = f(t)	I = f(t)
Amperometry: Although amperometry is also a voltammetric measurement, there is a difference in operation between them. In voltammetry, the current is measured over a controlled potential variation; on the contrary, in amperometry, the current is measured at a constant voltage applied to the cell. Amperometric sensors quantify the current produced by the oxidation or reduction of electroactive species in solution when the experiment is performed by applying a steady potential to the working electrode with respect to the reference electrode [24]. The current is linearly proportional to the concentration of the electroactive species.	E	I
Impedance: EIS is employed to investigate both bulk and interfacial electrical properties of the electrode systems, which can be used to determine quantitative parameters of EC processes. For example, if a chemical reaction (oxidation/reduction) or biorecognition event occurs at the modified surface, the interfacial properties change [21, 22]. Therefore, EIS furnishes the fingerprint of the interfacial region.	E = constant + f(t)	Electrical impedance (Z)

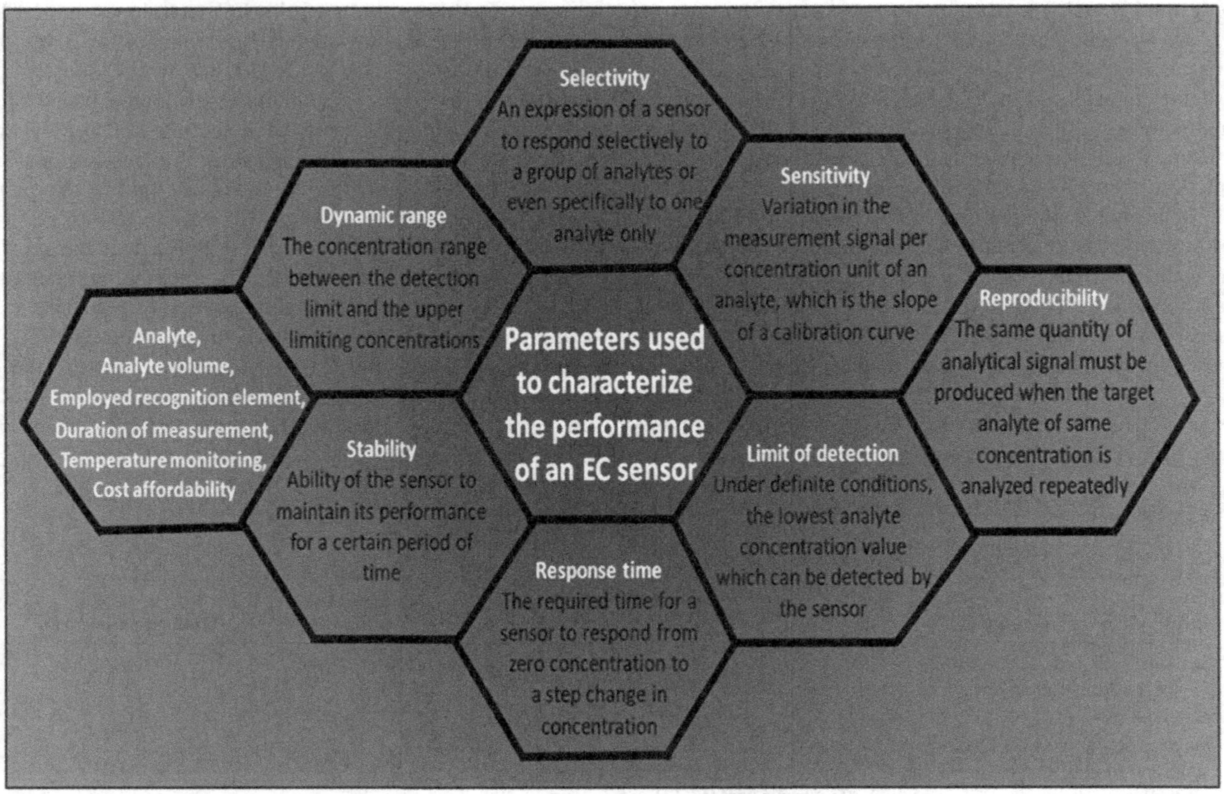

FIGURE 10.1 Static and dynamic parameters used to characterize the performance of an EC sensor.

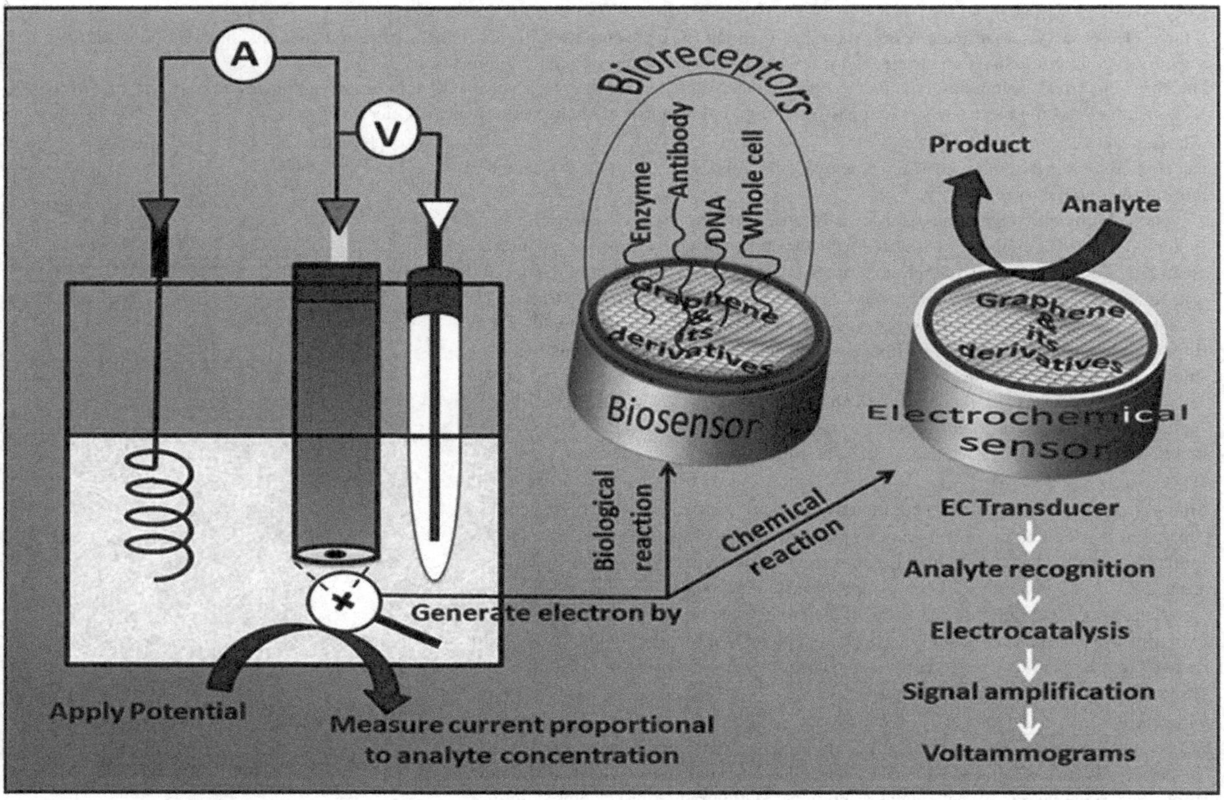

FIGURE 10.2 Scheme of three-electrode system and the analytical principle of EC sensors.

10.5 Graphene-Based EC Sensors for Dopamine

Parkinson's disease, Huntington's, and schizophrenia are characterized by the abnormalities of an important neurotransmitter, DOPA. The measurement of this catecholamine is considered as a biomarker to evaluate the dopaminergic nigrostriatal system of the brain. However, due to its very low physiological concentrations, DOPA sensing is a challenging task. In addition to that, uric acid (UA) and ascorbic acid (AA) are present in high concentrations that possibly interfere with DOPA detection. Because at similar potentials, they can all get oxidized and their signals get overlapped [52]. So we shall review the literature for the most recently developed DOPA sensors based on graphene and its derivatives.

Ahammad et al. [53] have best described the general mechanism of DOPA, AA, and UA simultaneous determination when graphene is used as a modifier. The NH_2 moiety on the DOPA becomes positively charged above pH 5.0; hence when GO, or rGO (negatively charged) are used as transducer elements, strong electrostatic interaction between the opposite charges, makes the pi stacking interactions stronger. DOPA bonds with graphene via pi-pi, pi-CH, and pi-N^+H_3 interactions, observed due to the HOMO-LUMO interactions of DOPA and graphene.

Additionally, DOPA forms hydrogen bonds with graphene. While on the other hand, AA has one five-membered ring, and UA has a conjugated pi-bonding five- and six-membered rings. These two molecules generally exist as anions above pH 7.0 due to their pKa values of ~4.0 and ~5.0. As a consequence, the pi-ring system in both molecules is hindered. Thus the pi-interaction in UA is weaker than in DOPA, while for AA, it is almost nonexistent. Due to their anionic forms, the two molecules also experience electrostatic repulsion when GO and rGO are used. Figure 10.3 displays the scheme of electrode-analyte interaction, while the same group discusses a detailed explanation of this mechanism in their previous publication [54].

An EC sensor using graphene quantum dots/multi-walled carbon nanotubes (GQDs-MWCNTs) composite was developed by Huang et al. for DOPA determination at physiological pH [55]. The modification of MWCNTs with graphene quantum dots enhanced the surface area of the sensor, and the good conductive properties of MWCNTs improved. Under optimal conditions, the differential pulse voltammetry (DPV) results exhibited a broad linear range of 0.005 to 100.0 μM with the detection limit of 0.87 μM (S/N = 3). The sensor also depicted excellent selectivity for DOPA among other analytes such as AA and UA. The selectivity was achieved because the modified GQDs-MWCNTs electrode had a cationic exchanger electrode surface that could selectively attract cationic DOPA, as well as many anionic groups that could attract cations. This electron transfer of DOPA, which is facilitated by the pi-pi superposition interaction between GQDs-MWCNTs and

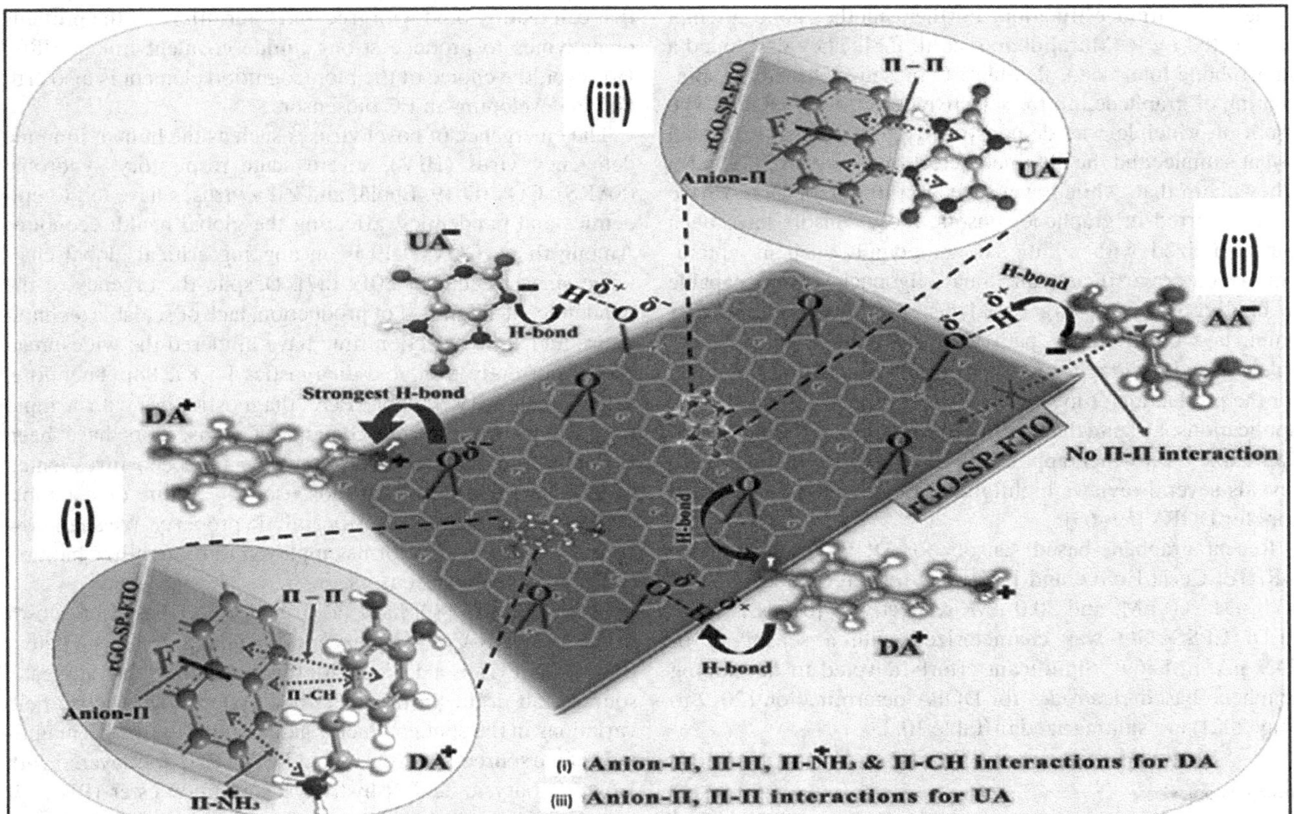

FIGURE 10.3 Schematic representation of the possible interaction of functionalized graphene with analytes: (i) DOPA, (ii) AA, and (iii) UA interaction [54].

DOPA, may prevent UA and AA from reaching the electrode surface. The practical applicability was tested in human serum samples, and it was the first example used to measure DOPA secreted from live cells with high accuracy. Another interesting contribution of this sensor was that it displayed good cell biocompatibility via the cell viability test.

Asif et al. created a self-assembled periodic superlattice material by combining positively charged semiconductive Zn-NiAl layered double hydroxide (LDH) sheets and negatively charged rGO layers [56]. In the composite material, both the exfoliated counterpart constituents were hybridized on a molecular scale in an alternate sequence. This hybridization endowed the superlattice with direct neighboring conductive rGO to semiconductive LDH channels, significantly improving electron transfer capability for EC reaction. Zn-NiAl LDH/rGO sensor also exhibited excellent synergistic effects, huge active surface sites, and electrocatalytic properties. This sensor demonstrated exceptional performance in the specific and sensitive detection of DOPA, UA, and AA at the same time. A low detection limit of 0.1 nM (S/N = 3) was obtained for DOPA using DPV. The sensor's practicability was assessed for real-time monitoring of extracellular DOPA efflux excreted by live cells and in the presence of an endogenous amount of UA from human urine samples.

Wei and colleagues [57] employed a two-step EC reduction to fabricate palladium nanoparticles/rGO for simultaneous determination of DOPA, AA, and UA. But this sensor was less superior in comparison to the previous work with respect to detection limit estimation. Further, another good attempt is reported by R. Muralidharan et al. [58]. They developed a smartphone-interfaced, flexible EC biosensor based on spin-coating of graphene ink for selective detection of DOPA. The proof-of-principle was demonstrated by testing an artificial sweat sample, and the obtained detection limit was 100.0 nM. They claim that, while lower detection limits for DOPA have been reported in graphene sensors, those sensors have been functionalized with additional compounds such as aptamers, which are artificial functional oligonucleic acids capable of binding ranges of target analytes. This work, on the other hand, lays out a simple path for developing DOPA sensors solely based on graphene with minimal processing, allowing for the production of low-cost sensors for point-of-care (POC) applications. Despite that, this sensor is yet to be validated and upgraded for further application. Meanwhile, the literature reveals several reviews highlighting the graphene-based sensors for DOPA [59–65].

Recent graphene-based sensors for DOPA such as MIP/GR/GE, Ceria-ErGO, and FTO/rGO rendered good LOD of 33.0 nM, 3.0 nM, and 70.0 nM, respectively [66–68], while PEDOT:PSS/rGO was characterized with a sensitivity of 24.9 μA/mM [69]. Significant efforts devoted to developing graphene-based electrodes for DOPA determination (2019 to May 2021) are summarized in Table 10.2.

10.6 Graphene-Based EC Biosensor

It is also known as a chemically modified electrode, and according to IUPAC [85], "an electrochemical biosensor is a self-contained integrated device capable of providing specific quantitative or semi-quantitative analytical information using a biological recognition element (bioreceptor) retained in direct spatial contact with an electrochemical transduction element". Graphene, a two-dimensional hexagonal pattern of carbon atoms, can be used as a supporting matrix to keep biomolecules' orientation and biological activity after immobilization [38]. The key process for the high performance of a biosensor is appropriate immobilization of the bioreceptor employing proper functional material for the electrode.

Graphene-based materials are considered to cater to an adequate microenvironment for biomolecule immobilization, promoting electron transfer between the immobilized biomolecules and the electrode substrate [38, 86]. When used alone, graphene has a low hydrophobicity and throughput; this limits its usage in biosensor applications. While GO and rGO solved the problem by increasing the hydrophilicity of the graphene layer and eliminating the oxygen groups of GO, they achieved significant electrical conductivity and ease of surface modification for biomolecule immobilization. Graphene has been considered an ideal support source for label-free biosensors due to its excellent electronic and mechanical properties. In addition, chemically modified graphene composed of functional groups and defects or vacancies act as a highly desirable solid support for the immobilization of enzymes and inorganic nanoparticles with enhanced stability and loading efficiency [38]. Novel graphene-based EC biosensors have been developed because possessing reactive functional groups that can readily bind with free $-NH_2$ terminals of the proteins or enzymes to produce strong amide covalent linkage [86]. However, the choice of the biorecognition element is also crucial in developing an EC biosensor.

The emergence of novel viruses such as the human immune deficiency virus (HIV), severe acute respiratory syndrome (SARS), COVID-19, Ebola, and Zika viruses have led to epidemics and pandemics, affecting the global health economy. Among them, COVID-19 is an ongoing critical global challenge since December 2019 [87]. Despite the urgency of the pandemic, the high cost of production, lack of scalable technologies, and slow detection time have hindered the widespread use of currently available diagnostics [4, 87, 88]. Therefore, a low-complexity, reliable POC diagnostic test with a rapid turn-around time is needed, and EC biosensors have been considered to fill this void. Graphene has been investigated to develop highly efficient biosensors by virtue of its stable EC behavior and good electrocatalytic property. We shall now consider graphene platforms employed to immobilize biomolecules to develop biosensors.

Seo and group [89] have fabricated a FET-based biosensor to detect SARS-CoV-2 by exploiting spike protein as a detection probe. FET is a device with three terminals, namely gate, source, and drain. In this electrode system, the electric field variations in the semiconductor gate region control the magnitude of the source-drain current [27]. The FET was layered with 1-pyrene butyric acid N-hydroxysuccinimide ester (PBASE) functionalized graphene sheets on which SARS-CoV-2 anti-spike protein antibodies were immobilized. PBASE acted as a probe linker between graphene-coated FET and antibodies. The performance of the FET was tested using the antigen

TABLE 10.2

EC Sensors Developed for the Sensitive Detection of DOPA

Graphene material	Method of electrode modification	Buffer and pH	Technique used	Dynamic range	LOD	Interferents	Validation in real sample	Ref.
GQD-MWCNTs	Drop-cast	PBS, pH 7.0	DPV	0.005–100.0 μM	0.87 nM	AA, UA	Serum, live cell	[55]
Zn-NiAl LDH/rGO	Drop-cast	PBS, pH 7.4	DPV	0.001–1.0 μM	0.1 nM	AA, UA	Urine, live cell	[56]
PdNPs/rGO	Electrodeposition	PBS, pH 7.2	DPV	15.0–42.0 μM	6.083 μM	AA, UA	Serum	[57]
Graphene ink	Spin-coat	PBS, pH 7.4	CV	U pto 1.0 mM	100.0 nM	AA, UA	Artificial sweat	[58]
MIP/MWCNTs/GA	Drop-cast, molecular imprinting	PBS, pH 6.0	DPV	0.005–20.0 μM	1.67 nM	AA, UA, FA, EP, VB_6, inorganic ions	Serum	[70]
HCONS@rGO@MWCNT	Drop-cast	PBS, pH 7.4	DPV	1.6–23.6 μM	0.012 μM	AA, UA, GLU, urea, L-lysine, tryptophan	Serum	[71]
Graphene/P_4SAc	Drop-cast	PBS, pH 7.0	CV	5.0–100.0 μM	0.26 μM	AA, UA	Injection	[72]
Pd-Au-P-PDDA/rGO	Drop-cast	PBS, pH 6.5	DPV	3.5–125.0 μM	0.70 μM	AA, UA, inorganic ions	Serum	[73]
GO-CMF/PdSPs	Drop-cast, sonication	PBS, pH 7.0	DPV	0.3–196.3 μM	0.023 μM	UA,AA, EP, CC, HQ, GLU	Injection, serum	[74]
MGH-600	Drop-cast	PBS, pH 7.0	SWV	0.5–50.0 μM	0.44 μM	AA	–	[75]
p-L-Trp/GN	Drop-cast, electropolymerization	PBS, pH 3.0	DPV	0.2–100.0 μM	0.06 μM	AA, cysteine, tyrosine, phenylalanine, inorganic ions	Injection	[76]
Pt-Ag/Gr	Drop-cast	PBS, pH 6.5	DPV	0.1–60.0 μM	0.012 μM	AA, UA, PAP, AC	Injection	[77]
N-rGO	Drop-cast	PBS, pH 6.0	DPV	100.0–3,000.0 μM	0.057 μM	AA, L-cystine	–	[78]
PVP-Gr-GCE	Drop-cast	PBS, pH 6.0	SDLSV	0.02–100.0 μM	0.002 μM	AA, UA	Urine	[79]
rGO/Pd@PPy NPs	Drop-cast	PBS, pH 3.0	DPV	38.0–1,647.0 μM	0.056 μM	AA, UA	Serum	[80]
GO/P(ANI-co-THI)	Drop-cast, electropolymerization	PBS, pH 7.0	DPV	0.002–0.5 mM	2.0 μM	AA, GLU, sucrose, citric acid, glycine	–	[81]
GNPs-Naf/SPE	Drop-cast	PBS, pH 7.4	DPV	0.02–30.0 μM	0.13 μM	APAP	Human urine	[82]
Pt NPs/OPPy/rGO	Electrochemical synthesis	PBS, pH 7.0	DPV	0.1–256.0 μM	41.0 nM	AA, 5-HT	Plasma	[83]
N-GA	Drop-cast	PBS, pH 6.0	SWV	1.0–100.0 μM	0.06 μM	AA	Artificial cerebrospinal fluid	[84]

Abbreviations: Zn-NiAl LDH/rGO – Zn-NiAl layered double hydroxide (LDH) and negatively charged layers of rGO, PBS – phosphate buffer solution, graphene/P_4SAc – graphene/azobenzene-perylene diimide derivative, Pd-Au-P-PDDA/rGO – phosphorous incorporated into bimetallic Pd-Au supported on poly(diallydimethyammonium chloride)-functionalized rGO, GO-CMF/PdSPs – Pd nanostructures on graphene oxide-cellulose microfiber, EP – epinephrine, CC – catechol, HQ – hydroquinone, GLU –glucose, MGH-600 – metal containing graphene hybrid carbonized at 600° C, p-L-Trp/GN/GCE – polymerized L-tryptophan on graphene composite, PAP – p-aminophenol, AC – acetaminophenol, PdNPs/rGO – Pd nanoparticles/rGO, N-rGO – nitrogen-doped-rGO, PVP-Gr-GCE – polyvinylpyrrolidone-graphene composite film, SDLSV – second-order derivative linear sweep voltammetry, rGO/Pd@PPy NPs – Pd nanoparticles supported on polypyrrole/rGO, GO/P(ANI-co-THI) – poly(aniline-co-thionine) modified graphene oxide, GNPs-Naf/SPE – graphene nanoplatelets-nafion nanocomposite, APAP – N-acetyl-p-aminophenol, Pt NPs/OPPy/rGO – Pt nanoparticles/over-oxidized polypyrrole nanofiber/rGO, 5-HT – 5-hydroxy tryptamine, N-GA – nitrogen-doped graphene aerogel, MIP/MWCNTs/GA – molecularly imprinted polymers/multi-walled carbon nanotubes spaced graphene aerogels; FA – folic acid, EP – epinephrine, VB_6 – vitamin B_6, HCONS@rGO@MWCNT – hexagonal Co_3O_4nanosheets@rGO@MWCNT-rGO-MWCNT.MIP/GR/GE – MIP/graphene/gold, Ceria-ErGO – ceria (CeO_2) nanoparticles decorated with electrochemically reduced graphene oxide, FTO/rGO – fluorine-doped tin oxide/rGO.

protein, cultured virus, and nasopharyngeal swab samples from COVID-19 patients. The biosensor detected spike protein antigen in both PBS and universal transport medium (UTM-employed to suspend nasopharyngeal swabs in clinical diagnosis). The tests demonstrated the sensitive detection of the SAR-CoV-2 virus in PBS and UTM with a LOD of 1.0 fg/mL and 100.0 fg/mL, respectively. In contrast, the cultured virus and nasopharyngeal swab samples produced a LOD of 1.6 10^1 pfu/mL and 2.42×10^2 copies/mL, respectively. While on the other hand, a company named Grapheal introduced this graphene-based rapid COVID-19 screening test for direct detection of viral particles from saliva samples.

Zhao et al. [90] have designed and assembled the super sandwich-type EC biosensor for SARS-CoV-2 through the following steps; at first, the capture probes (CP) labeled with thiol were immobilized on the surface of Au@Fe_3O_4 nanoparticles to result in CP/Au@Fe_3O_4 nanocomposite. Secondly, the host-guest complex, p-sulfocalix[8]-toluidine blue (SCX8-TB) was immobilized on rGO to give rise to Au@SCX8-TB-rGO-Label probe bioconjugate. Next, the sandwich structure of the CP-target-Label probe was produced, and finally, the auxiliary probe was introduced to form long concatamers. The (Au@SCX8-TB-rGO-Label probe-target/HT/CP/Au@Fe_3O_4)-modified screen-printed carbon electrode depicted high conductivity attributed to the rGO materials and enhanced the capability of signal molecule TB based on the supramolecular recognition of SCX8. This biosensor could detect RNA present in SAR-CoV-2, as the RNA target could bind specifically with the CP which is attached on the Au/Fe_3O_4. In this work, the common challenge pertaining to low analytical signals was addressed. Hence, with the utilization of nanocomposite and DPV, signal amplification was achieved. Another advancement in this work is a smartphone could read the TB signal.

Meanwhile, Yakoh and group [91] developed a label-free paper-based EC device and the mechanism of detection of SARS-CoV-2 antibody relied on the binding event of the antibody target with the SARS-CoV-2 S protein receptor-binding domain (RBD), which is attached to the detection zone of the graphene-based EC sensor. The results show an LOD of 0.11 ng/mL. Therefore, further improvement in lowering the detection limit is required. Table 10.3 represents the currently available biosensors for early detection of SARS-CoV-2 infection (during 2019–2021 May); however, they are still no match to the current gold standards, RT-PCR. Although they might present future use, for which further investigations must be performed.

Graphene has diverse applications in biosensors to detect several types of infections, including viruses such as Ebola, HIV, and Zika. A life-threatening epidemic disease is caused by the Ebola virus that invades most major organs and triggers multisystem failure in humans. The disease also presents a high mortality rate of up to 90%; due to its high fatality rate, early disease detection is important. Therefore, Jin et al. [97] have recently developed a FET biosensor, which is an effective way for direct electrical detection of inactivated Ebola virus (EBOV) using semiconducting rGO configured FET (Figure 10.4a).

TABLE 10.3
Reported Graphene-Based Biosensors for SARS-CoV-19 in Literature

Transducer	Method	Analyte	Detection medium	Analytical characteristics	Ref
FET biosensor	Label-free graphene-based FET functionalization with antibody	Spike protein	Nasopharyngeal swab	Range: 1.6×10^1–1.6×10^4 pfu/mL LOD: 100 fg/mL in UTM 2.42×10^2 copies/mL in nasopharyngeal swab	[89]
EC biosensor	A super sandwich-type sensor based on p-sulfocalix[8]arene functionalized graphene	ORF1ab	Throat and oral swab, sputum, saliva, whole blood, plasma, feces, urine	Range: 10^{-2}fM–1.0 pM LOD: 200 copies/mL	[90]
EC biosensor	Graphene oxide immobilization onto a paper substrate	Receptor-binding domain (RBD)	Human sera	Range: 1.0–1,000.0 ng/mL LOD: 0.11 ng/mL	[91]
EC biosensor	3D nanoprinting of electrodes coated by rGO	Spike protein RBD	Fetal bovine serum, rabbit serum	Range: 1.0 fM–1.0 nM LOD: 2.8 fM for S protein LOD: 16.9 fM for RBD	[92]
EC biosensor	Screen-printed electrodes coupled graphene oxide along with EDC-NHS and gold nanostars	S1 and S2 glycoprotein	Oropharyngeal/ nasopharyngeal swab, saliva, blood	LOD: 1.68×10^{-13}fg/mL	[93]
EC biosensor	Single-step electrodeposited AuNS onto LSG with a custom made portable potentiostat connected to a smartphone for POC diagnosis	Spike protein antibody	Blood serum	LOD: 2.9 ng/mL	[94]
EC biosensor	Multiplexed PBA-functionalized graphene platform	S1-IgG and S1-IgM	Serum, saliva	–	[95]
FET biosensor	Gr-FET immunosensor	Spike protein S1-RBD	–	LOD: 0.2 pM	[96]

Anti-EBOV glycoprotein antibodies were immobilized onto rGO modified FET through PBASE, a crosslinker molecule. Subsequently, Ebola in serum was incubated to allow detection, which was measured through a shift in Dirac voltage. The Dirac voltage is the peak voltage that is observed at the V-shaped curve of rGO-FET due to its ambipolar properties. The change in conductance occurs upon binding of the charged EBOV to the antibody immobilized on the rGO-based FET chip surface. The sensors presented good selectivity, which was confirmed by obtaining negligible conductance when non-specific viruses were applied to the FET sensor. The obtained detection range for EBOV concentration at rGO-FET was 2.4×10^{-12} g/mL to 1.2×10^{-7} g/mL, and the limit of detection was 2.4 pg/mL (S/N = 3). This sensing platform is attractive as it could directly detect the EBOV in whole spiked serum, making it possible for practical applications.

Meanwhile, HIV causes another major epidemic despite significant efforts made for its early diagnosis and treatment. Researchers have been developing simplified and cost-effective devices as diagnostics tools for HIV; among them, Gong and team [98] have fabricated a DNA sensor based on polyaniline/graphene (PAN/GN) nanocomposite for the detection of HIV-1. To fabricate the sensor, a single-stranded DNA (ssDNA) was immobilized on PAN/GN, and the negatively charged phosphate backbone of HIV-1 was bound to the PAN/GN surface through pi-pi interactions. The hybridization between ssDNA and the target HIV-1 resulted in double-stranded DNA, as represented in Figure 10.4b. Under optimized conditions the impedimetric study, values were linearly related to the logarithm of the concentration of HIV genes in the range 5.0×10^{-16} M – 1.0×10^{-10} M. The developed sensor presented good sensitivity of 1.0×10^{-16} M (S/N = 3) for HIV-1 gene detection. The authors have investigated the excellent selectivity of the proposed sensor, while Islam and team [99] have researched developing an EC sensor for HIV by amine-functionalized graphene (afG) conjugated with antibodies.

The graphene immunosensor was fabricated as depicted in Figure 10.5. First, exfoliation of graphene was conducted, followed by a source-drain contact pad formed by electron beam lithography. Next, 1-ethyl-3-(3-dimethylaminopropyl)carboii mide hydrochloride (EDC) and N-hydroxysulfosuccinimide (NHS) were used for the activation of carboxylic groups on the graphene surface, enhancing the formation of activated graphene-antibody complexes through amino groups. Subsequently, onto the fabricated graphene-FET, antibodies were drop-cast. Finally, biosensing was performed by adding antigens to the functionalized graphene. The interaction of the target antigen bound to the graphene surface with its specific antibody leads to a redistribution of local doping. This manifests as a change in resistance of the graphene channel, which was used to determine the concentration of the target analyte. Under optimized conditions the sensor depicted good linear range from 1.0 fg/mL to 1.0 µg/mL with a LOD of 100.0 fg/mL for p24. The electrical response to the added antigens was instantaneous; hence the analytical time of the device was extremely short. However, the practical application is not discussed in this work.

Shamsipur and team's work [100] represents an impedimetric DNA-based biosensor. The proposed composite was arrayed based on a conductive layer that consists of

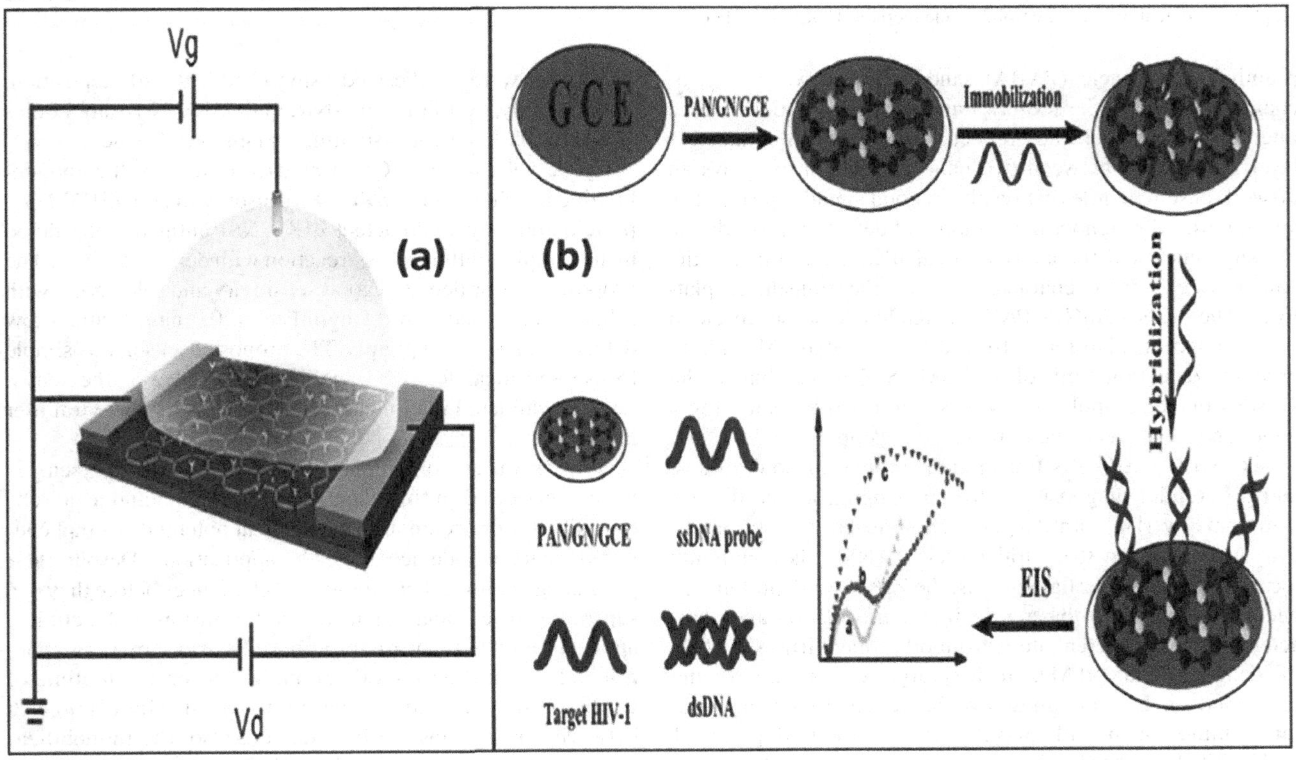

FIGURE 10.4 (a) Graphical representation of FET modified with rGO for Ebola virus detection [97]. (b) Schematic presentation of step-wise fabrication of the DNA sensor to detect HIV-1 gene through EIS [98].

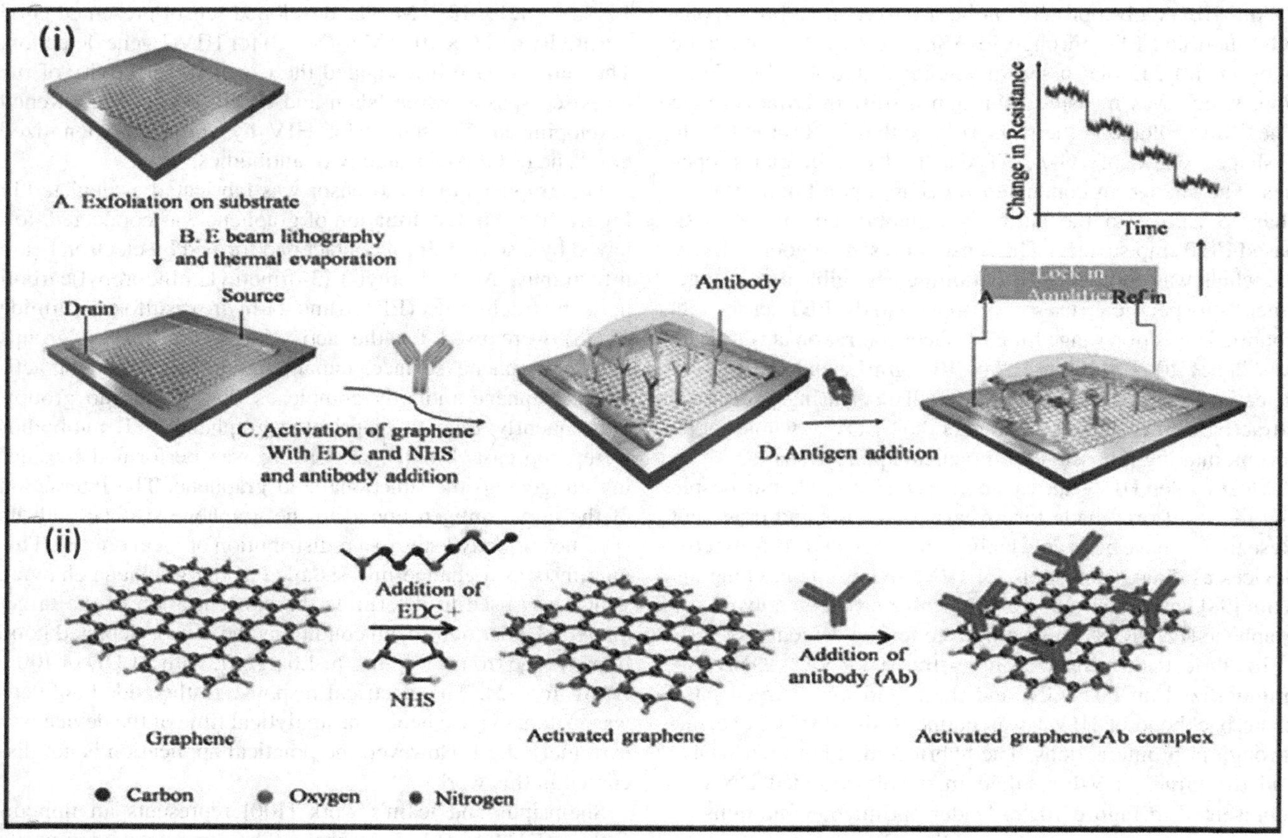

FIGURE 10.5 Schematic illustration of graphene immunosensor development [99]. (iA) Exfoliation of graphene, (iB) formation of source-drain contact pad, (iC) activation of graphene and antibody attachment, (iD) addition of antigen. (ii) Activation procedure for antibody conjugation to the graphene nanomaterial using carbodiimide method (EDC and NHS).

p-aminobenzoic acid (PABA) sandwiched between electrochemically rGO as the sub-layer and gold nanoparticles as the interfacial layer. The study reflects the loin PABA conductive layer sandwiched between two nanostructure layers plays an artwork-ensemble role that results in good signal repeatability and stability throughout incubation and detection procedures. Density functional theory (DFT) calculations confirmed the enhanced electronic characteristics of the transducer platform. The ssDNA/AuNPs/PABA/ERGO/GCE sensor revealed a wide concentration range from 0.1 fM to 10.0 nM with an excellent detection limit of 37.0 aM (S/N = 3). Finally, the sensor's practical applicability was confirmed by obtaining a good recovery percentage in real serum samples.

Zika virus (ZIKV) is transmitted through mosquito bites, and differential diagnosis is still challenging as the disease symptoms overlap with dengue, yellow fever, and chikungunya. Although non-structural protein-1 (NS-1) is commonly located in many other flaviviruses, the ZIKV-NS-1 protein has been substantiated to have a distinct conformation and characteristics that differentiate it from other flaviviruses such as the dengue. Faria and Mazon [101] targeted this property and developed an EC immunosensor based on ZnO nanostructures immobilized with anti-ZIKV-non-structural protein-1 (ZIKV-NS-1) antibodies. The seeding layer of the working electrode was composed of GO on which the zinc oxide nanorods (ZnO-NR) were synthesized using chemical bath deposition. Subsequently, on to the ZnO-NR, anti-ZIKV-NS-1 antibodies were immobilized via cystamine and glutaraldehyde.

Cyclic voltammetry (CV) was used to test for the antigen-binding in urine. The GO/Zn NRs sensor with anti-ZIKV-NS-1 protein antibody could detect ZIKV-NS1 antigen in undiluted urine samples without cross-reaction with dengue antigen. The proposed sensor depicted good sensitivity and selectivity, with a linear range between 0.1 ng/mL to 100.0 ng/mL and a low detection limit of 1.0 pg/mL. The proposed sensor was simple to use and ideal for POC applications. However, the sensor must be validated in clinical samples, as a very low viral titer is present.

A wide variety of graphene-based ultra-sensitive sensors were overviewed in this chapter, with the potential as a POC tool. These works demonstrate the great potential of graphene-based smart nanodevices for POC applications. Despite their promising features, there are several challenges before they can substitute current analysis methods. For instance, the clinical application of most of these proposed sensors requires more research and testing for validation. Moreover, the lifetime of EC biosensors and their capability to specifically distinguish between viruses must be investigated; also, the immobilization of nanomaterials and biological components is decisive in minimizing false positive or negative results.

10.7 Conclusions and Future Scope

Graphene and its composites have been considered excellent materials to develop EC sensors, as these sensors amplify signals, lower overpotential, and enable sensitive and selective determinations. To enhance sensitivity and lower the detection limit, a wide range of graphene and its derivatives have been utilized as promising candidates, which is creditable to their capability to act as an efficient transduction element and even immobilize more bioreceptor units. The scientific community has achieved significant progress in graphene-based EC and biosensors for biomedical applications. Despite substantiate progress, the ongoing research in the field of EC sensors has not yet overcome scientific challenges, such as the gap between a novel sensing concept and a premeditated real-world application. We foresee that an EC sensor depicting good sensitivity to GLU in 0.1 M NaOH may not be ideal for GLU monitoring under physiological conditions.

Subsequent advancement and realization of these sensors hinge on persistent academic and collaborative research endeavors. This can be achieved when researchers are encouraged to think beyond academic proof-of-concept studies with the view to improving these devices for personalized and real-time health monitoring. From the literature survey, it is understood that graphene-based EC sensors can be classified as in vitro and in vivo. Among the two, in vitro applications of EC sensors are much straight forward to clinically translate in comparison to in vivo sensors, as the former is applied to study biological processes via isolated molecules from biological fluids. Thus the requirement of authorization by health agencies is less arduous than those for in vivo applications. For instance, for the POC of the EC sensor, the device must demonstrate upgraded clinical specificity and sensitivity and provide a new function unavailable in the existing tests. Therefore, graphene-based EC sensors have been envisaged as promising next-generation sensors.

REFERENCES

1. Lisdat, F. (2020). Coupling biology to electrochemistry – future trends and needs. *Journal of Solid State Electrochemistry*, *24*(9), 2125–2127.
2. Popovtzer, R., Neufeld, T., Rishpon, J., & Shacham-Diamand, Y. (2006). Electrochemical detection of biological reactions using a novel nano-bio-chip array. *Sensors and Actuators B: Chemical*, *119*(2), 664–672.
3. Chokkareddy, R. (2018). *Fabrication of sensors for the sensitive electrochemical detection of anti-tuberculosis drugs* (Doctoral dissertation), Durban University of Technology, Durban, South Africa.
4. Suleman, S., Shukla, S. K., Malhotra, N., Bukkitgar, S. D., Shetti, N. P., Pilloton, R., … & Aminabhavi, T. M. (2021). Point of care detection of COVID-19: Advancement in biosensing and diagnostic methods. *Chemical Engineering Journal*, *414*, 128759.
5. Hartati, Y., Yusup, S., Fitrilawati, F., Wyantuti, S., Sofiatin, Y., & Gaffar, S. (2020). A voltammetric epithelial sodium channels immunosensor using screen printed carbon electrode modified with reduced graphene oxide. *Current Chemistry Letters*, *9*(4), 151–160.
6. Nabok, A., Abu-Ali, H., Takita, S., & Smith, D. P. (2021). Electrochemical detection of prostate cancer biomarker PCA3 using specific RNA-based Aptamer labelled with ferrocene. *Chemosensors*, *9*(4), 59.
7. Arvinte, A., & Sesay, A. M. (2017). Graphene applications in biosensors and diagnostics. *Biosensors and Nanotechnology: Applications in Health Care Diagnostics*, *1*, 297–326.
8. Shetti, N. P., Mishra, A., Basu, S., Mascarenhas, R. J., Kakarla, R. R., & Aminabhavi, T. M. (2020). Skin-patchable electrodes for biosensor applications: A review. *ACS Biomaterials Science & Engineering*, *6*(4), 1823–1835.
9. Manasa, G., Mascarenhas, R. J., Bhakta, A. K., & Mekhalif, Z. (2021). Nano-graphene-platelet/Brilliant-green composite coated carbon paste electrode interface for electrocatalytic oxidation of flavanone hesperidin. *Microchemical Journal*, *160*, 105768.
10. D'Souza, O. J., Mascarenhas, R. J., Satpati, A. K., & Basavaraja, B. M. (2019). A novel ZnO/reduced graphene oxide and Prussian blue modified carbon paste electrode for the sensitive determination of rutin. *Science China Chemistry*, *62*(2), 262–270.
11. D'Souza, O. J., Mascarenhas, R. J., Satpati, A. K., & Basavaraja, B. M. (2019). A novel ZnO/reduced graphene oxide and Prussian blue modified carbon paste electrode for the sensitive determination of rutin. *Science China Chemistry*, *62*(2), 262–270.
12. Talikoti, N. G., Devarushi, U. S., Tuwar, S. M., Shetti, N. P., & Malode, S. J. (2019). Electrochemical behavior of mefenamic acid at graphene oxide modified carbon paste electrode. *Materials Today: Proceedings*, *18*, 582–589.
13. Todakar, A., Shetti, N. P., Devarushi, U. S., & Tuwar, S. M. (2019). Electro oxidation and analytical applications of valacyclovir at reduced graphene oxide modified carbon paste electrode. *Materials Today: Proceedings*, *18*, 550–557.
14. Totaganti, A., Malode, S. J., Nayak, D. S., & Shetti, N. P. (2019). Voltammetry and analytical applications of hydrochlorothiazide at graphene oxide modified glassy carbon electrode. *Materials Today: Proceedings*, *18*, 542–549.
15. Bukkitgar, S. D., Shetti, N. P., Reddy, K. R., Saleh, T. A., & Aminabhavi, T. M. (2020). Ultrasonication and electrochemically-assisted synthesis of reduced graphene oxide nanosheets for electrochemical sensor applications. *FlatChem*, *23*, 100183.
16. D'Souza, O. J., Mascarenhas, R. J., Satpati, A. K., Aiman, L. V., & Mekhalif, Z. (2016). Electrocatalytic oxidation of l-tyrosine at carboxylic acid functionalized multi-walled carbon nanotubes modified carbon paste electrode. *Ionics*, *22*(3), 405–414.
17. D'Souza, O. J., Mascarenhas, R. J., Thomas, T., Namboothiri, I. N., Rajamathi, M., Martis, P., & Dalhalle, J. (2013). Electrochemical determination of L-tryptophan based on a multiwall carbon nanotube/Mg–Al layered double hydroxide modified carbon paste electrode as a sensor. *Journal of Electroanalytical Chemistry*, *704*, 220–226.
18. D'Souza, O. J., Mascarenhas, R. J., Satpati, A. K., Namboothiri, I. N., Detriche, S., Mekhalif, Z., & Delhalle, J. (2015). A multi-walled carbon nanotube/poly-2,

6-dichlorophenolindophenol film modified carbon paste electrode for the amperometric determination of l-tyrosine. *RSC Advances*, *5*(111), 91472–91481.
19. Thomas, T., Mascarenhas, R. J., Swamy, B. K., Martis, P., Mekhalif, Z., & Sherigara, B. S. (2013). Multi-walled carbon nanotube/poly (glycine) modified carbon paste electrode for the determination of dopamine in biological fluids and pharmaceuticals. *Colloids and Surfaces B: Biointerfaces*, *110*, 458–465.
20. Pradhan, P., Mascarenhas, R. J., Thomas, T., Namboothiri, I. N., D'Souza, O. J., & Mekhalif, Z. (2014). Electropolymerization of bromothymol blue on carbon paste electrode bulk modified with oxidized multiwall carbon nanotubes and its application in amperometric sensing of epinephrine in pharmaceutical and biological samples. *Journal of Electroanalytical Chemistry*, *732*, 30–37.
21. D'Souza, O. J., Mascarenhas, R. J., Satpati, A. K., Detriche, S., Mekhalif, Z., Delhalle, J., & Dhason, A. (2017). High electrocatalytic oxidation of folic acid at carbon paste electrode bulk modified with iron nanoparticle-decorated multiwalled carbon nanotubes and its application in food and pharmaceutical analysis. *Ionics*, *23*(1), 201–212.
22. Manasa, G., Raj, C., Satpati, A. K., & Mascarenhas, R. J. (2020). S(O)MWCNT/modified carbon paste–A non-enzymatic amperometric sensor for direct determination of 6-mercaptopurine in biological fluids. *Electroanalysis*, *32*(11), 2431–2441.
23. D'Souza, O. J., Mascarenhas, R. J., Satpati, A. K., Mane, V., & Mekhalif, Z. (2017). Application of a nanosensor based on MWCNT-sodium dodecyl sulphate modified electrode for the analysis of a novel drug, alpha-hydrazinonitroalkene in human blood serum. *Electroanalysis*, *29*(7), 1794–1804.
24. D'Souza, O. J., Mascarenhas, R. J., Thomas, T., Basavaraja, B. M., Saxena, A. K., Mukhopadhyay, K., & Roy, D. (2015). Platinum decorated multi-walled carbon nanotubes/Triton X-100 modified carbon paste electrode for the sensitive amperometric determination of paracetamol. *Journal of Electroanalytical Chemistry*, *739*, 49–57.
25. Shetti, N. P., Bukkitgar, S. D., Reddy, K. R., Reddy, C. V., & Aminabhavi, T. M. (2019). Nanostructured titanium oxide hybrids-based electrochemical biosensors for healthcare applications. *Colloids and Surfaces B: Biointerfaces*, *178*, 385–394.
26. Bukkitgar, S. D., Kumar, S., Singh, S., Singh, V., Reddy, K. R., Sadhu, V., ... & Naveen, S. (2020). Functional nanostructured metal oxides and its hybrid electrodes–Recent advancements in electrochemical biosensing applications. *Microchemical Journal*, *159*, 105522.
27. Shetti, N. P., Bukkitgar, S. D., Reddy, K. R., Reddy, C. V., & Aminabhavi, T. M. (2019). ZnO-based nanostructured electrodes for electrochemical sensors and biosensors in biomedical applications. *Biosensors and Bioelectronics*, *141*, 111417.
28. Kumar, S., Bukkitgar, S. D., Singh, S., Singh, V., Reddy, K. R., Shetti, N. P., ... & Naveen, S. (2019). Electrochemical sensors and biosensors based on graphene functionalized with metal oxide nanostructures for healthcare applications. *ChemistrySelect*, *4*(18), 5322–5337.
29. Thomas, T., Mascarenhas, R. J., D'Souza, O. J., Martis, P., Dalhalle, J., & Swamy, B. K. (2013). Multi-walled carbon nanotube modified carbon paste electrode as a sensor for the amperometric detection of L-tryptophan in biological samples. *Journal of Colloid and Interface Science*, *402*, 223–229.
30. Bhajanthri, N. K., Arumugam, V. K., Chokkareddy, R., & Redhi, G. G. (2016). Ionic liquid based high performance electrochemical sensor for ascorbic acid in various foods and pharmaceuticals. *Journal of Molecular Liquids*, *222*, 370–376.
31. Lee, J. H., Park, S. J., & Choi, J. W. (2019). Electrical property of graphene and its application to electrochemical biosensing. *Nanomaterials*, *9*(2), 297.
32. Saini, D. (2016). Synthesis and functionalization of graphene and application in electrochemical biosensing. *Nanotechnology Reviews*, *5*(4), 393–416.
33. Atta, N. F., Galal, A., & El-Ads, E. H. (2015). Graphene—A platform for sensor and biosensor applications. *Biosensors-Micro and Nanoscale Applications*, *9*, 38–84.
34. Shetti, N. P., Nayak, D. S., Reddy, K. R., & Aminabhvi, T. M. (2019). Graphene–Clay-based hybrid nanostructures for electrochemical sensors and biosensors. In Pandikumar, Alagarsamy & Perumal Rameshkumar (Eds.), *Graphene-based electrochemical sensors for biomolecules* (pp. 235–274). Elsevier.
35. Lebedev, A. A., Davydov, S. Y., Eliseyev, I. A., Roenkov, A. D., Avdeev, O., Lebedev, S. P., ... & Usikov, A. S. (2021). Graphene on SiC substrate as biosensor: Theoretical background, preparation, and characterization. *Materials*, *14*(3), 590.
36. Immanuel, S., Aparna, T. K., & Sivasubramanian, R. (2019). Graphene–metal oxide nanocomposite modified electrochemical sensors. In Pandikumar, Alagarsamy & Perumal Rameshkumar (Eds.), *Graphene-based electrochemical sensors for biomolecules* (pp. 113–138). Elsevier.
37. Viswanathan, S., Narayanan, T. N., Aran, K., Fink, K. D., Paredes, J., Ajayan, P. M., ... & Renugopalakrishanan, V. (2015). Graphene–protein field effect biosensors: Glucose sensing. *Materials Today*, *18*(9), 513–522.
38. Cho, I. H., Kim, D. H., & Park, S. (2020). Electrochemical biosensors: Perspective on functional nanomaterials for on-site analysis. *Biomaterials Research*, *24*(1), 1–12.
39. Manasa, G., Mascarenhas, R. J., Satpati, A. K., D'Souza, O. J., & Dhason, A. (2017). Facile preparation of poly (methylene blue) modified carbon paste electrode for the detection and quantification of catechin. *Materials Science and Engineering: C*, *73*, 552–561.
40. Manasa, G., Mascarenhas, R. J., Bhakta, A. K., & Mekhalif, Z. (2020). MWCNT/Nileblue heterostructured composite electrode for flavanone naringenin quantification in fruit juices. *Electroanalysis*, *32*(5), 939–948.
41. Manasa, G., Bhakta, A. K., Mekhalif, Z., & Mascarenhas, R. J. (2020). Bismuth-nanoparticles decorated multi-wall-carbon-nanotubes cast-coated on carbon paste electrode; an electrochemical sensor for sensitive determination of gallic acid at neutral pH. *Materials Science for Energy Technologies*, *3*, 174–182.

42. Manasa, G., Mascarenhas, R. J., Satpati, A. K., Basavaraja, B. M., & Kumar, S. (2018). An electrochemical bisphenol F sensor based on ZnO/G nano composite and CTAB surface modified carbon paste electrode architecture. *Colloids and Surfaces B: Biointerfaces*, *170*, 144–151.
43. Wu, S., He, Q., Tan, C., Wang, Y., & Zhang, H. (2013). Graphene-based electrochemical sensors. *Small*, *9*(8), 1160–1172.
44. Panchakarla, L. S., Subrahmanyam, K. S., Saha, S. K., Govindaraj, A., Krishnamurthy, H. R., Waghmare, U. V., & Rao, C. N. R. (2009). Synthesis, structure, and properties of boron-and nitrogen-doped graphene. *Advanced Materials*, *21*(46), 4726–4730.
45. Liu, H., Liu, Y., & Zhu, D. (2011). Chemical doping of graphene. *Journal of Materials Chemistry*, *21*(10), 3335–3345.
46. Wang, X., Sun, G., Routh, P., Kim, D. H., Huang, W., & Chen, P. (2014). Heteroatom-doped graphene materials: Syntheses, properties and applications. *Chemical Society Reviews*, *43*(20), 7067–7098.
47. Menon, S., Mathew, M. R., Sam, S., Keerthi, K., & Kumar, K. G. (2020). Recent advances and challenges in electrochemical biosensors for emerging and re-emerging infectious diseases. *Journal of Electroanalytical Chemistry*, *878*, 114596.
48. Baryeh, K., Takalkar, S., Lund, M., & Liu, G. (2017). Introduction to medical biosensors for point of care applications. In Narayan, Roger (Ed.), *Medical biosensors for point of care (POC) applications* (pp. 3–25). Woodhead Publishing.
49. Chokkareddy, R., Bhajanthri, N. K., Redhi, G. G., & Redhi, D. G. (2018). Ultra-sensitive electrochemical sensor for the determination of pyrazinamide. *Current Analytical Chemistry*, *14*(4), 391–398.
50. Bahadır, E. B., & Sezgintürk, M. K. (2016). A review on impedimetric biosensors. *Artificial Cells, Nanomedicine, and Biotechnology*, *44*(1), 248–262.
51. Chokkareddy, R., Thondavada, N., Bhajanthri, N. K., & Redhi, G. G. (2019). An amino functionalized magnetite nanoparticle and ionic liquid based electrochemical sensor for the detection of acetaminophen. *Analytical Methods*, *11*(48), 6204–6212.
52. Thomas, T., Mascarenhas, R. J., Swamy, B. K., Martis, P., Mekhalif, Z., & Sherigara, B. S. (2013). Multi-walled carbon nanotube/poly (glycine) modified carbon paste electrode for the determination of dopamine in biological fluids and pharmaceuticals. *Colloids and Surfaces B: Biointerfaces*, *110*, 458–465.
53. Ahammad, A. S., Islam, T., & Hasan, M. M. (2019). Graphene-based electrochemical sensors for biomedical applications. In Nurunnabi, Md & Jason McCarthy (Eds.),*Biomedical applications of graphene and 2D nanomaterials* (pp. 249–282). Elsevier.
54. Ahammad, A. S., Islam, T., Hasan, M. M., Mozumder, M. I., Karim, R., Odhikari, N., ... & Kim, D. M. (2018). Reduced graphene oxide screen-printed FTO as highly sensitive electrodes for simultaneous determination of dopamine and uric acid. *Journal of the Electrochemical Society*, *165*(5), B174.
55. Huang, Q., Lin, X., Tong, L., & Tong, Q. X. (2020). Graphene quantum dots/multiwalled carbon nanotubes composite-based electrochemical sensor for detecting dopamine release from living cells. *ACS Sustainable Chemistry & Engineering*, *8*(3), 1644–1650.
56. Asif, M., Aziz, A., Wang, H., Wang, Z., Wang, W., Ajmal, M., ... & Liu, H. (2019). Superlattice stacking by hybridizing layered double hydroxide nanosheets with layers of reduced graphene oxide for electrochemical simultaneous determination of dopamine, uric acid and ascorbic acid. *Microchimica Acta*, *186*(2), 61.
57. Wei, Y., Liu, Y., Xu, Z., Wang, S., Chen, B., Zhang, D., & Fang, Y. (2020). Simultaneous detection of ascorbic acid, dopamine, and uric acid using a novel electrochemical sensor based on palladium nanoparticles/reduced graphene oxide nanocomposite. *International Journal of Analytical Chemistry*, *2020*, Article ID 8812443.
58. Muralidharan, R., Chandrashekhar, V., Butler, D., & Ebrahimi, A. (2020). A smartphone-interfaced, flexible electrochemical biosensor based on graphene ink for selective detection of dopamine. *IEEE Sensors Journal*, *20*(22), 13204–13211.
59. Rahman, M. M., & Lee, J. J. (2019). Electrochemical dopamine sensors based on graphene. *Journal of Electrochemical Science and Technology*, *10*(2), 185–195.
60. Matsoso, B. J., Lerotholi, T., Coville, N. J., & Jones, G. (2019). Designing parameters of surfactant-free electrochemical sensors for dopamine and uric acid on nitrogen doped graphene films. *Johnson Matthey Technology Review*, *63*(2), 76–88.
61. Cernat, A., Ştefan, G., Tertiş, M., Cristea, C., & Simon, I. (2020). An overview of the detection of serotonin and dopamine with graphene-based sensors. *Bioelectrochemistry*, *136*, 107620.
62. Cho, Y. W., Park, J. H., Lee, K. H., Lee, T., Luo, Z., & Kim, T. H. (2020). Recent advances in nanomaterial-modified electrical platforms for the detection of dopamine in living cells. *Nano Convergence*, *7*(1), 1–14.
63. Eddin, F. B. K., & Fen, Y. W. (2020). Recent advances in electrochemical and optical sensing of dopamine. *Sensors (Basel, Switzerland)*, *20*(4), 1039.
64. Chokkareddy, R., Bhajanthri, N. K., & Redhi, G. G. (2017). A novel electrode architecture for monitoring rifampicin in various pharmaceuticals. *International Journal of Electrochemical Sciences*, *12*, 9190–9203.
65. Stanford, M. G., Yang, K., Chyan, Y., Kittrell, C., & Tour, J. M. (2019). Laser-induced graphene for flexible and embeddable gas sensors. *ACS Nano*, *13*(3), 3474–3482.
66. Liu, R., Li, J., Zhong, T., & Long, L. (2019). Graphene modified molecular imprinting electrochemical sensor for determining the content of dopamine. *Current Analytical Chemistry*, *15*(6), 628–634.
67. Deng, P., Feng, J., Xiao, J., Liu, J., Nie, X., Li, J., & He, Q. (2020). Highly sensitive voltammetric sensor for nanomolar dopamine detection based on facile electrochemical reduction of graphene oxide and ceria nanocomposite. *Journal of the Electrochemical Society*, *167*(14), 146511.

68. Homola, T., Lorencova, L., Parráková, L., Gemeiner, P., & Tkac, J. (2020). Graphene oxide sensors of high sensitivity fabricated using cold atmospheric-pressure hydrogen plasma for use in the detection of small organic molecules. *Journal of Applied Physics*, *128*(24), 243301.
69. Settu, K., Huang, Y. M., & Zhou, S. X. (2020). A facile approach for the electrochemical sensing of dopamine using paper-based PEDOT: PSS/RGO graphene biosensor. *ECS Journal of Solid State Science and Technology*, *9*(12), 121002.
70. Huang, B., Xiao, L., Dong, H., Zhang, X., Gan, W., Mahboob, S., … & Li, Y. (2017). Electrochemical sensing platform based on molecularly imprinted polymer decorated N, S co-doped activated graphene for ultrasensitive and selective determination of cyclophosphamide. *Talanta*, *164*, 601–607.
71. Zahed, M. A., Barman, S. C., Toyabur, R. M., Sharifuzzaman, M., Xuan, X., Nah, J., & Park, J. Y. (2019). Ex situ hybridized hexagonal cobalt oxide nanosheets and RGO@ MWCNT based nanocomposite for ultra-selective electrochemical detection of ascorbic acid, dopamine, and uric acid. *Journal of The Electrochemical Society*, *166*(6), B304.
72. Varol, T. Ö., Perk, B., Avci, O., Akpolat, O., Hakli, Ö., Xue, C., … & Anik, Ü. (2019). Fabrication of graphene/azobenzene-perylene diimide derivative modified electrochemical sensors for the dopamine detection based on full factorial experimental design. *Measurement*, *147*, 106867.
73. Zhang, M., & Fu, D. (2019). An electrochemical sensor for dopamine detection based on ternary Pd-Au-P composites supported on PDDA/RGO. *International Journal of Electrochemical Sciences*, *14*, 9909–9920.
74. Palanisamy, S., Velusamy, V., Ramaraj, S., Chen, S. W., Yang, T. C., Balu, S., & Banks, C. E. (2019). Facile synthesis of cellulose microfibers supported palladium nanospindles on graphene oxide for selective detection of dopamine in pharmaceutical and biological samples. *Materials Science and Engineering: C*, *98*, 256–265.
75. Vermisoglou, E. C., Jakubec, P., Malina, O., Kupka, V., Schneemann, A., Fischer, R. A., … & Otyepka, M. (2020). Hierarchical porous graphene–iron carbide hybrid derived from functionalized graphene-based metal–organic gel as efficient electrochemical dopamine sensor. *Frontiers in Chemistry*, *8*, 544.
76. Gong, Q. J., Han, H. X., Wang, Y. D., Yao, C. Z., Yang, H. Y., & Qiao, J. L. (2020). An electrochemical sensor for dopamine detection based on the electrode of a poly-tryptophan-functionalized graphene composite. *New Carbon Materials*, *35*(1), 34–41.
77. Anuar, N. S., Basirun, W. J., Shalauddin, M., & Akhter, S. (2020). A dopamine electrochemical sensor based on a platinum–silver graphene nanocomposite modified electrode. *RSC Advances*, *10*(29), 17336–17344.
78. Soni, R., Palit, K., Soni, M., Kumar, R., & Sharma, S. K. (2020). Highly sensitive electrochemical sensing of neurotransmitter dopamine from scalable UV irradiation-based nitrogen-doped reduced graphene oxide-modified electrode. *Bulletin of Materials Science*, *43*(1), 1–11.
79. Wu, Y., Deng, P., Tian, Y., Feng, J., Xiao, J., Li, J., … & He, Q. (2020). Simultaneous and sensitive determination of ascorbic acid, dopamine and uric acid via an electrochemical sensor based on PVP-graphene composite. *Journal of Nanobiotechnology*, *18*(1), 1–13.
80. Demirkan, B., Bozkurt, S., Cellat, K., Arıkan, K., Yılmaz, M., Şavk, A., … & Sen, F. (2020). Palladium supported on polypyrrole/reduced graphene oxide nanoparticles for simultaneous biosensing application of ascorbic acid, dopamine, and uric acid. *Scientific Reports*, *10*(1), 1–10.
81. Song, N. N., Wang, Y. Z., Yang, X. Y., Zong, H. L., Chen, Y. X., Ma, Z., & Chen, C. X. (2020). A novel electrochemical biosensor for the determination of dopamine and ascorbic acid based on graphene oxide/poly (aniline-co-thionine) nanocomposite. *Journal of Electroanalytical Chemistry*, *873*, 114352.
82. Krampa, F. D., Aniweh, Y., Kanyong, P., & Awandare, G. A. (2020). Graphene nanoplatelet-based sensor for the detection of dopamine and N-acetyl-p-aminophenol in urine. *Arabian Journal of Chemistry*, *13*(1), 3218–3225.
83. Kilele, J. C., Chokkareddy, R., & Redhi, G. G. (2021). Ultra-sensitive electrochemical sensor for fenitrothion pesticide residues in fruit samples using IL@ CoFe2O4NPs@ MWCNTs nanocomposite. *Microchemical Journal*, *164*, 106012.
84. Urbanová, V., Kment, Š., & Zbořil, R. (2020). Nitrogen-doped graphene aerogel for simultaneous detection of dopamine and ascorbic acid in artificial cerebrospinal fluid. *Journal of the Electrochemical Society*, *167*(11), 116521.
85. Thévenot, D. R., Toth, K., Durst, R. A., & Wilson, G. S. (2001). Electrochemical biosensors: Recommended definitions and classification. *Biosensors and Bioelectronics*, *16*(1–2), 121–131.
86. Chatterjee, S., Shetti, N. P., & Reddy, K. R. (2020). Fundamentals, recent advances, and perspectives of electrode materials for bioelectrochemical sensing applications. In Kanchi, Suvardhan & Deepali Sharma (Eds.), *Nanomaterials in diagnostic tools and devices* (pp. 557–589). Elsevier.
87. Shetti, N. P., Mishra, A., Bukkitgar, S. D., Basu, S., Narang, J., Raghava Reddy, K., & Aminabhavi, T. M. (2021). Conventional and nanotechnology-based sensing methods for SARS Coronavirus (2019-nCoV). *ACS Applied Bio Materials*, *4*(2), 1178–1190.
88. Bukkitgar, S. D., Shetti, N. P., & Aminabhavi, T. M. (2021). Electrochemical investigations for COVID-19 detection-A comparison with other viral detection methods. *Chemical Engineering Journal*, *420*, 127575.
89. Novodchuk, I., Kayaharman, M., Ausri, I. R., Karimi, R., Tang, X. S., Goldthorpe, I. A., … & Yavuz, M. (2021). An ultrasensitive heart-failure BNP biosensor using B/N co-doped graphene oxide gel FET. *Biosensors and Bioelectronics*, *180*, 113114.
90. Zhao, H., Liu, F., Xie, W., Zhou, T. C., OuYang, J., Jin, L., … & Li, C. P. (2021). Ultrasensitive supersandwich-type electrochemical sensor for SARS-CoV-2 from the infected COVID-19 patients using a smartphone. *Sensors and Actuators B: Chemical*, *327*, 128899.

91. Yakoh, A., Pimpitak, U., Rengpipat, S., Hirankarn, N., Chailapakul, O., & Chaiyo, S. (2021). Paper-based electrochemical biosensor for diagnosing COVID-19: Detection of SARS-CoV-2 antibodies and antigen. *Biosensors and Bioelectronics, 176*, 112912.
92. Ali, M. A., Hu, C., Jahan, S., Yuan, B., Saleh, M. S., Ju, E., ... & Panat, R. (2021). Sensing of COVID-19 antibodies in seconds via aerosol jet nanoprinted reduced-graphene-oxide-coated 3D electrodes. *Advanced Materials, 33*(7), 2006647.
93. Hashemi, S. A., Behbahan, N. G. G., Bahrani, S., Mousavi, S. M., Gholami, A., Ramakrishna, S., ... & Omidifar, N. (2021). Ultra-sensitive viral glycoprotein detection NanoSystem toward accurate tracing SARS-CoV-2 in biological/non-biological media. *Biosensors and Bioelectronics, 171*, 112731.
94. Beduk, T., Beduk, D., de Oliveira Filho, J. I., Zihnioglu, F., Cicek, C., Sertoz, R., ... & Timur, S. (2021). Rapid point-of-care COVID-19 diagnosis with a gold-nanoarchitecture-assisted laser-scribed graphene biosensor. *Analytical Chemistry, 93*(24), 8585–8594.
95. Torrente-Rodríguez, R. M., Lukas, H., Tu, J., Min, J., Yang, Y., Xu, C., ... & Gao, W. (2020). SARS-CoV-2 RapidPlex: A graphene-based multiplexed telemedicine platform for rapid and low-cost COVID-19 diagnosis and monitoring. *Matter, 3*(6), 1981–1998.
96. Zhang, X., Qi, Q., Jing, Q., Ao, S., Zhang, Z., Ding, M., ... & Fu, W. (2020). Electrical probing of COVID-19 spike protein receptor binding domain via a graphene field-effect transistor. *arXiv preprint arXiv:2003.12529*.
97. Jin, X., Zhang, H., Li, Y. T., Xiao, M. M., Zhang, Z. L., Pang, D. W., ... & Zhang, G. J. (2019). A field effect transistor modified with reduced graphene oxide for immunodetection of Ebola virus. *Microchimica Acta, 186*(4), 1–9.
98. Gong, Q., Han, H., Yang, H., Zhang, M., Sun, X., Liang, Y., ... & Qiao, J. (2019). Sensitive electrochemical DNA sensor for the detection of HIV based on a polyaniline/graphene nanocomposite. *Journal of Materiomics, 5*(2), 313–319.
99. Islam, S., Shukla, S., Bajpai, V. K., Han, Y. K., Huh, Y. S., Kumar, A., ... & Gandhi, S. (2019). A smart nanosensor for the detection of human immunodeficiency virus and associated cardiovascular and arthritis diseases using functionalized graphene-based transistors. *Biosensors and Bioelectronics, 126*, 792–799.
100. Shamsipur, M., Samandari, L., Taherpour, A. A., & Pashabadi, A. (2019). Sub-femtomolar detection of HIV-1 gene using DNA immobilized on composite platform reinforced by a conductive polymer sandwiched between two nanostructured layers: A solid signal-amplification strategy. *Analytica Chimica Acta, 1055*, 7–16.
101. Faria, A. M., & Mazon, T. (2019). Early diagnosis of Zika infection using a ZnO nanostructures-based rapid electrochemical biosensor. *Talanta, 203*, 153–160.

11

Latest Trends in Bioimaging Using Quantum Dots

Monalisa Mishra

CONTENTS

11.1 Introduction

Quantum dots (QDs) were first introduced by Ekimov and Onushenko in 1981 [1]. However, it took 17 years for scientists to introduce it into the field of bioimaging. Later, QDs were used in several studies since the photo-bleaching is less [2]. The wavelength of semiconductor QDs can be altered by varying their size. The surface property can be modified to make it hydrophobic and biocompatible. NIR-emitting QDs which are emitting in the near-infrared region (NIR) help to prevent interference from autofluorescence, single cells, hemoglobin and have less absorption coefficient toward water.

The size of a QD varies from 2 to 10 nm. Elements like Cd, Pd, Hg, Cu, Ag, Ga, In, S or Se are used as QDs. What makes a QD an ideal molecule to be used for biomedical imaging? Most bioimaging needs the near-infrared region since light scattering in this region is less and also the tissue absorbs less [3]. This need provoked researchers to use QDs for medical imaging. Besides this, the QDs have some unique properties like optical properties, high quantum yields, tunable light emission properties and good chemical and photostability. The QDs are excited at a particular wavelength and their emission is used for innumerable assays. The emission band is dependent on the size as well as thickness of the QDs. They are highly reactive and have large surface areas. Thus, they need to be stabilized before they are used in ambient conditions. The property of a stabilizer further depends on absorption, luminescence and environmental condition. Since most of the elements used in QDs are toxic, reduction of the toxicity of QDs will make it more suitable for biomedical applications [4].

11.2 Modification in Quantum Dots for Specific Labeling

QDs are widely used for labeling purposes. The small size of the QDs makes it easy to pass through the barriers. Often this property also causes the non-functioning of the living cells, tissue and organs. Thus, the toxicity of QDs often arises with the property of the nanoparticle [5]. The toxicity depends on more than a few factors like size, charge, concentration, associated functional group, mechanical stability and oxidation. In the case of CdSe QDs, cytotoxicity is associated with the release of Cd^{+2} ions [6]. To overcome this problem CdSe/ZnS or CdSe/Zns/TEOS QDs were developed [7]. Furthermore, to use a QD for a biological material it should be soluble in water. Thus, dispersion in water media is considered a parameter while developing the QDs [8]. However, some of the QDs are also soluble in lipids [9]. To achieve the desirable solubility, QDs are coated with either polymer or lipids. Besides solubility to achieve desired sensitivity and specificity, the QDs are functionalized at their surface. One example is a QD conjugated

DOI: 10.1201/9781003157823-11

with antibodies [10]. Such QDs can detect T and B lymphocytes of the HRS cells [11]. Fe_3O_4 with amino terminated QDs can enhance the fluorescence for in vivo and in vitro imaging [12]. QDs conjugated with molecules can recognize the target, thus precluding non-specific binding. To boost the specificity of the imaging many modifications were made in the ligand.

11.2.1 Application of Quantum Dots in Bioimaging

Bioimaging is a requisite tool to understand the structural, functional dynamic of a cellular process. It enables us to gain information about a specimen without any physical interaction. QDs are predominantly appropriate for imaging biomolecules like proteins at the surface of the cell or within the intracellular organelles of the cell. Using a confocal microscope having time-lapse option permits spatial as well as temporal information of protein and nanoparticles interactions [13]. With cutting-edge techniques, the protein–nanoparticle interaction can be studied in a time-dependent manner. Such techniques help to provide information regarding protein–protein as well as protein–nanoparticle interaction at normal and under pathological conditions. QDs conjugated with specific antibodies are capable of distinguishing different types of neuronal and glial cells without the use of secondary antibodies [14]. Antibodies of beta-tubulin (a maker for neuron-specific cytoskeletal protein) and glia-specific proteins are conjugated with streptavidin-conjugated QDs for labeling of primary cortical cultures. QDs are further modified to label many organelles like mitochondria and lysosomes [15]. To detect the signal of QDs equipment like light, fluorescence, confocal microscopes, X-ray, positron and magnetic resonance (MRI) is used for imaging. QDs are designed in several ways to obtain images using these techniques. Different types of QDs that are designed for (1) labeling of lipids, (2) labeling of neuronal cells, (3) in vitro imaging, (4) in vivo imaging, (5) ex vivo imaging and (6) transfection are described in the next section.

11.2.1.1 QDs as a Nanoprobe for Labeling of Lipids

Accumulation of lipid droplets is reported from several metabolic disorders. Consequently, many methods are adopted to detect the lipid droplets within the organism [16]. Numerous fluorescent probes like Nile red, BODIPY and Seoul-Fluor are used for the detection of lipid droplets. Moreover, all these conventional fluorophores undergo photo-bleaching. To overcome the issues associated with fluorescent probes QD-based fluorescent probes are developed. To design such QDs quite a few subcellular targeting properties are used [17]. For example, the incorporation of zwitterionic surface charge is introduced to the QDs. Likewise, QDs are designed in a way that they should be prevented from cellular endocytosis. To achieve this the surface of the QDs is modified chemically to reduce clathrin-mediated endocytosis and trafficking by endosome or lysosome [18]. Such QDs are capable of entering within the cell also via lipid-raft or via caveolae-mediated endocytosis. Once the QDs enter the cell they label organelles like mitochondria and Golgi apparatus and are transported toward the perinuclear region. Henceforth zwitterionic–lipophilic nanoprobes can efficiently label the lipids [19]. The zwitterionic–lipophilic nanoprobe interacts through the membrane of the cell and assists in the endocytosis of lipid-raft. The endocytosis step is trailed by subcellular trafficking deprived of endosomal/lysosomal entrapment. The nanoprobe with zwitterionic functional groups (octyl- or oleyl-functionalized LQD) has almost zero surface charge [17]. Accordingly, it is more stable and beneficial for functioning at physiological pH and stable toward continuous light exposure. The efficacy of the nanoprobe to label lipids was 10 times more as evidenced from the labeling of two different cell lines. These nanoprobes are used to deliver drugs to lipids to regulate the biogenesis process.

11.2.1.2 QDs for Imaging of Neurons

Neuronal cells are the most expensive cells of the body since the ability to regenerate is much less. An alteration or reduction of neuronal imbalance results in neurodegenerative disease [20]. The function of the neurons cannot be determined until and unless we know the spatial and temporal organization of the molecules associated with it. The molecular interaction of the neurons can be tracked in real time. To achieve this target extremely fluorescent and photostable QDs are used for live neuronal cell imaging [21]. An antibody against glycine receptors (GlyR), i.e. GlyR a1, is tagged with QDs (QD-GlyR) to label rat neuronal cells of the spinal cord [22]. Using these QDs lateral dynamics in the neuronal membranes can be calculated up to 20 min. Previously the life span was only five seconds. With the synaptic cleft, the diffusion coefficients were calculated larger compared to the bead-GlyRs as evidenced by microscopy research and technique 598 E. Using QDs the internalization as well as dimerization of epidermal growth factor receptors was revealed [23], which can provide erb/HER receptor-mediated signaling. Dorsal root ganglia (DRG) cultures are ideal to discover signal transduction associated with nerve growth and survival [24]. QDs conjugated with nerve growth factor (NGF) are used to track the uptake as well as retrograde transportation of NGF in rat DRG cultures. By introducing QDs to primary culture (cell bodies are detached from the axons), the kinetics of internalization and the internalization of QDs can be detected with high accuracy. Monovalent QDs, conjugated with streptavidin or antibody against carcinoembryonic antigens are used for labeling protein present in the cell surface [25]. This QD-nano-construct is used for labeling glutamate receptors present on the surface of the neurons. Labeling of glutamate receptors on synaptic spines is the region used to detect the abnormalities present in neurodegenerative disorders. Thus, it is considered a highly sensitive tool to detect the function of the neurons in a time-dependent manner. QD-protein conjugations are ideal to track any type of protein of interest since protein surface undergoes modification in many pathological conditions [26]. QDs-emitting NIR regions are of particular interest to detect the receptor present in the plasma membrane, since it helps to prevent tissue autofluorescence and allows live brain imaging for a longer time. NIR InGaP-QDs are employed for thick tissue (a thickness of around 6 mm) imaging [27].

11.2.2.3 QDS for In Vitro Imaging

Different types of QDs are often used to label cells for in vitro imaging. QDs like PbS and PbSe capped with carboxylic groups are used to tag cells. CdTe coated with 3-mercaptopropionic acid (3-PMA) is cast off to label *Salmonella typhimurium*. Acid-capped CdSe/ZnSQDs are used to label Hela cells. Hela cells can uptake the QDs by endocytosis. CdSe/ZnS coated with PEG or conjugated with peptides are localized in the cytosol and can target the nucleus. CdSe/L-cystein labels the serum albumin as well as cells [28]. PS-PEG-COOH conjugated with streptavidin and immunoglobulin G is used to label the surface receptor and subcellular structure present within the cell. Silica coated with Fe3O4 and TGA-capped CdTeQDs is used to label cells, imaging and magnetic separation [29]. Avidin -onjugated CdTeQDs are used for specific labeling of cells. CdTeQDs conjugated with biotinylated Annexin V can label specifically to phosphatidylserine. Henceforth this dye can bind specifically to apoptotic but not to necrotic cells [30]. Functionalized silica-coated CdTe QDs has more advantage since it not only labels the proteins but also prevent the leaking of Cd^{+2}. QDs covered with PEG grafted polyethyleneimine (PEI) can pass the cell membrane and disrupt the endosomal organelles [31]. Gd-doped ZnO QDs are less toxic and have improved fluorescence in the yellow range [31]. Near-IR QDs (QD800) through emission wavelength 800 nm are used to label the U14 cell line by endocytosis [32]. QDs aptamer-doxorubicin is used to sense prostrate-specific membrane antigen [33]. CdSe/ZnS with paramagnetic covering arginine–glycine–aspartic acid conjugation is used to bind specific proteins during imaging [34]. CdSe/CdS/ZnS used ligands like transferrin and anti-Claudin-4 are used for labeling pancreatic cancer cells.

11.2.2.4 QDS for In Vivo Imaging

It is a non-invasive way to label the living organism either morphologically or molecularly. Along with several conventional probes QDs are also used for in vivo imaging. Spherical QDs have the upper hand over the other shapes for in vivo imaging. However, the molecular interaction of different shaped QDs with the biological system is not yet known. Among different QDs, QDs with base CdTe or CdSe are widely used for ex vivo imaging.

Application of any probe for in vivo imaging is a challenging task since it has to be (1) less toxic, (2) high contrast, (3) highly sensitive and (4) extremely photostable. Several QDs satisfy all these characteristics and are employed for in vivo imaging [35–37]. CdSe/ZnSQDs are encapsulated with phospholipid block copolymer micelle and further functionalized with DNA. This QD is injected into a Xenopus embryo does not show toxicity and is highly stable in nature [38]. CdSe/Zns QDs are used in live mice to visualize blood vessels [37]. Coating of PEF further reduces the toxicity and accumulation of QDs in the liver cells. CuInS2/ZnScor/shell nanocrystal was injected into the tail vein of nude mice to check its toxicity [39]. QD800-labeled BcaCD885 cells (BcaCd885/QD800) are implanted in a tumor to detect the signal [40]. These QDs can give a signal even after 16 days of implantation. Consequently, they can be used for the detection of cancer cells. Five carboxyl QDs having the same size with different emission spectra are used for imaging lymphs of the tumor due to its bright, photostable and luminescent properties [41]. Polydentate phosphine-coated QDs are used for the operation of big animals to visualize the lymph node which permits the resection [42]. Capped INP@ZnSQDs are used for cellular imaging to distinguish the five separate lymphatic flows [43].

CdTe QDs are used to measure human prostate cells [44]. CdTe QDs are stabilized with L-glutathione in water media followed by linking with prostate-specific antigen. This help to bind human prostate cancer cell specifically. Moreover, QD labeling affects the growth of the cancer cell. This QD is more stable than FITC conjugated dyes.

11.3 Heavy Metal–Free QDs for Ex Vivo Imaging

Ex vivo is a term associated to label cells or tissues taken out from the body. Ex vivo labeling also needs a non-toxic, biocompatible mode and more photostable probe for labeling. The Indium and Gadolinium-based QDs are widely used for ex vivo imaging[45]. The Indium-based probe can reside in the lymph and has stable photoluminescence. Gadolinium-doped carbon dots are used for fluorescence as well as MRI [46, 47]. Nanoprobes (approximately 5.5 nm and almost spherical) prepared from citric acid have great special resolution plus are used to enhance fluorescence compared to carbon dots [48]. Several types of QDs that are designed for ex vivo imaging include:

- Graphene quantum dots
- Semiconductor quantum dots
- Near-infrared quantum dots
- Fluorescent jelly quantum dots
- PEG-coated biocompatible quantum dots

11.3.1 Graphene Quantum Dots (GQDs)

Graphene is widely used in several biomedical fields [49] besides its toxic reports from model organisms [50]. It is composed of one to many strata of graphene sheets having dimensions < 10 nm. GQDs possess the property of 2D graphene with extraordinary physicochemical properties. The physicochemical properties include a non-zero bandgap for the electronic plus optical industries [51]. Fluorescent GQDs have the upper hand over conventional organic and semiconductor QDs due to their outstanding photostability, nontoxicity, prolonged fluorescence property, small size, little cost and easy grounding methods. They are used for bioimaging and diagnosis [52, 53] plus cancer theranostics [54, 55]. GQDs are further employed in two-photon imaging and MRI plus dual-modal imaging. GQDs were employed as a fluorescent probe by Pan et al. to image in vitro as well as in vivo [56]. Redox subtle fluorescent probe founded on GQDs was designed to determine the proteins [57, 58]. GQDs were modified to detect the carbohydrate receptor [59]. Li et al. designed them to detect

H_2S by photoluminescence [60]. PEI-coated GQDs are used to image tumor cells in vitro [61]. Ding et al. established a GQD grounded theranostic nanoagent loaded through doxorubicin [62]. GQD-conjugated protein nanofiber (PNF-GQDs) is used to target RGD receptors [63]. PNF-GQDs can be taken up by the cell five times faster than the normal GQDs. FA GQDs alter electronic energy. Heteroatom-doped GQDs are used for cellular imaging. N-GQDs can detect formaldehyde optically due to the fluorescence turn on/off mechanism, which is activated by a redox-sensitive mechanism.

11.3.2 Semiconductor Quantum Dots

These are the alternative to organic fluorophore due to their photostability and brightness, less toxicity and sparing solubility in water [64–66]. Their molecular weight is 500 kDA in size and they affect the function and dynamics of the target molecule. They regulate the growth of the cell without altering its morphology. They function well at 37° C, which is the ideal temperature for imaging since at higher temperatures the media evaporates. Colloidal QDs (CQDs) are synthesized and used to label yeast and *Escherichia coli* [67]. The QD binding peptide which is encoded genetically (the glutathione molecules) is used to label cells. Heavy metals used in these CQDs have toxic effects if they remain unremoved from the body. Polymer-based Janus spheres have biotin tags on one side and are further labeled with green and red [68, 69]. QDs are also used in human umbilical vein endothelial cells (HUVECs). Coating of polyethylene glycol (PEG) homogenously to the Janus beads prevents non-specific adhesion [70, 71].

11.3.3 Near-Infrared Quantum Dots (NIR QDs)

NIR QDs are extensively used in the biomedical field especially in imaging. The wavelengths of NIR range from 650 to 900 nm and from 1,200 to 1,600 nm. This range is well known for its large penetration depth. Its photostability and biocompatibility make it an ideal bioluminescent probe [72, 73]. CdTe QD emits in the NIR region. However, most of the NIR QDs have heavy metals like Cd, Hg and Pb which restrict their application in the arena of biomedical imaging [74, 75]. Later, low-toxic NIR QDs were developed. This includes the family of I–III–VI, I–VI and II–V. Such QDs are synthesized either using organic or aqueous phases [76]. The QDs synthesized in the water media are non-toxic, biocompatible and economical. Thus, they can be applied straight for the investigation of biological systems. Most of the QDs (except CdTe and HgTe) synthesized in the aqueous phase have deprived luminescence properties [77]. Organic phase synthesized QDs are insoluble in water [78]. Thus, additional modifications are needed for their application in the field of biology. QDs are encapsulated with amphiphilic polymers or silica to improve the photoluminescence property [79, 80].

CdTe QDs are used for tumor imaging. Arginine–glycine–aspartic acid (RGD) peptide-labeled CdTe/ZnS QDs can label the cancer cells [81, 82]. RGD peptide attached to the outside of CdTe QDs radiate near the NIR fluorescence region. Later it was vaccinated to mice to transplant within U87MG tumors to distinguish the tumor borderline [83, 84]. DNA functionalized Zn^{2+} doped CdTe QDs is used to image live tumors. Doping of Zn2+ in presence of DNA further reduces its toxicity [85]. Biotoxicity of CuInS2/ ZnS and CdTeSe/CdZnS QDs can be detected by monitoring the inflammatory parameters within mice [39]. Intravenous injection of Ag2S QDs to mice can be removed from the body without any toxicity [86]. Ag2S QDs have a great signal-to-noise ratio, great resolution in imaging and thus are used to image blood-associated parameters. Parameters like the flow of blood, vessels of the lymph and formation of blood vessels can be monitored using these QDs. Ag2S QDs coated with antibody (anti-vascular endothelial growth factor) can target U87MG tumors [87]. Ag2S QDs linked to thiopropionic acid are used to image 4T1luc tumors in mice[88]. S-nitrosothiols with glutathione-stabilized Ag2S QDs were used for the regulated release of nitric oxide and for fluorescence imaging at the NIR range [89]. Ag2Se QDs having emission wavelengths at 1,300 nm are used for in vivo imaging.

11.3.4 Fluorescent Jelly Quantum Dots

To make the quantum dots biocompatible, numerous advancements were made. One of the methods is by fabricating with natural polymers. Amid the natural polymers, gelatin is used in many food products. Thus it is an indispensable polymer for pharmaceuticals and medical applications. The structure of gelatin allows us to use it in hydrogels, polymer conjugates, cocktails and in the synthesis of microparticles as well as nanoparticles. Gelatin is a non-toxic biopolymer derived from collagens. Collagens are composed of polypeptides, glycine, proline and 4-hydroxyproline. Gelatin-modified QDs are widely used these days [90, 91]. CdTe/CdSe QDs embedded in gelatin are synthesized using the aqueous phase. Such nanoparticles are more biocompatible since they prevent the release of cadmium to the surroundings [92].

11.3.5 PEG-Coated Biocompatible Quantum Dots

To make the nanoparticle biocompatible quite a lot of surface coating is made on the nanoparticle. They include polycaprolactone, polylactic acid, polyethylene glycol (PEG), polyglycolic acid, poly dl-lactideco-glycolide and poly alkyl cyanoacrylate. Among all, PEG is commonly used because of its biocompatible property for several cytological studies. Lipids-conjugated PEG is also biocompatible and its fluorescence property remains intact. The lipophilic property helps to enter within the cell easily. PEG-lipid conjugates further allow adding several functional groups like peptides, antibodies and aptamers to target a particular type of cells. In QD lipid-PEG several hydrophilic branches of PEG are in the outside to make it biocompatible [93, 94]. PEG-DSPE can retain in the vascular system for a longer time period by decreasing the clearance from reticuloendothelial cells [95, 96]. PEG-DSPE is coated with coated cadmium chalcogenide QDs to achieve different emissions [97, 98]. CdS has blue, CdSe has green and CdTe has red luminescence property.

11.4 QDs for Transfection

Transfection is a method to introduce the DNA in a cell to be used for several purposes. A number of viral and non-viral methods are used for transfection [99, 100]. Among them, QDs are broadly used for transfection or gene delivery purposes. QDs have several advantages over other gene delivery vectors. QDs are labeled with small interfering RNA (siRNA) via disulfide linkages or via noncovalent adsorption of siRNA to alter the surface charge [101, 102]. Labeling of nucleic acids is still essential for the investigation of the release of siRNA. The transfer of electrons among QDs and DNA persuades quenching via QD emission. Though the application of QDs for the investigation of gene delivery is greatly encouraging, numerous encounters still hinder the practice of QDs in vivo. For example, long-term inquiries of QDs in live cells are still challenging [103]. Grafted cationic polypeptides developed after cationized bovine serum albumin (cBSA-147) are used for stable gene transfection. QDs are cytotoxic at high concentrations, thus a number of approaches are made to develop noncytotoxic QDs. QDs are encapsulated using polycationic polyethylene oxide (PEO) or polypeptide copolymer to develop a steady coating on the QDs for multivalent interactions. Bovine serum albumin (BSA) is used for surface coating, to obtain decent solubility in saline buffer and underneath physiological circumstances. pH-responsive QDs are designed subsequently coated with a human serum albumin (HSA)-derived copolymer with grafted PEO chains through ligand exchange (HSA-QDs) [104]. HSA-QDs having negative charges are biocompatible, however, the cells cannot take it easily due to the repulsion of charge among the HSA-QDs and the membrane of the cell. A novel polypeptide copolymer, dcBSA-PEO (5000) 27-TA 26, has 26 thioctic acid groups lengthwise as the backbone of the polypeptide. This polymer was used to coat QDs (cBSA-QDs), and the respective QDs were used for imaging of the live cell. These QDs are competent enough to be taken up the cell, highly stable within the cell and less cytotoxic. A complex of cBSA-QDs with plasmid DNA depicts promising photo-responsive properties. Thus, these QDs are most appropriate to investigate the transportation and unpacking of DNA during the process of gene delivery. Furthermore, optimization of transfection reagents and DNA is a time-consuming approach. The photo-responsive behavior of cBSA-QDs to plasmid DNA makes it more effective for gene transfection. cBSA-QDs were detected within large-sized endosomes[103]. Henceforth, cBSA-QDs are used as a method for transfection. QDs along with allatostatin (an insect neuropeptide) can enhance the transfection efficiency and is well suitable for photodynamic therapy [105].

11.5 Future Perspective or Conclusion

QDs with their highly photostable property can solve the issues associated with in vivo and in vitro imaging in the future. Currently, the application of QDs is facing several challenges like stability and toxicity. Especially Cd-containing QDs release Cd^{+2} ion. Appropriate surface modification of such QDs will be helpful to reduce the toxicity and make them suitable for biomedical applications. To completely overcome this issue many non-cadmium QDs should develop for the desired application. Since QDs like $NaYF_4$ and $NaGaF_4$ are less toxic and more suitable for biomedical applications they should be substituted with other heavy metal QDs. The development of new QDs with less noxious properties will solve the toxicity issues associated with the application in certain biomedical fields.

Acknowledgments

MM laboratory is supported by grants from SERB/EMR/2017/003054, BT/PR21857/NNT/28/1238/2017 and Odisha DBT 3325/ST (BIO)-02/2017.

REFERENCES

1. Ornes, S. (2016). Core concept: Quantum dots. *Proceedings of the National Academy of Sciences, 113*(11), 2796–2797.
2. Martynenko, I. V., Litvin, A. P., Purcell-Milton, F., Baranov, A. V., Fedorov, A. V., & Gun'Ko, Y. K. (2017). Application of semiconductor quantum dots in bioimaging and biosensing. *Journal of Materials Chemistry B, 5*(33), 6701–6727.
3. Pandey, S., & Bodas, D. (2020). High-quality quantum dots for multiplexed bioimaging: A critical review. *Advances in Colloid and Interface Science, 278*, 102137.
4. Wu, P., & Yan, X. P. (2013). Doped quantum dots for chemo/biosensing and bioimaging. *Chemical Society Reviews, 42*(12), 5489–5521.
5. Yong, K. T., Law, W. C., Hu, R., Ye, L., Liu, L., Swihart, M. T., & Prasad, P. N. (2013). Nanotoxicity assessment of quantum dots: From cellular to primate studies. *Chemical Society Reviews, 42*(3), 1236–1250.
6. Wang, L., Nagesha, D. K., Selvarasah, S., Dokmeci, M. R., & Carrier, R. L. (2008). Toxicity of CdSe nanoparticles in Caco-2 cell cultures. *Journal of Nanobiotechnology, 6*(1), 1–15.
7. Alencar, L. D., Pilla, V., Andrade, A. A., Donatti, D. A., Vollet, D. R., & De Vicente, F. S. (2014). High fluorescence quantum efficiency of CdSe/ZnS quantum dots embedded in GPTS/TEOS-derived organic/silica hybrid colloids. *Chemical Physics Letters, 599*, 63–67.
8. Deng, D., Zhang, W., Chen, X., Liu, F., Zhang, J., Gu, Y., & Hong, J. (2009). Facile synthesis of high-quality, water-soluble, near-infrared-emitting PbS quantum dots. *European Journal of Inorganic Chemistry, 2009*(23), 3440–3446.
9. Al-Jamal, W. T., Al-Jamal, K. T., Tian, B., Lacerda, L., Bomans, P. H., Frederik, P. M., & Kostarelos, K. (2008). Lipid-quantum dot bilayer vesicles enhance tumor cell uptake and retention in vitro and in vivo. *ACS Nano, 2*(3), 408–418.
10. Kuo, Y. C., Wang, Q., Ruengruglikit, C., Yu, H., & Huang, Q. (2008). Antibody-conjugated CdTe quantum dots for *Escherichia coli* detection. *The Journal of Physical Chemistry C, 112*(13), 4818–4824.
11. Liu, J., Lau, S. K., Varma, V. A., Kairdolf, B. A., & Nie, S. (2010). Multiplexed detection and characterization of rare tumor cells in Hodgkin's lymphoma with multicolor quantum dots. *Analytical Chemistry, 82*(14), 6237–6243.
12. Liu, Z., Li, G., Xia, T., & Su, X. (2015). Ultrasensitive fluorescent nanosensor for arsenate assay and removal using oligonucleotide-functionalized CuInS2 quantum dot@ magnetic Fe3O4 nanoparticles composite. *Sensors and Actuators B: Chemical, 220*, 1205–1211.

13. Pinaud, F., Clarke, S., Sittner, A., & Dahan, M. (2010). Probing cellular events, one quantum dot at a time. *Nature Methods*, *7*(4), 275–285.
14. Pathak, P. K., & Hughes, S. (2009). Generation of entangled photon pairs from a single quantum dot embedded in a planar photonic-crystal cavity. *Physical Review B*, *79*(20), 205416.
15. Xiao, Y., Forry, S. P., Gao, X., Holbrook, R. D., Telford, W. G., & Tona, A. (2010). Dynamics and mechanisms of quantum dot nanoparticle cellular uptake. *Journal of Nanobiotechnology*, *8*(1), 1–9.
16. Nayak, N., & Mishra, M. (2019). Simple techniques to study multifaceted diabesity in the fly model. *Toxicology Mechanisms and Methods*, *29*(8), 549–560.
17. Mandal, S., & Jana, N. R. (2017). Quantum dot-based designed nanoprobe for imaging lipid droplet. *The Journal of Physical Chemistry C*, *121*(42), 23727–23735.
18. Anas, A., Okuda, T., Kawashima, N., Nakayama, K., Itoh, T., Ishikawa, M., & Biju, V. (2009). Clathrin-mediated endocytosis of quantum dot–peptide conjugates in living cells. *ACS Nano*, *3*(8), 2419–2429.
19. Chakraborty, A., & Jana, N. R. (2015). Clathrin to lipid raft-endocytosis via controlled surface chemistry and efficient perinuclear targeting of nanoparticle. *The Journal of Physical Chemistry Letters*, *6*(18), 3688–3697.
20. Nanda, R., Panda, P., & Mishra, M. (2020). Biomarker detection of Parkinson's disease: Therapy and treatment using nanomaterials. In Kanchi, Suvardhan & Deepali Sharma (Eds.), *Nanomaterials in Diagnostic Tools and Devices* (pp. 479–523). Elsevier.
21. Wang, T., & Xu, C. (2020). Three-photon neuronal imaging in deep mouse brain. *Optica*, *7*(8), 947–960.
22. Dahan, M., Levi, S., Luccardini, C., Rostaing, P., Riveau, B., & Triller, A. (2003). Diffusion dynamics of glycine receptors revealed by single-quantum dot tracking. *Science*, *302*(5644), 442–445.
23. Kawashima, N., Nakayama, K., Itoh, K., Itoh, T., Ishikawa, M., & Biju, V. (2010). Reversible dimerization of EGFR revealed by single-molecule fluorescence imaging using quantum dots. *Chemistry–A European Journal*, *16*(4), 1186–1192.
24. Ernsberger, U. (2009). Role of neurotrophin signalling in the differentiation of neurons from dorsal root ganglia and sympathetic ganglia. *Cell and Tissue Research*, *336*(3), 349–384.
25. Howarth, M., Liu, W., Puthenveetil, S., Zheng, Y., Marshall, L. F., Schmidt, M. M., ... & Ting, A. Y. (2008). Monovalent, reduced-size quantum dots for imaging receptors on living cells. *Nature Methods*, *5*(5), 397–399.
26. Maysinger, D., Boridy, S., & Hutter, E. (2010). Subcellular fate of nanodelivery systems. In *Organelle-Specific Pharmaceutical Nanotechnology* (pp. 93–121). John Wiley & Sons.
27. Sandros, M. G., Behrendt, M., Maysinger, D., & Tabrizian, M. (2007). InGaP@ ZnS-enriched chitosan nanoparticles: A versatile fluorescent probe for deep-tissue imaging. *Advanced Functional Materials*, *17*(18), 3724–3730.
28. Delehanty, J. B., Boeneman, K., Bradburne, C. E., Robertson, K., Bongard, J. E., & Medintz, I. L. (2010). Peptides for specific intracellular delivery and targeting of nanoparticles: Implications for developing nanoparticle-mediated drug delivery. *Therapeutic Delivery*, *1*(3), 411–433.
29. Ahmed, S. R., Dong, J., Yui, M., Kato, T., Lee, J., & Park, E. Y. (2013). Quantum dots incorporated magnetic nanoparticles for imaging colon carcinoma cells. *Journal of Nanobiotechnology*, *11*(1), 1–9.
30. Biju, V., Itoh, T., & Ishikawa, M. (2010). Delivering quantum dots to cells: Bioconjugated quantum dots for targeted and nonspecific extracellular and intracellular imaging. *Chemical Society Reviews*, *39*(8), 3031–3056.
31. Duan, H., & Nie, S. (2007). Cell-penetrating quantum dots based on multivalent and endosome-disrupting surface coatings. *Journal of the American Chemical Society*, *129*(11), 3333–3338.
32. Cao, Y. A., Yang, K., Li, Z., Zhao, C., Shi, C., & Yang, J. (2010). Near-infrared quantum-dot-based non-invasive in vivo imaging of squamous cell carcinoma U14. *Nanotechnology*, *21*(47), 475104.
33. Bagalkot, V., Zhang, L., Levy-Nissenbaum, E., Jon, S., Kantoff, P. W., Langer, R., & Farokhzad, O. C. (2007). Quantum dot–aptamer conjugates for synchronous cancer imaging, therapy, and sensing of drug delivery based on bi-fluorescence resonance energy transfer. *Nano Letters*, *7*(10), 3065–3070.
34. Cui, Y., Hu, Z., Zhang, C., & Liu, X. (2014). Simultaneously enhancing up-conversion fluorescence and red-shifting down-conversion luminescence of carbon dots by a simple hydrothermal process. *Journal of Materials Chemistry B*, *2*(40), 6947–6952.
35. Xing, Y., & Rao, J. (2008). Quantum dot bioconjugates for in vitro diagnostics & in vivo imaging. *Cancer Biomarkers*, *4*(6), 307–319.
36. Cai, W., Hsu, A. R., Li, Z. B., & Chen, X. (2007). Are quantum dots ready for in vivo imaging in human subjects? *Nanoscale Research Letters*, *2*(6), 265–281.
37. Chokkareddy, R., & Kanchi, S. (2020). Simultaneous detection of ethambutol and pyrazinamide with IL@ CoFe 2 O 4 NPs@ MWCNTs fabricated glassy carbon electrode. *Scientific Reports*, *10*(1), 1–10.
38. Stylianou, P., & Skourides, P. A. (2009). Imaging morphogenesis, in *Xenopus* with quantum dot nanocrystals. *Mechanisms of Development*, *126*(10), 828–841.
39. Zou, W., Li, L., Chen, Y., Chen, T., Yang, Z., Wang, J., ... & Wang, X. (2019). In vivo toxicity evaluation of PEGylated CuInS2/ZnS quantum dots in BALB/c mice. *Frontiers in Pharmacology*, *10*, 437.
40. Yang, K., Cao, Y. A., Shi, C., Li, Z. G., Zhang, F. J., Yang, J., & Zhao, C. (2010). Quantum dot-based visual in vivo imaging for oral squamous cell carcinoma in mice. *Oral Oncology*, *46*(12), 864–868.
41. Kobayashi, H., Hama, Y., Koyama, Y., Barrett, T., Regino, C. A., Urano, Y., & Choyke, P. L. (2007). Simultaneous multicolor imaging of five different lymphatic basins using quantum dots. *Nano Letters*, *7*(6), 1711–1716.
42. Kim, S., Lim, Y. T., Soltesz, E. G., De Grand, A. M., Lee, J., Nakayama, A., ... & Frangioni, J. V. (2004). Near-infrared fluorescent type II quantum dots for sentinel lymph node mapping. *Nature Biotechnology*, *22*(1), 93–97.
43. Wegner, K. D., & Hildebrandt, N. (2015). Quantum dots: Bright and versatile in vitro and in vivo fluorescence imaging biosensors. *Chemical Society Reviews*, *44*(14), 4792–4834.
44. Hu, P., Wang, X., Wei, L., Dai, R., Yuan, X., Huang, K., & Chen, P. (2019). Selective recognition of CdTe QDs and strand displacement signal amplification-assisted

label-free and homogeneous fluorescence assay of nucleic acid and protein. *Journal of Materials Chemistry B*, *7*(31), 4778–4783.
45. Matea, C. T., Mocan, T., Tabaran, F., Pop, T., Mosteanu, O., Puia, C., ... & Mocan, L. (2017). Quantum dots in imaging, drug delivery and sensor applications. *International Journal of Nanomedicine*, *12*, 5421.
46. Bourlinos, A. B., Bakandritsos, A., Kouloumpis, A., Gournis, D., Krysmann, M., Giannelis, E. P., ... & Zboril, R. (2012). Gd (III)-doped carbon dots as a dual fluorescent-MRI probe. *Journal of Materials Chemistry*, *22*(44), 23327–23330.
47. Huang, Y., Li, L., Zhang, D., Gan, L., Zhao, P., Zhang, Y., ... & Jia, C. (2020). Gadolinium-doped carbon quantum dots loaded magnetite nanoparticles as a bimodal nanoprobe for both fluorescence and magnetic resonance imaging. *Magnetic Resonance Imaging*, *68*, 113–120.
48. Shangguan, J., He, D., He, X., Wang, K., Xu, F., Liu, J., ... & Huang, J. (2016). Label-free carbon-dots-based ratiometric fluorescence pH nanoprobes for intracellular pH sensing. *Analytical Chemistry*, *88*(15), 7837–7843.
49. Priyadarsini, S., Mohanty, S., Mukherjee, S., Basu, S., & Mishra, M. (2018). Graphene and graphene oxide as nanomaterials for medicine and biology application. *Journal of Nanostructure in Chemistry*, *8*(2), 123–137.
50. Priyadarsini, S., Sahoo, S. K., Sahu, S., Mukherjee, S., Hota, G., & Mishra, M. (2019). Oral administration of graphene oxide nano-sheets induces oxidative stress, genotoxicity, and behavioral teratogenicity in *Drosophila melanogaster*. *Environmental Science and Pollution Research*, *26*(19), 19560–19574.
51. Chen, F., Gao, W., Qiu, X., Zhang, H., Liu, L., Liao, P., ... & Luo, Y. (2017). Graphene quantum dots in biomedical applications: Recent advances and future challenges. *Frontiers in Laboratory Medicine*, *1*(4), 192–199.
52. Li, K., Liu, W., Ni, Y., Li, D., Lin, D., Su, Z., & Wei, G. (2017). Technical synthesis and biomedical applications of graphene quantum dots. *Journal of Materials Chemistry B*, *5*(25), 4811–4826.
53. Zhang, R., & Ding, Z. (2018). Recent advances in graphene quantum dots as bioimaging probes. *Journal of Analysis and Testing*, *2*(1), 45–60.
54. Kumawat, M. K., Thakur, M., Bahadur, R., Kaku, T., Prabhuraj, R. S., Ninawe, A., & Srivastava, R. (2019). Preparation of graphene oxide-graphene quantum dots hybrid and its application in cancer theranostics. *Materials Science and Engineering: C*, *103*, 109774.
55. Schroeder, K. L., Goreham, R. V., & Nann, T. (2016). Graphene quantum dots for theranostics and bioimaging. *Pharmaceutical Research*, *33*(10), 2337–2357.
56. Pan, D., Zhang, J., Li, Z., & Wu, M. (2010). Hydrothermal route for cutting graphene sheets into blue-luminescent graphene quantum dots. *Advanced Materials*, *22*(6), 734–738.
57. Eivazzadeh-Keihan, R., Pashazadeh-Panahi, P., Baradaran, B., Maleki, A., Hejazi, M., Mokhtarzadeh, A., & de la Guardia, M. (2018). Recent advances on nanomaterial based electrochemical and optical aptasensors for detection of cancer biomarkers. *TrAC Trends in Analytical Chemistry*, *100*, 103–115.
58. Zhang, L., Ying, Y., Li, Y., & Fu, Y. (2020). Integration and synergy in protein-nanomaterial hybrids for biosensing: Strategies and in-field detection applications. *Biosensors and Bioelectronics*, *154*, 112036.
59. Chen, J., Than, A., Li, N., Ananthanarayanan, A., Zheng, X., Xi, F., ... & Chen, P. (2017). Sweet graphene quantum dots for imaging carbohydrate receptors in live cells. *FlatChem*, *5*, 25–32.
60. Li, D., Qin, L., Zhao, P., Zhang, Y., Liu, D., Liu, F., ... & Lu, G. (2018). Preparation and gas-sensing performances of ZnO/CuO rough nanotubular arrays for low-working temperature H_2S detection. *Sensors and Actuators B: Chemical*, *254*, 834–841.
61. Gao, T., Wang, X., Yang, L. Y., He, H., Ba, X. X., Zhao, J., ... & Liu, Y. (2017). Red, yellow, and blue luminescence by graphene quantum dots: Syntheses, mechanism, and cellular imaging. *ACS Applied Materials & Interfaces*, *9*(29), 24846–24856.
62. Ding, H., Zhang, F., Zhao, C., Lv, Y., Ma, G., Wei, W., & Tian, Z. (2017). Beyond a carrier: Graphene quantum dots as a probe for programmatically monitoring anti-cancer drug delivery, release, and response. *ACS Applied Materials & Interfaces*, *9*(33), 27396–27401.
63. Su, Z., Shen, H., Wang, H., Wang, J., Li, J., Nienhaus, G. U., ... & Wei, G. (2015). Motif-designed peptide nanofibers decorated with graphene quantum dots for simultaneous targeting and imaging of tumor cells. *Advanced Functional Materials*, *25*(34), 5472–5478.
64. Kairdolf, B. A., Smith, A. M., Stokes, T. H., Wang, M. D., Young, A. N., & Nie, S. (2013). Semiconductor quantum dots for bioimaging and biodiagnostic applications. *Annual Review of Analytical Chemistry*, *6*, 143–162.
65. Martynenko, I. V., Litvin, A. P., Purcell-Milton, F., Baranov, A. V., Fedorov, A. V., & Gun'Ko, Y. K. (2017). Application of semiconductor quantum dots in bioimaging and biosensing. *Journal of Materials Chemistry B*, *5*(33), 6701–6727.
66. Arya, H., Kaul, Z., Wadhwa, R., Taira, K., Hirano, T., & Kaul, S. C. (2005). Quantum dots in bio-imaging: Revolution by the small. *Biochemical and Biophysical Research Communications*, *329*(4), 1173–1177.
67. Gao, Z., Zhao, C. X., Li, Y. Y., & Yang, Y. L. (2019). Beer yeast-derived fluorescent carbon dots for photoinduced bactericidal functions and multicolor imaging of bacteria. *Applied Microbiology and Biotechnology*, *103*(11), 4585–4593.
68. Le, T. C., Zhai, J., Chiu, W. H., Tran, P. A., & Tran, N. (2019). Janus particles: Recent advances in the biomedical applications. *International Journal of Nanomedicine*, *14*, 6749.
69. Walther, A., & Muller, A. H. (2013). Janus particles: Synthesis, self-assembly, physical properties, and applications. *Chemical Reviews*, *113*(7), 5194–5261.
70. Liao, M., Liu, H., Guo, H., & Zhou, J. (2017). Mesoscopic structures of poly (carboxybetaine) block copolymer and poly (ethylene glycol) block copolymer in solutions. *Langmuir*, *33*(30), 7575–7582.
71. Posel, Z., & Posocco, P. (2019). Tuning the properties of nanogel surfaces by grafting charged alkylamine brushes. *Nanomaterials*, *9*(11), 1514.
72. Zhao, J., Zhong, D., & Zhou, S. (2018). NIR-I-to-NIR-II fluorescent nanomaterials for biomedical imaging and cancer therapy. *Journal of Materials Chemistry B*, *6*(3), 349–365.
73. Hong, G., Antaris, A. L., & Dai, H. (2017). Near-infrared fluorophores for biomedical imaging. *Nature Biomedical Engineering*, *1*(1), 1–22.

74. Chinnathambi, S., & Shirahata, N. (2019). Recent advances on fluorescent biomarkers of near-infrared quantum dots for in vitro and in vivo imaging. *Science and Technology of Advanced Materials*, *20*(1), 337–355.
75. Quek, C. H., & Leong, K. W. (2012). Near-infrared fluorescent nanoprobes for in vivo optical imaging. *Nanomaterials*, *2*(2), 92–112.
76. Ma, Y., Zhang, Y., & William, W. Y. (2019). Near infrared emitting quantum dots: Synthesis, luminescence properties and applications. *Journal of Materials Chemistry C*, *7*(44), 13662–13679.
77. Zhao, D., He, Z., Chan, W. H., & Choi, M. M. (2009). Synthesis and characterization of high-quality water-soluble near-infrared-emitting CdTe/CdS quantum dots capped by N-acetyl-L-cysteine via hydrothermal method. *The Journal of Physical Chemistry C*, *113*(4), 1293–1300.
78. Navarro, D. A., Watson, D. F., Aga, D. S., & Banerjee, S. (2009). Natural organic matter-mediated phase transfer of quantum dots in the aquatic environment. *Environmental Science & Technology*, *43*(3), 677–682.
79. Hu, X., & Gao, X. (2010). Silica–polymer dual layer-encapsulated quantum dots with remarkable stability. *ACS Nano*, *4*(10), 6080–6086.
80. Nida, D. L., Nitin, N., Yu, W. W., Colvin, V. L., & Richards-Kortum, R. (2007). Photostability of quantum dots with amphiphilic polymer-based passivation strategies. *Nanotechnology*, *19*(3), 035701.
81. Cai, W., & Chen, X. (2008). Preparation of peptide-conjugated quantum dots for tumor vasculature-targeted imaging. *Nature Protocols*, *3*(1), 89–96.
82. He, X., Gao, J., Gambhir, S. S., & Cheng, Z. (2010). Near-infrared fluorescent nanoprobes for cancer molecular imaging: Status and challenges. *Trends in Molecular Medicine*, *16*(12), 574–583.
83. Gao, J., Chen, K., Xie, R., Xie, J., Yan, Y., Cheng, Z., … & Chen, X. (2010). In vivo tumor-targeted fluorescence imaging using near-infrared non-cadmium quantum dots. *Bioconjugate Chemistry*, *21*(4), 604–609.
84. Cheng, Z., Wu, Y., Xiong, Z., Gambhir, S. S., & Chen, X. (2005). Near-infrared fluorescent RGD peptides for optical imaging of integrin αvβ3 expression in living mice. *Bioconjugate Chemistry*, *16*(6), 1433–1441.
85. Zhang, C., Ji, X., Zhang, Y., Zhou, G., Ke, X., Wang, H., … & He, Z. (2013). One-pot synthesized aptamer-functionalized CdTe: Zn2+ quantum dots for tumor-targeted fluorescence imaging in vitro and in vivo. *Analytical Chemistry*, *85*(12), 5843–5849.
86. Zhang, Y., Zhang, Y., Hong, G., He, W., Zhou, K., Yang, K., … & Wang, Q. (2013). Biodistribution, pharmacokinetics and toxicology of Ag2S near-infrared quantum dots in mice. *Biomaterials*, *34*(14), 3639–3646.
87. Zhang, H., Yee, D., & Wang, C. (2008). Quantum dots for cancer diagnosis and therapy: Biological and clinical perspectives. *Nanomedicine (Lond)*, *3*(1), 83–91.
88. Tang, R., Xue, J., Xu, B., Shen, D., Sudlow, G. P., & Achilefu, S. (2015). Tunable ultrasmall visible-to-extended near-infrared emitting silver sulfide quantum dots for integrin-targeted cancer imaging. *ACS Nano*, *9*(1), 220–230.
89. Tan, L., Wan, A., & Li, H. (2013). Conjugating S-nitrosothiols with glutathiose stabilized silver sulfide quantum dots for controlled nitric oxide release and near-infrared fluorescence imaging. *ACS Applied Materials & Interfaces*, *5*(21), 11163–11171.
90. Elzoghby, A. O. (2013). Gelatin-based nanoparticles as drug and gene delivery systems: Reviewing three decades of research. *Journal of Controlled Release*, *172*(3), 1075–1091.
91. Yasmin, R., Shah, M., Khan, S. A., & Ali, R. (2017). Gelatin nanoparticles: A potential candidate for medical applications. *Nanotechnology Reviews*, *6*(2), 191–207.
92. Girija Aswathy, R., Sivakumar, B., Brahatheeshwaran, D., Ukai, T., Yoshida, Y., Maekawa, T., & Kumar, S. D. (2011). Biocompatible fluorescent jelly quantum dots for bioimaging. *Materials Express*, *1*(4), 291–298.
93. Jokerst, J. V., Lobovkina, T., Zare, R. N., & Gambhir, S. S. (2011). Nanoparticle PEGylation for imaging and therapy. *Nanomedicine*, *6*(4), 715–728.
94. Daglar, B., Ozgur, E., Corman, M. E., Uzun, L., & Demirel, G. B. (2014). Polymeric nanocarriers for expected nanomedicine: Current challenges and future prospects. *RSC Advances*, *4*(89), 48639–48659.
95. Phillips, W. T., Klipper, R. W., Awasthi, V. D., Rudolph, A. S., Cliff, R., Kwasiborski, V., & Goins, B. A. (1999). Polyethylene glycol-modified liposome-encapsulated hemoglobin: A long circulating red cell substitute. *Journal of Pharmacology and Experimental Therapeutics*, *288*(2), 665–670.
96. Che, J., Okeke, C., Hu, Z. B., & Xu, J. (2015). DSPE-PEG: A distinctive component in drug delivery system. *Current Pharmaceutical Design*, *21*(12), 1598–1605.
97. Poulose, A. C., Veeranarayanan, S., Mohamed, M. S., Raveendran, S., Nagaoka, Y., Yoshida, Y., … & Kumar, D. S. (2012). PEG coated biocompatible cadmium chalcogenide quantum dots for targeted imaging of cancer cells. *Journal of Fluorescence*, *22*(3), 931–944.
98. Wang, Y., Wu, B., Yang, C., Liu, M., Sum, T. C., & Yong, K. T. (2016). Synthesis and characterization of Mn: ZnSe/ZnS/ZnMnS sandwiched QDs for multimodal imaging and theranostic applications. *Small*, *12*(4), 534–546.
99. Biju, V., Muraleedharan, D., Nakayama, K. I., Shinohara, Y., Itoh, T., Baba, Y., & Ishikawa, M. (2007). Quantum dot-insect neuropeptide conjugates for fluorescence imaging, transfection, and nucleus targeting of living cells. *Langmuir*, *23*(20), 10254–10261.
100. Li, Y., Jing, L., Yang, C., Qiao, R., & Gao, M. (2011). Quantum dot-antisense oligonucleotide conjugates for multifunctional gene transfection, mRNA regulation, and tracking of biological processes. *Biomaterials*, *32*(7), 1923–1931.
101. Derfus, A. M., Chen, A. A., Min, D. H., Ruoslahti, E., & Bhatia, S. N. (2007). Targeted quantum dot conjugates for siRNA delivery. *Bioconjugate Chemistry*, *18*(5), 1391–1396.
102. Bilan, R., Fleury, F., Nabiev, I., & Sukhanova, A. (2015). Quantum dot surface chemistry and functionalization for cell targeting and imaging. *Bioconjugate Chemistry*, *26*(4), 609–624.
103. Wu, Y., Eisele, K., Doroshenko, M., Algara-Siller, G., Kaiser, U., Koynov, K., & Weil, T. (2012). A quantum dot photoswitch for DNA detection, gene transfection, and live-cell imaging. *Small*, *8*(22), 3465–3475.
104. Wu, Y., Chakrabortty, S., Gropeanu, R. A., Wilhelmi, J., Xu, Y., Er, K. S., … & Weil, T. (2010). pH-responsive quantum dots via an albumin polymer surface coating. *Journal of the American Chemical Society*, *132*(14), 5012–5014.
105. Alzugaray, M. E., Hernández-Martínez, S., & Ronderos, J. R. (2016). Somatostatin signaling system as an ancestral mechanism: Myoregulatory activity of an allatostatin-C peptide in hydra. *Peptides*, *82*, 67–75.

12

Quantum Dots as a Versatile Tool for Bioimaging Applications

Shaik Baji Baba, Naresh Kumar Katari, Rajasekhar Chokkareddy and Gan G. Redhi

CONTENTS

12.1 Introduction

Many researchers are exploring the outstanding potentiality of QDs in biomedical research in medical imaging to develop sophisticated theranostic agents to indicate the potentiality of QDs in biomedical applications. Nanoprobes (nanoparticles, fluorescent proteins, and carbon/graphene-based nanomaterials) and semiconductor QDs have many applications themselves in bioimaging because of their size and optical properties [1]. The size of the QDs is smaller than the excitation Bohr radius. The small size property (1–10 nm) of these QDs is supreme candidate to identify and track the biomolecules inside the cells [2]. Moreover, carbon dots and red fluorescence emission are possible with QD semiconductors, which may be favored to color in biological imaging because of the autofluorescence property of cells in the range of 400–600 nm (like green, blue, and yellow). In targeting applications, QDs effectively target tumors or biomarkers and play a key role in cancer detection, monitoring, and diagnosis [1].

QDs, also known as semiconductor nanocrystals, have been introduced as a capable tool in life sciences due to their optical and physical properties, consisting of II-VI, III-V, or IV-VI elements in the periodic table [3]. QDs are remarkably stable at the time of excitation and contain typical absorption and emission spectra [4]. These nanoparticles are shown narrow emission peaks and the semiconductor QD fluorescence is brighter than the organic fluorescent dyes. Therefore, the visibility of particles is improved, and less laser intensity is required for the bioimaging process. These properties help biomedical researchers build a potential tool to express the importance of QDs in diagnosis and imaging. Many fluorescent materials (quantum dots, carbon dots, organic dyes, metal-organic frameworks) are designed, developed, and applied in several fields like biomedicine [5], sensing [6], photo-electrochemistry [7], and catalysis [8], etc. Compared to all QDs, CdS play a key role in developing the various fluorescent nanomaterials. QDs are now considered as potential candidates as luminescent probes and labels in life science applications such as disease diagnosis, biological imaging, and molecular histopathology [9]. Many studies have reported the use of QDs for sentinel lymph nodes for in vitro and in vivo imaging [10], tumor immune responses [11], tumor-specific receptors [12], and malignant tumor detectors [13]. Figure 12.1 shows the number of articles published using nanotechnology in biological applications during the last 10 years.

Many reviews addressed and explained various questions related to QDs, including recent developments in surface modification of QDs, cellular uptake mechanism, bioimaging, and structure–activity relationships. This review chapter has shown developments and applications of QDs in biomedical research, especially in bioimaging. This review gives an excellent understanding of why and how QDs are an important versatile tool in biomedical applications.

DOI: 10.1201/9781003157823-12

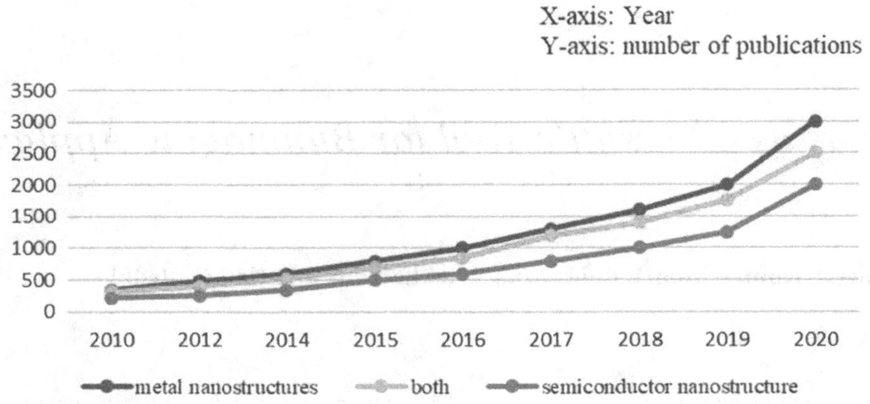

FIGURE 12.1 Number of publications between 2010 to 2020 reports the use of metallic and semiconductor QD nanostructure in biomedical applications. This data was assembled using the Medline database (approximately). It reveals the increasing applications of nanostructured materials in biological applications.

The basic structural phenomenon of QDs contains a semiconductor core, shell, and cap. The shell acts as a protective layer for the core and is responsible for capping ligands coating, which helps to form optical and conductive properties (Figure 12.2). Core-shell QDs contain many excellent application properties, but they also have a few disadvantages. The disadvantage of QDs core-shell depends on free metal ion leaching to aqueous solutions from the core-shell QDs. The core-shell QDs toxicity depends on various physicochemical and environmental features such as mechanical stability, size, capping material, concentration, charge, and functional groups.

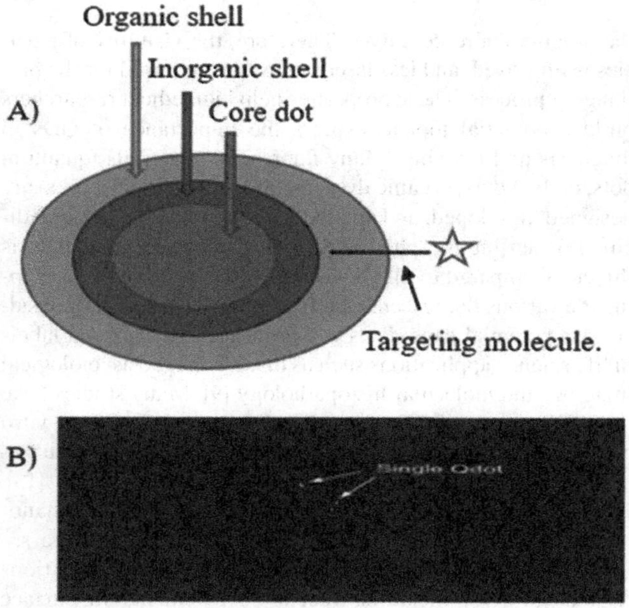

FIGURE 12.2 (A) Schematic representation of quantum dot targeted probes. The common current QD probes for imaging are consist of a semiconductor core with organic and inorganic shells. The inorganic cell increases the semiconductor QDs optical property. It decreases cytotoxicity after photodegradation. The organic core allows the quantum dots to interact with the biological system and provides a suitable functional group for coupling with the targeted molecule. (B) The optical image of ZnS-coated CdSe QDs.

Between the 1970s and 1980s, understanding the photophysical parameters of semiconductor constructions was essential in computer and electronics applications. Research and synthesis of semiconductor nanostructures, also known as quantum dots, began in the 1980s, and the applications of these materials play a significant role in electronic and biomedical industries. These QDs contain zero-dimensional electron systems, and the properties depend on the spatial detection of electrons. In electronic and computer applications, scientists are designed and synthesized nanometer-sized building blocks to construct very faster and small computer chips and high effective light-emitting diode systems. Henglein [14] and Rossetti et al. [15] designed and synthesized colloidal CdS QDs. CdS QDs were obtained by mixing cadmium and sulfide salts in an aqueous buffer solution. After this successful development, many researchers attempted to develop good optical properties of these QDs, increasing the monodispersity of particle and size tunability by varying the effect of different reaction conditions (for example, solvents, temperature, pH, salts) on their surface morphology and optoelectric properties [16]. These early primary research results guided the current state of the art in QDs synthesis using the organometallic synthetic process, which was developed by Murray et al. [17]. The "Greener" method uses organic stabilizing agents and metallic salts [18]. At present, these two synthetic methods (with small variations) are used to develop QDs with desired optical properties such as broad absorption, greater quantum yield, stability, and narrow fluorescence emission for biomedical applications. QDs like both group II-VI (e.g., CdSe, CdS, CdTe, and ZnSe) and group III-V (e.g., InP and InAs) semiconductor nanocrystals have been synthesized and revealed their properties [19, 20]. In the early 1990s, QDs were synthesized in aqueous solutions by adding stabilizing agents (like polyphosphate/thioglycerol). However, this process yields a deficient quality of QDs with poor fluorescence efficiency and significant size differences. In 1993, using high-temperature organometallic procedures, extremely luminescent CdSe QDs were developed by Bawendi et al. [21]. In this development,

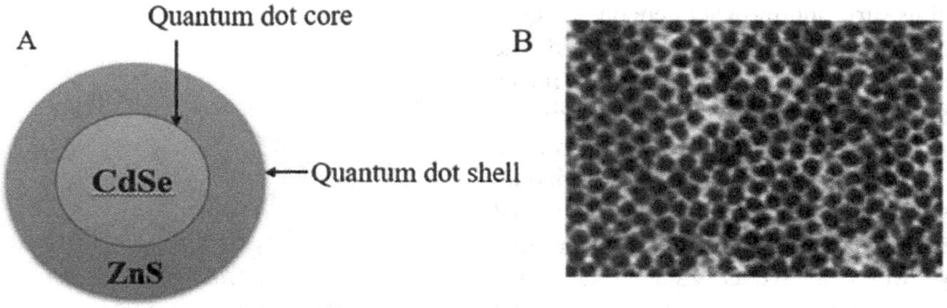

FIGURE 12.3 (A) Schematic representation of ZnS (inorganic shell) coated CdSe QDs. (B) Transmission electron microscopy image of ZnS-capped CdSe QDs. (Reprinted with permission from [22], copyright 1996, American Chemical Society.)

nanocrystals contain approximately clear crystal structure and size variations but the fluorescence quantum yields are very low (~10–15%).

The use of QDs in biology is shown to be tremendously important to cap or passivate the CdSe QDs with CdS or ZnS layers. The CdS/ZnS layers may improve the QDs fluorescence quantum yield and protect from photo-oxidation, which is more important for decreasing the cytotoxicity and increasing the photostability [22]. A ZnS capping layer is produced by dipping dimethylzinc and hexamethyldisilathiane (in tri-n-octylphosphine) slowly into the reaction vessel. The obtained ZnS shell thickness is based on the amount of dimethylzinc and hexamethyldisilathiane injected into the reaction vessel. Compare to CdSe, ZnS shells have large bandgap energy. The ZnS shell contains an approximate bond length to the CdSe, minimizing the crystal lattice strain. Figure 12.3 shows the schematic representation of CdSe/ZnS-capped QD shells with their TEM and optical fluorescence microscopy images [22].

12.2 Synthesis of QDs

Two main methods to synthesize QDs include "top-down" and "bottom-up". In the top-down method for the synthesis of quantum dots, a large semiconductor breaks down into small-sized material using beam lithography, etching, chemical oxidation, reactivation etching, wet chemical etching, and arc discharge [23]. The main disadvantage of this method is the QDs impurity and structural defects with designing [24]. In the bottom-top method, the synthesis of QDs by molecular precursors is broadly classified into vapor phase and wet chemical approaches. The wet chemical method using conventional precipitation approaches contains nucleation and partial growth of NPs with specific control properties for both single and mixture solutions. The nucleation approach was made by following ways such as heterogeneous, homogeneous, and secondary nucleation. Few examples of the wet chemical approaches are sol-gel [25], microemulsion [26], hot-solution thermal-decomposition [27], electrochemical approach [28], and microwaves [29]. A graphite rod with a one-step electrolysis method and a water-diffused phosphorus/sulfur co-doped GQDs are synthesized and used as a bright electrochemiluminescence signal maker by conjugating with a monoclonal antibody against okadaic acid and gradually it increases the electrochemiluminescence nature. In the vapor phase approach QD layers are grown like atom by atom on a smooth substrate. Therefore, QDs with self-assembly are grown by molecular beam epitaxy, aggregation, and sputtering types on a substrate without any forms [30]. Many QDs synthesis methods have difficulties with multistep, expensive use of toxic solvents requirement and need of surface passivation. The main advantage of the QDs colloidal chemical synthesis method is high quality and monodispersed nano-size particles [31]. This synthetic route rises from classical chemistry. To develop the monodispersed colloids, it needs a fast nuclear occurrence followed by nuclei with slow development. The different sizes of QDs can be obtained by optimizing the temperature in reactions, time, and using different types of solvents. In colloidal synthesis, solvent plays an important factor. Solvents like methanol, butanol, and hexane flocculate the QDs, whereas olive oils and n-octadecane prevent unsaturated compounds formation and improve reusability as dissolution media. The purification of QDs can be done using precipitation and centrifugation with different organic solvents which can be dispersed again into other organic solvents. Here, the polarity of the solvents may involve the precipitation and purification of QDs [32]. The usage of polymers in QDs synthesis increases the QDs stability and controls the growth of QDs. Another critical factor in the QDs development is temperature, which affects the QDs properties like photoluminescence intensity and wavelength. Using different synthetic temperatures in a cluster solution of QDs will result in different photoluminescence intensities and photoluminescence peak shifts attributed to the thermal trapping in electrons in the QDs interface trap state at all individual clusters. At the time of synthesis, the temperature-dependent bandgap shrinkage increases the nonradiative decay of QDs [33]. It should be noted that the synthesis of QDs via different approaches and solvents has different properties in terms of size, surface group, shape, edge state, etc., affecting their photoluminescence and physicochemical properties. Wide research on the QDs photoluminescence emission shows that it is affected by the size, edge configurations, and surface functional groups of quantum dots.

12.3 Optical Properties

The quantum dot optical properties (semiconductor nanocrystals) are obtained by the interactions between the electrons, holes, and corresponding surroundings. Semiconductor QDs absorbed the photons once the excitation energy exceeds the bandgap. Valence and conduction band systems exist in the semiconductor systems, and the energy difference between the two bands is called bandgap energy. In the entire process, electrons are promoted from the valence band to the conduction band. The UV-visible spectra measurements indicate that a greater number of energy states appear in QDs. The very first peak observes the very lowest excited energy state peak, called quantum confinement peak. The molar extinction coefficient is slowly increased to shorter wavelengths. This parameter is essential for biology applications because it permits instantaneous excitation of multicolor QDs with a single light-emitting source [34].

The other important biomedical parameter of QDs includes their constant absorption nature, which permits different QD emissions to be excited by using single wavelength and 20 nanoseconds fluorescent lifetime. It helps QDs to be used in time-resolved fluorescence bioimaging, stability against photobleaching, and allows them to have applications in biomonitoring, like protein tracking and intrinsic brightness, which gives a single QD to be imaged [35–37]. The comparison of optical properties of QDs with organic dye as well as biological applications is shown in Table 12.1 and Figure 12.4.

The main important features of QDs can be explained in three major properties as follows:

i) Using a single light-emitting source, the QDs can exhibit large spectral excitation to excite multicolor QDs and a thin sharp emission peak which decreases the spectral overlapping [38].
ii) The major change between the absorbed and emitted wavelengths of QDs is the Stokes shift, which permits collecting the full emission spectrum by separating the QDs fluorescence signal from the autofluorescence backgrounds and improving the detection sensitivity. This is an important factor for the tissue imaging process [39].
iii) Compared to organic dyes, quantum dots contain a long fluorescence lifetime due to QDs inorganic composition and bright emissions. The brightness of QDs is about 10–20 times greater than the single organic fluorescent compounds.

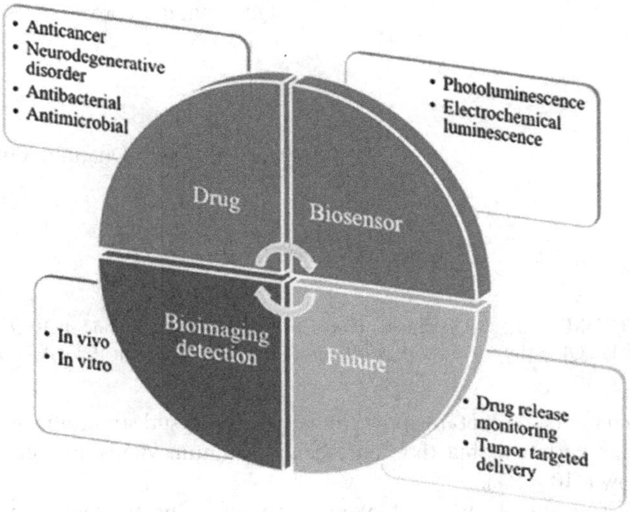

FIGURE 12.4 Various biological applications of QDs.

12.4 The Application of QDs to Cell Imaging

The potential applications of QDs are generally focused on cell imaging and labeling in vivo and in vitro research due to their higher fluorescent nature than the organic dyes or fluorescent proteins. The fluorescence emission wavelength should be maintained by size so that the color can be maintained by QDs size adjustment. From the recent reports, the fluorescence stability of labeled cells with QDs is approximately 400-fold greater than the Alexa 488 (organic pharmacophore) [33]. QDs applications in cell imaging are divided into three major categories such as cell staining, living cell tracking, and fluorescence probe. A few examples are shown in Table 12.2. Recent developments of QDs applications in cell imaging mainly focus the fluorescence and near-infrared luminescence QDs, which show an emission wavelength range from 700 to 900 nm. The light-emitting with this range contains its maximum possibility of tissue penetration and interfering of tissue autofluorescence at minimal (emission ranging from 400 to 600 nm) [40].

12.4.1 Cell Staining

QDs with no other particular functionalization or surface modifications can be used for cell staining. Kumawat and coworkers

TABLE 12.1
Comparison of Optical Properties of QDs with Organic Dye

Property	Quantum dot	Organic dye
Absorption spectra	Brad band	Red and blue two absorption bands
Coefficient of extinction	Large and the extinction range from 10^5–10^6M^{-1}cm^{-1} for the first excitonic absorption	Typically ranges from 10^3–10^5M^{-1}cm^{-1}
Emission spectrum	Narrow, symmetric, and it turns to NIR region with excellent quantum yield	Generally, at red wavelengths greater than 600 nm and boarder than QDs with low fluorescence quantum yield
Photostability	Great photostability	Less photostability
Water solubility	The surface of the QD can be modified by water-soluble and biocompatible nature	Poor water solubility and in few cases requires liposomal and micellar delivery
Metabolic degradation	High resistant	It depends on sensitizers

TABLE 12.2
QDs with Cell Imaging Applications

QD core/shell	Size	Modification	Receptor	Application	Ref.
CdTe/CdS	5.4 ± 1.8 nm	Human epidermal growth factor receptor 2 (HER2) receptors	HER2	Targeted cell imaging	[41]
CdSe/ZnS	13 nm	His8-tagged affirmers	Affimer targeted protein	For imaging targeted cancer cell	[42]
$AgInS_2$	3.1 nm	Polyethylenimine		For probing of water and glucose	[43]
ZnCuInSe/ZnS	3–4 nm			In vivo labeled red blood cell tracking	[44]
GQD	1–8 nm			There is no specific cell nucleus imaging	[45]
Nitrogen, sulfur-doped CQDs	6.5 ± 0.5 nm			There is no specific cancer cell detection or imaging	[46]
Nitrogen-doped GQDs	4–6 nm			No specific rBMSCs imaging	[47]

used synthesized graphene QDs (GQDs) in an aqueous solution using grape seed as the main precursor [48]. Without any capping or targeting agents, the developed GQDs with bright red fluorescence can slowly enter into the cell nucleus within six to eight hours and finally self-localize in the cell nucleus. At the initial four hours, GQDs were concentrated and situated around the nucleus and this nucleus was dyed with strong blue by 40,6-diamidino-2-phenylindole (DAPI, nuclear fluorescent dye). After 6 h, GQDs enter into the nucleus and produce solid and continuous red fluorescence, where DAPI stains the blue fluorescent. Hong and coworkers synthesized very ultra-small single-layered GQDs with blue fluorescence using trisodium citrate as a precursor and used them to measure the optical effectiveness of HeLa cells. This report also found that GQDs can easily enter into the nucleus and dye the cell with fluorescent blue. Therefore, GQDs are showing great ability to enter the cell nucleus for staining. The major issue of QDs in cell staining is they may enter the cell or maintain for a long time without any elimination [49]. Based on GQDs' or carbon QDs' (CQDs) great potential activity in the field of bioimaging, many researchers have made significant scientific efforts to develop their fluorescence property by doping with sulfur or nitrogen into QDs [50]. The metal ions can easily quench the fluorescence, and this is one of the major challenges in biological imaging using CQDs [51]. A kind of sulfur and nitrogen-doped GQDs are developed by Schroer et al. [51] which can attack the metal ion quenching.

12.4.2 Fluorescence Probe and Sensor

The surface of the quantum dots is modified with a specific capping or targeting agent that can be used as a fluorescence probe and quantitatively detects the particular cellular molecules. In 1998, the first time demonstrated the QDs applications in cell labeling. After this, many QD probe compounds have been developed because of the development of many ligands from chemical molecules to biological macromolecules. There is a high potentiality of QD probes in the detection of cancer cells. Grafting of folic acid on the Ag_2Te/SiO_2 QD surface to target cancer cells was developed by Jin et al. [52]. Folic acid is greatly expressed in a few tumor cells and they can be specifically recognized by folic acid. In tumor cells, the vascular endothelial growth factor (VEGF) is greatly expressed to accumulated angiogenesis. Panagiotopoulou et al. [53] designed and developed two types of molecular imprinted polymer-coated QDs that emit red and green fluorescence which helps to identify the glucuronic acid and N-acetylneuraminic acid correspondingly in human keratinocytes and it realizes the multicolor imaging of these cells. Hence, the fluorescent QDs combination with receptor-ligand mechanism has great significance in the specified cell line imaging and tumor diagnosis. The specific cellular imaging and quantitative detection with QD probes are further defined as biosensors based on cell fluorescence imaging. Zhao et al. showed that adding chlorin e6 on the silicon QD surface via electrostatic adsorption to obtain a Si QD-Ce6 fluorescence probe can detect and monitor the hydroxy radical production [54]. When •OH exists, the sensing of Si QDs is quenched, while chlorin e6 is continuing to emit fluorescence, as shown in Figure 12.5.

12.4.3 Living Cell Tracking

By considering the advantages of high photostability and non-photobleaching properties, QDs have been shown high potentiality in the area of living cell imaging and tracking. The Rowland team developed a voltage-sensitive QD probe for imaging the electrically active cells and the cell tracking profile with millisecond time resolution [56]. In this study, due to the electric field in the neuronal membrane droves, the QDs ionization and the photoluminescence of QDs were demolished as the action potential changings. Next, QDs can also use for tracking circulating tumor cells that have high significance to treat the metastatic potential of many cancers. Kuo et al. delivered a conjugated QD antibody into the living animal bloodstreams, monitored CTCs exceptionally in real time, and indicated its great impact in identifying CTC by using multi-photon microscopy [55]. Photostability and non-toxicity both are two essential parameters for the living cell tracings.

12.5 Quantum Dots for Multiplexed Bioimaging

The quantum confinement effect of quantum dots exhibits size-dependent emissions. When the radius of the QDs is less

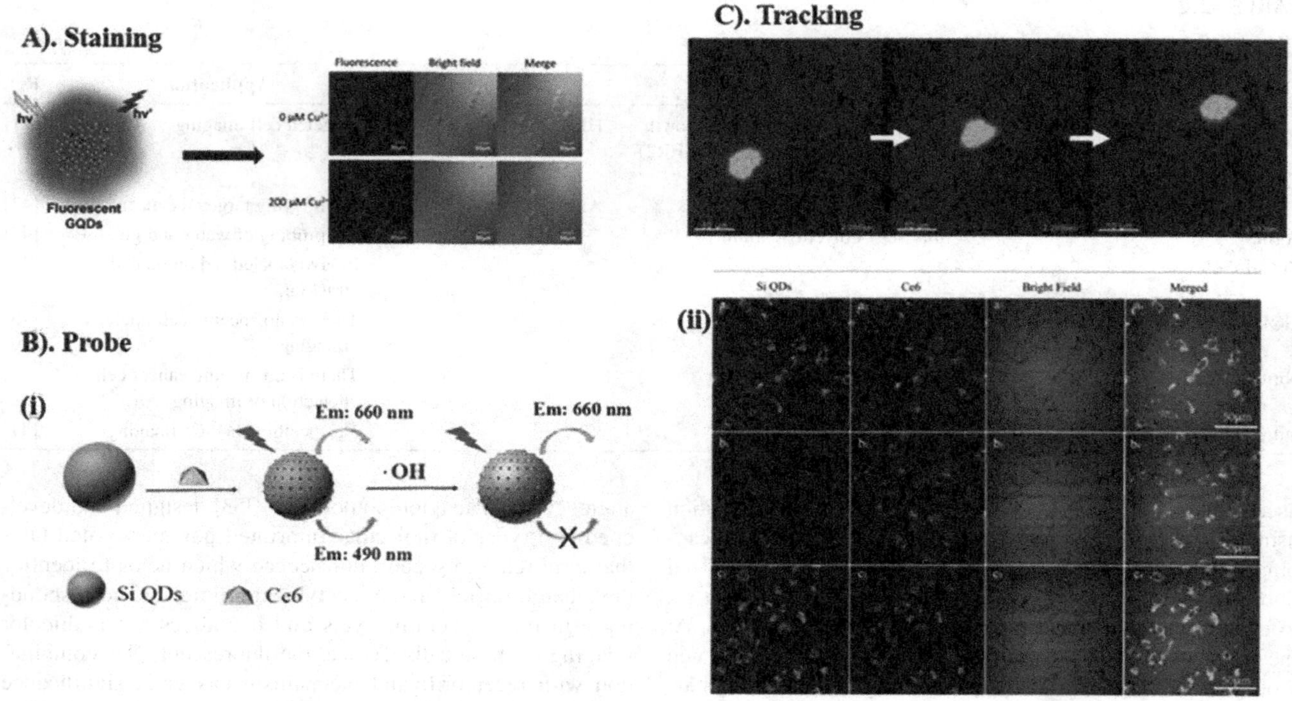

FIGURE 12.5 Examples of QDs in cell imaging. (A) Laser scanning confocal microscopy image of RAW 264.7 cells with Cu^{2+}. Reprinted with permission from [51], copyright 2019, American Chemical Society. (B) (i) Probe principle. (ii) Confocal fluorescence microscopy images of Si QDs-Ce6 complex probe: a) incubation of cells with 100μg/mL Si-Ce6 QDs for two hours at 38° C; b) with Si-Ce6 QDs after that treated with LPS (10 μg/mL) for one hour; c) incubation of cells with Si–Ce6 QDs after that treated with 0.1% DMSO and LPS (10 μg/mL) in sequence. Reprinted with permission from [54], copyright 2017, American Chemical Society. (C) Three consecutive images of CD133+CTC moving in the blood vessels (time interval: 33 ms). Red signals are from quantum dots and green signals are from tumor cells expressing green fluorescent protein (GFP). The movie can be found in Additional File 5: Movie M3. Tumor cells: BXPC3-GFP. Reproduced with permission from [55], copyright 2019, Springer Nature.

than the Bohr radius exciton, the energy levels are quantized [57, 58]. QDs can emit any light colors simply by changing the QDs size, making them an alternative to current fluorophores for multiplexed imaging. QDs can be conjugated with many biomolecules such as DNA, proteins, antibodies, etc. to target several organelles.

Multicolor CQDs (blue, yellow, and green) from Manilkara zapota fruits were developed by Bhamore et al. [59]. The developed multiple fluorescence emissive QDs used for the multicolor imaging of various bacterial cells.

Jeong et al. developed a multiplexed in vivo cell imaging using PbS (core)/CdS (shell) QDs encapsulates in the polymer matrix [60]. Polymer QDs emitting at 1,285 nm and 1,085 nm were developed and injected into a tumor-xenografted mouse and detected concurrently using CCD near IR at a single excitation (650 nm).

Kobayashi et al. performed simultaneous multicolor in vivo imaging using five different QDs with different emission spectra; the lymphatic imaging was performed in mice using distinct types of carboxyl QDs [61].

Five different streptavidin conjugates with commercially available QDs with different spectral imaging values (525, 565, 585, 605, 655, 705, and 810 nm) were used by Fountaine et al. for imaging the several tissue targets [62]. This study effectively confirms the application of QDs in multiplex imaging and helps researchers understand the cellular interactions that motivate certain diseases.

The multiplex imaging using exchange PAINT was developed by Werbin et al. [63]. Using this method, they constantly detect the five receptor tyrosine kinases such as EGFR, IGF-1R, ErbB2, Met, and ErbB3 at endogenous appearance levels in BT20 carcinoma cells. They also reported the changes of distributions in receptors following ligand stimulations. Finally, the author concludes with the use of PAINT for the analysis of distribution and membrane protein interactions.

Quantum confinement also allows the quantum dots to release many colors by size variations [64]. Therefore, this type of QDs can be used specifically in multiplexed bioimaging applications. Also, they exhibit more quantum efficiency, photochemical stability, and reproducibility in synthesis. Coating of QDs with a biocompatible molecule (biocompatibility) reduces its toxicity and allows the biomarker conjugation for multiplexed imaging. These properties have outstanding alternatives to new fluorophores in different imaging applications.

In the upcoming days, QDs can show the major importance and impact in biology research. The major importance in biology is the finding and map the biomolecule networks that identify the cell function and cell viability [65, 66]. This helps in many research areas such as molecular diagnostics, tissue engineering, and molecular biology. Active cell QDs can be designed for biomolecule color-coding due to their excellent photostability and brightness so that they can be imaged and monitored for a long time. The bioimage analysis can show (i) the molecular family's interactions with one another, (ii) the binding interaction time, and (iii) the rate of molecular assemblies.

12.6 In Vitro and In Vivo Imaging Applications of Quantum Dots

QDs were initially measured as potential optical probes to replace the organic fluorophores for biomedical applications due to their advantageous properties. Many researchers have established the application and use of QDs for cell labeling [67, 68], flow cytometry [69, 70], cell tracking migration [71, 72], pathogen detection [73, 74], fluorescence resonance energy transfer (FRET) sensors [75, 76], fluorescence in situ hybridization [77, 78], high-throughput screening of biomolecules [79, 80], and whole-animal contrast agents [81, 82]. These investigations demonstrated the importance of QDs in biological applications.

So many types of research have been introduced to applications of QDs for in vitro imaging over the past few decades. In vivo applications face many problems for researchers because they need to optimize several parameters associated with imaging living cells and tissues. These applications require several great optical properties exhibited by QDs but also require biocompatibility and less toxicity. Fluorescent imaging in vitro mainly belongs to three categories, cellular imaging, biomolecular tracking in cells, and tissue staining.

Hasan et al. [45] synthesized near-IR-emitting GQDs from reduced graphene oxide for in vitro and in vivo bioimaging applications. NIR fluorescence imaging of GQDs exhibits effective internalization in HeLa cell lines at 12 hours with other estimated excretion. GQDs with NIR 950 nm emission is largely biocompatible in vitro with no other primary toxic nature response in vivo. They outlined the new GQDs as NIR fluorescence imaging probes suitable for tracking therapeutic delivery in animal cells. Sun et al. [83] synthesized a novel fluorescent Ag-In-S/ZnS QDs by solution reaction after that dipped in poly(vinyl pyrrolidone) for in vivo applications with red emission for tumor drainage lymph node imaging (Figure 12.6). The developed QDs are greatly dispersed in an aqueous solution with 295 nm hydrodynamic radius for lymph node imaging. After an intratumoral injection to mice, within 10 mins, tumor drainage lymph node imaging was achieved.

Raji et al. [84] synthesized CQDs from banana peel waste using a hydrothermal process. The developed CQDs contain fine size distribution and average particle size, exhibit excellent water solubility, and excitation depending on the emission property. The synthesized CQDs have confirmed less toxicity for nematodes at 200 μg mL^{-1} due to their fluorescence stability and biocompatibility CQDs are used for bioimaging applications in nematodes.

Shu et al. [85] synthesized Zn doped Cu_2S QDs with intrinsic properties of thermodynamic therapy activity and NIR fluorescence imaging capability using the one-step synthesis method. In vitro and in vivo experiments reveal that Zn doped Cu_2S QDs can effectively inhibit the cancer cell growth rate without damaging other cell lines. Xin et al. [86] synthesized nitrogen-doped graphene QDs using graphite as a precursor. These QDs exhibits a green and blue fluorescence with different sizes and type of nitrogen doping. In vitro and in vivo imaging of these QDs shows a high and clear fluorescence property with less toxicity and high biocompatibility. The author concludes that the nitrogen-doped graphene QDs with more excellent fluorescent properties can be used for bioimaging applications

Li et al. [87] used amino acid 4-nitro-3-phenyl-L-alanine, as a triplet state quencher to link covalently to Cy5.5 dye and an amphiphilic polymer of poly (oligo [ethylene glycol] methyl ether meth-acrylate)-block-poly(2-amino-N4-[2-diisopropylamin oethyl]-L-aspartic acid). With the intramolecular electron transfer between the amino acid and Cy5.5, the amino acid with the protecting agent acts as a potential quencher. Using confocal laser scanning microscopy, the photostability of the obtained fluorophore was measured in HepG2 cell lines. Fluorophore coupling with copolymers for self-healing shows an essential pathway to developing its biological imaging applications to monitor the living cells.

Zhang et al. [88] synthesized a boron-dipyrromethene derivative fluorophore. The developed fluorophore exhibits less toxicity and 53% photoluminescence quantum yield. Next, fluorophore photostability was evaluated in MCF-7 cell lines by using a confocal fluorescence microscope. Results indicate that developed fluorophore is shown applications to track the endocytosis process activated by the bacterium and proved significant tracking lysosomes at different physical conditions.

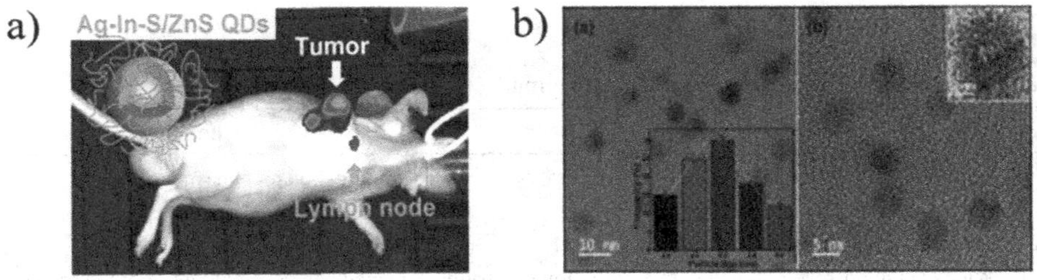

FIGURE 12.6 (a) In vivo tumor drainage lymph node imaging of Ag-In-S/ZnS QDs. (b) TEM image of synthesized QDs: a) particle size distribution graph of QDs; b) HRTEM image of synthesized QDs. Reprinted with permission from [83], Copyright 2021, American Chemical Society.

Zhen et al. [89] developed a series of red and NIR fluorophores using electron-withdrawing BDTO (benzo[1,2-b:4,5-b'] dithiophene 1,1,5,5-tetraoxide). The developed nanoparticles with a NIR fluorophore triphenylamine: BDTO, which exhibits great photostability in targeted cancer cell photodynamic ablation. They demonstrated that developed fluorophores can be used for two-photon fluorescence imaging in intraviral mouse brain vasculature.

Narasimhan et al. [90] reported high applications of graphene quantum dots for in vivo bioimaging. GQDs were loaded with polyacrylamide gel, and they can be implanted intravenously near the euthanized mouse's thoracic region. A deep red fluorescence was detected successfully, and results indicate that developed GQDs can effectively help in cell line imaging as an optical contrast agent and also in pre-clinical imaging.

Other investigators have reported the applications of 3–5 nm CdS QDs, green synthesized from tea leaves extract for the applications of A549 cancer cell in vitro imaging by Ponmurugan et al. [91]. The developed QDs are stable as analyzed by flow cytometry and concluded with high impact in intracellular fluorescence cell tracking.

Goreham et al. [92] published the water-soluble, less cytotoxic graphene oxide quantum dots, and folic acid functionalization. The obtained QDs have resulted as non-toxic to macrophage cells. For the bioimaging process, they used fluorescence lifetime imaging and multi-photon microscopy to detected HeCaT cell imaging. The results indicate that the synthesized QDs are effectively showing impact in QDs bioimaging applications.

Wu et al. [93] reported a diketopyrrolopyrrole conjugate porphyrin-based organic nanoparticles (NPs) for in vivo photoacoustic imaging and photo-thermal therapy. The obtained porphyrin NPs exhibited a significant photo-thermal effect compared to reported organic and inorganic nanoparticles. Conjugates have shown remarkable photo-thermal therapeutic action in HeLa tumors in mice.

Yang et al. [94] reported a synthesis of amphiphilic fluoroporphyrins NPs for photo-thermal and photodynamic synergistic cancer therapy. They performed in vitro testing using HeLa cells. The reported NPs were found as highly biocompatible and also used as potent theragnostic nano-agents for cancer diagnosis in in vivo treatment.

So many fluorophores have been reported for different applications in cellular bioimaging. The majority of them face some limitations in bioimaging via low biocompatibility, photostability, emission spectra, broad excitation, and the difficulties in surface modification for coupling of other biomarkers, etc. These limitations affect the applications of QDs in bioimaging, mainly in multiplex imaging applications. But QDs exhibits great potential applications because of their optical properties such as great biocompatibility, easy surface modification, UV single excitation and size tunability, etc. for bioimaging applications.

12.7 Challenges

Quantum dots show a great impact on cell imaging and at the same time, there are various problems in their practical application. The few advantages and challenges of QDs in bioimaging were listed in Table 12.3. First, because of intracellular vesicle transport mode, QDs may be ready to aggregate in the cell, resulting in uneven stating. There are very few active ways to transfer monodisperse QDs into living cells [95]. Another drawback is QDs can easily wrap up via biomacromolecules and encapsulates by endocytosis vesicles in cells, which affects the loss of its surface labeling effect [96]. In QDs, non-specified binding is also a major problem like interfering with electrostatic adherence [97]. The main important parameter ID application of QDs depends on the type of biomarker development.

12.8 Cytotoxicity

QDs possess their unique physicochemical properties and also exhibit their potential cytotoxicity. Hardman reported the toxicity of quantum dots as parameters of physicochemical and environmental features [97]. QD concentration, charge, size, outer coating like capping material and functional groups, photolytic, oxidative, and mechanical stability affect the important factors in the toxicity of QDs. The in vitro of QDs reveals few cytotoxicities in cells. In 2005, Lovic et al. [98] established that (i) cadmium telluride (CdTe) QDs can enter into the cells and is distributed in different subcellular sections; (ii) CdTe QDs may exert cytotoxicity categorized by various changes in the morphology of the cell nucleus as wee as reductions in the metabolic activity; (iii) the QDs-induced toxicity mechanism may be arbitrated by the other mechanisms than the formation of free radical. They also reveal that mercaptopropionic acid and cysteamine coated CdTe showed cytotoxicity to pheochromocytoma cells in the rat (PC12) culture at a concentration of 10 μg/mL and also CdTe uncoated QDs

TABLE 12.3

Advantages and Challenges of QDs for Cell Imaging

S. no.	Advantage	Challenges
1	Great fluorescence strength	Aggregation and uneven dyeing
2	High fluorescence stability	Nonspecific binding
3	Long fluorescence period	Easy to be covered up on cells
4	Wide excitation spectrum and narrow emission spectrum	Inadequate development of biometric coatings
5	Tunable fluorescence release	Cytotoxicity
6	Able for surface-functionalized modification	Low production and high cost

were exhibiting the cytotoxicity at 1 μg/mL. The cell apoptosis was measured by membrane blebbing, chromatin condensation, and symptomatic of apoptosis. QDs with a small positive charge (2.2 ± 0.1 nm) exhibits more cytotoxicity than the large positively charged QDs (5.2 ± 0.1 nm) at same concentrations, and the cytotoxicity was determined by using MTT assay. The size of the quantum dots may also affect the subcellular distribution, where the small cationic QDs localize to the nuclear section and large cationic QDs localize to the cytosol. CdTe QD-induced cytotoxicity reveals the partially dependence on the size of the QDs and QDs coating, intracellular reactions, and intracellular degradation of QDs into the metalloid ions.

Maysinger et al. [99] reported the mechanisms to cellular cytotoxicity of surface uncoated CdTe Qds in MCF-7 cell culture. The proposed investigation reveals that CdTe QDs induce cell death by nonclassical apoptosis initiated by the reactive oxygen species imaging in live cells. Voura et al. [100] performed the treating of B16F10 melanoma cells with dihydroxy lipoic acid capped ZnS/CdSe QDs. After 5 h of QDs injection, the tumor cell in the lung imaging and the results indicate that tumor cell labeling has no detectable effects on survival ability in the circulation of the endothelium. Next, QDs labeling with B16F10 cells has also shown no effect on tumor formation ability. Another study exposed the capping of CdSe/ZnS QDs with 11-mercaptoundecanoic acid and coated with sheep serum albumin and applied on Vero cells (0.24 mg/mL); results exhibit that QDs are not showing any effect on cell viability [101].

Apoptosis is one of the major cytotoxicities when QDs are treated with living organisms. Many researchers are trying to explain the mechanism of apoptosis caused by QDs use in cells. Reactive oxygen species (ROS) generation will exhibit in the oxidative stress that can mediate apoptosis. Yan et al. reported that CdTe QDs has planned to induce oxidative stress, then it can play a key role in QDs-mediated mitochondrial-dependent apoptosis in HUVECs cells [102]. Peynshaert et al. [103] reported that PEGylated-QDs were exhibiting more toxicity than MPA-coated QDs because of ROS production increasing and lysosomal damage, resulting in cytotoxicity and autophagy dysfunction. The cytotoxicity effect of carboxylic acid-coated QDs such as QD 565 and QD 655 on human keratinocytes [104] perceptibly inhibited keratinocytes cell viability in an equal concentration.

The mechanisms for QDs cytotoxicity are complicated and no one has shown the exact theory up to now. According to the previous literature, four types of mechanisms are influencing cytotoxicity, i.e., mitochondrial damage, synthetic materials toxicity in QDs, immune toxicity, and more remarkable ROS synthesis (Figure 12.7) [105–107]. This classification is also not clear due to the free Cd^{2+} and increased ROS synthesis, mitochondrial damage, and immune toxicity.

12.9 Future Prospects

For the past few years, QDs have been considered important in bio-nanotechnology and biomedicine because of their unique optical properties and emissions. Along with QDs, other nanostructured materials such as carbon nanoparticles and metal nanoparticles were added to the precursors to accelerate both fundamental and applied biochemical research. Like any other new field, there are also a few disadvantages in QDs, such as emission variations of QDs in the different solvents, high cytotoxicity, etc.

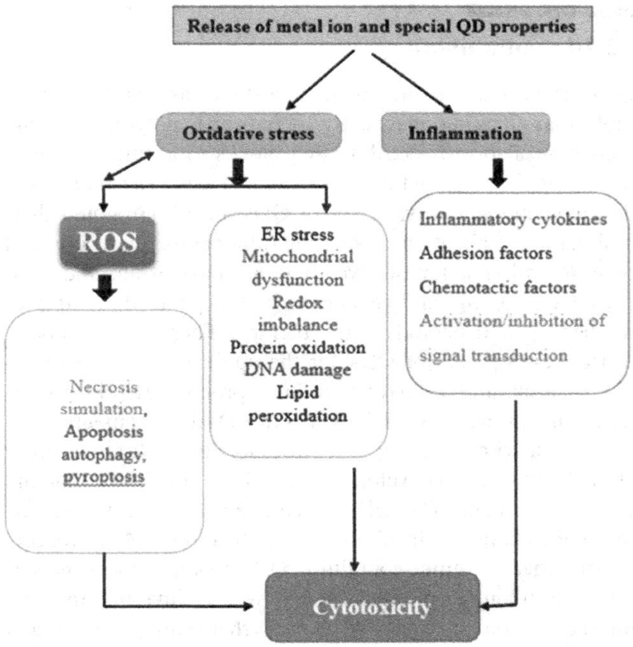

FIGURE 12.7 Possible molecular mechanism of QDs leading to cellular cytotoxicity.

The above-mentioned reports suggest that high photostability, board excitation wavelength, fluorescence emission wavelength, high fluorescent signals, and thin and tunable fluorescence QDs can be useful applications to in vitro and in vivo cell imaging. Research in QDs applications in bioimaging achieved significant success but also faced many problems. Among them, heavy metal elements are being used in the synthesis approach. Cadmium is a class A metal element, and its use is not tolerated in today's environment [108]. However, fluorescent emitter-doped QDs, which depend on transition metal ion-doped QDs without heavy metallic ions, can overcome the few problems of QDs in biological applications. A wide range of bandgap semiconductor QDs like Mn ion-doped ZnSe and ZnS have attracted significant attention as the novel luminescence nanocrystal generation [109, 110]. Mn-doped ZnS/ZnSe QDs prepared by aqueous synthesis can be most suitable for biomedical research. Reiss et al. [111] reported the in vivo imaging of cadmium-free QDs and results showed that no modification of the mice's vital functions can be detected and insignificant fluorescence during the time of death was observed. This study opens the way to QDs directions toward the cell visualization. The development of a new class of QDs, more environmentally eco-friendly QDs with greater photoluminescence quantum yield synthesized and studied their application on molecular imaging, tumor imaging, molecular detection, clinical diagnosis, immunofluorescence, etc. [69].

12.10 Conclusion

QDs have already gained their importance as a novel class of molecular imaging agents. By their proper polymer coating, QDs have also provided building blocks to accumulate multifunctional nanostructures and nanodevices. The interesting developments and advantages of QDs have highlighted their bioimaging applications, including tumor cells, stem cells, red or NIR emission for in vivo imaging, biosensing, and fluorescence imaging, and cancer, including photo-thermal chemotherapy, and photodynamic therapy. There is no problem in the development of QDs for the applications in biomedical research, and we have seen huge progress within a short duration. Taking the initial success in QDs developments for in vivo and in vitro imaging, it is believable that future researchers will continue to develop the new QDs for cell labeling and cancer treatment. The QDs biocompatibility factor remains the main challenge in biological applications. Many reports showed that cadmium-containing QDs contain short time stabilities and long-term breakdown in in vivo imaging, increasing their chronic toxicity. We believe that future research and developments of QDs synthesis and applications can replace the organic fluorescent dyes and be safely applied in biology, chemistry, physics, material science, and biology. Great achievements are seen so far with quantum dots regarding their applications in biological systems, such as cell imaging, cancer diagnosis, and cell labeling, and success is expected in other fields, such as nanotechnology and nanomedicine for biomedical applications.

Declaration of Competing Interest

The authors declare no competing financial interest.

Acknowledgment

The author gratefully thanks the GITAM Deemed University, Hyderabad, for supporting this work.

REFERENCES

1. Filali, S., Pirot, F., & Miossec, P. (2020). Biological applications and toxicity minimization of semiconductor quantum dots. *Trends in Biotechnology*, *38*(2), 163–177.
2. Himmelstoß, S. F., & Hirsch, T. (2019). A critical comparison of lanthanide based upconversion nanoparticles to fluorescent proteins, semiconductor quantum dots, and carbon dots for use in optical sensing and imaging. *Methods and Applications in Fluorescence*, *7*(2), 022002.
3. Michalel, X., Pinaud, F. F., Bentolila, L. A., Tsay, J. M., Doose, S., Li, J. J., & Weiss, S. Quantum dots for live cells. *Vivo Imaging, and Diagnostics. Science*, *307*(5709), 538–544.
4. Jaiswal, J. K., & Simon, S. M. (2004). Potentials and pitfalls of fluorescent quantum dots for biological imaging. *Trends in Cell Biology*, *14*(9), 497–504.
5. Namdari, P., Negahdari, B., & Eatemadi, A. (2017). Synthesis, properties and biomedical applications of carbon-based quantum dots: An updated review. *Biomedicine & Pharmacotherapy*, *87*, 209–222.
6. Long, R., Tang, C., Xu, J., Li, T., Tong, C., Guo, Y., … & Wang, D. (2019). Novel natural myricetin with AIE and ESIPT characteristics for selective detection and imaging of superoxide anions in vitro and in vivo. *Chemical Communications*, *55*(73), 10912–10915.
7. Shu, J., & Tang, D. (2017). Current advances in quantum-dots-based photoelectrochemical immunoassays. *Chemistry–An Asian Journal*, *12*(21), 2780–2789.
8. Han, M., Zhu, S., Lu, S., Song, Y., Feng, T., Tao, S., … & Yang, B. (2018). Recent progress on the photocatalysis of carbon dots: Classification, mechanism and applications. *Nano Today*, *19*, 201–218.
9. Jin, S., Hu, Y., Gu, Z., Liu, L., & Wu, H. C. (2011). Application of quantum dots in biological imaging. *Journal of Nanomaterials*, *2011*, Article ID 834139.
10. Takeda, M., Tada, H., Higuchi, H., Kobayashi, Y., Kobayashi, M., Sakurai, Y., … & Ohuchi, N. (2008). In vivo single molecular imaging and sentinel node navigation by nanotechnology for molecular targeting drug-delivery systems and tailor-made medicine. *Breast Cancer*, *15*(2), 145–152.
11. Sen, D., Deerinck, T. J., Ellisman, M. H., Parker, I., & Cahalan, M. D. (2008). Quantum dots for tracking dendritic cells and priming an immune response in vitro and in vivo. *PloS One*, *3*(9), e3290.
12. Gao, X., Cui, Y., Levenson, R. M., Chung, L. W., & Nie, S. (2004). In vivo cancer targeting and imaging with semiconductor quantum dots. *Nature Biotechnology*, *22*(8), 969–976.
13. Qi, L., & Gao, X. (2008). Quantum dot - Amphipol nanocomplex for intracellular delivery and real-time imaging of SiRNA. *ACS Nano*, *2*, 1403–1410.
14. Henglein, A. (1982). Photochemistry of colloidal cadmium sulfide. 2. Effects of adsorbed methyl viologen and of colloidal platinum. *The Journal of Physical Chemistry*, *86*(13), 2291–2293.
15. Ekimov, A. I., & Onushchenko, A. A. (1982). Quantum size effect in the optical-spectra of semiconductor micro-crystals. *Soviet Physics Semiconductors-USSR*, *16*(7), 775–778.
16. Fischer, C. H., Weller, H., Katsikas, L., & Henglein, A. (1989). Photochemistry of colloidal semiconductors. 30. HPLC investigation of small CdS particles. *Langmuir*, *5*(2), 429–432.
17. Bawendi, M. G. (1995). Synthesis and spectroscopy of II–VI quantum dots: An overview. *Confined Electrons and Photons*, *27*, 339–356.
18. Mekis, I., Talapin, D. V., Kornowski, A., Haase, M., & Weller, H. (2003). One-pot synthesis of highly luminescent CdSe/CdS core– shell nanocrystals via organometallic and "Greener" chemical approaches. *The Journal of Physical Chemistry B*, *107*(30), 7454–7462.
19. Henglein, A. (1992). Chemistry and optical properties of small metal particles in aqueous solution. *MRS Online Proceedings Library (OPL)*, *272*, 77–83.
20. Nirmal, M., & Brus, L. (1999). Luminescence photophysics in semiconductor nanocrystals. *Accounts of Chemical Research*, *32*(5), 407–414.

21. Murrag, C. B., Norris, D. B., & Bawendi, M. G. (1993). Synthesis and characterization of nearly monodisperse CdE (E= S, Se, Te) semiconductor nanocrystallite. *Journal of the American Chemical Society*, *115*(19), 8706–8715.
22. Hines, M., & Guyot-Sionnest, P. (1996). Synthesis and characterization of strongly luminescing ZnS-capped CdSe nanocrystals. *Journal of Physical Chemistry*, *100*(2), 468–471.
23. Scheer, H., Yang, X., & Zhao, K. H. (2015). Biliproteins and their applications in bioimaging. *Procedia Chemistry*, *14*, 176–185.
24. Guo, C., Cao, K., Zhang, Z., Xiong, Y., Chen, Y., & Wang, Y. (2020). ZnS quantum dots/gelatin nanocomposites with a thermo-responsive Sol–Gel transition property produced by a facile and green one-pot method. *ACS Sustainable Chemistry & Engineering*, *8*(11), 4346–4352.
25. Zhang, A., Chen, L., Wang, M., Li, J., Chen, L., Shi, R., … & Yang, P. (2020). Study on the luminescence stability of CdSe/CdxZn1-xS quantum dots during the silication process. *Journal of Luminescence*, *219*, 116907.
26. Lisensky, G., McFarland-Porter, R., Paquin, W., & Liu, K. (2020). Synthesis and analysis of zinc copper indium sulfide quantum dot nanoparticles. *Journal of Chemical Education*, *97*(3), 806–812.
27. Mollarasouli, F., Majidi, M. R., & Asadpour-Zeynali, K. (2019). Facile synthesis of ZnTe/Quinhydrone nanocomposite as a promising catalyst for electro-oxidation of ethanol in alkaline medium. *International Journal of Hydrogen Energy*, *44*(39), 22085–22097.
28. Peng, J., Zhao, Z., Zheng, M., Su, B., Chen, X., & Chen, X. (2020). Electrochemical synthesis of phosphorus and sulfur co-doped graphene quantum dots as efficient electrochemiluminescent immunomarkers for monitoring okadaic acid. *Sensors and Actuators B: Chemical*, *304*, 127383.
29. Thomas, D., Lee, H. O., Santiago, K. C., Pelzer, M., Kuti, A., Jenrette, E., & Bahoura, M. (2020). Rapid microwave synthesis of tunable cadmium selenide (CdSe) quantum dots for optoelectronic applications. *Journal of Nanomaterials*, *2020*, Article ID 5056875.
30. Kanatzidis, M. G., Wu, C. G., Marcy, H. O., & Kannewurf, C. R. (1989). Conductive-polymer bronzes. Intercalated polyaniline in vanadium oxide xerogels. *Journal of the American Chemical Society*, *111*(11), 4139–4141.
31. Schiffman, J. D., & Balakrishna, R. G. (2018). Quantum dots as fluorescent probes: Synthesis, surface chemistry, energy transfer mechanisms, and applications. *Sensors and Actuators B: Chemical*, *258*, 1191–1214.
32. Justino, C. I. L., Gomes, A. R., Freitas, A. C., Duarte, A. C., & Rocha-Santos, T. A. (2017). Graphene based sensors and biosensors. *TrAC Trends in Analytical Chemistry*, *91*, 53–66.
33. Mo, D., Hu, L., Zeng, G., Chen, G., Wan, J., Yu, Z., … & Cheng, M. (2017). Cadmium-containing quantum dots: Properties, applications, and toxicity. *Applied Microbiology and Biotechnology*, *101*(7), 2713–2733.
34. Jaiswal, J. K., Mattoussi, H., Mauro, J. M., & Simon, S. M. (2003). Long-term multiple color imaging of live cells using quantum dot bioconjugates. *Nature Biotechnology*, *21*(1), 47–51.
35. Dahan, M., Laurence, T., Pinaud, F., Chemla, D. S., Alivisatos, A. P., Sauer, M., & Weiss, S. (2001). Time-gated biological imaging by use of colloidal quantum dots. *Optics Letters*, *26*(11), 825–827.
36. Chan, W. C., Maxwell, D. J., Gao, X., Balley, R. E., Han, M., & Nie, S. (2002). Luminescent quantum dots for multiplexed biological detection and imaging. *Current Opinions in Biotetechnology*, *13*, 40–46.
37. Tope, S., Saudagar, S., Kale, N., & Bhise, K. (2014). Therapeutic application of quantum dots (QD). *Pharma Innovation*, *2*(12, Part A), 86.
38. Dos Santos, M. C., Algar, W. R., Medintz, I. L., & Hildebrandt, N. (2020). Quantum dots for Förster resonance energy transfer (FRET). *TrAC Trends in Analytical Chemistry*, *125*, 115819.
39. Pandurangan, D. K., & Mounika, K. S. (2012). Quantum dot aptamers-an emerging technology with wide scope in pharmacy. *International Journal of Pharmacy and Pharmaceutical Sciences*, *4*(3), 24–31.
40. Cao, Y. A., Yang, K., Li, Z., Zhao, C., Shi, C., & Yang, J. (2010). Near-infrared quantum-dot-based non-invasive in vivo imaging of squamous cell carcinoma U14. *Nanotechnology*, *21*(47), 475104.
41. Ag, D., Bongartz, R., Dogan, L. E., Seleci, M., Walter, J. G., Demirkol, D. O., … & Scheper, T. (2014). Biofunctional quantum dots as fluorescence probe for cell-specific targeting. *Colloids and Surfaces B: Biointerfaces*, *114*, 96–103.
42. Wang, W., Guo, Y., Tiede, C., Chen, S., Kopytynski, M., Kong, Y., … & Zhou, D. (2017). Ultraefficient cap-exchange protocol to compact biofunctional quantum dots for sensitive ratiometric biosensing and cell imaging. *ACS Applied Materials & Interfaces*, *9*(18), 15232–15244.
43. Wang, L., Kang, X., & Pan, D. (2017). Gram-scale synthesis of hydrophilic PEI-coated AgInS2 quantum dots and its application in hydrogen peroxide/glucose detection and cell imaging. *Inorganic Chemistry*, *56*(11), 6122–6130.
44. Pons, T., Bouccara, S., Loriette, V., Lequeux, N., Pezet, S., & Fragola, A. (2019). In vivo imaging of single tumor cells in fast-flowing bloodstream using near-infrared quantum dots and time-gated imaging. *ACS Nano*, *13*(3), 3125–3131.
45. Hasan, M. T., Lee, B. H., Lin, C. W., McDonald-Boyer, A., Gonzalez-Rodriguez, R., Vasireddy, S., … & Naumov, A. V. (2021). Near-infrared emitting graphene quantum dots synthesized from reduced graphene oxide for in vitro/in vivo/ex vivo bioimaging applications. *2D Materials*, *8*(3), 035013.
46. Shi, C., Qi, H., Ma, R., Sun, Z., Xiao, L., Wei, G., … & Guo, Z. (2019). N, S-self-doped carbon quantum dots from fungus fibers for sensing tetracyclines and for bioimaging cancer cells. *Materials Science and Engineering: C*, *105*, 110132.
47. Geng, H., Qiu, J., Zhu, H., & Liu, X. (2018). Achieving stem cell imaging and osteogenic differentiation by using nitrogen doped graphene quantum dots. *Journal of Materials Science: Materials in Medicine*, *29*(6), 1–13.
48. Kumawat, M. K., Thakur, M., Gurung, R. B., & Srivastava, R. (2017). Graphene quantum dots for cell proliferation, nucleus imaging, and photoluminescent sensing applications. *Scientific Reports*, *7*(1), 1–16.
49. Hong, G. L., Zhao, H. L., Deng, H. H., Yang, H. J., Peng, H. P., Liu, Y. H., & Chen, W. (2018). Fabrication of ultra-small monolayer graphene quantum dots by pyrolysis of trisodium citrate for fluorescent cell imaging. *International Journal of Nanomedicine*, *13*, 4807.

50. Lu, H., Li, W., Dong, H., & Wei, M. (2019). Graphene quantum dots for optical bioimaging. *Small, 15*(36), 1902136.
51. Schroer, Z. S., Wu, Y., Xing, Y., Wu, X., Liu, X., Wang, X., … & Chen, J. (2019). Nitrogen–sulfur-doped graphene quantum dots with metal ion-resistance for bioimaging. *ACS Applied Nano Materials*, *2*(11), 6858–6865.
52. Jin, H., Gui, R., Sun, J., & Wang, Y. (2016). Glycerol-regulated facile synthesis and targeted cell imaging of highly luminescent Ag2Te quantum dots with tunable near-infrared emission. *Colloids and Surfaces B: Biointerfaces*, *143*, 118–123.
53. Panagiotopoulou, M., Salinas, Y., Beyazit, S., Kunath, S., Duma, L., Prost, E., … & Haupt, K. (2016). Molecularly imprinted polymer coated quantum dots for multiplexed cell targeting and imaging. *Angewandte Chemie*, *128*(29), 8384–8388.
54. Zhao, Q., Zhang, R., Ye, D., Zhang, S., Chen, H., & Kong, J. (2017). Ratiometric fluorescent silicon quantum dots–Ce6 complex probe for the live cell imaging of highly reactive oxygen species. *ACS Applied Materials & Interfaces*, *9*(3), 2052–2058.
55. Kuo, C. W., Chueh, D. Y., & Chen, P. (2019). Real-time in vivo imaging of subpopulations of circulating tumor cells using antibody conjugated quantum dots. *Journal of Nanobiotechnology*, *17*(1), 1–10.
56. Rowland, C. E., Susumu, K., Stewart, M. H., Oh, E., Mäkinen, A. J., O'Shaughnessy, T. J., … & Erickson, J. S. (2015). Electric field modulation of semiconductor quantum dot photoluminescence: Insights into the design of robust voltage-sensitive cellular imaging probes. *Nano Letters*, *15*(10), 6848–6854.
57. Bawendi, M. G., Steigenvald, M. L., & Brus, L. E. (1990). The quantum mechanics of larger semiconductor clusters ("quantum dots"). *Annual Review of Physical Chemistry*, *41*, 477.
58. Reimann, S. M., & Manninen, M. (2002). Electronic structure of quantum dots. *Reviews of Modern Physics*, *74*(4), 1283.
59. Bhamore, J. R., Jha, S., Park, T. J., & Kailasa, S. K. (2019). Green synthesis of multi-color emissive carbon dots from *Manilkara zapota* fruits for bioimaging of bacterial and fungal cells. *Journal of Photochemistry and Photobiology B: Biology*, *191*, 150–155.
60. Jeong, S., Jung, Y., Bok, S., Ryu, Y. M., Lee, S., Kim, Y. E., … & Kim, S. (2018). Multiplexed in vivo imaging using size-controlled quantum dots in the second near-infrared window. *Advanced Healthcare Materials*, *7*(24), 1800695.
61. Kobayashi, H., Ogawa, M., Alford, R., Choyke, P. L., & Urano, Y. (2010). Review New strategies for fluorescent probe design in medical diagnostic imaging. *Chemicals Reviews*, *110*, 2620–2640.
62. Fountaine, T. J., Wincovitch, S. M., Geho, D. H., Garfield, S. H., & Pittaluga, S. (2006). Multispectral imaging of clinically relevant cellular targets in tonsil and lymphoid tissue using semiconductor quantum dots. *Modern Pathology*, *19*(9), 1181–1191.
63. Werbin, J. L., Avendaño, M. S., Becker, V., Jungmann, R., Yin, P., Danuser, G., & Sorger, P. K. (2017). Multiplexed exchange-PAINT imaging reveals ligand-dependent EGFR and Met interactions in the plasma membrane. *Scientific Reports*, *7*(1), 1–12.
64. Martynenko, I. V., Litvin, A. P., Purcell-Milton, F., Baranov, A. V., Fedorov, A. V., & Gun'Ko, Y. K. (2017). Application of semiconductor quantum dots in bioimaging and biosensing. *Journal of Materials Chemistry B*, *5*(33), 6701–6727.
65. Hood, L. Health, J. R., Phelps, M. E., & Lin, B. (2004). Systems biology and new technologies enable predictive and preventative medicine. *Science*, *306*, 640–643.
66. Walling, M. A., Novak, J. A., & Shepard, J. R. (2009). Quantum dots for live cell and in vivo imaging. *International Journal of Molecular Sciences*, *10*(2), 441–491.
67. Wu, X., Liu, H., Liu, J., Haley, K. N., Treadway, J. A., Larson, J. P., Ge, N., Peale, F., & Bruchez, M. P. (2003). Immunofluorescent labeling of cancer marker Her2 and other cellular targets with semiconductor quantum dots. *Nature Biotechnology*, *21*, 41.
68. Gao, X., Xing, Y., Chung, L. W., & Nie, S. (2007). Quantum dot nanotechnology for prostate cancer research. In Chung, L. W. K., Isaacs, W. B. & J. W. Simons (Eds.), *Prostate Cancer* (pp. 231–244). Humana Press.
69. Klostranec, J. M., & Chan, W. C. (2006). Quantum dots in biological and biomedical research: Recent progress and present challenges. *Advanced Materials*, *18*(15), 1953–1964.
70. Lagerholm, B. C., Wang, M., Ernst, L. A., Ly, D. H., Liu, H., Bruchez, M. P., & Waggoner, A. S. (2004). Multicolor coding of cells with cationic peptide coated quantum dots. *Nano Letters*, *4*(10), 2019–2022.
71. Pathak, S., Choi, S. K., Amheim, N., & Thompson, M. (2001). Hydroxylated quantum dots as luminescent probes for in situ hybridization. *Journal of the American Chemistry Society*, *123*, 4103–4104.
72. Parak, W. J., Boudreau, R., Le Gros, M., Gerion, D., Zanchet, D., Micheel, C. M., … & Larabell, C. (2002). Cell motility and metastatic potential studies based on quantum dot imaging of phagokinetic tracks. *Advanced Materials*, *14*(12), 882–885.
73. Zhu, L., Ang, S., & Liu, W. T. (2004). Quantum dots as a novel immunofluorescent detection system for *Cryptosporidium parvum* and *Giardia lamblia*. *Applied and Environmental Microbiology*, *70*(1), 597–598.
74. Edgar, R., McKinstry, M., Hwang, J., Oppenheim, A. B., Fekete, R. A., Giulian, G., … & Adhya, S. (2006). High-sensitivity bacterial detection using biotin-tagged phage and quantum-dot nanocomplexes. *Proceedings of the National Academy of Sciences*, *103*(13), 4841–4845.
75. Medintz, I. L., Stewart, M. H., Trammell, S. A., Susumu, K., Delehanty, J. B., Mei, B. C., … & Mattoussi, H. (2010). Quantum-dot/dopamine bioconjugates function as redox coupled assemblies for in vitro and intracellular pH sensing. *Nature Materials*, *9*(8), 676–684.
76. Clapp, A. R., Medintz, I. L., Mauro, M. J., Fisher, B. R., Bawendi, M. G., & Mattoussi, H. (2004). Fluorescence resonance energy transfer between quantum dot donors and dye-labeled protein acceptors. *Journal of the American Chemical Society*, *126*, 301.
77. Xiao, Y., & Barker, P. E. (2004). Semiconductor nanocrystal probes for human metaphase chromosomes. *Nucleic Acids Research*, *32*(3), e28–e28.
78. Medintz, I. L., Konnert, J. H., Clapp, A. R., Twigg, M. E., Mattoussi, H., Mauro, J. M., & Deschamps, J. R. (2004). Fluorescence resonance energy transfer-derived structure

of a quantum dot-protein bioconjugate nanoassembly. *Proceedings of the National Academy of Sciences of the United States of America*, *101*(26), 9612–9617.

79. Niebling, T., Zhang, F., Ali, Z., Parak, W. J., & Heimbrodt, W. (2009). Excitation dynamics in polymer-coated semiconductor quantum dots with integrated dye molecules: The role of reabsorption. *Journal of Applied Physics*, *106*(10), 104701.
80. Kim, S., Lim, Y. T., Soltesz, E. G., De Grand, A. M., Lee, J., Nakayama, A., … & Frangioni, J. V. (2004). Near-infrared fluorescent type II quantum dots for sentinel lymph node mapping. *Nature Biotechnology*, *22*, 93.
81. Bentolila, L. A., Ebenstein, Y., & Weiss, S. (2009). Quantum dots for in vivo small-animal imaging. *Journal of Nuclear Medicine*, *50*(4), 493–496.
82. Walker, L. S., & Abbas, A. K. (2002). The enemy within: Keeping self-reactive T cells at bay in the periphery. *Nature Reviews Immunology*, *2*(1), 11–19.
83. Sun, X., Shi, M., Zhang, C., Yuan, J., Yin, M., Du, S., … & Yang, S. T. (2021). Fluorescent Ag–In–S/ZnS quantum dots for tumor drainage lymph node imaging in vivo. *ACS Applied Nano Materials*, *4*(2), 1029–1037.
84. Atchudan, R., Edison, T. N. J. I., Shanmugam, M., Perumal, S., Somanathan, T., & Lee, Y. R. (2021). Sustainable synthesis of carbon quantum dots from banana peel waste using hydrothermal process for in vivo bioimaging. *Physica E: Low-dimensional Systems and Nanostructures*, *126*, 114417.
85. Li, S. L., Jiang, P., Hua, S., Jiang, F. L., & Liu, Y. (2021). Near-infrared Zn-doped Cu 2 S quantum dots: An ultrasmall theranostic agent for tumor cell imaging and chemodynamic therapy. *Nanoscale*, *13*(6), 3673–3685.
86. Xin, Q., Shah, H., Xie, W., Wang, Y., Jia, X., Nawaz, A., … & Gong, J. R. (2021). Preparation of blue-and green-emissive nitrogen-doped graphene quantum dots from graphite and their application in bioimaging. *Materials Science and Engineering: C*, *119*, 111642.
87. Li, T., Liu, L., Jing, T., Ruan, Z., Yuan, P., & Yan, L. (2018). Self-healing organic fluorophore of cyanine-conjugated amphiphilic polypeptide for near-infrared photostable bioimaging. *ACS Applied Materials & Interfaces*, *10*(17), 14517–14530.
88. Zhang, X., Wang, C., Han, Z., & Xiao, Y. (2014). A photostable near-infrared fluorescent tracker with pH-independent specificity to lysosomes for long time and multicolor imaging. *ACS Applied Materials & Interfaces*, *6*(23), 21669–21676.
89. Zhen, S., Wang, S., Li, S., Luo, W., Gao, M., Ng, L. G., … & Tang, B. Z. (2018). Efficient red/near-infrared fluorophores based on benzo [1, 2-b: 4, 5-b′] dithiophene 1, 1, 5, 5-tetraoxide for targeted photodynamic therapy and in vivo two-photon fluorescence bioimaging. *Advanced Functional Materials*, *28*(13), 1706945.
90. Narasimhan, A. K., Santra, T. S., Rao, M. R., & Krishnamurthi, G. (2017). Oxygenated graphene quantum dots (GQDs) synthesized using laser ablation for long-term real-time tracking and imaging. *RSC Advances*, *7*(85), 53822–53829.
91. Shivaji, K., Mani, S., Ponmurugan, P., De Castro, C. S., Davies, M. L., Balasubramanian, M. G., & Pitchaimuthu, S. (2018). Green-synthesis-derived CdS quantum dots using tea leaf extract: Antimicrobial, bioimaging, and therapeutic applications in lung cancer cells. *ACS Applied Nano Materials*, *1*, 1683–1693.
92. Goreham, R. V., Schroeder, K. L., Holmes, A., Bradley, S. J., & Nann, T. (2018). Demonstration of the lack of cytotoxicity of unmodified and folic acid modified graphene oxide quantum dots, and their application to fluorescence lifetime imaging of HaCaT cells. *Microchimica Acta*, *185*(2), 1–7.
93. Wu, F., Chen, L., Yue, L., Wang, K., Cheng, K., Chen, J., … & Zhang, T. (2019). Small-molecule porphyrin-based organic nanoparticles with remarkable photothermal conversion efficiency for in vivo photoacoustic imaging and photothermal therapy. *ACS applied Materials & Interfaces*, *11*(24), 21408–21416.
94. Yang, L., Li, H., Liu, D., Su, H., Wang, K., Liu, G., … & Wu, F. (2019). Organic small molecular nanoparticles based on self-assembly of amphiphilic fluoroporphyrins for photodynamic and photothermal synergistic cancer therapy. *Colloids and Surfaces B: Biointerfaces*, *182*, 110345.
95. Tasso, M., Singh, M. K., Giovanelli, E., Fragola, A., Loriette, V., Regairaz, M., … & Pons, T. (2015). Oriented bioconjugation of unmodified antibodies to quantum dots capped with copolymeric ligands as versatile cellular imaging tools. *ACS Applied Materials & Interfaces*, *7*(48), 26904–26913.
96. Smith, A. M., Duan, H., Mohs, A. M., & Nie, S. (2008). Bioconjugated quantum dots for in vivo molecular and cellular imaging. *Advanced Drug Delivery Reviews*, *60*(11), 1226–1240.
97. Tada, H., Higuchi, H., Wanatabe, T. M., & Ohuchi, N. (2007). In vivo real-time tracking of single quantum dots conjugated with monoclonal anti-HER2 antibody in tumors of mice. *Cancer Research*, *67*(3), 1138–1144.
98. Lovrić, J., Bazzi, H. S., Cuie, Y., Fortin, G. R., Winnik, F. M., & Maysinger, D. (2005). Differences in subcellular distribution and toxicity of green and red emitting CdTe quantum dots. *Journal of Molecular Medicine*, *83*(5), 377–385.
99. Winnik, J. L. S. C. F., & Maysinger, D. (2005). Unmodified cadmium telluride quantum dots induce reactive oxygen species formation leading to multiple organelle damage and cell death. *Chemical Biology*, *12*, 1227–1234.
100. Voura, E. B., Jaiswal, J. K., Mattoussi, H., & Simon, S. M. (2004). Tracking metastatic tumor cell extravasation with quantum dot nanocrystals and fluorescence emission-scanning microscopy. *Nature Medicine*, *10*, 993–998.
101. Hanaki, K. I., Momo, A., Oku, T., Komoto, A., Maenosono, S., Yamaguchi, Y., & Yamamoto, K. (2003). Semiconductor quantum dot/albumin complex is a long-life and highly photostable endosome marker. *Biochemical and Biophysical Research Communications*, *302*(3), 496–501.
102. Yan, M., Zhang, Y., Qin, H., Liu, K., Guo, M., Ge, Y., … & Zheng, X. (2016). Cytotoxicity of CdTe quantum dots in human umbilical vein endothelial cells: The involvement of cellular uptake and induction of pro-apoptotic endoplasmic reticulum stress. *International Journal of Nanomedicine*, *11*, 529.
103. Peynshaert, K., Soenen, S. J., Manshian, B. B., Doak, S. H., Braeckmans, K., De Smedt, S. C., & Remaut, K. (2017). Coating of quantum dots strongly defines their effect on lysosomal health and autophagy. *Acta Biomaterialia*, *48*, 195–205.

104. Lee, E. Y., Bae, H. C., Lee, H., Jang, Y., Park, Y. H., Kim, J. H., … & Son, S. W. (2017). Intracellular ROS levels determine the apoptotic potential of keratinocyte by quantum dot via blockade of AKT phosphorylation. *Experimental Dermatology*, *26*(11), 1046–1052.
105. Kang, M. K., Lee, G. H., Jung, K. H., Jung, J. C., Kim, H. K., Kim, Y. H., … & Chang, Y. (2016). Gadolinium nanoparticles conjugated with therapeutic bifunctional chelate as a potential t 1 theranostic magnetic resonance imaging agent. *Journal of Biomedical Nanotechnology*, *12*(5), 894–908.
106. Xiang-Yi, D. E. N. G., Ya-Li, F. E. N. G., Dong-Sheng, H. E., Zhang, Z. Y., De-Feng, L. I. U., & Ru-An, C. H. I. (2020). Synthesis of functionalized carbon quantum dots as fluorescent probes for detection of Cu2+. *Chinese Journal of Analytical Chemistry*, *48*(10), e20126–e20133.
107. Peng, X. (2009). An essay on synthetic chemistry of colloidal nanocrystals. *Nano Research*, *2*, 425–447.
108. Pradhan, N., & Peng, X. (2007). Efficient and color-tunable Mn-doped ZnSe nanocrystal emitters: Control of optical performance via greener synthetic chemistry. *Journal of the American Chemical Society*, *129*(11), 3339–3347.
109. Zhuang, J., Zhang, X., Wang, G., Li, D., Yang, W., & Li, T. (2003). Synthesis of water-soluble ZnS: Mn^{2+} nanocrystals by using mercaptopropionic acid as stabilizer. *Journal of Materials Chemistry*, *13*(7), 1853–1857.
110. Wang, C., Gao, X., Ma, Q., & Su, X. (2009). Aqueous synthesis of mercaptopropionic acid capped Mn^{2+}-doped ZnSe quantum dots. *Journal of Materials Chemistry*, *19*(38), 7016–7022.
111. Xia, C., Meeldijk, J. D., Gerritsen, H. C., & de Mello Donega, C. (2017). Highly luminescent water-dispersible NIR-emitting wurtzite $CuInS_2$/ZnS core/shell colloidal quantum dots. *Chemistry of Materials*, *29*(11), 4940–4951.

13

Nanodevices for Drug Delivery Systems

Kajal Karsauliya, Sheelendra Pratap Singh and Manu Sharma

CONTENTS

13.1 Introduction

The term "nanotechnology" involves manipulation and control over the engineered particles (nanoparticles) synthesized by chemical reactions, manipulating single-molecule or electron beam lithography. In recent years, nanotechnology proved effective and gained popularity in the area of current science and technology. Nanoparticles have a wide range of applications that includes information technology, aerospace, energy, medicinal, transportation, and defense as well [1]. According to the definition of the US National Nanotechnology Initiative the range of the nanoscale is 1–100 nm.

"Nanomedicine" comprises one of the significant parts of nanotechnology that introduce extremely specialized medical involvement at the molecular level providing a tool for prevention, diagnosis and treatment of diseases [2]. The technology of nanomedicine on the ground of developing innovative nanodevices has a humongous capability to reform therapeutics and diagnostics [3]. Usually, the size of the nanoparticles is like small nanospheres as their material has been developed at an atomic or molecular level [4]. The application of nanodevices in drug delivery constitutes a significant portion of nanomedicine. The nanosystems such as conjugates of drug-polymer and polymer micelles to microparticles ranging from 1 to100 μm are proved to be useful for developing clinically convenient drug delivery systems [5]. The nanoparticles are the base of nanotechnology that governs the area of nanomedicine comprising drug delivery, biosensors, and tissue engineering [6].

Currently, nanomedicines are highly acknowledged for the application of nanostructures as delivery agents by binding or encapsulating the therapeutic drugs to convey their precise delivery and release at targeted tissues in a controlled manner [7]. Various "nano" devices that are currently used for drug delivery purposes, however, are derived from traditional drug delivery such as nanocrystals, lipid-based nanoparticles, polymer micelles, dendrimers, and organic/inorganic nanoparticles systems that exist in the nanometer range [8]. The FDA-approved lipid systems such as liposomes and micelles as the first generation therapy based on nanoparticles incorporated with inorganic nanoparticles such as magnetic or gold nanoparticles [9]. Moreover, encapsulating the water-soluble drugs with nanoparticles prevents degradation in the gastrointestinal tract and aids in its delivery to the specific location. The oral bioavailability of nano-drugs is high due to their endocytic uptake mechanisms [8].

Metallic, organic, inorganic, nanocrystals, carbon nanotubes and fullerenes (carbon-based particles), nano-emulsions, nanoshells (quantum dots), and polymeric nanostructures, including dendrimers, micelles, and liposomes are the main categories of nanoparticles that are applied for designing the target-specific drug carriers [1]. Being incorporated with nanoparticles, the potency of the drugs with poor absorption and solubility has been enhanced [10]. Various synthetic (poly-L-lactic acid, poly[lactic-*co*-glycolic acid], polyethylene glycol, polyvinyl alcohol) and natural polymers such as chitosan and alginate have their application for the purpose of the nanofibrillation of nanoparticles due to the high bioactivity and biodegradable properties [11, 12]. The physical and chemical properties of the targeted drugs perform a major role in developing an ideal nano-drug delivery system [11].

DOI: 10.1201/9781003157823-13

Nanomaterials remain in the circulatory system for a long duration and control the release of a specified dose of concomitant drug reducing the unfavorable effects and variation in plasma [13]. Due to their smaller size, the nanoparticles promote cell uptake of the drug for efficient delivery with reduced side effects [14]. They provide ease in entering the tissue system and interact directly with diseased cells. The surface property of the nanomaterials performs an important part in procuring the information for their application in the treatment and diagnoses of diseases. The effectiveness of a drug carrier differs with the shape, size and other inherent biophysical/chemical characteristics. Nanomaterials offer a large surface area for drugs to get affixed at their surface either by covalent conjugation or noncovalent attachment such as selective absorption of chemotherapeutic drugs on the surface or interior of the nanomaterials. The solubility and biocompatibility of the nanoparticle drug carriers can be enhanced by polymer coatings [1]. However, the selection of an appropriate nanomaterial for drug delivery is totally based on the physicochemical characteristic of the drugs.

Recently, the application of nanoscience in collaboration with bioactive natural compounds is increasing rapidly. The natural compounds possess numerous medicinal properties such as curcumin (from turmeric) and caffeine (from cacao plant) have shown autophagy; on the other hand, cinnamaldehyde (from cinnamon), carvacrol (from aromatic plants such as oregano), and essential oils such as curcumin and eugenol that are obtained from clove oil, nutmeg, cinnamon, etc. are known to have antimicrobial properties [15–17]. Their properties such as bioavailability, specified dose, or controlled release at the targeted site can be improved by combining them with the nanostructures. For example, thymoquinone (a bioactive compound of *Nigella sativa*) after encapsulation in a lipid nanocarrier had shown six times increased bioavailability than the free thymoquinone [18].

The concept and potentiality of nanotechnology to orchestrate molecules with supramolecular structures to develop devices with predetermined functions are significant for drug delivery. The nanoparticles with great loading capability and extremely consistent size of particles could be manufactured by using nanotechnologies, such as nanopatterning [19]. The development of medically potent nanodevices having utility inside the living body is one of the eventual objectives of nanomedicine. It is visualized that nanodevices will be the fusion of biological molecules and synthetic polymers that can easily interact with DNA and proteins by entering into the cells and the organelles. In addition, newly discovered nanomedicine and nanodevices such as metallic nanoparticles, nanoshells, nanotubes, nanopores, nanowires, nanocantilevers, and quantum dots have shown their application in the field of various cancer treatments.

Consequently, nanotechnology proved to be beneficial for the treatment of chronic diseases by allowing specified, controlled, and targeted delivery of drugs. However, the toxicity associated with the nanostructure is of major concern which needs further research for safer and more effective application of nano-drug delivery systems. Based on the above description this chapter is focused on various drug delivery systems used for target-specific drug delivery, approach and applicability of nanomaterials, fundamental perspective for the development of nanodevices, natural compound-based nanomedicine, and the biomedical application of nanoparticles for diagnosis and treatment.

13.2 Nano-Drug Delivery Systems

The system of nanomaterials used for drug delivery (nano-drug delivery systems) is the category of nanomaterials capable of enhancing the solubility and stability of drugs, as well as cell or tissue uptake rate, and decreasing the enzyme degradation, consequently improving the safety and efficacy of drugs. As an efficacious method to improve drug delivery, nano-drug carriers have evolved as an area of high interest in the field of biomedicine and pharmacy. Depending on the composition of nanomaterials being used for constructing the nano-drug delivery system, the nanomaterials are classified into organic, inorganic, and composite materials, for example, liposomes, polymer micellar co-delivery system, dendritic macromolecules, metal nanomaterials, and inorganic non-metallic nanomaterials (Figure 13.1) [20].

13.2.1 Liposomes

Liposomes refer to the lipid vesicles made up of arranged phospholipid bilayers having a structure similar to a cell [21]. Liposomes have been proved to be very advantageous as drug carrier as they are safe, non-immunogenic, offer constant discharge of drugs, enhance action time of drugs, vary in vivo distribution of drugs, ameliorate drug treatment profile, and decrease side effects of drugs [22]. It is hard to develop the liposomes to integrate with hydrophilic and ionic molecules; however, they are compatible for encapsulation of hydrophobic drugs [23]. Modification of different lipid materials can be done to alter the surface chemistry, potential, and size of the particle. The positively charged cationic liposomes may result in cytotoxicity depending on their dose and inflammatory responses, and they may undergo non-specific interaction with negatively charged serum proteins in the form of complexes. Thus, the use of neutral lipids and pH-sensitive liposomes can solve the safety issues associated with charged liposomes [20].

Another important class of lipid-based nanoparticles is known as lipid nanoparticles (LNPs) – structurally similar to liposomes and used for the purpose of nucleic acid delivery [24]. The formation of micellar structures within the particle core differentiates them from the traditional liposomes. On the basis of formulation and synthesis criteria, the morphology of particles can be changed. The composition of LNPs comprises four main components – (i) cationic lipids that form a conjugate with negatively charged genetic compounds and support the endosomal escape, (ii) phospholipids responsible for providing support to the structure of the particle, (iii) cholesterol for membrane fusion and stability, and (iv) non-covalently or covalently bonded polyethylene glycol to lipids also provide better stability and circulation [25]. The LNPs have their applicability particularly in personalized genetic therapy as its synthesis is simple, small in size and stable in serum as well efficient for nucleic acid delivery [26]. An ideal platform

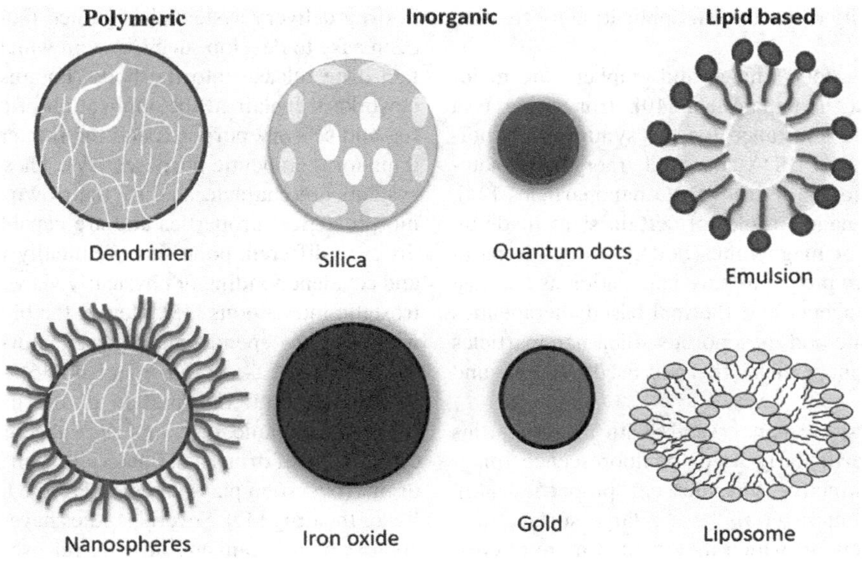

FIGURE 13.1 Types of nanomaterials.

for the delivery of nucleic acid therapies is ionizable LNPs. These ionizable LNPs at physiological pH possess charge near-neutral whereas in acidic endosomal compartments they become charged particles, initiating escape from endosomal vicinity for intracellular delivery [26, 27]. However, in spite of these advantages, higher uptake of LNPs in the liver and spleen can happen during bio-distribution due to the higher perfusion rate of these organs [28].

13.2.2 Polymer Micellar Co-Delivery System

Other carriers for drug delivery are polymeric nanoparticles that can be non-biodegradable or biodegradable in nature [29]. The synthetic polymer materials primarily comprise polyvinyl alcohol (PVA), polycaprolactone (PCL), polyvinyl imine, poly(lactic-co-glycolic acid) (PLGA), etc. [30]. These synthetic polymers are biocompatible, non-hazardous, and non-teratogenic. They have zero toxic effect on cells upon degradation along with oligomerization and final products, and also have stable synchronization with most drugs [31]. The peptides, polysaccharides, and cyclodextrin inclusion complexes are some of the natural polymers. Self-polymerization of amphiphilic block copolymers results in the formation of stable polymeric nanoparticles that can be used to entrap the insoluble drugs [32]. The stable polymer nanoparticles maintain uniform particle size and the controlled drug release behavior and can effectively provide protection into the gastrointestinal environment on oral administration [33]. The nanoscale size and large surface area facilitate the easy uptake of drugs by cells and improve bioavailability. However, there are certain drawbacks with some polymer nanoparticles. For instance, chitosan, a natural polymer, which is not compatible with biological fluids, causes particles to degrade with decreased work competency. This problem can be solved by taking structural changes into account. Conjugation of polyethylene glycol with chitosan has provided protection from the unique endocytic mechanism and macrophage phagocytosis [34]. Also, combining polypeptide into chitosan can refine the working efficiency of the natural compound [35].

13.2.3 Dendritic Macromolecules

The macromolecules are synthetic in nature with different shapes having branched structures. The spherical macromolecules have a tendency to organize in monodisperse space and mostly have application as nanocarriers to administer drugs with poor solubility [20]. Dendritic macromolecules with distinctive branched assembly are also monodispersing with controlled molecular weight. Moreover, various functional groups on the surface and hydrophobic packaging make them an excellent material for drug delivery [36]. Dendritic macromolecules are extensively used in the biomedical and pharmaceutical areas due to their exceptional biological characteristics, but their scientific application is limited due to the presence of a positive charge on the surface [20].

13.2.4 Inorganic Metallic/Non-Metallic Nanomaterials

Inorganic materials such as iron, gold, silver, and silica are useful in synthesizing nanomaterials for drug delivery purposes and imaging applications. Gold and silver nanomaterials of different shapes such as rods, capsules, cuboids, and wire are the most commonly used metal nanomaterials [37]. Apart from being used as a nano-contrast agent for surface-enhanced Raman spectroscopy and CT, gold nanomaterials have also been utilized for photothermic therapy of rheumatoid arthritis and tumors [20]. According to studies, silver nanomaterials have found their application as antibacterial, anti-infection, and anti-tumor [38]. Furthermore, hollow gold or silver nanostructures are physically loaded with some therapeutic drugs or are chemically attached to the surface of therapeutic agents to attain aimed drug delivery [39]. Conversely, the clearance of gold nanomaterials from humans is extremely slow, and the in vivo silver ion toxicity restricts their application for treating chronic diseases.

Iron oxide, quantum dots, silicon, and graphene are major inorganic non-metallic nanomaterials [40]. Iron oxide is a commonly investigated substance for the synthesis of inorganic nanoparticles and FDA-approved inorganic material comprising a majority of iron oxide nanoparticles [24]. Magnetic iron oxides nanoparticles of certain sizes made up of magnetite (Fe_3O_4) or maghemite (Fe_2O_3) with superparamagnetic properties are proven to have application as contrast agents, drug transfer agents and thermal-based therapeutics [41]. Calcium phosphate and mesoporous silica nanoparticles are other inorganic nanoparticles that are used for gene and drug delivery [42].

Quantum dots (QDs) are nanocrystals with the properties of semiconductors, particularly used for fluorescence imaging due to their distinctive incandescent properties [43]. Mesoporous silicon nanomaterials have a large surface area and porous structure due to which they gained more interest for disease therapy in recent years. These inorganic nanomaterials are applied to ameliorate the efficiency of drug delivery and genes transport in mammal cells by combining with different functional groups [44]. However, a major drawback of inorganic non-metallic nanomaterials is their bio-safety which makes them unsuitable for clinical application [45].

Inorganic nanoparticles due to their plasmonic, radioactive, or magnetic properties are distinctively certified for purposes, e.g., imaging, diagnostics, and photothermic therapies. The inorganic nanoparticle with good biocompatibility and stability can replace the organic materials that are unable to attain the required properties [41]. However, their clinical application is limited due to toxicity and low solubility, especially when formulations of heavy metals are used.

13.2.5 Composite Nanomaterials

Various composite nanomaterials having unique properties are explored for the development of drug carrier systems. For example, multifunctional novel drug delivery systems (NDDSs) can be synthesized by assimilation of metal or inorganic non-metallic nanomaterials with polymeric or lipid nanomaterials serving as both contrast agents and therapeutic agents' carriers. The physical and chemical properties, in vivo kinetic behavior, and biocompatibility of metallic or inorganic nanomaterials are enhanced by reframing them with organic material. The amalgamation of diverse metals and inorganic substances leads to the formation of NDDSs having special structures and multiple functions [20].

13.3 Drug Delivery Process

The designing of drugs has been evolved as a favorable trait for characterization of the innovation of new lead drugs depending on the information regarding the biological site for the target. Various studies and reviews have described the coherent designing of variant molecules that conjugate with nanocarriers and exhibit different mechanisms of drug release [46]. As well, in the last few years, the application of nanomaterials in drug delivery systems has gained much importance. It has been easy to develop such a system which promotes the modified drug release into the body. For instance, Chen and his coworkers explained the application of nanocarriers in imaging and sensory purposes and the influence of these nanosystems for therapeutic purposes [47]. These nano-drug delivery systems are characterized by their own physicochemical and morphological properties and are capable of accommodating drugs of different polarities chemically via hydrogen bonding and covalent bonding or physically via van der Waals and electrostatic interactions [48]. Hence, the biological interaction of nanocarriers depends on all these factors as well as the kinetics of drug release in the organism [46].

The structural information of the nanomaterials such as organic, inorganic, or composite substances and the nature of conjugation of drugs nanomaterials such as a core-shell system or matrix system plays an essential role to understand the profile of the drug [49]. Several studies have been published in the context of mechanisms of drug release involving nano-drug delivery systems. The mechanisms such as chemical reaction, diffusion, and stimuli-controlled release have explained the release of drugs from nanocarriers to the targeted site [50]. Kamaly and his coworkers reviewed controlled drug release systems with his study centered on controlled release of drugs from polymeric nanocarriers [51]. Even though different drug release profiles are associated with different nanomaterials, the present approaches have been developed for improving the site specificity of the nanostructures and lessening the immunogenic property either via functionalizing or coating them with several substances chemically, such as natural polysaccharides, antibodies, polymers, surfactants, cell membrane, peptides, etc. [52–58].

There are several cases where drugs have zero affinity for binding to a particular target or are unable to traverse some barriers such as the blood-cerebrospinal fluid barrier or the blood-brain barrier. Thus, for programmed drug delivery to a specific target the drugs are conjugated with ligand-modified nanocarriers to cross the cell membrane [59]. Nonetheless, the development of the ligand-based drug carrier systems is laborious [60]. The cell uptake of the nanoparticles occurs through a phagocytic or non-phagocytic mechanism like clathrin-mediated endocytosis or caveolae-mediated endocytosis. However, each delivery system has some specific physicochemical characteristics which make it challenging to predict the mode of action of such systems in the cell [61]. In a review by Salatin and Khosroushahi, endocytosis was highlighted as the main mechanism for the acceptance of polysaccharide nanoparticles loaded with active ingredients/drugs by the cell [62]. Contrarily, stimuli-responsive nano-drug delivery systems are able to control the profile of drug release under the influence of external aspects such as heat, light, pH, magnetic property, ionic force, and ultrasound, allowing better targeting of drugs and greater control over the dose. For instance, superparamagnetic iron oxide nanoparticles are initially stimulating an organized discharge of drug by employing the external magnetic field [8]. Ulbrich and his coworkers reviewed success made in the domain of nano-drug delivery systems particularly on nanoparticles of magnetic and polymeric origins and also discussed the impact of covalently or non-covalently bonded drugs in the treatment of cancer [63]. Furthermore, for NIR-triggered chemo/photo-thermal therapy, gold and iron oxide (Au/Fe_3O_4) have been also synthesized [64]. Therefore, hybrid nanomaterials are the most efficient tools in the field of nanomedicine as they offer different properties of individual systems in one system, thus assuring the enhanced performance of the material for its application in both therapeutic and diagnostic areas [3]. Besides, studies have increased regarding the synthesis of nanomaterials by chemical reactions that are environmentally safe using natural components such as plant extracts and microorganisms.

13.3.1 Targeting Mechanism for Nano-Drug Delivery System

Modulation of drug distribution to targeted cells or tissues can improve the therapeutic outcomes for nano-drug delivery systems through passive targeting and active targeting. Both active and passive targeting has reduced systemic exposure of drugs to healthy tissues and organ systems thus lowering the adverse effect of drugs [1].

Passive targeting involves the accumulation of nano-sized particles at the sites of tumor or inflammation depending on their size and enhanced permeability and retention (EPR) effect (in case of tumors) (Figure 13.2) [65]. The microvascular endothelial cell space of normal tissue is thick and inviolate, and it is not easy for NDDSs carrying drugs with large molecular weight to traverse through the vascular wall. However, tumors of soft-tissue and epithelial cell origin have abundant blood vessels and poor stability of the structure [66]. The NDDSs loaded with large molecular weight drugs can permeate selectively across the wall of vessels and persist in the tumor tissue. The EPR effect allows permeation and accumulation of nanoscale particles into tumor interstitials improving the bioavailability. The EPR effect on nanotech-based drugs has demonstrated evident improvements in efficacy [67]. Another passive targeted transport is shear-induced targeting. Cardiovascular diseases such as atherosclerosis (AS plaque), myocardial infarction, etc. involve the abnormal narrowing of blood vessels, and flow of blood increases due to plaque, causing an increase in fluid shear force [68]. Hence, physicochemical targeting can be achieved by designing blood fluid shear-sensitive nanoparticles depending on the variation of blood fluid shear force between AS plaque and normal blood vessels [68]. The drug-loaded nanodevices are structurally stable in normal blood vessels and deliver the drug under the influence of high blood fluid shear force via the blood circulation to the plaque utilizing the change in the configuration [20]. Magnetically directed nanoparticle involves a remarkable "pseudopassive" targeting process. Hypothetically, in this phenomenon, the magnetic nanoparticles are guided to the site of disease under the application of an external magnetic field [69]. Iron oxide particles, superparamagnetic iron oxide nanoparticles, ultra-small superparamagnetic iron oxide nanocarrier, and very small superparamagnetic iron oxide nanoparticles are some of the magnetically guided nanoparticles [20]. Studies have shown that the application of external magnetic field transfers particles from the cell-free layer devoid of red blood cells to the wall of the vessel [70].

Active targeting facilitates the delivery of drugs to diseased tissues by connecting through the biochemical moieties, such as monoclonal antibodies, which enunciate biomarkers (mutated cellular proteins and membrane receptors) that differentiate the diseased tissue from the surrounding healthy tissue (Figure 13.3) [71]. In endothelial cells, there is a surface glycoprotein called E-selectin that induces the inflammatory response through the connection of monocytes or lymphocytes and macrophages, eventually leading to the development of cardiovascular diseases [72].

The final targeting efficiency of both passive targeting and active targeting depends on the physical and biological characteristics of nanoparticles which include the size of the particle and distribution, types of unit to be targeted, surface chemistry, structural information and density [73]. The targeting efficiency is greatly affected by the stage of development, nature, and location of cardiovascular diseases and tumor, shear rate of the vascular wall, constitution of the blood, and type of its fluid, along with other characteristics [74]. Even though the utility of nano-drug delivery system has great potential to actively target the drugs for clinical diagnosis and therapy,

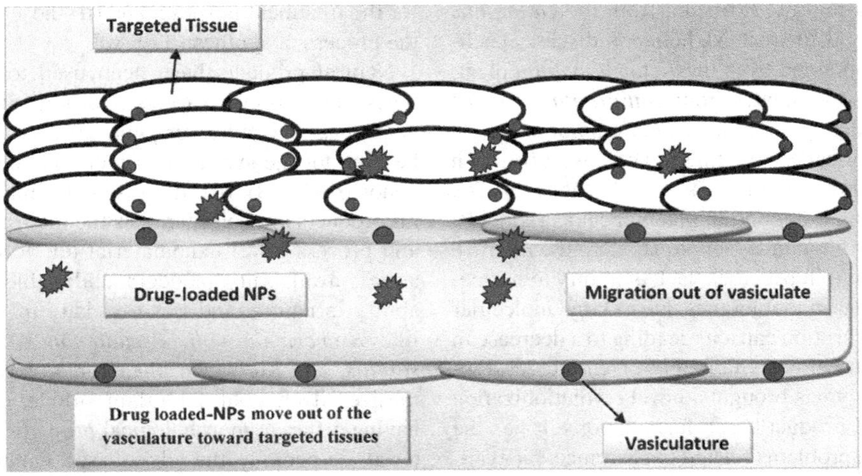

FIGURE 13.2 Diagrammatic representation of passive targeting.

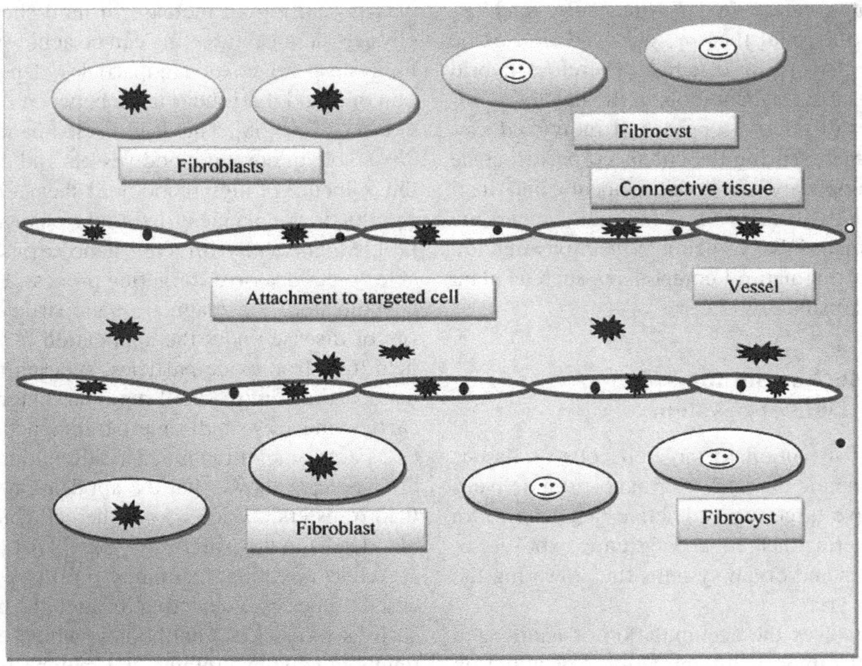

FIGURE 13.3 Diagrammatic representation of active targeting.

the development of such NDDSs is a challenge. There are two main challenges in the development of active targeting NDDSs – first is the restriction of finding a perfect target, and second is that the design and preparation of effective targeting nanosystems still have many gridlock problems [20].

13.3.2 Natural Product-Based Drug Delivery

There are several economically important natural compounds discovered in higher plants with medicinal properties and are already being marketed. Several commercial drugs are available in the market that are structurally composed of naturally occurring therapeutic agents. These include a traditional Chinese medicine plant known as *Artemisia annua* L., used to extract artemotil for malaria treatment; an acetylcholinesterase inhibitor called reminyl extracted from the *Galanthus woronowii* Losinsk used to treat Alzheimer's disease; paclitaxel and its analogs obtained from the *Taxus brevifolia* plant; vinblastine and vincristine isolated from *Catharanthus roseus*; camptothecin and its analogs obtained from *Camptotheca acuminata* Decne used for cancer treatment; and silymarin from *Silybum marianum* for the treatment of liver disease [75]. The bioactive molecules that are found in plants include alkaloids, flavonoids, phenolic compounds, saponins, steroids, tannins, terpenes, etc. However, in most cases, lack of ability to permeate through the lipid-based membranes due to large molecular sizes restricts their absorption capacity leading to a decrease in their bioavailability and efficacy [76]. The scientific development of nanotechnology has brought a novel revolutionization for developing natural product-based formulations. It has the ability to elucidate the problems related to clearance, bioavailability, efficiency, solubility, stability, and other factors that limit the application of such compounds in nanomedicine on a large scale [77]. The application of nanotechnology can overcome the barriers allowing usage of variant compounds and their combinations to prepare the same formulation, and can also modify the characteristics and behavior of a compound inside the biological vicinity. The association of NDDSs with the compound enhances the site specificity and bioavailability, prolongs the compound activity, and conglomerates molecules with varying degrees of hydrophilicity or lipophilicity [78]. Also, it has been proven that the combination of NDDSs with natural complexes may hinder the drug resistance process and is consequently necessary to discover new possibilities to treat various diseases having low reactions toward treatment from conventional approaches and modern medicine. There are two categories of natural product-based materials: (i) those directed to a particular site and released at that specified site for the treatment purpose, and (ii) those commonly applied in the process of synthesis [79, 80].

Natural products have been used to synthesize nanoparticles; for example, various microorganisms including algae, yeast, bacteria, fungi, or plant extracts have been reported to be used for the synthesis of metallic, metal oxide and sulfide nanoparticles [81, 82]. At present, the materials based on natural products are considered as the main component to prepare and process novel nanomaterial due to their properties such as easy availability, biodegradability, biocompatibility, renewability in nature, and less toxicity [83]. Numerous bacteria like *Escherichia coli*, *Pseudomonas aeruginosa*, *Bacillus subtilis*, and *Klebsiella pneumoniae* have been used to synthesize silver, gold, cadmium sulfide, and titanium dioxide having different morphological properties [84]. Such nanomaterials, especially the silver-based nanomaterials, are vastly analyzed in vitro for their antifungal, antibacterial, and cytotoxic nature as they possess higher potential than all-metal nanoparticles [85]. Gupta et al. prepared paclitaxel (Taxol) extracted from *Taxus brevifolia* loaded chitosan nanoparticles and employed them for actively targeting various types of cancer. Paclitaxel-loaded nanoparticles showed improved action with prolonged release, greater uptake via cells, and decreased hemolytic toxicity [86]. Chang and his coworkers developed a heparin/berberine (an alkaloid released from the barberry plant) conjugate to enhance the growth of suppressive *Helicobacter pylori* and in the meantime to lower the cytotoxic outcomes in affected cells [87].

13.3.3 Biomedical Application of Nanoparticles for Diagnosis and Treatment

Pharmaceutical industries are continuously looking for strategies to develop novel technologies and apply them for designing new advanced drugs and enhancing the efficacy of drugs that are currently in use. The accelerated development in the field of nanotechnology has allowed designing new formulations of drugs through different approaches with their utility as nanobiomaterials, nanopharmaceuticals, imaging and diagnosis (nanodiagnostics), and nanotheranostics (Table 13.1). The surface chemistry, size of the particle, shape, dispersion activity, stability, encapsulation ability, and biological actions are the factors that are used to categorize the delivery systems which are applied according to their requirements. The liposomes, polymeric nanoparticles, crystal nanoparticles, micelles, solid-lipid nanoparticles, superparamagnetic iron oxide nanoparticles, and dendrimers are the most usually studied nanocarriers for drug delivery purposes.

The combination of therapeutic and diagnostic fields is known as theranostic and has wide application in cancer therapy. The application of theranostic nanoparticles allows us to diagnose the disease, detect the stage and location of the disease, and provide information to deduce the possible treatment response [88]. As well, such nanomaterials have the ability to convey the therapeutic agent at the tumor site, providing the essential amount of the therapeutic agent either by molecular or external stimuli [89]. A biopolymer, chitosan has unique characteristics and is biocompatible. It is utilized to coat or encapsulate different kinds of nanoparticles allowing them to perform multiple functions making them capable to detect and diagnose of various diseases [90]. Near-infrared and magnetic resonance imaging (MRI) mechanisms are analytically applied to investigate the accretion of nanoparticles in tumor cells as a result of the penetrability and holding (EPR) effect. Yang and his coworkers synthesized nanoparticles of high efficiency and used light-mediated mechanisms to reveal colorectal cancer cells. Alginate and folic acid-amended chitosan physically result in the development of nanoparticles with increases in the release of 5-aminolevulinic acid (5-ALA) in the cellular lysosome. The findings have shown that the endocytosis (folate receptor-based endocytosis) of functionalized nanoparticles was voluntarily performed through the colorectal cancer cells. In accord with this research, for endoscopic fluorescent detection of colorectal cancer cells, the nanoparticles based on chitosan in conjugation with other substances like folic acid and alginate are proven to be stupendous paths for the specific delivery of 5-ALA to the cells [91]. Ryu and his coworkers created a spherical Cathepsin B (for the identification of metastasis in a cell) sensitive nanoprobe (CB-CNP) of diameter 280 nm. The combination of CB-CNP and fluorogenic peptide forms a conjugate with glycol chitosan nanoparticles (CNPs) that has the capability of targeting tumor cells and can potentially be applied to discriminate between metastatic cells and healthy cells by non-invasive imaging because the fluorescence power was completely quenched in biological condition [92]. Hyaluronic acid (HA) is a bio-polymeric material composed of negatively charged glycosaminoglycan amide mainly found in the extracellular matrix, can combine with CD44 receptor (found in cancerous cells) as a result of the receptor-linked interface, stimulating their use in the recognition and cure of cancer due to its biocompatibility [93]. Wang and his coworkers used dopamine-modified HA for coating the iron oxide nanoparticles surface providing a hydrophilic outer side and a hydrophobic inner area for the encapsulation of chemotherapeutic drug homocamptothecin. Choi and his coworkers also designed hyaluronic acid nanoparticles of various diameters varying the rate with which HA is subjected for hydrophobic substitution to decrease tumor expansion with decreased systemic harmfulness as they have an extraordinary capability to target tumors [94]. Baghbani and his coworkers synthesized stable perfuorohexane nano-droplets using alginate to deliver doxorubicin and also estimated its sensitivity for ultrasound and imaging [95]. Podgorna and his coworkers designed nanogels of gadolinium to load hydrophilic drugs and to be applied in MRI images as positive contrast agents due to their paramagnetic behavior. Ding and his coworkers used dextran nanoparticles in combination with redox-responsive chlorine 6 (C6) to encapsulate Fe_3O_4 nanoparticles for formulating a multifunctional composite nanoparticulate system for magnetic resonance (MR) and near-infrared (NIR) imaging [96].

13.4 Conclusion

The recent advancements in nanomedicines include technological development for drug delivery as well as novel diagnostic approaches. Originally, nanotechnology was applied to increase the bioavailability, solubility, absorption, and controlled release of drugs by developing a nano-drug delivery system. However, a number of uncertainties are associated with the invention of nano-drugs. The nano-drug delivery systems involve an effectual, specified, and controllable intracellular drug release, demonstrating distinctive benefits in the area of therapeutic and diagnostic. It can successfully decipher the problems regarding targeting at specific sites, localized and controlled drug release, and prolonged release, as well as decreasing the toxic effects. The integration of nanotechnology into drug delivery has offered a multifunctional and assimilated direction in the area of diagnosis and treatment. With the modernization of nanotechnology, the usage of nano-drug delivery systems will be encouraged with the introduction of novel techniques and methods for clinical diagnostic and therapeutic aspects. Around some 50 years ago, the synthesis of pharmacologically active compounds (berberine, curcumin, ellagic acid, resveratrol, curcumin, and

TABLE 13.1
Nanomedicine with Biomedical and Diagnostic Applications

Nano-drug delivery system	Application	Advantage	Ref.
Triiodobenzoate cholecalciferol nano-emulsion	Micro-computed tomography	Size of the nano-emulsions compatible with the cell and tissues interaction	[97]
5-fluorouracil-loaded chitosan nanoparticles	Clinical chemotherapy for the treatment of carcinoma of the colon and rectum	Prevent the side effects of 5-fluorouracil	[98]
Ibuprofen-amine-functionalized SBA-15	Treatment of inflammation, pain or rheumatism	Increases the drug loading capacity of ibuprofen and decreases its drug delivery rate	[99]
5-fluorouracil-grafted with chitosan-coated Fe_3O_4 magnetic nanoparticles	Anticancer drug for the treatment of stomach, liver, and intestinal cancer	Decreases the hydrophobic effect of drug	[100]
Doxorubicin encapsulated folate-conjugated starch nanoparticles	Antineoplastic agent	Improve the targeting function and decreases the toxic effect of doxorubicin	[101]
Azelaic acid grafted poly(lactide-co-glycolide) nanoparticles	Acne treatment	For control release of drug in follicular system	[102]
Doxorubicin encapsulated carboxylic functional group coated magnetic nanoparticles	Anticancer drug used in the treatment of various carcinoma	Decreases the toxicity of doxorubicin	[103]
Doxorubicin loaded poly(N-isopropyl acrylamide) co-acrylic acid nanoparticles	Antineoplastic agent for breast carcinoma	Chemotherapeutic drug screening applications	[104]
Epigallocatechin-3-gallate loaded PEGylated PLGA nanoparticles	Treatment of Alzheimer's disease	Enhancement of therapeutic effect of drug and its stability	[105]
Ellipticine encapsulated poly-(3-hydroxybutyr ate-co-3-hydroxyvalerate) based nanoparticles	Model anticancer drug in brain tumor	Enhance the bioavailability of ellipticine and its cytotoxic effect	[106]
Enalapril encapsulated mPEG3-PCL3 nanoparticles	Treatment of hypertension and congestive heart failure	Enhancement of hydrophilic nature of drug	[107]
Paclitaxel-loaded nanoparticles	Treatment of prostate cancer	Enhance the therapeutic efficacy of drug	[108]
Paclitaxel-rapamycin encapsulated glycerol monooleate coated magnetic nanoparticles	Antineoplastic agent	Enhances the therapeutic effect of drug	[109]
Curcumin loaded folate-modified-chitosan-nanoparticles	Treatment of breast cancer	Enhances the therapeutic efficacy of curcumin	[110]
Raloxifene HCl loaded nanoparticles	Estrogenic effect on bone and reduces the bone loss	Reduces the drug release rate so enhances the therapeutic rate	[111]
Bortezomib grafted micellar nanoparticles	Anticancer drug in treatment of mantle cell lymphoma and multiple myeloma	Enhances the therapeutic effect of drug by sustained release of drug	[112]
Doxycycline loaded poly (D,L-lactide-co-glycolide): poly (ε-caprolactone) nanoparticles	Broad-spectrum antibiotic	Increases the therapeutic efficacy of drug by sustained release	[113]
Lamivudine drug-loaded Poly(lactic acid)/chitosan nanoparticles	Antiretroviral drug for HIV-treatment	Control drug delivery of lamivudine	[114]
Naringenin-loaded nanoparticles	Natural bioflavonoid used as anticancer drug	Enhances the anticancer potential of naringenin	[115]
Latanoprost and curcumin loaded nanoparticles	Treatment of glaucoma by decreasing intraocular pressure	Increases trabecular and uveoscleral outflow	[116]

quercetin) from natural sources was not a preferred option. However, the application of nanotechnology to improve the efficiency of naturally occurring bioactive compounds has turned out to be a familiar characteristic. The usefulness of the natural products has been greatly enhanced by using nanocarriers designed with polymeric nanoparticles, crystal nanoparticles, liposomes, micelles, superparamagnetic iron oxide nanoparticles dendrimers and metallic components such as gold, silver, cadmium sulfide, and titanium dioxide. Even though in the future the regulatory mechanisms involved in assessing the safety/toxicity of the nanomedicines will be the cause of new innovation, nanomedicine has already reformed a potential pathway for the administration and discovery of drugs in biological systems. Advancements that have been made in the field of nanomedicine enable a high ability to diagnose diseases and also allow conjugating the diagnosis with therapy.

REFERENCES

1. McNeil, S. E. (2011) Unique benefits of nanotechnology to drug delivery and diagnostics, in *Characterization of nanoparticles intended for drug delivery*, Scott E. McNeil (Ed.), pp. 3–8, Springer.
2. Orive, G., Gascon, A. R., Hernández, R. M., Domínguez-Gil, A., and Pedraz, J. L. (2004) Techniques: New approaches to the delivery of biopharmaceuticals, *Trends in Pharmacological Sciences 25*, 382–387.
3. Jahangirian, H., Lemraski, E. G., Webster, T. J., Rafiee-Moghaddam, R., and Abdollahi, Y. (2017) A review of drug delivery systems based on nanotechnology and green chemistry: Green nanomedicine, *International Journal of Nanomedicine 12*, 2957.
4. Rudramurthy, G. R., Swamy, M. K., Sinniah, U. R., and Ghasemzadeh, A. (2016) Nanoparticles: Alternatives against drug-resistant pathogenic microbes, *Molecules 21*, 836.

5. Patra, J. K., and Baek, K.-H. (2014) Green nanobiotechnology: Factors affecting synthesis and characterization techniques, *Journal of Nanomaterials 2014*, Article ID 417305.
6. Joseph, R. R., and Venkatraman, S. S. (2017) Drug delivery to the eye: What benefits do nanocarriers offer?, *Nanomedicine 12*, 683–702.
7. Lam, P.-L., Wong, W.-Y., Bian, Z., Chui, C.-H., and Gambari, R. (2017) Recent advances in green nanoparticulate systems for drug delivery: Efficient delivery and safety concern, *Nanomedicine 12*, 357–385.
8. Patra, J. K., Das, G., Fraceto, L. F., Campos, E. V. R., del Pilar Rodriguez-Torres, M., Acosta-Torres, L. S., Diaz-Torres, L. A., Grillo, R., Swamy, M. K., and Sharma, S. (2018) Nano based drug delivery systems: Recent developments and future prospects, *Journal of Nanobiotechnology 16*, 1–33.
9. Shi, X., Sun, K., and Baker Jr, J. R. (2008) Spontaneous formation of functionalized dendrimer-stabilized gold nanoparticles, *The Journal of Physical Chemistry C 112*, 8251–8258.
10. Mirza, A. Z., and Siddiqui, F. A. (2014) Nanomedicine and drug delivery: A mini review, *International Nano Letters 4*, 94.
11. Watkins, R., Wu, L., Zhang, C., Davis, R. M., and Xu, B. (2015) Natural product-based nanomedicine: Recent advances and issues, *International Journal of Nanomedicine 10*, 6055.
12. Tan, Q., Liu, W., Guo, C., and Zhai, G. (2011) Preparation and evaluation of quercetin-loaded lecithin-chitosan nanoparticles for topical delivery, *International Journal of Nanomedicine 6*, 1621.
13. De Villiers, M. M., Aramwit, P., and Kwon, G. S. (2008) *Nanotechnology in drug delivery*, Springer Science & Business Media.
14. Kabanov, A. V., Lemieux, P., Vinogradov, S., and Alakhov, V. (2002) Pluronic® block copolymers: Novel functional molecules for gene therapy, *Advanced Drug Delivery Reviews 54*, 223–233.
15. Wang, N., and Feng, Y. (2015). Elaborating the role of natural products-induced autophagy in cancer treatment: Achievements and artifacts in the state of the art, *BioMed Research International 2015*, Article ID 934207.
16. Sharma, G., Raturi, K., Dang, S., Gupta, S., and Gabrani, R. (2014) Combinatorial antimicrobial effect of curcumin with selected phytochemicals on Staphylococcus epidermidis, *Journal of Asian Natural Products Research 16*, 535–541.
17. Ouattara, B., Simard, R. E., Holley, R. A., Piette, G. J.-P., and Bégin, A. (1997) Antibacterial activity of selected fatty acids and essential oils against six meat spoilage organisms, *International Journal of Food Microbiology 37*, 155–162.
18. Abdelwahab, S. I., Sheikh, B. Y., Taha, M. M. E., How, C. W., Abdullah, R., Yagoub, U., El-Sunousi, R., and Eid, E. E. (2013) Thymoquinone-loaded nanostructured lipid carriers: Preparation, gastroprotection, in vitro toxicity, and pharmacokinetic properties after extravascular administration, *International Journal of Nanomedicine 8*, 2163.
19. Park, K. (2007) Nanotechnology: What it can do for drug delivery, *Journal of Controlled Release 120*, 1.
20. Deng, Y., Zhang, X., Shen, H., He, Q., Wu, Z., Liao, W., and Yuan, M. (2020) Application of the nano-drug delivery system in treatment of cardiovascular diseases, *Frontiers in Bioengineering and Biotechnology 7*, 489.
21. Landesman-Milo, D., Goldsmith, M., Ben-Arye, S. L., Witenberg, B., Brown, E., Leibovitch, S., Azriel, S., Tabak, S., Morad, V., and Peer, D. (2013) Hyaluronan grafted lipid-based nanoparticles as RNAi carriers for cancer cells, *CancerLetters 334*, 221–227.
22. Yingchoncharoen, P., Kalinowski, D. S., and Richardson, D. R. (2016) Lipid-based drug delivery systems in cancer therapy: What is available and what is yet to come, *Pharmacological Reviews 68*, 701–787.
23. Chandrasekaran,S.,andKing,M.R.(2014)Microenvironment of tumor-draining lymph nodes: Opportunities for liposome-based targeted therapy, *International Journal of Molecular Sciences 15*, 20209–20239.
24. Mitchell, M. J., Billingsley, M. M., Haley, R. M., Wechsler, M. E., Peppas, N. A., and Langer, R. (2020) Engineering precision nanoparticles for drug delivery, *Nature Reviews Drug Discovery, 20*, 1–24.
25. Cheng, X., and Lee, R. J. (2016) The role of helper lipids in lipid nanoparticles (LNPs) designed for oligonucleotide delivery, *Advanced Drug Delivery Reviews 99*, 129–137.
26. Berraondo, P., Martini, P. G., Avila, M. A., and Fontanellas, A. (2019) Messenger RNA therapy for rare genetic metabolic diseases, *Gut 68*, 1323–1330.
27. Vhora, I., Lalani, R., Bhatt, P., Patil, S., and Misra, A. (2019) Lipid-nucleic acid nanoparticles of novel ionizable lipids for systemic BMP-9 gene delivery to bone-marrow mesenchymal stem cells for osteoinduction, *International Journal of Pharmaceutics 563*, 324–336.
28. Fenton, O. S., Olafson, K. N., Pillai, P. S., Mitchell, M. J., and Langer, R. (2018) Advances in biomaterials for drug delivery, *Advanced Materials 30*, 1705328.
29. Shi, C., He, Y., Ding, M., Wang, Y., and Zhong, J. (2019) Nanoimaging of food proteins by atomic force microscopy. Part I: Components, imaging modes, observation ways, and research types, *Trends in Food Science & Technology 87*, 3–13.
30. Wei, D., Qiao, R., Dao, J., Su, J., Jiang, C., Wang, X., Gao, M., and Zhong, J. (2018) Soybean lecithin-mediated nanoporous PLGA microspheres with highly entrapped and controlled released BMP-2 as a stem cell platform, *Small 14*, 1800063.
31. Zhao, L., Ding, J., Xiao, C., He, P., Tang, Z., Pang, X., Zhuang, X., and Chen, X. (2012) Glucose-sensitive polypeptide micelles for self-regulated insulin release at physiological pH, *Journal of Materials Chemistry 22*, 12319–12328.
32. Afsharzadeh, M., Hashemi, M., Mokhtarzadeh, A., Abnous, K., and Ramezani, M. (2018) Recent advances in co-delivery systems based on polymeric nanoparticle for cancer treatment, *Artificial Cells, Nanomedicine, and Biotechnology 46*, 1095–1110.
33. Wang, W., Ding, J., Xiao, C., Tang, Z., Li, D., Chen, J., Zhuang, X., and Chen, X. (2011) Synthesis of amphiphilic alternating polyesters with oligo (ethylene glycol) side chains and potential use for sustained release drug delivery, *Biomacromolecules 12*, 2466–2474.
34. Yang, C., Gao, S., Dagnæs-Hansen, F., Jakobsen, M., and Kjems, J. (2017) Impact of PEG chain length on the physical properties and bioactivity of PEGylated chitosan/siRNA nanoparticles in vitro and in vivo, *ACS Applied Materials & Interfaces 9*, 12203–12216.
35. Sun, P., Huang, W., Kang, L., Jin, M., Fan, B., Jin, H., Wang, Q.-M., and Gao, Z. (2017) siRNA-loaded poly (histidine-arginine) 6-modified chitosan nanoparticle with enhanced

cell-penetrating and endosomal escape capacities for suppressing breast tumor metastasis, *International Journal of Nanomedicine 12*, 3221.
36. Kesharwani, P., Gajbhiye, V., and Jain, N. K. (2012) A review of nanocarriers for the delivery of small interfering RNA, *Biomaterials 33*, 7138–7150.
37. Baeza, A., Ruiz-Molina, D., and Vallet-Regí, M. (2017) Recent advances in porous nanoparticles for drug delivery in antitumoral applications: Inorganic nanoparticles and nanoscale metal-organic frameworks, *Expert Opinion on Drug Delivery 14*, 783–796.
38. Di Pietro, P., Strano, G., Zuccarello, L., and Satriano, C. (2016) Gold and silver nanoparticles for applications in theranostics, *Current Topics in Medicinal Chemistry 16*, 3069–3102.
39. Liang, J.-J., Zhou, Y.-Y., Wu, J., and Ding, Y. (2014) Gold nanoparticle-based drug delivery platform for antineoplastic chemotherapy, *Current Drug Metabolism 15*, 620–631.
40. Khafaji, M., Zamani, M., Golizadeh, M., and Bavi, O. (2019) Inorganic nanomaterials for chemo/photothermal therapy: A promising horizon on effective cancer treatment, *Biophysical Reviews 11*, 335–352.
41. Arias, L. S., Pessan, J. P., Vieira, A. P. M., Lima, T. M. T. d., Delbem, A. C. B., and Monteiro, D. R. (2018) Iron oxide nanoparticles for biomedical applications: A perspective on synthesis, drugs, antimicrobial activity, and toxicity, *Antibiotics 7*, 46.
42. Huang, K.-W., Hsu, F.-F., Qiu, J. T., Chern, G.-J., Lee, Y.-A., Chang, C.-C., Huang, Y.-T., Sung, Y.-C., Chiang, C.-C., and Huang, R.-L. (2020) Highly efficient and tumor-selective nanoparticles for dual-targeted immunogene therapy against cancer, *Science Advances 6*, eaax5032.
43. Su, X., Chan, C., Shi, J., Tsang, M.-K., Pan, Y., Cheng, C., Gerile, O., and Yang, M. (2017) A graphene quantum dot@ Fe3O4@ SiO2 based nanoprobe for drug delivery sensing and dual-modal fluorescence and MRI imaging in cancer cells, *Biosensors and Bioelectronics 92*, 489–495.
44. Wang, W., Sun, X., Zhang, H., Yang, C., Liu, Y., Yang, W., Guo, C., and Wang, C. (2016) Controlled release hydrogen sulfide delivery system based on mesoporous silica nanoparticles protects graft endothelium from ischemia–reperfusion injury, *International Journal of Nanomedicine 11*, 3255.
45. Perioli, L., Pagano, C., and Ceccarini, M. R. (2019) Current highlights about the safety of inorganic nanomaterials in healthcare, *Current Medicinal Chemistry 26*, 2147–2165.
46. Wong, P. T., and Choi, S. K. (2015) Mechanisms of drug release in nanotherapeutic delivery systems, *Chemical Reviews 115*, 3388–3432.
47. Chen, G., Roy, I., Yang, C., and Prasad, P. N. (2016) Nanochemistry and nanomedicine for nanoparticle-based diagnostics and therapy, *Chemical Reviews 116*, 2826–2885.
48. Mattos, B. D., Rojas, O. J., and Magalhães, W. L. (2017) Biogenic silica nanoparticles loaded with neem bark extract as green, slow-release biocide, *Journal of Cleaner Production 142*, 4206–4213.
49. Siepmann, F., Herrmann, S., Winter, G., and Siepmann, J. (2008) A novel mathematical model quantifying drug release from lipid implants, *Journal of Controlled Release 128*, 233–240.
50. Ding, C., and Li, Z. (2017) A review of drug release mechanisms from nanocarrier systems, *Materials Science and Engineering: C 76*, 1440–1453.
51. Kamaly, N., Yameen, B., Wu, J., and Farokhzad, O. C. (2016) Degradable controlled-release polymers and polymeric nanoparticles: Mechanisms of controlling drug release, *Chemical Reviews 116*, 2602–2663.
52. Edgar, J. Y. C., and Wang, H. (2017) Introduction for design of nanoparticle based drug delivery systems, *Current Pharmaceutical Design 23*, 2108–2112.
53. Pelaz, B., del Pino, P., Maffre, P., Hartmann, R., Gallego, M., Rivera-Fernandez, S., de la Fuente, J. M., Nienhaus, G. U., and Parak, W. J. (2015) Surface functionalization of nanoparticles with polyethylene glycol: Effects on protein adsorption and cellular uptake, *ACS Nano 9*, 6996–7008.
54. Almalik, A., Benabdelkamel, H., Masood, A., Alanazi, I. O., Alradwan, I., Majrashi, M. A., Alfadda, A. A., Alghamdi, W. M., Alrabiah, H., and Tirelli, N. (2017) Hyaluronic acid coated chitosan nanoparticles reduced the immunogenicity of the formed protein corona, *Scientific Reports 7*, 1–9.
55. Kolhar, P., Anselmo, A. C., Gupta, V., Pant, K., Prabhakarpandian, B., Ruoslahti, E., and Mitragotri, S. (2013) Using shape effects to target antibody-coated nanoparticles to lung and brain endothelium, *Proceedings of the National Academy of Sciences 110*, 10753–10758.
56. Gao, W., and Zhang, L. (2015) Coating nanoparticles with cell membranes for targeted drug delivery, *Journal of Drug Targeting 23*, 619–626.
57. Müller, J., Bauer, K. N., Prozeller, D., Simon, J., Mailänder, V., Wurm, F. R., Winzen, S., and Landfester, K. (2017) Coating nanoparticles with tunable surfactants facilitates control over the protein corona, *Biomaterials 115*, 1–8.
58. Gao, H., Yang, Z., Zhang, S., Cao, S., Shen, S., Pang, Z., and Jiang, X. (2013) Ligand modified nanoparticles increases cell uptake, alters endocytosis and elevates glioma distribution and internalization, *Scientific Reports 3*, 1–9.
59. Jain, A., and Jain, S. K. (2015). Ligand-appended BBB-targeted nanocarriers (LABTNs), *Critical Reviews™ in Therapeutic Drug Carrier Systems 32*(2), 149–180.
60. Muro, S. (2012) Challenges in design and characterization of ligand-targeted drug delivery systems, *Journal of Controlled Release 164*, 125–137.
61. Kou, L., Sun, J., Zhai, Y., and He, Z. (2013) The endocytosis and intracellular fate of nanomedicines: Implication for rational design, *Asian Journal of Pharmaceutical Sciences 8*, 1–10.
62. Salatin, S., and Yari Khosroushahi, A. (2017) Overviews on the cellular uptake mechanism of polysaccharide colloidal nanoparticles, *Journal of Cellular and Molecular Medicine 21*, 1668–1686.
63. Ulbrich, K., Hola, K., Subr, V., Bakandritsos, A., Tucek, J., and Zboril, R. (2016) Targeted drug delivery with polymers and magnetic nanoparticles: Covalent and noncovalent approaches, release control, and clinical studies, *Chemical Reviews 116*, 5338–5431.
64. Chen, C.-W., Syu, W.-J., Huang, T.-C., Lee, Y.-C., Hsiao, J.-K., Huang, K.-Y., Yu, H.-P., Liao, M.-Y., and Lai, P.-S. (2017) Encapsulation of Au/Fe 3 O 4 nanoparticles into a polymer nanoarchitecture with combined near infrared-triggered

chemo-photothermal therapy based on intracellular secondary protein understanding, *Journal of Materials Chemistry B 5*, 5774–5782.
65. Fang, J., Sawa, T., and Maeda, H. (2004) Factors and mechanism of "EPR" effect and the enhanced antitumor effects of macromolecular drugs including SMANCS, *Polymer Drugs in the Clinical Stage 519*, 29–49.
66. Torchilin, V. (2011) Tumor delivery of macromolecular drugs based on the EPR effect, *Advanced Drug Delivery Reviews 63*, 131–135.
67. Fang, J., Nakamura, H., and Maeda, H. (2011) The EPR effect: Unique features of tumor blood vessels for drug delivery, factors involved, and limitations and augmentation of the effect, *Advanced Drug Delivery Reviews 63*, 136–151.
68. Korin, N., Kanapathipillai, M., Matthews, B. D., Crescente, M., Brill, A., Mammoto, T., Ghosh, K., Jurek, S., Bencherif, S. A., and Bhatta, D. (2012) Shear-activated nanotherapeutics for drug targeting to obstructed blood vessels, *Science 337*, 738–742.
69. Prijic, S., and Sersa, G. (2011) Magnetic nanoparticles as targeted delivery systems in oncology, *Radiology and Oncology 45*, 1–16.
70. Freund, J., and Shapiro, B. (2012) Transport of particles by magnetic forces and cellular blood flow in a model microvessel, *Physics of Fluids 24*, 051904.
71. Sahoo, S. K., and Labhasetwar, V. (2003) Nanotech approaches to drug delivery and imaging, *Drug Discovery Today 8*, 1112–1120.
72. Ma, S., Tian, X. Y., Zhang, Y., Mu, C., Shen, H., Bismuth, J., Pownall, H. J., Huang, Y., and Wong, W. T. (2016) E-selectin-targeting delivery of microRNAs by microparticles ameliorates endothelial inflammation and atherosclerosis, *Scientific Reports 6*, 1–11.
73. Morachis, J. M., Mahmoud, E. A., and Almutairi, A. (2012) Physical and chemical strategies for therapeutic delivery by using polymeric nanoparticles, *Pharmacological Reviews 64*, 505–519.
74. Charoenphol, P., Mocherla, S., Bouis, D., Namdee, K., Pinsky, D. J., and Eniola-Adefeso, O. (2011) Targeting therapeutics to the vascular wall in atherosclerosis–carrier size matters, *Atherosclerosis 217*, 364–370.
75. Atanasov, A. G., Waltenberger, B., Pferschy-Wenzig, E.-M., Linder, T., Wawrosch, C., Uhrin, P., Temml, V., Wang, L., Schwaiger, S., and Heiss, E. H. (2015) Discovery and resupply of pharmacologically active plant-derived natural products: A review, *Biotechnology Advances 33*, 1582–1614.
76. Yuan, H., Ma, Q., Ye, L., and Piao, G. (2016) The traditional medicine and modern medicine from natural products, *Molecules 21*, 559.
77. Namdari, M., Eatemadi, A., Soleimaninejad, M., and Hammed, A. T. (2017) A brief review on the application of nanoparticle enclosed herbal medicine for the treatment of infective endocarditis, *Biomedicine & Pharmacotherapy 87*, 321–331.
78. Bonifácio, B. V., da Silva, P. B., dos Santos Ramos, M. A., Negri, K. M. S., Bauab, T. M., and Chorilli, M. (2014) Nanotechnology-based drug delivery systems and herbal medicines: A review, *International Journal of Nanomedicine 9*, 1.
79. Ramana, K. V., Singhal, S. S., and Reddy, A. B. (2014) *Therapeutic potential of natural pharmacological agents in the treatment of human diseases*, Hindawi.
80. Guo, W. (2013) Green technology for nanoparticles in biomedical applications, in *Green biosynthesis of nanoparticles: Mechanisms and applications*, Rai, M., and Posten, C. (Eds.), pp. 1–12, CABI.
81. Boroumand Moghaddam, A., Namvar, F., Moniri, M., Azizi, S., and Mohamad, R. (2015) Nanoparticles biosynthesized by fungi and yeast: A review of their preparation, properties, and medical applications, *Molecules 20*, 16540–16565.
82. Metz, K. M., Sanders, S. E., Pender, J. P., Dix, M. R., Hinds, D. T., Quinn, S. J., Ward, A. D., Duffy, P., Cullen, R. J., and Colavita, P. E. (2015) Green synthesis of metal nanoparticles via natural extracts: The biogenic nanoparticle corona and its effects on reactivity, *ACS Sustainable Chemistry & Engineering 3*, 1610–1617.
83. Teleanu, D. M., Chircov, C., Grumezescu, A. M., Volceanov, A., and Teleanu, R. I. (2018) Impact of nanoparticles on brain health: An up to date overview, *Journal of Clinical Medicine 7*, 490.
84. Iravani, S. (2014) Bacteria in nanoparticle synthesis: Current status and future prospects, *International Scholarly Research Notices 2014*, Article ID 359316.
85. Franci, G., Falanga, A., Galdiero, S., Palomba, L., Rai, M., Morelli, G., and Galdiero, M. (2015) Silver nanoparticles as potential antibacterial agents, *Molecules 20*, 8856–8874.
86. Gupta, U., Sharma, S., Khan, I., Gothwal, A., Sharma, A. K., Singh, Y., Chourasia, M. K., and Kumar, V. (2017) Enhanced apoptotic and anticancer potential of paclitaxel loaded biodegradable nanoparticles based on chitosan, *International Journal of Biological Macromolecules 98*, 810–819.
87. Chang, C.-H., Huang, W.-Y., Lai, C.-H., Hsu, Y.-M., Yao, Y.-H., Chen, T.-Y., Wu, J.-Y., Peng, S.-F., and Lin, Y.-H. (2011) Development of novel nanoparticles shelled with heparin for berberine delivery to treat *Helicobacter pylori*, *Acta Biomaterialia 7*, 593–603.
88. Swierczewska, M., Han, H. S., Kim, K., Park, J., and Lee, S. (2016) Polysaccharide-based nanoparticles for theranostic nanomedicine, *Advanced Drug Delivery Reviews 99*, 70–84.
89. Chen, F., Hong, H., Goel, S., Graves, S. A., Orbay, H., Ehlerding, E. B., Shi, S., Theuer, C. P., Nickles, R. J., and Cai, W. (2015) In vivo tumor vasculature targeting of CuS@ MSN based theranostic nanomedicine, *ACS Nano 9*, 3926–3934.
90. Yhee, J. Y., Son, S., Kim, S. H., Park, K., Choi, K., and Kwon, I. C. (2014) Self-assembled glycol chitosan nanoparticles for disease-specific theranostics, *Journal of Controlled Release 193*, 202–213.
91. Yang, S.-J., Lin, F.-H., Tsai, H.-M., Lin, C.-F., Chin, H.-C., Wong, J.-M., and Shieh, M.-J. (2011) Alginate-folic acid-modified chitosan nanoparticles for photodynamic detection of intestinal neoplasms, *Biomaterials 32*, 2174–2182.
92. Ryu, J. H., Na, J. H., Ko, H. K., You, D. G., Park, S., Jun, E., Yeom, H. J., Seo, D. H., Park, J. H., and Jeong, S. Y. (2014) Non-invasive optical imaging of cathepsin B with activatable fluorogenic nanoprobes in various metastatic models, *Biomaterials 35*, 2302–2311.

93. Lapčík, L., Lapcik, L., De Smedt, S., Demeester, J., and Chabrecek, P. (1998) Hyaluronan: Preparation, structure, properties, and applications, *Chemical Reviews 98*, 2663–2684.
94. Wang, G., Gao, S., Tian, R., Miller-Kleinhenz, J., Qin, Z., Liu, T., Li, L., Zhang, F., Ma, Q., and Zhu, L. (2018) Theranostic hyaluronic acid–iron micellar nanoparticles for magnetic-field-enhanced in vivo cancer chemotherapy, *ChemMedChem 13*, 78–86.
95. Baghbani, F., Moztarzadeh, F., Mohandesi, J. A., Yazdian, F., and Mokhtari-Dizaji, M. (2016) Novel alginate-stabilized doxorubicin-loaded nanodroplets for ultrasounic theranosis of breast cancer, *International Journal of Biological Macromolecules 93*, 512–519.
96. Podgórna, K., Szczepanowicz, K., Piotrowski, M., Gajdošová, M., Štěpánek, F., and Warszyński, P. (2017) Gadolinium alginate nanogels for theranostic applications, *Colloids and Surfaces B: Biointerfaces 153*, 183–189.
97. Attia, M. F., Anton, N., Akasov, R., Chiper, M., Markvicheva, E., and Vandamme, T. F. (2016) Biodistribution and toxicity of X-ray iodinated contrast agent in nano-emulsions in function of their size, *Pharmaceutical Research 33*, 603–614.
98. Yang, H.-C., and Hon, M.-H. (2009) The effect of the molecular weight of chitosan nanoparticles and its application on drug delivery, *Microchemical Journal 92*, 87–91.
99. Ahmadi, E., Dehghannejad, N., Hashemikia, S., Ghasemnejad, M., and Tabebordbar, H. (2014) Synthesis and surface modification of mesoporous silica nanoparticles and its application as carriers for sustained drug delivery, *Drug Delivery 21*, 164–172.
100. Ding, Y., Shen, S. Z., Sun, H., Sun, K., Liu, F., Qi, Y., and Yan, J. (2015) Design and construction of polymerized-chitosan coated Fe3O4 magnetic nanoparticles and its application for hydrophobic drug delivery, *Materials Science and Engineering: C 48*, 487–498.
101. Xiao, S., Tong, C., Liu, X., Yu, D., Liu, Q., Xue, C., Tang, D., and Zhao, L. (2006) Preparation of folate-conjugated starch nanoparticles and its application to tumor-targeted drug delivery vector, *Chinese Science Bulletin 51*, 1693–1697.
102. Reis, C. P., Martinho, N., Rosado, C., Fernandes, A. S., and Roberto, A. (2014) Design of polymeric nanoparticles and its applications as drug delivery systems for acne treatment, *Drug Development and Industrial Pharmacy 40*, 409–417.
103. Hua, X., Yang, Q., Dong, Z., Zhang, J., Zhang, W., Wang, Q., Tan, S., and Smyth, H. D. C. (2017) Magnetically triggered drug release from nanoparticles and its applications in anti-tumor treatment, *Drug Delivery 24*, 511–518.
104. Kang, A., Seo, H. I., Chung, B. G., and Lee, S.-H. (2015) Concave microwell array-mediated three-dimensional tumor model for screening anticancer drug-loaded nanoparticles, *Nanomedicine: Nanotechnology, Biology and Medicine 11*, 1153–1161.
105. Cano, A., Ettcheto, M., Chang, J.-H., Barroso, E., Espina, M., Kühne, B. A., Barenys, M., Auladell, C., Folch, J., Souto, E. B., Camins, A., Turowski, P., and García, M. L. (2019) Dual-drug loaded nanoparticles of epigallocatechin-3-gallate (EGCG)/ascorbic acid enhance therapeutic efficacy of EGCG in a APPswe/PS1dE9 Alzheimer's disease mice model, *Journal of Controlled Release 301*, 62–75.
106. Masood, F., Chen, P., Yasin, T., Fatima, N., Hasan, F., and Hameed, A. (2013) Encapsulation of ellipticine in poly-(3-hydroxybutyrate-co-3-hydroxyvalerate) based nanoparticles and its in vitro application, *Materials Science and Engineering: C 33*, 1054–1060.
107. Danafar, H., Manjili, H. K., and Najafi, M. (2016) Study of copolymer composition on drug loading efficiency of enalapril in polymersomes and cytotoxicity of drug loaded nanoparticles, *Drug Res (Stuttg) 66*, 495–504.
108. Sahoo, S. K., Ma, W., and Labhasetwar, V. (2004) Efficacy of transferrin-conjugated paclitaxel-loaded nanoparticles in a murine model of prostate cancer, *International Journal of Cancer 112*, 335–340.
109. Dilnawaz, F., Singh, A., Mohanty, C., and Sahoo, S. K. (2010) Dual drug loaded superparamagnetic iron oxide nanoparticles for targeted cancer therapy, *Biomaterials 31*, 3694–3706.
110. Esfandiarpour-Boroujeni, S., Bagheri-Khoulenjani, S., Mirzadeh, H., and Amanpour, S. (2017) Fabrication and study of curcumin loaded nanoparticles based on folate-chitosan for breast cancer therapy application, *Carbohydrate Polymers 168*, 14–21.
111. Bikiaris, D., Karavelidis, V., and Karavas, E. (2009) Novel biodegradable polyesters. Synthesis and application as drug carriers for the preparation of raloxifene HCl loaded nanoparticles, *Molecules, 14*(7), 2410–2430.
112. Lee, A. L. Z., Voo, Z. X., Chin, W., Ono, R. J., Yang, C., Gao, S., Hedrick, J. L., and Yang, Y. Y. (2018) Injectable coacervate hydrogel for delivery of anticancer drug-loaded nanoparticles in vivo, *ACS Applied Materials & Interfaces 10*, 13274–13282.
113. Misra, R., Acharya, S., Dilnawaz, F., and Sahoo, S. K. (2009) Sustained antibacterial activity of doxycycline-loaded poly(D,L-lactide-co-glycolide) and poly(ε-caprolactone) nanoparticles, *Nanomedicine 4*, 519–530.
114. Dev, A., Binulal, N. S., Anitha, A., Nair, S. V., Furuike, T., Tamura, H., and Jayakumar, R. (2010) Preparation of poly(lactic acid)/chitosan nanoparticles for anti-HIV drug delivery applications, *Carbohydrate Polymers 80*, 833–838.
115. Krishnakumar, N., Sulfikkarali, N., RajendraPrasad, N., and Karthikeyan, S. (2011) Enhanced anticancer activity of naringenin-loaded nanoparticles in human cervical (HeLa) cancer cells, *Biomedicine & Preventive Nutrition 1*, 223–231.
116. Cheng, Y.-H., Ko, Y.-C., Chang, Y.-F., Huang, S.-H., and Liu, C. J.-l. (2019) Thermosensitive chitosan-gelatin-based hydrogel containing curcumin-loaded nanoparticles and latanoprost as a dual-drug delivery system for glaucoma treatment, *Experimental Eye Research 179*, 179–187.

14

Nanodevices for the Detection of Cancer Cells

Annu and Priya Chauhan

CONTENTS

DOI: 10.1201/9781003157823-14

14.1 Introduction

In the battle against cancer, the major fight is enduring based on its early detection. The early-phase detection of cancer considerably raises the possibility for its successful treatment. Vigorous checking of population may assist to recognize individuals who have the disease, including those who do not so far have the indications like breast or cervical cancer [1]. A major development in the diagnosis of all types of cancer in the past three decades is very challenging task. This prominent and incessant decrement in cancer mortality is attributed to the advance in avoidance, early-on diagnosis [2].

Early detection of cancer, particularly prior to cancer cells metastasizing, is serious for cancer judgment since early-stage cancers can be treated more efficiently. However, timely stage diagnosis has been tricky to attain in the majority of cases since clinical signs are inclined to demonstrate themselves at later stages. Consequently, non-invasive techniques are immediately required for early cancer detection [3].

In the past few years, nanodevices have been potentially investigated to detect cancer at early stages. Nano-biosensors are being extended to sense blood-borne biomarkers, viz. cancer-related proteins. These biosensors possess elevated sensitivity, specificity, robustness, speed and multiplexed investigation. Next-generation nanodevices linked with genetic examination could be proficient to promote diagnosis of a patient's cancer and potential treatments and disease course.

NPs are being built up to perform as molecular imaging agents for detection of the presence of cancer-appropriate genetic alterations or the functional distinctiveness of tumor cells. The rising area of nanotechnology engrosses scientists from numerous diverse fields, including physicists, chemists, engineers and biologists. Nanodevices are the construction of functional materials and devices during the treatment of such miniscule matters [4].

This chapter offers an extensive summary of the application of nanodevices for the early detection of cancer.

14.2 Application of Nanodevices for Recognition of Cancer Cells

Despite advances in our understanding of molecular biology, chemotherapy, radiotherapy and conventional surgical procedures, cancer is known as one of the leading causes of death in the world. Existing therapies for cancer, such as chemotherapy and radiotherapy, frequently lack tumor cell specificity, ensuing in severe toxic effects for cancer patients enduring such treatments. Thus, the significant goal in cancer therapy generally focuses on the development of treatment modalities that efficiently kill tumor cells without harming normal cells. Therefore, new strategies for targeted cancer therapy have been posed in great demand for effective cancer treatment. With the rapid advancement of nanotechnology, several nanostructured materials have been efficiently synthesized for the purpose of biomedical applications. Moreover, their diverse characteristics with multifunctional theranostic ability confirm favorable potential in cancer therapy [5]. Such nanomaterials can non-specifically accumulate in cancer tissue by the increased permeability and retention (EPR) effect, i.e., by passive targeting, although with limited dosage and selectivity [6]. Currently, though, the active, cell-specific targeting of nanomaterials has begun to signify a potentially powerful technology related to cancer treatment. Besides this, active targeting is attained by conjugating nanomaterials with targeting ligands that bind to overexpressed antigens or receptors on the target cells. Such specific binding to targeted cells leads to an enhanced accumulation of nanomaterials on target cells while reducing detrimental toxicity to non-target cells.

Over the past few years, the well-known aptamers have been known as a novel class of targeting ligands for diagnostic and therapeutic application purposes in cancer therapy [7, 8]. Aptamers are also called short, synthetic, single-stranded oligonucleotides which specifically bind to several molecular targets, involving small molecules, proteins, nucleic acids and even cells as well as tissues along with high affinity and specificity [9, 10]. They have been derived from an iterative process referred to as the systematic evolution of ligands through exponential enrichment and also exemplified as an exclusive class of molecules that are larger than small-molecule drugs but smaller compared to antibodies [11, 12]. Compared with traditional ligands, including antibodies, peptides and small molecules, aptamers possess advantages like low cost, low immunogenicity as well as toxicity, small size in order to assist solid tumor penetration and high affinity so as to bind with the target, all of which make aptamers ideal candidates for targeted cancer therapy [13, 14]. Although on uniting the intrinsic features of nanomaterials with the specific recognition ability of aptamers, aptamer-conjugated nanomaterials might offer a highly effective as well as less harmful method to meet the emergent demands for new strategies to fight against cancer [15–18]. Thus, this aptamer-targeted strategy validates high efficacy and also low side effects for cancer treatment, making aptamer-conjugated nanomaterials promising materials to be applicable in future cancer therapy.

14.2.1 Aptamer-Conjugated Nanomaterials for Specific Cell Recognition

The important way for effective cancer therapy is to differentiate cancer cells from normal cells. Sensitive and selective cancer cell detection by precise molecular recognition is greatly preferred for the development of targeted cancer therapy and also the potential efficacy of novel therapeutic modalities. Enthused by aptamer technology and nanotechnology, various strategies developed for specific cell recognition have been discussed. Based on aptamer-tethered DNA nanodevices (aptNDs), specific recognition and in situ self-assembly of aptNDs on target living cell surfaces have been accomplished. To develop aptNDs, aptamer sgc8, which may bind to target human protein tyrosine kinase 7, was chosen as a model. Protein tyrosine kinase 7 is overexpressed on the cell membrane of CCRF-CEM cells (human T-cell acute lymphocytic leukemia) but not on non-target Ramos cells. Additionally, two partially complementary hairpin monomers, i.e., M1 and M2, and also an aptamer probe were utilized so as to construct aptNDs by using either a hybridization chain reaction-based self-assembly upon initiation by an aptamer-tethered trigger probe or through cascading another hybridization of two partially complementary monomers begun by aptamer seed probes. Moreover, the aptND scan effectively anchors or in situ self-assembles onto the target cell surfaces. Either covalent chemical labeling of multiple copies of fluorophores or non-covalent physical association with multiple fluorogenic double-stranded DNA-intercalating fluorophores on each nanodevice provided increased fluorescence signals for efficient detection of cancer [19]. Herein, the high specificity of aptamers to target cells has led to selectivity improvement in the area of electrochemical and electrochemiluminescence detection of cancer cells [20–22]. By means of fluorescence and electrochemical methods, a signal amplification super-sandwich strategy has been constructed for highly selective and sensitive detection of cancer cells by using aptamer-DNA concatamer quantum dot probes. The developed supers and cyto-sensor showed high sensitivity, with a detection limit of 50 cells per ml [23]. Additionally, a novel cycle-amplifying technique utilizing a DNA device on magnetic beads was further employed in order to expand the sensitivity of the electrochemiluminescence assay of cancer cells [24]. Especially, a strategy expending an aptamer and RNA polymerase-based amplification has been also been developed for highly sensitive and selective cancer cell detection [25].

A no magnetic background has been exhibited virtually by most biological samples, thus the application of magnetic nanoparticles (MNPs) may lead to ultrasensitive detection. On the basis of a magnetic relaxation switch technique and a self-amplifying proximity assay using the change of spin–spin relaxation time (DT2) of the surrounding water protons, Bamrungsap et al. developed aptamer-conjugated magnetic nanoparticles (ACMNPs) for the detection of cancer cells. The ACMNPs capitalize on the ability of the sgc8 aptamer to specifically bind to target cancer cells, also the large surface areas of MNPs, so as to accommodate multiple aptamer-binding events. However, the ACMNPs may detect as few as 10 cancer cells in 250 ml of sample. Additionally, their specificity as well as sensitivity were also confirmed by detection in cell mixtures and complex biological media, comprising fatal bovine serum, human plasma and whole blood. Likewise, using an array of ACMNPs, different cell types were distinguished by using pattern recognition, and there by producing a cellular molecular profile which allows clinicians to exactly recognize cancer cells at the molecular as well as the single-cell level [26]. Alternatively, another study demonstrates a DNA aptamer-polyethylene glycol (PEG)-lipid composite in order to modify cell surfaces for specific cell recognition. Aptamer TD05, which may selectively bind to IgG receptors on the surface of Ramos cells, a B-cell lymphoma cell line and sgc8 aptamer were utilized for the testing purposes. Leukemia cell lines were applied so as to validate that aptamers attached to the cell surface might act as targeting ligands, which may specifically identify their target cells. Moreover, the potential of such a probe was explored in adoptive cell therapy. Immune-effector cells modified by the probe confirmed enhanced affinity, though remaining cytotoxic to target cancer cells. Surface modification of living cells by the aptamer-PEG-lipid offers an efficient approach for the recognition of cells as well as suggesting significant potential in cell-based therapy [27]. Double aptamer-conjugated gold manganese oxide (Au@MnO) hybrid nanoflowers were also utilized as a multifunctional platform in order to specifically target CCRF-CEM cells, and also to capture ATP molecules from cell lysate. Additionally, such sgc8 aptamer- and ATP aptamer-modified nanoflowers have been used as an effective ionization substrate for the purpose of laser desorption/ionization, leading to highly selective detection and for analysis of metabolites from cancer cells. Single-platform nanoflower conjugates comprising MnO and Au components give an ideal all-in-one system for selective binding to the target molecule and for laser desorption ionization-mass spectrometry as an ionization substrate [28]. These merits, together with the simple preparation of aptamer-conjugated nanomaterials, make such strategies highly promising to effectively diagnose and target cancer therapy.

In the recent era, nanotechnology has led to numerous promising results with its applications for the diagnosis and treatment of cancer, involving drug delivery, gene therapy, detection and diagnosis, drug carriage, biomarker mapping, targeted therapy and molecular imaging. Nanotechnology has been engaged for the development purpose of nanomaterials, like gold nanoparticles and quantum dots, that are useful for cancer diagnosis at the molecular level. Nanotechnology-based molecular diagnostics, for instance, the development of biomarkers, may accurately as well as rapidly detect cancers. Treatments based on nanotechnology, like the development of nanoscale drug delivery, might confirm accurate cancerous tissue targeting with the least side effects. Because of their biological nature, nanomaterials may simply cross cell barriers. Over the years, nanomaterials have been applied in the treatment of tumors, as they offer active and passive targeting. Even though several drugs may be used to treat cancers, and thus the sensitivity of the drugs usually leads to insufficient results and also might have distinct side effects and could lead to damage to the healthy cells. In view of that, various studies have observed different forms of nanomaterials, like liposomes, polymers, molecules and antibodies, with the conclusion that a combination of these nanomaterials in cancer drug design may accomplish a balance between enhancing efficacy as well as reducing the toxicity of drugs. Conversely, because of the potential toxicity of nanomaterials, it still requires a lot of advancement to be done on them before their willing acceptance in the clinic for cancer management.

14.2.2 Nanotechnology in Cancer Diagnosis

Genetic mutations might cause changes in the synthesis of certain biomolecules which leads to uncontrolled cell proliferation and ultimately cancerous tissues [29]. Classification of cancers may be either benign or malignant where benign tumors are confined to the origin of cancer while malignant tumors actively shed cells invading surrounding tissues and distant organs. Thus, cancer diagnostic and therapeutic strategies are targeted on early detection and inhibition of cancerous cell growth as well as their spread. Noteworthy among the early diagnostic tools for cancers involve the use of positron emission tomography (PET), magnetic resonance imaging (MRI), computed tomography (CT) and ultrasound [30]. Such imaging systems are limited by their inadequate provision of relevant clinical information about various types of cancer and its stage. Therefore, making it challenging to acquire a full assessment of the disease state based on which an optimum therapy can be delivered [31, 32].

14.2.3 Tools Based on Nanotechnology to Be Used in Cancer Diagnosis

In recent research, nanotechnology may endorse cancer imaging at the tissue, cell and molecular levels [33]. This might be achieved by the capacity of nanotechnology applications to explore the tumor's environment. For example, pH response to fluorescent nanoprobes may help to detect fibroblast activated protein on the cell membrane of tumor-related fibroblasts [34]. Herein, some nanotechnology-based spatial and temporal techniques have been discussed that can help accurately track living cells and monitor dynamic cellular events in tumors.

14.2.3.1 Near-Infrared (NIR) Quantum Dots

The lack of capability to pierce objects confines the utilization of visible spectral imaging. Quantum dots that emit fluorescence in the near-infrared spectrum (i.e., 700–1,000 nanometers) have been developed so as to overcome such problems, which makes them highly suitable for imaging colorectal cancer, liver cancer, pancreatic cancer and lymphoma [35–37]. A second near-infrared (NIR) window (NIR-II, 900–1,700 nm) with higher tissue penetration depth, higher spatial and temporal resolution has also been designed to aid cancer imaging. Similarly, the development of a silver-rich Ag2Te quantum dots (QDs) covering a sulfur source has been reported to allow visualization of better spatial resolution images over a wide infrared range [38].

14.2.3.2 Nanoshells

Alternatively, the commonly used nanotechnology application is the utilization of nanoshells. They are dielectric cores between 10 and 300 nanometers in size, generally made up of silicon as well as coated with a thin metal shell (usually gold) [39, 40]. Such nanoshells work by converting plasma-mediated electrical energy into light energy which may be flexibly tuned optically by UV-infrared emission/absorption arrays. Nanoshells are desirable as their imaging is devoid of heavy metal toxicity [41] despite the fact their uses are limited by their large sizes.

14.2.3.3 Colloidal Gold Nanoparticles

Gold nanoparticles (AuNPs) are known as a good contrast agent due to their small size, good biocompatibility and high atomic number. Several research studies suggested that AuNPs work by both active and passive means to target cells.

Usually, the principle of passive targeting is directed by a congregation of the gold nanoparticles to enhance imaging due to the permeability tension effect (EPR) in tumor tissues [42]. On the other hand, Active targeting is mediated by the coupling of AuNPs with tumor-specific targeted drugs, like EGFR monoclonal antibodies, in order to attain AuNP active targeting of tumor cells. As the energy exceeds 80 kev, the mass attenuation rate of gold becomes higher as compared to alternative elements such as iodine, indicating greater prospect gold nanoparticles [43]. Such findings have essential implications for early diagnosis, with the technique permitting tumors as small as a few millimeters in diameter to be detected in the body [44].

14.2.4 Recognition of Circulating Tumor Cells

Approximately 90% of deaths from solid tumors are accredited to metastasis [45]. In the course of metastatic dissemination, a cancer cell from the primary tumor first conquers the surrounding tissue and then moves in the microvasculature of the blood (intravasation) and lymph systems, followed by survival and translocation by the bloodstream to micro-vessels in distant tissues, consequent exit from the bloodstream (extravasation) and survival in the microenvironment of distant tissues, presenting an appropriate foreign microenvironment for the development of a macroscopic secondary tumor [46]. The early detection of metastatic cancer cells within the bloodstream, referred to as circulating tumor cells (CTCs), may potentially affect cancer prognosis as well as diagnosis. As a portion of a liquid biopsy, CTCs have been studied broadly because of their potential applications. The detection of CTC may help us in understanding the molecular organization of a tumor in a slightly intrusive manner. Nonetheless, CTCs show relatively low abundance as well as heterogeneity, presenting technical challenges for CTC isolation and characterization. Recently, researchers have generally focused on the application of nanotechnologies for the sensitive detection of CTCs.

These technologies might help in characterizing cells and molecules, thus possessing broad clinical applications, like the detection of disease at an early stage as well as evaluation of the treatment response and disease development. As confirmed in many studies, it is possible for cell pseudopodia to develop on surfaces with nanostructure, hence increasing the local topographical interactions among the cancer cells and nanostructured substrates, which is beneficial to CTC enrichment. For CTC detection, nanomaterials have an important advantage in their large surface-to-volume ratio, enabling adsorption of high-efficiency targeting ligands with the ability to recognize specific molecules on cancer cells; consequently, CTC isolation indicates high specificity and recovery, thus the detection sensitivity is increased [47].

Various types of nanomaterials have been reported, for instance, magnetic nanoparticles (MNPs), gold nanoparticles (AuNPs), quantum dots (QDs), nanowires, nanopillars, silicon nanopillars, carbon nanotubes, dendrimers, graphene oxide and polymers, for CTC detection.

It has been suggested that such nanomaterials might progress the sensitivity as well as the specificity of CTC capture devices and possess the potential to ease cancer diagnosis and prognosis. In the area of nano-biotechnology, MNPs are known as mature nanomaterials which can bind to cells and have long been utilized for in vitro separation with the help of an external magnetic field [48]. Antibody-functionalized

MNPs, viz. immunomagnetic nanoparticles, are often applied in the biomedical field. For CTC detection, immunomagnetic technologies generally specifically target EpCAM expressing CTCs with anti-EpCAM functionalized MNPs. Moreover, to perform single-cell transcriptional profiling of CTCs purified from breast cancer patients, Powell et al. [49] utilized Mag Sweeper, which is an immunomagnetic enrichment device that may isolate tumor cells from unfractionated blood. Mag Sweeper is known to serve as a magnetic cell sorting system that uses magnetic rods covered by a sheath to sweep across capture wells and also attract target cells labeled with magnetic nanoparticles [50]. It can be applied to achieve high purity CTCs from patient blood, though preserving their capacity in order to initiate tumors and metastasize, enabling robust analysis of single CTCs. Thereby, using the system, the authors effectively purified CTCs from 70% of patients with primary and metastatic breast cancer and also performed direct measurement of the gene expression in individual CTCs. QDs are being characterized by using unique optical properties that might increase their expediency in cancer cell detection [51]. Because of their high quantum yields, QDs are proven to be helpful in the detection of materials with low abundance. Thus, to enhance QD electrical characteristics, Pang et al. [52] hybridized ZnO NDs and g-C3N4 QDs so as to afford higher photoelectron transfer as well as separation efficiency. As they possess exceptional advantages, ZnO NDs and g-C3N4 QDs adored the prolonged application, and a photocatalyzed renewable self-powered cytosensing device was presented based on ZnO NDs@g-C3N4 QDs. By means of conjugation to the membrane PTK7-specific aptamer Sgc8c, the device was applied in order to detect CCRF-CEM cells (human acute lymphoblastic leukemia cells), that expresses PTK7. The results suggested that the device shows better performance in terms of detection range, detection limit, selectivity and reproducibility. Hence, it captured only CCRF-CEM cells (500 cell/mL) while no other cell types, such as HL-60, K562 and HeLa cells. Wu et al. [53] described a strategy for semiconducting polymer dots (PDs) functionalization by entrapment of heterogeneous polymer chains into a single dot, expedited by hydrophobic interactions, at the time of nanoparticle formation. A few amphiphilic polymers with functional groups for successive covalent conjugation of biomolecules, like streptavidin and immunoglobulin G (IgG), suggested co-condensation with most semiconducting polymers for modification as well as functionalization of a nanoparticle surface. The PDs bioconjugates showed the ability to label cellular targets in an efficient and specific manner, involving a cell surface marker on human breast cancer cells, with no requirement to detect nonspecific binding. For the differentiation of different types of cells and cancer states, the authors utilized AuNPs capped with ligands of different hydrophobicity and were coated with green fluorescent protein (GFP). As the capping ligands applied suggested different chemical structures, each AuNP-GFP complex was linked to cancer cells to a different degree, considering cell membrane composition differences [54]. Magnetic bio targeting multifunctional nano-bioprobes (MBMNs) were utilized to detect and also isolate a small subset of malignant cells from normal cells. CoFe2O4@BaTiO3 magnetoelectric nanoparticles illustrated different cancer cells from each other and from their normal counterparts by using a magnetoelectric effect [55].

14.2.5 Detection through Cell Surface Protein Recognition

The chief approach to detect cancer cells is based on the binding of nanoparticle probes conjugated with moieties (protein, short peptides, antibodies, oligonucleotide aptamers) to surface markers on cancer cells, and also on those entering cells and detecting genetic content. For the detection of cancer cells, like CTCs, capture or isolation is the first and highly important stage. Though the cell's physical properties, for instance, size, deformability and density, are sometimes used, capture is mostly based on the affinity of cell surface molecules on CTCs detected with materials like antibodies or aptamers. Exclusive surface proteins on CTCs are the primary targets. Since several studies have shown that EpCAM is highly expressed on CTCs from many human malignancies, EpCAM can be utilized as a cell surface biomarker. Thus, anti-EpCAM molecules are frequently applied to the screening of CTCs. CTCs experiencing EMT may cause inefficient positive sorting based on EpCAM expression. Consequently, an alternative approach is to find supplemental or replacement markers for EpCAM. Various cell surface markers, like vimentin, androgen receptor, glycan, major vault protein (MVP), as well as fibroblast activation protein α (FAPα), have been considered for the detection of CTCs. Thus, majority of such markers are only specific to definite cells, and several markers do not exist after CTCs experienced EMT as well. More mesenchymal CTCs may be seen in the metastatic stages of cancer, thereby looking for proper EMT markers to assess prognosis and metastasis in cancer patients is important.

14.2.6 Detection Based on mRNA

Additionally, to the detection of extracellular nucleic acids, nanoparticles have also been proposed as intracellular nucleic acid sensors. Seferos et al. [56] suggested that it is possible to utilize novel gold nanoparticle probes fabricated by oligonucleotides hybridized to complements labeled with a fluorophore as transfection agents and cellular "nanoflares" to detect mRNA in living cells. Nanoflares may overcome different challenges in the formation of effective as well as sensitive intracellular probes and express a large signal-to-noise ratio and also sensitivity to changes in the number of RNA transcripts in cells. Nanoflares that indicate high orientation, dense oligonucleotide coating and which might enter cells without the requirement for cytotoxic transfection agents [57] are highly useful for detecting intracellular mRNA.

Meanwhile, researchers have proposed nanoflares for simultaneous intracellular detection of various mRNA transcripts. In such multiplexed nanoflare studies, AuNPs functionalized with two to three DNA recognition strands and thus later hybridized with short complementary reporter strands were produced as nanoflares. For instance, the application of multiplexed nanoflares to detect survive in addition to actin has been examined for normalizing nanoflare fluorescence differences in cellular uptake. Hence, the technique is known to be comparable with conventional qRT-PCR for the quantification of intracellular mRNA however, can be performed at the single live cell level. While, in a few cases, the nanoflare platform was expanded in order to quantify intracellular RNA as well as to detect spatiotemporal localization in living cells [58]. Hereon, β-act in targeting nanoflares has been incubated with HeLa cells and offered a perceptibly different intracellular distribution, demonstrating strong colocalization with mitochondria. Furthermore, smart flares were used for studying purposes of melanoma tumor cell heterogeneity [59]. Such smart flares have the ability to quantify genomic expression at the single-cell level, thereby expanding our knowledge of cancer as well as metastasis. Scrutinizing the heterogeneity of cancer cells is essential for recognizing novel biomarkers for early cancer diagnosis. Halo et al. [60] proposed nanoflares, which were used to capture live circulating breast cancer cells. These nanoflares may detect target mRNA in model metastatic breast cancer cell (MBC) lines in human blood and showed high recovery and fidelity reaching 99%. They have also utilized nanoflares together with later cultured mammo spheres to reimplant the regained live recurrent breast cancer cells into whole human blood, where only 100 live cancer cells may be detected per mL of blood. Based on the nanoflare technology, it was likely to simultaneously isolate as well as characterize intracellular live cancer cells from whole blood. The authors confirmed the ability of nanoflares to collect CTCs for future culture and study. Moreover, nanoflares subsidize the technology of combining intracellular markers with cell-surface markers for dually identifying putative CTCs. The combined method is likely to increase the function of more platforms to specifically detect CTCs and subpopulations of CTCs. It was suggested that nanoflares offer the first gene-based approach in order to detect, isolate and characterize live cancer cells in the blood and are probable to contribute to cancer diagnosis, prognosis and prediction, as well as personalized treatment. Lee et al. reported a method relying on elegant plasmonic nanoparticle network structure, creating a plasmon-coupled dimer possessing the ability to detect single mRNA variants [61]. The method has been applied for the detection and quantification of BRCA1 mRNA splice variants in vitro and in vivo. The method is powerful and can successfully detect, quantify and differentiate between different BRCA1 splice variants with single-copy sensitivity, thus laying a foundation for quantitative, single-cell genetic profiling in the future.

14.2.7 Nanotechnology for In Vivo Imaging

Furthermore, to cancer diagnosis by ex vivo detection of cancer cells and biomarkers in liquid biopsy samples, identifying cancerous tissues in the body has several advantages in diagnosing as well as treating cancer. A proper nanoparticle probe to detect cancer tissue must exhibit a long circulation time, be specific to tumor tissue as well as present low toxicity to adjacent healthy tissue [62]. Recently, related studies have generally engrossed on nanoparticle probe accumulation in tumor tissue to diagnose cancer in animal models, usually mouse models. Nanoparticle probes might especially accumulate in tumor tissues by active or positive targeting, thus permitting imaging and diagnosis of cancer in vivo [63]. Interactions among nanoparticles and blood proteins, uptake and clearance by the reticuloendothelial system (RES), dissemination into solid tumors and optimized active (vs. passive) targeting for diagnosis of cancer establish the main clinical application barriers. Auspiciously, various developments related to such aspects have been attained.

14.2.7.1 Passive Targeting

Passive targeting signifies the favored extravasation capacity of 10–150 nm nanoparticles from the bloodstream into tumor tissue. As the tight junction mong endothelial cells in new blood vessels in tumors do not generate appropriately, nanoparticles might favorably accumulate in tumor tissue [64]. Thus, this form of passive nanoparticle entry into the tumor microenvironment is known as the enhanced permeability and retention (EPR) effect, which was detected around 30 years ago by studying macromolecule transport into tumor tissues [65]. QDs are characterized by obvious photostability, tunable emission and high quantum yield, contributing to their extensive application in tumor tissue imaging through passive accumulation dependent on the EPR effect. Hong et al. [66] reported the application of a novel NIR-II fluorophore, six-armed PEG-Ag2S QDs, for imaging of subcutaneous xenograft4T1 murine tumors. They examined how the NIR-II signal was distributed within the mice for a long period of time (up to 24 h post-injection [p.i.]) and observed that the NIR-II fluorescence of 6PEGAg2S QDs enhanced firmly in the tumor region and also reduced in the skin and other organs in the range of 30 min p.i. to 24 h p.i. The in vivo QD pharmacokinetics showed extraordinary accumulation of6PEG-Ag2S QDs in tumors (> 10% ID/gram, where % ID/gram signifies the concentration of the probe relative to the injected dose [ID] percentage per gram of tissue) through the EPR effect. They declare that imaging with these NIR-II QDs offered deep inner organ registration, dynamic tumor contrast and quick detection of the tumor. Researchers have also applied AuNPs to in vivo tumor imaging by passive targeting. Lai et al. [67] suggested that mercaptoundecanoic acid-coated AuNPs may identify as well as track primary glioma cells at the inoculation sites in mouse brains. Moreover, such particles detected tumor-associated microvasculature in detail. However, in a few cases, chitosan nanoparticles have been utilized for in vivo imaging by applying the EPR effect. Nam et al. [68] proposed a tumor-targeting nanoparticle for use as an underlying multimodal imaging probe through optical/MR (MR: magnetic resonance) dual imaging based on self-assembled glycol chitosan. Based on chemical modification and conjugation, they proposed stable chitosan nanoparticles labeled with Cy5.5 and encapsulated by Gd (III) (Cy5.5CNP-Gd [III]). The Cy5.5-CNP-Gd (III) were spherical in nature, with a size of approximately 350 nm. According to cellular experiments, Cy5.5-CNP-Gd (III) was taken up in an actual manner, also the distribution in the cytoplasm was observed. Subsequently, after administration using the tail vein of tumor-bearing mice, the nanoparticles localized in large numbers within the tumor, and further were detected through non-invasive NIR fluorescence together with an MR imaging system. The authors suggested that the exceptional characteristics of the glycol chitosan nanoparticles, like blood stability, deformability, as well as rapid cellular uptake, might prominently affect the in vivo tumor-targeting ability, which depends on the EPR effect. Further, their results showed that Cy5.5-CNP-Gd (III) can actually be utilized as an optical/MR dual imaging agent so as to detect and treat cancer. Although nanoparticle size as well as its shape affect the EPR effect. Consequently, such factors should be regarded at the time of designing nanoparticle probes for high tumor accumulation. Nanoparticles with a size of less than 10 nm may be rapidly eradicated by the kidneys, reducing their localization in tumor tissue [69]. Anisotropic particles showed an augmentation in circulation time, perhaps as anisotropic nanoparticles are known to be less expected to infuse endothelial gaps in the liver in the range of hundreds of nanometers to tens of micrometers. Silica-coated QDs of various thicknesses were utilized to explore the impact of nanoparticle size on tumor tissue accumulation [70]. The 12 nm QDs pierced the tumor tissue with the least hindrance, as the 60 nm QDs extravasated while remaining in 10 µm blood vessels. By contrast, the 120 nm QDs suggested no substantial extravasation. Once the nanoparticles interact with a biological fluid, their surface will become covered with a "corona" of biological macromolecules [71]. As serum proteins adsorb onto a nanoparticle surface (opsonization), the in vivo trafficking, uptake, as well as clearance of nanoparticles are prominently changed. Using PEG to coat a nanoparticle surface diminishes nonspecific adsorption of serum proteins and also reduces protein corona formation, which enhances the circulating time of the nanoparticle. PEGylation of several nanoparticles, like AuNPs and QDs, resulting in a longer circulation time in the blood, also slows accumulation in the liver and spleen [72]. However, it is estimated that on the one hand nanotechnology-based imaging might progress the specificity as well as the sensitivity of cancer diagnosis, also decreasing toxicity on the other hand. Garrigue et al. [73] currently reported that harnessing nanoparticles and the "enhanced permeation and retention" (EPR) effect assisted them to develop an innovative nanosystem for the purpose of positron emission tomography (PET) imaging. The system embraces a self-assembling amphiphilic dendrimer which helps in retaining different PET reporting units at terminals. Such dendrimers have the ability to self-assemble into small uniform nano micelles that accumulated in tumors, permitting effective PET imaging. Because of the dendrimeric multivalence combined with the passive tumor targeting facilitated by EPR, the nanosystem showed improved imaging sensitivity as well as stronger specificity, with PET signal ratios which enhanced approximately by 14-fold as compared to the clinical gold standard 2-fluorodeoxyglucose ([18F] FDG). Likewise, the dendrimer exhibited an exceptional safety profile and virtuous pharmacokinetics for PET imaging. Thus, the authors believe that the study contributes to the development of dendrimer nanosystems for effective and promising cancer imaging.

14.2.7.2 Active Targeting

Furthermore, tumor imaging with the help of nanoparticle accumulation by means of passive targeting relying on the EPR effect, researchers have employed many studies on recognition of receptors on the cell surface for active targeting of tumor tissues. Generally, such approaches enhance the number of nanoparticles conveyed to tumor tissue in each unit time, thus increasing the sensitivity demonstrated by in vivo tumor detection methods [74]. For the detection purposes of tumors at an early stage with high contrast imaging, active tumor targeting attained an improved result as compared to passive targeting which relies on the EPR effect. Levenson and Nie proposed antibody-conjugated QDs to target PSMA for active tumor targeting. The in vivo imaging results for three types of QD surface modifications were observed: (1) COOH groups, (2) PEG groups and (3) PEG plus PSMA Ab (PEG-PSMA Ab). On the basis of histological examinations, the COOH probe did not show any tumor signals, whereas only weak tumor signals have been detected with the PEG probe (passive targeting), while the PEG-PSMA Ab-conjugated probe (active targeting) demonstrated intense signals. Thus, the evaluation suggested the conclusion of highly efficient as well as much more rapid active targeting of tumors with a tumor-specific ligand as compared to passive targeting in terms of tumor permeation, retention and uptake [75]. However, a current study revealed the frequent utilization of peptides to active targeting of cancerous tissues in vivo. The RGD peptide is analyzed by a receptor (integrin $\alpha v\beta 3$) on the cell surface involved in cancer angiogenesis and metastasis and additionally has been applied to the targeting of tumor tissue in vivo for diagnosis [76]. In one of the studies, an iRGD-mediated and enzyme-induced precise targeting gold nanoparticle system (iRGD/AuNPs-A&C) was proposed by simply co-administering tumor-homing penetration peptide iRGD with a legumain-responsive aggregable gold nanoparticle. Intravenously injected compounds coupled to iRGD were bound to tumor vessels and after that being spread to extravascular tumor parenchyma, whereas traditional RGD peptides only transported cargo into blood vessels. iRGD homes to tumors by using three steps: the RGD motif reveals a mediating effect on the binding to αv integrins on the tumor endothelium, and further on, a proteolytic cleavage executes a binding motif for neuropilin-1, that helps in penetration into the cells. Conjugation to iRGD contributed to an obvious enhancement in the sensitivity of the tumor imaging agents as well as in the activity of the anti-tumor drug.

14.3 Application of Nanodevices in Delivery of Anticancer Drugs

The use of new agents in cancer therapy has significantly enriched patient survival, however, there are still various biological barriers that may alienate drug delivery to target cells and tissues, viz. unfavorable blood half-life and physiologic behavior with high off-target effects as well as effective clearance from the human organism [77–79]. Likewise, in cancer, there is a small subset of cancer cells – cancer stem cells (CSC) – that, like normal stem cells, can self-renew, contribute to rising heterogeneous populations of daughter cells and also proliferate widely [80]. Moreover, standard chemotherapy is directed against quickly dividing cells, the bulk of non-stem cells of a tumor, and thereby CSC frequently seems to be relatively refractory to those agents. The expansion of side effects in normal tissues (e.g., nephrotoxicity, neurotoxicity, cardiotoxicity, etc.) and multidrug resistance (MDR) mechanisms by cancer cells leads to a decrease in drug concentration at the target location, a deprived accumulation in the tumor with a subsequent decrease in efficacy which might be linked to patient deterioration [81–86]. In order to overcome such types of issues and still progress the efficiency of chemotherapeutic agents as less toxic and more target-specific therapies toward cancer cells are in great demand, i.e., novel drugs, drug delivery systems (DDSs) and also gene delivery systems. Several new medications have been developed to efficiently treat complicated conditions, while at the same time a few of them produce severe side effects in which the benefit does not always outweigh the risk. In contrast, a few drugs have been confirmed to be highly effective in vitro, however, cannot endure the endogenous enzymes found within the gastrointestinal (GI) tract (if taken orally), rendering them nearly insignificant in vivo. Even though improbable progress has been made to identify drug targets, as well as designing and making a better drug molecule, there is still room to improve the drug delivery systems and also targeting. In the past few decades, nanotechnology, especially the manufacturing of nanoparticles, has found extraordinary attention in broad areas of science. The application of nanoparticles has reformed how drugs are formulated as well as delivered. Nanotechnology is known as a multidisciplinary scientific field spanning engineering and manufacturing principles at the molecular level. By means of the application of nanotechnology to medicine, nanoparticles have been created to mimic or modify biological processes. Nanoparticles are also called solid, colloidal particles whose size ranges from 10 nm to <1,000 nm; though, for nanomedical application, the preferential size is less than 200 nm. As one of the highly significant areas of study has been in the construction of nanoparticle drug delivery systems [87].

"Nanotechnology" may also be known as the manipulation of matter on an atomic, molecular and supramolecular scale including the design, production, characterization and utilization of various nanoscale materials in several key areas proposing novel technological advances mostly in the field of medicine (so-called nanomedicine) [88]. The development as well as the optimization of drug delivery-based methods in nanoparticles involves the early detection of cancer cells and/or specific tumor biomarkers, and the improvement of the efficacy of the treatments applied [89].

The development of nanoparticles suggested it as an effective target-specific strategy for cancer treatment, acting as nanocarriers and also as active agents as well [90]. Over the last few decades, several types of nanoparticles have been developed based on different components, comprising carbon, silica oxides, metal oxides, nanocrystals, lipids, polymers, dendrimers and quantum dots, together with an enhanced variety of newly developed materials. Such nanomaterials have shown the ability to offer a high degree of biocompatibility before and after conjugation to biomolecules for specific functions to translate into nanomedicines and clinical practice. Indeed, the protection from adsorption to plasma proteins and/or deprivation by circulating nucleases permits for enhanced availability of effector molecules at sites of interest. However, this is further increased by the substantial reduction in clearance from the organism where conjugation to nanoparticles confers. The modulation of pharmacokinetic as well as pharmacodynamics parameters creates a key factor for amending the approach of administration which is generally neglected compared to the ability of therapeutic nanoconjugates to offer the possibility of improved targeting (active and/or passive) and cell uptake. While considering nanoparticles for therapeutics one should also assess the effect on cellular metabolism and also the fate which might be acquired through optimal conjugation with (bio)molecules of interest. DDSs have been known to improve the properties of free drugs by enhancing their in vivo stability and also biodistribution, solubility and even modulation of pharmacokinetics, endorsing the transport and even more essential the release of higher doses of the drug in the target site in order to be effective. DDSs may be created by direct conjugation with the drugs, later surface modifications might lead to aim proved delivery for such systems, endorsing a targeted delivery to specific types of cells and accomplishing cell compartments like nucleus and mitochondria. Insofar as drug delivery is concerned, the essential nanoparticle platforms are liposomes, polymer conjugates, metallic nanoparticles (for example, AuNPs), polymeric micelles, dendrimers, nanoshells and protein as well as nucleic acid-based nanoparticles.

Due to the huge impact of such a lethal disease, researchers worldwide have been engrossed in emerging several novel carrier systems to deliver anti-cancerous agents to the target sites without generating detrimental effects on healthy tissues. By the nature of cancer, the malignant tumor area and its progression stage, the outcome of therapy is determined. Several strategies, such as immunotherapy, hormone therapy, surgery, chemotherapy and radiotherapy, are applied to treat cancer. Among these, chemotherapy is known as the primary choice for the abolition of cancerous cells. Though the conventional delivery systems for chemotherapy possess a few limitations, such as toxic side effects of the chemotherapeutic agents on fast-growing healthy cells and other side effects like nausea, vomiting, fatigue, hair loss and even death in extreme cases. Alternatively, a key obstacle in managing the tumors is the development of multidrug resistance (MDR). The cancerous cell produces resistance as well as escaping from the effects of chemotherapeutic drugs. To overcome such limitations of chemotherapy, the researchers are emerging novel nanotechnology-based advances in drug delivery systems for substantial developments in onco-therapy through the delivery of anticancer agents at higher concentrations to targeted sites. However, nanocarriers have been developed and also exploited as a new tool for cancer treatment to decrease numerous precincts of conventional drug delivery systems. Researchers are actively working to expand the treatment options for the management of malignant tumors. Keeping this in view, nanotechnology-based drug carriers have been trialed and considered as useful cancer management applications that have later led to nano-drug carriers (10–100 nm) approaching as different therapeutics for cancer treatment. The potential use as well as efficacy of several nano-drug carriers as anticancer agents is greatly higher compared to normal ones. These nano carriers propose numerous benefits over conventional drug delivery systems such as increased plasma half-life, improved biodistribution and also targeted delivery of a drug to tumor microenvironment by endothelial layers. Such nanocarrier-based drug delivery systems are being utilized against various types of cancerous tumors comprising tissues.

Different nanocarrier preparations have been intended to mend therapeutic efficacy as well as diminish toxic side effects. Nanocarriers prospect a continuous, direct and controlled release of the drug to the malignant cells selectively with augmented drug localization and cellular uptake. Nanoparticles may be automated in order to identify malignant cells and provide discriminating and also precise drug transfer, evading contact through the normal cells. Additionally, various drawbacks, including systemic toxicity, low oral bioavailability, decreased solubility, narrow therapeutic indices and chemoresistance, have been overcome with the advancement of nanocarrier-based DDS [91]. Nanotechnology has become an immense field for the purpose of biomedical applications and moreover nanocarriers like polymeric nanoparticles, micelles, dendrimers, solid lipid nanoparticles, quantum dots and magnetic nanoparticles with significant physical as well as chemical properties along with the nano-sized effect. The resultant of different cancer research studies has been usually accredited that nanoparticle-based drug delivery systems (DDS) with improved bioavailability and least side effects hold propitious anticancer effects as compared with free drugs.

Different types of nanocarriers have been efficaciously implemented in various nanomedicines. However, their transformation to medical oncology remains insufficient. Engineered nanoparticles hold advanced applications including gene delivery, 18 site-specific targeted drug delivery systems and agents in magnetic resonance imaging to increase diagnostic possibilities, or to design and also develop new imaging methods. Nanocarriers offer unique properties like quantum effects, a high ratio of surface to volume and a capability to carry therapeutically active compounds to the targeted site because of their nano size.

Numerous novel nanocarrier-mediated drug delivery systems in order to deliver the chemotherapeutic agents at targeted sites have recently come into practice. A few important nanocarriers include polymeric nanoparticles, liposomes, polymeric micelles, carbon nanotubes, dendrimers, solid lipid nanoparticles, magnetic nanoparticles and quantum dots.

14.4 Nanoparticle-Based Drug Formulations

Several reasons are involved for using nanoparticles as therapeutic and diagnostic agents, and advancement of drug delivery, which is very essential and highly needed as well. One of them is that, nowadays, traditional drugs existing for oral or injectable administration are not always manufactured as the optimal formulation for every product. Products that contain proteins or nucleic acids require a highly innovative type of carrier system to enhance their efficacy and to protect them from unwanted degradation [92]. It is prominent that the efficiency of various drug delivery systems is directly associated with particle size (excluding intravenous and solution). Because of their small size and large surface area, drug nanoparticles demonstrate enhanced solubility and thereby increased bioavailability, additional ability so as to cross the blood-brain barrier (BBB), entering into the pulmonary system and might be absorbed by the tight junctions of endothelial cells of the skin. In particular, the nanoparticles made up from natural and synthetic polymers (biodegradable and nonbiodegradable) have received huge attention as they may be customized for targeted delivery of drugs, progress bioavailability and also provide a controlled release of medication from a single dose; by adaptation of the system, it may prevent endogenous enzymes from degrading the drug. Furthermore, the development of new drug delivery systems is offering alternative advantages for pharmaceutical sales to branch out. Pioneering drug delivery is driving the pharmaceutical companies in order to generate novel formulations of existing drugs. Although such new formulations will be highly beneficial to the patients, and will also create a powerful market force, driving the development of even highly effective delivery approaches.

14.5 Characteristics of Nanoparticle Drug Formulations

Defining exactly what an ideal nanoparticle-based drug delivery system is made of and also understanding how the body handles the exogenous particulate matter are warranted. Nanoparticles may enter the human body by using three main routes, direct injection, inhalation and oral intake. After being entered into the systemic circulation, particle-protein interaction is the first known phenomenon taking place prior to distribution into the distant organs. Absorption from the blood capillaries permits the lymphatic system to distribute and also eliminate the particles. Such a system includes three chief functions, two of which concern drug delivery. The first, fluid recovery, includes the filtering of fluids through the lymphatic system from blood capillaries. Whereas the second encompasses immunity and may be the most relevant to this topic. As the system mends the excess fluid, it also picks up foreign cells as well as the chemicals from the tissues. As the fluids are filtered back into the blood, the lymph nodes identify any foreign matter passing through. If something is acknowledged as foreign, macrophages will help in engulfing and clearing it from the body. This may tend to be the struggle with nanoparticle-based drug delivery; thus, clearance might be influenced by the size and also surface characteristics of particles, which will be explained in the following subsections.

Nanoparticles should possess the ability to endure in the bloodstream for a significant time without being eradicated for effective delivery of drugs to the targeted tumor tissue. Conventional surface particles as well as non-modified nanoparticles are generally caught in the circulation by the reticuloendothelial system, including the liver and the spleen, based on their size and also surface characteristics [93]. The outcome of injected nanoparticles may be controlled by regulating their size and surface characteristics as well.

14.5.1 Size of Particle

The shape as well as the size of nanoparticles affects cells in the body, thereby dictating their distribution, toxicity and also targeting ability. Most significantly, nanoparticles may cross the BBB providing sustained delivery of medication for diseases that were difficult to treat earlier. Despite being possible to reach new targets, such a technique may be manipulated to govern drug distribution. It has been described that 100 nm nanoparticles demonstrated a 2.5-fold greater uptake as compared to 1 mm diameter particles and a six-fold greater uptake than 101 m particles. It has been conferred how essential the nanoparticle drug delivery systems are, but these systems would be of no use if the drug is not released or released efficiently. As particle size gets smaller, their surface area to volume ratio gets larger. This would imply that more of the drug is closer to the surface of the particle compared to a larger molecule. Being at or near the surface would lead to faster drug release. It would be beneficial to create nanoparticle systems that have a large surface area to volume ratio; however, toxicity must always be monitored. As mentioned earlier, the size of the nanoparticle determines biological fate. Remember that the vascular and lymph systems are responsible for the filtering and clearance of foreign matter and chemicals. This is yet another factor that must be engineered into the ideal nanoparticle system. It has been shown that particles 200 nm or larger tend to activate the lymphatic system and are removed from circulation quicker. Thus, from the literature evaluation and discussion so far, it is clear as though the optimum size for a nanoparticle is approximately 100 nm. However, this size is highly accurate for the particle to pass through the BBB, enough drug delivery because of high surface area to volume ratio, also evading abrupt clearance by the lymphatic system.

14.5.2 Surface Properties

Surface characteristics of nanoparticles have been considered as an essential factor for defining their lifespan and destiny at the time of circulation related to their capture by macrophages. Preferably, nanoparticles must possess a hydrophilic surface so that they may escape macrophage capture [94]. However, this can be acquired by using two methods, the first one involves coating the surface of nanoparticles with a hydrophilic polymer, like PEG, and the second protects them from opsonization by repelling plasma proteins; conversely, nanoparticles can be produced from block copolymers with hydrophilic and hydrophobic domains. It has been prominent how size might affect the performance of nanoparticle-based drug formulations; though manipulation of surface characteristics is another opportunity to form the ideal system. In order to produce an optimum nanoparticle drug delivery system, the integration of suitable targeting ligands, surface curvature as well as reactivity is essential so as to address the prevention of aggregation, stability and receptor binding and also ensuing pharmacological effects of the drug. Firstly, the clearance of nanosystems should be addressed. Since nanoparticles may be acknowledged by the lymphatic system, they are being subjected to the body's natural immune response to foreign matter. It is very clear, the more hydrophobic a nanoparticle is, the more expected it is to be cleared because of the higher binding of blood components. As hydrophobic nanoparticles are being cleared certainly, logically it appears to assume that making their surface hydrophilic might enhance their time in circulation. Indeed, coating the nanoparticles by using polymers or surfactants or creating copolymers like polyethylene glycol (PEG) reduces the opsonization; polyethylene oxide, polyethyleneglycol, polyoxamer, poloxamine and polysorbate 80 have been proven to be valuable. PEG is known to be hydrophilic in nature and a comparatively inert polymer as incorporated in the nanoparticle surface impedes the binding of plasma proteins (opsonization), thereby preventing considerable loss of the given dose. PEGylated nanoparticles are frequently referred to as "stealth" nanoparticles, as without opsonization, they remain undetected by the reticuloendothelial system (RES). Mostly, nanoparticles like dendrimers, quantum dots and micelles are particularly susceptible to aggregation. Various strategies have been engaged to prevent aggregation and call for particles coating as well along with capping agents and also modifying the zeta potential (surface charges). Generally, such approaches and theories may be précised into one idea: the size of the particle should be large enough to avoid leakage into blood capillaries, however not too large to become susceptible to clearance of macrophage. Thus, by manipulating the surface, the extent of aggregation as well as clearance might be controlled.

14.5.3 Drug Loading and Release

The size and surface properties of nanoparticles have been explored in order to optimize bioavailability, reduce clearance and enhance stability. Through controlling such characteristics, it is possible to get the drug into tissues in the body that may have been unapproachable before. Nevertheless, there is no significance to this practice if the drug cannot then be released from the nanoparticle matrix. The drug released from the nanoparticle-based formulation relies on various factors such as pH, temperature, drug solubility, desorption of the surface-bound or adsorbed drug, drug diffusion through the nanoparticle matrix, nanoparticle matrix swelling and erosion, and the combination of erosion and diffusion processes. Based on the type of nanoparticle being utilized, the lease of the drug will vary. The developed polymeric nanoparticles may be called nanocapsules or nanospheres based on their composition. Nanospheres are referred to as homogeneous systems such that the polymer chains organize in like fashion to surfactants in micelle formation, even though nanocapsules are heterogeneous systems, such that the drug within a reservoir is comprised of the polymer (like vesicle). With respect to nanospheres, which are matrix systems where the drug is physically and evenly isolated, the drug is released by the erosion of the matrix. There is a quick burst of drug release related to an uncertainly bound drug to the large surface area of the nanoparticle followed by a sustained release. In contrast, if nanocapsules are being used, the release is thus controlled by drug diffusion by the polymeric layer and thereby, drug infusibility by using that polymer is the determining factor of its deliverability. Hence, if there are ionic interactions between the drug and polymer, they will produce complexes that might obstruct the release of the drug from the capsule. Therefore, this can be avoided through the addition of other auxiliary agents like polyethylene oxide-propylene oxide (PEO-PPO), thereby diminishing the interactions between the drug and capsule matrix permitting for greater release of drug to target tissues.

14.5.4 Passive and Active Targeting

Nanocarriers come across several barriers in their route to the target, like mucosal barriers and nonspecific uptake [95]. To report the challenges in targeting tumors with nanotechnology, it is important to combine the rational design of nanocarriers with the fundamental understanding of tumor biology. Common features of tumors include poor lymphatic drainage and leaky blood vessels. While free drugs might diffuse nonspecifically, a nanocarrier may escape into the tumor tissues through the leaky vessels by the increased permeability and retention effect (EPR effect). There is rapid and defective angiogenesis due to which there is enhanced permeability of blood vessels within tumor cells. Moreover, the dysfunctional lymphatic drainage in tumors also helps in retaining the accumulated nanocarriers and permits them to release drugs into the locality of the tumor cells. Experiments using liposomes of various mean sizes propose that the threshold vesicle size for extravasation into tumors is 400 nm, while other studies have suggested that particles having diameters < 200 nm are highly effective. Relying on clinical therapy passive targeting methods suffers from numerous limitations. Fewer drugs cannot diffuse proficiently, thus targeting cells within the tumor is not every time possible, and the random nature of the method makes it difficult in order to control the process due to this lack of control multi-drug resistance (MDR) might persuade – a situation where chemotherapy treatments fail patients indicates the resistance of cancer cells toward one or more drugs.

14.5.5 Targeted Drug Delivery

Subsequently, identifying the significance of nanoparticle manipulation to attain a prosperous drug delivery system, the subsequent logical step involves the development of targeted drug delivery. The nanoparticles may breach the inflamed or damaged tissue because of larger epithelial junctions. Such penetration might occur passively or actively. Active targeting occurs as the drug carrier system is conjugated to a tissue or cell-specific ligand, whereas passive targeting is when the nanoparticle reaches the target organ because of the leaky junctions. An ultimate nanoparticle drug delivery system must be able to reach, recognized, bind and deliver its load to specific pathologic tissues, as well as minimize or avoid damage induced by the drug to healthy tissues. However, coating specific targeting ligand(s) on the surface of nanoparticles is known as the highly common strategy. Such targeting ligands might be in the form of small molecules, peptides, antibodies, designed proteins and nucleic acid aptamers. Small organic molecules are usually employed targeting agents because of the relative ease of preparation, stability and control of conjugation chemistry. These targeting ligands may not possess desired specificity as well as affinity. Biotin (vitamin H), because of its very high affinity for streptavidin has been extensively applied for conjugation with nanoparticles [96]. Folic acid (vitamin B9) possess an extreme affinity for endogenous folate receptors and thereby has been examined for targeting various types of cancers where folate receptors are greatly expressed. Likewise, numerous other task-specific carbohydrates, short peptides, antibodies as well as small molecules have been designed and employed. Alternatively, a useful discovery in order to aid in the targeted delivery of drugs is liposomes. As they mimic the cell membrane, one may design specific lipid monomer to tailor physicochemical properties like size and charge also might integrate surface targeting ligands as discussed above. Such system over additional advantage, since liposomal composition is alike to the targeted cell membrane, an increased lipid-lipid exchange occurs. This prompts up the convective flux of lipophilic drugs from the liposomal lipid layer into the targeted cell membrane.

14.6 Application of Nanoparticle Technology

Recently, chemotherapy is generally aimed at rescinding all quickly dividing cells. The downside of this therapy is that the body's other rapidly proliferating cells, such as in the hair follicles as well as in the intestinal epithelium, are also being killed off, leaving the patient to cope with life-changing side effects. The expansion of nanoparticles has provided a novel opportunity for chemotherapy. With briskly designed nanoparticles, targeted drug delivery at the tumor site or a certain group of cells greatly avoids the toxic effects on other normal tissues and organs [97]. There have been various systems tested to provide this type of therapy. Micelles and liposomes provide other options for the delivery of chemotherapeutic agents. Moreover, micelles have also been proven to be a great way for making insoluble drugs soluble because of their hydrophobic core and hydrophilic shell. However, if the micelle's surface is further PEGylated, it enhances the capability of the nanocarriers to get through the fenestrated vasculature of tumors and inflamed tissue by passive transport, thereby resulting in higher drug concentration in tumors. As of now, various polymeric micelles comprising anticancer drugs, NK012, NK105, NK911, NC-6004 and SP1049C are under clinical trials whereas one such system, Genexol PM (paclitaxel) is approved for breast cancer patients. Conversely, dendrimers are known as highly branched macromolecules with various functional groups available for the attachment of drug, targeting and imaging agents as well as their absorption, distribution, metabolism and elimination (ADME) profile based upon several structural features. A polyfunctional dendrimer system has been proposed for effective localization (folic acid), imaging (fluorescein) and for the delivery of the anticancer drug methotrexate in vitro.

Nanoparticle therapeutics relying on dendrimers might increase the therapeutic index of cytotoxic drugs by engaging biocompatible components, and the surface derivatization with PEGylation, acetylation, glycosylation and different amino acids. Though there are various other forms of nanoparticles that have exposed promise in cancer treatment, currently one of the highly used systems is the carbon nanotubes. Carbon nanotubes (CNTs) are referred to as an allotropic form of carbon with a cylindrical framework that has been deepened on a number of sheets in concentric cylinders, and thus may be classified as single-walled carbon nanotubes (SWCNTs) and multiwalled carbon nanotubes (MWCNTs). Meanwhile, carbon nanotubes possess a highly hydrophobic hollow interior and water-insoluble drugs might easily be loaded to them. The large surface area permits for outer surface functionalization and may be done specifically for a particular cancer receptor and for contrast agents. Finally, Buckminsterfullerene C60 (spherical molecule) and its derivatives as well have been assessed for the treatment of cancer. Whereas, fullerene C60 might bind up to six electrons, thus acting as an exceptional scavenger of reactive oxygen species (ROS) as well. It has been proposed that fullerene nanocrystals (nano-C60) may increase the cytotoxicity of chemotherapeutic agents and thereby nano-C60 adjunct chemotherapy may be further evaluated. Another study used the complex of fullereneC60 with doxorubicin and observed the tumor volumes of the treated rats (C60+Dox) to be 1.4 times lower as compared to the control group (untreated rats). In addition, the mechanism of action of C60+Dox complex is supposed to be through its direct action on tumor cells and immune-modulating effect.

14.6.1 Nanoparticles: Particles Having Unique Properties to Be Considered as Delivery Vehicles

NP-based drug delivery systems have made an astonishing difference in the site-specific release of drugs particularly chemotherapeutic agents, owing to their physical as well as chemical characteristics and biological attributes [98]. Several studies in this rousing area have been conducted, numerous formulations are released in the market that have been now routinely utilized in clinics. NPs are known as a material with overall dimensions in the nanoscale range. In the current decades, various types of NPs and microparticles have been synthesized as well as proposed to be applied as contrast agents for diagnostics and imaging purposes and also for drug delivery, for instance, in cancer therapy. The mechanism involved by which NPs move into the cell has essential implications not only for their providence but also for their influence on biological systems. Numerous research discussed about the potential risks related to exposure of NP, while more recently the concept that even sublethal doses of NPs might prompt a cell response has been proposed. NPs should no longer be viewed only as simple carriers for biomedical applications purposes, but they can also play a dynamic role in intervening biological effects. NPs offer the inimitable possibility to overcome cellular barriers so as to enhance the delivery of several drugs and drug candidates, involving promising therapeutic biomacromolecules like nucleic acids, antisense oligonucleotides, small interfering ribonucleic acid (siRNA) and plasmid DNA, that may only employ their function once inside the cells, or else may not be delivered. However, polar molecules cannot invade the lipid bilayer of the plasma membrane or other biological membranes (blood-brain, air-blood, gastrointestinal barriers). Through such NPs, these therapeutic agents can not only be delivered site specifically while there is the possibility to load NPs with a high concentration of the desired drug. In carrying a large payload, nanocarriers might auspiciously modulate biodistribution as well as pharmacokinetic profiles of the drug formulations. They may be also applied as carriers for contrast agents in vivo magnetic resonance imaging or, again, as an all-in-one system. In some NP applications, the first aim is to avoid clearance by the reticuloendothelial system, thereby prolonging the circulation time in the blood and enhancing the bioavailability at the target site. To one side from size, the shape is regarded as one of the primary parameters that require great attention. The majority of NPs developed for drug delivery have a spherical shape, while other forms like cube-shaped, cylindrical, ellipsoids and disks have currently been suggested as new drug nanocarriers. The rigidity of NP is another significant factor that influences the entry pathway. Whereas other characteristics suggest that interaction between NPs and serum protein convinces the formation of a protein corona which may rapidly cover the entire nanoparticle surface. Generally, these include poly(ethylene glycol), polysaccharides (such as dextran), 266 poly(N-vinylpyrrolidone), polyvinyl alcohol, poly(2-methyl-2-oxazoline), poly(2-ethyl-2-oxazoline), poly(2-methacryloyloxyethyl phosphorylcholine) and polysulfobetaine methacrylate. The application of stimuli-responsive nanocarriers provides a remarkable opportunity for drug as well as gene delivery in the optimization of therapies. The example of a biological stimulus that may be exploited to target drugs and genetic material is known to be pH [22]. Cellular components like the cytoplasm, endosomes, lysosomes, endoplasmic reticulum, Golgi apparatus, mitochondria and nuclei have been known to maintain their own characteristic pH values, which range from 4.5 in the lysosome to about 8.0 in the mitochondria. Furthermore, pH value is prominently affected by diseases: the hypoxic environment in cancer leads to an enhancement in the production of lactic acid and hydrolysis of ATP, as both contribute to acidification.

Indeed, various solid tumors possess lower extracellular pH (pH 6.5) as compared to the surrounding tissues (pH 7.5). Thereby selecting the right material composition, it is likely to be possible to engineer nanocarriers that might exploit such pH differences and permit the release of the delivered drugs or genes to the selected target site. pH-sensitive poly(β-amino ester), known as a biodegradable cationic polymer, in an acidic microenvironment experiences a quick dissolution as well as releasing its content all at once, hence it can represent a virtuous scaffold in order to deliver anticancer drugs. Several other strategies include the presence of acid-sensitive spacers such as poly (vinylpyrrolidone-codimethyl maleic anhydride) among the drug and also the polymer, which enables, after endocytosis, drug release in endosomes or lysosomes of tumor cells. In recent scenarios, NPs are called promising vehicles for anti-tumor drug delivery that are designed to be pH-responsive, experiencing physicochemical changes to release enclosed drugs at acidic pH conditions. Thus, if the target is not the lysosome or in general the acidic compartments of the cells, the low pH environment and numerous lysosomal enzymes result in the degradation of endocytosed components, thereby the loss of the therapeutic effect. However, this happens if there are specific mechanisms for the payload in order to escape out of the lysosomes as well as maximize the efficiency of several treatments [99, 100].

14.7 Types of Nanoparticle Carriers

14.7.1 Liposomes

A liposome is known as an artificially developed vesicle comprised of a lipid bilayer. Liposomes are frequently composed of phosphatidylcholine-enriched phospholipids that may also contain mixed lipid chains with surfactant properties like egg phosphatidylethanolamine. There are some major types of liposomes i.e., multilamellar vesicle (MLV), the small unilamellar vesicle (SUV), the large unilamellar vesicle (LUV) and the cochleate vesicle. Moreover, a liposome encapsulates a region of aqueous solution within a hydrophobic membrane; dissolved hydrophilic solutes cannot readily pass through the lipids. Such hydrophobic chemicals may be dissolved into the membrane, thus in such a way liposomes might carry hydrophobic molecules as well as hydrophilic molecules. However, liposomes have gained huge attention in various scientific disciplines. Since the mid-1970s, the cytotoxic drug incorporation into the liposomes has been reported. Previous reports have shown the groundwork for selecting therapeutic agents pliable to incorporation by liposomes and determining optimal liposome size as well as net charge needed for effective drug delivery. Thus, liposomes were primarily utilized as carriers for lipophilic cytostatic agents, while they were rapidly found to be appropriate for both hydrophilic and hydrophobic drugs. Further, the role of liposomes which acts as drug delivery systems was established for their potentials so as to modify drug pharmacokinetics as well as distribution, allowing an enhanced proportion of the cytostatic agents to be delivered within the tumor tissue and consequently decrease the exposure of normal tissues to the cytostatic agents.

14.7.2 Bionanocapsules

A bio-nanocapsule (BNC) called a nanoparticle (about 70 nm in diameter) consists of the L protein of the hepatitis B virus (HBV), surface antigen (HBsAg) and a lipid bilayer. As of the presence of L protein, BNC can specifically transport genes or drugs to hepatocytes.

14.7.3 Gold Nanoparticles

Gold nanoparticles are of great interest as potential diagnostic as well as therapeutic agents in vivo, as they might be used as X-ray contrast agents, radiation enhancers, laser and radiofrequency thermotherapy enhancers. However, if tumors can be loaded with the help of gold, this may lead to a higher dose to the cancerous tissue as compared to the dose received by normal tissue at the time of radiotherapy treatment. Such nanoparticles are generally non-toxic possessing potentially good biocompatibility: the LD50 of this material is around 3.2 g (Au) per kg of body weight.

14.7.4 Polymeric Nanoparticles

Additionally, polymer-based nanoparticles demonstrated pronounced promise in pre-clinical studies, also polymer-based delivery systems are extremely biocompatible. Moreover, their structures may be modified to engineer multifunctional nanoparticles by using desired shapes, sizes and surface modifications. Polymers can be applied by isolation from their natural sources like chitosan which is produced from chitin or they might be synthesized in the desired structures, for instance, poly-lactic-co-glycolic acid (PLGA). PLGA, chitosan, human serum albumin, hyaluronic acid and cyclodextrins have been extensively utilized in pre-clinical studies for the purpose of drug delivery.

14.7.5 Chitosan Nanoparticles

Chitosan nanoparticles are considered one of the most popular polymeric delivery systems. Chitosan is a linear polysaccharide comprised mainly of randomly distributed (1-4)-linked D-glucosamine (deacetylated unit) and N-acetyl-D-glucosamine (acetylated unit). It is usually made by treating shrimp and other crustacean shells with the help of alkali sodium hydroxide. Chitosan possesses several significant biological properties such as biocompatibility, bioactivity and biodegradability with reactive chemical groups including OH and NH2. Consequently, chitosan as well as its derivatives have been extensively applied in the fields of pharmacy and biotechnology.

14.7.6 PLGA Nanoparticles

PLGA or poly(lactic-co-glycolic) acid is known as a copolymer that is used in a host of FDA-approved therapeutic devices. PLGA has proven to be efficacious as a biodegradable polymer as it undergoes hydrolysis in the body to produce the original monomers, lactic acid and glycolic acid. Thus, these two monomers are by-products of several metabolic pathways in the body under normal physiological conditions. Meanwhile, the body efficiently deals with the two monomers, there is insignificant systemic toxicity related to using PLGA for drug delivery or biomaterial applications.

14.7.7 Cyclodextrin Nanoparticles

Cyclodextrins are referred as cyclical sugar molecules with a hydrophilic exterior as well as a hydrophobic cavity interior. Though high aqueous solubility and the ability in order to encapsulate hydrophobic moieties inside their cavity by the formation of inclusion complexes enable cyclodextrins to increase the solubility, stability, as well as bioavailability of hydrophobic small-molecule drugs.

14.7.8 Polymeric Micelles

In general, polymeric micelles are known to be formed from the self-assembly of amphiphilic-block copolymers whose sizes range between 10 and 100 nm. They consist of a hydrophobic core as well as a hydrophilic corona. Micelles may increase the bioavailability of hydrophobic drugs to confer protection of the drugs under the effect of biological surroundings. Polymeric micelle formulations might be used for both passive and active targeting in anticancer therapy. For instance, Genexol PM is recently under investigation as a paclitaxel-loaded polymeric micelle formulation for treating breast, lung and pancreatic cancers.

14.7.9 Dendrimers

Dendrimers are 100–200 Å diametric, hyperbranched particles comprised of a core, branching chains and functionalized terminal groups. As the name comes from the Greek word "dendron" which refers to tree-like structures. Whereas the symmetric scaffold structure of dendrimers has been found to be an appropriate carrier for several drugs and siRNA, enhancing the solubility and bioavailability of poorly soluble agents.

14.7.10 Inorganic Nanoparticles

Inorganic nanoparticles made up of materials like silica and aluminum have also been used in cancer therapy. Thus, silica can be made multiporous (which is then called mesoporous silica) to carry more drugs as compared to that a simple sphere does. A layered double hydroxide nanoparticle has been made by layering a hydroxyl (–OH) group on inorganic materials in order to carry out drugs [101].

14.8 Therapeutic Application for Cancer Cells

In recent scenarios, principally two main hurdles are involved in cancer therapy, i.e., diagnosis of cancer at an advanced stage and the inability to deliver therapeutic chemicals to the targeted tumor site without consequent damage to healthy cells. Formerly, Paul Ehrlich defined the immunological work as the "magic bullet" paradigm that allows the selective delivery of therapeutics to cancer cells at its very early stage [102]. Thus, the proposed concept can be proved to be real by using the advancement of proteomics, genomics and nanotechnology. However, by the application of such techniques, researchers have identified various biomarkers related to tumorigenesis and progression. Moreover, novel screening technology also approves the identification of peptide sequences, antibodies and as well as nucleic acid aptamers along with high affinity toward cancer-specific biomarkers. Although nanotechnology is a multidisciplinary science comprising of biochemistry, physics and material science, which has found a wide range of applications in the biomedicine field. It is usually defined as fabrication and application of the man-made material, i.e., nanomaterials (NMs) (size range of 1–100 nm) in targeted drug delivery, biosensing, imaging, photo-thermal therapy (PTT), photodynamic therapy (PDT), etc. Nanoparticle (NP) is a widely used term to define distinct shapes and sizes of nano vector structures that possess unique physicochemical properties. Therefore, by using their distinctive properties, NPs exhibit the ability to revolutionize the way in which several diseases like cancer are presently being diagnosed as well as treated. Various nanoprobes coupled with respective ligands and chemotherapeutic drugs have been developed that may interact with biological systems and therefore might sense as well as monitor biological events with high efficiency along with therapeutic applications. Moreover, in in vivo execution, such nanoprobes are administered as well as accumulated at the tumor site by ligand-biomarker interaction to generate signals for diagnostic purpose. Additionally, because of their intrinsic property the NPs being accumulated might also be employed in cancer therapy by using PTT and PDT. However, it is also known to be equally beneficial in in vitro analysis of urine, blood, saliva and tears. Due to the high surface-to-volume ratio, magnetic and electrical properties, diversity of shape and size either solid or hollow, desirable chemical composition with preferred surface chemistry, NPs have the ability to afford such potential [103]. NPs have the capability to overcome the chemical as well as biological barriers consenting for augmented diagnosis and therapy with higher biocompatibility and low invasiveness within the human body also. Despite all these benefits, auspicious results of pre-clinical trials cannot be completely deciphered into real-life scenarios. Primarily, this is due to the challenges linked with its synthesis on a large scale, reproducibility, toxicology behavior and also imprecisely evaluated safety hazards. Thus, the long-term safety assessment of NPs needs to be primarily evaluated as it may persuade toxicity if being used for a longer time period, like bio-persistent and extended utilization of NPs in pre-clinical trials for pulmonary drug delivery prompts fibrosis in hypertensive rats. Likewise, in vivo studies on mice suggested that intranasal delivery of copper oxide (CuO) NPs induces pulmonary toxicity as well as fibrosis in lung tissues. Apart from such toxic hazardous effects, various NPs exhibit intrinsic enzyme mimetic and fluorescence properties, preferably allowing them to be applied in diagnosis as well as in treatment of different diseases such as cancer. NPs like gold NPs (AuNPs) and iron oxide (IONPs) have been well known to possess peroxidase mimetic activity and therefore may be utilized for visual detection of cancer cells and in therapy. Peroxidase-like activity of NPs catalyzes the decomposition of hydrogen peroxide (H_2O_2) to produce hydroxyl radical (·OH) leading to cytotoxic effect in cancer cells. On the other hand, quantum dots (QDs) suggests a prodigious advantage over traditional organic fluorescent dyes that can be engaged in cancer molecular imaging as well as in targeted therapy. Moreover, the fluorescence of quantum dots (QDs) and upconversion NPs (UCNPs) upon near-infrared (NIR) activation assists their detection in deep tissues making them suitable for in vivo imaging with a high signal-to-background ratio.

14.9 Conclusion

Currently, nanotechnology has been proven to be very useful as a drug delivery approach. As in nanomedicine formulation research, the development of nano dosage forms (polymeric NPs and nanocapsules, liposomes, solid lipid NPs, phytosomes and nano-emulsion, etc.) possesses a number of advantages for delivery systems, such as enhancement of solubility and bioavailability, protection from toxicity, enhancement of pharmacological activity, enhancement of stability, improving tissue macrophages distribution, sustained delivery, protection from physical and chemical degradation, etc. [104]. Because of the lack of drug availability, adverse side effects and drug resistance, conventional therapy has been proven to be unsuccessful in achieving proper treatment. During the past few decades, nanomedicine has suggested significant progress in improving cancer treatment. In 1959, after the instigation of nanotechnology, the area of nanomedicine has developed rapidly and recently approached solutions to different challenges successfully. Before the 1970s, the so-called term liposome was known to be described and the drug-encapsulated liposome was developed. However, from there itself the progression of nanomedicine passed through several achievements, starting from gold NPs, polymeric NPs, quantum dots, fullerenes, etc., to the clinically permitted nanomedicines for chemotherapy. The primacy of developing nanomaterial for cancer treatment involves: (i) multifunctionality, (ii) enhanced potency and multivalency, (iii) increased selectivity for targets, (iv) theranostic potential, (v) altered pharmacokinetics, (vi) controlled synthesis, (vii) controlled agent release and kinetics, (viii) novel properties and interactions, (ix) lack of immunogenicity and (x) enhanced physical stability.

14.9.1 Cancer Treatments Using Nanotechnology

There are several common examples belonging to nanotechnology platforms for the therapeutic purpose of cancer such as polymeric NPs, liposomes, dendrimers, nanoshells, carbon nanotubes and superparamagnetic NPs. With small size to various structural and physicochemical features, such nanotechnology platforms may enter tumor vasculature through enhanced permeability and retention effect (EPR). The application of cancer-specific targeting residues (e.g., antibodies, ligands and lectins) can also achieve tumor cell targeting [105]. Nano-scaled dimensions have the ability to improve the delivery of drug molecules and underrate the side effects of drugs and drug carriers; they can be used as a treatment of conditions such as various stages of tumors, cancer, microbial infections, gene therapy and chronic hyperglycemia. The chief superiorities of this model of treatment are that such particles may act as a contrast medium, i.e., enhancing the visibility of blood vessels for several diagnostic purposes and biosensors, cell culture, magnetic nanoparticles and enhancing renewal and growth of tissue. Moreover, in the field of tumor imaging nanostructures have been highly cherished. Due to the EPR (enhanced permeability and retention effects), the small size of nanoparticles can invade tumors leaking into abnormal tumor blood vessels as well as accumulate in tumor tissues. Conversely, in pharmaceutical development, nanotechnology has been proven to be the most essential milestone. In humans, nanoparticles exhibit superlative activity against cancer in vitro. Nanotubes possess great insistence and strength, producing heat and destroying the surrounding tumor cells, thus absorbing near-infrared (IR) light rays. Indium-111 radio nucleus, a labeled carbon nanotube, has been known to be inspected to destroy tumor cells. The SWNT (single-walled carbon nanotube) is being investigated as a novel transporting agent in vitro. SWNTs can efficiently transport several bimolecular substances within human cells such as therapeutic agents, amino acids, proteins, DNA and intervening RNA, through endocytosis. Carbon nanotubes possess the intrinsic near-infrared (NIR) light absorption property that is useful in rescinding cancer slots and NIR photoluminescence property, thus applicable for cell visualization and exploring the cell in vitro. Though physicians can perform precise molecular as well as cellular intervention by using nano-robotic devices. Medical nano-robots have been highly recommended for the purpose of gerontological applications in pharmaceutical studies, clinical science and modern dentistry. However, it can also be applied for reoccurring atherosclerosis, boosting breathing ability, permitting near-immediate homeostasis, boosting the immune system, substituting sequences of DNA in cells, healing brain damage and fixing gross cellular distress that may have occurred because of unalterable processes or through storage of biological tissues at low temperature. Active and passive are known as two models which have been applied for targeting nano-scaled particles to specific sites of tumors. Active refers to connecting the ligands to tumor-specific nanoparticles. The basic and characteristic mass of nanoparticles and the typical ability of the tumor vascular system are the benefits of passive targeting [106–109].

REFERENCES

1. Mukerjee, A., Ranjan, A. P., Vishwanatha, J. K., (2012). Combinatorial nanoparticles for cancer diagnosis and therapy. *Current Medicinal Chemistry*, 19, 3714–3721.
2. Barbas, A. S., Mi, J., Clary, B. M., White, R. R., (2010). Aptamer applications for targeted cancer therapy. *Future Oncology*, 6, 1117–1126.
3. Bamrungsap, S., Zhao, Z., Chen, T., Wang, L., Li, C., Fu, T., Tan, W., (2012). Nanotechnology in therapeutics: A focus on nanoparticles as a drug delivery system. *Nanomedicine*, 7, 1253–1271.
4. Tan, W., Wang, H., Chen, Y., Zhang, X., Zhu, H., Yang, C., Yang, R., Liu, C., (2011). Molecular aptamers for drug delivery. *Trends in Biotechnology*, 29, 634–640.
5. Gu, F. X., Karnik, R., Wang, A. Z., Alexis, F., Levy-Nissenbaum, E., Hong, S., Langer, R. S., Farokhzad, O. C., (2007). Targeted nanoparticles for cancer therapy. *Nanotoday*, 2, 14–21.
6. Hu, M., Zhang, K., (2013). The application of aptamers in cancer research: An up-to-date review. *Future Oncology*, 9, 369–376.
7. Shangguan, D., Li, Y., Tang, Z., Cao, Z. C., Chen, H. W., Mallikaratchy, P., Sefah, K., Yang, C. J., Tan, W., (2006). Aptamers evolved from live cells as effective molecular probes for cancer study. *Proceedings of the National Academy Science United States of America*, 103, 11838–11843.
8. Sefah, K., Bae, K. M., Phillips, J. A., Siemann, D. W., Su, Z., McClellan, S., Vieweg, J., Tan, W., (2013). Cell-based selection provides novel molecular probes for cancer stem cells. *International Journal of Cancer*, 132, 2578–2588.
9. Sefah, K., Tang, Z. W., Shangguan, D. H., Chen, H., Lopez-Colon, D., Li, Y., Parekh, P., Martin, J., Meng, L., Phillips, J. A., Kim, Y. M., Tan, W., (2009). Molecular recognition of acute myeloid leukemia using aptamers. *Leukemia*, 23, 235–244.
10. Fang, X., Tan, W., (2010). Aptamers generated from cell-SELEX for molecular medicine: A chemical biology approach. *Accounts of Chemical Research*, 43, 48–57.
11. Ni, X., Castanares, M., Mukherjee, A., Lupold, S. E., (2011). Nucleic acid aptamers: Clinical applications and promising new horizons. *Current Medicinal Chemistry*, 18, 4206–4214.
12. Chang, Y. M., Donovan, M. J., Tan, W., (2013). Using aptamers for cancer biomarker discovery. *Journal of Nucleic Acids*, 817350.
13. Zhang, Y., Hong, H., Cai, W., (2011). Tumor-targeted drug delivery with aptamers. *Current Medicinal Chemistry*, 18, 4185–4194.

14. Li, X., Zhao, Q., Qiu, L., (2013). Smart ligand: Aptamer-mediated targeted delivery of chemotherapeutic drugs and siRNA for cancer therapy. *Journal of Control Release*, 171, 152–162.
15. Wang, H., Yang, R., Yang, H. L., Tan, W. H., (2009). Nucleic acid conjugated nanomaterials for enhanced molecular recognition. *American Chemical Society Nanomaterials*, 3, 2451–2460.
16. Stadler, N., Chi, C., Lelie, D. V. D., Gang, O., (2010). DNA-incorporating nanomaterials in biotechnological applications. *Nanomedicine*, 5, 319–334.
17. Lee, J. H., Yigit, M. V., Mazumdar, D., Lu, Y., (2010). Molecular diagnostic and drug delivery agents based on aptamer-nanomaterial conjugates. *Advanced Drug Delivery Reviews*, 62, 592–605.
18. Chen, T., Shukoor, M. I., Chen, Y., Yuan, Q., Zhu, Z., Zhao, Z., Gulbakan, B., Tan, W., (2011). Aptamer-conjugated nanomaterials for bioanalysis and biotechnology applications. *Nanoscale*, 3, 546–556.
19. Zhu, G., Zhang, S., Song, E., Zheng, J., Hu, R., Fang, X., Tan, W., (2013). Building fluorescent DNA nanodevices on target living cell surfaces. *Angewandte Chemie International Edition*, 52, 5490–5496.
20. Ding, C., Ge, Y., Zhang, S., (2010). Electrochemical and electrochemiluminescence determination of cancer cells based on aptamers and magnetic beads. *Chemistry- A European Journal*, 16, 10707–10714.
21. Wu, M., Yuan, D., Xu, J., Chen, H., (2013). Sensitive electrochemiluminescence biosensor based on Au-ITO hybrid bipolar electrode amplification system for cell surface protein detection. *Analytical Chemistry*, 85, 11960–11965.
22. Yan, M., Sun, G., Liu, F., Lu, J., Yu, J., Song, X., (2013). An aptasensor for sensitive detection of human breast cancer cells by using porous GO/Au composites and porous PtFe alloy as effective sensing platform and signal amplification labels. *Analytica Chimica Acta*, 798, 33–39.
23. Liu, H., Xu, S., He, Z., Deng, A., Zhu, J., (2013). Super sandwich cytosensor for selective and ultrasensitive detection of cancer cells using aptamer-DNA concatamer-quantum dots probes. *Analytical Chemistry*, 85, 3385–3392.
24. Jie, G., Wang, L., Yuan, J., Zhang, S., (2011). Versatile electrochemiluminescence assays for cancer cells based on dendrimer/CdSe-ZnS-quantum dot nanoclusters. *Analytical Chemistry*, 83, 3873–3880.
25. Zhao, J., Zhang, L., Chen, C., Jiang, J., Yu, R., (2012). A novel sensing platform using aptamer and RNA polymerase-based amplification for detection of cancer cells. *Analytica Chimica Acta*, 745, 106–111.
26. Bamrungsap, S., Chen, T., Shukoor, M. I., Chen, Z., Sefah, K., Chen, Y., Tan, W., (2012). Pattern recognition of cancer cells using aptamer-conjugated magnetic nanoparticles. *American Chemical Society Nanomaterials*, 6, 3974–3981.
27. Xiong, X., Liu, H., Zhao, Z., Altman, M. B., Lopez-Colon, D., Yang, C. J., Chang, L. J., Liu, C., Tan, W., (2013). DNA aptamer-mediated cell targeting. *Angewandte Chemie International Edition*, 52, 1472–1476.
28. Ocsoy, I., Gulbakan, B., Shukoor, M. I., Xiong, X., Chen, T., Powell, D. H. & Tan, W., (2013). Aptamer-conjugated multifunctional nanoflowers as a platform for targeting, capture, and detection in laser desorption ionization mass spectrometry. *American Chemical Society Nanomaterials*, 7, 417–427.
29. Chaturvedi, V. K., Singh, A., Singh, V. K., Singh, M. P., (2019). Cancer nanotechnology: A new revolution for cancer diagnosis and therapy. *Current Drug Metabolism*, 20, 416–429.
30. Kim, D., Jeong, Y. Y., Jon, S. J. A. N., (2010). A drug-loaded aptamer-gold nanoparticle bioconjugate for combined CT imaging and therapy of prostate cancer. *American Chemical Society Nanomaterials*, 4, 3689–3696.
31. Akhter, S., Ahmad, I., Ahmad, M. Z., Ramazani, F., Singh, A., Rahman, Z., (2013). Nanomedicines as cancer therapeutics: Current status. *Current Cancer Drug Targets*, 13, 362–378.
32. Wang, J., Sui, M., Fan, W., (2010). Nanoparticles for tumor targeted therapies and their pharmacokinetics. *Current Drug Metabolism*, 11, 129–141.
33. Garrigue, P., Tang, J., Ding, L., Bouhlel, A., Tintaru, A., Laurini, E., (2018). Self-assembling supramolecular dendrimer nanosystem for PET imaging of tumors. *Proceedings of the National Academy of Sciences of the United States of America*, 115, 11454–11459.
34. Ji, T., Zhao, Y., Wang, J., Zheng, X., Tian, Y., Zhao, Y., (2013). Tumor fibroblast specific activation of a hybrid ferritin nanocage-based optical probe for tumor microenvironment imaging. *Nano-Micro Small*, 9, 2427–2431.
35. Parungo, C. P., Ohnishi, S., De Grand, A. M., Laurence, R. G., Soltesz, E. G., Colson, Y. L., (2004). In vivo optical imaging of pleural space drainage to lymph nodes of prognostic significance. *Annals of Surgical Oncology*, 11, 1085–1092.
36. Gao, X., Cui, Y., Levenson, R. M., Chung, L. W., Nie, S., (2004). In vivo cancer targeting and imaging with semiconductor quantum dots. *Nature Biotechnology*, 22, 969–976.
37. Dubertret, B., Skourides, P., Norris, D. J., Noireaux, V., Brivanlou, A. H., Libchaber, A., (2002). In vivo imaging of quantum dots encapsulated in phospholipid micelles. *Science (New York, NY)*, 298, 1759–1762.
38. Zhang, Y., Yang, H., An, X., Wang, Z., Yang, X., Yu, M., (2020). Controlled synthesis of Ag(2) Te@Ag(2) S core-shell quantum dots with enhanced and tunable fluorescence in the second near-infrared window. *Small (Weinheiman der Bergstrasse, Germany)*, 16, 2001–2003.
39. Hirsch, L. R., Stafford, R. J., Bankson, J. A., Sershen, S. R., Rivera, B., Price, R. E., (2003). Nanoshell-mediated near-infrared thermal therapy of tumors under magnetic resonance guidance. *Proceedings of the National Academy of Sciences of the United States of America*, 100, 13549–13554.
40. Loo, C., Lin, A., Hirsch, L., Lee, M. H., Barton, J., Halas, N., (2004). Nanoshell-enabled photonics-based imaging and therapy of cancer. *Technology in Cancer Research & Treatment*, 3, 33–40.
41. Nunes, T., Pons, T., Hou, X., Van Do, K., Caron, B., Rigal, M., (2019). Pulsed-laser irradiation of multifunctional gold nanoshells to overcome trastuzumab resistance in HER2-overexpressing breast cancer. *Journal of Experimental & Clinical Cancer Research*, 38, 306.
42. Fu, N., Hu, Y., Shi, S., Ren, S., Liu, W., Su, S., (2018). Au nanoparticles on two-dimensional MoS(2) nanosheets as a photoanode for efficient photoelectrochemical miRNA detection. *The Analyst*, 143, 1705–1712.
43. Fu, F., Li L, Luo, Q., Li, Q., Guo, T., Yu, M., (2018). Selective and sensitive detection of lysozyme based on plasmon resonance light-scattering of hydrolyzed peptidoglycan stabilized-gold nanoparticles. *The Analyst*, 143, 1133–1140.

44. Shrivas, K., Nirmalkar, N., Thakur, S. S., Deb, M. K., Shinde, S. S., Shankar, R., (2018). Sucrose capped gold nanoparticles as a plasmonic chemical sensor based on non-covalent interactions: Application for selective detection of vitamins B(1) and B(6) in brown and white rice food samples. *Food Chemistry*, 250, 14–21.
45. Gupta, G. P., Massague, J., (2006). Cancer metastasis. *Cell*, 127 (4), 679–695.
46. Chaffer, C. L., Weinberg, R. A., (2011). A perspective on cancer cell metastasis. *Science*, 331 (6024), 1559–1564.
47. Huang, Q., Yin, W., Chen, X., Wang, Y., Li, Z., Du, S., Wang, L., Shi, C., (2018). Nanotechnology-based strategies for early cancer diagnosis using circulating tumor cells as a liquid biopsy. *Nanotheranostics*, 2 (1), 21–41.
48. Akbarzadeh, A., Samiei, M., Davaran, S., (2012). Magnetic nanoparticles: Preparation, physical, properties, and applications in biomedicine. *Nanoscale Research Letters*, 7, 144.
49. Powell, A. A., Talasaz, A. H., Zhang, H., Coram, M. A., Reddy, A., Deng. G., Telli, M. L., Advani, R. H., Carlson, R. W., Mollick, J. A., Sheth, S., Kurian, A. W., Ford, J. M., Stockdale, F. E., Quake, S. R., Pease, R. F., Mindrinos, M. N., Bhanot, G., Dairkee, S. H., Davis, R. W., Jeffrey, S. S., (2012). Single cell profiling of circulating tumor cells: Transcriptional heterogeneity and diversity from breast cancer cell lines. *Plos One*, 7 (5), 337–388.
50. Talasaz, A. H., Powell, A. A., Huber, D. E., Berbee, J. G., Roh, K. H., Yu, W., Xiao, W., Davis, M. M., Pease, R. F., Mindrinos, M. N., Jeffrey, S. S., Davis, R. W., (2009). Isolating highly enriched populations of circulating epithelial cells and other rare cells from blood using a magnetic sweeper device. *Proceedings of National Academy Sciences of United States of America*, 106 (10), 3970–3975.
51. Peng, Y. Y., Hsieh, T. E., Hsu, C. H., (2009). The conductive property of ZnO QDs-SiO2 and ZnOQDs-SiOx Ny nanocomposite films. *Journal of Nanoscience & Nanotechnology*, 9 (8), 4892–4900.
52. Pang, X., Cui, C., Su, M., Wang, Y., Wei, Q., Tan, W., (2018). Construction of self-powered cytosensing device based on ZnO nanodisks@g-C3N4 quantum dots and application in the detection of CCRF-CEM cells. *Nano Energy*, 46, 101–109.
53. Wu, C., Schneider, T., Zeigler, M., Yu, J., Schiro, P. G., Burnham, D. R., McNeill, J. D., Chiu, D. T., (2010). Bioconjugation of ultrabright semiconducting polymer dots for specific cellular targeting. *Journal of American Chemical Society*, 132 (43), 15410–15417.
54. Bajaj, A., Miranda, O. R., Kim, I. B., Phillips, R. L., Jerry, D. J., Bunz, U. H., Rotello, V. M., (2009). Detection and differentiation of normal, cancerous, and metastatic cells using nanoparticle-polymer sensor arrays. *Proceedings of National Academy of Sciences of the United States of America*, 106 (27), 10912–10916.
55. Nagesetti, A., Rodzinski, A., Stimphil, E., Stewart, T., Khanal, C., Wang, P., Guduru, R., Liang, P., Agoulnik, I., Horstmyer, J., Khizroev, S., (2017). Erratum multiferroic coreshell magnetoelectric nanoparticles as NMR sensitive nanoprobes for cancer cell detection. *Scientific Reports*, 7, 1610.
56. Seferos, D. S., Giljohann, D. A., Hill, H. D., Prigodich, A. E., Mirkin, C. A., (2007). Nanoflares: Probes for transfection and MRNA detection in living cells. *Journal of American Chemical Society*.129 (50), 15477–15479.
57. Choi, C. H. J., Hao, L., Narayan, S. P., Auyeung, E., Mirkin, C. A., (2013). Mechanism for the endocytosis of spherical nucleic acid nanoparticle conjugates. *Proceedings of National Academy Sciences of the United States of America*, 110 (19), 7625–7630.
58. Briley, W. E., Bondy, M. H., Randeria, P. S., Dupper, T. J., Mirkin, C. A., (2015). Quantification and real-time tracking of RNA in live cells using sticky flares. *Proceedings of National Academy of Sciences of the United States of America*, 112 (31), 9591–9595.
59. Seftor, E. A., Seftor, R., Weldon, D., Kirsammer, G. T., Margaryan, N. V., Gilgur, A., Hendrix, M., (2014). Melanoma tumor cell hetrogeneity: A molecular approach to study subpopulations expressing the embryonic morphogen nodal. *Seminars in Oncology*, 41 (2), 259–266.
60. Halo, T. L., McMahon, K. M., Angeloni, N. L., Xu, Y., Wang, W., Chinen, A. B., Malin, D., Strekalova, E., Cryns, V. L., Cheng, C., Mirkin, C. A., Thaxton, C. S., (2014). Nanoflares for the detection, isolation, and culture of live tumor cells from human blood. *Proceedings of National Academy of Sciences of the United States of America*, 111 (48), 17104–17109.
61. Lee, K., Cui, Y., Lee, L. P., Irudayaraj, J., (2014). Quantitative imaging of single MRNA splice variants in living cells. *Nature Nanotechnology*, 9 (6), 474–480.
62. Chinen, A. B., Guan, C. M., Ferrer, J. R., Barnaby, S. N., Merkel, T. J., Mirkin, C. A., (2015). Nanoparticles probes for the detection of cancer biomarkers, cells, and tissues by fluorescence. *Chemical Reviews*, 115 (19), 10530–10574.
63. Bertrand, N., Wu, J., Xu, X., Kamaly, N., Farokhzad, O. C., (2014). Cancer nanotechnology: The impact of passive and active trageting in the era of modern cancer biology. *Advanced Drug Delivery Reviews*, 66 (2), 25.
64. Golombek, S. K., May, J. N., Theek, B., Appold, L., Drude, N., Kiessling, F., Lammers, T., (2018). Tumor targeting via EPR: Strategies to enhance patient responses. *Advanced Drug Delivery Reviews*, 130, 17–38.
65. Matsumura, Y., Maeda, H., (1986). A new concept for macromolecular therapeutics in cancer chemotherapy: Mechanism of tumoritropic accumulation of proteins and the antitumor agent smancs. *Cancer Research*, 46, 6387–6392.
66. Hong, G., Robinson, J. T., Zhang, Y., Diao, S., Antaris, A. L., Wang, Q., Dai, H., (2012). In vivo fluorescence imaging with Ag2S quantum dots in the second near-infrared region. *Angewandte Chemie International Edition*, 51 (39), 9818–9821.
67. Lai, S. F., Ko, B. H., Chien, C. C., Chang, C. J., Yang, S. M., Chen, H. H., Petibois, C., Hueng, D. Y., Ka, S. M., Chen, A., Margaritondo, G., Hwu, Y., (2015). Gold nanoparticles as multimodality imaging agents for rain gliomas. *Journal of Nanobiotechnology*, 13, 85.
68. Nam, T., Park, S., Lee, S. Y., Park, K., Choi, K., Song, I. C., Han, M. H., Leary, J. J., Yuk, S. A., Kwon, I. C., Kim, K., Jeong, S. Y., (2010). Tumor targeting chitosan nanoparticles for dual-modality optical/MR cancer imaging. *Bioconjugate Chemistry*, 21 (4), 578–82.
69. Venturoli, D., Rippe, B., (2005). Ficoll and dextran vs. globular proteins as probes for testing glomerular permselectivity: Effects of molecular size, shape, charge and deformability. *American Journal of Physiology Renal Physiology*, 288, F605–F613.

70. Popovic, Z., Liu, W., Chauhan, V. P., Lee, J., Wong, C., Greytak, A. B., Insin, N., Nocera, D. G., Fukumura, D., Jain, R. K., Bawendi, M. G., (2010). A nanoparticle size series for in vivo fluorescence imaging. *Angewandte Chemie International Edition*, 49 (46), 8649–52.
71. Lundqvist, M., Stigler, J., Cedervall, T., Berggard, T., Flanagan, M. B., Lynch, I., Elia, G., Dawson, K., Lundqvist, M., Stigler, J., Cedervall, T., Berggard, T., Flanagan, M. B., Lynch, I., Elia, G., Dawson, K., (2011). The evolution of the protein corona around nanoparticles: a test study. *ACS Nano*, 5, 7503.
72. Schipper, M. L., Iyer, G., Koh, A. L., Cheng, Z., Ebenstein, Y., Aharoni, A., Keren, S., Bentolila, L. A., Li, J., Rao, J., Chen, X., Banin, U., Wu, A. M., Sinclair, R., Weiss, S., Gambhir, S. S., (2009). Particle size, surface coating, and PEGylation influence the biodistribution of quantum dots in living mice. *Nano Micro Small*, 5, 126.
73. Garrigue, P., Tang, J., Ding, L., Bouhlel, A., Tintaru, A., Laurini, E., Huang, Y., Lyu, Z., Zhang, M., Fernandez, S., Balasse, L., Lan, W., Mas, E., Marson, D., Weng, Y., Liu, X., Giorgio, S., Iovanna, J., Pricl, S., Guillet, B., Peng, L., (2018). Self-assembling nanotechnology for cancer personalized medicine. *Chemical Engineering Transactions*, 115, 11454.
74. Bertrand, N., Wu, J., Xu, X., Kamaly, N., Farokhzad, O. C., (2014). Cancer nanotechnology: The impact of passive and active targeting in the era of modern cancer biology. *Advances in Drug Delivery Review*, 66, 2.
75. Gao, X., Cui, Y., Levenson, R. M., Chung, L. W., Nie, S., (2004). In vivo cancer targeting and imaging with semiconductor quantum dots. *Nature Biotechnology*, 22, 969.
76. Yang, Y., Chen, Q., Li, S., Ma, W., Yao, G., Ren, F., Cai, Z., Zhao, P., Liao, G., Xiong, J., Yu, Z., (2018). Novel chemically synthesized, alpha-mangostin-loaded nano-particles, enhanced cell death through multiple pathways against malignant glioma. *Journal Biomedical Nanotechnology*, 14, 1396.
77. Blanco, E., Hsiao, A., Mann, A. P., Landry, M. G., Meric-Bernstam, F., & Ferrari, M., (2011). Nanomedicine in cancer therapy: Innovative trends and prospects. *Cancer Science*, 102 (7), 1247–1252.
78. Conde, J., Doria, G., & Baptista, P., (2012). Noble metal nanoparticles applications in cancer. *Journal of Drug Delivery*, 2012 (2012), 1–12.
79. Sanna, A., Pala, N., & Sechi, M., (2014). Targeted therapy using nanotechnology: Focus on cancer. *International Journal of Nanomedicine*, 9, 467–483.
80. Clarke, M. F., Dick, J. E., Dirks, P. B., Eaves, C. J., Jamieson, C. H. M., Jones, D. L., Visvader, J., Weissman, I. L., Wahl, G. M., (2006). Cancer stem cells-perspectives on current status and future, directions: AACR workshop on cancer stem cells. *Cancer Research*, 66 (19), 9339–9344.
81. Jordan, C. T., Guzman, M. L., Noble, M., (2006). Mechanisms of disease: Cancer stem cells. *The New England Journal of Medicine*, 355 (12), 1253–1261.
82. Fernandes, A. R., Baptista, P. V., (2014). Nanotechnology for cancer diagnostics and therapy – An update on novel molecular players. *Current Cancer Therapy Reviews*, 9 (3), 1–9.
83. Brigger, I., Dubernet, C., Couvreur. P., (2012). Nanoparticles in cancer therapy and diagnosis. *Advanced Drug Delivery Reviews*, 64, 24–36.
84. Oerlemans, C., Bult, W., Bos, M., Storm, G., Nijsen, J. F. W., Hennink, W. E., (2010). Polymeric micelles in anticancer therapy: Targeting, imaging and triggered release. *Pharmaceutical Research*, 27 (12), 2569–2589.
85. Larsen, A. K., Escargueil, A. E., & Skladanowski, A., (2000). Resistance mechanisms associated with altered intracellular distribution of anticancer agents. *Pharmacology & Therapeutics*, 85 (3), 217–229.
86. Conde, J., de la Fuente, J. M., & Baptista, P. V., (2013). Nanomaterials for reversion of multidrug resistance in cancer: A new hope for an old idea. *Frontiers and Pharmacology*, 4 (134), 1–5.
87. Syed, A., Rizvia,A., & Saleh, A. M., (2018). Applications of nanoparticle systems in drug delivery technology, *Saudi Pharmaceutical Journal*, 26 (2018), 64–70.
88. Bishwajit, K., Sutradharan, Amin, M. L., (2014). Nanotechnology in cancer drug delivery and selective targeting. *ISRN Nanotechnology*, 2014, 1–12.
89. Sanna, A., Pala, N., Sechi, M., (2014). Targeted therapy using nanotechnology: Focus on cancer. *International Journal of Nanomedicine*, 9, 467–483.
90. Donga, P., Rakesha, K. P., Manukumar, H. M., Mohammede, Y. H. E., Karthik, C. S., Sumathif, S., Mallud, P., LiQina, H., (2019). Innovative nano-carriers in anticancer drug delivery-a comprehensive review. *Bioorganic Chemistry*, 85, 325–336.
91. Acharya, S., Sahoo, S. K., (2011). PLGA nanoparticles containing various anti cancer agents and tumour delivery by EP Reff ect. *Advances in Drug Delivery Review*, 63, 170–183.
92. Alimoradi, H., Greish, K., Barzegar-Fallah, A., Alshaibani, L., Pittalà,V., (2018). Nitric oxide-releasing nanoparticles improve doxorubicin anticancer activity. *International Journal of Nanomedicine*, 13, 7771–7787.
93. Almeida, P. V., Shahbazi, M. A., Mäkilä, E., Kaasalainen, M., Salonen, J., Hirvonen, J., et al., (2014). Amine-modified hyaluronic acid-functionalized porous silicon nanoparticles for targeting breast cancer tumors. *Nanoscale*, 6, 10377–10387.
94. Edis, Z., Wang, Z., Waqas, M. K., Ijaz, M., & Ijaz, M., (2021). Nanocarriers-mediated drug delivery systems for anticancer agents: An overview and perspectives. *International Journal of Nanomedicine*, 16, 1313–1330.
95. Dadwal, A., Baldi, A., Narang, R. K., (2018). Nanoparticles as carriers for drug delivery in cancer. *Artificial Cells, Nanomedicine, and Biotechnology*, 46 (S2), S295–S305.
96. Li, Z., Tan, S., Li, S., Shen, Q., Wang, K., (2017). Cancer drug delivery in the nano era: An overview and perspectives (Review). *Oncology Reports*, 38, 611–624.
97. American Cancer Society I (2018). *Cancer facts & figures 2018*. Atlanta: American Cancer Society.
98. Anders, C. K., Adamo, B., Karginova, O., (2013). Pharmacokinetics and efficacy of PEGylated liposomal doxorubicin in an intracranial model of breast cancer. *PLoS ONE*, 8, e61359.
99. Arvizo, R., Bhattacharya, R., & Mukherjee, P., (2010). Gold nanoparticles: Opportunities and challenges in nanomedicine. *Expert Opinion in Drug Delivery*, 7, 753–763.
100. Patra, J. K., Das, G., Fraceto, L. F., Campos, E. V. R., Rodriguez Torres, M. P., Laura Acosta Torres, L. S., Diaz Torres, L. A., Grillo,E., Swamy, M. K., Sharma, S., Habtemariam, S., & Shin, H., (2018). Nano based drug delivery systems: Recent developments and future prospects. *Journal of Nanobiotechnology*, 16 (71), 1–33.

101. Zhu, Y., & Liao L., (2015). Applications of nanoparticles for anticancer drug delivery: A review. *Journal of Nanoscience and Nanotechnology*, 15, 4753–4773.
102. Singh, R., (2019). Nanotechnology based therapeutic application in cancer diagnosis and therapy. *Biotechnology,* 9 (415), 1–29.
103. Copeland, G., Lake, A., Firth, R., (2018). *Cancer in North America: 2010– 2014. Volume one: Combined cancer incidence for the United States, Canada and North America.* USA: North American Association of Central Cancer Registries Inc.
104. Zhang, C., Wang, S., Xiao, J., (2013). Sentinel lymph node mapping by a near-infrared fluorescent heptamethine dye. *Biomaterials*, 31 (7), 1911–1917.
105. Munyendo, W. L., Lv H, Benza-Ingoula H., Baraza, L D., & Zhou, J., (2012). Cell penetrating peptides in the delivery of biopharmaceuticals. *Biomolecules*, 2 (2), 187–202.
106. Chen, T., Shukoor, M. I., & Wang, R., (2011). Smart multifunctional nanostructure for targeted cancer chemotherapy and magnetic resonance imaging. *ACS Nano*, 5 (10), 7866–7873.
107. Schweiger, C., Hartmann R., Zhang, F., Parak, W. J. Thomas, K., & Gil, P. R., (2012). Quantification of the internalization patterns of superparamagnetic iron oxide nanoparticles with opposite charge. *Journal of Nanobiotechnology*, 10, 1–28.
108. Shapero, K., Fenaroli, F., Lynch, I., Cottell, D. C., Salvati, A., Dawson, K. A., (2011). Time and space resolved uptake study of silica nanoparticles by human cells. *Molecular BioSystems*, 7 (8), 371–378.
109. Biswas, A. K., Islam, M. R., Choudhury, Z. S., Mostafa, A., Kadir, M. F., (2014). Nanotechnology based approaches in cancer therapeutics. *Advances in Natural Sciences: Nanoscience and Nanotechnology*, 5 (4), 043001.

15

Nanomaterial-Modified Pencil Graphite Electrode as a Multiplexed Low-Cost Point-of-Care Device

Mansi Gandhi, Nandimalla Vishnu and Naresh Kumar Katari

CONTENTS

15.1 Introduction

A pencil is an implement used for drawing, shading and the scribing of words. It is constructed of a narrow and solid pigment core in the protective casing that prevents it from breaking and helps to keep its mechanical strength intact. The pencil marks are resistant to water/moisture, most chemicals, radiation (UV) and even remain undamaged during aging. With innovations, varieties from colored to grease to clay to charcoal are available based on the needs of consumers.

Due to the electrical resistance and mechanical strength of graphite, pencil-based electrodes display lower background noise and current values with respect to other working electrodes. They have good reproducibility, higher sensitivity and an adjustable electro-active surface area that permit the analysis of low concentration and low sample volumes without any pre-concentration step.

Many reports have come up using graphite-reinforced carbon, commonly employed as mechanical pencil lead for electrochemical analyses. The commercial availability of graphite leads to mechanical pencils having a low content of heavy metal impurities with high quality, tailored by the manufacturers, thus enhancing the electrochemical characteristics of carbon-based disposable sensors. To the best of our knowledge, no paper has given an account of the holistic approach of pencil types and their usage as electrode counterparts with its relevant characterization review.

The initial phase of employing pencils as a substitute in electrochemical units emerged in 1996 in abrasive stripping voltammetry [1]. Later after two years, a vibrating electrode based on a mechanical graphite pencil was reported for simultaneous potentiometric stripping determination for metal traces [2]. Further to add, pencil electrodes were set up for the immobilization and characterization of different electro-active species using voltammetry on micro-particles [3]. Later on, after a few technical modifications, the lead corrosion and degradation determination in archeological samples were extended using pencil electrodes [4]. Wherein, the stripping peaks were estimated for patina metals, based on the potential shift of the reduction peak, and signals due to PbO_2 reduction were accounted for [5].

Pencil graphite is used as a working counterpart besides being cheap and easy to use, as shown in Figure 15.1. The electrode surface doesn't require cleaning or surface polishing protocols with the window of allowing different types of voltammetric techniques. It has been extended for quantifying different analytical systems with a wide range of samples enabling reproducible signals. The obtained well-defined voltammetric signals have helped in yielding a robust electrode with a viable, renewable and economical tool. The most interesting part of a pencil is the electrical resistance lower than 5 ohms irrespective of the producer or the manufacturer, thus making it one of the most prominent electrode materials.

The economical characteristics of the pencil graphite electrode (PGE) found in recent years have attracted widespread application in various fields. It has been applied for electroanalysis of bare and modified forms of pencil electrodes for the determination of various samples ranging from simple matrices like water to complex systems of biological fluids like human urine and serum samples. It has even been extended for complex formation based on the intrinsic metals that are present in the electrode due to manufacturing defects. Their

DOI: 10.1201/9781003157823-15

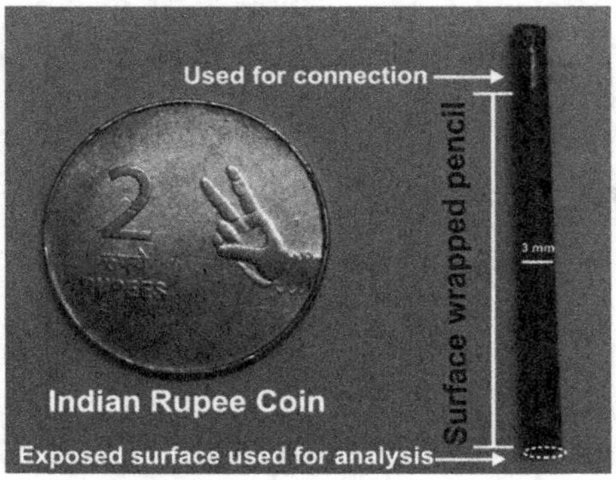

FIGURE 15.1 Representation of the pencil electrode which is wrapped to ensure the fixed exposed surface area for analysis. Further, its economic efficiency has been highlighted.

usage has even been extended as a high-performance supercapacitor material, in which 6H* displays ultrahigh areal capacitance [6]. This review represents a detailed version of the material quality and characteristics, design and characterization, along with applications of crude and chemically modified PGEs. Following is the overview about their usage in real-life scenarios and their bottlenecks and finally we conclude with future prospects too.

15.2 Material Constituent and Quality

Graphite consists of both metallic and non-metallic dimensions being utilized as an electrode material. Nearly 5% of graphite has been used for the making of pencils all over the world [7]. The pencil graphite consists of fine powder in an inorganic resin or organic matrix of a high polymer example cellulose [7, 8]. The pencil graphite electrodes or PGEs are composite materials consisting of graphite (~65%), clay (~30%) and a binder wax (resins of high polymer ~5%) [9]. According to the European letter scale, the graphite scale is marked with the letter "H" for "hardness" and "B" for "blackness/boldness" and the numbers indicating the degree of hardness or blackness. The H type signifies the percentage of graphite content in the pencil lead, while the B represents the darkness of the mark that it produces on the paper. The proportional/optimal mixture of graphite and clay makes the pencil give the perfect impression. Notably, the greasy material or the graphite which is very soft gives a dark mark, while on the other hand the clay is very hard and leaves a light gray mark when rubbed against the paper [10]. They are currently 24 degrees on the pencil lead types in total ranging from 10H (hardest) to 12B (softest) based on altering the proportion, as shown in Figure 15.2. Similarly, the perfect mixture, i.e. HB, corresponds to an equal percentage of clay and graphite. The clay is an important constituent having an influence on the chemical (e.g. ion exchange) and structural properties (i.e. degree of disorder or surface morphology of the lead electrodes).

The electrode material of this type is constituted with easily available commercial pencil graphite leads. The PGEs are cheap, eco-friendly and user-friendly and can be employed as disposable electrodes avoiding the time-consuming and cumbersome protocols of electrode preparation and electrode cleanings during experimentations [11]. The work using simple electrodes can be handled with non-professionals as no sophisticated handling is required in this setup.

The electrochemical characteristics of a pencil lead depend on surface properties which can be changed/improved either by pretreatment or amperometric pre-anodization or chemical modification of electrode surface. The electrochemical pretreatment PGE presented improved electrocatalytic activity with higher sensitivity and a ten-fold increment in the voltammetric response with higher selectivity of peak separation and better adsorption capacity of organic compounds [12–18]. A brief comparative analyte detection has been tabulated accounting for their detection in real samples as in Table 15.1.

In comparison with other electrodes, the pencil is represented only for writing purposes offering the advantage of a user-friendly approach and toward a sustainable development initiative. It is non-toxic, largely available for purchase all over the world and its simplicity has made it a disposable and readily available electrode type for the research community.

15.3 Design and Characterization

The design of a commonly used pencil commercially used includes a mechanical pencil holder where the lead is generally a bit outside. In order to obtain a minimum constant electro-active surface area, a constant length is always introduced into the solution in an upright position [17, 30–36]. Many authors reported the direct connection of the electrode led to the potentiostat cable or the mounting on an improvised electrode clip [37].

An interesting article by Pokpas et al. reported sealing the epoxy of graphite lead into a syringe acting as a holder [38, 39]. The one end of the lead remained outside the holder, whereas on the other end a copper wire was attached [38]. On the other hand, Prasad et al.'s lead graphite disk was prepared by introducing the pencil lead into a Teflon tube, a PVC tube, the tip of a micropipette or covering the electrode using a Teflon tape or a parafilm wax [40–46].

Numerous reports on the performance of PGEs have been compared based on electroanalysis of bioanalytes. The pencil lead exhibited a well-defined reduction peak of O_3 molecule in protic solvent presenting a better signal compared to conventional electrodes but the determination of boron is nearly sextuple with a lower potential range if PGEs are considered [47]. Kwade et al. in 2016 reported a Bi-GCE chemically modified electrode showed a better signal for the determination of heavy metal ions using the square wave voltammetric technique [21]. On the contrary, Bond et al. demonstrated better performance of the "low tech" renewable lead electrode compared to "high tech" GCE for in situ mercury film-coated electrodes [48]. Successively, another article for the determination of carbon-based electrodes vs. graphite leads has been elaborately compared for melanin-type polymer sensors for the determination of dopamine. Wherein the amperometric studies show that pencil graphite was not suitable for the determination probably due

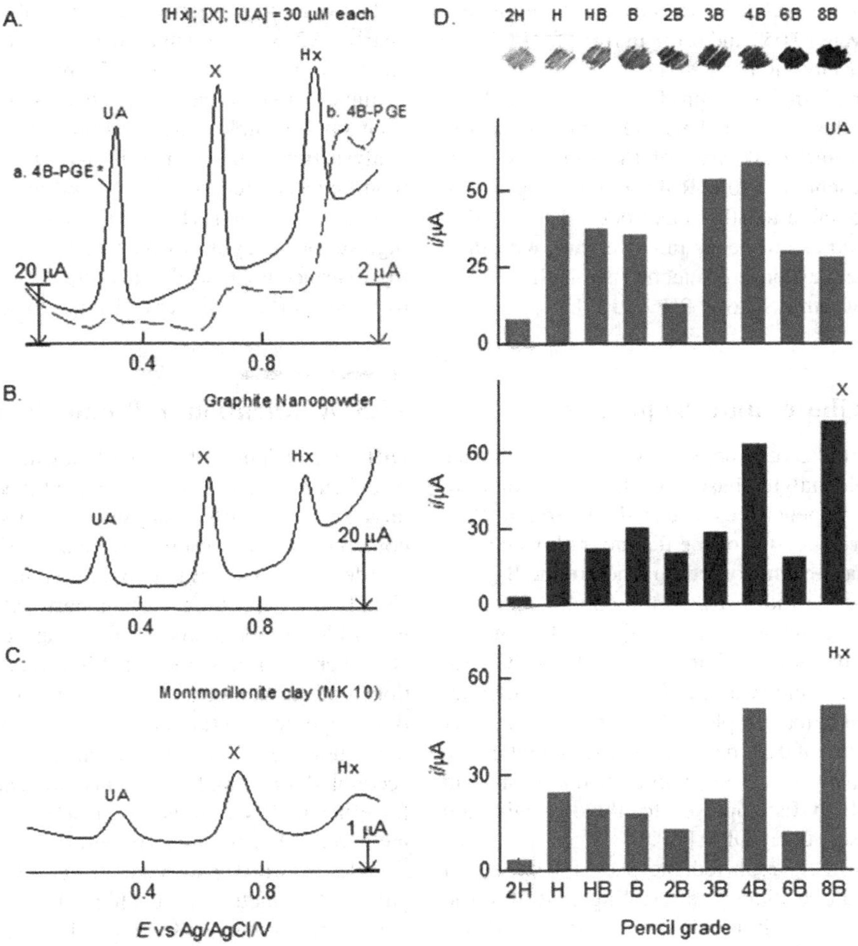

FIGURE 15.2 Illustration of the variation in response characteristics of different grades and their pre-anodization response for the simultaneous detection of three analytes in fish samples using the differential pulse voltammetric technique [19].

TABLE 15.1

Use of Pencil Graphite Sensing Platform for the Determination of Various Analytes in Real Samples

S. no.	Real sample	Technique	Analyte	Limit of detection	Linear range	Ref.
1.	Pharmaceuticals	Differential pulse voltammetry	Acyclovir	0.3 µM	1–100 µM	[20]
2.	Tap water	Amperometric i-t	Ozone	0.6 µM	30–400 µM	[21]
3.	Rocks and earthworm	Square wave anodic adsorption stripping voltammetry	Uranium	< 0.00008 ng/mL	< 0.000007 ng/mL	[22]
4.	Tobacco shells	Differential pulse voltammetry	Silver	0.015–0.25 µM	0.01 µM	[23]
5.	Willow barks	Square wave voltammetry	Salicylic acid	0.1–80 µg/mL	1.7 ng/mL	[24]
6.	Biological fluids	Square wave anodic adsorption stripping voltammetry	Itraconazole	10.6–127 ng/mL	8.7 µM	[25]
7.	Human urine	Differential pulse voltammetry	Hydroxyurea	10–1000 µM	7.89 µM	[26]
8.	Human mother milk	Differential pulse voltammetry	Dantrolene sodium	0.395–2.955 µg/mL	0.09 µM	[27]
9.	Serum	Square wave cathodic adsorption stripping voltammetry	Flutamide irinotecan	0.398–6.36 µM	0.0155 µM	[28]
10.	Dietary supplementary	Differential pulse voltammetry	Chlorogenic acid	0.1–500 µM	0.0714 µM	[29]

to the inherent high electro-active surface. But interestingly, the author mentions that electron transfer kinetics of wooden PGEs behave similarly to GCEs [49]. Later the applicability of a lead sensor has been extended for antioxidant activity determination in fruits and veggies, hot beverages such as tea and coffee, which showed quite intense peaks compared to traditional GCEs [50]. Further, the higher sensitivity of pencil electrodes has been concluded in comparison to carbon paste and glassy electrodes by Rizk and co-workers [51]. Herein, the limits of detection obtained for sodium ranelate were 0.17 (PGE),

0.24 (CPE) and 0.39 μg/mL (GCE), respectively, while for dantrolene sodium they were (0.052 and 0.158 μg/mL) for PGE and (0.095 and 0.28 μg/mL) for GCE.

The base for determining the reproducibility of results is the relative standard deviation or the RSD value of the signals, peak potentials and peak area of the redox systems. Researchers have presented $< 10\%$ RSD while using a lead pencil which is comparable to other electrodes. The stability of the response for lead electrodes is in correlation with their grade quality. A better response of electrochemically treated pencil electrodes shows an RSD of $< 2\%$ [52, 53].

The relatively large surface area of the pencil electrode is nearly ~0.255 cm^2 compared to 0.0951 cm^2 for carbon paste electrode compared to 0.0707 cm^2 for glassy carbon. Hence a higher area of electrode ensures a better response with even small sample concentrations/volumes, thus reducing the analysis time. They can easily have miniaturized form and be modified as better chemically modified electrodes.

The higher electrochemical reactivity of PGEs can be due to high surface irregularity morphology which helps in enhancing their surface area, while edge plane sites result in better electron transport and higher resistance to surface passivation [64].

15.4 Inbuilt Attributes and Properties

The use of different grades of pencils is expected to influence signals for various bioanalytes based on the nature of lead. But contradictorily, the pencil electrodes show similar electrochemical signals irrespective of the distinct redox couples. But various articles have been worked on wherein the B grade has been extended for the indirect determination of boron in the presence of iron, showing the best signal [54]. Another interesting report accounts use of pre-anodized 4B PGE for the determination of tea quality assays based on the polyphenolic content available in tea samples [55]. Contrarily, another report involved the use of HB pencils for the polyphenolic acid content determination [32, 33], while Vishnu et al. had reported the use of 6B PGEs employed for the determination of phenols and m-cresols using DPV [13].

Gong and his coworkers reported the use of B-type soft pencils consisting of large diameters enabling easy electron transfer generating higher peak signals ensuring defined determination while smaller diameter H-type electrodes provide better reversibility, being more adequate for qualitative investigations. HB2 graphite lead has electron transfer rates similar to glassy carbon electrodes [50].

Various polyphenolic content of sea buckthorn was estimated using a DPV technique for flavonoid detection and their interaction with copper ions. The results were quite useful in understanding the role of flavonoids in decreasing copper-induced oxidation of substrates [56].

Pencil lead is not as rigid as GCE but neither is it as soft as carbon paste, and it is quite stable compared to others and helps in eliminating the chances of electrode fouling or poisoning constraints [57]. In addition to this, the working potential window bar can be extended to both more positive and more negative sides compared to both Au and Pt electrodes [58]. Conversely, authors have added another important point of low background current using the PGEs compared to other traditional setup systems [20, 26, 33, 38, 40, 59–62]. The non-graphitic component of the pencil does not get any background signal, but the different types of pencil lead with different compositions of polymers generate a high analyte signal and relatively low noise, i.e. low signal-to-noise ratio [31, 55, 63].

An essential attribute for clean, adsorption-free, polishing and pretreated electrodes before initiating any experiment is one of the prima facie for electrochemical setup. Using low-cost pencil electrodes helps in ensuring a new surface every time as they are quite cheap and disposal.

15.5 Application in Biomedical Platforms

Different attributes of electrochemical and economical aspects of different pencil electrodes have hiked in recent times with large applicability for analysis of various inorganic and organic compounds from various different matrices [65, 66].

Adenine and adenine-copper complexes represented higher signals on PGE than carbon paste along with peak shifts toward lower potentials and the electrode discharging appears at higher potentials, thus enabling more sensitive determinations indicating an easier electrode process and a larger potential range for voltammetric measurements [34]. Similarly, the simultaneous detection of adenine and guanine has been accounted for using MoS_2 grown on pencil graphite electrodes for efficient DNA studies [67], while the end product of the metabolic cycle was determined using a 4B-PGE* in real fish samples in pH 7.0 phosphate buffer solution using differential pulse voltammetric approach, as shown in Figure 15.2 [19].

A huge amount of literature has been extended to DNA-modified PGEs, as referred to in Table 15.2 and Table 15.3. One such report has been published using Cibacron Blue modified pencil electrodes obtained using CV, and passive adsorption was developed to investigate the dye affinity to serum albumin. Using cyclic voltammetry, the peak current decreased with linear increments in the BSA concentration [70]. Even pencil surfaces can be altered using metal-based systems such as thin films, nanostructures for various detection of bioactives, as shown in Figure 15.3 [38, 41, 68, 69].

David et al. in 2013 reported an in situ generated electrochemical Hg-film used for potentiometric stripping of copper, cadmium, zinc and lead ions from river sediments of Prut river for a period of three years [71, 72]. Slater in 2013 and Aziz et al. reported detection of reactive oxygen species using a Pd nanoparticle modified graphite lead after selectively improving the peak separation compared to traditional graphite lead [30, 62, 73]. Different methodologies for pencil electrodes with nanomaterials such as metal nanoparticles, metal oxides, metal complex nanostructures and nanotubular structures with conducting polymers and the electrochemical applications of such modified electrochemical setup have been recently reviewed by Akanda et al. [62].

The redox systems namely $[Fe(CN)_6]^{3-/4-}$ and $[Ru(NH_3)_6]^{2+/3+}$ with different grades of pencils been accounted for by Paulo et al. Based on their experiments, they suggested that harder graphite lead are much more appropriate as electrode systems regardless of their manufacturer [78].

TABLE 15.2

The Modification of Pencil Electrodes with Different Enzyme Modifiers Has Been Critically Tabulated with Their Technique, Analyte Detection, Real Sample and the Linear Range of Their Analysis

S. no.	Electrode modifier	Technique	Analyte	Real sample	Linear range	Ref
1.	RuO_2-rGO	Differential pulse voltammetry	Insulin	Urine	0.0008–0.02	[74]
2.	Molecular imprinted polymer	Solid phase microextraction; differential pulse anodic stripping voltammetry	D and L aspartic acid	Blood serum	0.6–10	[75]
3.	Electrochemically synthesized molecular imprinted polymer	Electrochemical micro-solid phase extraction followed by differential pulse cathodic stripping voltammetry	D and L methionine	Human blood plasma	0.03–30 ng/mL	[74]
4.	N-one molecularly imprinted polymer-MWCNT	Micro-solid phase extraction; differential pulse anodic stripping voltammetry	Epinephrine	Cerebrospinal fluid	0.005–8 ng/mL	[76]
5.	rGO	Electro-membrane; micro-solid phase extraction followed by linear sweep voltammetry (LSV)	Tramadol	Urine	0.01–0.50 μg/mL	[77]

TABLE 15.3

Use of Pencil Graphite Sensing Platform for the Determination of Various Analytes in Real Samples

S. no.	Electrode modifier enzyme	Analyte	Pencil grades	Critical comments	Ref.
1.	Uricase, horseradish peroxidase	Uric acid and H_2O_2	HB	Glutaraldehyde as a cross-linking agent	[79]
2.	Alcohol dehydrogenase	Ethanol mediator-NADH	Pencil lead	Used for electro-polymerization of pyrocatechol	[80]
3.	Ascorbate oxidase	Ascorbic acid	HB	Using MWCNT, fullerenes and enzymes modified using polyurethane	[81]
4.	Lipase, glycerol kinase, glycerol 3-phosphate oxidase	Triglycerides	6B	Electrostatic interaction is the major reason for enzyme immobilization	[82]
5.	Hemoglobin	Nitrite	H	A multi-layered pencil electrode modification has been done using carboxylated multiwalled-carbon nanotube/copper nanoparticle/polyaniline which leads to slower electron transfer	[83]
6.	Glucose oxidase	Glucose	2B	Increase in the edge plane due to increased disorder on electrode surface thereby leading to faster charge kinetics	[82]
7.	Glucose dehydrogenase	Glucose mediators; NADH	2B	Electrode preparation via precipitation of quantum dots; involving glutardehyde as a cross-linking agent	[84]
8.	Alkaline phosphatase	mir21 RNA, substrate-α-NAP	HB	COOH-group functionalized system has been used for immobilization	[85]
9.	Horseradish peroxidase	H_2O_2	6H	Comparatively, 6H portrayed the reversible redox system	[80]
10.	Laccase	Oxygen mediator-ABTS	Simple pencil	Adsorption of MWCNT followed by electro-polymerization	[86]

An interesting report on pencil electrochemical genosensors based on colloidal gold nanoparticle has been employed for Leiden mutation (Factor V) detection and even their homozygous and heterozygous mutations were compared based on peak currents of Au signals [87]. In extension to this, the uric acid determination in urine and serum samples have been determined using an electrochemically treated graphite electrode in pH 3 PBS but showed interference from dopamine and ascorbic acid [16]. Contrary, Ali et al. have reported highly selective poly(diallyl dimethylammonium chloride) dispersed in MWCNT, layer by layer to detect the DNA damage induced by radicals generated by sulfite autoxidation using cyclic voltammetry and impedance spectroscopy techniques, as shown in Figure 15.4 [88]. The first serum insulin voltammetric for diagnosis of type-1 and type-2 immunosensor has been studied on edge-plane pyrolytic graphite electrodes by Vini and co-workers [89]. Alvaro et al. have initiated a self-powered biosensor for glucose detection involving pencils as transducers for biofuel cell applications [90]. In continuation many such systems have been studied for Cu/CuO_X nanoparticles, Cu-Al layered double hydroxide chicken feet yellow membrane, a thin film of Preyssler nanocapsules, nitrogen-doped graphene nanoflowers and even 3D-printed lab-in a-pencil graphite for microfluidic sensing [91–99].

Quite interesting literature recounts a galvanostatic pulse voltammetry study and cyclic voltammetry for the development of metal complex nanostructures multilayers; especially copper films and copper particles modified graphite leads and extended them for electrocatalytic activity toward nitrate reduction [101, 102]. Copper hexacyanoferrate modified the

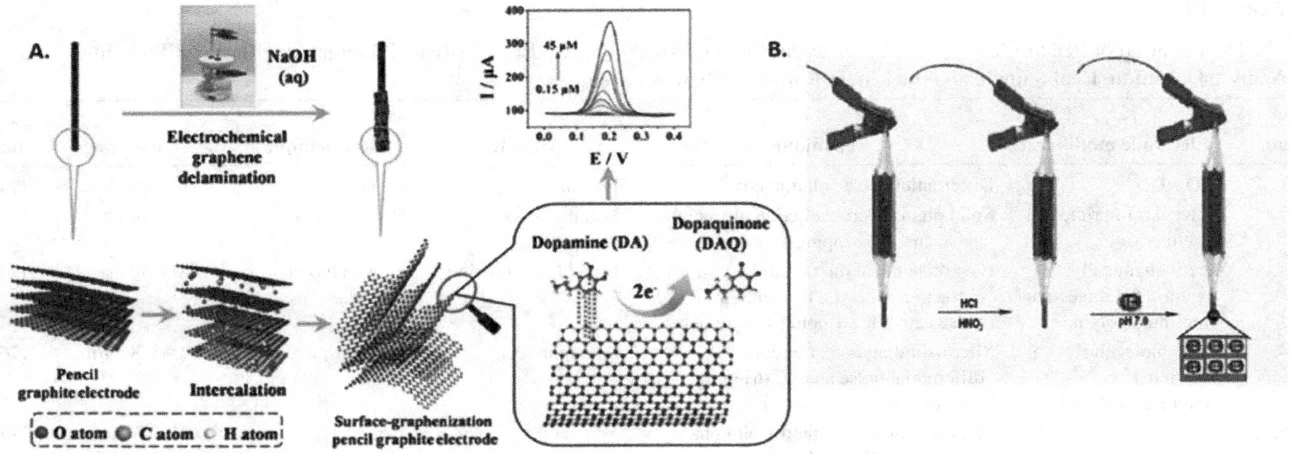

FIGURE 15.3 Schematic illustration of pencil electrodes (A) based on graphenization of PGE (NaOH treated anodically) for dopamine detection (Copyright [68]), (B) using chemisorption immobilization of cholesterol oxidase (Copyright [69]).

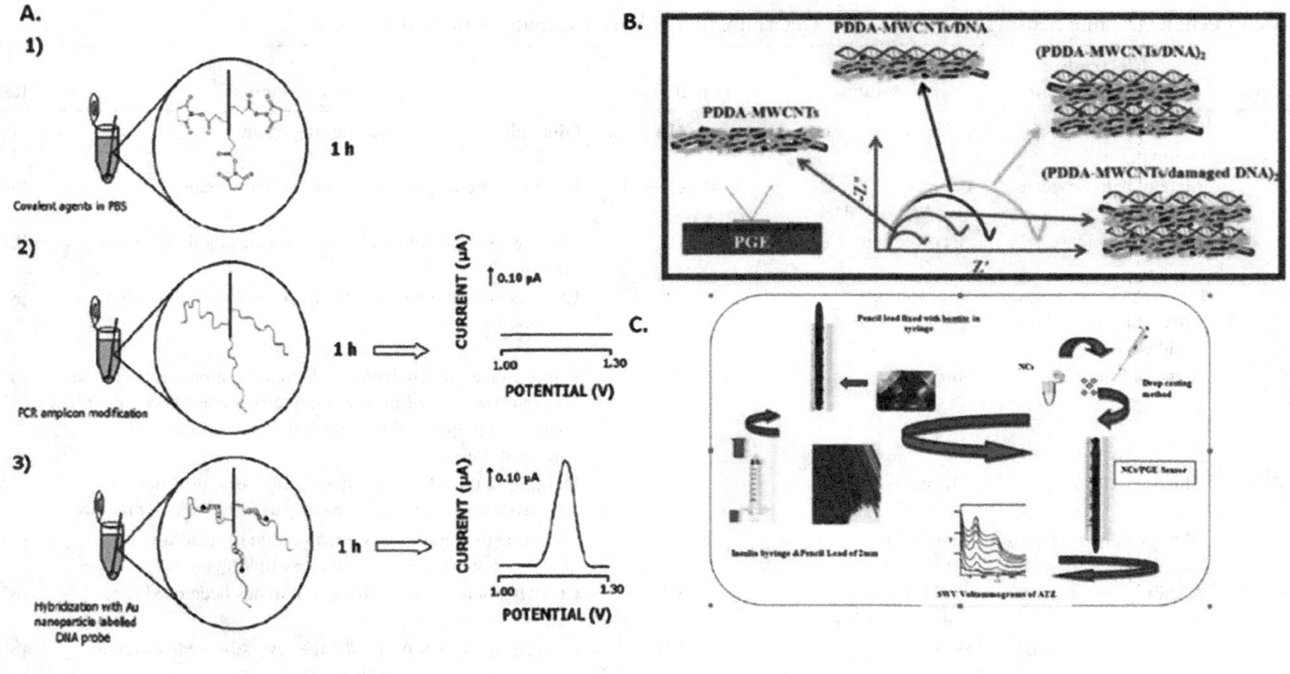

FIGURE 15.4 Schematic illustration of pencil electrodes for (A) detection of Factor V Leiden mutation as electrochemical genosensor platform (Copyright [87]), (B) DNA based biosensor to study catalytic effect of transition metals on autoxidation of sulfide (Copyright [88]), (C) modifications using bontite (Copyright [100]).

electrochemical platform and further, $K_3\{Fe(CN)_6\}$ solution [103]. Moreover, the harder pencil lead presents higher voltammetric signals, and separation of the peak potentials is closer to the theoretical value for reversible systems. There is not much difference in the voltammetric response of benchmark electrochemical redox systems, namely $[Fe(CN)_6]^{3-/4-}$ and $[Ru(NH_3)_6]^{2+/3+}$ with different grades of pencils.

A report involves clozapine analysis in human plasma using a microsystem consisting of a micropipette tip with a pencil as a working electrode (12 μm dia), i.e. an HB grade pencil insulated by an electrogenerated o-phenylenediamine [104, 105]. It led to initiating a flexible paper-based electrochemical platform, where paper acts as a substrate with electrodes being drawn using graphite pencils [106–109]. This concept is relatively cheaper compared to high-tech lithographic techniques [110].

15.5.1 Usage/Applicability in Real-Life Scenarios

PGEs have gained tremendous applicability in electroanalysis of very different types of analytes from various matrices. A careful literature search related different types of carbon-based electrodes and their different application areas of these electrodes in which PGEs account for a huge weightage.

The pencil graphite gives a smaller background current generated by the clays and polymers contained in the lead but nevertheless, they are still lower than the conventional electrode systems especially GCE, Au and Pt, but quite higher than boron-doped diamond electrodes [30, 31]. The history includes the use of graphite pencils as electrodes with low-weighted pencils for analysis of environmental samples [73]. Its economic cost is so minute that it is nearly 5,000 times less than any other electrode type, as the cost of each glassy carbon electrode is US$ 200 while a pencil costs just US$ 0.10.

15.6 Bottlenecks

The biggest bottleneck with these pencil electrodes is that they are produced by different manufacturers, with different quality schemes. The strongest point of its economic advantage leads to falling of its quality constraint most of the time [111, 112]. Due to different interactions between any analyte and the ingredients of pencil graphite electrode and the different electro-active species, they exhibit different voltammetric behavior of electrodes of the same hardness as produced by different manufacturers [78]. They can't be reused due to adsorption complications. Adding to this, comparing the voltammogram recorded on PGEs with those obtained on GCEs, some authors attribute higher peak intensities using PGEs and even a high specific surface area with a good electrolytic effect of pencil electrode on the electrode process for various analytes like acyclovir, eugenol, etc. [20, 60].

15.7 Conclusion and Future Prospects

This chapter has mainly been intended to review the various prospects of pencil graphite systems from their initial days of usage and the developments associated with them. Further, their usage in day-to-day applications has helped us in understanding the approach in an easy and economical route. Proper choice of the analyte, surface modification of PGEs and detection pathways have been exquisitely discussed in this chapter.

Acknowledgment

Mansi Gandhi thanks the Indian Council of Medical Research (ICMR) for the award of her SRF (2019-4952).

REFERENCES

1. Blum, D., Leyffer, W., & Holze, R. (1996). Pencil-leads as new electrodes for abrasive stripping voltammetry. *Electroanalysis*, *8*(3), 296–297.
2. Kakizaki, T., & Hasebe, K. (1998). Potentiometric stripping determination of heavy metals using a graphite-reinforcement carbon vibrating electrode. *Fresenius' Journal of Analytical Chemistry*, *360*(2), 175–178.
3. Perdicakis, M., Aubriet, H., & Walcarius, A. (2004). Use of a commercially available wood-free resin pencil as convenient electrode for the "voltammetry of microparticles" technique. *Electroanalysis*, *16*(24), 2042–2050.
4. Doménech-Carbó, A., Doménech-Carbó, M. T., & Peiró-Ronda, M. A. (2011). "One-touch" voltammetry of microparticles for the identification of corrosion products in archaeological lead. *Electroanalysis*, *23*(6), 1391–1400.
5. Doménech-Carbó, A., Doménech-Carbó, M. T., Peiró-Ronda, M. A., & Osete-Cortina, L. (2011). Electrochemistry and authentication of archaeological lead using voltammetry of microparticles: Application to the Tossal de Sant Miquel Iberian plate. *Archaeometry*, *53*(6), 1193–1211.
6. Vishnu, N., Gopalakrishnan, A., & Badhulika, S. (2018). Impact of intrinsic iron on electrochemical oxidation of pencil graphite and its application as supercapacitors. *Electrochimica Acta*, *269*, 274–281.
7. Keeling, J. (2017). Graphite: Properties, uses and South Australian resources. *3*, 28–41.
8. David, I. G., Popa, D., & Buleandra, M. (2017). Pencil graphite electrodes: A versatile tool in electroanalysis. *Journal of Analytical Methods in Chemistry*, *2017*(Cv), 1905968.
9. Das, N., Neog, K., Pathak, R., & Sarmah, S. (2019). Study of structural, elemental and thermal properties of pencil graphites and their applications as electrodes. *Bulletin of Physics Projects*, *4*, 21–24.
10. Pollock, S. (2016). From clay to stone: Material practices and writing in third millennium mesopotamia, in *Materiality of Writing in Early Mesopotamia*, Eds: Deutsche Forschungsgemeinschaft Edited by: Thomas E. Balke and Christina Tsouparopoulou, De Gruyter.
11. Foster, C. W., Brownson, D. A. C., Souza, P. R. De, Bernalte, E., Iniesta, J., & Banks, C. E. (2016). Pencil it in: pencil drawn electrochemical sensing platforms. *Analyst*, *141*, 4055–4064.
12. Alipour, E., Majidi, M. R., Saadatirad, A., Golabi, S. M., & Alizadeh, A. M. (2013). Simultaneous determination of dopamine and uric acid in biological samples on the pretreated pencil graphite electrode. *Electrochimica Acta*, *91*, 36–42.
13. Vishnu, N., & Kumar, A. S. (2015). A preanodized 6B-pencil graphite as an efficient electrochemical sensor for monophenolic preservatives (phenol and meta-cresol) in insulin formulations. *Analytical Methods*, *7*(5), 1943–1950.
14. Asofiei, I., Calinescu, I., Trifan, A., David, I. G., & Gavrila, A. I. (2016). Microwave-assisted batch extraction of polyphenols from sea buckthorn leaves. *Chemical Engineering Communications*, *203*(12), 1547–1553.
15. Dilgin, Y., Kizilkaya, B., Ertek, B., Eren, N., & Dilgin, D. G. (2012). Amperometric determination of sulfide based on its electrocatalytic oxidation at a pencil graphite electrode modified with quercetin. *Talanta*, *89*, 490–495.
16. Özcan, A., & Şahin, Y. (2010). Preparation of selective and sensitive electrochemically treated pencil graphite electrodes for the determination of uric acid in urine and blood serum. *Biosensors and Bioelectronics*, *25*(11), 2497–2502.
17. Özcan, A., & Şahin, Y. (2009). Selective and sensitive voltammetric determination of dopamine in blood by electrochemically treated pencil graphite electrodes. *Electroanalysis*, *21*(21), 2363–2370.
18. Özcan, A., & Şahin, Y. (2011). A novel approach for the determination of paracetamol based on the reduction of N-acetyl-p-benzoquinoneimine formed on the electrochemically treated pencil graphite electrode. *Analytica Chimica Acta*, *685*(1), 9–14.
19. Vishnu, N., Gandhi, M., Rajagopal, D., & Kumar, A. S. (2017). Pencil graphite as an elegant electrochemical sensor for separation-free and simultaneous sensing of hypoxanthine, xanthine and uric acid in fish samples. *Analytical Methods*, *9*(15), 2265–2274.

20. Dilgin, D. G., & Karakaya, S. (2016). Differential pulse voltammetric determination of acyclovir in pharmaceutical preparations using a pencil graphite electrode. *Materials Science and Engineering C*, *63*, 570–576.
21. Kawde, A. N. (2016). Electroanalytical determination of heavy metals in drinking waters in the eastern province of Saudi Arabia. *Desalination and Water Treatment*, *57*(33), 15697–15705.
22. Ly, S. Y., Lee, J. H., & Jung, D. H. (2010). Radioactive uranium measurement in vivo using a handheld interfaced analyzer. *Environmental Toxicology and Chemistry*, *29*(5), 1025–1030.
23. Krizkova, S., Krystofova, O., Trnkova, L., Hubalek, J., Adam, V., Beklova, M., … & Kizek, R. (2009). Silver(I) ions ultrasensitive detection at carbon electrodes-analysis of waters, tobacco cells and fish tissues. *Sensors*, *9*(9), 6934–6950.
24. Petrek, J., Havel, L., Petrlova, J., Adam, V., Potesil, D., Babula, P., & Kizek, R. (2007). Analysis of salicylic acid in willow barks and branches by an electrochemical method. *Russian Journal of Plant Physiology*, *54*(4), 553–558.
25. Shalaby, A., Hassan, W. S., Hendawy, H. A. M., & Ibrahim, A. M. (2016). Electrochemical oxidation behavior of itraconazole at different electrodes and its anodic stripping determination in pharmaceuticals and biological fluids. *Journal of Electroanalytical Chemistry*, *763*, 51–62.
26. Naik, K. M., Ashi, C. R., & Nandibewoor, S. T. (2015). Anodic voltammetric behavior of hydroxyurea and its electroanalytical determination in pharmaceutical dosage form and urine. *Journal of Electroanalytical Chemistry*, *755*, 109–114.
27. Omran, M. A. (2016). Highly sensitive voltammetric determination of dantrolene sodium in pure form, pharmaceuticals, human breast milk and urine at pencil graphite and glassy carbon electrodes. *Indo American Journal of Pharmaceutical Sciences*, *3*(10), 1210–1222.
28. Temerk, Y. M., Ibrahim, H., & Schuhmann, W. (2016). Square wave cathodic adsorptive stripping voltammetric determination of the anticancer drugs flutamide and irinotecan in biological fluids using renewable pencil graphite electrodes. *Electroanalysis*, *28*(2), 372–379.
29. David, I. G., Popa, D. E., Buleandra, M., Moldovan, Z., Iorgulescu, E. E., & Badea, I. A. (2016). Cheap pencil graphite electrodes for rapid voltammetric determination of chlorogenic acid in dietary supplements. *Analytical Methods*, *8*(35), 6537–6544.
30. Aziz, M. A., & Kawde, A. N. (2013). Nanomolar amperometric sensing of hydrogen peroxide using a graphite pencil electrode modified with palladium nanoparticles. *Microchimica Acta*, *180*(9–10), 837–843.
31. Yardim, Y., & Şentürk, Z. (2013). Electrochemical evaluation and adsorptive stripping voltammetric determination of capsaicin or dihydrocapsaicin on a disposable pencil graphite electrode. *Talanta*, *112*, 11–19.
32. David, I. G., Buleandră, M., Popa, D. E., Bîzgan, A. M. C., Moldovan, Z., Badea, I. A., … & Basaga, H. (2016). Voltammetric determination of polyphenolic content as rosmarinic acid equivalent in tea samples using pencil graphite electrodes. *Journal of Food Science and Technology*, *53*(6), 2589–2596.
33. David, I. G., Bizgan, A. M. C., Popa, D. E., Buleandra, M., Moldovan, Z., Badea, I. A., … & Ciucu, A. A. (2015). Rapid determination of total polyphenolic content in tea samples based on caffeic acid voltammetric behaviour on a disposable graphite electrode. *Food Chemistry*, *173*, 1059–1065.
34. Aladag, N., Trnkova, L., Kourilova, A., Ozsoz, M., & Jelen, F. (2010). Voltammetric study of aminopurines on pencil graphite electrode in the presence of copper ions. *Electroanalysis*, *22*(15), 1675–1681.
35. David, I. G., Florea, M. A., Cracea, O. G., Popa, D. E., Buleandra, M., Iorgulescu, E. E., … & Ciucu, A. A. (2015). Voltammetric determination of B1 and B6 vitamins using a pencil graphite electrode. *Chemical Papers*, *69*(7), 901–910. https://doi.org/10.1515/chempap-2015-0096
36. David, I. G., Popa, D. E., Calin, A. A., Buleandra, M., & Iorgulescu, E. E. (2016). Voltammetric determination of famotidine on a disposable pencil graphite electrode. *Turkish Journal of Chemistry*, *40*(1), 125–135.
37. Ang, J. Q., & Li, S. F. Y. (2012). Novel sensor for simultaneous determination of K + and Na + using Prussian blue pencil graphite electrode. *Sensors and Actuators, B: Chemical*, *173*, 914–918.
38. Pokpas, K., Jahed, N., Tovide, O., Baker, P. G., & Iwuoha, E. I. (2014). Nafion-graphene nanocomposite in situ plated bismuth-film electrodes on pencil graphite substrates for the determination of trace heavy metals by anodic stripping voltammetry. *International Journal of Electrochemical Science*, *9*(9), 5092–5115.
39. Buratti, S., Scampicchio, M., Giovanelli, G., & Mannino, S. (2008). A low-cost and low-tech electrochemical flow system for the evaluation of total phenolic content and antioxidant power of tea infusions. *Talanta*, *75*(1), 312–316.
40. Prasad, B. B., Madhuri, R., Tiwari, M. P., & Sharma, P. S. (2010). Electrochemical sensor for folic acid based on a hyperbranched molecularly imprinted polymer-immobilized sol-gel-modified pencil graphite electrode. *Sensors and Actuators, B: Chemical*, *146*(1), 321–330.
41. Rehacek, V., Hotovy, I., Vojs, M., & Mika, F. (2008). Bismuth film electrodes for heavy metals determination. *Microsystem Technologies*, *14*(4–5), 491–498.
42. Prasad, B. B., Madhuri, R., Tiwari, M. P., & Sharma, P. S. (2010). Imprinting molecular recognition sites on multiwalled carbon nanotubes surface for electrochemical detection of insulin in real samples. *Electrochimica Acta*, *55*(28), 9146–9156.
43. Prasad, B. B., & Pandey, I. (2013). Metal incorporated molecularly imprinted polymer-based electrochemical sensor for enantio-selective analysis of pyroglutamic acid isomers. *Sensors and Actuators, B: Chemical*, *186*, 407–416.
44. Prasad, B. B., Kumar, D., Madhuri, R., & Tiwari, M. P. (2011). Metal ion mediated imprinting for electrochemical enantioselective sensing of l-histidine at trace level. *Biosensors and Bioelectronics*, *28*(1), 117–126.
45. Cantalapiedra, A., Gismera, M. J., Procopio, J. R., & Sevilla, M. T. (2015). Electrochemical sensor based on polystyrene sulfonate-carbon nanopowders composite for Cu (II) determination. *Talanta*, *139*, 111–116.
46. Rad, A. S. (2011). Vitamin C determination in human plasma using an electro-activated pencil graphite electrode. *Arabian Journal for Science and Engineering*, *36*(1), 21–28.

47. Aoki, K., Kobayashi, A., & Kato, N. (1990). Amperometric determination of ozone in water at disposable graphite reinforcement carbon electrodes. *Electroanalysis*, *2*(1), 31–34.
48. Bond, A. M., Mahon, P. J., Schiewe, J., & Vicente-Beckett, V. (1997). An inexpensive and renewable pencil: Electrode for use in field-based stripping voltammetry. *Analytica Chimica Acta*, *345*(1–3), 67–74.
49. Rubianes, M. D., & Rivas, G. A. (2003). Amperometric quantification of dopamine using different carbon electrodes modified with a melanin-type polymer. *Analytical Letters*, *36*(2), 329–345.
50. Kariuki, J., Ervin, E., & Olafson, C. (2015). Development of a novel, low-cost, disposable wooden pencil graphite electrode for use in the determination of antioxidants and other biological compounds. *Sensors (Switzerland)*, *15*(8), 18887–18900.
51. Rizk, M., Abou El-Alamin, M. M., Hendawy, H. A. M., & Moawad, M. I. (2016). Highly sensitive differential pulse and square wave voltammetric methods for determination of strontium ranelate in bulk and pharmaceutical dosage form. *Electroanalysis*, *28*(4), 770–777.
52. Koyun, O., Gorduk, S., Arvas, M. B., & Sahin, Y. (2018). Electrochemically treated pencil graphite electrodes prepared in one step for the electrochemical determination of paracetamol. *Russian Journal of Electrochemistry*, *54*(11), 796–808.
53. Dossi, N., Toniolo, R., Terzi, F., Impellizzieri, F., & Bontempelli, G. (2014). Pencil leads doped with electrochemically deposited Ag and AgCl for drawing reference electrodes on paper-based electrochemical devices. *Electrochimica Acta*, *146*, 518–524.
54. Liv, L., & Nakiboğlu, N. (2016). Simple and rapid voltammetric determination of boron in water and steel samples using a pencil graphite electrode. *Turkish Journal of Chemistry*, *40*(3), 412–421.
55. Vishnu, N., Gandhi, M., Badhulika, S., & Kumar, A. S. (2018). Tea quality testing using 6B pencil lead as an electrochemical sensor. *Analytical Methods*, *10*(20), 2327–2336.
56. Vestergaard, M., Kerman, K., & Tamiya, E. (2005). An electrochemical approach for detecting copper-chelating properties of flavonoids using disposable pencil graphite electrodes: Possible implications in copper-mediated illnesses. *Analytica Chimica Acta*, *538*(1–2), 273–281.
57. Yang, X., Kirsch, J., Fergus, J., & Simonian, A. (2013). Modeling analysis of electrode fouling during electrolysis of phenolic compounds. *Electrochimica Acta*, *94*, 259–268.
58. Sweeney, J., Hausen, F., Hayes, R., Webber, G. B., Endres, F., Rutland, M. W., ... & Atkin, R. (2012). Control of nanoscale friction on gold in an ionic liquid by a potential-dependent ionic lubricant layer. *Physical Review Letters*, *109*(15), 1–5.
59. Gowda, J. I., & Nandibewoor, S. T. (2014). Electrochemical characterization and determination of paclitaxel drug using graphite pencil electrode. *Electrochimica Acta*, *116*, 326–333.
60. Santhiago, M., Henry, C. S., & Kubota, L. T. (2014). Low cost, simple three dimensional electrochemical paper-based analytical device for determination of p-nitrophenol. *Electrochimica Acta*, *130*, 771–777.
61. Abdul Aziz, M., & Kawde, A. N. (2013). Gold nanoparticle-modified graphite pencil electrode for the high-sensitivity detection of hydrazine. *Talanta*, *115*, 214–221.
62. Akanda, M. R., Sohail, M., Aziz, M. A., & Kawde, A. N. (2016). Recent advances in nanomaterial-modified pencil graphite electrodes for electroanalysis. *Electroanalysis*, *28*(3), 408–424.
63. Gholivand, M. B., Khodadadian, M., & Bahrami, G. (2015). Molecularly imprinted polymer preconcentration and flow injection amperometric determination of 4-nitrophenol in water. *Analytical Letters*, *48*(18), 2856–2869.
64. Arvas, M. B., Gorduk, O., Gencten, M., & Sahin, Y. (2021). Differential pulse voltammetric (DPV) determination of phosphomolybdenum complexes by a poly (vinyl chloride) coated molybdenum blue modified pencil graphite electrode (PVC-MB-PGE) differential pulse voltammetric (DPV) determination of phosphomoly. *Analytical Letters*, *54*, 492–511.
65. V. A., Sajid, M., & Kawde, A. (2016). One-pot microwave-assisted in situ reduction of Ag+ and Au3+ ions by Citrus limon extract and their carbon-dots based nanohybrids: a potential nano-bioprobe for cancer cellular imaging. *RSC Advances*, *6*, 103482–103490.
66. Fan, X., Xu, Y., Sheng, T., Zhao, D., Yuan, H., Liu, F., ... & Lu, J. (2019). Amperometric sensor for dopamine based on surface-graphenization pencil graphite electrode prepared by in-situ electrochemical delamination. *Microchimica Acta*, *186*(5), 5–12.
67. Chauhan, N., Narang, J., & Pundir, C. S. (2010). Amperometric determination of serum cholesterol with pencil graphite rod. *American Journal of Analytical Chemistry*, *1*(2), 41–46.
68. Vishnu, N., & Badhulika, S. (2019). Single step grown MoS 2 on pencil graphite as an electrochemical sensor for guanine and adenine: A novel and low cost electrode for DNA studies. *Biosensors and Bioelectronics*, 124–125, 122–128.
69. Kuralay, F., Yilmaz, E., Uzun, L., & Denizli, A. (2013). Cibacron blue F3GA modified disposable pencil graphite electrode for the investigation of affinity binding to bovine serum albumin. *Colloids and Surfaces B: Biointerfaces*, *110*, 270–274.
70. David, I. G., Matache, M. L., Radu, G. L., & Ciucu, A. A. (2013). Cheap in situ voltammetric copper determination from freshwater samples. *E3S Web of Conferences*, *1*(April), 37004.
71. Bund, A. (1996). A simple and versatile PSA system for heavy metal determinations. *Fresenius' Journal of Analytical Chemistry*, *356*(1), 27–30.
72. Kawde, A. N., Baig, N., & Sajid, M. (2016). Graphite pencil electrodes as electrochemical sensors for environmental analysis: A review of features, developments, and applications. *RSC Advances*, *6*(94), 91325–91340.
73. Ensafi, A. A., Khoddami, E., Rezaei, B., & Jafari-Asl, M. (2015). A supported liquid membrane for microextraction of insulin, and its determination with a pencil graphite electrode modified with RuO2-graphene oxide. *Microchimica Acta*, *182*(9–10), 1599–1607.
74. Prasad, B. B., Srivastava, A., & Tiwari, M. P. (2013). Highly sensitive and selective hyphenated technique (molecularly imprinted polymer solid-phase microextraction-molecularly

imprinted polymer sensor) for ultra trace analysis of aspartic acid enantiomers. *Journal of Chromatography A*, *1283*, 9–19.

75. Prasad, B. B., Srivastava, A., Pandey, I., & Tiwari, M. P. (2013). Electrochemically grown imprinted polybenzidine nanofilm on multiwalled carbon nanotubes anchored pencil graphite fibers for enantioselective micro-solid phase extraction coupled with ultratrace sensing of d- and l-methionine. *Journal of Chromatography B: Analytical Technologies in the Biomedical and Life Sciences*, *912*, 65–74.
76. Fakhari, A. R., Sahragard, A., Ahmar, H., & Tabani, H. (2015). A novel platform sensing based on combination of electromembrane-assisted solid phase microextraction with linear sweep voltammetry for the determination of tramadol. *Journal of Electroanalytical Chemistry*, *747*, 12–19.
77. Tavares, P. H. C. P., & Barbeira, P. J. S. (2008). Influence of pencil lead hardness on voltammetric response of graphite reinforcement carbon electrodes. *Journal of Applied Electrochemistry*, *38*(6), 827–832.
78. Tsai, W. C., & Wen, S. T. (2006). Determination of uric acid in serum by a mediated amperometric biosensor. *Analytical Letters*, *39*(5), 891–901.
79. Zhu, J., Wu, X. Y., Shan, D., Yuan, P. X., & Zhang, X. J. (2014). Sensitive electrochemical detection of NADH and ethanol at low potential based on pyrocatechol violet electrodeposited on single walled carbon nanotubes-modified pencil graphite electrode. *Talanta*, *130*, 96–102.
80. Barberis, A., Spissu, Y., Fadda, A., Azara, E., Bazzu, G., Marceddu, S., … & Serra, P. A. (2015). Simultaneous amperometric detection of ascorbic acid and antioxidant capacity in orange, blueberry and kiwi juice, by a telemetric system coupled with a fullerene- or nanotubes-modified ascorbate subtractive biosensor. *Biosensors and Bioelectronics*, *67*, 214–223.
81. Narwal, V., & Pundir, C. S. (2017). An improved amperometric triglyceride biosensor based on co-immobilization of nanoparticles of lipase, glycerol kinase and glycerol 3-phosphate oxidase onto pencil graphite electrode. *Enzyme and Microbial Technology*, *100*, 11–16.
82. Batra, B., Lata, S., Sharma, M., & Pundir, C. S. (2013). An acrylamide biosensor based on immobilization of hemoglobin onto multiwalled carbon nanotube/copper nanoparticles/polyaniline hybrid film. *Analytical Biochemistry*, *433*(2), 210–217.
83. Ertek, B., Akgül, C., & Dilgin, Y. (2016). Photoelectrochemical glucose biosensor based on a dehydrogenase enzyme and NAD^+/NADH redox couple using a quantum dot modified pencil graphite electrode. *RSC Advances*, *6*(24), 20058–20066.
84. Down, M. P., Foster, C. W., Ji, X., & Banks, C. E. (2016). Pencil drawn paper based supercapacitors. *RSC Advances*, *6*(84), 81130–81141.
85. Kashyap, D., Kim, C., Kim, S. Y., Kim, Y. H., Kim, G. M., Dwivedi, P. K., … & Goel, S. (2015). Multi walled carbon nanotube and polyaniline coated pencil graphite based biocathode for enzymatic biofuel cell. *International Journal of Hydrogen Energy*, *40*(30), 9515–9522.
86. Ozsoz, M., Erdem, A., Kerman, K., Ozkan, D., Tugrul, B., Topcuoglu, N., … & Taylan, M. (2003). Electrochemical genosensor based on colloidal gold nanoparticles for the detection of Factor V Leiden mutation using disposable pencil graphite electrodes. *Analytical Chemistry*, *75*(9), 2181–2187.
87. Ensafi, A. A., Heydari-Bafrooei, E., & Rezaei, B. (2013). DNA-based biosensor for comparative study of catalytic effect of transition metals on autoxidation of sulfite. *Analytical Chemistry*, *85*(2), 991–997.
88. Singh, V., & Krishnan, S. (2015). Voltammetric immunosensor assembled on carbon-pyrenyl nanostructures for clinical diagnosis of type of diabetes. *Analytical Chemistry*, *87*(5), 2648–2654.
89. Torrinha, Á., Tavares, M., Delerue-Matos, C., & Morais, S. (2021). A self-powered biosensor for glucose detection using modified pencil graphite electrodes as transducers. *Chemical Engineering Journal*, *426*(May), 131835.
90. Bahadori, Y., & Razmi, H. (2021). Design of an electrochemical platform for the determination of diclofenac sodium utilizing a graphenized pencil graphite electrode modified with a Cu-Al layered double hydroxide/chicken feet yellow membrane. *New Journal of Chemistry*, *45*(32), 14616–14625.
91. Bahrami, E.; Amini, R.; Vardak, S. (2021). Electrochemical detection of dopamine via pencil graphite electrodes modified by Cu/CuxO nanoparticles. *Journal of Alloys and Compounds*, *855*.
92. Rouhani, M., & Soleymanpour, A. (2021). Ultrasensitive electrochemical determination of trace ceftizoxime using a thin film of Preyssler nanocapsules on pencil graphite electrode surface modified with reduced graphene oxide. *Microchemical Journal*, *165*, 106160.
93. Senel, M., & Alachkar, A. (2021). Lab-in-a-pencil graphite: A 3D-printed microfluidic sensing platform for real-time measurement of antipsychotic clozapine level. *Lab on a Chip*, *21*(2), 405–411.
94. Findik, M., Bingol, H., & Erdem, A. (2021). Hybrid nanoflowers modified pencil graphite electrodes developed for electrochemical monitoring of interaction between mitomycin C and DNA. *Talanta*, *222*, 121647.
95. Erden, S. (2021). Square wave anodic adsorptive stripping voltammetric determination of cefpodoxime proxetil by using pencil graphite electrode. *Monatshefte Fur Chemie*, *152*(6), 587–592.
96. Hassan Oghli, A., & Soleymanpour, A. (2021). Ultrasensitive electrochemical sensor for simultaneous determination of sumatriptan and paroxetine using molecular imprinted polymer/sol-gel/polyoxometalate/rGO modified pencil graphite electrode. *Sensors and Actuators, B: Chemical*, *344*(May), 130215.
97. Samukaite-Bubniene, U., Valiūnienė, A., Bucinskas, V., Genys, P., Ratautaite, V., Ramanaviciene, A., … & Ramanavicius, A. (2021). Towards supercapacitors: Cyclic voltammetry and fast Fourier transform electrochemical impedance spectroscopy based evaluation of polypyrrole electrochemically deposited on the pencil graphite electrode. *Colloids and Surfaces A: Physicochemical and Engineering Aspects*, *610*, 125750.
98. Nagarajan, S., & Vairamuthu, R. (2021). Electrochemical detection of riboflavin using tin-chitosan modified pencil graphite electrode. *Journal of Electroanalytical Chemistry*, *891*, 115235.

99. Annu, Sharma, S., Jain, R., & Raja, A. N. (2020). Review—Pencil graphite electrode: An emerging sensing material. *Journal of the Electrochemical Society*, *167*(3), 37501.
100. Majidi, M. R., Asadpour-Zeynali, K., & Hafezi, B. (2011). Fabrication of nanostructured copper thin films at disposable pencil graphite electrode and its application to elecrocatalytic reduction of nitrate. *International Journal of Electrochemical Science*, *6*(1), 162–170.
101. Alam, M. M., Hasnat, M. A., Rashed, M. A., Uddin, S. M. N., Rahman, M. M., Amertharaj, S., ... & Mohamed, N. (2015). Nitrate detection activity of Cu particles deposited on pencil graphite by fast scan cyclic voltammetry. *Journal of Analytical Chemistry*, *70*(1), 60–66.
102. Majidi, M. R., Asadpour-Zeynali, K., & Hafezi, B. (2010). Sensing L-cysteine in urine using a pencil graphite electrode modified with a copper hexacyanoferrate nanostructure. *Microchimica Acta*, *169*(3), 283–288.
103. Rouhollahi, A., Kouchaki, M., & Seidi, S. (2016). Electrically stimulated liquid phase microextraction combined with differential pulse voltammetry: A new and efficient design for in situ determination of clozapine from complicated matrices. *RSC Advances*, *6*(16), 12943–12952.
104. Hsieh, B. C., Cheng, T. J., Shih, S. H., & Chen, R. L. C. (2011). Pencil lead microelectrode and the application on cell dielectrophoresis. *Electrochimica Acta*, *56*(27), 9916–9920.
105. Honeychurch, K. C. (2015). The voltammetric behaviour of lead at a hand drawn pencil electrode and its trace determination in water by stripping voltammetry. *Analytical Methods*, *7*(6), 2437–2443.
106. Li, W., Qian, D., Li, Y., Bao, N., Gu, H., & Yu, C. (2016). Fully-drawn pencil-on-paper sensors for electroanalysis of dopamine. *Journal of Electroanalytical Chemistry*, *769*, 72–79.
107. Foster, C. W., Brownson, D. A. C., Ruas De Souza, A. P., Bernalte, E., Iniesta, J., Bertotti, M., & Banks, C. E. (2016). Pencil it in: Pencil drawn electrochemical sensing platforms. *Analyst*, 141, 4055–4064.
108. Dossi, N., Toniolo, R., Impellizzieri, F., & Bontempelli, G. (2014). Doped pencil leads for drawing modified electrodes on paper-based electrochemical devices. *Journal of Electroanalytical Chemistry*, 722–723, 90–94.
109. He, J., Luo, M., Hu, L., Zhou, Y., Jiang, S., Song, H., ... & Tang, J. (2014). Flexible lead sulfide colloidal quantum dot photodetector using pencil graphite electrodes on paper substrates. *Journal of Alloys and Compounds*, *596*, 73–78.
110. Anderson, R. C., Laband, D. N., Hansen, E. N., & Knowles, C. D. (2005). Price premiums in the mist. *Forest Products Journal*, *55*(6), 19–22.
111. Voice, E. H. (2016). *Transactions of the Newcomen Society the History of the Manufacture of Pencils. Transactions of the Newcomen Society*, *27*(1), 131–141.
112. Chokkareddy, R., & Redhi, G. G. (2019). Recent sensing technologies for first line anti-tuberculosis drugs in pharmaceutical dosages and biological fluids: A review. *Sensor Letters*, *17*(11), 833–858.

16

An Outbreak of Oxidative Stress in Pathogenesis of Alzheimer's Disease

Sourbh Suren Garg, Poojith Nuthalapati, Sruchi Devi, Atulika Sharma, Debasis Sahu and Jeena Gupta

CONTENTS

16.1 Introduction

Oxygen is vital for every single creature present on this planet. Violence to the metabolism and the generation of reactive oxygen species (ROS) result in an indefinite number of neural ailments [1]. The chemical reactions and the central role of free radicals in the body make free radicals a highly attractive topic in biology. The excess of free radicals increases the level of oxidative stress (OS) which ultimately increases the chances of progression of chronic [2] and neural diseases [3]. The aging factor is considered as the main connection between OS and neuronal diseases [4]. Women are expected to have the maximum risk of getting affected with Alzheimer's disease (AD). The decline in levels of estrogen causes menopause which might be a possible factor for developing the disease in women [5]. Oxidative damage (OD) and dysfunctioning of mitochondria are commonly observed during OS. The functioning of mitochondrial complex IV is reduced in the hippocampus as a model of AD [6]. Increased levels of free radicals cause an increase in OS which ultimately ends in several neural and chronic disorders. Additionally, an increased level of OS directly contributes to the OD in mitochondria.

The β-amyloid is a toxic peptide, known to be involved in the spread of AD. ROS helps to increase the productivity of β-peptides in the body [7]. In neocortical neurons, the association of β-amyloid with activated hydrogen peroxide results in OS. Moreover, the microglia cells are known to associate with dopaminergic neurons by dysregulating the functioning of NADPH in Parkinson's disease [8]. Dementia is the primary consequence of AD in old age [9]. AD is rated as the fourth leading ailment to cause death in the world. Cholinergic and amyloid are two important terms that are used to illustrate the mechanism of pathogenesis of AD. Acetylcholine (Ach) is a cerebral neurotransmitter whose esterification is catalyzed by acetylcholinesterase (AChE). The levels of AChE are found to increase from 40% to 90% in AD [10].

Lipid peroxidation (LPO) is delineated as the oxidation of the lipid membrane. The end products of LPO are malondialdehyde, 4-hydroxynonenal, and F2-isoprostanes [11]. In the cerebrospinal fluid of the brain, levels of these end products are increased in patients suffering from AD [12]. The upregulation of ROS causes OS, which declines the functioning of antioxidant enzymes in AD, i.e., superoxide dismutase and catalase. OD in lipids, proteins, and abnormal functioning of antioxidant enzymes are observed in the synapses of an Alzheimer's patient, which suggests that OS has a direct role in Alzheimer's-related synaptic loss. OS is considered a chief mediator for the progression of AD [13]. All the abbreviations used in this manuscript have been listed in Table 16.1.

DOI: 10.1201/9781003157823-16

TABLE 16.1
List of Abbreviations

S. no.	Abbreviation	Full name
1	OS	Oxidative stress
2	AD	Alzheimer's disease
3	ROS	Reactive oxygen species
4	BACE1	Beta-site APP cleaving enzyme
5	SP	Senile plaques
6	OD	Oxidative damage
7	4-HNE	4-hydroxynonenal
8	AChE	Acetyl cholinesterase
9	BChE	Butyl cholinesterase
10	LPO	Lipid peroxidation
11	LRP1	Low-density lipoprotein receptor protein-1
12	NCT	Nicastrin
13	PS	Presenilin
14	H_2O_2	Hydrogen peroxide
15	APP	Amyloid precursor protein
16	PS	Presenilin
17	APH-1	Anterior pharynx defective-1

16.2 Sources of Free Radicals

Both endogenous and exogenous sources are the leading producers of free radicals. Mitochondria, peroxisomes, and endoplasmic reticulum are counted as the main cell organelles for endogenous free radicals in the body. On the contrary, ultraviolet rays, air pollution, water pollution, and heavy metals like copper and zinc are viewed as exogenous sources of free radicals [14]. A list of free radicals with their contribution to the progression of various diseases has been provided in Table 16.2.

16.2.1 Mitochondria as a Site of Free Radical Generation

Mitochondria, also known as the powerhouse of the cell, are the major site for the generation of free radicals in the body. Free radicals such as superoxide radicals are generated in the body during the cascade of electron transport [15]. Complex I and complex III are major sites in mitochondria to generate superoxide radicals [16]. NADPH dehydrogenase and ubiquinone cytochrome C reductase are other names for complex I and complex III, respectively. Briefly, there is the transport of electrons either from complex I or complex II to coenzyme Q. With this transfer, a reduced form of coenzyme (QH2) is generated which regenerates the coenzyme Q via semiquinone anion (•Q–). This reaction further continues with the transfer of electrons from semiquinone anion to oxygen leading to the formation of superoxide radicals in the mitochondria [17]. The generation of superoxide radicals in the body is responsible for a number of ailments. Mitochondrial superoxide dismutase is an enzyme responsible for the catalyzation of superoxide anion to other free radicals such as hydrogen peroxide [18] (H_2O_2).

16.2.2 Peroxisomes as a Site of Free Radical Generation

Peroxisome, a site where H_2O_2 is produced during the β-oxidation of fatty acids [19]. Additionally, O2•–, OH•, and NO• are produced from the peroxisomes. Various enzymes such as acyl CoA oxidase, urate oxidase, D-aspartate oxidase, D-amino-acid-oxidase, and xanthine oxidase are majorly responsible for the generation of ROS in the peroxisomes [14].

16.2.3 Endoplasmic Reticulum as a Site of Free Radical Generation

Cytochrome P450 and diamine oxidase are two reticulum enzymes, known as a source of the generation of ROS [20]. In addition to these, the reaction of transferring electrons from dithiols to oxygen involves the action of Erop1. And Erop1 produces H_2O_2 when acting on thiol substrates in the presence of oxygen [14].

16.3 Hallmarks of AD

Senile plaques (SP), neurofibrillary tangles (NFTs), and neuronal loss are the main hallmarks of AD (Figure 16.1). SP consists of the beta-amyloid peptide (Abeta). The proteolytic cleavage of the transmembrane amyloid precursor protein (APP) generates the beta-amyloid peptides [21].

The deposition of amyloid plaques in the cortex of the brain is considered an influential trademark of AD [22]. This deposition was first explained by Aloïs Alzheimer. These plaques are termed the SP. These plaques are known to contain amyloid-β in an aggregated form. And, later this peptide forms the β-sheets which are rich in fibrils [23]. The combination of amyloid-beta proteins and some metal ions such as iron, copper, and zinc comprise the extracellular SP [24]. Intracellular

TABLE 16.2
Role of Free Radicals in Progression of Disease

S. no.	Free radicals	Symbol	Role in disease
1	Superoxide radical	$O_2^{\bullet}$	It is produced in mitochondria and causes damage to biomolecules.
2	Hydroxyl radical	$OH^{\bullet}$	It causes severe damage to the biomolecules of a cell.
3	Singlet oxygen	$^1O_2$	It is produced in vivo by the activation of neutrophils and eosinophils and causes damage to DNA.
4	Hydrogen peroxide	H_2O_2	It is produced in the peroxisome and it penetrates the cell membranes and causes damage to DNA.
5	Peroxyl radical	$ROO^{\bullet}$	It initiates fatty acid peroxidation and promotes tumor development.
6	Ozone	O_3	It causes lipid peroxidation and chromosomal aberrations.
7	Hypochlorous acid	HOCl	It is produced at the site of inflammation and mainly contributes to inflammation-induced ailments.

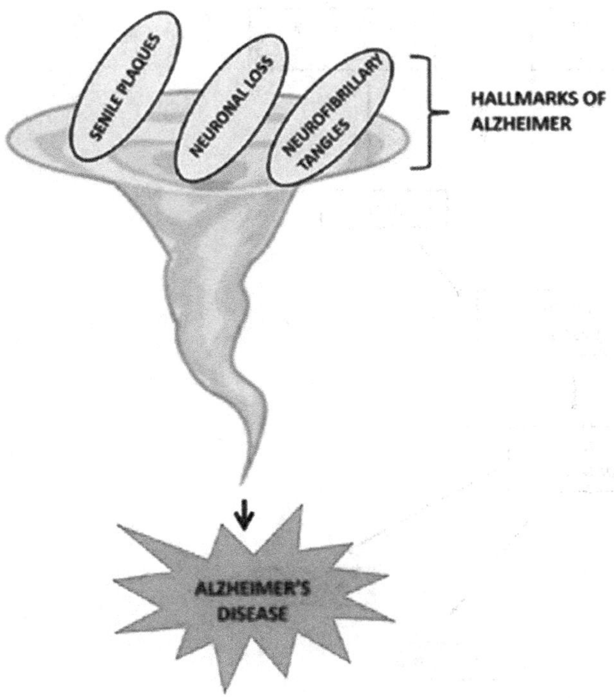

FIGURE 16.1 Notable hallmarks in Alzheimer's disease. Senile plaques, intracellular neurofibrillary tangles, and neuronal loss are the main hallmarks of Alzheimer's disease.

neurofibrillary tangles and SP are the trademarks of AD. The attachment of copper as redox-active metal with amyloid-β (Aβ) catalyzes the reaction and increases the generation of ROS, especially the hydroxyl radical [25]. This radical provides oxidative and cellular damage to amyloid-β peptides and lipids, etc. [26], as depicted in Figure 16.2.

The existence of intracellular neurofibrillary tangles in the brain is possibly a hallmark of AD [27]. The event of hyperphosphorylation of tau protein contributes to the pathogenesis of Alzheimer's [28]. The interaction of tubulin with proteins attached to microtubules results in the stabilization of microtubules [29]. It has been noted that the cascade of p38 MAPK gets activated due to the presence of amyloid-β. This activation results in the aberrant phosphorylation of tau proteins in AD [30]. This aberrant phosphorylation event causes the aggregation of neurons in neurofibrillary tangles. However, aggregation of neurons makes the microtubule unstable and causes the loss of neuron functioning [31], as depicted in Figure 16.3.

16.4 Role of Cholesterol in AD

Cholesterol is a sterol having a molecular formula of $C_{17}H_{62}O_{23}$ and, present in the cell membranes [32]. The abundance of cholesterols in microdomains of cell membranes is called lipid raft [33]. The binding of APP with these rafts commands the β-secretase to insert APP in a monolayer of phospholipids which accumulate the Aβ1–42 peptides through a mechanism called the amyloidogenic pathway [34]. The mechanism of propagation of Aβ is favored by esterified cholesterols. The elimination of excess cholesterols in the brain can be achieved by an oxidation reaction. Oxysterols are liberated as an end product of oxidation. The oxidation of cholesterols is helpful to prevent their accumulation in the brain. Cholesterol is the major component of the brain. The biochemical event of cholesterol 24-hydroxylase lays the formation of 24-hydroxycholesterol, which has the potential to cross the blood-brain barrier [35]. The brain is the major site for their formation. In astrocytes, the liver-X-receptor controlled pathway is responsible for mediating the efflux of apoE cholesterols. Together, oxysterol and efflux of apoE upregulate the cholesterol homeostasis in the brain which directly links with the progression of AD [36].

16.5 Molecular Link of OS with Abeta-Induced Toxicity

OS is considered a chief mediator for the progression of AD. Beta-secretase and gamma-secretase help in the formation of Abeta followed by the cleavage of APP [37]. Beta-secretase and gamma-secretase are two membrane proteases. Beta-secretase is also called the beta-site APP cleaving enzyme (BACE1). Gamma-secretase is a meshwork of multi-proteins that structurally contains presenilin (PS), nicastrin (NCT), and anterior pharynx defective-1 (APH-1). In addition to PS, NCT, and APH-1, gamma-secretase also consists of the presenilin enhancer protein-2 [38] (PEN-2). The 99 amino acids and APP C-terminal fragments are released after the cleavage of APP by BACE-1 at the N-terminal end [39]. Gamma-secretase recognizes this cleavage and signals the transmembrane domain for further cleavage. Thus, the cleavage by gamma-secretase secretes the Abeta peptides [40]. Large numbers of peptides having different lengths are formed via cleavage through these secretases such as Abeta40 and Abeta42 [41]. The 40-amino acid form of peptide is called Abeta40 and the 42-amino acid form of peptide is called Abeta42. Out of these freshly formed peptides, Abeta42 is the more toxic form of the peptide [42]. The rapid self-aggregation of peptides into soluble oligomers contributes to the toxicity of Abeta42. Alpha-secretase is another enzyme that is responsible for the cleavage of APP [43]. The declines in the clearance rate of Abeta peptides lead to their accumulation in the brain which ultimately provokes the pathway of cell-signaling [44]. In addition to cell-signaling, their accumulation also causes the degeneration in synaptic clefts and abnormalities in neurons which cause the neuronal death [45], as the mechanism is depicted in Figure 16.4.

16.6 Proteins Involved in AD

One of the major target proteins of OS in AD is tau protein [46]. For example, the sudden change in the conformations of tau protein is observed with 4-hydroxynonenal [47] (4-HNE). These change in protein conformations are preferably occurred due to amyloid β which ultimately lead to the formation of neurofibrillary tangles for the progression of AD. In addition to amyloid-β, the nitration reaction of tau protein also results in the formation of neurofibrillary tangles by inducing a conformational change [48]. The nitrated tau proteins are known

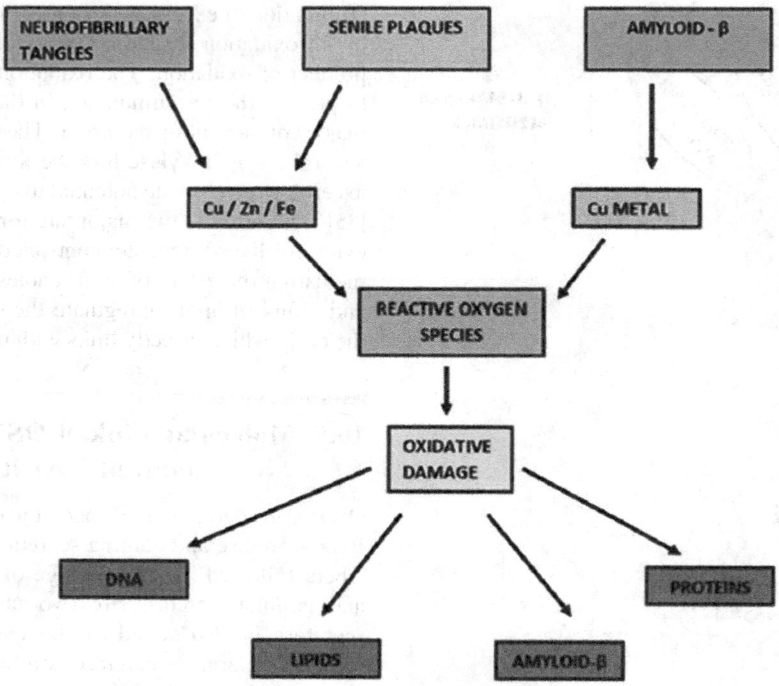

FIGURE 16.2 Reaction of senile plaques, amyloid-β, and neurofibrillary tangles with redox-active-metals in Alzheimer's disease. The attachment of neurofibrillary tangles, senile plaques, and amyloid-β with copper, zinc, and iron as redox-active metals supports the generation of reactive oxygen species. ROS further cause oxidative damage to macromolecules such as DNA, lipids, amyloid-β, and proteins, which supports the progression of Alzheimer's disease.

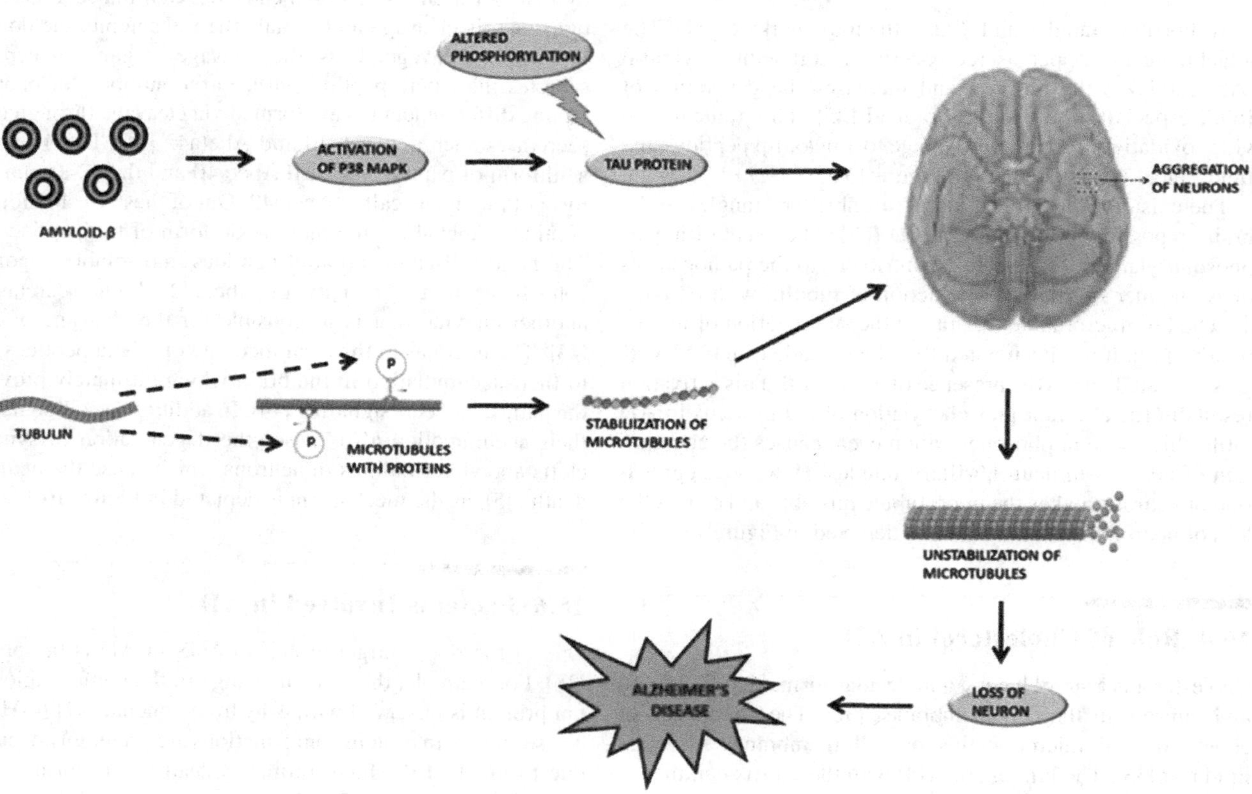

FIGURE 16.3 Aggregation of neurons in the brain in Alzheimer's disease. The amyloid-β activates the p38 MAPK which causes the altered phosphorylation in tau proteins. This alteration results in the accumulation of neurons in the cortex of the brain. Meanwhile, the interaction of tubulin with protein present on microtubules stabilizes its structure. However, the accumulation of neurons destabilizes the microtubular structure and hence increases the chances of the pathogenesis of Alzheimer's disease.

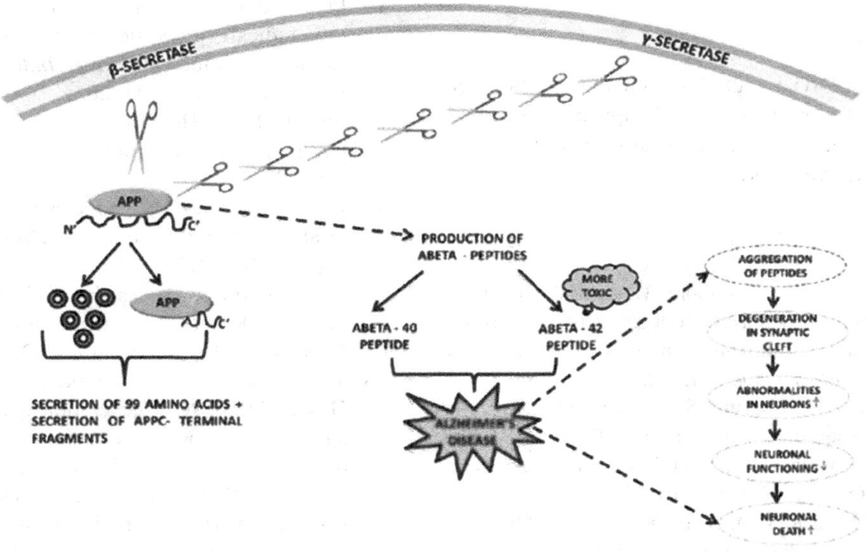

FIGURE 16.4 Molecular mechanism of peptides produced from β and γ-secretase in Alzheimer's disease. β and γ-secretase are two membrane proteases known to cleave the APP. Cleavage of APP produces the peptides responsible for the occurrence of Alzheimer's disease. 99-amino acids and APP with C-terminal fragments are produced after the cleavage of APP by β-secretase. Abeta peptides (Abeta 40 and 42) are the end products of cleavage of APP by γ-secretase. These toxic peptides cause Alzheimer's disease.

to associate with neurofibrillary tangles before the maturation event of tau proteins which ultimately bring AD to early stages [49].

The tau protein prevents the neuronal genomic DNA, cytoplasmic RNA, and nuclear RNA from the harmful consequences of damage induced by ROS [50]. The oxidation event in DNA and RNA was increased with the alteration in tau proteins. The undefined alterations in tau proteins lead to damage in DNA by increased oxidation of RNA and DNA [51]. The damage in DNA is repaired by a mechanism called the base excision repair system. Several repair proteins are known to repair the DNA. But oxidation of DNA and RNA inactivates those proteins which are responsible for the repair system [52]. It has been observed that the breakage in strands of DNA favors the generation of neurofibrillary tangles and causes neuronal disorders [53].

The OS might be considered as an influential factor for the clearance of amyloid-β. Several reports showed that amyloid-β helps in the oxidation of low-density lipoprotein receptor protein (LRP1) and results in the aggregation of amyloid-β, which is a neurotoxic peptide [54]. Moreover, LRP1 is a notable protein known to have multifunctionality. The amyloid-β is released out from the brain to blood capillaries by crossing the blood-brain barrier [55]. And, LRP1 is responsible for the release of amyloid-β from the brain. The functioning of LRP1 was declined in AD [56]. The oxidation of LRP1 via amyloid-β decreases the clearance rate of amyloid-β [57]. LRP1 oxidation has been evidenced by the presence of 4-HNE-LRP1 oxidation. As a result, the alternations in the clearance rate of amyloid-β favor the pathogenesis of AD by depositing the amyloid-β in the brain.

16.7 Lethal Consequences of AD

Loss of memory, imbalance of redox signaling, increased OS, and death of neurons are considered lethal consequences of AD. Moreover, AD is directly linked to OS which is associated with the pathogenesis of other ailments.

16.8 Conclusion

In summary, free radicals generated from the cell organelles are the largest mediator for the pathogenesis of oxidative stress in a number of diseases. However, various mechanisms such as the dysfunctioning of mitochondria and alteration in transition metals participate in the production of reactive oxygen species. The abnormal accumulation of Abeta and tau proteins in the brain supports the promotion of redox imbalance. However, the exact mechanism of redox imbalance still remains unclear in Alzheimer's. The therapeutic approach of removal of reactive oxygen species might be able to diminish the rate of the molecular event involved in the pathogenesis of Alzheimer's disease.

Credit Author's Statement

Jeena Gupta has conceptualized and designed the manuscript. Sourbh Suren Garg and Poojith Nuthalapati have done the literature review, manuscript writing, referencing, and diagrammatical work. Together, Sourbh Suren Garg, Debasis Sahu, Atulika Sharma, and Jeena Gupta have contributed to reviewing the manuscript.

Declaration of Competing Interest

We wish to confirm that there is no conflict of interest associated with this publication. Further, there has been no significant financial support for this work that could influence its outcome.

Acknowledgments

The authors thank the Division of Research and Development (DRD), Lovely Professional University, Phagwara, for providing us with space, time, and a healthy atmosphere to make possible this collaborative work.

REFERENCES

1. Uttara, B., Singh, A. V., Zamboni, P., & Mahajan, R. T. (2009). Oxidative stress and neurodegenerative diseases: a review of upstream and downstream antioxidant therapeutic options. *Current Neuropharmacology*, *7*(1), 65–74.
2. Garg, S. S., Gupta, J., Sharma, S., & Sahu, D. (2020). An insight into the therapeutic applications of coumarin compounds and their mechanisms of action. *European Journal of Pharmaceutical Sciences*, *152*, 105424.
3. Barber, S. C., Mead, R. J., & Shaw, P. J. (2006). Oxidative stress in ALS: a mechanism of neurodegeneration and a therapeutic target. *Biochimica et Biophysica Acta (BBA)-Molecular Basis of Disease*, *1762*(11–12), 1051–1067.
4. Cenini, G., Lloret, A., & Cascella, R. (2020). Oxidative stress and mitochondrial damage in neurodegenerative diseases: from molecular mechanisms to targeted therapies. *Oxidative Medicine and Cellular Longevity*, 2020, Article ID 1270256.
5. Cheignon, C., Tomas, M., Bonnefont-Rousselot, D., Faller, P., Hureau, C., & Collin, F. (2018). Oxidative stress and the amyloid beta peptide in Alzheimer's disease. *Redox Biology*, *14*, 450–464.
6. Fukui, H., Diaz, F., Garcia, S., & Moraes, C. T. (2007). Cytochrome c oxidase deficiency in neurons decreases both oxidative stress and amyloid formation in a mouse model of Alzheimer's disease. *Proceedings of the National Academy of Sciences*, *104*(35), 14163–14168.
7. Collin, F. (2019). Chemical basis of reactive oxygen species reactivity and involvement in neurodegenerative diseases. *International Journal of Molecular Sciences*, *20*(10), 2407.
8. Peterson, L. J., & Flood, P. M. (2012). Oxidative stress and microglial cells in Parkinson's disease. *Mediators of Inflammation*, *2012*, Article ID 401264.
9. Qiu, C., Kivipelto, M., & von Strauss, E. (2009). Epidemiology of Alzheimer's disease: occurrence, determinants, and strategies toward intervention. *Dialogues in Clinical Neuroscience*, *11*(2), 111.
10. Chokkareddy, R., Thondavada, N., Kabane, B., & Redhi, G. G. (2020). Nanotechnology-based devices in the treatment for Alzheimer's disease. In *Nanomaterials in Diagnostic Tools and Devices* (pp. 241–256). Elsevier.
11. Ito, F., Sono, Y., & Ito, T. (2019). Measurement and clinical significance of lipid peroxidation as a biomarker of oxidative stress: oxidative stress in diabetes, atherosclerosis, and chronic inflammation. *Antioxidants*, *8*(3), 72.
12. Vida, C., Martinez de Toda, I., Garrido, A., Carro, E., Molina, J. A., & De la Fuente, M. (2018). Impairment of several immune functions and redox state in blood cells of Alzheimer's disease patients. Relevant role of neutrophils in oxidative stress. *Frontiers in Immunology*, *8*, 1974.
13. Lee, K. H., Cha, M., & Lee, B. H. (2020). Neuroprotective effect of antioxidants in the brain. *International Journal of Molecular Sciences*, *21*(19), 7152.
14. Phaniendra, A., Jestadi, D. B., & Periyasamy, L. (2015). Free radicals: properties, sources, targets, and their implication in various diseases. *Indian Journal of Clinical Biochemistry*, *30*(1), 11–26.
15. Finkel, T., & Holbrook, N. J. (2000). Oxidants, oxidative stress and the biology of ageing. *Nature*, *408*(6809), 239–247.
16. Brand, M. D. (2010). The sites and topology of mitochondrial superoxide production. *Experimental Gerontology*, *45*(7–8), 466–472.
17. Cadenas, E., & Davies, K. J. (2000). Mitochondrial free radical generation, oxidative stress, and aging. *Free Radical Biology and Medicine*, *29*(3–4), 222–230.
18. Thondavada, N., Chokkareddy, R., Naidu, N. V., & Redhi, G. G. (2020). New generation quantum dots as contrast agent in imaging. In *Nanomaterials in Diagnostic Tools and Devices* (pp. 417–437). Elsevier.
19. Elsner, M., Gehrmann, W., & Lenzen, S. (2011). Peroxisome-generated hydrogen peroxide as important mediator of lipotoxicity in insulin-producing cells. *Diabetes*, *60*(1), 200–208.
20. Snezhkina, A. V., Kudryavtseva, A. V., Kardymon, O. L., Savvateeva, M. V., Melnikova, N. V., Krasnov, G. S., & Dmitriev, A. A. (2019). ROS generation and antioxidant defense systems in normal and malignant cells. *Oxidative Medicine and Cellular Longevity*, *2019*, Article ID 6175804.
21. O'brien, R. J., & Wong, P. C. (2011). Amyloid precursor protein processing and Alzheimer's disease. *Annual Review of Neuroscience*, *34*, 185–204.
22. Murphy, M. P., & LeVine III, H. (2010). Alzheimer's disease and the amyloid-β peptide. *Journal of Alzheimer's Disease*, *19*(1), 311–323.
23. Chen, G. F., Xu, T. H., Yan, Y., Zhou, Y. R., Jiang, Y., Melcher, K., & Xu, H. E. (2017). Amyloid beta: structure, biology and structure-based therapeutic development. *Acta Pharmacologica Sinica*, *38*(9), 1205–1235.
24. Maynard, C. J., Bush, A. I., Masters, C. L., Cappai, R., & Li, Q. X. (2005). Metals and amyloid-β in Alzheimer's disease. *International Journal of Experimental Pathology*, *86*(3), 147–159.
25. Smith, D. G., Cappai, R., & Barnham, K. J. (2007). The redox chemistry of the Alzheimer's disease amyloid β peptide. *Biochimica et Biophysica Acta (BBA)-Biomembranes*, *1768*(8), 1976–1990.
26. Borra, S. K., Mahendra, J., & Gurumurthy, P. (2014). Effect of curcumin against oxidation of biomolecules by hydroxyl radicals. *Journal of Clinical and Diagnostic Research: JCDR*, *8*(10), CC01.
27. Armstrong, R. A. (2009). The molecular biology of senile plaques and neurofibrillary tangles in Alzheimer's disease. *Folia Neuropathologica*, *47*(4), 289–99.
28. Naseri, N. N., Wang, H., Guo, J., Sharma, M., & Luo, W. (2019). The complexity of tau in Alzheimer's disease. *Neuroscience Letters*, *705*, 183–194.
29. Kadavath, H., Hofele, R. V., Biernat, J., Kumar, S., Tepper, K., Urlaub, H., … & Zweckstetter, M. (2015). Tau stabilizes microtubules by binding at the interface between tubulin heterodimers. *Proceedings of the National Academy of Sciences*, *112*(24), 7501–7506.
30. Corrêa, S. A., & Eales, K. L. (2012). The role of p38 MAPK and its substrates in neuronal plasticity and neurodegenerative disease. *Journal of Signal Transduction*, *2012*, Article ID 649079.

31. Hahn, I., Voelzmann, A., Liew, Y. T., Costa-Gomes, B., & Prokop, A. (2019). The model of local axon homeostasis-explaining the role and regulation of microtubule bundles in axon maintenance and pathology. *Neural Development, 14*(1), 1–28.
32. Delhom, R., Nelson, A., Laux, V., Haertlein, M., Knecht, W., Fragneto, G., & Wacklin-Knecht, H. P. (2020). The antifungal mechanism of amphotericin B elucidated in ergosterol and cholesterol-containing membranes using neutron reflectometry. *Nanomaterials, 10*(12), 2439.
33. Magee, A. I., & Parmryd, I. (2003). Detergent-resistant membranes and the protein composition of lipid rafts. *Genome Biology, 4*(11), 1–4.
34. Chow, V. W., Mattson, M. P., Wong, P. C., & Gleichmann, M. (2010). An overview of APP processing enzymes and products. *Neuromolecular Medicine, 12*(1), 1–12.
35. Russell, D. W., Halford, R. W., Ramirez, D. M., Shah, R., & Kotti, T. (2009). Cholesterol 24-hydroxylase: an enzyme of cholesterol turnover in the brain. *Annual Review of Biochemistry, 78*, 1017–1040.
36. Staurenghi, E., Cerrato, V., Gamba, P., Testa, G., Giannelli, S., Leoni, V., ... & Leonarduzzi, G. (2021). Oxysterols present in Alzheimer's disease brain induce synaptotoxicity by activating astrocytes: a major role for lipocalin-2. *Redox Biology, 39*, 101837.
37. Zhang, X., & Song, W. (2013). The role of APP and BACE1 trafficking in APP processing and amyloid-β generation. *Alzheimer's Research & Therapy, 5*(5), 1–8.
38. Steiner, H. (2004). Uncovering γ-secretase. *Current Alzheimer Research, 1*(3), 175–181.
39. Nhan, H. S., Chiang, K., & Koo, E. H. (2015). The multifaceted nature of amyloid precursor protein and its proteolytic fragments: friends and foes. *Acta Neuropathologica, 129*(1), 1–19.
40. Xu, X. (2009). γ-Secretase catalyzes sequential cleavages of the AβPP transmembrane domain. *Journal of Alzheimer's Disease, 16*(2), 211–224.
41. Siegel, G., Gerber, H., Koch, P., Bruestle, O., Fraering, P. C., & Rajendran, L. (2017). The Alzheimer's disease γ-secretase generates higher 42: 40 ratios for β-amyloid than for p3 peptides. *Cell Reports, 19*(10), 1967–1976.
42. Yan, Y., & Wang, C. (2006). Aβ42 is more rigid than Aβ40 at the C terminus: implications for Aβ aggregation and toxicity. *Journal of Molecular Biology, 364*(5), 853–862.
43. Postina, R. (2008). A closer look at α-secretase. *Current Alzheimer Research, 5*(2), 179–186.
44. Prasansuklab, A., & Tencomnao, T. (2013). Amyloidosis in Alzheimer's disease: the toxicity of amyloid beta (Aβ), mechanisms of its accumulation and implications of medicinal plants for therapy. *Evidence-Based Complementary and Alternative Medicine, 2013*,. 413808.
45. Carter, J., & Lippa, C. F. (2001). β-Amyloid, neuronal death and Alzheimer's disease. *Current Molecular Medicine, 1*(6), 733–737.
46. Teixeira, J. P., de Castro, A. A., Soares, F. V., da Cunha, E. F., & Ramalho, T. C. (2019). Future therapeutic perspectives into the Alzheimer's disease targeting the oxidative stress hypothesis. *Molecules, 24*(23), 4410.
47. Castro, J. P., Jung, T., Grune, T., & Siems, W. (2017). 4-Hydroxynonenal (HNE) modified proteins in metabolic diseases. *Free Radical Biology and Medicine, 111*, 309–315.
48. Horiguchi, T., Uryu, K., Giasson, B. I., Ischiropoulos, H., LightFoot, R., Bellmann, C., ... & Trojanowski, J. Q. (2003). Nitration of tau protein is linked to neurodegeneration in tauopathies. *American Journal of Pathology, 163*(3), 1021–1031.
49. Zhang, Y. J., Gendron, T. F., Xu, Y. F., Ko, L. W., Yen, S. H., & Petrucelli, L. (2010). Phosphorylation regulates proteasomal-mediated degradation and solubility of TAR DNA binding protein-43 C-terminal fragments. *Molecular Neurodegeneration, 5*(1), 1–13.
50. Violet, M., Delattre, L., Tardivel, M., Sultan, A., Chauderlier, A., Caillierez, R., ... & Galas, M. C. (2014). A major role for Tau in neuronal DNA and RNA protection in vivo under physiological and hyperthermic conditions. *Frontiers in Cellular Neuroscience, 8*, 84.
51. Koren, S. A., Galvis-Escobar, S., & Abisambra, J. F. (2020). Tau-mediated dysregulation of RNA: evidence for a common molecular mechanism of toxicity in frontotemporal dementia and other tauopathies. *Neurobiology of Disease, 141*, 104939.
52. Kwiatkowski, D., Czarny, P., Toma, M., Jurkowska, N., Sliwinska, A., Drzewoski, J., ... & Sliwinski, T. (2016). Associations between DNA damage, DNA base excision repair gene variability and Alzheimer's disease risk. *Dementia and Geriatric Cognitive Disorders, 41*(3–4), 152–171.
53. Lin, X., Kapoor, A., Gu, Y., Chow, M. J., Peng, J., Zhao, K., & Tang, D. (2020). Contributions of DNA damage to Alzheimer's disease. *International Journal of Molecular Sciences, 21*(5), 1666.
54. Kanekiyo, T., & Bu, G. (2014). The low-density lipoprotein receptor-related protein 1 and amyloid-β clearance in Alzheimer's disease. *Frontiers in Aging Neuroscience, 6*, 93.
55. Zenaro, E., Piacentino, G., & Constantin, G. (2017). The blood-brain barrier in Alzheimer's disease. *Neurobiology of Disease, 107*, 41–56.
56. Van Gool, B., Storck, S. E., Reekmans, S. M., Lechat, B., Gordts, P. L., Pradier, L., ... & Roebroek, A. J. (2019). LRP1 has a predominant role in production over clearance of Aβ in a mouse model of Alzheimer's disease. *Molecular Neurobiology, 56*(10), 7234–7245.
57. Ramanathan, A., Nelson, A. R., Sagare, A. P., & Zlokovic, B. V. (2015). Impaired vascular-mediated clearance of brain amyloid beta in Alzheimer's disease: the role, regulation and restoration of LRP1. *Frontiers in Aging Neuroscience, 7*, 136.

17

Applications of Nanotechnology and Nanodevices for the Early-Stage Detection of Cancer Cells

Shaik Baji Baba, Moses Kigozi, Naresh Kumar Katari and Vishnu Nandimalla

CONTENTS

17.1 Introduction

Cancer treatment and diagnosis are very interesting because of the disease's widespread rate, mortality rate, and recurrence after treatment [1]. According to the World Health Organization, cancer is a leading cause of death worldwide, accounting for nearly 10 million deaths in 2020, as shown in Table 17.1. The WHO reports from 2020 to 2021, 19.3 million new cancer reports have been notified and 10 million deaths have been reported [2]. Out of the total, approximately 70% of deaths from cancer occur in low- and middle-income countries. American Chemical Society (ACS) researchers estimated that 1.9 million new cancer cases would be reported in the United States in 2021, and more than 600,000 patients would be dead because of cancer [3]. According to the National Cancer Institute reports, different types of cancers have been reported in the past few years, and among all breast cancer, prostate cancer, and lung cancer are causing the highest death rates. Cancer is a disease characterized by uncontrolled cell proliferation growth that spreads from an early focal point to other body parts, eventually causing death [4].

Several therapies have been introduced to detect early-stage cancer cells, but developing the most advanced techniques to detect cancer is necessary. For this reason, it is crucial to identify earlier detection and diagnosis of cancer to decrease disease spreading and mortality rate. Currently, nanotechnology-based diagnostic methods are developed as a promising tool for real-time and cost-effective detection and diagnosis of cancer [5]. Nanotechnology and nanoparticles have led to numerous promising results with their many applications in cancer detection and diagnoses, such as gene therapy, drug delivery, targeted therapy, molecular imaging, biomarker mapping, and drug carriage, to mention but a few [6]. Also, nanotechnology has been applied to develop various nanomaterials such as silver, gold, platinum, etc., as well as nanoparticles and quantum dots, which are used for bio-imaging and early-stage cancer detection and diagnosis.

Molecular diagnostic-based nanomaterials, such as developing and introducing nano biomarkers, can easily and quickly detect cancer cell lines [6]. Nano therapies, such as creating a nanoparticle-based drug delivery system, can precisely target the cancer tissues with fewer side effects [7, 8]. Nanomaterials can easily cross the cell barrier because of their physicochemical properties and biological nature [9]. Nanomaterials have been extensively used in tumor treatment for several years because of their passive and active targeting profile.

Many generic drugs are used for cancer treatment, but these drugs' sensitivity leads to various side effects and damaging the healthy cells. Many studies reveal that different forms of nanomaterials, such as polymers, antibodies, liposomes, and molecules with a combination of various nanomaterials or nanoparticles in the development of cancer drugs can balance drug efficiency and decrease the cytotoxicity of drugs [10].

DOI: 10.1201/9781003157823-17

TABLE 17.1

The Most Common in Terms of New Cases of Cancer and Most Common Causes of Cancer Death in 2020

S. no.	Type of cancer	Number of cases	Number of deaths
1	Breast	2.26 million	685,000
2	Lung	2.21 million	1.80 million
3	Colon and rectum	1.93 million	935,000
4	Prostate	1.41 million	375,304
5	Skin (non-melanoma)	1.20 million	~100,000
6	Stomach	1.09 million	769,000

Source: Ref. [2].

Several cancer-screening methods are available in the market, such as (i) the Papanicolaou test for cervical screening, which is used to detect the potentially precancerous and cancerous processes in the cervix or colon; (ii) prostate-specific antigen level detection in men's blood samples to detect and identify prostate cancer; (iii) for colon cancer, occult blood detection; (iv) endoscopy, (v) X-ray, ultrasound imaging; (vi) CT scanning; and (vii) MRI imaging to detect various cancers [1]. These traditional detection and diagnostic methods are not satisfying at very early-stage cancer detection, and also some of these screening methods are very costly and burdensome to many people. Therefore, there is an urgent need to develop nanotechnology-based selective and sensitive sources for early-stage cancer detection and diagnosis.

Nanotechnology provides a great innovative tool that sheds more excellent light on the average cell life cycle, molecular processing point, and changes occurring within cells that correlate with cancer development [11]. It needs to learn and obtain a large amount of information from small nanoparticles. With the rapid growth in nanotechnology, this chapter highlights the application of nanoparticles and nanodevices for early-stage detection and diagnosis of cancer cells. We also provide our perspective on challenges in the use of nanotechnology for cancer diagnosis.

Nanodevices can quickly detect cancer cells and identify cancer growth, and they provide target base delivery of anticancer and contrast agents to tumor cells. Nowadays, nanotechnology plays a vital role as a potential tool in cancer diagnosis and it also helps to detect molecular changes. Nanotechnology shows a high impact on challenges in cancer detection and therapy and also has significant advantages over conventional treatments. Tremendous research is going on the design and development of nanodevices capable of detecting the early-stage cancer cells, identifying their location within the human body, and delivering chemotherapeutic agents against cancerous cells. Different nanodevices are used in cancer detection and diagnosis, drug delivery and gene delivery, DNA detection, etc. (Table 17.2) [12].

The main area in which nanodevices and nanomedicine are being developed in cancer includes:

i) The early-stage detection of tumors (analysis of cancer-related makers and developing target contrast agents to improve the tumor area resolution compared to other normal tissues).
ii) Cancer therapy (developed nanodevices that can quickly release chemotherapeutic agents).

17.2 The Aim and Objective of This Chapter

The world of material science, nano-engineering, and nanotechnology is highly contrasted with a high scientific approach, adroitness, and keenness [13]. Science and technology only have an answer to overcome the various main issues in the world such as health, pollution, education, etc. The applications of nanodevices in cancer detection and diagnosis research are focused on in this chapter. Nanomedicine, a term encompassing the pharmaceutical and biomedicines of nanometer particles, involves connecting the various unique physicochemical and biological applications in an essentially innovative way [14].

17.3 Application Areas of Nanodevices

Nanodevices are the small innovative materials of human scientific growth. Nanodevices and their applications are viewing a new beginning in human scientific developments. Science, engineering, and technology are only showing significant impact globally and growing very fast [14, 15]. The achievement of nanodevices application on human health depends today on the efforts of scientists and engineers. Nanodevices include nanotubes, dendrimers, nanopores, quantum dots and nanoshells (Figure 17.1) [17]. To date, many potential nanodevices have been developed to improve cancer detection, treatment, and diagnosis [18]. Nanoparticles are small in size, but our human body can easily clear them too quickly, but they are very effective at imaging and detecting. Large particles may massively accumulate in active organs and exhibit toxicity. However, nanoscale devices can promptly enter into cells and organs, interacting with proteins and DNA. Medical devices and tools are developed via nanotechnology to detect various diseases in small tissues and cells. Currently, cancer detection and diagnosis commonly depend on changes in cells and tissues [19]. Early-stage cancer detection needs to be the future of science and development. Among all the nanodevices, carbon nanotubes (CNTs) are commonly used for cancer detection, and they mainly depend on the changes occurring in DNA structure. Nanoshells (NS) and nanopores (NPs) depend on genetic code and its sequences for cancer detection. Quantum dots (QDs) mainly rely on light emission, if any structural changes. Dendrimers transmit the molecules and identify the cells, and they kill the cell [20].

In the academic field, a number of research papers have been published on the application of nanotechnology and nanomaterials in healthcare and pharmaceutical science. A PubMed search on nanotechnology found 115,886 articles from 2000 to 2021 (Figure 17.2) related to different nanotechnology applications [21]. For the applications of nanotechnology, especially in medicine, pharmacy, and healthcare, the search database shows a total of 18,594, 8,234 and 1,217 articles from 2000 to 2021. These results indicate that an increasing number of articles are published every year. According to BCC research

TABLE 17.2

Types of Nanodevices Used in Clinical Applications

S. no.	Nanodevice	Monographs	Clinical applications
1	Nanoparticles	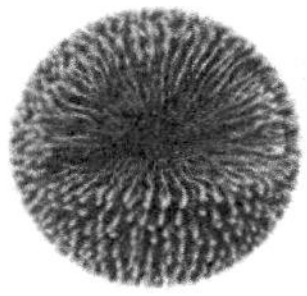	Target drug delivery MRI, USG image contrast agents, and reporters of apoptosis
2	Nanocrystals	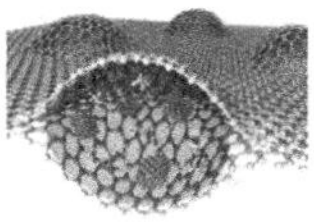	Improved formulation for poorly soluble drugs
3	Quantum dots	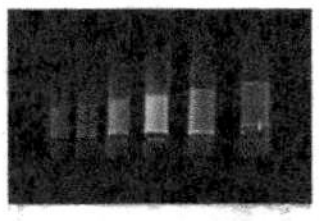	Optical detection of genes and proteins in the animal model, cell assays visualization of tumor and lymph node in the human
4	Nanowires	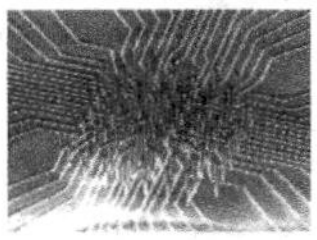	High thoughtful screening, disease protein biomarker detection, and SNPs gene expression analysis
5	Nanoshell	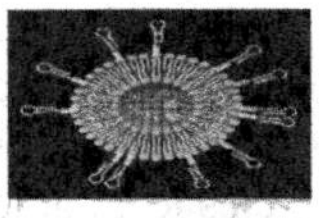	Tumor-specific imaging
6	Carbon nanotubes	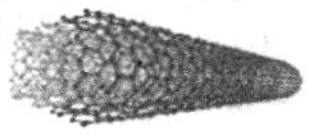	SNPs protein biomarker detection
7	Dendrimers	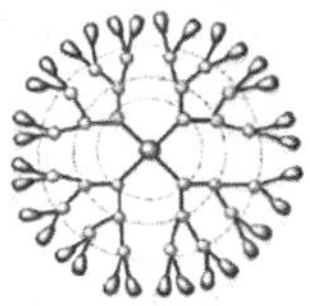	Image contrast agents

MRI: magnetic resonance imaging, SNP: single nucleotide polymorphisms.

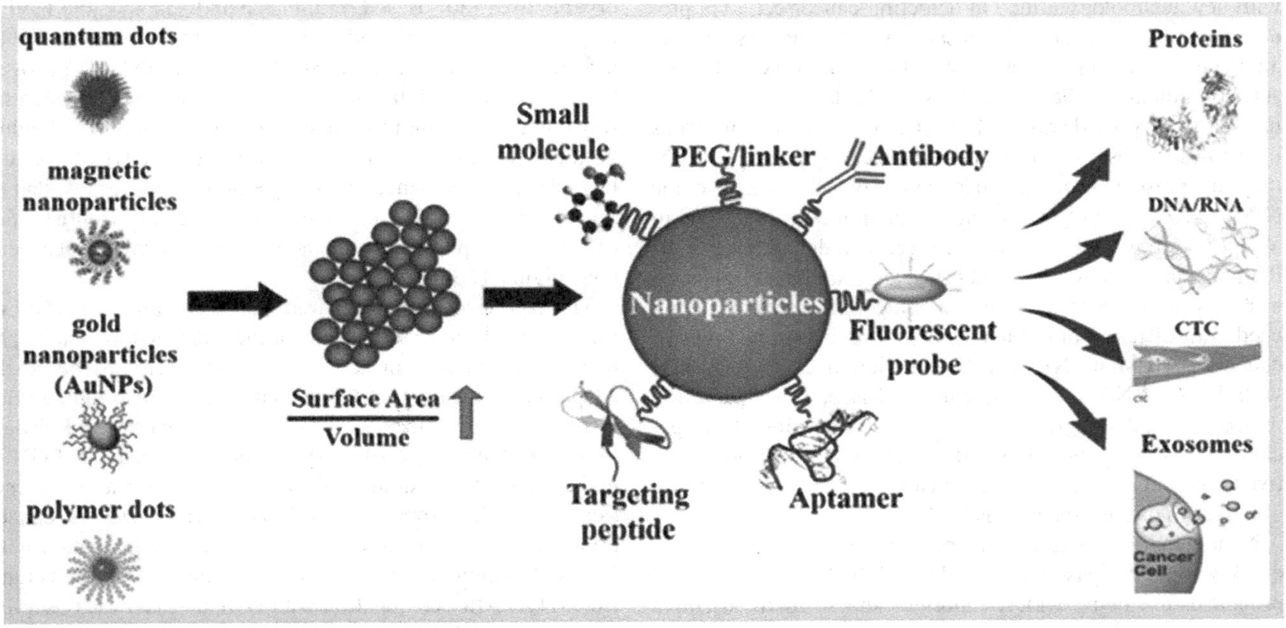

FIGURE 17.1 Nanotechnology improves cancer detection and diagnosis [16].

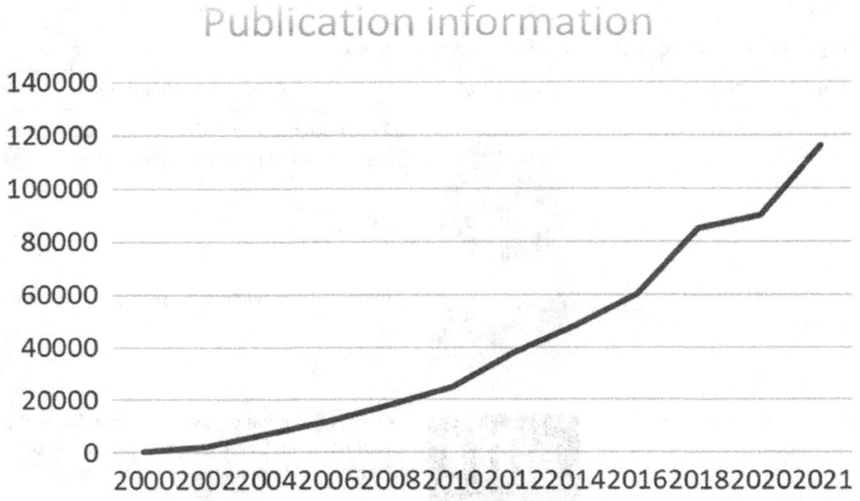

FIGURE 17.2 Trends in nanotechnology in medical and pharmaceutical fields. From Information obtained from PubMed, accessed 11.01.19.

reports, the global market range for nanodevices and nanomachines should increase from US$ 736.1 million in 2018 to US$ 1.3 billion in 2023 and also US$ 2.7 billion in 2028, at a compound annual growth rate (CAGR) of 11.6% from 2018 to 2023 and 16.0% from 2023 to 2028. Also, the worldwide market for nanosensors should grow from US$ 560.0 million in 2018 to US$ 976.5 million in 2023 and then to US$ 1.6 billion in 2028 at a CAGR of 11.8% from 2018 to 2023 and 10.7% from 2023 to 2028 [22].

17.3.1 Role of Nanotechnology and Nanodevices in Cancer Detection

Cancer cell detection and diagnosis have importance using a nanotechnology-based approach. It makes a specific method with new technology aimed at detecting cancer cells via prediction, prevention, and diagnosis – nanotechnology playing an essential role in precisely targeted drug delivery and early-stage treatment and diagnosis of disease [23]. Nanodevices and nanoparticles provide fast and sensitive detection of cancerous molecules to observe and identify molecular changes even if they are present in a small percentage of cells. Metal oxide NPs (e.g., ZnO, AgO, etc.) coated with antibodies that bind to receptors are used to produce stronger signals in cancer cells than normal cells by MRI and CT scans. Nanomaterials such as USPIO and super magnetic iron oxide (SPIO) are commonly used with different brand names to examine various types of cancers [24]. Many NPs can be functionalized concurrently with DNA, RNA, peptides, carbohydrates, other targeting molecules, and imaging agents. NPs can specifically target cancer cells and cancer biomarkers, allowing for more sensitive treatment and monitoring the cancer therapy progress and obliteration of the cancer cells [25].

Nanotechnology-based therapies have been developing quickly over the past few decades. With this, nanomaterial properties are being widely reported, and so many attempts are made to fabricate the suitable nanomaterial. Nanomaterials have been widely used for sensitive and specific biomarker detection because of their unique optical, mechanical, physicochemical, and magnetic properties. Nanomaterials applicable to sensing cancer biomarkers vary from inorganic nanoparticles (e.g., gold), magnetic nanodevices, quantum dots, nanowires, nanoshells, and carbon nanotubes [26, 27].

17.3.2 Gold Nanoparticles

Gold nanoparticles (GNPs) have gained their attention in bioimaging and cancer cell detections because of their unique optical properties [28, 29]. GNPs with firm surface plasmon-enhanced scattering and absorption allowed them to develop GNPs as more powerful contrasting agents and imaging labels. GNPs have good dispersion and absorption bands than organic dyes [30]. Based on the size and shape of the GNPs, they can easily absorb and scatter the light from the visible to near-infrared (NIR) region [31]. Moreover, GNPs are proven to have biocompatibility, less cytotoxicity, and photobleaching resistance according to human cell experiments [32]. Among various inorganic and organic nanoparticles, GNPs have surface plasmon resonance and unique optical properties; due to this nature, it has become the first choice for scientists and researchers, especially in the medical and pharmaceutical sectors (Figure 17.3).

The core size of the prepared GNPs contains a 1–150 nm range, making it easier to control their dispersion. The negative charge on the surface of the GNPs helps to be easily adaptable. It means that GNPs can easily functionalize with different biomolecules such as ligands, genes, and drugs. Moreover, the biocompatibility and non-toxic nature of GNPs make them an essential applicant for their use as drug carriers [34]. Methotrexate (MTX) is a drug used for cancer treatment for the past few decades, but upon conjugation with GNPs, it exhibited more cytotoxicity against tumor cells than free MTX. After conjugation with GNPs, MTX was observed to accumulate fast in the tumor cells and at high levels [35].

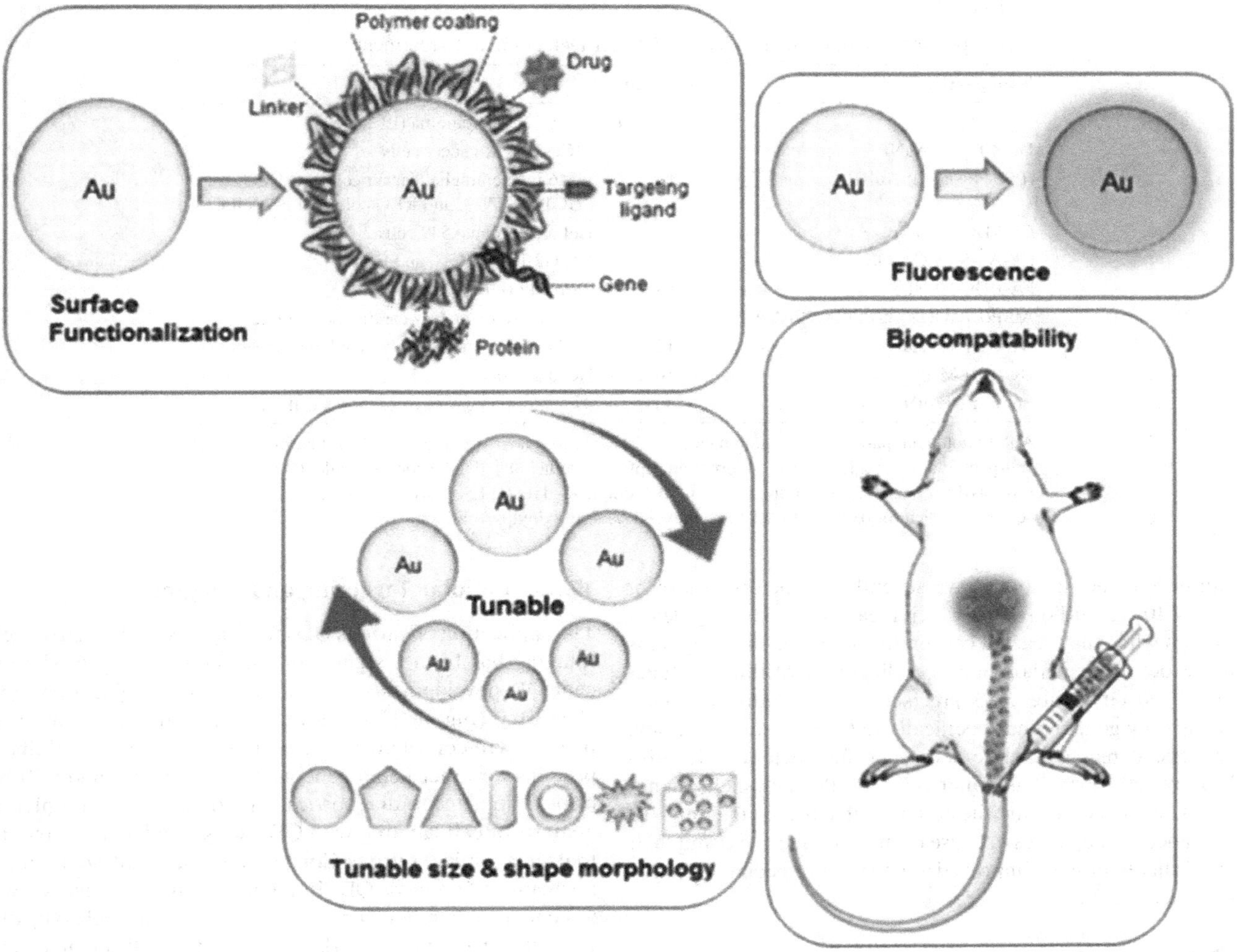

FIGURE 17.3 Important properties of gold nanoparticles [33].

Another example, doxorubicin (DOX), when it bounds to GNPs through an acid linker, exhibited increased toxicity against the multidrug-resistant MCF-7/ADR breast cancer cell line [36]. Many studies are reported on the application of GNPs on cancer detection, diagnosis and drug delivery (Table 17.3).

17.3.3 Gold Nanoparticles in Photo-Thermal Therapy and Photo-Imaging

GNP-based selective and sensitive photo-thermal and photo-imaging applications are still under development. Photo-thermal therapy is described as photon-medicated induction of the localized therapeutic temperature that stimulates hyperthermic physiological responses. Metal nanoparticles such as GNPs are used in this therapy, and they can exhibit surface plasmon resonance. It converts light into heat efficiently [37]. The morphology of GNPs used in photo-thermal therapy generally contains rod or shell shape. When it is released into a biological environment like other materials, the cellular uptake can be limited. Near-IR light is preferable for photo-thermal therapy due to the little absorption by tissue at 650–900 nm wavelength [38], where this range is sufficient to induce cytotoxicity damage [39]. Other studies have shown photo-thermal therapy where GNPs were also used at maximum absorption of 795 nm and branched GNPs surface modification with nanobodies that can efficiently kill the cancerous cells without harming healthy cells [40].

The Food and Drug Administration (FDA) approved many of the most successful GNPs, and PEGylated gold nanoparticles are one of the most challenging and ongoing human pilot studies. They have shown enhanced absorption and accumulation at the targeted tumor site in the near-IR region [41]. Reports state that more than 60% of the mice are treated with a single dose of PEGylated gold NPs administration.

Photo-imaging is an updated and advanced method to help detect early-stage cancer cells and guide doctors for care treatment. One of the doctors' significant problems and challenges is identifying the tumor ending and health tissue clear pictures. At the time of operations, physicians can't remember and decide at what level the tumor has been removed and its effect on healthy tissues. Computed tomography (CT) and magnetic resonance imaging (MRI) scans are limited. These

TABLE 17.3

Developments of Gold Nanoparticles in Cancer Detection and Treatment

Nanoparticle	Size (nm)	Cell lines
MTX-AuNP	8–80	Lewis lung carcinoma (LL2) cells
DOX-Hyd@AuNP	30	MCF-7/ADR cancer cells
(Pt(R, R-dach))-AuNP	26.7	A546 lung epithelial cancer cell line, HCT116, HCT15, HT29, and RKO colon cancer cell lines
CPP-DOX-AuNP	25	HeLa cells and A549 cells
FA-Au-SMCC-DOX		HepG2-R, C0045C, and HDF
FA-BHC-AuNP	20–60	Vero and HeLa
Au-P(LA-DOX)-b-PEG-OH/FA NP	34	4T1 mouse mammary carcinoma cell line
DOX@PVP-AuNP	12	A549, H460, and H520 human lung cancer cells
DOX-BLM-PEG-AuNP	10	HeLa cell lines
EpCam-RPAuN	48	4T1 mouse mammary carcinoma cell line

AuNP: gold nanoparticle, AuN: gold nanocage, BHC: berberine hydrochloride, BLM: bleomycin, CPP: cell-penetrating peptides, DOX: doxorubicin, EpCam: epithelial cell adhesion molecule, FA: folic acid, Hyd: hydrazone, MTX: methotrexate, PEG: polyethylene glycol, PLA: poly-L-asparate, Pt(R, R-dach): active ingredient of oxaliplatin, PTX: pacitaxel, PVP: polyvinylpyrrolidone.

can only detect tumors above several millimeters in size or nearly 10–12 million cells, which means that they only detect tumors when they reach a certain threshold. Photo-imaging is a new detection method in cancer diagnosis. Millions of functionalized GNPs are spots precisely injected into the tumor, where they go and bind specifically to the cancerous cells and scatters. It makes it more accessible for doctors to identify healthy cells as well as tumor cells. GNPs (nanoshells, nanocages, and nanorods) are the best available photo-imaging NPs for cancer therapeutics because of their bio-inertness and their capability to provide improved temporal and special resolution in imaging [42].

17.3.4 Quantum Dots

Quantum dots (QDs) are one of the most emerging fields in nanotechnology. Scientists and researchers have widely reported them because of their advantages and potentiality in bio-imaging, especially in cancer cell imaging [43]. QDs-based nanodevices are very helpful in building a biomedical imaging platform for cancer behaviors. QDs are semiconducting, light-emitting nanocrystals that have developed as powerful biomolecular imaging agents. The main advantages of QDs include size, shape, better signal brightness, tunable light emission, photobleaching resistance, etc. [44]. Also, QDs with different colors can be excited simultaneously by a single light source with slight spectral overlapping, and it can provide more advantages for multiplexed detection of targeted cells [45, 46]. The potential QDs application in cancer imaging and cancer detection are summarized in Table 17.4.

17.3.5 Quantum Dot Applications in Cancer Imaging and Cancer Detection

Recent developments of QDs in the biology and therapeutic area have already gained much scope for future researchers. Current achievements, challenges, and future advances of QDs in cancer cell imaging and diagnosis developments are summarized in Table 17.5.

17.3.6 Cellular Targeting and Imaging

The application of various QDs for selective and sensitive cell imaging has been a significant advantage owing to significant QD developments: synthesis, surface conjugation, and chemistry (Figure 17.4). However, these developments and benefits will contribute to detecting various cancer cell lines, proteins, or other heterogeneous tumor samples and will be critical in cancer diagnostics and treatment. Multiplexed analysis of cell labeling with QDs has gained many potential tools in a clinical setting. Compared to the traditional immunochemistry process, QD, immunostaining has been shown to be more accurate and precise at low protein levels [48, 49]. In 1998, Chan and co-workers reported the QDs conjugated to transferrin (a membrane translocating protein) should cause endocytosis of QDs via living cancerous cell lines in medium, and fluorescence of QDs should retain within the cell line and could be utilized as intracellular labeling for living cells [50].

17.3.7 In Vivo Targeting and Imaging

The long-term physicochemical, photostability, and optical properties of QDs make them an ideal candidate for in vivo targeting and imaging [51]. A phospholipid micelle-coated QD was first reported and used for in vivo cell imaging. Since then, many studies have been reported and gained a lot of advantages from their unique optical and fluorescent properties for imaging and diagnosis purposes [52]. A photostability property is one of the main advantages of QDs for in vivo imaging applications. It allows images to record over a long time, existing with fluorescent materials such as dyes or proteins because of resistance in photobleaching. Nie et al. developed QD-based nanoparticle probes for cancer cell imaging and targeting in live animals. The first encapsulated QDs with ABC triblock copolymer and linked them with amphiphilic polymer to cancer tumor-targeting ligands. Next, they applied this luminescent probe with active and passive tumor imaging to achieve high selective, sensitive

TABLE 17.4

Applications of Quantum Dot Conjugates in Cancer Detection and Diagnosis

Application	Conjugates	Target	Potential
Imaging	Transferrin-QD conjugates	Mouse heart and femur up to 0.8 mm deep beneath the skin	QD conjugates in the near-IR allow for greater visualization depth
	DHLA-QD conjugates	Interstitial fluids in rats, where the QD conjugates exited the blood vessels	Potential to interrogate the delivery mechanism of QDs to tumor cells
Vasculature imaging	Tri-peptide-QD conjugates	Human glioblastoma and human breast cells	Help differentiate cancer based on integrin expression levels
	Biotinylated fibrinogen-QD conjugates	Robust biocompatibility, longer stay in circulation without toxicity	Non-invasive visualization of blood vessel development over time
Tracking	Anti-HER2 monoclonal antibodies-QDs conjugates	Visualization of the nanoparticles in blood vessels serving tumor cells in mice	Track biomolecules and understand sub-cellular movements; drug delivery
	PEG-coated QDs	A right dorsal flank of mice, monitor biodistribution by ICP-MS	Determination of non-specific accumulation of the nanoparticles into the organs involved with immune response
	EGF-QDs conjugates	Continuous observation of protein diffusion on the cellular membrane, even after the proteins were internalized	Specific recognition and tracking of plasma membrane antigens
	Antibody-QDs conjugates	Antibody fragment specific for glycine receptors on the membranes of living neurons	Tracking of single receptors
Cellular imaging	QDs with heterofunctional, biocompatible, charge tunable surface coatings	Utility of the alloyed shell and surface coating	Live-cell labeling
	Phospholipid micelle encapsulated QDs	Internalization in human pancreatic cancer cells	Functionalizing-dependent cell uptake
	QD-coated substrate	Non-specific endocytosis	Cell motility and metastatic potential studies
	Tat-peptide-QDs conjugates	Cellular uptake and intracellular transport of nanoparticles in live cells	Development of nanoparticle probes for intracellular targeting and imaging
Targeting	Folate-QD conjugates	Folate receptors in mouse lymphoma cells	Folate is critical for cell growth, may be important to cancer diagnosis
	Antibody-QD conjugates	Imaging of the retinal vasculature, spatial resolution down to the level of single cells	Cancer growth and metastasis
	A ternary system composed of a QD, an aptamer, and doxorubicin	Gradual release of DOX was found to "turn on" the fluorescence of both QDs and DOX, sensing the release of the drug	In vivo targeted imaging, therapy and sensing of drug release

TABLE 17.5

Summary of QDs in Cancer Diagnosis and Treatment

Applications of QD in cancer	Current achievements	Challenges	Future goals
SLN mapping	Map SLNs in pigs and mice		SLN mapping in cancers that involve visceral organs
Detecting primary tumor and metastasis	Detect primary tumor in xenograft models in mice	RES uptake	Metastasis detection, especially micro-metastasis
Molecular target identification for targeted therapy	Reach in vivo tumor vasculature	RES uptake; difficulty in extravasating to reach tumor cells in vivo	Reach tumor cell lines; calculate the level of molecular targets quantitatively and predict response post-therapy

RES: reticuloendothelial system; SLN: sentinel lymph node.
Source: Ref. [47].

multi-color capabilities. The main advantage of this study includes the enhanced permeability and retention of tumor sites and antibody diagnosing cancer-specific cell surface biomarkers [49] (Figure 17.5).

Gao et al. reported the synthesis of ultrasmall NIR non-cadmium quantum dots, InAs/InP/ZnSe QDs coated with mercaptopropionic acid with 800 nm emission wavelength, and applied these QDs for in vivo and ex vivo imaging [54]. Pang et al. developed the NIR Ag_2Se QDs with tunable fluorescence application. They found that the NIR fluorescence of these QDs easily penetrates the abdominal cavity of a living nude mouse and represents their potent applications in vivo imaging [55]. They integrated polyisoprene-block-poly(ethylene oxide) ligand with Cds/CdSe/ZnS QDs and iron oxide (Fe_3O_4) nanocrystals. Then they conjugated with antigen-related cell adhesion molecule (CEACAM) specific monoclonal antibody T84.1 and found in vivo and in vitro cell targeting with magnetic resonance and fluorescence imaging [56].

These advantages of QDs have successfully encouraged their applications for fluorescent biosensing ranging

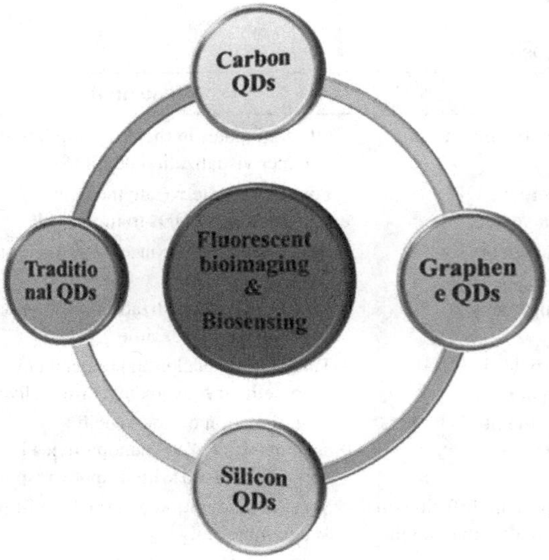

FIGURE 17.4 Different types of QDs used in imaging and sensing.

for in vivo targeting and imaging in the past few decades. Biocompatibility of novel emerging QDs and QD-based fluorescence imaging is even more important and promising in cancer research.

17.3.8 Nanowires

Nanowires (NWs) have their unique optical properties of selectivity and specificity to design molecular markers of malignant cells. NWs are laid down across a microfluidic channel and also, they allow cells or particles to flow through it. NWs are coated with a probe such as an antibody or oligonucleotide, and a short stretch of DNA can be used to diagnose. Protein binding with antibodies will change the NWs' electrical conductance, and it can be measured via a detector. Therefore, proteins produced by tumor cells can be detected, and early-stage tumor detection and diagnosis can be achieved [57]. Nanowire-based sensors can be configured from high-performance field-effect nanowire transistors [58, 59] by linking recognition groups to the nanowire surface (Figure 17.6). Silicon NWs with their native oxide coating makes the receptor linkages exactly and chemical modification of silicon oxide or glass surfaces from research on planar chemical and biological arrays [61].

Silicon nanowire field-effect transmitters have been highly used for single-stranded DNA detection [62]. The binding with polyanionic macromolecules to p-type nanowire surfaces enhances the conductance. DNA target molecule recognition was carried out using complementary single-stranded sequences of peptide nucleic acids (PNAs) (Figure 17.7) [60]. Here PNA is used as the receptor for DNA detection, and it is a potential general biomarker and therapeutic target for cancer detraction and diagnosis.

The advantages and developments of silicon NW sensor devices for cancer cell protein maker detection have been carried out by attaching the monoclonal antibodies to one nanowire element followed by device fabrication. The linkage chemistry between both is similar to that previously labeled protein microarrays, and it involves three main steps [63]. Initially, aldehyde propyltrimethoxysilane (APTMS) is conjugated with the surface of the oxygen plasma cleaned silicon NW to present terminal aldehydes at the surface of the nanowires. Next, coupling between the aldehyde functional group with the monoclonal antibodies and, finally, blocking unreacted free aldehyde groups via reaction with ethanolamine.

17.3.9 Nanoshells

Another important nanotechnology application in cancer cell detection and imaging is the use of nanoshells. Generally, nanoshells are dielectric cores between 10 and 300 nm in

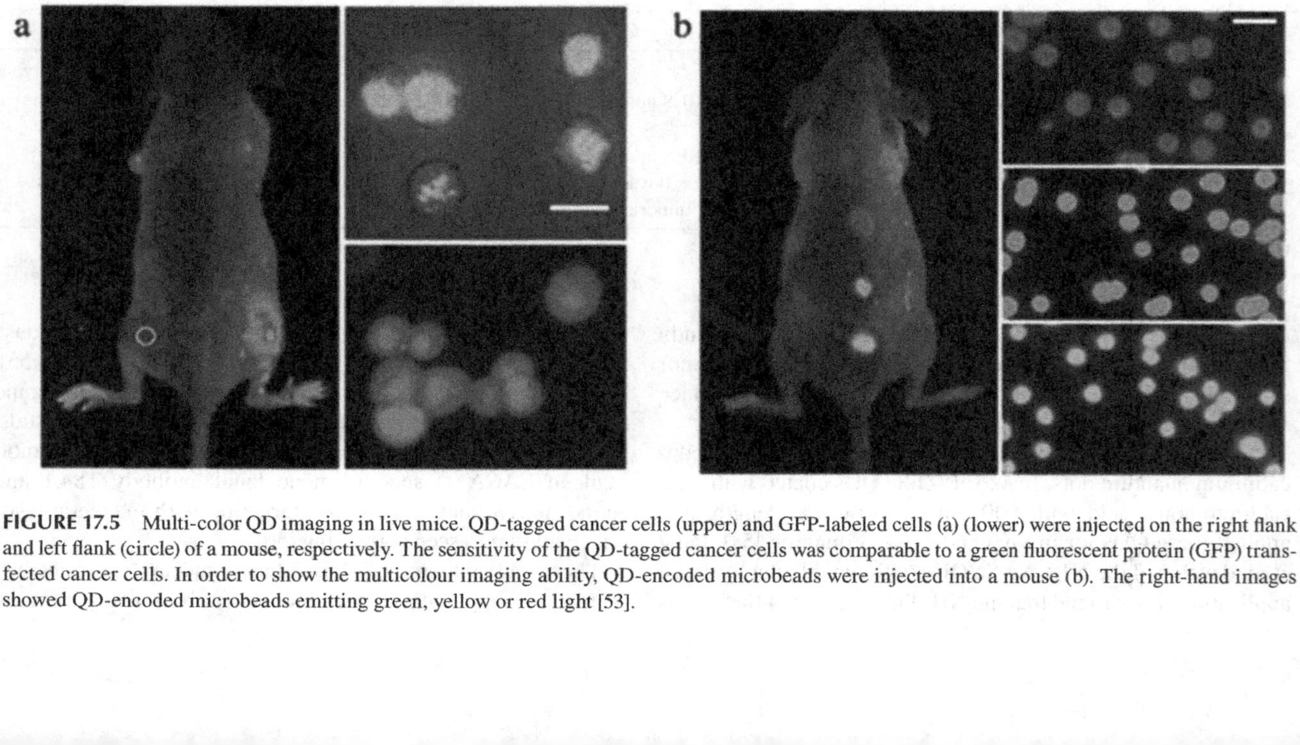

FIGURE 17.5 Multi-color QD imaging in live mice. QD-tagged cancer cells (upper) and GFP-labeled cells (a) (lower) were injected on the right flank and left flank (circle) of a mouse, respectively. The sensitivity of the QD-tagged cancer cells was comparable to a green fluorescent protein (GFP) transfected cancer cells. In order to show the multicolour imaging ability, QD-encoded microbeads were injected into a mouse (b). The right-hand images showed QD-encoded microbeads emitting green, yellow or red light [53].

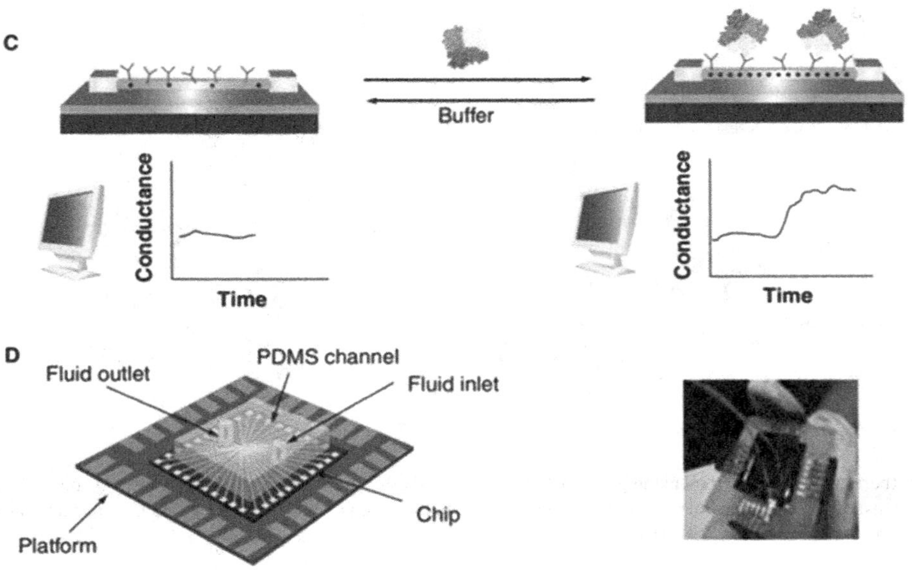

FIGURE 17.6 (C) Schematic of a nanowire device configured as a sensor with antibody receptors (shown top of the arrows) and binding a protein with net negative charge yields an increase in the conductance. (D) Schematic and photograph of a prototype nanowire sensor biochip with integrated microfluidic sample delivery [60].

particle size, typically silicon and coated with a thin metal shell (e.g., gold) [64, 65]. These nanoshells work via converting plasma-mediated electrical energy into light energy and can be flexibly tuned optically through UV-infrared emission/absorption arrays. Nanoshells are required because their imaging avoids heavy metal toxicity, but their uses are limited because of their size [66].

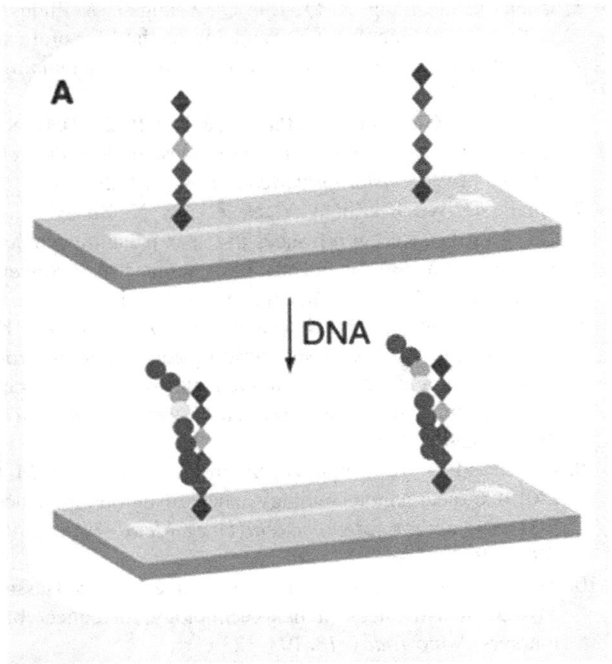

FIGURE 17.7 Schematic of silicon nanowire sensor surface modified with PNA receptor before and after duplex formation with target DNA [62].

Hirsch et al. reported that the nanoshells' surfaces could be easily functionalized for tumor-targeting applications [67]. Nanoshells increase chemical sensing more than 10 billion times and are 10,000 times more sensitive and effective at Raman scattering compared to traditional methods. Antibodies can be attached with NS and get them into incredibly particular and targeted cancer cells (example, breast adenocarcinoma cells overexpressing human epidermal growth factor receptor-2 [HPGF-2]) via in vitro [68]. First, antibodies will attach to PEG, and this antibody-PEG complex is further linked with the nanoshell surface via a sulfur-containing group present in the PEG linker.

17.3.10 Photo-Thermal Ablation Therapy

Nanoshells can be used for photo-thermal ablation of tumor tissue or cells, and it was confirmed in both murine models in vivo and human breast carcinoma cells in vitro [65, 69]. Nanoshells' spherical and layered nanoparticles consist of dielectric silica (SiO2) core coated with a thin metal shell (Figure 17.8). These nanoshells can safely be given into animal models, and they will specifically accumulate in cancer protein sites because of their unique size properties. Cancer cells easily bind with nanoshells through active targeting. Applying NIR light captivated by the nanoshells creates intense heating inside the tumor cells and selectively kills the tumor cells without harming any other healthy tissues or cells.

The efficiency of nanoshell-mediated photo-thermal ablation therapy has also been evaluated in several other in vivo studies. Early research involved directly injecting nanoshell suspensions into tumor-targeting sites, and MRI thermal imaging was used to screen temperature conditions during NIR-induced heating [71, 72]. These reports confirm the rapid heating of nanoshells-loaded cells upon exposure to the NIR light. This early-stage research also provided the informative relationship

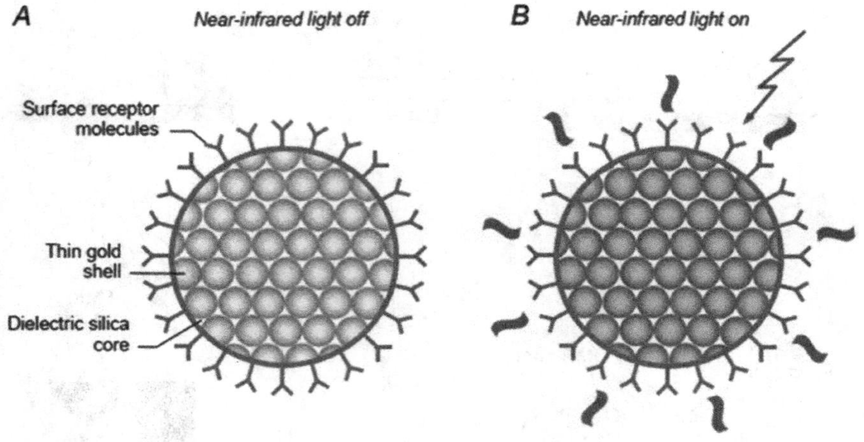

FIGURE 17.8 Photo-thermal ablation therapy using gold-coated silica nanoshells. Surface receptor molecules, e.g. antibodies, are used for targeting (A). Once accumulated inside a tumor, near-infrared light is used to activate the gold nanoparticles. The gold nanoparticles absorb near-infrared light turning it into heat which is lethal to cancer cells (B) [70].

between nanoshell dosage, light intensity, and illumination time with the ultimate heat profile and tissue damage results. However, the majority of the research shows direct injection into the tumor site may not be feasible and may be toxic.

This challenge calls for an alternative approach is to inject nanoshells intravenously and allow them to circulate and accumulate at the tumor point before performing NIR treatment. Here, NS size is critical to the success of this approach. Several research efforts have been demonstrated that particle size with 60–400 nm range will extravasate and accumulate in many types of tumors by passive mechanism states as the "high permeability and retention" [67]. This effect has been attributed to the highly proliferative vasculature within neoplastic tumors.

17.4 Conclusion

The application of material science and nanotechnology has gained more scope in healthcare in the last few decades and created a disease screening and treatment trend. It has challenged clinical applications, including designing practical diagnostic tools for earlier disease detection, drug delivery, and precision therapy. Primarily, nanotechnology provides innovative exposure to the density of medicine and surgery. With the advent and popularity of minimally invasive surgeries and interventional techniques, nanotechnology offers a future platform for further development of drugs for cancer diagnosis. The multidisciplinary field of nanotechnology holds the promise of delivering a scientific breakthrough and may move very fast from concept to reality. Severe limitations were identified in their applications for in vivo studies. However, the time the drug delivery or the microsurgery is performed due to nanorobots/nanodevices is not that far. Several examples of nanodevices, nanorobots, and nanomachines with applications on targeted drug delivery and sensing and diagnosis were discussed critically, insisting on type of motion and propulsion and their composition depending on the desired application.

REFERENCES

1. Choi, Y. E., Kwak, J. W., & Park, J. W. (2010). Nanotechnology for early cancer detection. *Sensors*, *10*(1), 428–455.
2. WHO cancer reports 2020–2021. https://www.who.int/news-room/fact-sheets/detail/cancer.
3. Siegal, R., Miller, K. D., & Jemal, A. (2014). Cancer statistics, 2012. *CA: A Cancer Journal for Clinicians*, *64*(1), 9–29.
4. Jin, C., Wang, K., Oppong-Gyebi, A., & Hu, J. (2020). Application of nanotechnology in cancer diagnosis and therapy-a mini-review. *International Journal of Medical Sciences*, *17*(18), 2964.
5. Chen, X. J., Zhang, X. Q., Liu, Q., Zhang, J., & Zhou, G. (2018). Nanotechnology: a promising method for oral cancer detection and diagnosis. *Journal of Nanobiotechnology*, *16*(1), 1–17.
6. Tran, S., DeGiovanni, P. J., Piel, B., & Rai, P. (2017). Cancer nanomedicine: a review of recent success in drug delivery. *Clinical and Translational Medicine*, *6*(1), 1–21.
7. Huang, D., Wu, K., Zhang, Y., Ni, Z., Zhu, X., Zhu, C., … & Hu, J. (2019). Recent advances in tissue plasminogen activator-based nanothrombolysis for ischemic stroke. *Reviews on Advanced Materials Science*, *58*(1), 159–170.
8. Hu, J., Huang, S., Zhu, L., Huang, W., Zhao, Y., Jin, K., & ZhuGe, Q. (2018). Tissue plasminogen activator-porous magnetic microrods for targeted thrombolytic therapy after ischemic stroke. *ACS Applied Materials & Interfaces*, *10*(39), 32988–32997.
9. Chaturvedi, V. K., Singh, A., Singh, V. K., & Singh, M. P. (2019). Cancer nanotechnology: a new revolution for cancer diagnosis and therapy. *Current Drug Metabolism*, *20*(6), 416–429.
10. Ye, F., Zhao, Y., El-Sayed, R., Muhammed, M., & Hassan, M. (2018). Advances in nanotechnology for cancer biomarkers. *Nano Today*, *18*, 103–123.
11. Ravindran, R. (2011). Nanotechnology in cancer diagnosis and treatment: an overview. *Journal of Oral and Maxillofacial Pathology*, *2*, 101–106.

12. Hussain, C. M., & Palit, S. (2017). Nanomaterials, eco-materials and wide vision of material science. In *Springer Handbook of Ecomaterials* (pp. 1–29), 1st edn. Springer.
13. Palit, S., & Hussain, C. M. (2018). Engineered nanomaterial for industrial use. In *Handbook of Nanomaterials for Industrial Applications* (pp. 3–12). Elsevier.
14. Palit, S. (2018). Recent advances in the application of engineered nanomaterials in the environment industry: a critical overview and a vision for the future. *Handbook of Nanomaterials for Industrial Applications* (pp. 883–893). Elsevier
15. Pala, N. (2016). Nanodevices. *Nanomaterials and Nanotechnology*, *6*, 1–2.
16. Zhang, Y., Li, M., Gao, X., Chen, Y., & Liu, T. (2019). Nanotechnology in cancer diagnosis: progress, challenges and opportunities. *Journal of Hematology & Oncology*, *12*(1), 1–13.
17. Palit, S., & Hussain, C. M. (2020). Nanodevices applications and recent advancements in nanotechnology and the global pharmaceutical industry. In *Nanomaterials in Diagnostic Tools and Devices* (pp. 395–415). Elsevier.
18. Tertis, M., Cernat, A., Mirel, S., & Cristea, C. (2021). Nanodevices for pharmaceutical and biomedical applications. *Analytical Letters*, *54*(1–2), 98–123.
19. Schöning, M. J., Abouzar, M. H., Wagner, T., Näther, N., Rolka, D., Yoshinobu, T., ... & Poghossian, A. (2006). A semiconductor-based field-effect platform for (bio-) chemical and physical sensors: from capacitive EIS sensors and LAPS over ISFETs to nano-scale devices. *MRS Online Proceedings Library (OPL)*, ISBN: 9781604234084. Boston, MA, United States.
20. El-Gazzar, N. S., & Enan, G. (2020). Advances in phage inspired nanoscience based therapy. In *Nanobioscience*, Saxena, S. K., & Khurana, S. P., Eds. Springer.
21. https://pubmed.ncbi.nlm.nih.gov/?term=nanotechnology&filter=years.2000-2022.
22. McWilliams, A. (2018). Nanodevices and nanomachines: the global market. https://www.bccresearch.com/market-research/nanotechnology/ nanodevices-and-nanomachines-market-report.html (accessed 11-09-2021).
23. Verma, M., Sheoran, P., & Chaudhury, A. (2018). Application of nanotechnology for cancer treatment. In *Advances in Animal Biotechnology and its Applications* (pp. 161–178). Springer.
24. Mousa, S. A., & Bharali, D. J. (2011). Nanotechnology-based detection and targeted therapy in cancer: nano-bio paradigms and applications. *Cancers*, *3*(3), 2888–2903.
25. Jaishree, V., & Gupta, P. D. (2012). Nanotechnology: a revolution in cancer diagnosis. *Indian Journal of Clinical Biochemistry*, *27*(3), 214–220.
26. Zhang, X., Guo, Q., & Cui, D. (2009). Recent advances in nanotechnology applied to biosensors. *Sensors*, *9*(2), 1033–1053.
27. Misra, R., Acharya, S., & Sahoo, S. K. (2010). Cancer nanotechnology: application of nanotechnology in cancer therapy. *Drug Discovery Today*, *15*(19–20), 842–850.
28. Jain, S., Hirst, D. G., & O'sullivan, J. M. (2012). Gold nanoparticles as novel agents for cancer therapy. *The British Journal of Radiology*, *85*(1010), 101–113.
29. Huang, X., Jain, P. K., El-Sayed, I. H., & El-Sayed, M. A. (2007). Gold nanoparticles: interesting optical properties and recent applications in cancer diagnostics and therapy. *Nanomedicine*, *2*(5), 1–5.
30. Jain, P. K., Lee, K. S., El-Sayed, I. H., & El-Sayed, M. A. (2006). Calculated absorption and scattering properties of gold nanoparticles of different size, shape, and composition: applications in biological imaging and biomedicine. *The Journal of Physical Chemistry B*, *110*(14), 7238–7248.
31. Kim, B., Tripp, S. L., & Wei, A. (2001). Tuning the optical properties of large gold nanoparticle arrays. *MRS Online Proceedings Library (OPL). ChemPhysChem*, *2* (12), 743–745.
32. Connor, E. E., Mwamuka, J., Gole, A., Murphy, C. J., & Wyatt, M. D. (2005). Gold nanoparticles are taken up by human cells but do not cause acute cytotoxicity. *Small*, *1*(3), 325–327.
33. Singh, P., Pandit, S., Mokkapati, V. R. S. S., Garg, A., Ravikumar, V., & Mijakovic, I. (2018). Gold nanoparticles in diagnostics and therapeutics for human cancer. *International Journal of Molecular Sciences*, *19*(7), 1979.
34. Ajnai, G., Chiu, A., Kan, T., Cheng, C. C., Tsai, T. H., & Chang, J. (2014). Trends of gold nanoparticle-based drug delivery system in cancer therapy. *Journal of Experimental & Clinical Medicine*, *6*(6), 172–178.
35. Chen, Y. H., Tsai, C. Y., Huang, P. Y., Chang, M. Y., Cheng, P. C., Chou, C. H., ... & Wu, C. L. (2007). Methotrexate conjugated to gold nanoparticles inhibits tumor growth in a syngeneic lung tumor model. *Molecular Pharmaceutics*, *4*(5), 713–722.
36. Wang, F., Wang, Y. C., Dou, S., Xiong, M. H., Sun, T. M., & Wang, J. (2011). Doxorubicin-tethered responsive gold nanoparticles facilitate intracellular drug delivery for overcoming multidrug resistance in cancer cells. *ACS Nano*, *5*(5), 3679–3692.
37. Dreaden, E. C., Austin, L. A., Mackey, M. A., & El-Sayed, M. A. (2012). Size matters: gold nanoparticles in targeted cancer drug delivery. *Therapeutic Delivery*, *3*(4), 457–478.
38. Weissleder, R. (2001). A clearer vision for in vivo imaging. *Nature Biotechnology*, *19*(4), 316–317.
39. Huff, T. B., Tong, L., Zhao, Y., Hansen, M. N., Cheng, J. X., & Wei, A. (2007). Hyperthermic effects of gold nanorods on tumor cells. *Nanomedicine* (Lond), *2*(1), 125–132.
40. El-Sayed, I. H., Huang, X., & El-Sayed, M. A. (2006). Selective laser photo-thermal therapy of epithelial carcinoma using anti-EGFR antibody conjugated gold nanoparticles. *Cancer Letters*, *239*(1), 129–135.
41. Dickerson, E. B., Dreaden, E. C., Huang, X., El-Sayed, I. H., Chu, H., Pushpanketh, S., ... & El-Sayed, M. A. (2008). Gold nanorod assisted near-infrared plasmonic photothermal therapy (PPTT) of squamous cell carcinoma in mice. *Cancer Letters*, *269*(1), 57–66.
42. Menon, J. U., Jadeja, P., Tambe, P., Vu, K., Yuan, B., & Nguyen, K. T. (2013). Nanomaterials for photo-based diagnostic and therapeutic applications. *Theranostics*, *3*(3), 152.
43. Liu, B., Jiang, B., Zheng, Z., & Liu, T. (2019). Semiconductor quantum dots in tumor research. *Journal of Luminescence*, *209*, 61–68.
44. Kobayashi, H., Hama, Y., Koyama, Y., Barrett, T., Regino, C. A., Urano, Y., & Choyke, P. L. (2007). Simultaneous multicolor imaging of five different lymphatic basins using quantum dots. *Nano Letters*, *7*(6), 1711–1716.

45. Morgan, N. Y., English, S., Chen, W., Chernomordik, V., Russo, A., Smith, P. D., & Gandjbakhche, A. (2005). Real time in vivo non-invasive optical imaging using near-infrared fluorescent quantum dots1. *Academic Radiology*, *12*(3), 313–323.
46. Santra, S., Xu, J., Wang, K., & Tand, W. (2004). Luminescent nanoparticle probes for bioimaging. *Journal of Nanoscience and Nanotechnology*, *4*(6), 590–599.
47. Zhang, H., Yee, D., & Wang, C. (2008). Quantum dots for cancer diagnosis and therapy: biological and clinical perspectives. *Nanomedicine* (Lond), *3*(1), 83–91.
48. Gao, X., Yang, L., Petros, J. A., Marshall, F. F., Simons, J. W., & Nie, S. (2005). In vivo molecular and cellular imaging with quantum dots. *Current Opinion in Biotechnology*, *16*(1), 63–72.
49. Gao, X., Cui, Y., Levenson, R. M., Chung, L. W., & Nie, S. (2004). In vivo cancer targeting and imaging with semiconductor quantum dots. *Nature Biotechnology*, *22*(8), 969–976.
50. Chan, W. C., & Nie, S. (1998). Quantum dot bioconjugates for ultrasensitive nonisotopic detection. *Science*, *281*(5385), 2016–2018.
51. Michalet, X., Pinaud, F. F., Bentolila, L. A., Tsay, J. M., Doose, S. J. J. L., Li, J. J., … & Weiss, S. (2005). Quantum dots for live cells, in vivo imaging, and diagnostics. *Science*, *307*(5709), 538–544.
52. Dubertret, B., Skourides, P., Norris, D. J., Noireaux, V., Brivanlou, A. H., & Libchaber, A. (2002). In vivo imaging of quantum dots encapsulated in phospholipid micelles. *Science*, *298*(5599), 1759–1762.
53. Li, J., & Zhu, J. J. (2013). Quantum dots for fluorescent biosensing and bio-imaging applications. *Analyst*, *138*(9), 2506–2515.
54. Gao, J., Chen, K., Xie, R., Xie, J., Lee, S., Cheng, Z., … & Chen, X. (2010). Ultrasmall near-infrared non-cadmium quantum dots for in vivo tumor imaging. *Small*, *6*(2), 256–261.
55. Gu, Y. P., Cui, R., Zhang, Z. L., Xie, Z. X., & Pang, D. W. (2012). Ultrasmall near-infrared Ag2Se quantum dots with tunable fluorescence for in vivo imaging. *Journal of the American Chemical Society*, *134*(1), 79–82.
56. Pöselt, E., Schmidtke, C., Fischer, S., Peldschus, K., Salamon, J., Kloust, H., … & Weller, H. (2012). Tailor-made quantum dot and iron oxide-based contrast agents for in vitro and in vivo tumor imaging. *ACS Nano*, *6*(4), 3346–3355.
57. Barani, M., Bilal, M., Sabir, F., Rahdar, A., & Kyzas, G. Z. (2020). Nanotechnology in ovarian cancer: diagnosis and treatment. *Life Sciences*, 118914.
58. Cui, Y., Wei, Q., Park, H., & Lieber, C. M. (2001). Nanowire nanosensors for highly sensitive and selective detection of biological and chemical species. *Science*, *293*(5533), 1289–1292.
59. Wang, W. U., Chen, C., Lin, K. H., Fang, Y., & Lieber, C. M. (2005). Label-free detection of small-molecule–protein interactions by using nanowire nanosensors. *Proceedings of the National Academy of Sciences*, *102*(9), 3208–3212.
60. Jensen, K. K., Ørum, H., Nielsen, P. E., & Nordén, B. (1997). Kinetics for hybridization of peptide nucleic acids (PNA) with DNA and RNA studied with the BIAcore technique. *Biochemistry*, *36*(16), 5072–5077.
61. Patolsky, F., Zheng, G., & Lieber, C. M. (2006). Nanowire sensors for medicine and the life sciences. *Nanomedicine* (Lond), *1*(1), 51–65.
62. Li, Z., Chen, Y., Li, X., Kamins, T. I., Nauka, K., & Williams, R. S. (2004). Sequence-specific label-free DNA sensors based on silicon nanowires. *Nano Letters*, *4*(2), 245–247.
63. Arenkov, P., Kukhtin, A., Gemmell, A., Voloshchuk, S., Chupeeva, V., & Mirzabekov, A. (2000). Protein microchips: use for immunoassay and enzymatic reactions. *Analytical Biochemistry*, *278*(2), 123–131.
64. Hirsch, L. R., Stafford, R. J., Bankson, J. A., Sershen, S. R., Rivera, B., Price, R. E., … & West, J. L. (2003). Nanoshell-mediated near-infrared thermal therapy of tumors under magnetic resonance guidance. *Proceedings of the National Academy of Sciences*, *100*(23), 13549–13554.
65. Loo, C., Lin, A., Hirsch, L., Lee, M. H., Barton, J., Halas, N., … & Drezek, R. (2004). Nanoshell-enabled photonics-based imaging and therapy of cancer. *Technology in Cancer Research & Treatment*, *3*(1), 33–40.
66. Nunes, T., Pons, T., Hou, X., Van Do, K., Caron, B., Rigal, M., … & Bousquet, G. (2019). Pulsed-laser irradiation of multifunctional gold nanoshells to overcome trastuzumab resistance in HER2-overexpressing breast cancer. *Journal of Experimental & Clinical Cancer Research*, *38*(1), 1–13.
67. Hirsch, L. R., Gobin, A. M., Lowery, A. R., Tam, F., Drezek, R. A., Halas, N. J., & West, J. L. (2006). Metal nanoshells. *Annals of Biomedical Engineering*, *34*(1), 15–22.
68. Loo, C., Lowery, A., Halas, N., West, J., & Drezek, R. (2005). Immunotargeted nanoshells for integrated cancer imaging and therapy. *Nano Letters*, *5*(4), 709–711.
69. O'Neal, D. P., Hirsch, L. R., Halas, N. J., Payne, J. D., & West, J. L. (2004). Photo-thermal tumor ablation in mice using near infrared-absorbing nanoparticles. *Cancer Letters*, *209*(2), 171–176.
70. Roszek, B., De Jong, W. H., & Geertsma, R. E. (2005). Nanotechnology in medical applications: state-of-the-art in materials and devices, Department of Pharmaceutical Affairs and Medical Technology of the Dutch Ministry of Health, Welfare and Sports, within the framework of project V/265001, Support for Policy on Medical Technology.
71. Park, J., Estrada, A., Sharp, K., Sang, K., Schwartz, J. A., Smith, D. K., … & Tunnell, J. W. (2008). Two-photon-induced photoluminescence imaging of tumors using near-infrared excited gold nanoshells. *Optics Express*, *16*(3), 1590–1599.
72. Karunakaran, C., Bhargava, K., & Benjamin, R. (2015). *Biosensors and Bioelectronics*. Elsevier.

18

Nanoparticles: The Promising Future of Advanced Diagnosis and Treatment of Neurological Disorders

Poojith Nuthalapati, Sudharshan Asaithambi, Malavika Kumar and Dinesh Reddy

CONTENTS

DOI: 10.1201/9781003157823-18

18.1 Introduction

Nanotechnology is one of the leading fields of science that deals with matter in the nano-size range. Nanotechnology products like liposomes, exosomes, nanotubes, nanorods, nano-emulsions, etc. are used in effective drug delivery to the brain. The evolution of this branch has led to development in various fields, including medicine, especially in neurology. The neurological problems due to etiologies of unknown origin are rising daily, and the most important of them are multiple sclerosis (MS), Parkinson's disease (PD), Huntington's disease (HD), and Alzheimer's disease (AD). There are currently various technologies that are employed to treat the diseases effectively, but most of them fail to cross the blood-brain barrier (BBB) and show their fullest activity in the brain. There were no advanced technologies to tackle this issue till the advent of nanotechnology. This stood as a promising technology as the drug molecules are delivered through nano-sized particles that can cross the BBB. The use of nanoparticles in neurology addresses the limitations of current therapies. Polymeric nanoparticle technology (PNT), magnetic iron-oxide nanotechnology (MFN), exosomes and liposomes (E&L), and gold nanoparticles (AuNP) are various nanoparticle strategies that deliver the drug candidates that cross the BBB to the brain to treat the diseases effectively. Imaging using fluorescence, Raman scattering, persistent luminescence, and photoacoustics is used as a tool for bio-imaging. There are only a few existing clinical studies in this sector; therefore, further research is needed. The noninvasive process of introducing nanoparticles into the human body should also be based on mucosal, sublingual, nasal, or buccal channels. The toxicity of the drugs transmitted by nanoparticles must be limited with special focus.

18.2 Neurological Disorders and Nanoparticles

Multiple sclerosis (MS), Parkinson's disease (PD), Huntington's disease (HD), and Alzheimer's disease are the most common neurological problems caused by unknown etiologies. The blood-brain barrier (BBB) adequately protects the brain from the entrance of drugs and other pathological microorganisms. This quality also causes hindrances to the drugs that target the brain for activity [1].

Nanosomes have a number of advantages, including targeted action, high loading capacity, reduced toxicity, and increased therapeutic efficacy. They provide good bioavailability and enhanced pharmacokinetics of the drug. Hence, nanoscience has become the topic of research in the present stage, especially treating psychiatric and neurodegenerative disorders [2]. There are other disorders like leukodystrophy, lysosomal storage diseases, and various kinds of cancers of the brain that need drugs to cross the BBB for their effective delivery and activity, which demands targeted action.

Various nanoparticle strategies deliver drug candidates that cross the BBB to the brain to treat diseases effectively. They are as follows.

18.2.1 Polymeric Nanoparticle Technology (PNT)

Polymers encapsulate the drug molecules and deliver them across the BBB. There are two kinds of polymers, namely natural and synthetic polymers. Examples of natural polymers are chitosan, gelatin, and sodium alginate. Examples of synthetic polymers include polymers of lactic acid, caprolactone, cyanoacrylate, and glycolic acid derivatives. The advantages of this type of delivery are that they are stable and free-flowing in the bloodstream and can cross the BBB more effectively than liposomes [3].

18.2.2 Magnetic Iron-Oxide Nanotechnology (MFN)

The magnetic nanoparticles are efficient in sensing and separating the molecules that usually target the drugs and delivering the drug to the specified site in the brain. Iron is used to prepare nanoparticles in a variety of ways, including microemulsification, hydrothermal synthesis, sonochemical synthesis, laser-assisted pyrolysis, electricity-assisted synthesis, and pyro-decomposition. The iron molecules are carboxylated using polysaccharides and surface modification. The polymers used are chitosan, anti-ferritin, chlorotoxin, etc. Coating the nanoparticles with gold metal provides more stability and prevents oxidation in the presence of water [4].

18.2.3 Exosomes and Liposomes (E and L)

Liposomes in nanoscales are spherical and are biphasic with lipid layers, and carry drug molecules to the delivery site through the BBB. They have a hydrophilic core and a hydrophobic outer layer. The preparation is done with the nanoliposome film rehydration method. The polymers used in the preparation of the exosomes are chitosan, mannose, polyethylene glycol (PEG), interleukins, and aptamers [5].

18.2.4 Gold Nanoparticles (AuNP)

Gold nanoparticles are used in a variety of medicinal and neurological disease treatments. They are synthesized by the citrate reduction method using gold tetrachloride in an acid medium. This is a low-cost option that requires only mild conditions. Gold nanoparticles target effective cancers and readily cross the BBB with drug molecules. The brain tumors are targeted using AuNP attaching and are finally ablated using lasers or radiation. Gold nanoshells and nanocages are gaining significance in delivering drugs to the brain, as in Table 18.1. The limitation of this delivery vehicle is cytotoxicity due to the long-term usage of the metal gold [6].

18.3 Diagnostic Bio-Barcoding of Enzymes

There has been an increasing demand for the proper and specific analysis of drug molecules and the causative enzymes of neuro-degeneration. The ultrasensitive nanoparticle technique seemed promising; however, the technology has been slightly improved in the name of bio-barcoding (BBC) with

TABLE 18.1
Drug Candidates Delivered via Nanocarriers

S. no.	Disease	Type of nanoparticle	Drug candidate
1	Alzheimer's disease	E&L	BACE
		PNT	Curcumin
		AuNP	Polyoxometalate
2	Parkinson's disease	E&L	Dopamine
			Levodopa
		PNT	Nicotine
			Ropinirole
		AuNP	Synuclein
		MFN	Rhodamine
3	Malignancy	E&L	Doxorubicin
			Coumarin
		PNT	Paclitaxel
			Docetaxel
		AuNP	Phthalocyanine

the introduction of gold nanoparticles [7]. Other techniques for estimating brain enzymes, such as polymerase chain reaction (PCR) and western blots, are available; however, BBC offers benefits over these.

BBC was developed after extensive research and is several times more sensitive than ELISA and other comparatively advanced and sensitive methods. For diagnostic purposes, this technique employs nanogold technology. BBC refers to the identification or amplification of a specific moiety or functional group that may be utilized to attach nanoparticles and improve the sensitivity of the analysis. In the first step, the magnetic probe molecules are attached to the targeted brain enzyme. A magnetic probe is a gold nanoparticle that is attached to the anti-target protein and barcoded DNA strands. The second step is to apply a magnetic field to the medium so that the field forms a sandwich between the probes. Then a test sample is also run similarly to target a specific protein in the enzyme present in the brain tissue sample or blood. After the run, DNA is dissociated from the nanoparticle and the content of the enzyme is estimated using various methods that amplify the recording and detection of a bio-labeled sample [8]. There are various methods of detection of barcoded separated proteins; they are discussed as follows.

18.3.1 Fluorescent Labeling to Detect Cellular Abnormalities

Pericytes have an important role in the control of blood through the BBB. Fluorescent labeling of the pericytes was done in the central nervous system (CNS) to identify live cells and immune-chemistry [9]. Also, the enteric nervous system was successfully labeled using fluorescent stains and used for detection and imaging of cells and cell lines [10]. This approach is used to identify DNA samples that have been barcoded with stains and are frequently identified by measuring the signals by fluorescence detectors. It's ideal for improving detection stability in small runs of samples. This method is commonly used to test toxins such as ricin, bluetongue virus, and salmonella. Combining barcoding with fluorescence is an immature approach since the detection is well adapted and detected using other methods [11].

18.3.2 Biosensors to Detect Cognitive Decline and Neurotransmitters

Neurotransmitters are the common messengers of the CNS and their imbalance can cause neurological and cognitive disorders, such as Alzheimer's disease, Parkinson's disease, and schizophrenia. Bio-sensing systems could help detect the neurotransmitters that are indicative of neurological disorders [12]. Glutamate is one of those neurotransmitters in the brain that can be detected via bio-sensing using an implantable micro biosensor [13]. This method uses gold nanoparticles to amplify the signals and magnetic probes to split and extract the sample proteins from the tissue sample. The technique is applicable for simple runs that are short and portable. They have very low signal strength detection and high response rates. Platelet antigens in human blood, *E. coli*, and antigen A are detected using this method effectively. The repeatability of the results is not certain and completely depends on the signal strength of the binding of gold nanoparticles to the magnetic field applied [14].

18.3.3 Colorimetric Method to Analyze Inflammatory Mediators

This method involves estimations coded with gold nanoparticles that are usually formed into a colored solution which can be amplified and analyzed according to the changes in color. This method is simple and inexpensive. The estimations are highly responsive to color changes, but they often don't work with transparent or discolored solutions. Brain inflammatory mediators like cytokines and interleukins are estimated using this method, signifying the extent of neurodegenerative diseases like Alzheimer's and Parkinson's diseases.

18.3.4 Polymerase Chain Reaction (PCR) Method

The PCR method is widely used to diagnose various CNS disorders, genetic and autoimmune disorders, malignant neoplasms, and various kinds of infections. PCR can detect DNA and RNA changes in the tissue [15]. This method is used to effectively diagnose brain cancer and is effective in food testing. It functions based on the antigen-antibody testing protocols by PCR amplification and in time with the pairing of the fluorescence and the PCR method. HCV proteins and enterotoxins are often estimated using this method. The major disadvantages of this technique are the high cost of the machines and that the results are not stored for a long time.

18.3.5 Biochips to Detect Changes in the Brain

Micro-extrusion-based chips are used to align the axonal relationships and organization of cellular content [16]. The chip-based amplification or barcoding is done by ligating a chip on the surface of the probe nanoparticles to the DNA sequences. These are stained using silver or other conductive metals and the analysis is performed by scanning the barcodes on the

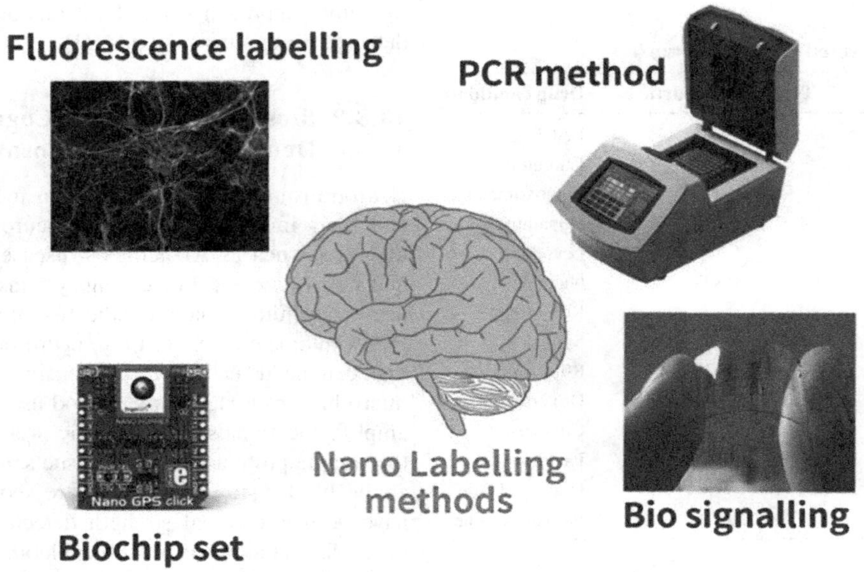

FIGURE 18.1 CNS barcoding and labeling methods using nanotechnology.

DNA sequences. It is a rapid method of estimation and the use of silver as an amplifier enables the detection of the slightest of signals. Anthrax, HIV, and human immunoglobulins are estimated effectively using this method. The system is not convenient for detecting the sensitive signals of DNA and related macromolecules as shown in Figure 18.1.

18.4 Applications of Nanotechnology in CNS Disorders

Nanotechnology has revolutionized the field of treating various neurological diseases and has produced a variety of novel methods that have demonstrated promise for treating neurodegenerative disorders, including Alzheimer's disease, Parkinson's disease, stroke, Huntington's disease, and brain tumors. Molecules have been nanoengineered to cross the BBB, target a particular cell or signaling pathway, and serve as a carrier for gene transmission. Due to nanotechnology's expanded usage in supplying therapeutic drugs and radiocontrast agents, imaging agents, and more, researchers are researching how they can utilize nanotechnology to distribute such imaging agents and other products. The nanotechnology-based therapy for CNS disorders when a particular stimulation or disease-related activation of glial cells is experienced, such as when viral infection or infection with other disease-causing organisms occurs, which promotes neuroinflammation and the progression from neuroinflammation to neural balance disruption and systemic imbalance between pro-inflammatory (IL-1β, TNF-α, and IL-6) and anti-inflammatory (IL-4, IL-10, etc.,) cytokines ensues. Here, the authors illustrate in a schematic representation how the ability of nano-formulations to localize and selectively engage molecules is key to prevent an unwanted, adverse effect on the central nervous system while also restoring balance in the overall system through reducing neuroinflammation and glial cells.

18.4.1 Epilepsy

Epilepsy, a central nervous system (CNS) condition, is described as a spike in brain electrical activity that can be either limited to a particular region or expanded across the brain, resulting in partial or widespread seizures [17]. While some treatment approaches, intending to reduce seizure incidence and intensity while minimally impacting the brain and other tissues of the body, have nearly failed due to their inability to reach the BBB, drug tolerance, and recurrence of disease after drug discontinuation, other treatment strategies, aimed at directly mitigating the toxic effects of the medications and at the same time declining seizure frequency and severity, are nearing failure because of the above problems. Although numerous techniques such as nano-based strategy, prodrugs, efflux pump inhibition, the opening of the BBB by hyperosmolar solution, and direct drug delivery to the ventricles and cortex have been evolved for the treatment of epilepsy, among these, the nanotechnological approach has shown tremendous potential for overcoming all the major obstacles in the treatment of epilepsy, especially concerning the crossing of the BBB and supplying targeted quantities of drugs at their therapeutic concentration. An *in vitro* analysis in which samples of either solid lipid nanoparticles loaded with carbamazepine or PLGA nanoparticles loaded with β-carotene were found to exhibit more promise for the anticonvulsant effect than *in vitro* emulsified loaded carbamazepine and polysorbate-80 coated PLGA nanoparticles [18]. Rats administered with a liposomal muscimol formulation showed that it had poor histological effects but effectively reduced focal seizures while inducing relatively minor cellular changes.

Moreover, mice injected with amiloride-loaded liposomes demonstrated that the formulation displayed higher anticonvulsant potency than the free drug. An experiment was performed on a rat model in which it was discovered that injection of ethosuximide-loaded chitosan nanocapsules under the skin leads to a decrease in the spike-wave discharge. Because of their capability of reliably generating a successful release of the medication, these nano-formulations can be turned into depot drug delivery systems that distribute antiepileptic medications for prolonged periods.

18.4.2 Alzheimer's Disease

Alzheimer's disease is a steadily progressing neurodegenerative condition, and one of the primary causes of dementia. Several pieces of evidence state that its occurrence and prevalence rate is much more in elderly people. Plaque deposits of the protein called amyloid-β (Aβ) and tau protein hyperphosphorylation are both observed in individuals with Alzheimer's disease. Degeneration of the nervous tissue occurs over time in patients with Alzheimer's disease. Traditional treatment strategies have fallen short of entirely curing the disease, allowing for the use of orally administered medicines such as tacrine, rivastigmine, and others, but they are no longer capable of doing so, suggesting that the use of nanotechnology to treat AD may become feasible.

A broad variety of nano-formulations have been developed to be effective for people with Alzheimer's disease. An experiment demonstrated that phospholipid-stabilized nano-micelles (also known as PEG-stabilized nano-micelles) made up of phospholipids decrease Aβ aggregation and delay Aβ-induced neurotoxicity in the SHSY-5Y human neuroblastoma cell line *in vitro*. After undertaking an *in vitro* analysis, it was discovered that the phytochemical curcumin could minimize Aβ oligomerization and cytotoxicity. However, bioavailability was found to be subpar when injected into mice. In a nano-liposomal formulation, a substance's ability to suppress Aβ aggregation was retained, while at the same time, its bioavailability was improved [19]. Another technique for AD treatment includes chelating agents, which improve the number of metal ions, including copper, and thus add to the pathology of AD. When added to a model system composed of pre-existing Aβ aggregates, the nanoparticles that carry a copper chelator known as d-penicillamine could permeate the BBB and dissolve the pre-existing Aβ aggregates.

The main component of oxidative disruption in Alzheimer's disease is apparent in using antioxidants to treat the disease. An exceptional number of fullerene compounds, active free-radical scavengers, have been shown to exhibit neuroprotective activity against glutamate receptors caused by excitotoxicity. On the other hand, there is not enough data to demonstrate if fullerene has a preventive function against Alzheimer's disease in either *in vitro* or *in vivo* models, but when it prevents the fibrillization of Aβ peptides, the increase in cognitive function that occurs after intraventricular administration suggests it is advantageous in AD care. Often, there is a pronounced deficit of acetylcholine (ACh) neurotransmitters, which is another significant feature in AD pathology. Since direct injection of free ACh has a slightly higher decomposition rate in blood, ACh direct injection is inadequate for treating the imbalance in ACh. Thus, a nanotechnological technique has been used to distribute ACh to the brain to reach an equilibrium degree of ACh. The ACh loaded into carbon nanotubes has been shown to substantially boost important cognitive functions to the degree previous to the onset of Alzheimer's disease in a kainic-acid mediated mouse model instead of free ACh [20].

When compared to untreated poly (n-butyl cyanoacrylate) NPs, those coated with polysorbate 80 and loaded with the anti-AD medication rivastigmine exhibited a 3.82-fold increase in drug concentration in the brain. This is in line with the findings of Borchard et al., who used a bovine model to show that polysorbate 80 is a very effective drug delivery agent across the BBB [21]. The potential of nanoparticles (NP) to improve biodistribution of commercially available drugs for AD has been explored in a series of studies, and the possibility of further improving uptake via functionalization of NPs with surfactants like polysorbate 80 has also been explored. A study by Sun et al. investigated the role of polysorbate 80 coating on NPs in their uptake into the brain, and found that it interacts with the brain micro-vessel endothelial cells, and leads to increased uptake of AD drugs [22]. Javed et al. coated gold nanoparticles with beta casein and administered them intracardially into zebrafish which was induced with neurotoxicity from Aβ injected into the cerebroventricular spaces. The authors reported that the administration of βCas gold NPs mitigated the toxicity of Aβ in zebra fish, and displayed an exceptional capability to rescue the animal from AD-like symptoms [23].

18.4.3 Parkinson's Disease

Parkinson's disease is the second most prominent neurodegenerative condition worldwide, after Alzheimer's disease. It is seen in the absence of dopaminergic neurons in the substantia nigra of the midbrain and the generation of α-synuclein aggregates (Lewy bodies). While now known to occur in the brain, pathological trends that arise from neurodegeneration in PD may potentially be due to impairment and depletion of several brain regions, including neurotransmitter concentrations and their metabolites, proteins that accumulate besides Lewy bodies, and others [24]. The symptoms of PD are mainly motor in nature. They include tardive dyskinesia, motor issues like tremors, slurred speech, writing problems, and non-motor issues like cognitive, mental, and autonomic problems. Owing to the drawbacks, the existing therapies for PD cannot prevent or alleviate the severity of the condition; instead, they only serve to improve the symptoms. On top of this, the inadequacy of medications for crossing the BBB is a difficulty in treating PD. To meet the sector's demands, more efficient care modalities must be produced in greater quantities. The nanotechnological method can tackle this issue since it has demonstrated notable efficacy in reversing or avoiding disease conditions, encouraging functional recovery of injured neurons, providing neuroprotection, and enabling the distribution of drugs that transcend the BBB [25].

Since PD often arises due to gene distribution issues, it is thought that this method will be used extensively in treating the disease. Viral vectors are commonly used in traditional gene delivery experiments. They are still problematic in terms of toxicity and immunogenicity, whereas the nanotechnological solution does not have these issues. Polyethylenimine nanogels complexes and PEG-based nanogels demonstrated the successful crossing of the blood-brain barrier *in vitro*.

Their intravenous administration through the blood-brain barrier with injection increases transmission efficiency, particularly when the nanogels are functionalized with insulin or Tf molecules. These results have been reported in a previous study on the 6-hydroxydopamine (6-OHDA) model of PD in rats in which enzyme activity in the striatum was found to be completely restored after a single intravenous administration of tyrosine hydroxylase-encoding plasmids and liposomes conjugated with Tf receptor antibody that is filled with PEG molecules. Recent research has reported that nerve growth factor (NGF) bound PBCA nanoparticles and L-Dopa encapsulated nanoparticles can permeate the blood-brain barrier and help relieve the symptoms of PD [26]. The length of mPEG PLGA nanoparticles (mPEG PLGA nanoparticles) was in the size range of 70 nm. It was noted that the Schisantherin A (SA) was used against Parkinson's disease (PD) in zebrafish larvae where SA was encapsulated in a nanoparticle formulation, which provided SA with an extended circulation in the bloodstream, which subsequently induced SA to have a higher brain uptake and was believed to be potentially efficacious for the treatment of PD. However, nanoparticulate delivery of SA was more efficient than SA suspension. The SA-NPs produced significant neuroprotective effects in zebrafish and a cell culture model of PD [27].

In the rotenone-induced PD animal model, rolinirole-loaded PLGA NPs were efficient in preventing neurodegeneration. Permeation tests in the rat model, both *in vitro* and *ex vitro*, demonstrated the sustained and extended release of RP as well as substantial permeability. The same authors also demonstrated that the PD model utilized in their investigation restored biochemical abnormalities, indicating that the NPs employed in their study had potential for additional research [28]. Trapani et al. developed dopamine-encapsulated polymeric chitosan NPs and tested them *in vitro* on Madin Darby canine kidney II-multi-drug resistance gene-1 (MDCKII-MDR1) cell lines [29]. The same researchers discovered an increase in striatal dopamine concentrations as well as a significant capacity for transportation across the BBB. In vivo testing on a rat model was also carried out in the same study, in which rats were acutely treated with the formed NPs and microdialysate samples were collected. Pillay et al. found that three days after treatment, there was a peak in dopamine release, followed by sustained dopamine release [30].

18.4.4 Huntington's Disease

Huntington's disease is caused by the destruction of some neurons in the striatum and other regions of the brain that gradually leads to gradual sensory, neurological, and psychological problems. Huntington's disease (HD) is a genetic disease. It is a trinucleotide repeat disorder with CAG repeats. Precisely, it is caused by a mutation in exon 1 of the huntingtin gene. Because of this mutation, there is an expansion of a polyglutamine (polyQ) and the misfolding and accumulation of huntingtin protein (HTT) in the brain [31]. Many studies affirm the role of astrocytes in HD. Studies that evaluate the brain tissue of HD patients and mouse models of HD indicate that HTT levels are abnormally elevated in astrocytes of the striatum, which contributes to the pathology and premature death of HD-like disease [32].

There are symptomatic and preventive therapeutic options for HD, but none effectively cure the disease. Tetrabenazine is a drug approved by the FDA for use in the symptomatic treatment of HD. When an elevated level of oxidative stress has been observed in combination with neurodegenerative diseases such as HD, a link has been confirmed in the literature [33]. Since this is the case, antioxidant therapy may be used in HD to further avoid oxidative stress. Fullerenols can remove free radicals and thus help reduce the oxidative burden on the cell. To conclude, it is said that the findings indicate antagonistic activity on glutamate receptors, and therefore, they could be suitable for neuroprotective purposes. In addition to treating individuals with elevated blood pressure, nitrendipine has been found to reduce the occurrence of dementia in patients with Alzheimer's disease (up to 50% in two years). Due to the hydrophilicity of this compound, there are permeability problems and therefore low BBB penetration. In a sample, the solid lipid nanoparticles (SLNs) of nitrendipine were made, and then a comparison was done to decide if bulk drug and nano formulation methods work the same. It was shown that higher absorption of the drug was possible when it was encapsulated in SLNs. The results of this research indicated that encapsulated cyclodextrin nanoparticles (CNPs) with short-interfering RNA (siRNA) are capable of reducing HTT mRNA expression both *in vivo* and *in vitro* [34].

18.4.5 Multiple Sclerosis (MS)

Multiple sclerosis is an autoimmune and inflammatory condition that develops over time. It is a multifactorial disorder that affects the nervous system, leading to an immune reaction by the body's cells. Demyelination of myelinated axons results from many pathogenic pathways, which contributes to the propagation of nerve impulses at a sluggish rate. All the data points out that microglia, astrocytes, and T cells play a cooperative role in the onset of demyelination [35]. Diagnosis is made based on clinical results, magnetic resonance imaging (MRI), and an analysis of cerebrospinal fluid (CSF). The prevalence rate of MS varies between two and 150 per 100,000 people, and females are more likely to be affected than males. An MS assault disrupts the brain barrier in the brain or spinal cord where peripheral T lymphocytes gain entry to the brain and eventually induces disruption to the myelin, which contributes to demyelination.

With nanotechnology's advances in the last few years, new methods for diagnosing and treating of multiple sclerosis have also been developed. The latest study has shown that therapeutic therapy can transform patient services and the treatment of multiple sclerosis. Through the association of carbon nanotubes with stem cells, new lines of research into cell activity are being made open, and new possibilities are being uncovered. It offers other strategies for treating the symptoms of MS pathology that are not linked to genetic modulation of signaling pathways *in vivo* [36]. To evaluate sustained *in situ* delivery of CNTF, researchers used microcapsules filled with CNTF and administered them by injection into the bloodstream. The result was found that this formulation had no effect on the immune system and was free of cytotoxicity. In addition to its role in therapeutics, the nano-based approach has applications in diagnosing multiple sclerosis. CSF or the diseased brain's biomarkers can be identified with the DNA-carrier gold NP-based barcode assay, a highly sensitive assay. Because of the strong link between the identification of MS using radio methods and the effective diagnostic assay, this can be a very helpful diagnostic tool for MS. Because of new advances in nanotechnology, nanotools are now possible to use diagnose and treat the different phases of multiple sclerosis.

18.5 Nanoparticles in Detection of Neurological Cancers

Since nanoparticles are potent in minimizing side effects and toxicity, they have their application in cancer diagnosis and treatment. They serve as an alternative source for multidrug-resistant cancers and easily pass the blood-brain barrier (BBB). As a result, they have a promising future for diagnosing and treating cancer [37]. Cancer treatment methodologies use them as carriers of anti-cancer drugs, gene delivery directly into the tumor, and diagnosing techniques employ them as contrasting agents for imaging. A wide range of nanoparticles was synthesized using organic, lipid, and glycan compounds and synthetic polymers [37].

In light of the benefits discussed above, the use of nanoparticles to diagnose or detect tumor and cancers in the brain is gaining traction among existing diagnostic facilities. There are various mechanisms by which cancer can be detected using nanoparticles. They are as follows.

18.5.1 Detection of Extracellular Biomarkers of Cancer

A CNS biomarker is a molecule that can be detected in a sample of tissue, blood, or any other specimen, which directly indicates cancer in the nervous system. They can be a variety of cell membrane proteins, simple carbohydrates, DNA and RNA of the tumor cells, etc. Usually, nanoparticles are specific for detecting the type of tumor/cancer affinity for a particular biomarker. Gold nanoparticles (Au-NPs), polymer dots (PD), and quantum dots (QD) are the most commonly used diagnosing agents for cancer detection [38].

18.5.2 Proteins as Biomarkers

Various proteins are used as biomarkers to indicate cancers, like PSA, AFP, CEA, etc. Usually, assessment of proteins as biomarkers is done using a sandwich method, which employs a capturing antibody, a secondary antibody, a secondary capture antibody, and a biomarker. The detection of secondary antibodies is achieved by staining or a fluorescence assay. The receptors identify peptides such as arginine-glycine-aspartame on the cell membrane that is undergoing metastasis or cancer. So, such peptides are identified using gold nanoparticles with legumain responsive gold nanoparticles that have high penetration into T1 mammary gland cancers [39].

Aptamers are single-stranded DNA/RNA pairs that rely on the conjugation of nanoparticles. Cy5-PLA/aptamer is commonly used to diagnose tumors. Upconverting nanophosphors (UCNP) is a novel bioluminescent biomarker that can absorb the infrared range of light used in detecting phospholipase, which is directly linked to cancer cell progression [40].

18.5.3 Detection of Micro-RNA (miR) as a Biomarker

There were novel nanoparticles developed to recognize the miR-141 biomarker that is sensitive and promising in detecting cancer. The nanoparticles of cadmium or zinc are functionalized using nucleic acid bonded with telomerase to recognize the miR sequence. After forming a bond with the miR, the nanoparticle cleaves into a fluorescent complex detected by chemiluminescence detectors [41].

18.5.4 Detection of Extracellular Vesicles (EV)

Generally, like the intracellular vesicles, the EV also encodes the information about the miR, DNA, and mRNA raised from the cancerous mother cell. This cellular, molecular detail was analyzed and detected by a novel nanoparticle technology called nanopore magnets. Machine learning algorithms detect RNA-sequencing and cancer biomarkers, and promising results are achieved using nanopore magnets to detect cancers. The usual nanoparticles that assist in this process are made from silver (AuNP-A650) [42].

18.5.5 Circulating DNA (ctDNA) as Biomarkers

Tumor-marked DNA fragments are circulated in the blood called ctDNA, which is primary cancer cells. Detection of such DNA fragments in the blood and the aberrations caused are detected by nanocluster silver probes developed using fluorescent indicators. They are used to detect the gene sequences responsible for or the result of cancers in the body that are freely circulated, as shown in Figure 18.2. The novel AgNC probes are bound to the fragmented DNA and are usually indicated by fluorescent detections ranging from higher to lower depending on the cancer cells [42].

18.6 Detection of Cancer Cells in the Direct Method

18.6.1 Detection of Circulating Cells

Metastasis is often the result of primary circulating cancer cells in the blood reaching various sites to cause secondary cancers. There are very limited or no means to diagnose or predict metastasis before the occurrence, so nanoparticles assist us in this respect. Gold nanoparticles (AuNP), nanoparticle-based magnets (NPM), nanowires, pillars and tubes, quantum dots (QD), and graphene oxide nanoparticles are used to detect cells directly in the bloodstream [43]. The Magsweeper is another technology developed as a magnetic immune device that is useful for isolating cancer cells. This device is used to

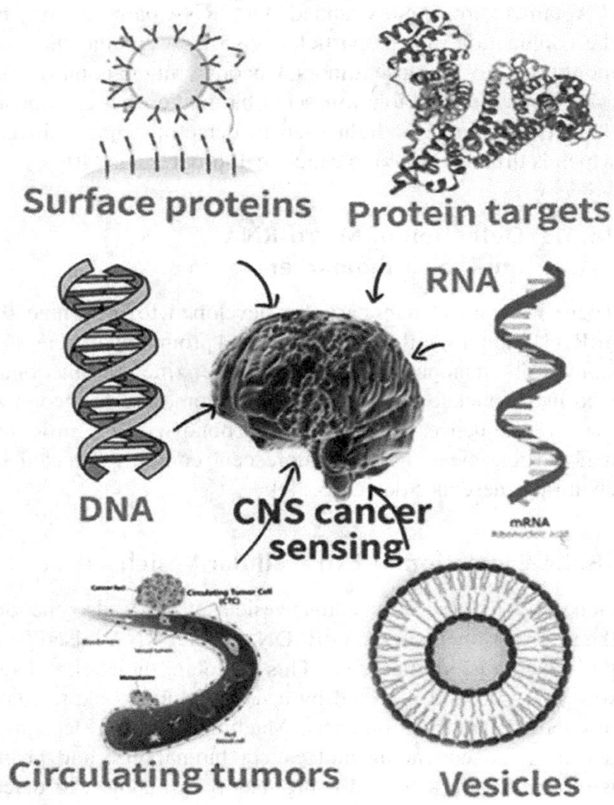

FIGURE 18.2 Detection of brain cancer using nanotechnology.

attract and trap magnetic labeled nanoparticles that are linked to cancer cells, as tabulated in Table 18.2.

18.6.2 Detection of Cells via Surface Protein Detection

The nanoparticles investigate and bind to surface moieties like proteins, commonly short peptide molecules, oligonucleotide aptamer molecules, antibodies, etc. Cell surface markers like androgen receptors, glycan proteins, vimentin, major vault proteins, and fibroblast activator proteins have also been investigated to estimate cancer prognosis.

18.6.3 Detection by Targeting the Tumors by Imaging

Cancer detection *in vitro* is a simple and cost-effective procedure that is gaining popularity. Blood samples and biopsy specimens of tumor tissue can be used to detect cancer cells. Imaging systems were created to detect changes in tissue caused by cancer. When nanoparticles are added to the system, they induce some changes that can be imaged directly by penetrating the cancer cells and the endothelium by targeting the tumor tissue.

18.6.4 Passive Targeting

The nanoparticles are linked to the extravasation capacity of the bloodstream before being transferred into tumor cells in this approach. Tumor cells do not develop in blood vessels because nanoparticles accumulate in the tissue's tight endothelial junctions. This passive entrance into the cell's space can be studied by understanding the macromolecular transport in the tissue of the tumor cells. QD and AgNP are the most common nanoparticles that are used to detect cancer cells using the photo-imaging technique.

18.6.5 Active Targeting

Researchers have discovered active receptors that can transport nanoparticles to cellular levels, in addition to passive transport and targeting of nanoparticles at the junction of cellular spaces. This is, in turn, time-dependent, and the sensitivity is usually high for these methods. Antibody-quantum dots conjugates attack the carboxylic acid antibody moieties of active tumor cells. Gold nanoparticles are also one of the best types transported in this system. Specific ligand-gated permeation can enhance tumor permeation and retention and reuptake for better stability and detection [44].

18.7 Ongoing Clinical Trials

The ongoing clinical trials using nanotechnology were tabulated in Table 18.3.

There is an ongoing clinical trial titled "A Phase II Study Evaluating AGuIX® Nanoparticles in Combination with Stereotactic Radiation for Brain Metastases", which targets the treatment and detection of brain metastasis. The Paucicentric phase II clinical trial was conducted in a single-arm in a flashing-high design that aims to detect the potency and efficacy of a nanoparticle adjuvant drug, AGulX. AGulX is defined as activation and guided irradiation of X-rays along with chelation of gadolinium with polysiloxane-based NP. These are combined with magnetic resonance properties to improve kidney tumor permeability and drug clearance [45].

TABLE 18.2
Detection of Cancers Using Nanoparticles

S. no.	Type of nanoparticle	Target	Type of targeting
1	Magnetic nanoparticles	Antibody	Direct circulating cells
2	Gold nanoparticles	AptamerAntibody	Active targeting
3	Quantum dots	Aptamer	Passive targeting, biomarkers
4	Nanorod arrays	DNA aptamer	Biomarkers
5	Nanofibers	Antibody	Biomarker proteins
6	Polymer dots	Antibody	Extracellular biomarkers
7	Carbon nanotubes	Antibody	Direct circulating cells
8	Nanoparticle-coated silicon bead	Antibody	Extracellular vesicles

TABLE 18.3

Ongoing Clinical Trials Using Nanotechnology

S. no.	Clinical trial	Technology applied	Brain cancer targets	NCT code
1	Phase 1	EGFR(V)-EDV-Dox with EnGeneIC EDV™	Glioblastoma, grade IV – astrocytoma	NCT02766699
2	Phase 2	AGuIX®	Multiple brain metastases	NCT03818386
3	Phase 1	89Zr-cRGDY ultrasmall silica particle	Brain metastases	NCT03465618
4	Phase 1	MTX110 convection-enhanced delivery	Diffuse midline gliomas	NCT04264143
5	Phase 2	ABI-009 Nab-Rapamycin in Bevacizumab-resistant subjects	Glioblastoma and gliomas	NCT03463265
6	Phase 1	124I-labeled cRGDY silica nanomolecular particle	Metastatic melanoma to the brain	NCT01266096
7	Phase 2	AGuIX® gadolinium with polysiloxane-based NP	Oligo brain metastases	NCT04094077

TherAguix SAS conducted the clinical trial titled "Radiotherapy of Multiple Brain Metastases Using AGuIX® Gadolinium-Chelated Polysiloxane Based Nanoparticles: A Prospective Randomized Phase II Clinical Trial" in patients with brain metastasis who were undergoing radiation therapy. The survival of those patients was estimated at six months, and the novel approach increased their life span significantly. AGuIX, developed by Aguix, was used in this study to investigate its radiosensitivity and multimodal imaging capability. This study focuses on the multiplicity of brain tumors, radiotherapy resistance due to tissue changes, and the lack of anti-cancer drug distribution during metastatic disease. The drug's effect was significant in rodents where the study was conducted in more than eight models of *in vivo* testing [46].

A clinical trial named "A Phase 2, Open-Label Study of ABI-009 (Nab-Rapamycin) in Bevacizumab-Nave Subjects with Progressive High-Grade Glioma Following Prior Therapy and Subjects with Newly Diagnosed Glioblastoma" is currently underway to investigate the effect of nanotechnology-based rapamycin and a few other drugs on progressive high-grade glioblastoma and glioma. These drugs were coupled with albumin in the form of nanoparticles and tested in subjects with gliomas. Rapamycin nanoparticles are paired with various drugs like temozolomide, lomustine, marizomib, and adjuvant radiotherapy [47].

A clinical trial was designed to study the maximum tolerable doses of nanoparticle design of panobinostat drug-water soluble form and gadolinium with the title "A Phase I Study Examining the Feasibility of Intermittent Convection-Enhanced Delivery (CED) of MTX110 for the Treatment of Children with Newly Diagnosed Diffuse Midline Gliomas". The scope of the study is to use the drug effectively against newer gliomas and to deliver the pump into the tumor tissue directly for nine days with a single dose. The drugs are delivered using a convection-enhanced delivery system [48].

In collaboration with Engenic, John Hopkins University designed a study to evaluate the safety and tolerable capacity of doxorubicin delivered through a nanoparticle technology named EGFR. The title of the study was "A Phase 1 Study to Evaluate the Safety, Tolerability, and Immunogenicity of EGFR (Vectibix® Sequence)-Targeted EnGeneIC Dream Vectors Containing Doxorubicin (EGFR (V)-EDV-Dox) in Subjects with Recurrent Glioblastoma Multiforme (GBM)". The formulation was tested to prove the body's immune responses to fight glioblastoma in multiple forms. The drug is attached to specific antibodies that are loaded and packaged into nano cells. The nano cells are derived from bacteria, and when they are given to the patient, they attach to the cancer cells. The body's immune system attacks the bacterial antigen and kills the cancer cells along with it [49].

18.8 Bioimaging

Imaging based on molecular markers has proven to be the most effective tool for investigating and identifying biological information and revealing the dynamics of disease pathways. It often becomes important to assess and view the response directly due to certain therapies [50]. By integrating the imaging and diagnostic systems, we can unveil the real-time pharmacokinetics and pharmacodynamics of the drug molecule. X-rays, fluorescence, radionuclides, spectroscopies, and magnetic resonance are currently used to image or analyze biological products. Optical imaging helps all the technologies to better evaluate and analyze the sample in real-time. Nanoparticles have been proven to have numerous applications in science and, especially, in medicine to treat and diagnose diseases. The main advantages of nanoparticles for imaging are size variations, morphological dissimilarities, and biocompatibility [51]. Nanoparticles can trigger optical signals and have the potential to aid in bio-imaging. As a result, current research interests revolve around nanoparticles to effectively use them in the methods mentioned earlier of analysis and bio-imaging. The nanoparticles interact with the cellular matter inside the body. That may also include cancer or tumor cells. This nature can be used to detect the growth and development of cancers, neurological disorders, and degeneration. This also helps to increase the specificity and selectivity of the organelle for the nanoparticles that help in bio-imaging.

18.8.1 Nanoparticles in Bioimaging

Nanoparticles due to their advanced properties can be effectively used in bioimaging. They are classified into two categories:

i) Organic nanoparticles
ii) Inorganic nanoparticles

Organic nanoparticles include quantum dots, metal nanoclusters, silica-based nanoparticles, graphene, carbon dots, etc.; inorganic nanoparticles include metallic nanoparticles with gold, silver, zinc, and other inorganic metals. They have unique properties of surface plasmon resonance and Raman scattering with fluorescence, unlike inorganic particles having photo fluorescence and ligand-based luminescence.

18.8.1.1 Imaging Using Fluorescence

Nano cages made of gold and viral vectors used for engulfing the DNA and other drugs are commonly used in this method. Paclitaxel inserted into nanoparticles of human serum albumin used for cancer treatment. Hyaluronic acid is delivered using miRNA using fluorescent AuNP complexes [52].

18.8.1.2 Raman Scattering

Nanoparticles having the Raman scattering works in the IR range of light. They are optical probes that have photo-thermal illumination. Gold nanorods are used along with multifunctional polymers for conduction to achieve the Raman scattering effect. Similarly, silver nanoshells are used in the medical arena along with serum albumin to enhance biocompatibility [53].

18.8.1.3 Imaging Using Persistent Luminescence

Luminescence exists for about seven to eight days which commonly employs metal nanoparticles like zinc. $ZnGa_2O_4:Cr_3^+$ possesses a strong and long-lasting luminescence in the UV range of excitation. They are charged and aligned to any drug candidate and can be visualized using detectors [54].

18.8.1.4 Imaging Using Photoacoustics

This is a method of visualizing photoacoustic response in the biomolecules and cell organelles. Perylene-3,4,9,10-tetracarboxylic diimide is one of those molecules designed to display properties of acrotism and is employed along with electronic devices that can track the responses. Semiconducting polymers and carbon dots are examples of this technique [55].

18.9 Tissue Engineering in Neurology with Nano-Scaffolds

Tissue engineering, also known as regenerative medicine, generates new tissues from a small number of cells with the assistance of scaffolds, either *in vivo* or *in vitro* [56]. Healthy cells are sourced and placed on the surface of a suitable scaffold, where they are cultured in a nurturing environment. The scaffold is made of nanofibers designed to mimic the extracellular matrix, which plays an important role in providing physical and biochemical support for the cells. Because of its nano-sized fibers, the scaffold is also known as a nano-scaffold. The scaffold's structure aids in cell attachment, biochemical factor regulation, and cell migration. The structure has a high porosity to allow for cell and nutrient migration, and it is frequently made of a biodegradable material that disintegrates after the tissue or organ is built. Materials are selected based on physical properties such as size, strength, and biodegradability, and a combination of materials can be used to achieve desired properties.

Depending on the material, different techniques are used to build the scaffold. Electrospinning is a popular technique that involves dissolving nanomaterials in a solvent and placing them in a syringe. As the solution is pumped from the syringe, a high voltage is applied at the same time, evaporating the solvent and producing wavy nanofibers that are collected by the neutrally charged collector. The physical properties of the fiber, such as size, porosity, and uniformity, can be controlled by adjusting the electrospinning environment parameters such as distance between the needle and collector, flow rate, needle diameter [57], solution concentration, viscosity, conductivity [58], molecular weight [59], and environmental temperature.

18.10 Neuro Knitting

Nano neuro knitting is a new technology that uses nano scaffolding techniques to repair nervous system tissues [60]. Nano neuro knitting, which is currently being investigated in various research projects, has been shown to allow partial reinnervation in damaged areas of the nervous system via interactions between potentially regenerative axons and peptide scaffolds [60]. This interaction has been shown to result in sufficient axon density renewal to restore functionality. While nano neuro knitting shows promise, the uncertainty of the effects on human subjects necessitates additional research before clinical trials. Researchers have been able to investigate clinically relevant applications involving the promotion of tissue regeneration at sites of acute damage thanks to nanotechnology scaffolds. These methods are used specifically in nano neuro knitting to repair nervous system tissues [60].

18.11 Future Prospects

As per the discussion in the article, nanoparticles and nanotechnology have been an important technology in the medical field and especially in neurology in treating neurodegenerative diseases and other types of cancers that occur in the nervous system, as shown in Figure 18.3.

The use of nanoscale materials and agents to treat CNS problems is already available on the market in cell cultures and animal systems. The possibility of the use of nanoparticles for novel treatment techniques and therapies for tissue regeneration and understanding and assessing the responses of cells to neural signals and the development of artificial nano neural networks is under discussion now [61]. The major hurdle to treating and delivering drugs to the nervous system is the BBB. It is understood that any disease should cross the BBB easily to show its full activity in the brain for any disease. The body's natural protection system poses a problem here, and nanotechnology answers all the problems effectively.

Various nano delivery systems are designed for enhanced efficacy and clinical applications. However, the drug's effectiveness is dependent on effective delivery, which is complicated by factors such as toxicity and aggregation. There is another problem with the clearance of the drug before showing its activity. To properly and safely deliver the drug to the

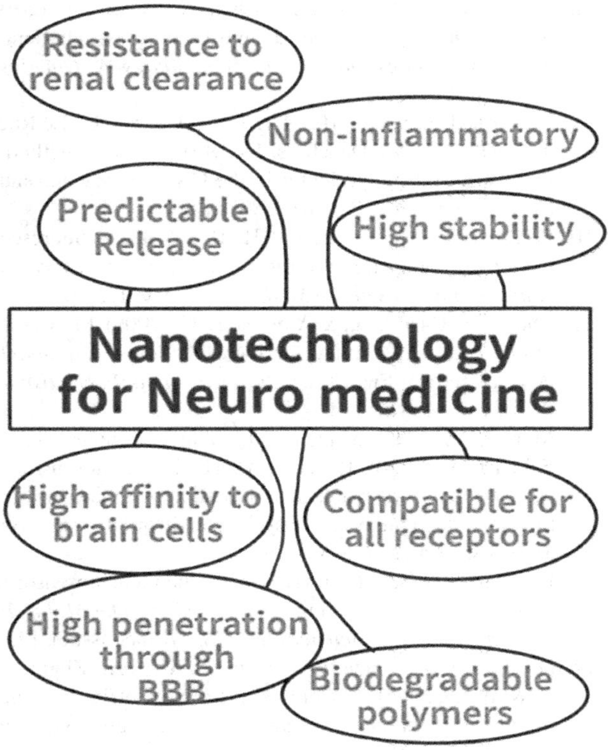

FIGURE 18.3 Advantages of nanotechnology applied to CNS.

targeted site, a thorough understanding of the mechanism of drug release from nano delivery systems and formulations is required. When nanotechnology offers a significant benefit for the treatment of neurodegenerative diseases, the following parameters must be thoroughly studied and adapted for effective and safer treatment methods.

i) The nanoparticles and the carrier size are to be maintained under 100 nm and biodegradable.
ii) The morphology and shape of the nanocarrier are to be maintained in sync with the receptors and the sites of binding.
iii) The production cost of the nanoparticles is to be maintained low and the productivity is to be enhanced to an industrial scale of production.
iv) The nanocarriers are said to be biocompatible and nontoxic, as well as anti-inflammatory and anti-thrombotic.
v) Ease of penetration through biological membranes like the BBB can be achieved effectively.
vi) The produced nanoparticles are stable and resist immediate renal clearance before the drug is delivered to the targeted site.
vii) The nanocarriers must be designed in such a way that the drug is delivered or released smoothly and predictably.
viii) The developed nanoparticles should have a high and specific affinity for receptors in the brain or nervous system, to deliver the drug to the targeted site.

18.11.1 NEMS: Nanoelectromechanical Devices

Nanoelectromechanical devices, or NEMS, are nanometer-scale devices that can perform electrical or mechanical work and can be used as sensors, actuators, or pumps. Because of their small size, these devices can monitor and measure physiological conditions at the cellular level investigated. NEMS as a microsurgical instrument is known as a nano knife to operate at the cellular level in peripheral neurosurgery [62]. Rasmeisel Jr. et al. used a nano knife to perform irreversible electroporation in canines to assess the procedure's safety, feasibility, and efficacy. They concluded that irreversible electroporation with nano knives is best suited for the controlled destruction of target cells in situations where conventional surgery is not possible [63].

18.11.2 Artificial Intelligence in Nanotechnology

Machine learning and artificial intelligence techniques, particularly computer vision, have advanced rapidly in the last decade because of deep neural networks. Computer vision is the study of inferring information from image data, such as object recognition, classification, and objection detection. These advancements have created new opportunities in neurology and nanotechnology [64].

It is critical to design nano-scaffolds with the desired structure and strength for successful cell growth and tissue generation. Eichholz et al. investigated the effect of scaffold architectures on stem cell behavior and concluded that precise scaffold construction with extracellular matrix architectures specific to the tissue drives stems cells without the need for biochemical signals [65]. This necessitates meticulous control over the nanofiber creation and layering processes. However, controlling the electrospinning process is a challenging task as it is highly stochastic, dynamic, and the structure is hard to observe.

Automation and artificial intelligence are increasingly being used to improve manufacturing processes. Tan et al. describe how multi-nozzles, angled multi-nozzles, and rotating nozzles can be used to control scaffolding using robotics [66]. Tourlomousis et al. created various scaffolds using the melt electro writing technique and modeled the cell shapes using machine learning and metrology [67]. Similarly, Shi et al. demonstrated that machine learning techniques such as deep neural networks with multi-objective optimization improve the speed and stability of bio-printing [68].

One of the most important applications of nanomaterials is drug delivery. However, locating a suitable nanomaterial for the task at hand is not easy. Recently, modeling the behaviors of nanomaterials suitable for targeted drug delivery with high cellular uptake has been done. Fourches et al. and Winkler et al. used data mining techniques to predict cellular uptake of cross-linked iron-oxide nanoparticles [69, 70].

Toxicity is still a major concern with nanomaterials and determines the viability of medical applications. Toxicity is assessed by cellular parameters such as oxidative stress, inflammatory response, genotoxicity, and cell viability. Jones et al. [71] published a review of studies that used data mining methods to assess the toxicity of metal oxide nanomaterials along with the other literature reports [72–77].

18.12 Conclusion

Alzheimer's disease, Parkinson's disease, myocardial infarction, and brain cancer are all serious health issues. These are the major causes of disability and mortality for most of the population in the world. Nanotechnology is the most advanced and current strategy for treating neurological problems. Nanotechnology products like liposomes, exosomes, nanotubes, nanorods, nano-emulsions, etc. are effectively used in drug delivery to the brain. They are efficient in delivering a high therapeutic concentration of drugs to the target site. The nanocarriers with the enhanced ability to carry and deliver drugs across the BBB into the brain are used to target the genes corresponding to cancers and other degenerations in the nervous tissue. Neuro knitting, which uses nanoparticles to repair damaged nervous tissue, is gaining popularity and may have applications in neurodegenerative disorders. Only a few clinical studies have been reported with success in this field, hence further research is required. Deep neural networks have paved a new path for the development of machine learning and artificial intelligence techniques that are helpful in bio-printing and the designing of medically applicable nanorobots. In addition, the emphasis should be on noninvasive methods of introducing nanoparticles into the human body, such as mucosal, sublingual, nasal, or buccal routes. Special emphasis has to be given on limiting the toxicity of the drugs delivered via nanoparticles.

REFERENCES

1. Kanwar, J. R., Sun, X., Punj, V., Sriramoju, B., Mohan, R. R., Zhou, S. F., … & Kanwar, R. K. (2012). Nanoparticles in the treatment and diagnosis of neurological disorders: untamed dragon with fire power to heal. *Nanomedicine: Nanotechnology, Biology and Medicine*, *8*(4), 399–414.
2. Chhabra, R., Tosi, G., & M Grabrucker, A. (2015). Emerging use of nanotechnology in the treatment of neurological disorders. *Current Pharmaceutical Design*, *21*(22), 3111–3130.
3. Nasir, A., Kausar, A., & Younus, A. (2015). A review on preparation, properties and applications of polymeric nanoparticle-based materials. *Polymer-Plastics Technology and Engineering*, *54*(4), 325–341.
4. Sood, A., Arora, V., Shah, J., Kotnala, R. K., & Jain, T. K. (2017). Multifunctional gold coated iron oxide core-shell nanoparticles stabilized using thiolated sodium alginate for biomedical applications. *Materials Science and Engineering: C*, *80*, 274–281.
5. Sercombe, L., Veerati, T., Moheimani, F., Wu, S. Y., Sood, A. K., & Hua, S. (2015). Advances and challenges of liposome assisted drug delivery. *Frontiers in Pharmacology*, *6*, 286.
6. Mody, V. V., Siwale, R., Singh, A., & Mody, H. R. (2010). Introduction to metallic nanoparticles. *Journal of Pharmacy and Bioallied Sciences*, *2*(4), 282.
7. Thaxton, C. S., Hill, H. D., Georganopoulou, D. G., Stoeva, S. I., & Mirkin, C. A. (2005). A bio-bar-code assay based upon dithiothreitol-induced oligonucleotide release. *Analytical Chemistry*, *77*(24), 8174–8178.
8. Zhang, Y., Li, M., Gao, X., Chen, Y., & Liu, T. (2019). Nanotechnology in cancer diagnosis: progress, challenges and opportunities. *Journal of Hematology & Oncology*, *12*(1), 1–13.
9. Edwards, I. J., Singh, M., Morris, S., Osborne, L., Le Ruez, T., Fuad, M., … & Deuchars, J. (2013). A simple method to fluorescently label pericytes in the CNS and skeletal muscle. *Microvascular Research*, *89*, 164–168.
10. Powley, T. L., & Berthoud, H. R. (1991). A fluorescent labeling strategy for staining the enteric nervous system. *Journal of Neuroscience Methods*, *36*(1), 9–15.
11. Zhang, D., Carr, D. J., & Alocilja, E. C. (2009). Fluorescent bio-barcode DNA assay for the detection of *Salmonella enterica* serovar Enteritidis. *Biosensors and Bioelectronics*, *24*(5), 1377–1381.
12. Mobed, A., Hasanzadeh, M., Ahmadalipour, A., & Fakhari, A. (2020). Recent advances in the biosensing of neurotransmitters: material and method overviews towards the biomedical analysis of psychiatric disorders. *Analytical Methods*, *12*(4), 557–575.
13. Hamdan, S. K. (2014). In vivo electrochemical biosensor for brain glutamate detection: a mini review. *The Malaysian Journal of Medical Sciences: MJMS*, *21*(Spec Issue), 12.
14. Du, P., Jin, M., Chen, G., Zhang, C., Cui, X., Zhang, Y., … & Wang, J. (2017). Competitive colorimetric triazophos immunoassay employing magnetic microspheres and multi-labeled gold nanoparticles along with enzymatic signal enhancement. *Microchimica Acta*, *184*(10), 3705–3712.
15. Yang, S., & Rothman, R. E. (2004). PCR-based diagnostics for infectious diseases: uses, limitations, and future applications in acute-care settings. *The Lancet Infectious Diseases*, *4*(6), 337–348.
16. Gu, Z., Fu, J., Lin, H., & He, Y. (2020). Development of 3D bioprinting: from printing methods to biomedical applications. *Asian Journal of Pharmaceutical Sciences*, *15*(5), 529–557.
17. R. Jabir, N., Tabrez, S., Firoz, C. K., Kashif Zaidi, S., S Baeesa, S., Hua Gan, S., … & Amjad Kamal, M. (2015). A synopsis of nano-technological approaches toward anti-epilepsy therapy: present and future research implications. *Current Drug Metabolism*, *16*(5), 336–345.
18. Samia, O., Hanan, R., & Kamal, E. T. (2012). Carbamazepine mucoadhesive nanoemulgel (MNEG) as brain targeting delivery system via the olfactory mucosa. *Drug Delivery*, *19*(1), 58–67.
19. Taylor, M., Moore, S., Mourtas, S., Niarakis, A., Re, F., Zona, C., … & Allsop, D. (2011). Effect of curcumin-associated and lipid ligand-functionalized nanoliposomes on aggregation of the Alzheimer's Aβ peptide. *Nanomedicine: Nanotechnology, Biology and Medicine*, *7*(5), 541–550.
20. Yang, Z., Zhang, Y., Yang, Y., Sun, L., Han, D., Li, H., & Wang, C. (2010). Pharmacological and toxicological target organelles and safe use of single-walled carbon nanotubes as drug carriers in treating Alzheimer disease. *Nanomedicine: Nanotechnology, Biology and Medicine*, *6*(3), 427–441.
21. Wilson, B., Samanta, M. K., Santhi, K., Kumar, K. P. S., Paramakrishnan, N., & Suresh, B. (2008). Poly (n-butylcyanoacrylate) nanoparticles coated with polysorbate 80 for the targeted delivery of rivastigmine into the brain to treat Alzheimer's disease. *Brain Research*, *1200*, 159–168.

22. Sun, W., Xie, C., Wang, H., & Hu, Y. (2004). Specific role of polysorbate 80 coating on the targeting of nanoparticles to the brain. *Biomaterials*, *25*(15), 3065–3071.
23. Javed, I., Peng, G., Xing, Y., Yu, T., Zhao, M., Kakinen, A., ... & Lin, S. (2019). Inhibition of amyloid beta toxicity in zebrafish with a chaperone-gold nanoparticle dual strategy. *Nature Communications*, *10*(1), 1–14.
24. Kumar, A., & Singh, A. (2015). A review on Alzheimer's disease pathophysiology and its management: an update. *Pharmacological Reports*, *67*(2), 195–203.
25. Soursou, G., Alexiou, A., Md Ashraf, G., Ali Siyal, A., Mushtaq, G., & Kamal, A. M. (2015). Applications of nanotechnology in diagnostics and therapeutics of Alzheimer's and Parkinson's disease. *Current Drug Metabolism*, *16*(8), 705–712.
26. Mohanraj, K., Sethuraman, S., & Krishnan, U. M. (2013). Development of poly (butylene succinate) microspheres for delivery of levodopa in the treatment of parkinson's disease. *Journal of Biomedical Materials Research Part B: Applied Biomaterials*, *101*(5), 840–847.
27. Chen, T., Li, C., Li, Y., Yi, X., Wang, R., Lee, S. M. Y., & Zheng, Y. (2017). Small-sized mPEG–PLGA nanoparticles of Schisantherin A with sustained release for enhanced brain uptake and anti-parkinsonian activity. *ACS Applied Materials & Interfaces*, *9*(11), 9516–9527.
28. Dudhipala, N., & Gorre, T. (2020). Neuroprotective effect of ropinirole lipid nanoparticles enriched hydrogel for parkinson's disease: in vitro, ex vivo, pharmacokinetic and pharmacodynamic evaluation. *Pharmaceutics*, *12*(5), 448.
29. Toropova, A. P., Toropov, A. A., Rallo, R., Leszczynska, D., & Leszczynski, J. (2015). Optimal descriptor as a translator of eclectic data into prediction of cytotoxicity for metal oxide nanoparticles under different conditions. *Ecotoxicology and Environmental Safety*, *112*, 39–45.
30. Thondavada, N., Chokkareddy, R., Naidu, N. V., & Redhi, G. G. (2020). New generation quantum dots as contrast agent in imaging. In *Nanomaterials in Diagnostic Tools and Devices* (pp. 417–437). Elsevier.
31. Bates, G. P., Dorsey, R., Gusella, J. F., Hayden, M. R., Kay, C., Leavitt, B. R., ... & Tabrizi, S. J. (2015). Huntington disease. *Nature Reviews Disease Primers*, *1*(1), 1–21.
32. Dobson, J. (2006). Gene therapy progress and prospects: magnetic nanoparticle-based gene delivery. *Gene therapy*, *13*(4), 283–287.
33. Liu, Z., Zhou, T., Ziegler, A. C., Dimitrion, P., & Zuo, L. (2017). Oxidative stress in neurodegenerative diseases: from molecular mechanisms to clinical applications. *Oxidative Medicine and Cellular Longevity*, *2017*, Article ID 2525967.
34. Godinho, B. M., Ogier, J. R., Darcy, R., O'Driscoll, C. M., & Cryan, J. F. (2013). Self-assembling modified β-cyclodextrin nanoparticles as neuronal siRNA delivery vectors: focus on Huntington's disease. *Molecular Pharmaceutics*, *10*(2), 640–649.
35. Niranjan, T., Chokkareddy, R., Redhi, G. G., & Naidu, N. V. (2020). Ionic liquids as gas sensors and biosensors. In *Green Sustainable Process for Chemical and Environmental Engineering and Science* (pp. 319–342). Elsevier.
36. Nunes, A., Al-Jamal, K. T., & Kostarelos, K. (2012). Therapeutics, imaging and toxicity of nanomaterials in the central nervous system. *Journal of Controlled Release*, *161*(2), 290–306.
37. Aghebati-Maleki, A., Dolati, S., Ahmadi, M., Baghbanzhadeh, A., Asadi, M., Fotouhi, A., ... & Aghebati-Maleki, L. (2020). Nanoparticles and cancer therapy: perspectives for application of nanoparticles in the treatment of cancers. *Journal of Cellular Physiology*, *235*(3), 1962–1972.
38. Zhang, H., Lv, J., & Jia, Z. (2017). Efficient fluorescence resonance energy transfer between quantum dots and gold nanoparticles based on porous silicon photonic crystal for DNA detection. *Sensors*, *17*(5), 1078.
39. Puig-Saus, C., Rojas, L. A., Laborda, E., Figueras, A., Alba, R., Fillat, C., & Alemany, R. (2014). iRGD tumor-penetrating peptide-modified oncolytic adenovirus shows enhanced tumor transduction, intratumoral dissemination and antitumor efficacy. *Gene Therapy*, *21*(8), 767–774.
40. Sharipov, M., Tawfik, S. M., Gerelkhuu, Z., Huy, B. T., & Lee, Y. I. (2017). Phospholipase A2-responsive phosphate micelle-loaded UCNPs for bioimaging of prostate cancer cells. *Scientific Reports*, *7*(1), 1–9.
41. Jou, A. F. J., Lu, C. H., Ou, Y. C., Wang, S. S., Hsu, S. L., Willner, I., & Ho, J. A. A. (2015). Diagnosing the miR-141 prostate cancer biomarker using nucleic acid-functionalized CdSe/ZnS QDs and telomerase. *Chemical Science*, *6*(1), 659–665.
42. Jiang, Y., Shi, M., Liu, Y., Wan, S., Cui, C., Zhang, L., & Tan, W. (2017). Aptamer/AuNP biosensor for colorimetric profiling of exosomal proteins. *Angewandte Chemie International Edition*, *56*(39), 11916–11920.
43. Huang, Q., Wang, Y., Chen, X., Wang, Y., Li, Z., Du, S., ... & Chen, S. (2018). Nanotechnology-based strategies for early cancer diagnosis using circulating tumor cells as a liquid biopsy. *Nanotheranostics*, *2*(1), 21.
44. Yang, Y., Chen, Q., Li, S., Ma, W., Yao, G., Ren, F., ... & Yu, Z. (2018). iRGD-mediated and enzyme-induced precise targeting and retention of gold nanoparticles for the enhanced imaging and treatment of breast cancer. *Journal of Biomedical Nanotechnology*, *14*(8), 1396–1408.
45. Thondavada, N., Chokkareddy, R., Naidu, N. V., & Redhi, G. G. (2019). Environmental science and engineering applications of polymer and nanocellulose-based nanocomposites. In *Composites for Environmental Engineering* (pp. 135–178). Scrivener Publishing LLC.
46. Verry, C., Sancey, L., Dufort, S., Le Duc, G., Mendoza, C., Lux, F., ... & Balosso, J. (2019). Treatment of multiple brain metastases using gadolinium nanoparticles and radiotherapy: NANO-RAD, a phase I study protocol. *BMJ Open*, *9*(2), e023591.
47. Habibi, N., Quevedo, D. F., Gregory, J. V., & Lahann, J. (2020). Emerging methods in therapeutics using multifunctional nanoparticles. *Wiley Interdisciplinary Reviews: Nanomedicine and Nanobiotechnology*, *12*(4), e1625.
48. Zacharoulis, S., Szalontay, L., Higgins, D., Englander, Z., Jin, Z., Garvin, J., ... & Bruce, J. (2020). Ddel-07. A phase i study examining the feasibility of intermittent convection-enhanced delivery (ced) of mtx110 for the treatment of children with newly diagnosed diffuse midline gliomas. *Neuro-Oncology*, *22*(Suppl 3), iii284.
49. Chokkareddy, R., Thondavada, N., Thakur, S., & Kanchi, S. (2019). Recent trends in sensors for health and agricultural applications. In *Advanced Biosensors for Health Care Applications* (pp. 341–355). Elsevier.

50. Fan, W., Yung, B., Huang, P., & Chen, X. (2017). Nanotechnology for multimodal synergistic cancer therapy. *Chemical Reviews*, *117*(22), 13566–13638.
51. Wang, M., & Thanou, M. (2010). Targeting nanoparticles to cancer. *Pharmacological Research*, *62*(2), 90–99.
52. Wolfbeis, O. S. (2015). An overview of nanoparticles commonly used in fluorescent bioimaging. *Chemical Society Reviews*, *44*(14), 4743–4768.
53. Premasiri, W. R., Chen, Y., Fore, J., Brodeur, A., & Ziegler, L. D. (2018). SERS biomedical applications: diagnostics, forensics, and metabolomics. In *Frontiers and Advances in Molecular Spectroscopy* (pp. 327–367). Elsevier.
54. Shi, J., Song, J., Song, B., & Lu, W. F. (2019). Multi-objective optimization design through machine learning for drop-on-demand bioprinting. *Engineering*, *5*(3), 586–593.
55. Lee, C., Kwon, W., Beack, S., Lee, D., Park, Y., Kim, H., ... & Kim, C. (2016). Biodegradable nitrogen-doped carbon nanodots for non-invasive photoacoustic imaging and photothermal therapy. *Theranostics*, *6*(12), 2196.
56. Chan, B. P., & Leong, K. W. (2008). Scaffolding in tissue engineering: general approaches and tissue-specific considerations. *European Spine Journal*, *17*(4, Supplement 4), 467–479.
57. Mathen, P., Rowe, L., Mackey, M., Smart, D., Tofilon, P., & Camphausen, K. (2020). Radiosensitizers in the temozolomide era for newly diagnosed glioblastoma. *Neuro-Oncology Practice*, *7*(3), 268–276.
58. Angammana, C. J., & Jayaram, S. H. (2011). Analysis of the effects of solution conductivity on electrospinning process and fiber morphology. *IEEE Transactions on Industry Applications*, *47*(3), 1109–1117.
59. Koski, A., Yim, K., & Shivkumar, S. J. M. L. (2004). Effect of molecular weight on fibrous PVA produced by electrospinning. *Materials Letters*, *58*(3–4), 493–497.
60. Ellis-Behnke, R. G., Liang, Y. X., You, S. W., Tay, D. K., Zhang, S., So, K. F., & Schneider, G. E. (2006). Nano neuro knitting: peptide nanofiber scaffold for brain repair and axon regeneration with functional return of vision. *Proceedings of the National Academy of Sciences*, *103*(13), 5054–5059.
61. Wang, K., Zhu, X., Yu, E., Desai, P., Wang, H., Zhang, C. L., ... & Hu, J. (2020). Therapeutic nanomaterials for neurological diseases and cancer therapy. *Journal of Nanomaterials*, *2020,* Article ID 2047379.
62. Chang, W. C., Hawkes, E. A., Kliot, M., & Sretavan, D. W. (2007). In vivo use of a nanoknife for axon microsurgery. *Neurosurgery*, *61*(4), 683–692.
63. Rossmeisl, J. H., Garcia, P. A., Pancotto, T. E., Robertson, J. L., Henao-Guerrero, N., Neal, R. E., ... & Davalos, R. V. (2015). Safety and feasibility of the NanoKnife system for irreversible electroporation ablative treatment of canine spontaneous intracranial gliomas. *Journal of Neurosurgery*, *123*(4), 1008–1025.
64. Krizhevsky, A., Sutskever, I., & Hinton, G. E. (2012). Imagenet classification with deep convolutional neural networks. *Advances in Neural Information Processing Systems*, *25*, 1097–1105.
65. Eichholz, K. F., & Hoey, D. A. (2018). Mediating human stem cell behaviour via defined fibrous architectures by melt electrospinning writing. *Acta Biomaterialia*, *75*, 140–151.
66. Tan, R., Yang, X., & Shen, Y. (2017). Robot-aided electrospinning toward intelligent biomedical engineering. *Robotics and Biomimetics*, *4*(1), 1–13.
67. Tourlomousis, F., Jia, C., Karydis, T., Mershin, A., Wang, H., Kalyon, D. M., & Chang, R. C. (2019). Machine learning metrology of cell confinement in melt electrowritten three-dimensional biomaterial substrates. *Microsystems & Nanoengineering*, *5*(1), 1–19.
68. Shi, J., Sun, X., Zheng, S., Li, J., Fu, X., & Zhang, H. (2018). A new near-infrared persistent luminescence nanoparticle as a multifunctional nanoplatform for multimodal imaging and cancer therapy. *Biomaterials*, *152*, 15–23.
69. Fourches, D., Pu, D., Tassa, C., Weissleder, R., Shaw, S. Y., Mumper, R. J., & Tropsha, A. (2010). Quantitative nanostructure: activity relationship modeling. *ACS Nano*, *4*(10), 5703–5712.
70. Sizochenko, N., & Leszczynski, J. (2016). Review of current and emerging approaches for quantitative nanostructure-activity relationship modeling: the case of inorganic nanoparticles. *Journal of Nanotoxicology and Nanomedicine (JNN)*, *1*(1), 1–16.
71. Jones, D. E., Ghandehari, H., & Facelli, J. C. (2016). A review of the applications of data mining and machine learning for the prediction of biomedical properties of nanoparticles. *Computer METHODS and Programs in Biomedicine*, *132*, 93–103.
72. Sayes, C., & Ivanov, I. (2010). Comparative study of predictive computational models for nanoparticle-induced cytotoxicity. *Risk Analysis: An International Journal*, *30*(11), 1723–1734.
73. Puzyn, T., Rasulev, B., Gajewicz, A., Hu, X., Dasari, T. P., Michalkova, A., ... & Leszczynski, J. (2011). Using nano-QSAR to predict the cytotoxicity of metal oxide nanoparticles. *Nature Nanotechnology*, *6*(3), 175–178.
74. Liu, R., Rallo, R., George, S., Ji, Z., Nair, S., Nel, A. E., & Cohen, Y. (2011). Classification NanoSAR development for cytotoxicity of metal oxide nanoparticles. *Small*, *7*(8), 1118–1126.
75. Horev-Azaria, L., Baldi, G., Beno, D., Bonacchi, D., Golla-Schindler, U., Kirkpatrick, J. C., ... & Korenstein, R. (2013). Predictive toxicology of cobalt ferrite nanoparticles: comparative in-vitro study of different cellular models using methods of knowledge discovery from data. *Particle and Fibre Toxicology*, *10*(1), 1–17.
76. Awasthi, R., Roseblade, A., Hansbro, P. M., Rathbone, M. J., Dua, K., & Bebawy, M. (2018). Nanoparticles in cancer treatment: opportunities and obstacles. *Current Drug Targets*, *19*(14), 1696–1709.
77. Liu, X., Tang, K., Harper, S., Harper, B., Steevens, J. A., & Xu, R. (2013). Predictive modeling of nanomaterial exposure effects in biological systems. *International Journal of Nanomedicine*, *8*(Suppl 1), 31.

19

Advances in Regenerative Medicine and Nano-Based Biomaterials

K. Ganesh Kadiyala, P.S. Brahmanandam, Rajya Lakshmi Chavakula and Naresh Kumar Katari

CONTENTS

19.1 Introduction to Regenerative Medicine

Regenerative medicine is a branch related to molecular biology and tissue engineering which involves the process of regenerating the structure and functions of damaged human cells, tissues or organs. The main motto of regenerative medicine is to find a way to cure previously untreatable injuries and diseases [1]. It is one of the modern and most interesting fields in medical science. Encouraging outcomes of the research recommend the enormous medical potential and the progressive impacts on the economy and healthcare system. Most human tissues do not regenerate naturally; this explains why regenerative medicine engineering has a lot of importance today with promising alternative treatments. Here, the principle is simple; cells are gathered and introduced with or without alteration of their biological properties into the injured tissues. After reaching their maturity, these cells or tissues can be grafted. Regenerative medicine has applications in various health issues like cardiac insufficiency, blood vessels, osteoarthritis, diabetes, liver diseases, etc., which could be considered. The concept of regenerative medicine is an evolving multidisciplinary field connecting medicine, biology, chemistry, mechanics and engineering to transform the way to advance health and quality of life by reestablishing, maintaining or enhancing organ and tissue functions.

19.1.1 Advantages of Regenerative Medicine

Regenerative medicine has surfaced due to the following reasons:

i) It can offer encouraging impacts on the health of the public by providing treatments for various diseases like spinal cord injury, Alzheimer's disease (the degradation of nerve cells), diabetes, arthritis (joint inflammation), chronic diseases and Parkinson's disease [2]. Moreover, it can address hereditary abnormalities such as the normal function of an organ that is initially absent in children [3]. If possible, it can replace old-style organ transplantation which, in the face of remarkable improvements, is not the finest treatment. Patients have to use immunosuppressant medicines continuously if they undergo transplant to avoid the body cells from disagreeing with the foreign organs or tissues [4]. This can pose risks of cancer or further diseases as the immune system is destabilized, enabling pathogens (e.g., bacteria, virus, etc.) and cancerous cells that could contaminate the patient. Therefore, regenerative medicine is considered a new effective option. Since the cells are

DOI: 10.1201/9781003157823-19

extracted from the patient, the possibility of rejection and the usage of immunosuppressant drugs are considerably reduced [5].

ii) The great shortage of organ donors adds another reason why regenerative medicine should be endorsed. As per the World Health Organization (WHO), 15 million people died of heart-related diseases in 2005 and by 2010 it is assessed that 600,000 people all over the world will need replacement heart valves. However, it was reported that 17.9 million people die every year as per WHO 2021 report. Undoubtedly, the existing demand for donor organs far surpasses the supply. Millions of patients worldwide are suffering or have died from various diseases that can be cured by the replacement of impaired organs [6]. In a few cases where organs are available, replacement cannot be carried out due to the recipient's body not accepting the donor's organ biologically [7]. Furthermore, this scarcity of organs also leads to illegal organ dealing on an international level [8]. The WHO states that there is a significant upsurge in kidney trafficking in the past few years. Thus, regenerative medicine can be a better solution to the scarce supply of donor organs.

iii) The success of this regenerative medicine may be beneficial to save the economy. The potentiality of regenerative medicine to reform the healthcare system, particularly for aged care, is assuming significance. It means that the government can reduce the budget allotment for medical purposes. Even though the early investment for regenerative medicine is high, the total cost of immunosuppressant medicine, checkup and various operations is estimated to be lower compared to that for organ transplantation [9]). Besides, the progress of regenerative medicine generates employment and the formation of native regenerative medicine healthcare organizations. It can attract a huge number of medical tourists, therefore potentially boosting up the local economy [6].

19.1.2 Disadvantages

The major barrier to the development of regenerative medicine is the cost. Since the cost for the development of cell-based products and their clinical trials is very huge, it needs huge investment both from the private and public sectors to continue the research [6]. Private companies have started funding a vast amount in the past few years as they are in competition as these technologies have huge potential [10]. Oftentimes, investment in regenerative medicine is risky and may result in loss of profits. Few biotechnology firms earn no profit after investing in regenerative medicine [11]. Thus, the cost of regenerative medicine is considered justified by many [10].

Meanwhile, regenerative medicine and tissue engineering emerged as an industry about two decades ago. A number of therapies have received FDA (Food and Drug Administration) approval and are commercially available (Table 19.1). The distribution of therapeutic cells that unswervingly contribute to the structure and function of formed new tissues is a principle model of regenerative medicine to date [11–13].

19.2 Common Biomaterials Used in Regenerative Medicine

Biomaterial is a substance that has been engineered to work together with biological systems for a medical purpose like a therapeutic (augment, treat, replace or repair a tissue function of the body) or a diagnostic one. The study of biomaterials is called biomaterials engineering or biomaterials science.

Biomaterials can be classified as bioactive ceramics, polymeric biomaterials and composites.

19.2.1 Bioactive Ceramics

Bioactive ceramics are of natural or synthetic origin, for example, coralline, bioactive glasses, calcium silicate, HA, tricalcium phosphate (TCP) and biphasic calcium phosphate (BCP). It is reported that the BCP scaffolds with multi-scale porosity and composition of HA (87%) and β-TCP (13%) show decent bioactivity and a positive effect on the growth of the bone when imbedded in porcine mandibular defects [15].

TABLE 19.1

FDA-Approved Regenerative Medicine

Category	Name	Biological agent	Approved use
Biologics	LaViv	Autologous fibroblasts	Improving nasolabial fold appearance
	Carticel	Autologous chondrocytes	Cartilage defects from acute or repetitive trauma
	Apligraf GINTUIT	Allogeneic cultured keratinocytes and fibroblasts in bovine collagen	Topical mucogingival conditions, leg and diabetic foot ulcers
	Cord blood	Hematopoietic stem and progenitor cells	Hematopoietic and immunological reconstitution after myeloablative treatment
Cell-based medical devices	Dermagraft	Allogenic fibroblasts	Diabetic foot ulcer
	Celution	Cell extraction	Transfer of autologous adipose stem cells
Biopharmaceuticals	GEM 125	PDGF-BB, tricalcium phosphate	Periodontal defects
	Regranex	PDGF-BB	Lower extremity diabetic ulcers
	Infuse bone graft	BMP-2	Tibia fracture and nonunion, and lower spine fusion
	Inductos	BMP-7	
	Osteogenic protein-1		Tibia nonunion

Source: [14].

Ceramic biomaterials have some advantages, for example, good biocompatibility as well as resistance to corrosion and compression, but their fragility and low tensile strength need to be improved [16].

19.2.2 Polymeric Biomaterials

The second category of biomaterials consists of natural polymers, such as hyaluronic acid, chitosan (CTS), fibrin, collagen (Col) and synthetic polymers like poly (vinyl phosphonic acid) (PVPA), polycaprolactone (PCL), polylactic acid (PLA) and polyglycolic acid (PGA). The structure and biochemical properties of naturally available polymers are far closer to those of the organic matrix of natural bone, but natural polymers have some disagreeable performances like poor thermal stability [17]. In the same way, synthetic polymers (PLA and PGA) are not ideal candidates (structurally) as polymeric biomaterials for bone tissue regrowth owing to the compressive strength and low osteoconductive [18]. Furthermore, an important class of polymeric biomaterials is represented by hydrogels (ECM analogs). Natural hydrogels (polymers are based on natural sources) include gelatin, Col, alginate and agarose. Synthetic hydrogels are made up using synthetic polymers such as polyamides, polyvinyl alcohol, and polyethylene glycol [19]. These synthetic polymers can be usually made up of cross-linking macromers (polymerization of monomers or self-assembly of small molecules) [20]. Usually, enhancing the mass concentration or cross-linking density of polymers is a reasonable way to improve stiffness and strength. Raucci's team prepared two kinds of gelatin-based scaffolds over surface modification by HA nanoparticles and adornment with BMP-2, respectively [21]. The scaffolds coated with inorganic contents show better cell attachment and early osteogenic differentiation in less time, whereas the ones altered with the BMP-2 peptide altered the biological response at a longer time.

19.2.3 Composites

The third category of biomaterials is composites. Composites consist of a blend of two or more materials with dissimilar properties in the form of co-polymers, polymer-ceramic composites, or polymer-polymer blends, such as poly (lactic-co-glycolic acid) (PLGA), PLA-HA, PCL/PVPA, and TS-calcium phosphate scaffolds, etc. [22, 23]. The composites combine the advantages of the above scaffolds and display decent mechanical hardness and load-bearing abilities as well as idyllic biocompatibility. Lai et al., prepared composite scaffolds of β-TCP and PLGA through 3D printing to upsurge mechanical stability and enhance tissue interactions [24]. The TCP/PLGA scaffold was stated to exhibit good biodegradability, biocompatibility, osteoconductivity and in vitro and in vivo studies. Besides, the addition of HA or β-TCP was validated to improve the physical strength of hydrogels and improve osteogenic differentiation and bone formation in vivo [25].

19.3 Biomedical Applications of New Classes of Scaffolds

The traditional biomaterials can be fabricated into a preferred three-dimensional scaffold which acts as a frame for the initial cell growth and their proliferation. By using the following technologies, these biomaterials can be fabricated into scaffolds:

i) Fiber networking
ii) Solvent casting in combination with particulate leaching
iii) Phase separation in combination with freeze-drying
iv) Solid freeform fabrication [26]

On the other hand, these scaffold fabrication technologies have some limitations including, lack of interconnectivity, desired pore size, etc., and, therefore, new scaffold fabrication technologies are evolving during recent times.

19.4 Hydrogels as Tissue Engineering Scaffolds

It is a colloidal gel wherein water is the dispersion medium. They emerge as significant scaffolding biomaterials as they look like native tissue. The aqueous nature of the hydrogels nearly resembles the cells in the body. Hydrogel possesses a noble porosity that allows the exchange of nutrients and waste. Natural and synthetic polymers can be fabricated into hydrogel scaffolds and then used for tissue regeneration [27]. The synthetic hydrogel scaffolds used in regenerative medicine are poly (vinyl alcohol) (PVA), polyurethanes (PU), poly (ethylene oxide) (PEO), poly (N-isopropyl acrylamide) (PNIPAAm), poly (propylene fumarate-co-ethylene glycol) (P[PF-co-EG]) and poly (acrylic acid) (PAA). Agarose, chitosan, alginate, collagen, gelatin, hyaluronic acid and fibrin are naturally derived polymers involved in tissue engineering scaffolds [28].

19.5 Cryogels as Tissue Engineering Scaffolds

These are the gel matrices manufactured by the process of cryogelation, which synthesizes the cryogel scaffolds at a sub-zero temperature from natural or synthetic polymers deprived of the use of inorganic solvents. The scaffolds generated by the route of cryogelation have benefits over other types of scaffolds. Cryogel scaffolds can be fabricated in various formats like sheets, disks, and monoliths with variable measurements. In the course of the process of cryogelation utmost of the solvent gets frozen but part of the solvent is left unfrozen, where monomeric or polymeric precursors undergo chemical reactions. The chemical reaction in the liquid microphase leads to the formation of gel that is transformed into a porous scaffold on melting the frozen part which acts as porogen. Cryogels retain continuously interconnected pores up to 200 mm, which provides a surface for the cell division of most of the cell types, making cryogel matrices fit for tissue engineering applications [29]. The prospective use of cryogel matrices has been explored extensively in the areas of tissue engineering or regenerative medicine, therapeutic protein production, bioseparation, etc. In tissue engineering, cryogels have shown potential for the restoration of cartilage, bone, liver, skin, etc. [30–32]. Cryogel scaffolds have the potential to be applied for other biomedical applications.

19.6 Application of Biomaterials in Regenerative Medicine

i) Cardiovascular devices and implants
ii) Artificial RBC cell substitutes
iii) Extracorporeal artificial organs
iv) Orthopedic applications/dental implants
v) Cartilage implants
vi) Surgical sutures/surgical adhesives
vii) Interactions of biomaterials with stem cells
viii) Regenerative medicine and tissue engineering anticipate making a functional tissue or organ. Here the discussion is about the advances in biomaterials and awareness of stem cell biology and its applicability in the field of regenerative medicine
ix) Stem cell niche
x) Sometimes there is an injury of deeper layers that have stem cell niches at the time of tissue damage. In those cases, biomaterials could be a valuable tool for the restoration of the niche functionality [33]. In contrast to the usual two-dimensional (2D), culture systems, a three-dimensional (3D) scaffold-based model would assist the spatial distribution of various cells resulting in the structural organization which may be similar to the in vitro tissue organization [34]

19.6.1 Bone Tissue

Both therapies for bone regeneration, autograft and allograft, have numerous drawbacks, so substitute methods are being explored. One such approach is the separation and growth of the mesenchymal stem cells (MSCs) from the patient and their seeding onto the porous three-dimensional scaffolds. In the in vitro cultivation, as the stem cells are bare to signaling molecules brought in the media, MSCs discriminate toward osteogenic lineage. After that, this engineered tissue can then be implanted at the defect location to regenerate the new bone as the scaffold degrades.

19.6.2 Nervous Tissue

Treatment of the nervous tissue mostly in spinal cord injuries needs new medical therapies because axons do not have the regenerative capacity in the natural environment. The existing strategy is the use of nerve autografts, but due to the confines like donor site morbidity, etc., this nerve autografts strategy does not offer a promising way out for the repair. In recent times it has been revealed that the design of surface topography may encourage stem cell differentiation toward the neural lineage. In this route amorphous hydrocarbons (a-C:H) groove topographies with the width-spacing ridges ranging from 80 to 40 μm, 40 to 30 μm, 30 to 20 μm, and depth of 24 nm were used as a single mechanotransducer stimulus to produce neural cells from IBM-MSCs in vitro [35].

19.6.3 Skeletal and Cardiac Muscles

Painful injuries can disturb muscle contraction by triggering damage to the skeletal muscles and peripheral nerves. Normal healing may result in scar tissue development. So, the use of three-dimensional (3D) scaffolds will trigger muscle cell orientation, elongation, striation, and fusion. Electrospun chitosan microfibers have been used as unique biomaterials for the repairing of muscle. Traditional porous scaffolds may be insufficient as they do not reproduce the characteristic myocardial environment. Considering the importance of topography in this progression, one approach is to simulate the microenvironment of the natural tissue. In this route, microfabricated scaffolds were generated with a soft lithography method using a bioartificial blend, based on gelatin and alginate and a unique poly (N-isopropyl acrylamide) based copolymer. This scaffold exhibited anisotropic mechanical properties which resembled the natural tissue [36].

19.6.4 Inorganic RG

By designing responsive biomaterials with suitable biophysical and biochemical characteristics, the cellular response can be controlled to narrate tissue healing. Recently, inorganic biomaterials have been shown to control cellular responses together with cell–cell and cell–matrix interactions. Besides, ions released from these mineral-based inorganic biomaterials play a dynamic role in cellular function by triggering specific genes or biochemical pathways. The fundamental properties of inorganic biomaterials, for example, the release of bioactive ions (e.g., Ca, Zn, Mg, Sr, B, Fe, Cu, Cr, Co, Si, Mo, Mn, Au, Ag, V, La, and Eu), can be the key source to induce phenotypic changes in cells [37]. Biophysical characteristics of biomaterials, such as topography, charge, size, electrostatic interactions, and stiffness can be modulated by the addition of inorganic micro and nanoparticles to polymeric networks. They have also been shown to play an important role in their biological response. Moreover, understanding physical and chemical properties with regard to cellular function is a key to developing fine-tuned regenerative medicine in therapeutics, as is shown in Table 19.2.

19.7 Toxicity of Biomaterials

The MTT (3-[4, 5-dimethylthiazol-2-yl]-2,5-diphenyl tetrazolium bromide) cytotoxicity assay is the quantitative approach recommended, as it can accurately count the metabolic activity of $\geq$ 950 cells. It is a colorimetric assay that detects the metabolic profile over the analysis of the number of viable cells. MTT is a water-soluble tetrazolium salt, which becomes an insoluble purple formazan product after the cleavage of the tetrazolium ring by succinate dehydrogenase, found within the cells of mitochondria. The product formazan is impermeable to the cell membranes and therefore it gathers in healthy cells. The water-insoluble product formazan can be dissolved in a solvent, like dimethyl sulfoxide. The absorbance of the resulting solution at 540 nm is directly associated with both

TABLE 19.2

Comparisons between Various Inorganic Nano-Based Biomaterials to Verify Their Applications, Advantages and Disadvantages

S. no.	Nanomaterial type	Applications	Advantages	Disadvantages
NPs of metallic substances				
1	Aluminum oxide	Fuel cells, polymers, paints, coatings, textiles, biomaterials, etc. (www.futuremarket sinc.com).	Aluminum oxide NPs, at concentrations of 10, 50, 100, 200 and 400 μg/mL possess no significant toxic effect on sustainability of mammalian cells [38].	Disturb the cell viability, alter mitochondrial function, upsurge oxidative stress and also modify tight junction protein expression of the blood-brain barrier (BBB) [39].
2	Gold	Gold NPs have the capability of easy functionalization; binding to amine and thiol groups. All these features possessed by gold NPs made a way for surface modification, and they are being explored as drug carriers in cancer and thermal therapy, and as contrast agents [40].	Gold NPs are considered to be comparatively safe, as their core is inert and non-toxic. In one experimental study, several gold NPs (4, 12, and 18 nm) with different capping agents have been investigated for any cytotoxicity against leukemia cell lines. The results of this report suggest that spherical gold NPs enter the cell and are non-toxic to cellular function. The cytotoxicity was evaluated by MTT assay [41].	However, there are some other reports suggesting that cytotoxicity associated with gold NPs depends on dose, side chain (cationic) and the stabilizer used [42, 43]. Cytotoxicity of gold NPs is dependent on the type of toxicity assay, cell line, and physical/chemical properties. The variation in toxicity with respect to different cell lines has been observed in human lung and liver cancer cell lines [44].
3	Copper oxide	Used in semiconductors, antimicrobial reagents, heat transfer fluids and intrauterine contraceptive devices [45].	–	Nano-copper has resulted in severe impairment in liver, kidney, and spleen in experimental animals. In one in vitro study, copper oxide NPs (50 nm) have been reported as being genotoxic and cytotoxic along with disturbing cell membrane integrity and inducing oxidative stress [46].
4	Silver	Its NPs are being used in a wide range of commercial products. Silver NPs are used in the form of wound dressings, coating of surgical instruments and prostheses [47].	They enter the human body in different ways and accumulate in different organs, crossing the BBB and reaching the brain.	In comparison to others, these NPs have shown more toxicity in terms of cell viability by generating reactive oxygen species (ROS) and lactate dehydrogenase (LDH) leakage [48].
5	Zinc oxide	Zinc oxide NPs have many applications in paints, wave filters, UV detectors, gas sensors, sunscreens, and many personal care products [49, 50].	–	Long-lasting exposure to zinc oxide NPs (300 mg/kg) resulted in oxidative DNA damage along with altering various enzymes of the liver.
6	Iron oxide	Iron oxide NPs have been used in biomedical, drug delivery and diagnostic fields.	Iron oxide NPs having magnetic properties have been observed to accumulate in the liver, spleen, lungs and brain after inhalation, showing their ability to cross BBB [51].	These NPs exert their toxic effect in the form of cell lysis, inflammation, and disturbing blood coagulation system [52]. Moreover, in in vitro studies, it has been reported that reduced cell viability is the most common toxic effect of iron oxide NPs.
7	Titanium oxide	–	Titanium oxide is chemically an inert compound.	Studies have shown that NPs of titanium dioxide possess some toxic health effects in in vivo studies including DNA damage as well as genotoxicity and lung inflammation [53, 54].

(*Continued*)

TABLE 19.2 (CONTINUED)

Comparisons between Various Inorganic Nano-Based Biomaterials to Verify Their Applications, Advantages and Disadvantages

S. no.	Nanomaterial type	Applications	Advantages	Disadvantages
NPs of non-metallic substances				
1	Carbon-based nanomaterials	The carbon-based nanomaterials, such as CNTs (carbon nanotubes), fullerenes, single- and multi-walled carbon nanotubes are the most widely used nanomaterials [55].	Carbon-based nanomaterials have been reported in literature as cytotoxic agents. It is reported that have carbon-based nanomaterials show size-dependent cytotoxicity [56].	However, accumulation of single-walled CNTs in the liver has caused disturbance in certain biochemical parameters in the form of LDH, aspartate transaminases, alanine transaminase, glutathione [57].
2	Silica	Usage of silica NPs has many benefits in drug delivery systems.	Silica NPs have been reported as simply functionalized drug carriers [58].	Nanosilica was a highly biocompatible material in drug delivery systems, but according to recent reports, NPs of silica cause the generation of ROS and subsequent oxidative stress [59].
NPs of polymeric materials		Polymeric NPs have the potential to be used in targeted drug delivery in cancer chemotherapy.	These NPs are also engaged in encapsulation of various molecules to advance nanomedicine providing sustained release and good biocompatibility with cells and tissues [60].	Recently, one report proposed that surface coating prompts the toxicity of polymeric NPs towards human-like macrophages [61].

TABLE 19.3

Biomaterials Tested on Different Types of Cells and Their Evaluation

Biomaterials tested	Cell type/cell line used	Toxicity observed
AuNPs	Human proximal tubule cells	Size-dependent toxicity was observed. AuNPs of size 80 nm were found to be more cyto-/genotoxic as compared to 40 nm AuNPs [63].
	Hypoxic MDA-MB-231 breast cancer	Uptake of AuNPs occurred in hypoxic conditions, causing radiosensitization in moderate, but not extreme hypoxia in a breast cancer cell line [64].
	Metastatic breast cancer cells and fibrosarcoma cells	Distribution of gold nanostar was five times greater in metastatic breast cancer cells in comparison to fibrosarcoma tumors [65].
Ficus religiosa AgNPs	A549 and Hep2 cells	Results exhibited deposition of AgNPs in liver, brain, and lungs on day 29 with respective concentrations of 4.77, 3.94 and 3.043 μg/g tissue, although complete elimination of silver was observed during washout period [66].
Tannic acid modified AgNPs	VK2-E6/E7 cells	Tannic acid-modified AgNPs with sizes of more than 30 nm were toxic to VK2-E6/E7 [67].
Ficus carica L modified AgNPs	MCF7 cell lines	Potential toxicity on MCF7 cell lines at size range 55–90 nm as revealed by *Ficus carica* L. modified AgNPs and chemical synthesized AgNPs both.
SiO_2NPs combined with lead acetate	A549 cells	Noncytotoxic concentration of SiO_2NPs exposure alone did not induce apoptosis in A549 cells, but changes were noticed when combined with lead acetate.

the number of living cells and their metabolic activity [62]. The MTT assay has advantages over other assays, it enables quantitative analysis without using radioactive labeling, and it is quick and reproducible. The various biomaterials tested on different types of cells and their evaluation are represented in Table 19.3.

19.8 Conclusion

So far, regenerative medicine has been directed to new, and a number of pathologies are being used to treat through FDA-approved therapies. Significant research has supported the fabrication of sophisticated grafts that exploit properties of scaffolding biomaterials and cell operation technologies for repairing tissue and controlling cell behavior. These scaffolds can be made to fit the patient's anatomy and be fabricated with considerable control over the three-dimensional positioning of cells. Approaches are being developed to enhance graft integration with the host vasculature and nervous system. New cell sources for a replacement that addresses the partial cell supply that hampered many preceding efforts are also being developed. A lot of issues will be vital for the progression of regenerative medicine as a field. The making of microenvironments, frequently modeled on several stem cell niches that offer specific cues, like morphogens and physical properties, will likely be a key to endorsing ideal regenerative responses from therapeutic cells. Generating a pro-regeneration environment inside the patient may dramatically progress outcomes of regenerative medicine plans in general. A better understanding of the immune system's role in regeneration could aid this goal and also a better understanding of how age and state of the disease affect regeneration will likely also be vital for advancing the field in various situations [68–70]. In conclusion, three-dimensional human tissue culture models of disease may permit testing of regenerative medicine methodologies in human biology, as compared to the animal models now used in preclinical studies. Augmented precision of disease models may advance the efficacy of regenerative medicine approaches and improve the translation to the clinic of promising methods [71]. Based on the advantages and vast applications of regenerative medicine, it can be used in nanomaterials to treat various types of health issues where there is a role for regenerative medicine.

REFERENCES

1. Greenwood, Heather L., Singer, Peter A., Downey, Gregory P., Martin, Douglas K., Thorsteinsdóttir, Halla, Daar, Abdallah S. (2006). Regenerative Medicine and the Developing World. *PLoS Med*, 3(9), e381.
2. Anthony, A., Darrell, J. I., Marsha, M., & Sunil, S. (2010). Wound healing versus regeneration: Role of the tissue environment in regenerative medicine, *MRS Bulletin*, 35(8), 597–606
3. J. Mawson, & K. Dunnil (2008). The scope of regenerative medicine biology essay, *Free Book Summary*, 3.
4. *Qualitative Research Through Case Studies* (2001). Max Travers Publications, 201–208.
5. Amishi, P. J., Elizabeth A. S., Anastasia, K., Ling, W., & Lois, G. (2010). Examining the protective effects of mindfulness training on working memory capacity and affective experience, *Emotion*, 10(1), 54–64.
6. Paul, K. (2006). History of regenerative medicine: Looking backwards to move forwards, *Regenerative Medicine*, 1(5), 653–669.
7. Fishman, J. A., & Rubin, R. H. (1998). Infection in organ-transplant recipients, *The New England Journal of Medicine: Research & Review*, 339, 1244–1246.
8. Yosuke, S. (2007). The state of the international organ trade: A provisional picture based on integration of available information, *Bulletin of the World Health Organization*, 85(12), 955–962.
9. James, Alexandra, & Mannon, Roslyn B. (2015). The Cost of Transplant Immunosuppressant Therapy: Is This Sustainable? *Current Transplantation Report*, 2(2), 113–121.
10. Chris M., & Peter D. (2008). A brief definition of regenerative medicine, *Regenerative Medicine*, 3(1), 1–5.
11. Gary, P. (2006). Profiting from innovation and the intellectual property revolution, *Research Policy*, 35(8), 1122–1130.
12. Buckler, L. (2011). Opportunities in regenerative medicine, *Bio Process International Magazine*, 14–18.
13. Fisher, M. B., & Mauck, R. L. (2013). Tissue engineering and regenerative medicine: Recent innovations and the transition to translation, *Tissue Engineering Part B: Reviews*, 19(1), 1–13.
14. Angelo, S. M., & David, J. M. (2015). Regenerative medicine: Current therapies and future directions, *PNAS*, 112, 14452–14459.
15. Rustom, L. E., Thomas, B., Siyu, L., Isabelle, P. P., Brett, W. N., Yan, L., Mark, D. M., Catherine, P., & Amy J. W. J. (2016). Micropore-induced capillarity enhances bone distribution in vivo in biphasic calcium phosphate scaffolds, *Acta Biomaterialia*, 44, 144–154.
16. Kaur, G., Om, P., Kulvir, S., Dan, H., Brian, S., & Gary, P. (2014). A review of bioactive glasses: Their structure, properties, fabrication and apatite formation, *Journal of Biomedical Materials Research*, 102, 254–274.
17. Thrivikraman, G., Athirasala, A., Twohig, C., & Bertassoni, L. E. (2017). Biomaterials for craniofacial bone regeneration, *Dental Clinics of North America*, 61, 835–856.
18. Iaquinta, M. R., Mazzoni, E., Manfrini, M., Agostino, A., Trevisiol, L., Nocini, R., Trombelli, L., Barbanti-Brodano, G., Martini, F., & Mauro, Tognon, (2019). Innovative biomaterials for bone regrowth, *International Journal of Molecular Sciences*, 20(618), 1–17.
19. Zhu, J., & Marchant, R. E. (2014). Design properties of hydrogel tissue-engineering scaffolds, *Expert Review of Medical Devices*, 8, 607–626.
20. Neves, S. C., Moroni, L., Barrias, C. C., & Granja, P. L. (2019). Leveling up hydrogels: Hybridsystems in tissue engineering, *Trends in Biotechnology*, S0167–7799(19), 30230–30236.
21. Raucci, M. G., D'Amora, U., Ronca, A., Demitri, C., & Ambrosio, L. (2019). Bioactivationroutes of gelatin-based scaffolds to enhance at nanoscale level bone tissue regeneration, *Frontiers in Bioengineering and Biotechnology*, 7(27). 1–11.
22. Hasan, A., Byambaa, B., Morshed, M., Cheikh, M. I., Shakoor R. A., Mustafy, T., & Marei, H. E. (2018). Advances in osteobiologic materials for bone substitutes, *Journal of Tissue Engineering and Regenerative Medicine*, 12, 1448–1468.
23. Kaur, G., Kumar, V., Biano, F., Mauro J. C., Pickrelle, G., Evans, I., & Bretcanu, O. (2019). Mechanical properties of bioactive glasses, ceramics, glassceramics and composites: State-of-the-art review and future challenges, *Materials Science and Engineering: C, Materials for Biological Applications*, 104, 109895.
24. Lai, Y., Cao H., Wang, X., Chen, S., Zhang, M., Wang N., Yao Z., Dai, Y., Xie, X., Zhang, P., Yao, X., & Qin, L. (2018). Porous composite scaffold incorporating osteogenic phytomoleculeicariin for promoting skeletal regeneration in challenging osteonecrotic bone in rabbits, *Biomaterials*, 153, 1–13.
25. Kim, H. D., Amirthalingam, S., Kim S. L., Lee S. S., Rangasamy, J., & Hwang N. S. (2017). Biomimetic materials and fabrication approaches for bone tissueengineering, *Advanced Healthcare Materials*, 6(23), 1–18.
26. Sachlos, E., & Czernuszka, J. T. (2003). Making tissue engineering scaffoldswork: Review on the application of solid freeform fabricationtechnology to the production of tissue engineering scaffolds, *European Cells & Materials*, 5, 29–40.
27. Patel, A., & Kibret, M. (2011). Hydrogel biomaterials, in *Biomedical Engineering-frontiers and Challenges*, Rezai, R. F. (Ed). InTech Publishers, 275–285.
28. Peppas, N., Hilt, J., Khademhosseini, A., & Langer, R. (2006). Hydrogels inbiology and medicine: From molecular principles to bionanotechnology, *Advanced Materials*, 18, 1345–1360.
29. Kumar, A., Mishra, R., Reinwald, Y., & Bhat, S. (2010). Cryogels: Freezing unveiled by thawing, *Material Today*, 13, 42–44.
30. Bhat, S., Tripathi, A., & Kumar, A. (2010). Super macro porus chitosanagarose-gelatin cryogels: In vitro characterization and in vivoassessment for cartilage tissue engineering, *Journal of The Royal Society Interface*, 8, 540–554.

31. Tripathi, A., Kathuria, N., & Kumar, A. (2009). Elastic and microporous agarose-gelatin cryogels with isotropic and anisotropic porosityfor tissue engineering, *Journal of Biomedical Materials Research A*, 90, 680–694.
32. Kumar, A., & Srivastava, A. (2010). Cell separation using cryogel-based affinity chromatography, *Nature Protocols*, 5, 1737–1747.
33. Lutolf, M. P., Doyonnas, R., Havenstrite, K., Koleckar, K., & Blau, H. M. (2009). Perturbation of single hematopoietic stem cell fates in artificialniches, *Integrative Biology*, 1, 59–69.
34. Di Maggio, N., Piccinini, E., Jaworski, M., Trumpp, A., Wendt, D. J., & Martin, I. (2011). Toward modeling the bone marrow niche usingscaffold-based 3D culture systems, *Biomaterials*, 32, 321–329.
35. D'Angelo, F., Armentano, I., Mattioli, S., Crispoltoni, L., Tiribuzi, R., & Cerulli, G. G. (2010). Micropatterned hydrogenated amorphous carbonguides mesenchymal stem cells towards neuronal differentiation, *European Cells & Materials*, 20, 231–244.
36. Rossellini, E., Vozzi, G., Barbani, N., Giusti, P., & Cristallini, C. (2010). Three dimensional microfabricated scaffolds with cardiac extracellular matrix-like architecture, *The International Journal of Artificial Organs*, 33, 885–894.
37. Anna, M. B., & Akhilesh, K. G. (2020). Inorganic biomaterials for regenerative medicine, *ACS Applied Materials & Interfaces*, 12(5), 5319–5344.
38. Radziun, E., Wilczyńska, D. J., Książek, I., Nowak, K., Anuszewska, E. L., Kunicki, A., Olszyna, A., & Zabkowski, T. (2011). Assessment of the cytotoxicity of aluminium oxide nanoparticles on selected mammaliancells, *Toxicology in Vitro*, 25(8), 1694–1700.
39. Chen, L., Yokel, R. A., Hennig, B., & Toborek, M. (2008). Manufactured aluminum oxide nanoparticles decreaseexpression of tight junction proteins in brainvasculature, *Journal of Neuroimmuno Pharmacology*, 3(4), 286–295.
40. Jain, S., Hirst, D. G., & O'sullivan, J. M. (2012). Gold nanoparticles as novel agents for cancer therapy, *The British Journal of Radiology*, 85(1010), 101–113.
41. Connor, E. E., Mwamuka, J., Gole, A., Murphy, C. J., & Wyatt, M. D. (2005). Gold nanoparticles are taken up by human cellsbut do not cause acute cytotoxicity, *Small*, 1(3), 325–327.
42. Boisselier, E., & Astruc, D. (2009). Gold nanoparticles in nanomedicine: Preparations, imaging, diagnostics, therapies and toxicity, *Chemical Society Reviews*, 38(6), 1759–1782.
43. Goodman, C. M., Mccusker, C. D., Yilmaz, T., & Rotello, V. M. (2004). Toxicity of gold nanoparticles functionalized with cationic and anionic side chains, *Bioconjugate Chemistry*, 15(4), 897–900.
44. Patra, H. K., Banerjee, S., Chaudhuri, U., Lahiri, P., & Dasgupta, A. K. (2007). Cell selective response to gold nanoparticles, *Nanomedicine: Nanotechnology, Biology and Medicine*, 3(2), 111–119.
45. Aruoja, V., Dubourguier, H. C., Kasemets, K., & Kahru, A. (2009). Toxicity of nanoparticles of CuO, ZnO and TiO_2 to microalgae Pseudo kirchneriellasubcapitata, *Science of the Total Environment*, 407(4), 1461–1468.
46. Ahamed, M., Siddiqui, M. A., Akhtar, M. J., Ahmad, I., Pant, A. B., & Alhadlaq, H. A. (2010). Genotoxic potential of copper oxidenanoparticles in human lung epithelial cells, *Biochemical and Biophysical Research Communication*, 396(2), 578–583.
47. Chen, X., & Schluesener, H. J. (2008). Nanosilver: A nanoproduct in medical application, *Toxicology Letters*, 176(1), 1–12.
48. Hussain, S. M., Hess, K. L., Gearhart, J. M., Geiss, K. T., & Schlager, J. J. (2005). In vitro toxicity of nanoparticles in BRL 3Arat liver cells, *Toxicology in Vitro*, 19(7), 975–983.
49. Huang, G. G., Wang, C. T., Tang, H. T., Huang, Y. S., & Yang, J. (2006). ZnO nanoparticle-modified infrared internal reflection elements for selective detection of volatile organic compounds, *Analytical Chemistry*, 78(7), 2397–2404.
50. Huang, C. C., Aronstam, R. S., Chen, D. R., & Huang, Y. W. (2010). Oxidative stress, calcium homeostasis, and altered gene expression in human lung epithelial cells exposed to ZnO nanoparticles, *Toxicology in Vitro*, 24(1), 45–55.
51. Liu, G., Gao, J., Ai, H., & Chen, X. (2013). Applications and potentialtoxicity of magnetic iron oxide nanoparticles, *Small*, 9(9–10), 1533–1545.
52. Zhu, M. T., Feng, W. Y., Wang, B., Wang, T. C., Gu, Y. Q, Wang, M., Wang, Y., Ouyang, H., Zhao, Y. L., & Chai, Z. F. (2008). Comparative study of pulmonary responses to nano- and submicron-sized ferric oxide in rats, *Toxicology*, 247(2–3), 102–111.
53. Trouiller, B., Reliene, R., Westbrook, A., Solaimani, P., & Schiestl, R. H. (2009). Titanium dioxide nanoparticles induce DNA damage and genetic instability in vivo in mice, *Cancer Research*, 69(22), 8784–8789.
54. Liu, R., Yin, L., Pu, Y., Liang, G., Zhang, J., Su, Y., Xiao, Z., & Ye, B. (2009). Pulmonary toxicity induced by three forms of titanium dioxide nanoparticles via intra-tracheal instillation in rats, *Progress in Natural Science*, 19(5), 573–579.
55. Huczko, A. (2001). Synthesis of aligned carbon nanotubes, *Journal of Applied Physics*, 74, 617–638.
56. Magrez, A., Kasas, S., Salicio, V., Pasquier, N., Seo, J. W., Celio, M., Catsicas, S., Schwaller, B., & Forro, L. (2006). Cellular toxicity of carbon-based nanomaterials, *Nano Letters*, 6(6), 1121–1125.
57. Yang, S. T., Wang, X., Jia, G., Gu, Y., Wang, T., Nie, H., Ge, C., Wang, H., & Liu, Y. (2008). Long-term accumulation and lowtoxicity of single-walled carbon nanotubes inintravenously exposed mice, *Toxicology Letters*, 181(7), 182–189.
58. Wilczewska, A. Z, Niemirowicz, K., Markiewicz, K. H., & Car, H. (2012). Nanoparticles as drug delivery systems, *Pharmacological Reports*, 64(5), 1020–1037.
59. Park, E. J., & Park, K. (2009). Oxidative stress and pro-inflammatory responses induced by silica nanoparticles in vivo and invitro, *Toxicological Letters*, 184(1), 18–25.
60. Panyam, J., & Labhasetwar, V. (2003). Biodegradable nanoparticles for drug and gene delivery to cells and tissue, *Advanced Drug Delivery Reviews*, 55(3), 329–347.
61. Grabowski, N., Hillaireau, H., Vergnaud, J., Tsapis, N., Pallardy, M., Kerdine-Rö me, S., & Fattal, E. (2015). Surface coatingmediates the toxicity of polymeric nanoparticlestowards human-like macrophages, *International Journal of Pharmseutics*, 482(1–2), 75–83.

62. Fotakis, G., & Timbrell J. A. (2006), In vitro cytotoxicity assays: Comparison of LDH, neutral red, MTT and protein assay in hepatoma cell lines following exposure to cadmium chloride, *Toxicology Letters*, 160(2), 171.
63. Ortega, M., Riviere, J., Choi, K., & Monteiro-Riviere, N. (2017). Biocorona formation on gold nano particles modulates human proximal tubule kidney cell uptake, cytotoxicity and gene expression, *Toxicology in Vitro*, 42, 150–160.
64. Jain, S., Coulter, J. A., Butterworth, K. T., Hounsell, A. R., McMahon, S. J., Hyland, W. B., Muir, M. F., Dickson, G. R., Prise, K. M., & Currell, F. M. (2014). Gold nanoparticle cellular uptake, toxicity and radio sensitisation in hypoxic conditions, *Radiotherapy and Oncology*, 110, 342–347.
65. Dam, D. H. M., Culver, K. S., Kandela, I., Lee, R. C., Chandra, K., Lee, H., Mantis, C., Ugolkov, A., Mazar, A. P., & Odom, T.W. (2015). Biodistribution and in vivo toxicity of aptamer loaded gold nanostars, *Nanomedicine*, 11, 671–679.
66. Nakkala, J. R., Mata, R., & Sadras, S. R. (2017). Green synthesized nano silver: Synthesis, physicochemical profiling, antibacterial, anticancer activities and biological in vivo toxicity, *Journal Colloid Interface Science*, 499, 33–45.
67. Orlowski, P., Soliwoda, K., Tomaszewska, E., Bien, K., Fruba, A., Gniadek, M., Labedz, O., Nowak, Z., Celichowski, G., & Grobelny, J. (2016). Toxicity of tannic acid-modified silver nanoparticles in keratinocytes: Potential for immunomodulatory applications, *Toxicology In Vitro*, 35, 43–54.
68. Oh, J., Lee, Y. D., & Wagers, A. J. (2014). Stem cell aging: Mechanisms, regulators and therapeutic opportunities, *Nature Medicine*, 20(8), 870–880.
69. Eming, S. A, Martin, P., & Tomic-Canic, M. (2014). Wound repair andregeneration: Mechanisms, signaling, and translation, *Science Translational Medicine*, 6(265), 1–16.
70. Scales, B. S., & Huffnagle, G. B. (2013). The microbiome in woundrepair and tissue fibrosis, *The Journal of Pathology*, 229(2), 323–331.
71. Bhatia, S. N., & Ingber, D. E. (2014). Microfluidic organs-on-chips, *Nature Biotechnology*, 32(8), 760–772.

20

Magnetic Nanocomposites and Their Biomedical Applications

Rajasekhar Chokkareddy, Raghavendra Vemuri, Nookaraju Muralasetti and Gan G. Redhi

CONTENTS

20.1 Introduction

"Nanotechnology" means the technology that performs on a nanoscale level by size and shape at the nanometer scale for design, production, characterization, and application. Renowned physicist Richard Feynman first described nanotechnology in a meeting held at the American Physical Society (APS), California Institute of Technology (CIT), in 1959, entitled "There Is Plenty of Room at the Bottom: An Invitation to Enter a New Field of Physics". The term "nano" comes from the Greek word "*nano*", which means "dwarf". One nanometer is designated as 1 nm and is equal to 10^{-9} m, which means one nanometer in length is approximately equivalent to the width of six carbon atoms or 10 water molecules [1, 2]. Development

DOI: 10.1201/9781003157823-20

of nanoscience and investigation of nanostructure was initiated around 1980. The field of nanotechnology is a versatile field of technology and proved to be a boon for many areas such as material technology [3], information technology [4], cellular and molecular biology [5], biotechnology [6], manufacturing [7], nano-electronics [8], communications [9], and robotics [10].

Nanomaterials or nanostructured materials at least one dimension of which lies in the 1–100 nm range include nanoparticles, quantum dots, nanorods, nanowire, and nanorings. The nanoparticles or nanomaterials are used in many applications, as shown in Figure 20.1 [11]. Thin films and bulk materials are also constructed from nanoscale building blocks or nanoscale structures [12]. Richard Feynman explored the manipulating possibility of material at an individual atomic or molecular level. Norio Taniguchi, a scientist at the University of Tokyo in 1974, first introduced nanotechnology. He referred to nanotechnology as an engineer, the materials precisely at the nanometer level. The research on nanoscience development and nanostructure investigation was started around 1980 with the invention of the scanning tunneling microscope (STM). Then the nanostructure solids concept was suggested [13, 14]. The nanostructure size of the single sugar molecule is nearly 1 nm measured by Albert Einstein during the study from his doctoral degree then-experimental work diffusion data of sugar in water [15]. A non-nano-crystalline matrix of one material filled with the nanoparticles of another material is present in some nanocomposite materials. The addition of nanoparticles enhanced improvement in mechanical strength, toughness, and electrical conductivity. Magnetic nanoparticles are a set of nanoparticles that can be influenced by a magnetic field. Magnetic nanoparticles have promising applications for magnetic hyperthermia, magnetic resonance imaging (MRI), drug delivery, and magnetic resonance tomography (MRT).

20.1.1 Introduction to Nanostructured Materials

Structured components with at least 100 nm are called nanostructured materials. The bulk properties of materials often change dramatically with nano ingredients. As an example, metals having grain sizes around 10 nm are nearly seven orders stronger and more demanding than metals having grain sizes around 100 nm. The fundamental reasons for the characteristics of materials to be different at the nanoscale are: nanomaterials have particle size made to fall; a considerable part of atoms is seen at the surface than in the interior. As an example, a 30 nm particle has 5% of its atoms on its surface, at 10 nm 20%, and at 3 nm 50% resides on the exterior. Quantum effects at the nanoscale can begin to dominate the behavior of matter, particularly at the lower end, affecting the optical, electrical, and magnetic properties of materials.

20.1.2 Morphology of Nanomaterials

An imperative characteristic of nanomaterials is the surface morphology which is primarily based on preparation approaches and leaves an intense effect on material features for fruitful exploitation in various purposes. Morphology

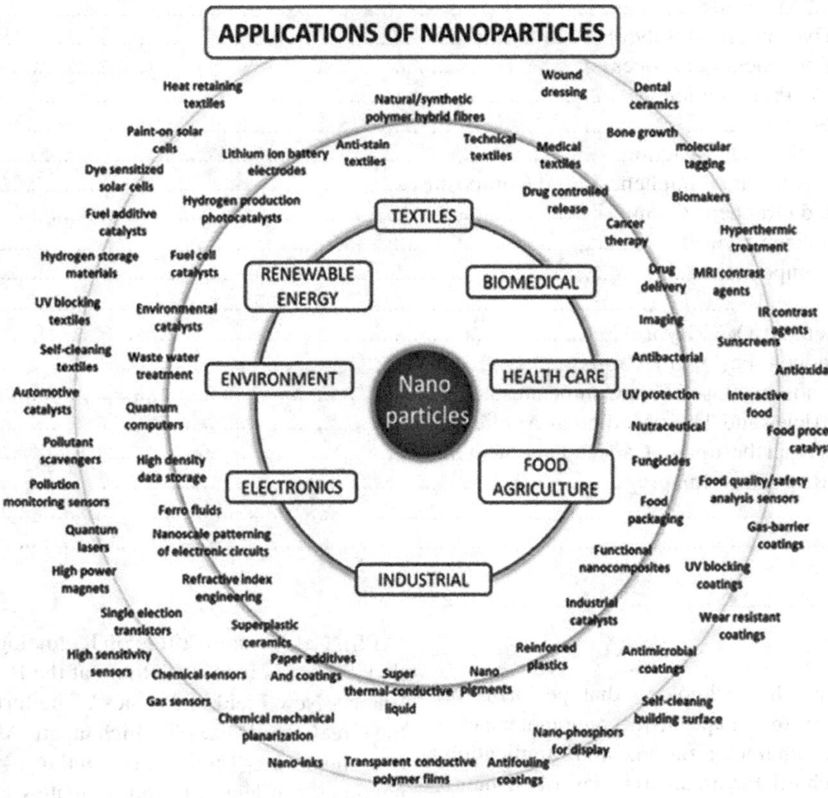

FIGURE 20.1 Applications of nanoparticles in various fields [11].

exploration of nano compounds is of primary significance soon after the preparation of the nanomaterials as it reveals a lot of informative statistics like size, shape, structure, and the surface of the class of the prepared sample. The nanomaterials possess features that are nowhere close to counterparts in their bulk form. The characteristics for dimensions of a solid enact a significant part in influencing its features, and it is recognized beyond imagination that with sizable change in the dimensions, the materials show tremendous change relative to their bulkier counterparts.

20.1.3 Classification of Nanomaterials

Nanomaterials are classified based on their dimensions.

(i) Materials having nanoscale one-dimensionally (1D) and grown in one dimension, like nano-wires and nanotubes.
(ii) Materials having nanoscale two-dimensionally (2D) with extension in other directions as layers, like thin films and surface coatings.
(iii) Materials having nanoscale three-dimensionally (3D) are particles; for instance, quantum dots, precipitates, and colloids are miniature particles of semiconductor materials. Nano-crystalline materials with nanometer-sized grains too come under this class.

20.1.3.1 Carbon Nanotubes (CNT)

They are formed by graphite sheets and then rolled up to make a tube. The dimensions are variable, and they exist as concentric nanotubes leading to the distinction between single-walled carbon nanotubes (SWCNT) and multi-walled carbon nanotubes (MWCNTs).

20.1.3.2 Carbon Black

It is one of the most widely used carbon nanomaterials in the current trend because it has various applications in the field of the painting industry.

20.1.3.3 Fullerenes

They are a class of carbon compounds that resemble a cage-like appearance composed of fused, pentagonal, and hexagonal sp2 carbon rings. Like most structures of carbon, they are insoluble in water. For reasons unknown there seem to be a considerable decline in research about fullerenes.

20.1.3.4 Nanocomposites

Nanocomposite materials can be classified in many ways. A non-nano-crystalline matrix of one material filled with the nanoparticles of another material is present in some nanocomposite materials. Nanocomposites also exist in this type of composite material, and the size of all constituent material grains is in the nanometer range.

20.1.3.5 Nano-Polymers

A polymer having nanoparticles widespread in the polymer matrix is called a "nano-polymer". The study focuses on nanopolymers and polymer nanocomposites. The nanostructure determines the essential modifications in the properties.

20.1.3.6 Nano-Ceramics

Nano-ceramic is one class of nanomaterial that comprise ceramic materials that are considered in the analysis of oxide and non-oxide ceramic materials, silicates, and hard metals like these material group composites.

20.2 Classification of Nanoparticles

Nanoparticles can be generally classified into two types.

20.2.1 Engineered Nanoparticles

They can be intentionally designed and created with physical properties, and their applications pertain to the below-mentioned businesses:

(i) Pharma
(ii) Chemical mechanical polishing
(iii) Quantum dots
(iv) Bio-detection and labeling
(v) Ceramic tiles
(vi) Food
(vii) Cosmetic and skin care

20.2.2 Non-Engineered Nanoparticles

These are non-deliberately bred or created naturally, such as nanoparticles in an atmosphere created through ignition. Physical properties play an essential role in this type of nanoparticle.

Some industries with non-engineered nanoparticles are:

(i) Environmental detection
(ii) Environmental monitoring
(iii) Controlled environments

20.3 Nanotechnology Applications

Nanoparticles that have changed their properties resulting from quantum confinements have drawn considerable interest and are currently being investigated [16–19]. By now, nanotechnology has started bringing innovations in day-to-day merchandise like cell phones, golf clubs, garments, and sunscreens. In another 10 years down the line, nanotechnology is very much likely to be a part of drug therapies, water filters, power lines, computers, fuel cells, and a variety of other purposes. Nanotechnology is pervading the

TABLE 20.1

Different Terms for Nanostructures and Their Commercial Impact

Near-term (1–5 years)	Mid-term (5–10 years)	Long-term (20+ years)
Long-lasting rechargeable batteries	New targeted drug therapies	New molecular electronics
Improved chemical and biological sensors	Enhanced medical imaging	New optical information processing
Point-of-care medical diagnostic devices	High-efficiency, cost-effective solar cells	New neural prosthetics for health care

commercial realm with its need overgrowing [20–22]. Table 20.1 shows the fields where nanotechnology is likely to stamp its authority commercially.

20.3.1 Synthesis Methods of Nanomaterials

Nanoparticles can be synthesized mainly in "top-down or "bottom-up" approaches in which the particles are fragmented into lower dimensions or aggregated to the bulk or higher dimensional form in the way of investigation of their properties (Figure 20.2) [24–27]. The synthesis processes are classified as physical methods and chemical methods [28–31]. Among several ways, the precipitation method is an appropriate, cheap, and easy method for nanoparticle synthesis working at low temperatures.

20.3.1.1 Chemical Precipitation

In this method, the knowledge of chemical reagents and reactions is essential. This method is more appropriate for metal oxide synthesis. In this process, care should be taken for the choice of vessels required for the reaction along with reactants, time of response, their amount, temperature and pH values are also vital in this process [32–34]. Also, the method of operation of every step in the synthesis process of nanomaterials makes a difference in their resulting properties.

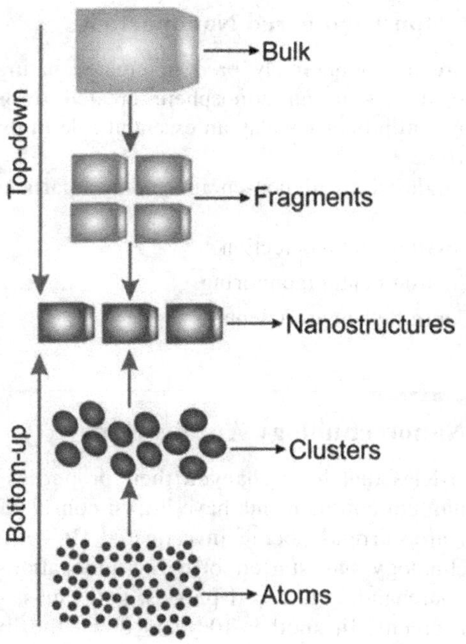

FIGURE 20.2 Top-down and bottom-up approaches for the fabrication of metal oxide nanostructures [23].

20.3.1.2 Surfactant and Capping Agent-Assisted Process

The surfactant and capping agents are the next essential component in the nanoparticle growth process which plays important properties in size and crystalline nature. They can alter the free energy of the system [35–37]. These agents can also control the morphology of the nanoparticles. The capping inorganic molecules on organic results in new and different useful photophysical and photochemical nature.

20.3.2 Synthesis of Materials

During the past few decades, research on nanoparticles has gained tremendous momentum due to the improved procurement techniques, which enhance precise control on shape and size. Numerous techniques are available to synthesize various types of nanoparticles in powders, clusters, colloids, tubes, rods, thin film, and wires. The usual conventional methods to synthesize different types of nanoparticles are optimized to produce novel nano products. The synthesis techniques that are to be employed mainly depend on the type of material, its nanostructure, namely zero, one- or two-dimensional materials, quality and size, etc. Designing and characterizing advanced materials paves the way for many improvised interdisciplinary studies. Thus, to develop novel advanced and functional materials, the nanoparticles' size and shape's uniformity play a key role. The nanoparticles' samples should be monodispersed in shape, size, surface chemistry, and internal structure. The synthesis of nanomaterials can be broadly classified into two types: the "top-down" approach and the "bottom-up" approach [23]. The "top-down" technique uses the conventional microfabrication methods that use tools controlled externally to reduce the bulk material dimension. This can be done by milling, cutting, and shaping the materials to form the required shape and order. The lithographic technique is the most common technique and laser beam processing. The molecules' chemical properties are exploited in a "bottom-up" approach, making them self-assemble into some productive conformation. Examples of the "bottom-up" approach are chemical vapor deposition, chemical synthesis, laser-induced assembly, self-assembly, growth, film deposition, and colloidal aggregation. The fact is that neither the top-down nor the bottom-up approaches are superior to each other. However, very small-sized particles are usually achieved using the bottom-up approach, a very cost-effective technique. The bottom-up

TABLE 20.2

Most Commonly Used Techniques for the Synthesis of Nanoparticles

Chemical methods	Physical methods
Sol-gel synthesis	Chemical vapor deposition
Co-precipitation	Arc discharge process
Solvo/hydrothermal synthesis	Inert gas condensation
Microemulsion	Plasma synthesis
Sonochemical method	High energy ball milling
Reverse micellar method	Aerosol synthesis
Microwave method	Electrodeposition
Thermal decomposition	Laser ablation
Solid-state reaction method	High temperature/grinding

method helps synthesize nanoparticles belonging to the various inorganic materials like oxides, metals, and chalcogenides. It involves the condensation of molecular entities in the gas phase. This approach is most sought after in the synthesis of nanoparticles. The basic principle in synthesizing nanoparticles is producing many nuclei and inhibiting the grains' growth and agglomeration. The various synthesis techniques like the solvo/hydrothermal, chemical co-precipitation process, sonochemical process, and sol-gel method are discussed briefly in this chapter (Table 20.2).

20.3.2.1 Hydrothermal/Solvothermal Synthesis

The most common technique of synthesizing nanoparticles is by using the hydro/solvothermal method [38–40]. The solvent is mixed with specific metal precursors. It is poured into an autoclave maintained at a very high temperature and pressure inside an oven to aid the crystal's growth and the assembling process. The force produced in the autoclave from the solvent's vapors also increases the solvent's boiling point. Inorganic nano samples are more easily synthesized at very low pressures and temperatures due to these metal salts' high reactivity. It is possible to fine-tune the reaction parameters such as pH, time, pressure, temperature, concentration to obtain the right nucleation rate and optimum particle size. Usually, solvents such as 1-butanol, methanol, ethanol, and toluene are employed. This technique is the most sought-after and is the most versatile method for procuring nanoparticles with a very narrow size distribution. The hydro/solvothermal process is mostly used to produce nanostructured metals, inorganic materials, and metal oxides, despite its yield being much less. The high pressures and temperatures employed in the present technique bring about an increase in the solids' solubility and hence accelerate the reaction occurring in the solid. In the case of magnetic nanoparticles, the solvothermal method is more preferred. Synthesized magnetic nanocrystals are having sizes such as 12 nm and 60 nm. These crystals were found to be water-soluble. The nanoparticles were prepared by refluxing a solution of 2-pyrrolidone of $FeCl_3$ proposed a generalized hydrothermal technique for synthesizing various nanoparticles via the liquid-solution reactions. Different magnetic nano microspheres that were highly functional and used in multiple bio-applications were using the hydrothermal reduction process. The solvothermal method of synthesis has its advantages as compared to the rest of the chemical methods. Firstly, the crystal size and morphology can be easily controlled by altering the conditions of synthesis such as temperature, pressure, pH, and the additives' nature. Secondly, the nanoparticles can be directly synthesized in the specified crystalline phase itself at much lower temperatures.

20.3.2.2 Sonochemical Process

Sono-chemistry [41] refers to the sound energy technique through which physical and chemical changes are induced within a medium. In this process, constant sonication aids the chemical reactions required for the synthesis of the nanoparticle. Consequently, the rate of the particle solvent dispersion and breaking down of the agglomerates are significantly enhanced. Ultrasound waves in the frequency of 20 kHz–10 MHz are used. This frequency range consists of three sub-regions: (i) high power-low frequency ultrasound (20–100 kHz); (ii) medium power-intermediate frequency ultrasound (100 kHz–2 MHz); and (iii) low power-high frequency ultrasound (2–10 MHz). The 2 kHz–2 MHz frequency range is usually made by the sonochemical processes. Frequencies that are above 3 MHz are mostly applicable in the field of non-destructive testing and medical imaging. The sonication process employed for sono-crystallization generally has three major effects:

- (i) The shock waves that are produced during the bubble implosion hinder the process of agglomeration.
- (ii) The liquid undergoes localized transient heating once the bubble collapses.
- (iii) The acoustic cavitation facilitates perfect mixing conditions.

All these factors contribute to particle size reduction and increase the homogeneity based on controlling the local nuclei population. Inorganic materials belonging to the nanostructured regime with high catalytic activities can be prepared by a process where the volatile organometallic precursors are sonochemically deposited. These materials are extensively used in industrial fields.

20.3.2.3 Co-Precipitation

This method of preparation is the commonest reaction that is used to prepare various kinds of nanoparticles [42, 43]. According to the LaMer diagram, in homogenous precipitations, nucleation is initiated as soon as the concentration increases beyond the saturation point. Monodispersed particles are prepared by increasing the rate of nucleation. The low surface energy raises the driving force required for the nucleation process because many nuclei are formed. Thus the nanoparticles are created through the process of precipitation as a result of three separate processes that occur sequentially: (i) nucleation, (ii) growth, and (iii) agglomeration. There are other precipitation methods and the co-precipitation method: the acid-base and the redox precipitation methods. Co-precipitation involves three main processes, namely:

inclusion, occlusion, and adsorption. Inclusion occurs when the impurity occupies a lattice site within the carrier crystal structure, resulting in a crystallographic defect. This usually happens when the impurity charge and its ionic radius are the same as that of the carrier. When an impurity is very weakly bound to the surface of the residue, it is called adsorbate. The physical trapping of a contaminant that is adsorbed during the growth process leads to what is called the process of occlusion. The co-precipitation technique is mostly used for the preparation of ceramic oxide powders. When the system is composite, this method is usually limited to cations with similar chemical properties.

20.3.2.4 Sol-Gel

In this method [44], the "sol" gradually evolves when the gel kind of a phase system comprising the liquid and a solid phase is formed [45]. In colloids, the particles' density may require removing a major quantity of the fluid initially to emphasize the gel kind of properties. The porosity of the gel ultimately decides the rate of the drying of the solvent. The changes imposed on the structure in the processing phase influence the microstructure of the prepared samples. This phase is then followed by the thermal treatment, which can also be called the firing process. Once the ignition process is initiated, the dried gel self-combusts till the entire gel is burned. This reaction self-propagates and can sustain for about 1–5 seconds within which the desired product is formed. Homogenous samples can be produced in a very short time without the need for high-temperature furnaces. The grain growth, structural stability, mechanical properties, and densification are all enhanced during the final sintering. Auto-combustion reaction provides the energy needed for the response between the component oxides, and hence an external energy source is not required. The sol-gel technique is also known by many other names like the self-propagating synthesis and auto-combustion synthesis, and was originally attempted in Russia by Merzhanov. This process was successfully developed to speed up complex oxide material procurement like ferrites and high-temperature superconductors. The sol-gel technique is mostly preferred over the other methods because of inexpensive precursors, low external energy consumption, a simple synthesis process, and a voluminous homogenous nano-sized yield.

20.3.2.5 Solid-State Reaction

The mixture of the raw materials is ground and sintered at 600°C–1,400°C range of temperatures for the desired time to complete the chemical reaction [46, 47]. Oxide synthesis materials are using a solid-state reaction technique as it is found to be a simple technique. The solid-state reaction synthesis method requires simple equipment, low cost, easily available raw materials, and straightforward synthesis procedures. This method's major drawbacks are non-uniform homogeneous composition, irregular morphology, uncontrollable particle growth, and longer heating time required. Due to structural and compositional variation, the materials produced by this method may not have consistent and attractive electrochemical performance.

The pellet preparation procedures for all samples are also the same. The pellet of each sample is prepared from the calcined powder as an active material and polyvinyl alcohol (PVA) as a binder. The proportion of binder to calcined powder is optimized for better results. The calcined powder is initially ground in an agate mortar for about 30 min. Further, the obtained powder is mixed with PVA and then ground for about 40 min. The binder added powder is then pressed at a pressure of six tons for 5 min in the hydraulic press using a die set pressure technique to form circular disk-shaped pellets. Before pressing the powder in pellet form, uniform tapping of the die set (filled with powders) is carefully adapted. The pellets are then sintered at 1,150°C for 4 h in air at heating and cooling rates of 5°C/min. The sintered pellets' surface layers are carefully polished by fine emery paper to make their faces smooth and parallel. The size of the pellets is around 10 mm in diameter and 1.1–1.3 mm in thickness. After polishing, the pellets are coated with silver paste on the opposite faces, which act as electrodes.

20.4 Magnetic Nanomaterials and Graphene-Based Composites

Graphene is made up of a single layer of carbon atoms and has been extensively studied in the past decade due to its well-ordered structure and electronic properties, which have been widely applicable in electronic devices. Synthesizing, characterizing and fabricating metal nanoparticles and other inorganic/organic species on its surface can enhance the magnetic and electronic properties of graphite. The entire nanocomposite material can have the same magnetic properties as of particles themselves are magnetic, thus opening up new possibilities and applications for graphene-based nanocomposite materials. There are many efforts on magnetic graphene and graphene oxide nanocomposites that are yet to be explored and fully reviewed.

20.4.1 Metal-Based Graphene Composites

The periodic table consists of many kinds of magnetic metals. Composites of graphite can only describe by impregnating a few typical metals with paramagnetism, such as iron, cobalt, or nickel.

In the theoretical study of the interaction between metal and graphene floor, the investigation of the effects of setting magnetic atoms subsequent to the vacancies inside the graphene layer [48, 49] and the explanation of the electronic and magnetic properties of the graphene ferromagnetic interface [50] are specifically applicable. In precise, it changed into determined that the high-spin C_{60}CrG nanostructure is stronger than its low-spin analogs. Structures containing unique metals can be of one type. For instance, C_{60}Ti stands symmetrically on the graphene floor, while C_{60}M fragments are bent in other metals and are not the same in every composite cloth (Figure 20.3). Importantly, adding C_{60} to the MG floor is extra famous than decorating graphene with C_{60} metal complexes. DFT studies have additionally been conducted on

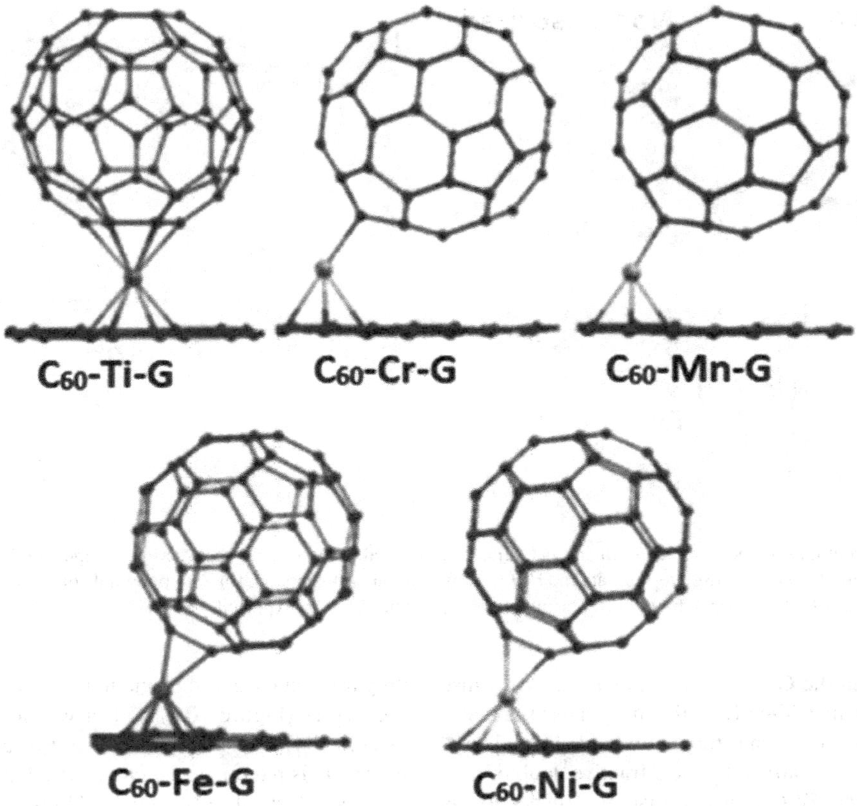

FIGURE 20.3 C_{60}-M-G nanostructures.

Co-G composites [51] or a single Co atom included into chair graphene nanoribbons with double vacancies [52]. In this case, it has been found that metal dopants introduce magnetic properties into the composite material formed.

The synthesis method of cobalt graphene nanocomposite involves the high-temperature decomposition of the starting materials. The magnetic cobalt-G nanocomposites have been prepared by means of self-assembly of cobalt carbide-based totally organometallic framework ZIF67 and graphene oxide [53]. This composite based totally on cobalt and reduced graphene oxide (RGO) is used as a catalyst for the activation of peroxymonosulfate (PMS) all through the decolorization of acid yellow dyes in water. The acquired regeneration performance remained at 97.6% after 50 cycles, indicating its effective and stable catalytic pastime. Another example is the combination of spontaneous combustion and sol-gel methods, which cause Co-G nanocomposites made from graphene oxide, cobalt nitrate, and citric acid as precursors [54]. First, the sol-gel is prepared, and then due to the effect of the generated reducing agents H_2 and CH_4, it spontaneously ignites in an Ar atmosphere at 300°C. In the formed nanocomposite, Co nanoparticles (about 10 nm in diameter) are uniformly distributed on the surface of the graphene. The same technique can be used to charge metals like nickel, copper, silver, and bismuth onto the surface of graphene. Pyrolysis is also used to produce cobalt phthalocyanine (Co-Pc) for organometallic/graphene composites [55]. Due to the interaction of cobalt and its oxides under pyrolysis or oxidation conditions, Co-Pc can disperse cobalt and its oxides on graphene sheets. The cobalt oxide-graphene nanocomposite exhibits excellent storage performance toward lithium, including good recycle performance and profoundly reversible capacity. Some people have proposed that copper oxide, ferric oxide, and other metals or metal oxide-based graphene composites can also be produced in this way.

20.4.2 Fe_2O_3-Graphene Hybrids

Iron can be present in the graphene composite in elemental form or in the form of iron core/oxide shell nanoparticles, iron oxide, etc. The ratio of iron nanomorphs in different oxidation states depends particularly on the oxygen-containing groups on the surface of the graphene, the use of reducing agents, and other conditions. Since these carbon materials are rich in oxygen-containing functional groups, magnetic iron-containing nanoparticles were loaded on to a graphene oxide layer, and their growth mechanism was studied [56]. During the heat treatment under reducing conditions, magnetic nanoparticles are partially converted to iron by the loss of these functional groups attached to carbon and the metal nanoparticles change the lattice structure and intrinsic function of graphene oxide; this effect depends on the amount of precursor. The influence of pH value is also very important, and several non-covalent magnetic GO-based materials made of Fe_2O_3 particles and magnetic surfactants have been studied [57]. The pH adjustment is used to effectively charge the repulsion or attraction between

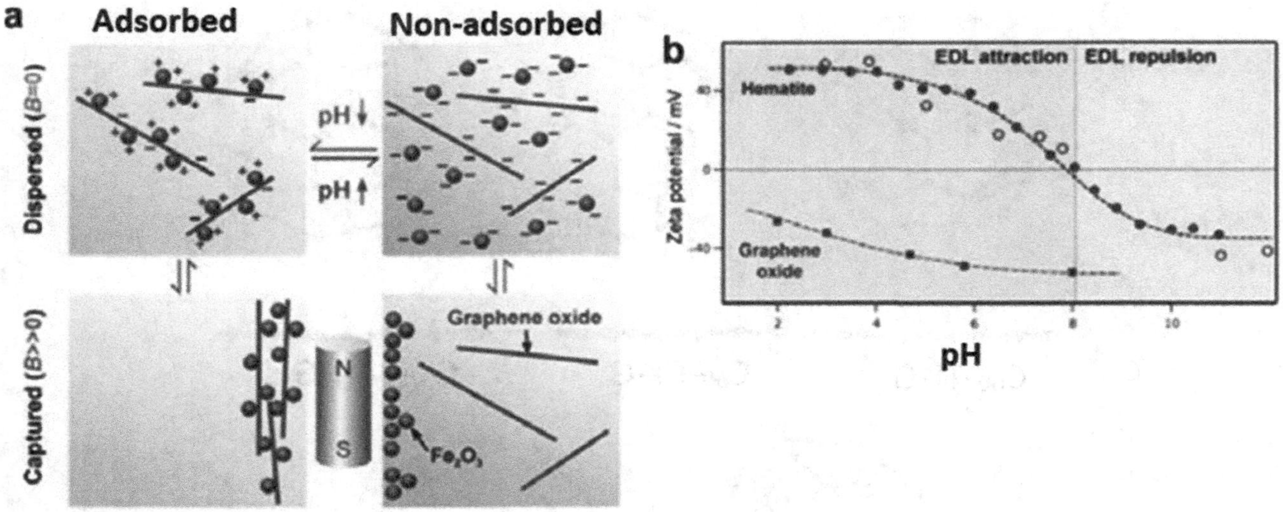

FIGURE 20.4 (a) Conceptual scheme of the experiment: pH adjustment is used to affect charge attraction or repulsion between the GO sheets and Fe_2O_3 particles. A magnetic field can be used to separate the Fe_2O_3 from solution or dispersion. (b) Zeta potentials of graphene oxide and Fe_2O_3, demonstrating the pH ranges at which electrical double-layer attraction or repulsion would be expected.

the Fe_2O_3 particles and the GO layer. Each material will cause GO co-flocculation at an acidic pH, which may cause the material to be captured by an external magnetic field. The adsorption of GO at low pH is explained by the attractive double-layer force between GO and Fe_2O_3. In contrast, the dispersion is stable at higher pH values due to the same charge repulsion. An amazing effect was found using Fe_2O_3 nanoparticles: low concentrations lead to GO flocculation, while high concentrations lead to re-stabilization, which can be explained by the effective overload of the GO surface (Figure 20.4). These systems have been found to remove model nanomaterial gold nanoparticles from water. The major synthesis methods of iron oxide-graphene nanohybrid are usually similar to the methods above for metals or Fe_3O_4, and these methods will have an insight below. For example, by high-pressure pyrolysis of ferrocene and pure grapheme, iron oxide nanoparticles wrapped in a permeable carbon layer consisting of multilayer graphene are synthesized [58]. The ferrocene precursor provides carbon and iron to form carbon-coated iron oxide, while graphene acts as a large surface area anchor to obtain small iron oxide nanoparticles. The material is used to improve the electrochemical performance of iron oxide-based electrodes on lithium-ion batteries. Similarly, a solvent-free thermal decomposition method was used to prepare iron oxide/graphene nanocomposites using iron oleate precursors (Figure 20.4) [59]. It was found that highly monodispersed γFe_2O_3 nanoparticles are in close contact with graphene. This nanomaterial can be used as a potentially valuable negative electrode material for high-rate lithium-ion batteries (Figure 20.5).

20.5 Fe_3O_4-Graphene Composites

20.5.1 Fe_3O_4/G Aerogels

Among various mixed-valence iron oxide @ graphene (or GO) composite materials, compared with the above-mentioned magnetic graphene hybrids, aerogels have received great attention. Graphene and other aerogel methods of synthesizing them are well known [60]. The introduction of magnetic components may bring a wider range of unusual properties and potential uses, that is, to simply remove contaminants such as crude oil. The composition of these nanocomposites can be simple (i.e., Fe_3O_4@G) or contain additional components, such as polymers. Therefore, 3D-N-doped graphene aerogel (NGA) supported Fe_3O_4 nanoparticles (Figure 20.6) (Fe_3O_4 @ N gas) are considered to be effective cathode catalysts for

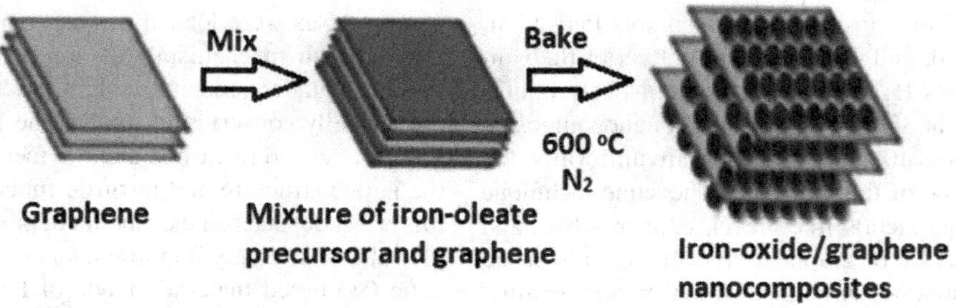

FIGURE 20.5 Schematic representation of the direct preparation of iron oxide/graphene nanocomposites by the solventless thermal decomposition method.

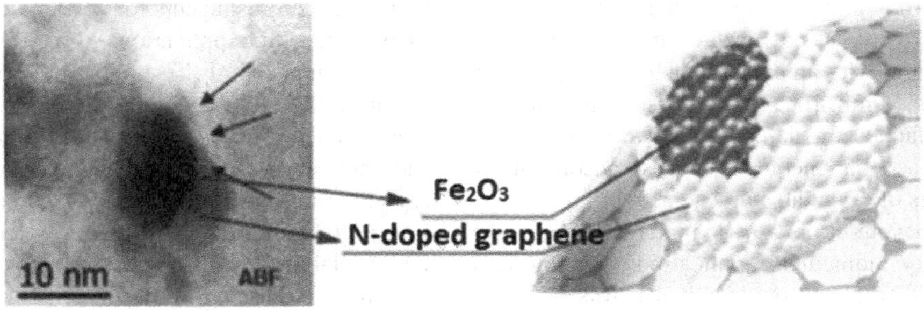

FIGURE 20.6 Nano-scaled Fe_2O_3 particles surrounded by N-doped graphene layers.

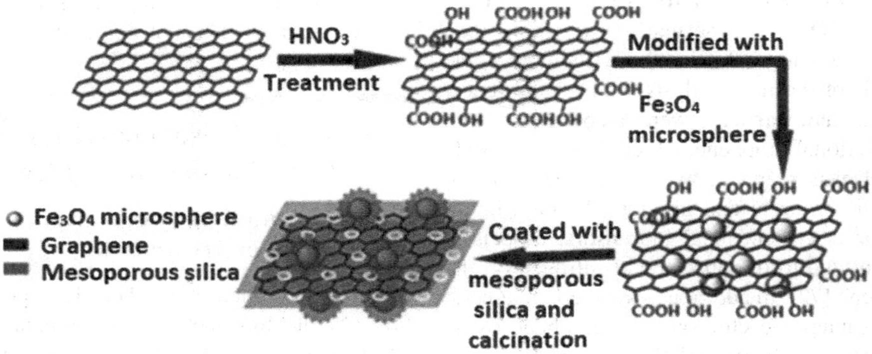

FIGURE 20.7 The workflow of synthesis of double-sided magnetic graphene/$mSiO_2$.

oxygen reduction reactions (ORR). These mixtures are produced through a combination of hydrothermal self-assembly, freeze-drying, and heat treatment. This product exhibits excellent electrocatalytic activity for ORR in alkaline electrolytes, higher current density, higher electron transfer number (~4), lower H_2O_2 yield, and better durability. 3D graphene aerogel with Fe_3O_4 nanoparticles (Fe_3O_4/GA) is the lightest magnetic elastomer ever, with a density of about 5.8 mg cm^3, prepared by self-organizing hydrothermal graphene and decorated with Fe_3O_4 nanoparticles [61]. With up to 52% reversible magnetic field-induced elongation and elongation-related resistance, they can be used to monitor the compression/elongation of materials.

Among the more complex aerogels, a hydrophobic graphene-aerogel/Fe_3O_4/polystyrene composite with a networked graphene structure (very low density, 0.005 g cm-3, corresponding to 99.7% volume porosity) has been prepared by the solvothermal method [62]. It is found that porous Fe_3O_4 nanoparticles can partially replace ethylenediamine-assisted cross-linking and the connection between graphene sheets. These composite materials are used for crude oil repair. After ten water-oil separation cycles, the capacity can reach 40 times its own weight.

20.5.2 Bicomponent Fe_3O_4/G Hybrids

With regard to the non-aerogel types of Fe_3O_4 @G hybrids, the hydrothermal [63] and the solvents interrupted synthesis [64] are also widely used, although other methods are commonly used, like deposition of the atomic layer, not only for Fe_3O_4 @ graphene, but also for Ni @ graphene composite [65]. Magnetic graphic foam with porous and hierarchical structures based on magnetic nanoparticles was made by gaseous reduction in a hydrothermal system and is used for the adsorption of oil and organic solvents, thus serving oil-type cleaning [66]. Fe_3O_4 nanoparticles on foam suspend various morphologies and cubic structures, while the reduction grade graphene oxide can be controlled under mild conditions. Various ratios of iron oxide to graphene are important for different applications [67]. The following morphologies were observed: uniform spherical homogeneous Fe_3O_4 nanoparticles without agglomeration over the graphene plates and a uniform plate-like shape of produced graphene. Increasing the Fe_3O_4 nanoparticles on the surface of the graft sheet proved to reduce the adsorption capacity while magnetization has been increased.

20.5.3 Multicomponent Fe_3O_4/G Hybrids

More complex systems containing Fe_3O_4@G and organic matter are common. Therefore, coating a layer of neutral silica material on each side of magnetic graphene, under the conditions of a surfactant-assisted sol-gel process, with an additional calcination process, leads to the production of neutral nanocomposites (Figure 20.7). Duplexity of magnetic graphene (G@SiO_2) with high specific surface and large pore volume [68]. The composite materials are applied for specific size and detection of peptides in human urine, protein digestion solutions, and standard peptide mixtures. One more important thing is the G@mSiO_2C_{18} material (surface area of 315 cm^2g^{-1} and pore size of 3.3 nm) with an expanded sheet morphology, prepared by coating medium silica layers. Adsorbent onto graphene through a surfactant-mediated condensing sol-gel process, and is applied as a magnetic solid-phase adsorbent for interactive increments of phthalates in water [69].

Some reports are devoted to magnetic materials containing chitosan with primary biomedical applications. Therefore, a mixture of iron oxide/graphene oxide/chitosan was obtained by evaporating the solution [70]. Among other things, the incorporation of 0.5% Fe_3O_4 and 1% by weight of graphene oxide improves the tensile strength and Young's modulus of the composite material by approximately 28% and 74%, respectively, compared to chitosan. Furthermore, it has been observed that Fe_3O_4-GO-CS is thermally less stable than GO-CS compounds and that graphite is thermally more stable than GO on the basis of thermal analysis. In a related report [71], Fe_3O_4 magnetic nanoparticles were incorporated into water-dispersing functional graphene, which is biocompatible with chitosan, which was prepared by one-step ball milling of chitosan-carboxylic chitosan and graphite. It could be an excellent catalyst for electrochemical biosensors, especially for glucose detection, due to the presence of nitrogen on the surface of the graphene [72]. In addition, a synthetic magnetic biosorbent based on magnetic chitosan and graphene oxide (MCGO) was produced [73]. Its methyl blue adsorption capacity after four uses was approximately 90% of the initial saturation adsorption capacity. Due to the ion exchange mechanism associated adsorption of methyl blue to MCGO is highly dependent on ionic strength and pH (Table 20.3).

20.5.4 Carbon Nanotube-Based Iron Composites

Superparamagnetic iron oxide nanoparticles (SPIONs) are known to have excellent adsorbent properties, wastewater treatment and in medicine [82]. Fe_2O_3 nanoparticles can be employed for heavy metal removal pertaining to their high specific surface area and tunable properties [83]. They can also be used as contrast agents for MRI, as drug delivery agents or for hyperthermia applications [84]. It is of greater interest to make a composite of these two materials for attaining beneficial properties. The conductivity for CNTs and the superparamagnetism for SPIONs are valuable for numerous applications. The applications include magnetic solid-phase extractions of dyes and pharmaceuticals [85]. Decorating carbon nanotubes with SPIONs is very essential to improve the reactivity and the recyclability of CNTs pertaining to their amphiphilic character [86]. Additionally, the decoration affects the magnetic and electric properties of this material [87–90]. A great step to improve CNTs' potential as a drug delivery vehicle was to combine them with MNPs. MWCNTs were functionalized with polyacrylic acid and then mediated with magnetite nanoparticles (Fe_3O_4) employing the co-precipitation step method in the presence of Fe^{2+} and Fe^{3+} [91].

20.6 Magnetic Nanoparticles and Their Medicinal Applications

20.6.1 Magnetic Hyperthermia in Cancer Treatment

In cancer treatment hyperthermia is previously known and a favorable healing method. In combination with radiotherapy or chemotherapy hyperthermia can improve the cell toxic effects. For cancer treatment, the use of local hyperthermia solely is investigated. The heating of tumor tissues up to over 40–45°C should lead to the devastation of tumors [92]. A problem in cancer treatment with hyperthermia is to restrict the heating effect of the tissue in the tumor area. Therefore, magnetic nanoparticles and nanocomposites were studied for local hyperthermia. For this method, iron oxide nanoparticles and nanocomposites are in the attention of research because of their compatibility in biological entities [93, 94]. Under the influence of an irregular magnetic area, magnetic nanoparticles and nanocomposites can be heated by hysteresis loss. In addition, one opportunity to focus these particles in the area

TABLE 20.3
Applications of Fe_3O_4-G Hybrids

Nanocomposite	Description/application
Graphene oxide, in the form of Fe_3O_4/GO magnetic particles	Adsorbent for wastewater treatment
Aerogel Fe_3O_4/grapheme	Removal of As from water [74]
Fe_3O_4 @ graphene	Electrochemical absorption for both inorganic and As species [75]
Magnetic support cyclodextrin graphene oxide	Removal of heavy metals from industrial effluents [76]
Polyethyleneimine-modified neutral magnetic silica (PEI) and graphene oxide	Simultaneous removal by synergistic adsorption of heavy metal ions and humic acids [77]
Magnetic graphene (Fe_3O_4@PDDA-GO@DNA) based on Fe_3O_4@PDDA polycation core-shell	Removal of brominated diphenyl ethers in water. Can be reusable for at least 20 times toward processing [78, 79]
Fe_3O_4@G composite disk	High-performance Li-ion battery – anodic material [80]
Porous graphene composite containing graphene scaffolds, Fe_3O_4 nanoparticles @TiO_2	Selective activity and magnetic separation of phosphopeptides target, exhibiting improved activity compared with commercial TiO_2 affinity materials [81]
Graphitic Fe_3O_4	Effective removal of bacteria. This is because it has the ability to eliminate a wide range of pathogens, including bacteria such as *E. faecium*, *E. coli*, *E. faecalis* and *S. aureus*, with up to 94% removal efficiency

of interest is the universal use and the detail that the nanoparticles are phagocytosed by the cells of the particular tissue. Hyperthermia advantages were also described for many illnesses such as syphilitic paralysis and gonococcal illnesses and are also estimated for HIV/AIDS infections [95]. Recent modalities for cancer hyperthermia could be categorized according to the nature of the heating basis and the heated target, from the whole body to the tumoral cell level. The heating is mainly classified into three classes: contactless applicator (microwave, ultrasound, infrared devices, and radiofrequency); contact with extremely heated liquid; and introduced heating source (antennas, probes, mediators, and laser fibers) (Figure 20.8a–c) [97]. Among most new hyperthermia devices, those that are based on focused electromagnetic or ultrasound radiation are commercially accessible. In addition, according to researchers from Korea, it is possible to use nanoparticles and nanocomposites as remote-controlled magnetic death switches to kill cancer cells from Korea. J. Cheon and J. S. Shin of Yonsei University, in Seoul, and their co-workers have attached zinc-doped iron oxide magnetic nanoparticles to an antibody, which targets a receptor on colon cancer cells. The cancer cells death receptor DR4 is targeted by the antibody when triggered by a magnetic field and programmed cell death or apoptosis is produced [98]. Magnetic nanoparticles enable the control of cell signaling paths. Cell signaling allows the exchange of information between cells and underpins growth metabolism, difference, and several other processes. It also allows cells to trigger apoptosis to make sure tissues do not grow out of control or cells breakdown.

In addition, adding a magnetic nanoparticle are suitable remote control of the signaling process using a magnetic area. This allows monitoring the position of magnetic nanoparticles in the body and exact malignant tissues are targeted. The research team confirmed this method using zebrafish and exhibited that the death switch can function at the micrometer scale. Shin emphasized that this is the first time an in vivo magnetic nano-switch has been revealed and been verified feasible. Johannsen and co-workers have presented data concerning the first medical use of this form of local hyperthermia with magnetic nanoparticles in cancer analysis of human patients [99]. J. Kolosnjaj-Tabi et al. displayed the outstanding tumor regression in mouse epidermoid carcinoma xenograft model using polyethylene glycol-coated magnetite nanoparticles after magnetic hyperthermia [100]. Hayashi and co-workers [101] have revealed improved accumulation and enlarged magnetic relativity using folic acid conjugated superparamagnetic nanoparticles. Furthermore, the mice were placed in an external magnetic field ($f = 230$ kHz; $Hf = 1.8 \times 10^9$ A/m•s and $H = 8$ kA/m) producing heat to the local tumor tissues (≈6°C higher than nearby tissues) causing a notable tumor decrease and higher survivability (Figure 20.9a–d). In addition, magnetic hyperthermia is useful for the precise release of cytotoxic agents in cancer cells using a heat-labile coating. Hu and co-workers have recently exhibited the controlled release of dual drugs (Dox and paclitaxel) from heat-sensitive polyvinyl alcohol (PVA) coated SPIONs using an outside magnetic field [102]. Furthermore, antibody conjugation with magnetic nanoparticles improved the effect of hyperthermia because of the anti-cancer effects of the antibody and selectivity of the cancer cells.

Examples include anti-FGFR1 aptamer-tagged magnetic nanoparticles for improved magnetic hyperthermia and antibody-conjugated magnetic nanoparticles for enhanced anticancer properties of cryptotanshinone [103, 104]. Kim and co-workers have reported a novel technique to improve the cytotoxicity of magnetic nanoparticles by selectively aiming for a thermally susceptible subcellular organelle (i.e., mitochondria) (Figure 20.10). The selective distribution of photothermal nanoparticles to a subcellular organelle can realize a higher photo-thermal healing efficiency in cancer analysis with negligible side effects. Recent results in improving the radiation-to-heat alteration efficiency using magnetic nanomaterials have attained a great achievement, but the improvement of magnetic nanoparticles based on novel methods and procedures to further enhance the heat transformation efficiency is still vital. In addition, cancer cells are more delicate to ionizing radiation and chemotherapeutic agents. Therefore, a magnet-mediated hyperthermia joint with either radiation or systemic chemotherapy could attain synergistic anti-tumor influence and quickly become a clinical reality for treating malignancy. Moreover, magnetic nanoparticles are already being used as difference agents in medicinal imaging, to visualize tumors

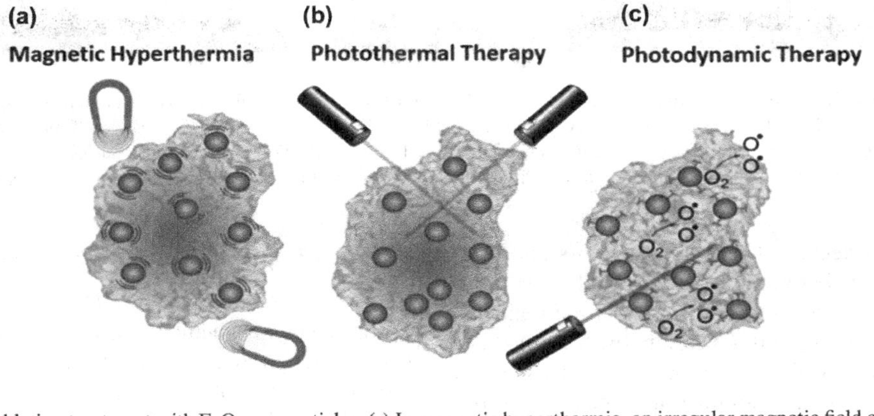

FIGURE 20.8 Tumor ablation treatment with FeO nanoparticles. (a) In magnetic hyperthermia, an irregular magnetic field causes FeO nanoparticles to produce heat, encouraging tumor necrosis. (b) In photo-thermal ablation, light absorbed by nanoparticles is transformed into thermal energy initiating cell death in the area. (c) For photodynamic treatment, photosensitizing agents involved in nanoparticles are stimulated by an outside light source to create single oxygen types that are cytotoxic to cells [96].

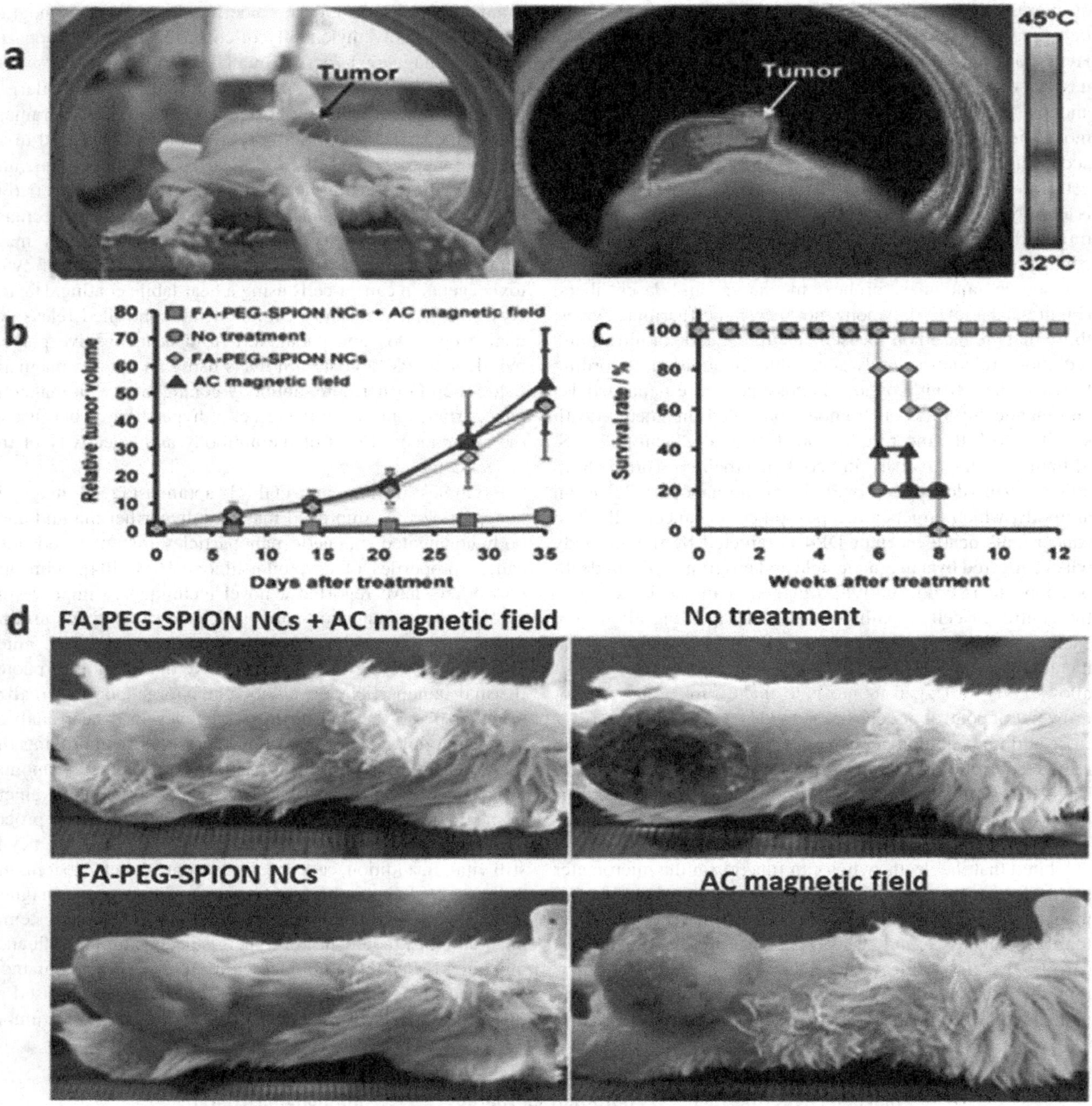

FIGURE 20.9 (a) Image (left) and thermal image (right) of a mouse 24 h after venous dose of folic acid conjugated pegylated superparamagnetic iron oxide nano-conjugates (FA-PEG-SPION NCs) under an AC magnetic field with H=8 kA/m and f=230 kHz. (b) Tumor-growth performance and (c) existence period of mice without cure and treated by venous injection of FA-PEG-SPION NCs, use of an alternating current (AC) magnetic field, and use of an AC magnetic field 24 h after venous injection of FA-PEG-SPION NCs (n=5). (d) Image of mice 35 days after treatment [102].

more exactly, with esteem not only to the limits and range but also to differentiate between the inactive and active areas. Therewith, treatment such as radiotherapy can be deliberate and passed out more simply and professionally. Nanomedicine is being used to improve the observing of the illness development so that the patient's administration can be familiar quickly, if essential. This section will focus more carefully on drug delivery and overcoming drug confrontation as means of the development of cancer therapy.

20.6.2 Magnetic Resonance Imaging

Magnetic resonance imaging (MRI) is one of the best powerful methods in medicinal imaging, due to its radiation-free nature and non-invasiveness. Related to other medical imaging procedures, MRI offers numerous advantages related to image elasticity and high three-dimensional resolution, good contrast in soft tissues and also the capability to provide data associated with blood movement and blood vessels [105]. In addition,

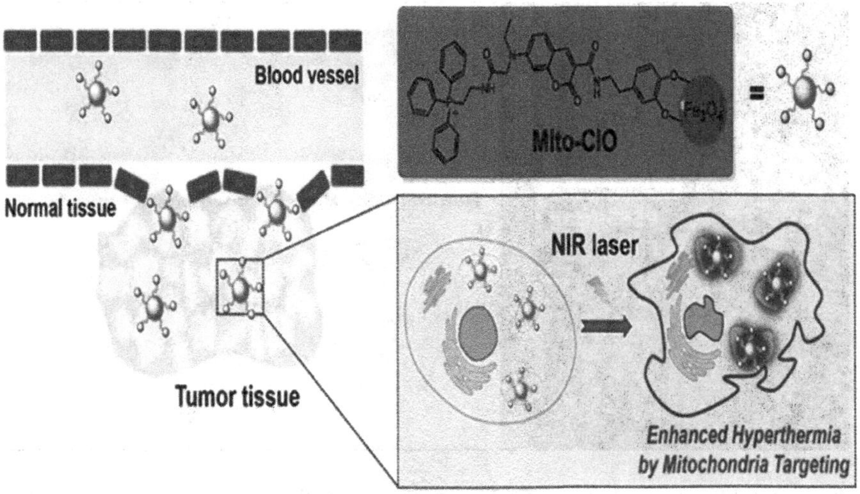

FIGURE 20.10 Improved hyperthermia by using mitochondria-aiming FeO nanoparticles.

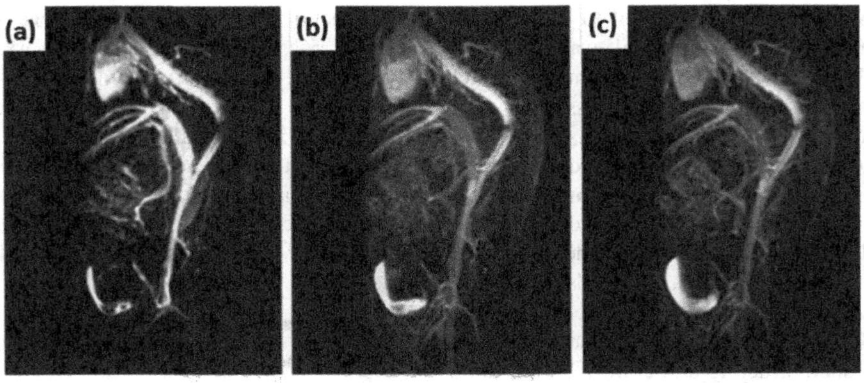

FIGURE 20.11 T1-weighted MR angiography, at 7 T, of a rat (a) 4 min; (b) 12 min; and (c) 20 min after the injection of ZES-SPIONs.

the main disadvantage is low sensitivity. Though, in current years, numerous types of contrast agents have been established to expand the MRI sensitivity and improve the info existing in the images, specifically by using magnetic nanoparticles and magnetic ions [106]. Many authors have also encouraged the use of FeO nanoparticles as T_1 contrast agents, later in medical practice, normally active contrast agents are gadolinium (Gd) complexes, which, as stated, pose health risks [107, 108]. Wei and co-workers have examined zwitterion (ZES)-coated superparamagnetic iron oxide nanoparticles (SPIONs) (ZES-SPIONs), possessing inorganic cores with a size of ~3 nm and an ultrathin hydrophilic shell (~1 nm). These nanoparticles were exposed to an existing a r_2/r_1 ratio equal to 2.0, i.e., a value lower than that related with other SPION-based positive difference agents, although this was in a factor of 2 to that showed by Gd-based chelates [109]. Also, in vivo MRI was done on rats injected with ZES-SPIONs to assess their preclinical potential as positive contrast agents for MR angiography and MRI (Figure 20.11). Such tests exposed a difference power that was suitably high for use in the measured uses.

An effective renal consent of the ZES-SPIONs was detected, and by calculating their r_2/r_1 ratio again after emission, the authors confirmed that the MR contrast power of the nanoparticles was mostly unchanged under physical conditions. Yin and co-workers have exhibited strong T_1 contrast improvement (brighter contrast) from SPIONs (diameters from 11 nm to 22 nm) as detected in the ultra-low field (ULF) MRI at 0.13 mT. They have attained a high longitudinal relaxivity for 18 nm SPION solutions, $r_1 = 615\ s^{-1}\ mM^{-1}$, which is two orders of magnitude larger than typical viable Gd-based T_1 contrast agents operating at high fields (1.5 T and 3 T). The significantly improved r_1 value at ultra-low areas is attributed to the coupling of proton spins with SPION magnetic fluctuations (Brownian and Néel) related with a low occurrence peak in the unreal part of AC susceptibility (χ"). SPION-based T_1-weighted ULF MRI has the use of improved signal, smaller imaging times, and FeO-based nontoxic biocompatible agents [110]. Corr and co-workers have studied suspensions collected of direct chains of magnetite nanoparticles which were modified through a procedure relating the cross-linking of adjacent nanoparticles with poly-electrolytic complexes. Later, a magnetic field was applied, leading to the reorganization of such nano-architectures into collateral arrangements. The relaxivity of the found nanostructures was evaluated through field-cycling NMR at 37°C, showing a significant drop in the reduction times for each field used. Furthermore, MR images of living rats to which such

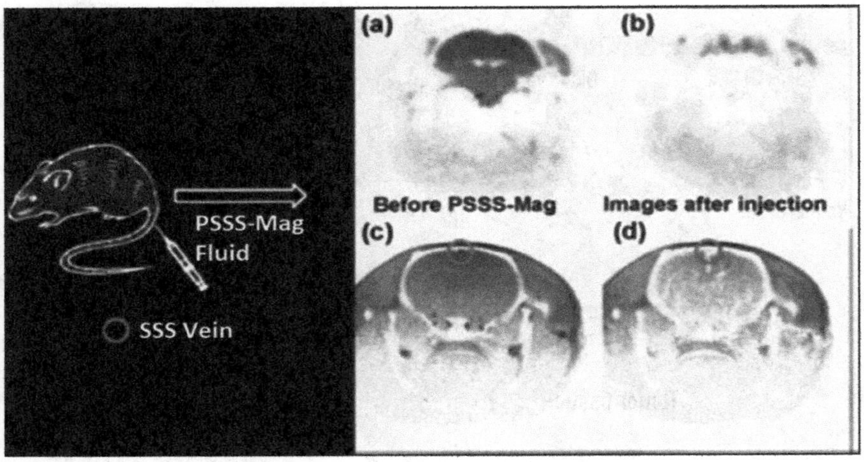

FIGURE 20.12 Echo planar image (EPI) of rat's brain (a) before and (b) as PSSS-Magl (Fe-polysodium-4-styrene sulfonate ratio 1:2) travels across the organ; fast low angle shot (FLASH) picture of a rat's brain (c) before and (d) as PSSS-Magl travels across the organ.

nanostructures were managed and obtained to calculate the influence that they had upon the brain. Not only was satisfactory biological compatibility for these nano-architectures detected, but in vivo MRI images also exhibited their possibility as T_1-contrast agents, since they covered brain areas, as displayed in Figure 20.12 [111]. Kozlova and co-workers have described that the control of the difference of an MRI scan can be attained by regulating the nanoparticle concentration in the controlled suspension, or by altering the number of magnetic nanoparticles in one carrier. In the same work, it was also stated that the core-shell structure of the nanoparticle influences the contrast effectiveness. The authors modified polymeric submicron particles with a core-shell structure and confirmed the development of T_1 and T_2 gradient contrast when magnetite nanoparticles were combined only in the carrier shell. The found result was related to the magnetic relations that were identified by the package density of the magnetite nanoparticles in the carrier [112]. Efremova and co-workers have described the modification of octahedral-shaped Fe_3O_4 magnetite nanoparticles with a diameter of 25 nm grown on 9 nm gold nanospheres. The nano-architectures showed magnetic features similar to those of bulk materials. Also, it was confirmed through in vivo and in vitro assays that the non-pure structures, because of their shape anisotropy, exist higher sloping relaxivity for MRI related to conventional contrast attractive agents [113]. Liu and co-workers have linked a post-processing susceptibility-gradient mapping (SGM) method to IRON and the White Marker (dephased process) positive contrast arrangements. SGM more evidently defines glioma tumors labeled with SPIONs than either the IRON or White Marker sequences in mice. However, further investigations are required to compare the limit of detection of the many positive contrast procedures with those of the traditional spin echo procedures [114]. Carvalho and co-workers have reported a novel dehydrodipeptides consisting of exact amino acids were prepared and investigated. SPIONs were combined into the hydrogels and the follow-on structures were completely categorized in order to know their novel properties. Carvalho has positively confirmed that self-build hydrogels sustain their magnetic features when SPIONs are combined in the structure, regardless of the existence of a diamagnetic module that rises from the hydrogel network. Dual T_1/T_2 MRI contrast agents were found, and it was detected that the SPIONs, with the use of an irregular magnetic field, were able to produce an increase in the local temperature that was enough to trigger a phase change of exact hydrogels; in this situation, the leak of SPIONs from the hydrogel mixtures upon the use of an irregular magnetic field could be an effective step in the improvement of more powerful theragnostic methods [115].

20.7 Conclusion

In brief, magnetic nanoparticles and nanocomposites have wide range of biomedical uses, with more applications for these magnetic nanoparticles being discovered. One of the limiting factors that these nanoparticles face in medical treatment is finding precise accurate targeting of areas within the body, be it for targeting of a magnetic resonance imaging and cancer therapy involving the magnetic nanoparticles and nanocomposites directly. Moreover, microfluidic methods are one of the extents that are steadily backed up along the entire path of magnetic nanocomposite-based biomedicine, contributing good clarifications and reproducibility from magnetic nanoparticle preparations to analysis, passing through evolving tests in drug progress and cell culture between others. Stimulating proofs-of-impression in a huge variation of biomedical regions have revealed the full possibility of this research part, demonstrating a significant development in the entire biomedical area.

REFERENCES

1. Mansoori, G. A. (2005). *Principles of Nanotechnology: Molecular-based Study of Condensed Matter in Small Systems* (p. 31) World Scientific Publishing Co.
2. Toumey, C. (2009). Plenty of room, plenty of history. *Nature Nanotechnology*, *4*(12), 783–784.
3. Navrotsky, A. (2003). Materials and nanotechnology. *Mrs Bulletin*, *28*(2), 92–94.

4. Waser, R. (Ed.). (2008). *Nanotechnology: Volume 4: Information Technology II*. Wiley.
5. Bogunia-Kubik, K., & Sugisaka, M. (2002). From molecular biology to nanotechnology and nanomedicine. *Biosystems*, *65*(2–3), 123–138.
6. Mazzola, L. (2003). Commercializing nanotechnology. *Nature Biotechnology*, *21*(10), 1137–1143.
7. Henini, M. (1998). Nanotechnology: Growing in a shrinking world. *III-Vs Review*, *11*(4), 30–34.
8. Geller, M., Hopfer, F., & Bimberg, D. (2008). Nanostructures for nanoelectronics: No potential for room temperature applications. *Microelectronics Journal*, *39*(3–4), 302–306.
9. Minoli, D. (2005). *Nanotechnology Applications to Telecommunications and Networking*. Wiley.
10. Smith, L. M. (2010). Molecular robots on the move. *Nature*, *465*(7295), 167–168.
11. Jordan, C. C., Kaiser, I., & Moore, V. C. (2014). 2013 nanotechnology patent literature review: Graphitic carbon-based nanotechnology and energy applications are on the rise. *Nanotechnology Law & Business*, *11*, 111.
12. Afanasiev, P. (2004). Elemental sulfur and sulfur-rich compounds II. In R. Steudel (Technische Universität Berlin) (Ed.), *Topics in Current Chemistry* (vol. 231, 248 pp). Springer-Verlag. ISBN 3-540-40378-7.
13. Binnig, G., Rohrer, H., Gerber, C., & Weibel, E. (1982). Surface studies by scanning tunneling microscopy. *Physical Review Letters*, *49*(1), 57.
14. Mecking, H. (1979). Deformation of polycrystals. In *Strength of Metals and Alloys* (pp. 1573–1594). Pergamon.
15. Rogers, B., Adams, J., & Pennathur, S. (2013). *Nanotechnology: The Whole Story*. CRC Press.
16. Kharisov, B. I., Dias, H. R., Kharissova, O. V., Vázquez, A., Pena, Y., & Gomez, I. (2014). Solubilization, dispersion and stabilization of magnetic nanoparticles in water and non-aqueous solvents: Recent trends. *RSC Advances*, *4*(85), 45354–45381.
17. Binnig, G., Quate, C. F., & Gerber, C. (1986). Atomic force microscope. *Physical Review Letters*, *56*(9), 930.
18. Kroto, H. W., Heath, J. R., O'Brien, S. C., Curl, R. F., & Smalley, R. E. (1985). C 60: Buckminsterfullerene. *Nature*, *318*(6042), 162–163.
19. Iijima, S., & Ajayan, P. M. (1992). Smallest carbon nanotube. *Nature*, *358*, 23–23.
20. Jortner, J. (1992). Cluster size effects. *Zeitschrift für Physik D Atoms, Molecules and Clusters*, *24*(3), 247–275.
21. Fang, X., Bando, Y., Gautam, U. K., Zhai, T., Zeng, H., Xu, X., & Golberg, D. (2009). ZnO and ZnS nanostructures: Ultraviolet-light emitters, lasers, and sensors. *Critical Reviews in Solid State and Materials Sciences*, *34*(3–4), 190–223.
22. Cao, G. (2004). *Nanostructures & Nanomaterials: Synthesis, Properties & Applications*. Imperial College Press.
23. Galstyan, V., Bhandari, M. P., Sberveglieri, V., Sberveglieri, G., & Comini, E. (2018). Metal oxide nanostructures in food applications: Quality control and packaging. *Chemosensors*, *6*(2), 16.
24. Lee, Y. S. (2008). *Self-assembly and Nanotechnology: A Force Balance Approach*. John Wiley & Sons.
25. Craig, P. J. (2007). Ulrich Muller. In *Inorganic Structural Chemistry* (vol. 2006, 268 pp). Wiley–VCH. ISBN 13: 978-0-470-01864-4 (hardback)/10: 0-470-01864-X.
26. Baraton, M. I. (Ed.). (2003). *Synthesis, Functionalization and Surface Treatment of Nanoparticles* (Vol. 9). American Scientific Publishers.
27. Gleiter, H. (1991). Nanocrystalline materials. *Advanced Structural and Functional Materials*, 1, 1–37.
28. Levins, C. G., & Schafmeister, C. E. (2005). The synthesis of curved and linear structures from a minimal set of monomers. *Journal of Organic Chemistry*, *70*(22), 9002–9008.
29. Das, S., Gates, A. J., Abdu, H. A., Rose, G. S., Picconatto, C. A., & Ellenbogen, J. C. (2007). Designs for ultra-tiny, special-purpose nanoelectronic circuits. *IEEE Transactions on Circuits and Systems I: Regular Papers*, *54*(11), 2528–2540.
30. Zhao, S., Hong, R., Luo, Z., Lu, H., & Yan, B. (2011). Carbon nanostructures production by AC arc discharge plasma process at atmospheric pressure. *Journal of Nanomaterials*, *2011*, 281–292.
31. Takacs, L. (2002). Self-sustaining reactions induced by ball milling. *Progress in Materials Science*, *47*(4), 355–414.
32. Hainfeld, J. F., Slatkin, D. N., Focella, T. M., & Smilowitz, H. M. (2006). Gold nanoparticles: A new X-ray contrast agent. *British Journal of Radiology*, *79*(939), 248–253.
33. Saito, T., Ohshima, S., Xu, W. C., Ago, H., Yumura, M., & Iijima, S. (2005). Size control of metal nanoparticle catalysts for the gas-phase synthesis of single-walled carbon nanotubes. *Journal of Physical Chemistry B*, *109*(21), 10647–10652.
34. Yang, X., Yang, M., Pang, B., Vara, M., & Xia, Y. (2015). Gold nanomaterials at work in biomedicine. *Chemical Reviews*, *115*(19), 10410–10488.
35. Chakraborty, S. (2018). *Multifunctional Nanocomposites in Coating Technology: Synthesis, Characterization, Formulation, Evaluation and Application*. Northern Illinois University.
36. Rao, C. R., Kulkarni, G. U., Thomas, P. J., & Edwards, P. P. (2000). Metal nanoparticles and their assemblies. *Chemical Society Reviews*, *29*(1), 27–35.
37. Hong, R. Y., Li, J. H., Chen, L. L., Liu, D. Q., Li, H. Z., Zheng, Y., & Ding, J. J. P. T. (2009). Synthesis, surface modification and photocatalytic property of ZnO nanoparticles. *Powder Technology*, *189*(3), 426–432.
38. Lester, E., Dunne, P., Chen, Y., & Al-Atta, A. (2018). The engineering of continuous hydrothermal/solvothermal synthesis of nanomaterials. *Supercritical and Other High-pressure Solvent Systems* (pp. 416–448). Green Chemistry Series, RSC.
39. Gersten, B. (2005). Solvothermal synthesis of nanoparticles. *Chemfiles*, *5*, 11–12.
40. Dunne, P. W., Lester, E., Starkey, C., Clark, I., Chen, Y., & Munn, A. S. (2018). The chemistry of continuous hydrothermal/solvothermal synthesis of nanomaterials. In *Supercritical and Other High-pressure Solvent Systems* (pp. 449–475). Green Chemistry Series, RSC.
41. Gedanken, A. (2004). Using sonochemistry for the fabrication of nanomaterials. *Ultrasonics Sonochemistry*, *11*(2), 47–55.
42. Kandpal, N. D., Sah, N., Loshali, R., Joshi, R., & Prasad, J. (201). Co-precipitation method of synthesis and characterization of iron oxide nanoparticles. *Journal of Scientific & Industrial Research*, *73*, 87–90.

43. Sagadevan, S., Chowdhury, Z. Z., & Rafique, R. F. (2018). Preparation and characterization of nickel ferrite nanoparticles via co-precipitation method. *Materials Research*, *21*(2), e20160533.
44. Chen, D. H., & He, X. R. (2001). Synthesis of nickel ferrite nanoparticles by sol-gel method. *Materials Research Bulletin*, *36*(7–8), 1369–1377.
45. Andrade, A. L., Souza, D. M., Pereira, M. C., Fabris, J. D., & Domingues, R. Z. (2009). Synthesis and characterization of magnetic nanoparticles coated with silica through a sol-gel approach. *Cerâmica*, *55*, 420–424.
46. Hasany, S. F., Ahmed, I., Rajan, J., & Rehman, A. (2012). Systematic review of the preparation techniques of iron oxide magnetic nanoparticles. *Nanosci. Nanotechnol*, *2*(6), 148–158.
47. Karami, H. (2010). Synthesis and characterization of iron oxide nanoparticles by solid state chemical reaction method. *Journal of Cluster Science*, *21*(1), 11–20.
48. Hu, F. M., Gubernatis, J. E., Lin, H. Q., Li, Y. C., & Nieminen, R. M. (2012). Behavior of a magnetic impurity in graphene in the presence of a vacancy. *Physical Review B*, *85*(11), 115442.
49. Hu, F. M., Ma, T., Lin, H. Q., & Gubernatis, J. E. (2011). Magnetic impurities in graphene. *Physical Review B*, *84*(7), 075414.
50. Weser, M., Voloshina, E. N., Horn, K., & Dedkov, Y. S. (2011). Electronic structure and magnetic properties of the graphene/Fe/Ni (111) intercalation-like system. *Physical Chemistry Chemical Physics*, *13*(16), 7534–7539.
51. Raji, A. T., & Lombardi, E. B. (2015). Stability, magnetic and electronic properties of cobalt–vacancy defect pairs in graphene: A first-principles study. *Physica B: Condensed Matter*, *464*, 28–37.
52. Li, B., Xu, D., Zhao, J., & Zeng, H. (2015). First principles study of electronic and magnetic properties of co-doped armchair graphene nanoribbons. *Journal of Nanomaterials*, *2015*, Article ID 538180.
53. Lin, K. Y. A., Hsu, F. K., & Lee, W. D. (2015). Magnetic cobalt–graphene nanocomposite derived from self-assembly of MOFs with graphene oxide as an activator for peroxymonosulfate. *Journal of Materials Chemistry A*, *3*(18), 9480–9490.
54. Zhang, L., Huang, Y., Zhang, Y., Ma, Y., & Chen, Y. (2013). Sol–gel autocombustion synthesis of graphene/cobalt magnetic nanocomposites. *Journal of Nanoscience and Nanotechnology*, *13*(2), 1129–1131.
55. Yang, S., Cui, G., Pang, S., Cao, Q., Kolb, U., Feng, X., … & Müllen, K. (2010). Fabrication of cobalt and cobalt Oxide/Graphene composites: Towards high-performance anode materials for lithium ion batteries. *ChemSusChem: Chemistry & Sustainability Energy & Materials*, *3*(2), 236–239.
56. Wang, Y., He, Q., Qu, H., Zhang, X., Guo, J., Zhu, J., … & Guo, Z. (2014). Magnetic graphene oxide nanocomposites: Nanoparticles growth mechanism and property analysis. *Journal of Materials Chemistry C*, *2*(44), 9478–9488.
57. McCoy, T. M., Brown, P., Eastoe, J., & Tabor, R. F. (2015). Noncovalent magnetic control and reversible recovery of graphene oxide using iron oxide and magnetic surfactants. *ACS Applied Materials & Interfaces*, *7*(3), 2124–2133.
58. Sun, Z., Madej, E., Wiktor, C., Sinev, I., Fischer, R. A., van Tendeloo, G., … & Ventosa, E. (2015). One-pot synthesis of carbon-coated nanostructured iron oxide on few-layer graphene for lithium-ion batteries. *Chemistry: A European Journal*, *21*(45), 16154–16161.
59. Jang, B., Chae, O. B., Park, S. K., Ha, J., Oh, S. M., Na, H. B., & Piao, Y. (2013). Solventless synthesis of an iron-oxide/graphene nanocomposite and its application as an anode in high-rate Li-ion batteries. *Journal of Materials Chemistry A*, *1*(48), 15442–15446.
60. Jung, S. M., Jung, H. Y., Dresselhaus, M. S., Jung, Y. J., & Kong, J. (2012). A facile route for 3D aerogels from nanostructured 1D and 2D materials. *Scientific Reports*, *2*(1), 1–6.
61. Wu, Z. S., Yang, S., Sun, Y., Parvez, K., Feng, X., & Müllen, K. (2012). 3D nitrogen-doped graphene aerogel-supported Fe3O4 nanoparticles as efficient electrocatalysts for the oxygen reduction reaction. *Journal of the American Chemical Society*, *134*(22), 9082–9085.
62. Xu, X., Li, H., Zhang, Q., Hu, H., Zhao, Z., Li, J., … & Gogotsi, Y. (2015). Self-sensing, ultralight, and conductive 3D graphene/iron oxide aerogel elastomer deformable in a magnetic field. *ACS Nano*, *9*(4), 3969–3977.
63. Zhou, S., Jiang, W., Wang, T., & Lu, Y. (2015). Highly hydrophobic, compressible, and magnetic polystyrene/Fe3O4/graphene aerogel composite for oil–water separation. *Industrial & Engineering Chemistry Research*, *54*(20), 5460–5467.
64. Ren, L., Huang, S., Fan, W., & Liu, T. (2011). One-step preparation of hierarchical superparamagnetic iron oxide/graphene composites via hydrothermal method. *Applied Surface Science*, *258*(3), 1132–1138.
65. Fu, C., Zhao, G., Zhang, H., & Li, S. (2014). A facile route to controllable synthesis of Fe3O4/graphene composites and their application in lithium-ion batteries. *International Journal of Electrochemical Sciences*, *9*(1), 46–60.
66. Wang, G., Gao, Z., Wan, G., Lin, S., Yang, P., & Qin, Y. (2014). Supported high-density magnetic nanoparticles on graphene by atomic layer deposition used as efficient synergistic microwave absorbers. *Nanoresearch*, *7*(5), 704–716.
67. Yang, S., Chen, L., Mu, L., & Ma, P. C. (2014). Magnetic graphene foam for efficient adsorption of oil and organic solvents. *Journal of Colloid and interface Science*, *430*, 337–344.
68. Farghali, M. A., El-Din, T. A. S., Al-Enizi, A. M., & El Bahnasawy, R. M. (2015). Graphene/magnetite nanocomposite for potential environmental application. *International Journal of Electrochemical Sciences*, *10*(1), 529–537.
69. Yin, P., Sun, N., Deng, C., Li, Y., Zhang, X., & Yang, P. (2013). Facile preparation of magnetic graphene double-sided mesoporous composites for the selective enrichment and analysis of endogenous peptides. *Proteomics*, *13*(15), 2243–2250.
70. Huang, D., Wang, X., Deng, C., Song, G., Cheng, H., & Zhang, X. (2014). Facile preparation of raisin-bread sandwich-structured magnetic graphene/mesoporous silica composites with C18-modified pore-walls for efficient enrichment of phthalates in environmental water. *Journal of Chromatography A*, *1325*, 65–71.

71. Yadav, M., Rhee, K. Y., Park, S. J., & Hui, D. (2014). Mechanical properties of Fe3O4/GO/chitosan composites. *Composites Part B: Engineering*, *66*, 89–96.
72. Zhang, W., Li, X., Zou, R., Wu, H., Shi, H., Yu, S., & Liu, Y. (2015). Multifunctional glucose biosensors from Fe 3 O 4 nanoparticles modified chitosan/graphene nanocomposites. *Scientific Reports*, *5*(1), 1–9.
73. Ou, J., Wang, F., Huang, Y., Li, D., Jiang, Y., Qin, Q. H., ... & Zhang, T. (2014). Fabrication and cyto-compatibility of Fe3O4/SiO2/graphene–CdTe QDs/CS nanocomposites for drug delivery. *Colloids and Surfaces B: Biointerfaces*, *117*, 466–472.
74. Fan, L., Luo, C., Li, X., Lu, F., Qiu, H., & Sun, M. (2012). Fabrication of novel magnetic chitosan grafted with graphene oxide to enhance adsorption properties for methyl blue. *Journal of Hazardous Materials*, *215*, 272–279.
75. Zhan, S., Zhu, D., Ma, S., Yu, W., Jia, Y., Li, Y., ... & Shen, Z. (2015). Highly efficient removal of pathogenic bacteria with magnetic graphene composite. *ACS applied Materials & Interfaces*, *7*(7), 4290–4298.
76. Ye, Y., Yin, D., Wang, B., & Zhang, Q. (2015). Synthesis of three-dimensional Fe3O4/graphene aerogels for the removal of arsenic ions from water. *Journal of Nanomaterials*, *2015*, Article ID 864864.
77. Mishra, A. K., & Ramaprabhu, S. (2012). Ultrahigh arsenic sorption using iron oxide-graphene nanocomposite supercapacitor assembly. *Journal of Applied Physics*, *112*(10), 104315.
78. Liu, S., Wang, H., Chai, L., & Li, M. (2016). Effects of single-and multi-organic acid ligands on adsorption of copper by Fe3O4/graphene oxide-supported DCTA. *Journal of Colloid and Interface Science*, *478*, 288–295.
79. Yang, X., Li, J., Wen, T., Ren, X., Huang, Y., & Wang, X. (2013). Adsorption of naphthalene and its derivatives on magnetic graphene composites and the mechanism investigation. *Colloids and Surfaces A: Physicochemical and Engineering Aspects*, *422*, 118–125.
80. Gan, N., Zhang, J., Lin, S., Long, N., Li, T., & Cao, Y. (2014). A novel magnetic graphene oxide composite absorbent for removing trace residues of polybrominated diphenyl ethers in water. *Materials*, *7*(8), 6028–6044.
81. Liu, Z., Chen, K., Davis, C., Sherlock, S., Cao, Q., Chen, X., & Dai, H. (2008). Drug delivery with carbon nanotubes for in vivo cancer treatment. *Cancer Research*, *68*(16), 6652–6660.
82. Chokkareddy, R., Bhajanthri, N. K., Kabane, B., & Redhi, G. G. (2018). Bio-Sensing Performance of Magnetite Nanocomposite for Biomedical Applications. *Nanomaterials: Biomedical, Environmental, and Engineering Applications*, 166–192.
83. Jin, H., Qian, Y., Dai, Y., Qiao, S., Huang, C., Lu, L., ... & Zhang, Z. (2016). Magnetic enrichment of dendritic cell vaccine in lymph node with fluorescent-magnetic nanoparticles enhanced cancer immunotherapy. *Theranostics*, *6*(11), 2000.
84. Wang, H., Cao, L., Yan, S., Huang, N., & Xiao, Z. (2009). An efficient method for decoration of the multiwalled carbon nanotubes with nearly monodispersed magnetite nanoparticles. *Materials Science and Engineering: B*, *164*(3), 191–194.
85. Chokkareddy, R., Bhajanthri, N. K., & Redhi, G. G. (2017). A novel electrode architecture for monitoring rifampicin in various pharmaceuticals. *International Journal of Electrochemical Sciences*, *12*, 9190–9203.
86. Kodama, R. H. (1999). Magnetic nanoparticles. *Journal of Magnetism and Magnetic Materials*, *200*(1–3), 359–372.
87. Pankhurst, Q. A., Connolly, J., Jones, S. K., & Dobson, J. (2003). Applications of magnetic nanoparticles in biomedicine. *Journal of Physics D: Applied Physics*, *36*(13), R167.
88. Gubin, S. P., Koksharov, Y. A., Khomutov, G. B., & Yurkov, G. Y. (2005). Magnetic nanoparticles: Preparation, structure and properties. *Russian Chemical Reviews*, *74*(6), 489.
89. Shubayev, V. I., Pisanic II, T. R., & Jin, S. (2009). Magnetic nanoparticles for theragnostics. *Advanced Drug Delivery Reviews*, *61*(6), 467–477.
90. Lu, A. H., Salabas, E. E., & Schüth, F. (2007). Magnetic nanoparticles: Synthesis, protection, functionalization, and application. *Angewandte Chemie International Edition*, *46*(8), 1222–1244.
91. Hyeon, T. (2003). Chemical synthesis of magnetic nanoparticles. *Chemical Communications*, 8, 927–934.
92. Lee, C. S., Lee, H., & Westervelt, R. M. (2001). Microelectromagnets for the control of magnetic nanoparticles. *Applied Physics Letters*, *79*(20), 3308–3310.
93. Berry, C. C., & Curtis, A. S. (2003). Functionalisation of magnetic nanoparticles for applications in biomedicine. *Journal of Physics D: Applied Physics*, *36*(13), R198.
94. Chokkareddy, R., Bhajanthri, N. K., Redhi, G. G., & Redhi, D. G. (2018). Ultra-sensitive electrochemical sensor for the determination of pyrazinamide. *Current Analytical Chemistry*, *14*(4), 391–398.
95. Hude, I., Sasse, S., Engert, A., & Bröckelmann, P. J. (2017). The emerging role of immune checkpoint inhibition in malignant lymphoma. *Haematologica*, *102*(1), 30.
96. Choi, J. S., Jun, Y. W., Yeon, S. I., Kim, H. C., Shin, J. S., & Cheon, J. (2006). Biocompatible heterostructured nanoparticles for multimodal biological detection. *Journal of the American Chemical Society*, *128*(50), 15982–15983.
97. Xiang, J., Xu, L., Gong, H., Zhu, W., Wang, C., Xu, J., ... & Liu, Z. (2015). Antigen-loaded upconversion nanoparticles for dendritic cell stimulation, tracking, and vaccination in dendritic cell-based immunotherapy. *ACS Nano*, *9*(6), 6401–6411.
98. Revia, R. A., & Zhang, M. (2016). Magnetite nanoparticles for cancer diagnosis, treatment, and treatment monitoring: Recent advances. *Materials Today*, *19*(3), 157–168.
99. Johannsen, M., Gneveckow, U., Eckelt, L., Feussner, A., Waldöfner, N., Scholz, R., ... & Jordan, A. (2005). Clinical hyperthermia of prostate cancer using magnetic nanoparticles: Presentation of a new interstitial technique. *International Journal of Hyperthermia*, *21*(7), 637–647.
100. Kolosnjaj-Tabi, J., Di Corato, R., Lartigue, L., Marangon, I., Guardia, P., Silva, A. K., ... & Gazeau, F. (2014). Heat-generating iron oxide nanocubes: Subtle "destructurators" of the tumoral microenvironment. *ACS Nano*, *8*(5), 4268–4283.
101. Hayashi, K., Nakamura, M., Sakamoto, W., Yogo, T., Miki, H., Ozaki, S., ... & Ishimura, K. (2013). Superparamagnetic nanoparticle clusters for cancer theranostics combining magnetic resonance imaging and hyperthermia treatment. *Theranostics*, *3*(6), 366.

102. Hu, S. H., Liao, B. J., Chiang, C. S., Chen, P. J., Chen, I. W., & Chen, S. Y. (2012). Core-shell nanocapsules stabilized by single-component polymer and nanoparticles for magneto-chemotherapy/hyperthermia with multiple drugs. *Advanced Materials*, *24*(27), 3627–3632.
103. Jurek, P. M., Zabłocki, K., Waśko, U., Mazurek, M. P., Otlewski, J., & Jeleń, F. (2017). Anti-FGFR1 aptamer-tagged superparamagnetic conjugates for anticancer hyperthermia therapy. *International Journal of Nanomedicine*, *12*, 2941.
104. Ota, S., Yamazaki, N., Tomitaka, A., Yamada, T., & Takemura, Y. (2014). Hyperthermia using antibody-conjugated magnetic nanoparticles and its enhanced effect with cryptotanshinone. *Nanomaterials*, *4*(2), 319–330.
105. Yousaf, T., Dervenoulas, G., & Politis, M. (2018). Advances in MRI methodology. *International Review of Neurobiology*, *141*, 31–76.
106. Studholme, C. (2011). Mapping fetal brain development in utero using magnetic resonance imaging: The Big Bang of brain mapping. *Annual Review of Biomedical Engineering*, *13*, 345–368.
107. Dai, Z. (Ed.). (2016). *Advances in Nanotheranostics II: Cancer Theranostic Nanomedicine* (Vol. 7). Springer.
108. Garcia, J., Liu, S. Z., & Louie, A. Y. (2017). Biological effects of MRI contrast agents: Gadolinium retention, potential mechanisms and a role for phosphorus. *Philosophical Transactions of the Royal Society A: Mathematical, Physical and Engineering Sciences*, *375*(2107), 20170180.
109. Wei, H., Bruns, O. T., Kaul, M. G., Hansen, E. C., Barch, M., Wiśniowska, A., ... & Bawendi, M. G. (2017). Exceedingly small iron oxide nanoparticles as positive MRI contrast agents. *Proceedings of the National Academy of Sciences*, *114*(9), 2325–2330.
110. Yin, X., Russek, S. E., Zabow, G., Sun, F., Mohapatra, J., Keenan, K. E., ... & Moreland, J. (2018). Large T 1 contrast enhancement using superparamagnetic nanoparticles in ultra-low field MRI. *Scientific Reports*, *8*(1), 1–10.
111. Corr, S. A., Byrne, S. J., Tekoriute, R., Meledandri, C. J., Brougham, D. F., Lynch, M., ... & Gun'ko, Y. K. (2008). Linear assemblies of magnetic nanoparticles as MRI contrast agents. *Journal of the American Chemical Society*, *130*(13), 4214–4215.
112. Kozlova, A. A., German, S. V., Atkin, V. S., Zyev, V. V., Astle, M. A., Bratashov, D. N., ... & Gorin, D. A. (2020). Magnetic composite submicron carriers with structure-dependent MRI contrast. *Inorganics*, *8*(2), 11.
113. Efremova, M. V., Naumenko, V. A., Spasova, M., Garanina, A. S., Abakumov, M. A., Blokhina, A. D., ... & Wiedwald, U. (2018). Magnetite-Gold nanohybrids as ideal all-in-one platforms for theranostics. *Scientific Reports*, *8*(1), 1–19.
114. Liu, W., Dahnke, H., Jordan, E. K., Schaeffter, T., & Frank, J. A. (2008). In vivo MRI using positive-contrast techniques in detection of cells labeled with superparamagnetic iron oxide nanoparticles. *NMR in Biomedicine: An International Journal Devoted to the Development and Application of Magnetic Resonance In vivo*, *21*(3), 242–250.
115. Carvalho, A., Gallo, J., Pereira, D. M., Valentão, P., Andrade, P. B., Hilliou, L., ... & Martins, J. A. (2019). Magnetic dehydrodipeptide-based self-assembled hydrogels for theragnostic applications. *Nanomaterials*, *9*(4), 541.

21

Ultrathin Graphene Structure, Fabrication and Characterization for Clinical Diagnosis Applications

Ganesh Gollavelli and Yong-Chien Ling

CONTENTS

21.1 Introduction

Graphene is a constituent element of carbon ($_{12}C^6$), placed sixth in the periodic table with a mass of 12. Carbon has rich chemistry and physics and its study has won many Nobel Prizes [1–4]. Graphene is a basic unit of all carbon allotropes such as 0D fullerenes, 1D CNTs and 3D graphite/diamonds/graphene ribbons [5]. Usually, carbon could form single, double and triple bonds with the hybridizations of sp^3, sp^2 and sp. The natures of C=C bonds are the measure of the strength of carbon material. The carbon in graphene undergoes sp^2 hybridization with the C=C bond, length of 1.42 A° with very high conductivity due to the outer plane Л electrons [6]. The series of sp^2 carbons join together and produce a hexagonal array of rings in a 2D manner and layer kind of structure known to be a graphene [7–9].

From a stereochemistry perspective, it has a zigzag and armed chair arrangement of C in the edges of graphene [9–11]. As a result, the surface chemistry, interface and edges associated properties, conductivity and other characteristics might vary (Figure 21.1) [12]. Graphene was well known to be a single layer of graphite or a layer of a pencil tip. After 2010 the 2D structure gained a lot of attention from the world by the pioneering work of Andre Geim and Konstantin Novoselov [2, 5]. These Manchester University researchers explored the intriguing properties of this wonder material and proved it as the best conductor of electricity [13]. Later, many researchers have revealed its structural, electrical, optical, thermal and other properties and produced thousands of articles [14, 15]. Various graphene-based platforms have been established in the field of energy, catalysis, medicine, etc. [16–18].

In the view of modern physicists, graphene has a hexagonal arrangement of carbon atoms in distinctive positions in a unit cell. The energy levels of the C atoms are in a cone shape and have zero bandgap at the K point like metals for single-layer graphene. The energy levels vary with the number of layers and dopants. The layer usually represents an AB-kind Bernal-Stacking. The hexagonal single ring of graphene is represented with K and K' for successive atomic points (Π bond or ortho, para positions) and M for the center of interatomic bonding space (σ bond), between K and K'. The center of the ring is represented with r (gamma); see Figure 21.2 [19–21].

DOI: 10.1201/9781003157823-21

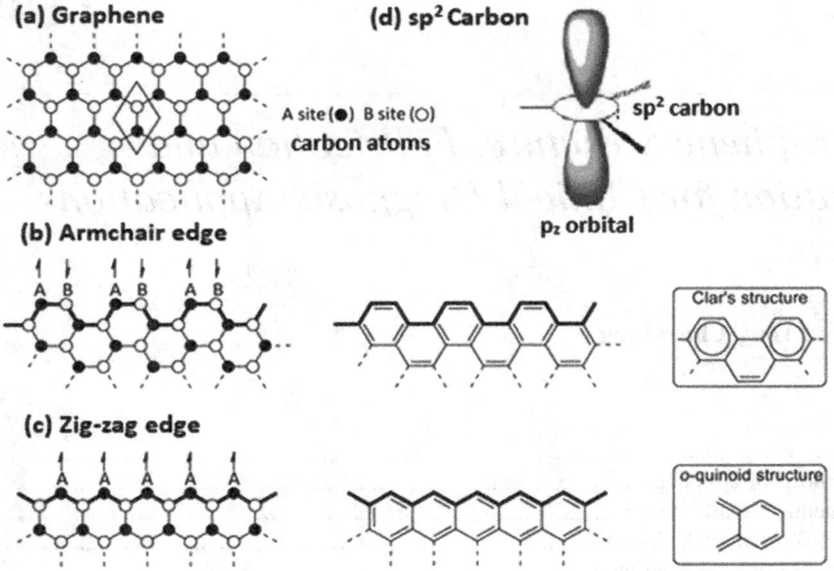

FIGURE 21.1 (a) Chemical structure of graphene with alternative carbon atoms in hexagonal lattice sites at A and B positions in dark and open circles. (b) Armchair edge represents A and B carbon sites toward outer edge with Clar's structure and (c) zigzag arrangement has a carbon atom site exposed to the outer edges with o-quinoid structure. (d) A single carbon atom has been shown with sp^2 hybridization in plane and p_z orbital perpendicular to it. Image reprinted from [10], [11] and [13].

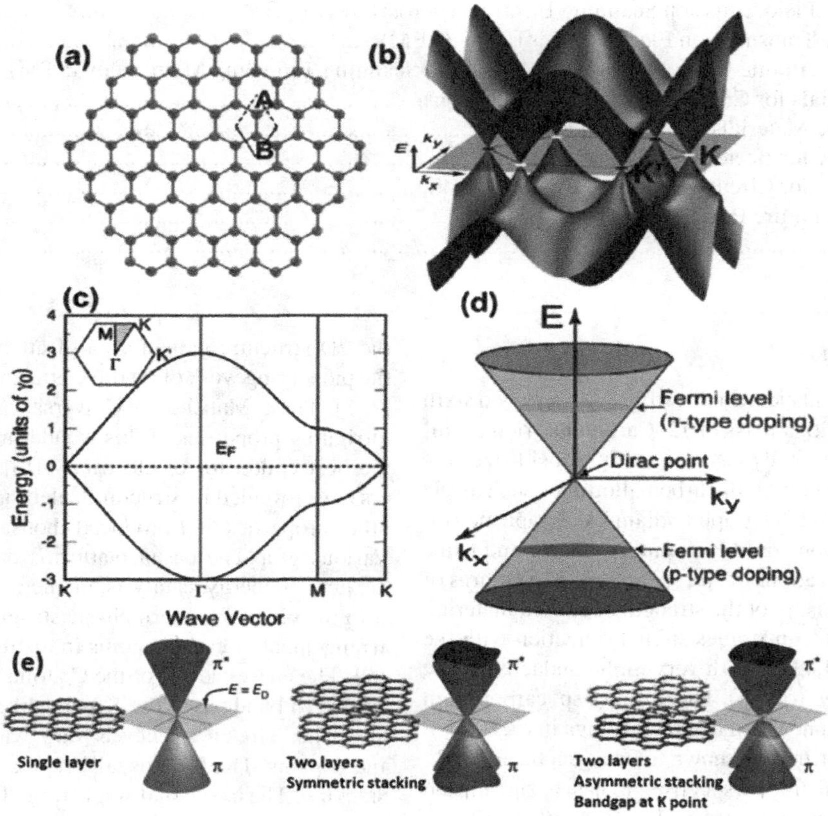

FIGURE 21.2 Electronic structure of graphene: (a) Graphene with two atoms (A and B) per unit cell in hexagonal honey comb lattice. (b) The graphene's 3D band structure. (c) Band structure of single-layer graphene between K and K'. (d) Cone-shaped energy band structure with Dirac point, and the Fermi level represents the nature of (n / p) doping. (e) Variations in band structure and band gap with number of layers and their symmetric and asymmetric staking. Images reprinted from [19], [20] and [21].

There are several ultrathin graphene synthesis methods that have been reported for electrode materials [22]. The exceptional physical, electrical and thermal properties of graphene make it a good candidate for electrode materials for ultrathin, flexible supercapacitors and field emission transistors (FET) [22, 23]. Usually, electrode materials must possess ultrahigh surface area, conductance and resistance to corrosion to the electrolyte [24]. Electrodes are the key components in the batteries, capacitors and in diagnosis sensors [25]. The lifetime of the electrode materials is very important for the efficiency of batteries/supercapacitors [26, 27]. There are many carbon electrode materials that have been successfully demonstrated as electrodes such as graphite, mesoporous carbon, CNTS and recently graphene and reduced graphene oxide (RGO) as they possess high surface area, pore structure and pore size distribution [28].

Here graphene has the superiority in the aforementioned features (Table 21.1) to fulfill the requirements of the best flexible/thin electrode sensor material for clinical diagnosis.

As the demands of health concerns are growing high due to the recent pandemic and empowerment of society, clinical diagnosis biosensors have a demanding role. Biosensor diagnosis technology was first developed in 1962 by Led and Clark to sense oxygen. Then onwards the research advancements made it a household diagnosis kit like glucose sensors, heart rate sensors and COVID-19 diagnosis sensors and till now up to 14 of its variations have been discovered. There is every concern to improve the quality, efficiency, sensitivity and price of diagnosis biosensors to avail to all, and it provides the disease results on time by self-check and prior to a visit to the healthcare professional, thereby minimizing the hospital visits. Hence, clinical biosensors could help to preserve health in our hands [28].

The invention is priceless; hence, avoiding the damage due to fracture, wrinkles and scratches of valuable devices is indispensable [29]. Moreover, the ease of its use and interface to the body and any external objects is a concern and demands the flexibility of the electronic devices such as displays, smart electronics, medical devices, flexible energy storage architecture, etc. which may transform our lives completely [23]. Among a plethora of carbon materials, graphene has revealed an outstanding character to fabricate flexible electrodes due to high tensile strength and Young's modulus for supercapacitors and diagnosis biosensors [30]. The ease of tailor-made morphology fabrication of graphene is an added advantage for the application of choice, as in Figure 21.3.

21.2 Design and Synthesis of Graphene

Designing of graphene depends on the choice of application. There are several graphene synthesis methodologies: for supercapacitor electrode materials, large-area flexible electronics and composites, drug delivery platforms, antimicrobial agents, toxic gas and pollutant adsorbents and diagnosis biosensors [31, 32]. Here we would like to focus on ultrathin graphene design and fabrication. However, from a general perspective, the most important synthesis methods of graphene from top-down to bottom-up approaches are listed in Figure 21.4. In the top-down process, the bulk graphite peeled off by micromechanical cleavage or oxidized followed by reduction or soaked with liquid solvents to separate stacked graphene layers into individual sheets. In contrast to it, in the bottom-up process individual carbon atoms join together and form a hexagonal network of carbons to a big sheet of graphene on the top of SiC in the epitaxial growth method and on Cu in the chemical vapor deposition (CVD). Here methane is used as a carbon source at elevated temperatures in CVD [33].

21.2.1 Experimental Details

The following top-down and bottom-up processes are the most important methods for producing graphene.

21.2.1.1 Top-Down Approach

21.2.1.1.1 Scotch Tape/Micromechanical Cleavage Method

This is the unique method to produce high-quality graphene, adopted in 2004, that was nominated for and won the Nobel in 2010. This method of production is simple, economic and easy to make at the laboratory level or even at home. Although it is labor-intensive, less skill is required to obtain high-quality large surface area graphene. The methodology usually requires Scotch tape which is easily available at book stores or

TABLE 21.1

Graphene Physical and Chemical Properties

Property	Value	Comparison with other materials
Young's module	1.1 T Pa	–
Tensile strength	125 G Pa	The specific strength is 100× greater than steel
Electron mobility (25° C)	2×10^5 cm²/V s	140× higher than Si
Thermal conductivity(25° C)	5×10^3 W/m K	10× higher than Cu
Light transmittance	97.7%	Alternative to ITO (indium-tin-oxide) and FTO (fluorine-doped tin oxide) films
Surface area	2,630 m²/g	2× larger than CNTs
Permeability	Impermeable to liquid/gases and permeable to protons	The geometric pore size is smaller than the diameter of He and H_2
Super capacitance	550 F/g	–

Source: Ref. [22].

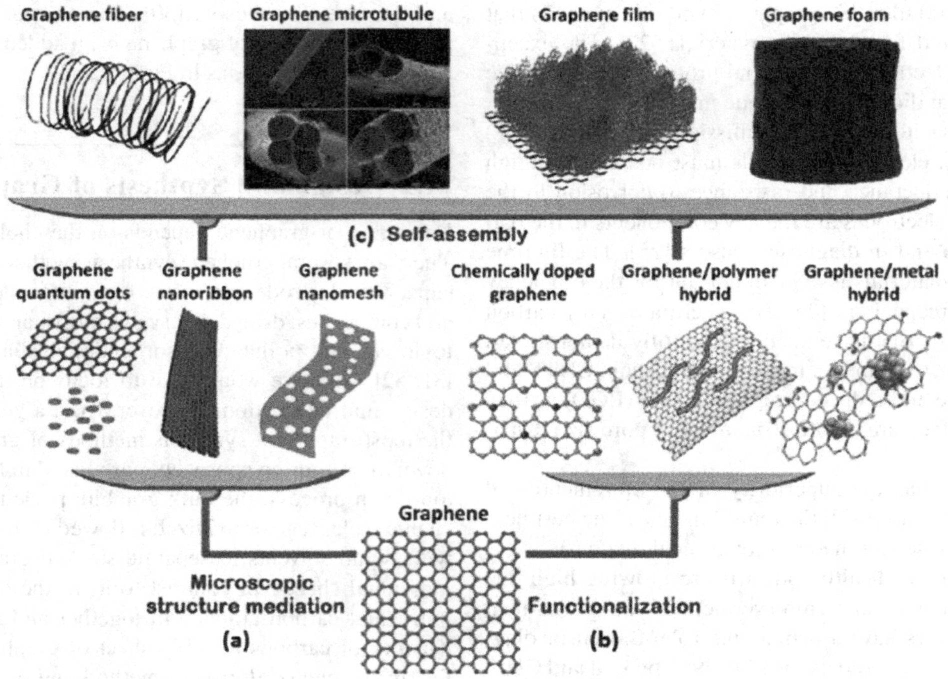

FIGURE 21.3 Types of graphene and its morphology: (a) Represents tailoring the graphene sheet into narrowing its shape as 0D quantum dots, 3D nanoribbons and porous nanomesh. (b) For external decoration of graphene with doping foreign atoms and functionalization of polymers etc. (c) is the self-assembly of graphene A and B structures. Image reprinted from [23].

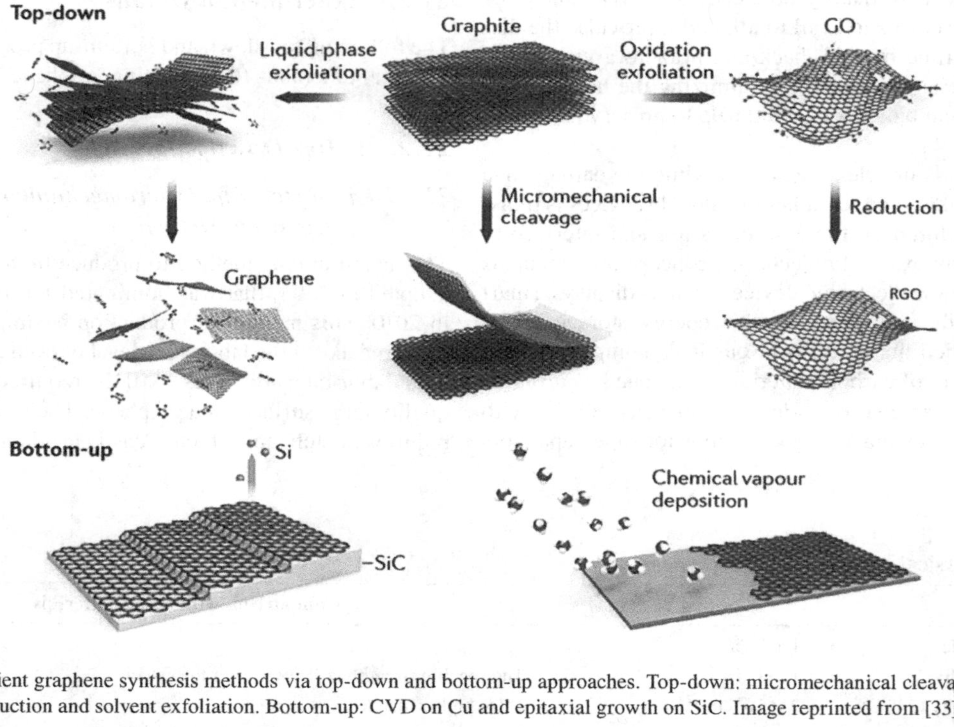

FIGURE 21.4 Salient graphene synthesis methods via top-down and bottom-up approaches. Top-down: micromechanical cleavage, graphite oxidation followed by reduction and solvent exfoliation. Bottom-up: CVD on Cu and epitaxial growth on SiC. Image reprinted from [33].

electrical departments and a solid graphite material. The preparation method was just to press the graphite with adhesive tape and peel it off, and then we can observe a thin gray material on the tape as seen in Figures 21.5 and 21.6. Then graphite carbon-containing tape is repeatedly stuck to itself and peeled away to obtain micron-size fine single-atom graphene layers. We can produce millimeters of graphene within a few hours by adopting this strategy [34–39]. Later on, many strategies have been reported after this piece of work as mass production is impossible with this method.

21.2.1.1.2 Oxidation/Liquid-Phase Exfoliation (LPE)

Oxidation and LPE of graphite are the best solution-phase methods to produce bulk quantity in an economic

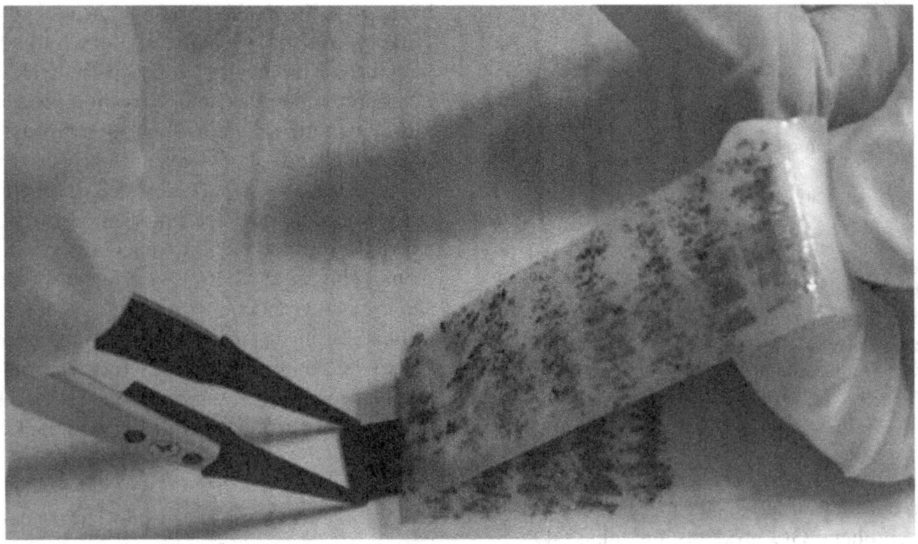

FIGURE 21.5 Scotch tape peeling method to produce graphene from solid graphite. Image reprinted from [37].

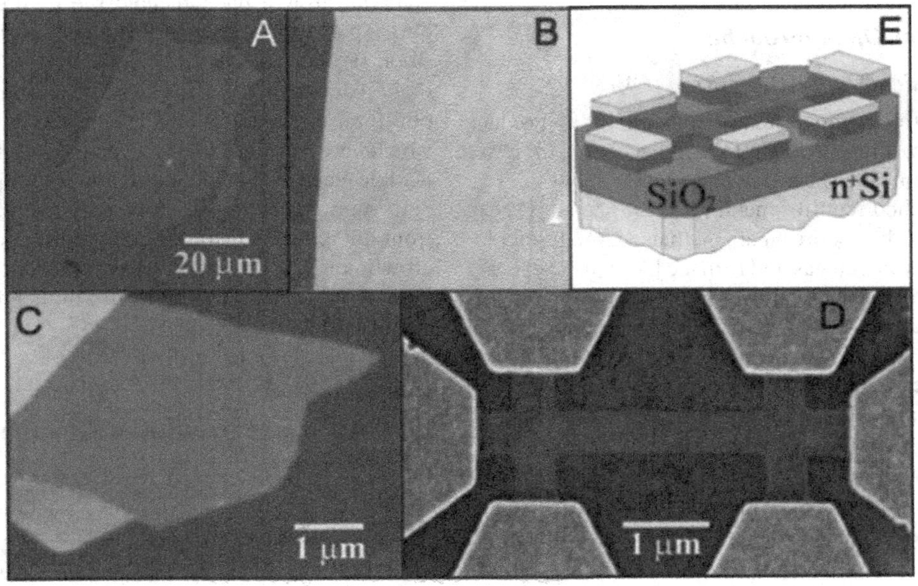

FIGURE 21.6 The first graphene layers removed from graphite by peeling method. (A) Optical image of multilayer graphene. (B) Atomic force microscope images (AFM) of graphene thin flakes at edges. (C) AFM image of single-layer graphene. (D) Experimental device fabricated using few-layer graphene and its scanning electron microscope image. (E) Device schematic view in (D). Image reprinted from [34].

process [40–43]. Direct LPE can produce high-quality graphene [44]. However, the graphene obtained in oxidation processes has to sacrifice optical and electronic properties due to the crystal defects and oxygen functional groups from sp^3 carbons [45]. In this method of obtaining graphene, the bulk graphite is oxidized and reduced chemically to obtain RGO [46, 47]. In the LPE the 3D graphite is soaked in organic solvents, ionic liquids and surfactants to inter-chelate between the 3D layers in graphite and diminish the physical forces/Π–Π interactions between the layers and separated into individual 2D graphene layers [48].

Hummer's method is the well-known oxidation method to prepare bulk graphene oxide (GO) and thereby obtain RGO/graphene [40]. Later many modified oxidation methods have been reported successfully by changing the oxidizing agent and other reaction parameters [46, 49]. Shear exfoliation of graphene is one of the best methods reported to obtain graphene solution in suitable solvents for bulk quantity industrial application [50]. Water-based green exfoliation was also reported successfully with the assistance of vapor pretreatment of graphite [51]. It was suggested that the solvents used in this method are non-toxic and environmentally friendly. Liquid-phase electrochemical [52, 53], microwave and sonication exfoliation also fall in this category, where electric force/charge is the driving force in the prior technique. In the later cases, microwave power and sonication jets are the factors to

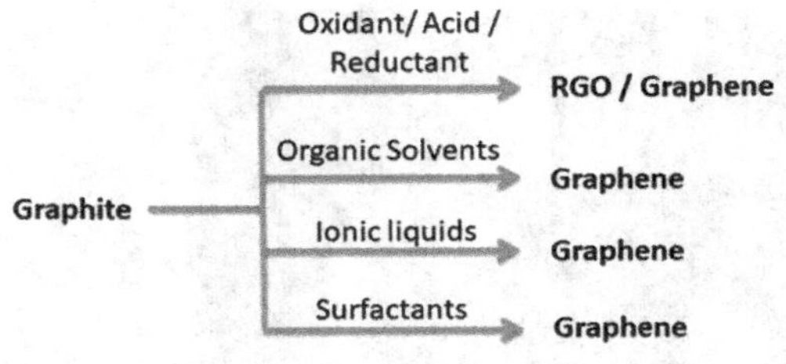

FIGURE 21.7 Synthesis of bulk quantity reduced graphene and graphene/functionalized graphene from bulk graphite from oxidation ($KMnO_4/H_2SO_4/H_3PO_4$) followed by reduction (hydrazine) or immersing the graphite with organic solvents NMP (N-methyl-2-pyrrolidone), ionic liquids (1-butyl-3-methyl-imidazolium bis[trifluoro-methane-sulfonyl] imide [{Bmim}-{Tf(2)N}]) and surfactants (sodium dodecylbenzene sulfonate).

exfoliate stacked layers of graphite into graphene [54, 55]. This method of solution exfoliation is versatile as one can produce many varieties of graphene/GO products for inks, QDs, biological, environmental, electronic and optical clinical diagnosis sensor applications [47], as shown in Figure 21.7.

21.2.1.2 Bottom-Up Approaches

21.2.1.2.1 Chemical Vapor Deposition (CVD)

CVD is one of the sophisticated best methods to produce large-area (in cm) high-quality graphene with no or fewer defects with controllable sheet size and sheet number in large quantity. This method requires metal foil (Ex: Cu, Ni, Pt, Co, Fe, Ta, Ru, etc.), carbon source (e.g., methane, CO_2, hexachlorobenzene, etc.), hydrogen gas and furnace [56–58]. The heavy atomic weight carbon from methane vapors deposit on the surface of Cu around 800–1,000° C and the light atomic weight hydrogen leaves the surface and the copper does not chemically interact much with it. The gaseous carbon atoms on the Cu surface form a stable atomic thick network by following C–C terravalency in a hexagonal framework. If the metals interact strongly with carbon it may form metal carbides and can disturb the hexagonal symmetry thereby sacrificing quality in terms of layers, size, defects, wrinkles, conductivity, etc. The carbon film growth on the metal surface can be explained by catalyst theory [59] and metal-induced crystallization [60]. Usually, transition metals are known to be good catalysts and help the gas molecules to adsorb on their surface and lower the reaction temperature and activation energies. Among all metals Cu can provide a low-energy pathway; as a result, it is the most widely adopted material for CVD-grown graphene above other metals [56, 61]. Thirty-inch graphene films for transparent electrodes have been fabricated by roll-to-roll production using Cu foil and the Cu etched away to recover the graphene from it (Figure 21.8) [62]. Very recently, Li et al. [63] prepared super-clean graphene with > 99% clean regions by avoiding all impurities during the manufacturing process. Several authors have worked on this strategy by changing the carbon source, proton source, time, temperature and metals to produce different kinds of graphene for diversified applications [37, 64, 65].

21.2.1.2.2 SiC Epitaxial Growth Method

Other than CVD, surface graphitization can also be done in the bottom-up process by sublimating the Si from the surface of SiC by heating of 4H-SiC/6H-SiC crystals > 1,200° C, where the remaining carbon atoms graphitize on the SiC surface via epitaxial growth phenomenon; see Figure 21.4 for reference [66]. Apart from CVD and SiC methods, physical vapor deposition (PVD) requires less temperature [67]. There are a few more manufacturing techniques such as laser-assisted deposition [68], unzipping of CNTS [69] and graphite evaporation which have been reported for quality graphene fabrication for flexible electronic devices in many other applications [22].

In summary, Figure 21.9 portrays the graphene production from micromechanical cleavage (mg production, 2004), CVD growth on Cu (2006), liquid-phase exfoliation (2008) and 30-inch film (2015) to tons (2019) of its production and applications [70].

21.3 Characterization of Graphene

Graphene could be characterized by following several spectroscopic and microscopic methods as detailed in the following sections. Spectroscopic methods are non-imaging characterization methods that provide the chemical compositions of the

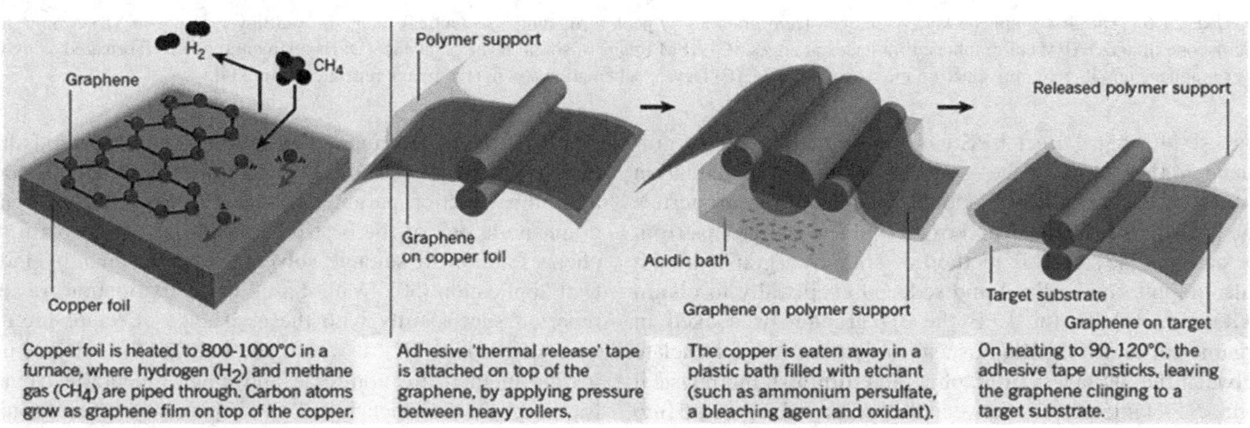

FIGURE 21.8 Synthesis of graphene from CVD, the yellow base is a Cu foil and the hexagonal carbon network deposition on Cu at 1,000° C from methane vapors. The roll-to-roll production and description of fabrication have been provided on the right side of the image. Image reprinted from [34].

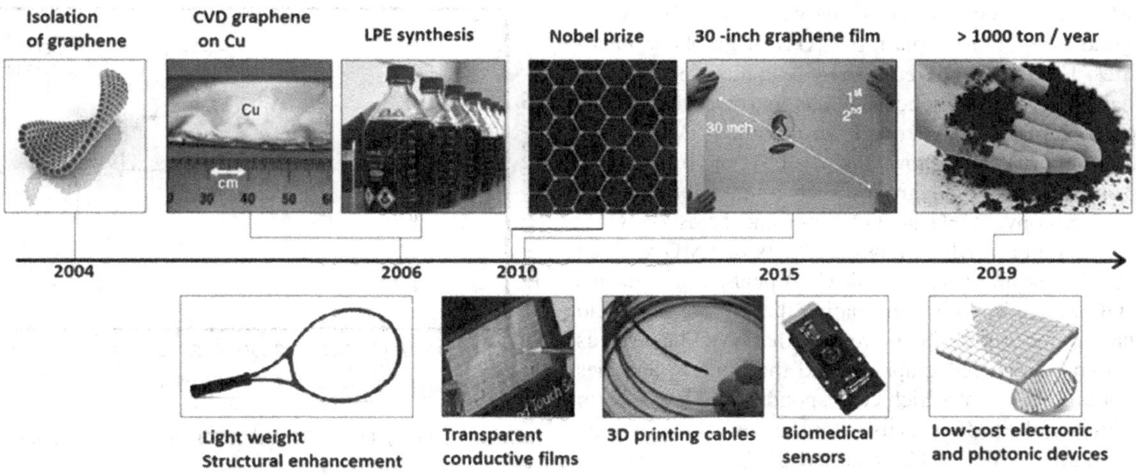

FIGURE 21.9 The 15-year voyage of graphene from mg to tons. Production and commercialization from 2004 to 2019 and its applications. Image reprinted from [70].

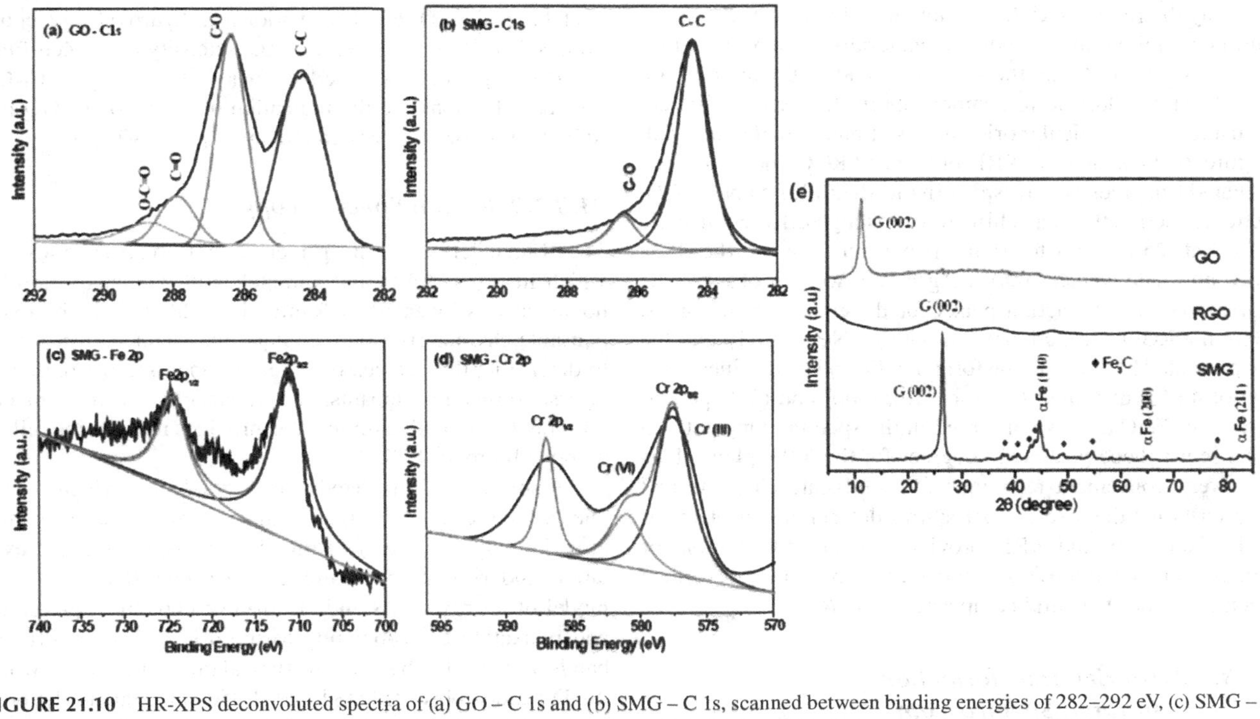

FIGURE 21.10 HR-XPS deconvoluted spectra of (a) GO – C 1s and (b) SMG – C 1s, scanned between binding energies of 282–292 eV, (c) SMG – Fe 2p (700–740 eV) and SMG – Cr 2p, scanned between 570–596 eV. (e) XRD of GO, RGO and SMG. Image reprinted from [71].

unknown nanomaterials synthesized in the lab. Here we discussed the most important spectroscopic techniques such as X-ray, FT-IR (Fourier transformation infrared) and Raman to determine surface chemical and elemental composition, functionality number of layers and defects of graphene.

21.3.1 Spectroscopic Characterization

21.3.1.1 X-Ray Photo Electron Spectroscopy (XPS)

XPS is the most powerful tool to determine the surface chemical composition and oxidation states of the nanomaterials up to 10 nm in depth. It involves measuring the binding energies of electrons ejected from the surface upon X-ray irradiation. As we know, graphene has only one chemical element "C" and has a uniform chemical environment of the C=C network; this tool is not prominent until it is functionalized with heteroatoms. Hence, if we can consider other graphene materials such as GO and RGO or functionalized graphene it could be an excellent tool to determine the chemical composition and nature of heteroatoms on its surface. Usually, the survey scan run between the ranges of expected binding energies can provide the elements present on the surface of graphene. From this, we can assess the nature of elements presented and their amount by the intensity of peaks. Figure 21.10 shows HR-XPS of GO, smart magnetic reduced graphene oxide (SMG) and

Cr adsorbed SMG. This is one of the works done by us and a very good example to explain the XPS of graphene nanocomposites. Here we prepared GO by oxidation method and functionalized with magnetic (Fe) nanoparticles by in situ reduction. Later we demonstrated the obtained SMG for Cr adsorption. Figure 21.10a shows that the C 1s of GO contains various oxygen functional groups such as alcohol (C–O at 286.3 eV), carbonyl (C=O at 287.9 eV) and acids (O–C=O at 288.7 eV). Figure 21.10b shows the C 1s of SMG with an intense C-C peak and a very weak C–O peak, which means that the GO was reduced by eliminating the carbonyl and acid functional groups. Figure 21.10c reveals that SMG has iron as a heteroatom on reduced graphene and the binding energies are 711.24 and 724.6 eV which correspond to the Fe_2O_3 form of iron. Further, the SMG was used for the adsorption of toxic Cr (VI). Figure 21.10d for Cr adsorbed SMG where one can see that some of adsorbed Cr (VI) reduced to Cr (III). Such XPS is highly advantageous to investigate the graphene surfaces after functionalization with different heteroatoms and dopants [71–73].

X-ray diffraction (XRD) is another technique for the structural determination of solid-state materials. Like XPS, X-rays are the source to target the substance, and diffracted rays are the structure illustrating components in the form of diffraction angles for particular orientations of planes in the material. Figure 21.10e shows the XRD of GO and RGO and SMG. The 2Θ at ~11 degrees is corresponding to the (002) plane of GO. After reduction the peak shifts to words high diffraction angle of 2Θ at ~25 degrees for RGO. If we functionalize the RGO with any metals such as Au, Ag or Fe, we can observe the corresponding diffraction peaks for different crystal planes. Here the RGO is functionalized with Fe NPs (SMG), and the Fe presented here was in the form of α Fe and determined from 2Θ of 44.63, 65.0 and 82.0 for (110), (200) and (211) planes. Traces of Fe_3C are also observed in the spectra along with an intense graphene peak at 25 degrees for the (002) plane. This is a very convenient technique for solid-state chemists and physicists for the size and structural determination of materials. Both XPS and XRD provide valuable information for graphene nanocomposites and catalysts to know the surface chemistry and structural composition [71, 74].

21.3.1.2 Fourier Transformation Infrared Spectroscopy

FTIR is a complementary technique to Raman spectroscopy, where IR mode active vibrations are Raman inactive. The information obtained in this method is corresponding to the stretchings and vibrations of bonds present in the chemical compounds. Every chemical compound has its distinctive chemical functional groups and a particular frequency of IR radiation is necessary for its activation. 500–4,000 cm^{-1} is the IR region to record the vibrational transitions of the functional groups in graphene-related materials. As we discussed in the earlier section GO contains –OH, –C–O–C–, –CHO, –COOH and –C=C– aromatic functional groups. After reduction in RGO, most of the oxygen functional groups leave the surface and we are left with the aromatic –C=C– skeleton. The intensity of aromatic stretching is corresponding to the quality of graphene obtained after reduction. Figure 21.11 is the FTIR of GO and RGO, where the GO shows the intense stretching frequencies of the abovementioned oxygen functional groups at 3,450 cm^{-1} for –OH, 1,732 cm^{-1} for C=O, 1,619 cm^{-1} for –C=C– skeletal vibrations and 1,389 to 1,046 cm^{-1} for (C–O) stretching and (O–H) deformations of hydroxyl and epoxy groups. The drastic decrease in the intensity of oxygen functional groups and increased intensity of aromatic (–C=C–) skeletal vibrations are the indication of reduction of GO into RGO by microwave-assisted reduction [71, 75–77].

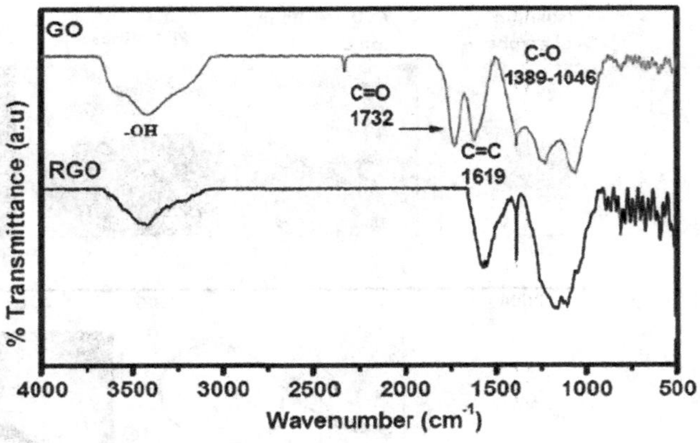

FIGURE 21.11 FTIR of GO and RGO. Image reprinted from [71].

21.3.1.3 Raman Spectroscopy

The Raman effect is the principle of Raman spectroscopy which involves stokes and anti-stokes of phonons. The phenomenon was invented by Raman in 1928. This is the indispensable characterization technique for graphene researchers to determine the defects, disorder, number and orientation of layers, strains and dopants. As graphene is the fundamental building block of all carbons, this provides insights into all sp^2 carbon allotropes [78].

Raman has four bands for graphene which occur due to the change in the symmetry modes at Γ (which are $= A_{2u} + B_{2g} + E_{1u} + E_{2g}$) during phonon displacement or excitation and relaxation. Figure 21.12a shows the Dirac cone model of energy levels and electron or phonon displacement and relaxation corresponding to the origin of four different bands. Figure 21.12b shows the typical graphene spectra where the D band is observed at ~1,320–1,350 cm^{-1}, due to the presence of defects and breathing modes of six-atom rings. The G peak at ~1,580–1,605 cm^{-1} corresponds to the high-frequency E_{2g} phonon at Γ and it is mostly observed in sp^2 carbons. The D' peak is observed at ~1,602–1,625 cm^{-1} by connecting two points belonging to the same cone around K/K'. The 2D/G' peak at ~2,640–2,680 cm^{-1} is the D peak overtone and no defect is required for its activation. Figure 21.12c is the elongated regions of the G' band and its deconvulsion to estimate the number of layers of graphene to graphite [77–80]. There are many trials that happen to produce high-quality graphene as defects are inevitable. Since graphene discovery, it was considered a challenge to produce high-quality, ultrathin, defect-free graphene. Albeit, Li Lin et al. in 2019 achieved the challenge to produce 99% super-clean graphene by CVD [63]. It was known that the transfer process (polymer assisted) of graphene to the target substrate can create defects

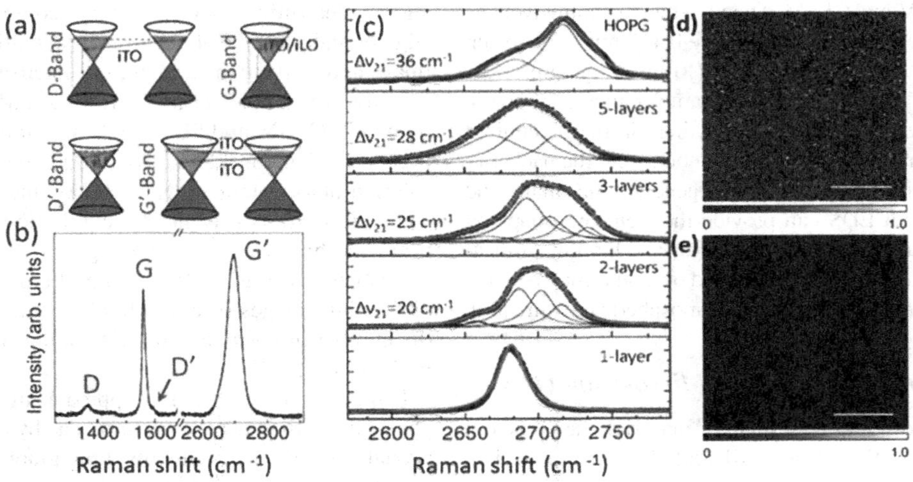

FIGURE 21.12 Raman spectra of graphene. (a) Dirac cone model to represent electron/phonon excitations and relaxation for corresponding Raman bands. (b) Raman spectra of single-layer graphene with characteristic D, G, D' and G' bands. (c) Expanded deconvulsion regions of G' band for different layers (1–5) of graphene and graphite (HOPG). (d) Intensity ratios (A_D/A_G) of G band and D band of graphene transfer assisted by PMMA and (e) flattened graphene. Image reprinted from [79] and [81].

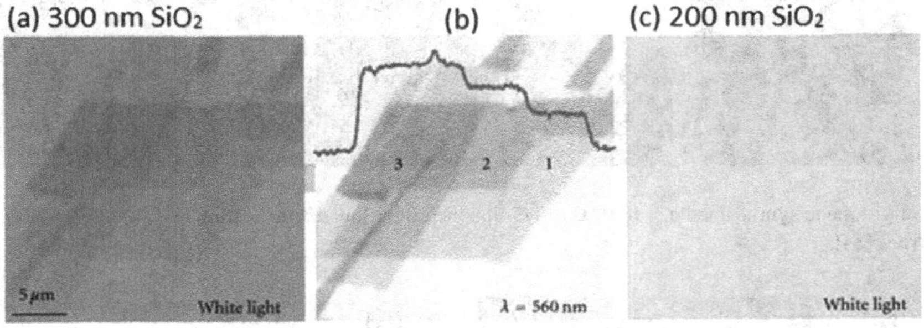

FIGURE 21.13 Graphene OM imaging on Si/SiO_2 substrate with different contrast and wavelengths. Image reprinted from [82].

by damaging the graphene surface. Very recently, this problem was also addressed successfully by Lukas Madauß et al. This method states that a swift technique can hydrophobize the graphene and increase its mechanical stability and charge carrier density. In Figures 21.12d and e, the Raman intensity ratios (A_D/A_G) of D and G band show that the graphene transferred through the regular PMMA (poly[methyl methacrylate]) method has more defects than the proposed latest shift technique [81]. Overall, the issues related to bulk production of graphene with controllable size, thickness, sheet number, high-quality and best transfer techniques to the substrate and its characterization methods have been successfully addressed here and in the next sections.

21.3.2 Microscopic Characterization

The important microscopic techniques to visualize the morphology, number of layers, wrinkles, edges, defects, grains and lattice structure of graphene are as follows.

21.3.2.1 Optical Microscope (OM)

OM is the simple and basic imaging technique and origin of fluorescence, confocal laser scanning microscopy (CLSM) and all electron microscopes. The graphene exfoliated from graphite by the micromechanical technique can quickly be identified by this method of imaging by mounting on the Si/SiO_2 substrate. The number of layers of graphene from monolayer, bilayer to multilayer can be identified by observing the contrast, and the quality of image contrast depends on the thickness of the substrate and wavelength of the light used, as shown in Figure 21.13a–c. However, its high optical transparency and ultra-low atomic thickness made it to be invisible in OM sometimes. The precise estimation of layers, number of defects cannot be identified very clearly. So one cannot completely rely on this technique, and further high-resolution imaging techniques are needed after preliminary evaluation under OM [34, 82].

Recently, CLSM has been reported for rapid optical characterization of one to five layers of exfoliated graphene on Si/SiO_2 and could able to distinguish epitaxial graphene (one to three layers) on SiC substrates. The CLSM has shown a very good correlation with OM, atomic force microscopy (AFM), scanning probe electron microscopy (SEM) and Raman techniques [83].

21.3.2.2 Field Emission Scanning Electron Microscopy (FESEM)

FESEM falls under electron microscope imaging techniques, which help to identify the morphology and sizes of conducting samples. Unlike OM and CLSM, it needs a high operating voltage of about 0.5–40 keV and a beam size of ~0.4–5 nm in diameter. Hence we can obtain better resolution (10 nm) large-area electron micrographs. Typically, graphene samples mount on the Si wafer and coat the surface with conducting plasma if required and analyze the sample in the SEM chamber. Here the backscattered electrons from the sample would generate the image and SEM connected with EDS can provide the elemental mapping in the selected area. Figure 21.14 is the low and high magnification FE-SEM image of graphene prepared by isothermal methane decomposition over the iron catalyst at atmospheric pressure [84].

21.3.2.3 Transmission Electron Microscopy (TEM)

TEM is another powerful technique of material chemists and physicists to image the ultra-small and thin nanomaterials and to determine their elemental compositions, grain and edge boundaries. TEM needs a very small sample on grids to image and analyze the transmitted electrons that pass through an ultrathin sample. TEM provides a much higher resolution than OM, CLSM and SEM, and can attain atomic resolution. Figure 21.15 is the TEM of the graphene exfoliated from the solution-phase method. Here NPM and other organic solvents are used as solvents for exfoliation. Figure 21.15a and b are the bright-field TEM images of single and multilayer graphenes and Figure 21.15c, d and e are the corresponding diffraction images of a and b which can help to distinguish the number of layers based on the intensity of Miller-Bravais indices (*hkil*) [42].

In the case of the bottom-up approach, in the CVD-grown graphene, the carbon nucleation starts one by one and forms a hexagonal network and eventually a big graphene sheet. However,

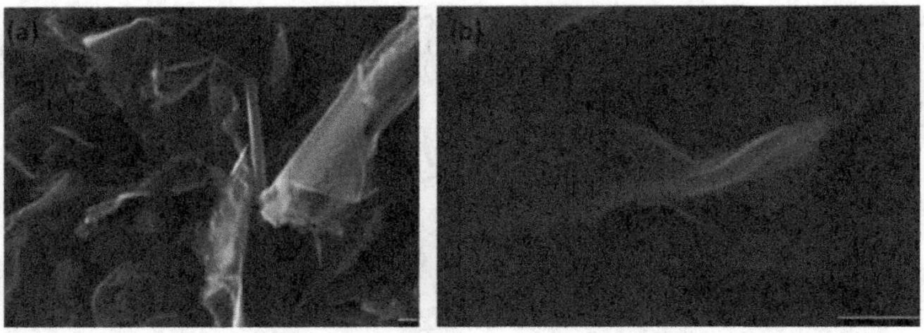

FIGURE 21.14 FESEM of graphene synthesized at 1,000° C. (a) Graphene under low magnification and (b) at high magnification; the scale bar was 100 nm. Image reprinted from [84].

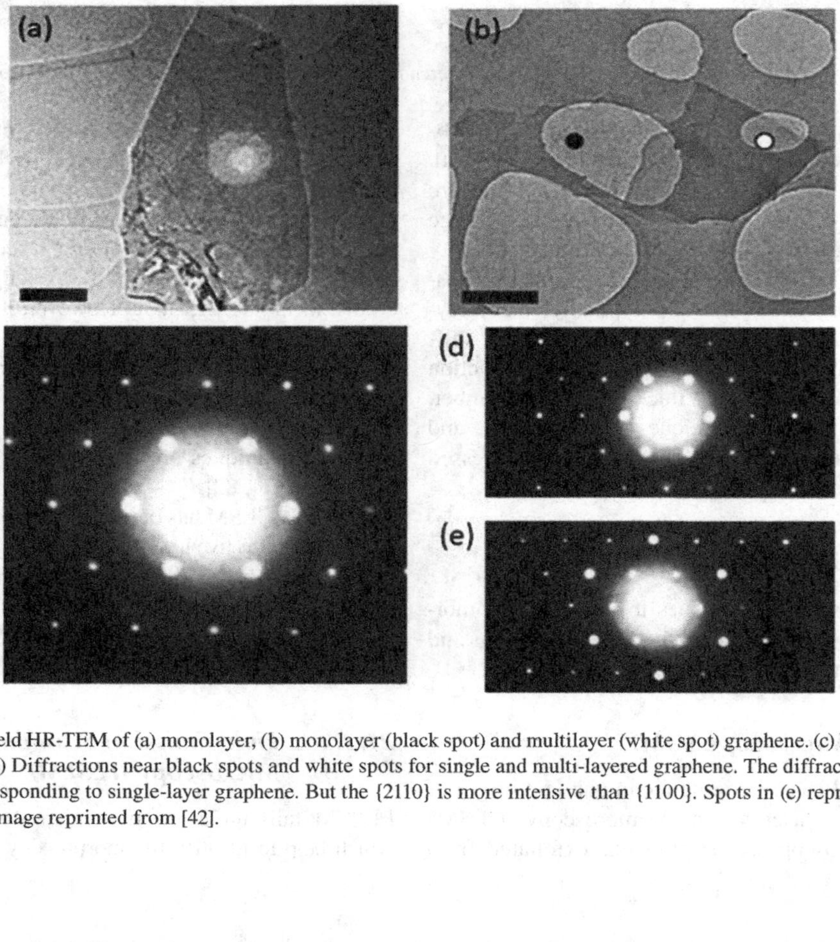

FIGURE 21.15 Bright-field HR-TEM of (a) monolayer, (b) monolayer (black spot) and multilayer (white spot) graphene. (c) Normal – electron diffraction pattern of (a). (d and e) Diffractions near black spots and white spots for single and multi-layered graphene. The diffractions of (d) are close to (c) so; both patterns are corresponding to single-layer graphene. But the {2110} is more intensive than {1100}. Spots in (e) represent multilayer graphene around white spot in (b). Image reprinted from [42].

sometimes the perfect hexagons may not form and there is a mismatch at the grain boundaries with a series of pentagons and heptagons along with hexagons. The size of grain and grain boundaries can be explained and lattice and atomic defects can be measured with aberration-corrected annular dark-field scanning transmission electron microscopy (ADF-STEM).

Figure 21.16a represents SEM of large-area single-layered graphene membranes on TEM grids covered by 90%, and Figure 21.16b for defect-free crystal lattice image of ADF-STEM which looks like a honeycomb structure with perfect hexagonal arrangement of carbon atoms in single-layered graphene. Whereas Figure 21.16c shows dislocations and the grain miss orientations of 27° at the tilt boundary. Figure 16d clearly highlights the stitching of two crystals with repeated pentagons, heptagons and elongated disordered hexagons [85].

21.3.2.4 Atomic Force Microscopy (AFM) and Scanning Tunneling Microscopy (STM)

AFM and STM are the probing methods to scan the surface of nanomaterials with atomically thin tips and gained a lot of attention in the surface sciences. The high-end AFM can image the surface topography with 0.3 nm resolution and microns in lateral dimensions.

It's a non-invasive and non-contaminating characterization technique to measure very thin and ultra-small nanomaterials like flat 2D graphene and QDs. Figure 21.17a shows the AFM image of graphene prepared by continuous mechanical exfoliation by the three-roll mill method revealing a sheet-like 2D structure with the height of 1 nm, as shown in Figure 21.17b [86]. Figure 21.17c shows the tapping mode AFM image of graphene prepared by the CVD method. The image contains 1, 2, 3, 4 and 5 layers of graphene on Si/SiO_2 substrate and the image can clearly portray the number of layers and wrinkles in it compared with OM and CLSM in the particular imaging work published by Vishal Panchal et al. [83]. Figure 21.17d shows the STM topography image of 0.35 monolayer of graphene grown on single crystal Cu (111) by thermal decomposition of ethylene. In Figure 21.17e, the image reveals atomically resolved single-layer graphene, its lattice parameters and Moire pattern and the honeycomb structure [87]. Elena Stolyarova et al. [36] examined the direct exfoliated graphene from graphite crystal, and the STM results are similar to the CVD-grown graphene by Li Gao [87]. The atomically resolved STM image (Figure 21.17f) is similar to the ADF-STEM image (Figure 21.16b) discussed in the TEM section.

In summary of characterization techniques, Figure 21.18a–f details the overall indispensable spectroscopic and imaging techniques and correlation among them. In (d), the AFM image more clearly describes the number of layers and wrinkles, whereas the CLSM can also do the same (b and c), but not the OM (a). The Raman (e and f) correlation graphs describe the increase in the intensity of G peak and G/2D peaks corresponding to the number of layers in the images shown (a and b). In (g and h) are the CLSM contrast and AFM height (nm) corresponding to the number of layers in the images (c and d). In conclusion of this part, AFM and Raman techniques are more valuable imaging and spectroscopic techniques for graphene researchers than other techniques. Well-characterized, ultra-pure defect-free graphene is highly important to sense the targets at a single atomic or molecular level for clinical diagnosis.

21.4 Graphene Materials for Clinical Diagnosis Applications

A single atomic layer of graphene offers a great advantage to detect viruses, bacteria and cancer down to the atom due to its

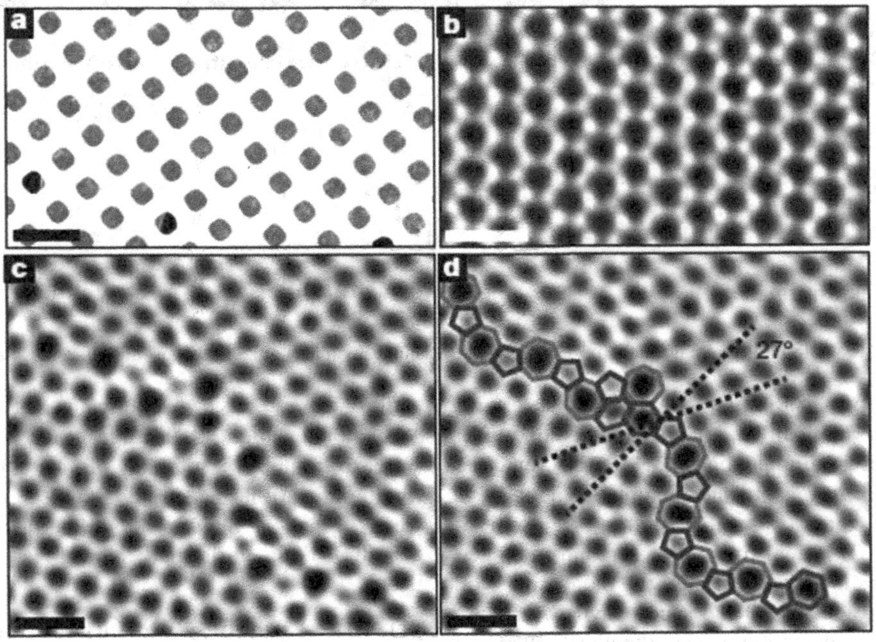

FIGURE 21.16 (a) SEM and (b) atomic resolution ADF-STEM of CVD-grown graphene and (c and d) are its grains or dislocations. Image reprinted from [85].

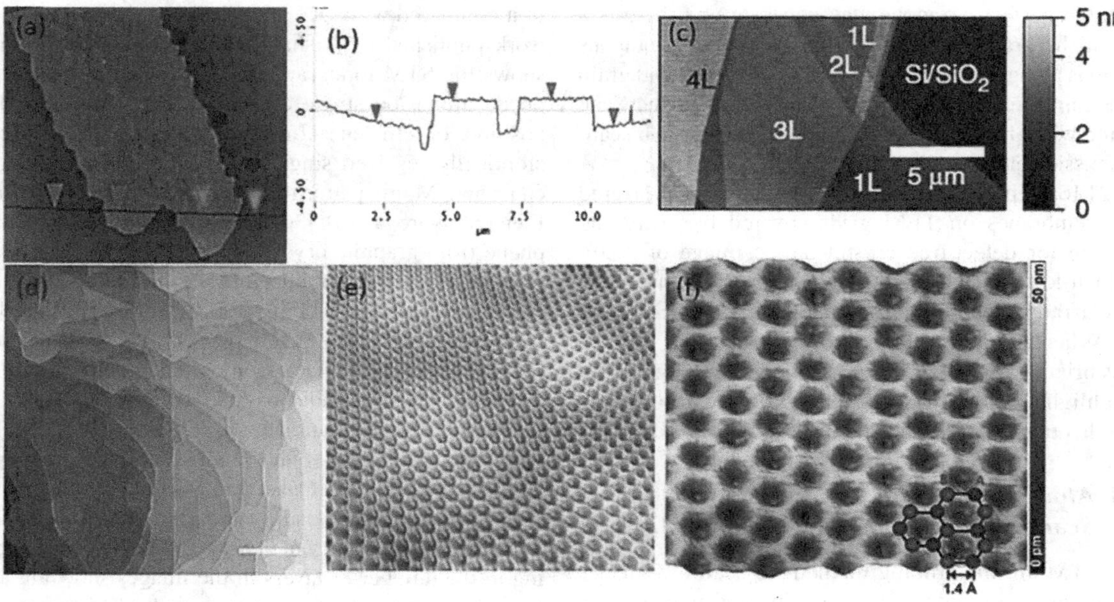

FIGURE 21.17 (a) AFM of graphene and its (b) height profiles and (c) number of layers and (d) STM topography of monolayer graphene (scale bar 200 nm); (e and f) Moire pattern and the honeycomb structure of atomically resolved graphene layers and lattice. The A and B spheres in (f) indicate the carbon atoms of the unit cell of graphene are separated with 1.4 A. Tunneling parameters are I = 10 pA and U = 50mV. Images (a and b) reprinted from [86], image (c) from [83], images (d and e) from [87] and image (f) from R. Wiesendanger et al., lab website, University of Hamburg [88].

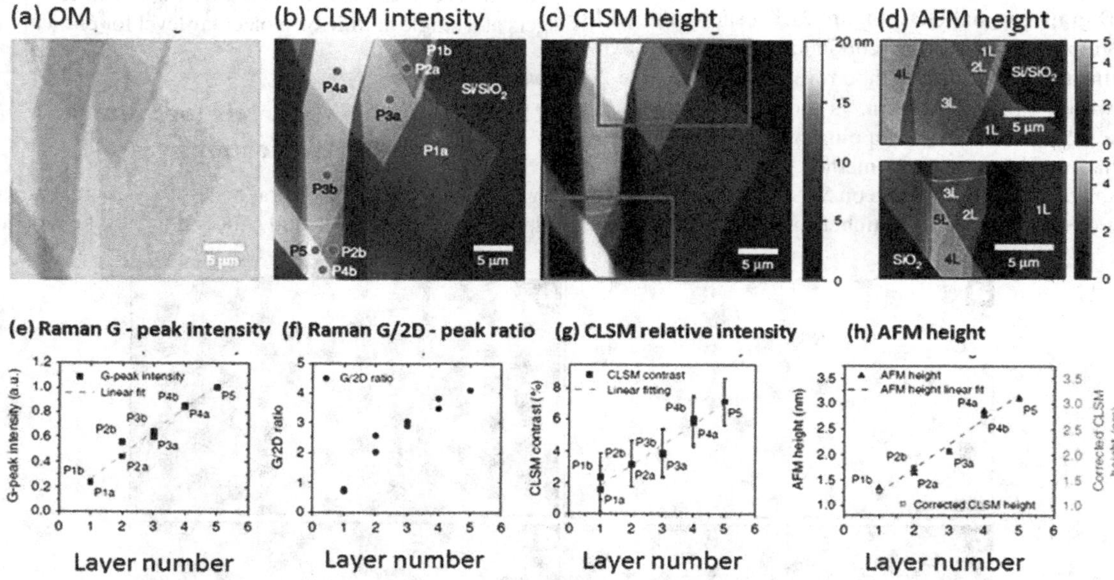

FIGURE 21.18 Correlation of graphene imaging with OM, CLSM, AFM and Raman for number of layers and thickness. Image reprinted from [83].

ultrahigh conductivity, sensitivity and surface area. Usually, the biomolecule detection can be done by functionalizing the graphene with corresponding peptides, DNA, RNA, enzymes and antigen-antibody. In recent times, nanotechnology-based graphene clinical diagnosis was anticipated to offer point-of-care detection (POCD) in real time with ease and in an economic way. Here I would like to discuss the pioneering work done based on the graphene POCD of coronavirus, bacteria, cancer and single-atom detection. Due to the ultrahigh optical and electrical properties of graphene, it is highly attractive in making the optical and electrical diagnosis sensors, which are affordable and easier to fabricate than other expensive mass detection techniques. The large surface area of graphene could offer to immobilize a variety of molecules in it without sacrificing much of its sensitivity and electrical conductivity. As a result, graphene materials could be one of the best materials for clinical diagnosis applications [89].

21.4.1 Graphene Materials for Virus Diagnosis

A virus is the smallest biofraction in the biological order compared with bacteria and human cancer cells, and making any device that can detect such small protein content is highly challenging. In very recent times, an unknown respiratory infectious disease created a disaster worldwide and its source and confirmation of the cause have made science mad. During such times POC diagnosis kits are highly indispensable. Several attempts have been made to identify the traces of disease based on reverse transcription-polymerase chain reaction (RTQPCR) and the limit of detection (LOD) is a starting hurdle. Due to the unique features of graphene, it has been made possible to detect the viruses like SARS-CoV-2 with very low LOD of 1.6×10^1 pfu/mL and 2.42×10^2 copies/mL in the culture medium and clinical samples. Here the authors Giwan Seo et al. prepared a graphene-based FET sensor designed by coating with a targeted antibody against SARS-CoV-2 spike protein. The graphene FET was tested against cultured virus and nasopharyngeal swabs from 19 infected COVID-19 patients [90]. This could be a very great attempt to make rapid and accurate POCD diagnosis kits based on graphene FET, as shown in Figure 21.19. Many other graphene-based viral sensors have been developed to detect the Zika virus, influenza, HIV, hepatitis C, dengue, etc. [91].

21.4.2 Graphene for Bacterial Diagnosis

Bacteria are other versatile contaminants after viruses that cause many contagious diseases. Early detection of its source with the highest accuracy is one of the challenges, and it is very important to trace out its presence in food, environment and biological fluids like saliva, blood and so on [92]. However, very limited work has been reported based on graphene bacterial sensors. Manu et al. have been fabricated a passive interfacing wireless sensor based on biotransferrable graphene. The graphene has printed on water-soluble silk thin-film substrates followed by functionalization of antimicrobial peptides (AMP) to detect the targeted bacteria, as shown in Figure 21.20a–f. The bio-interfacing wireless sensor has great capability to detect a single bacterial cell which offers great potential for point-of-care contamination detection [93].

21.4.3 Graphene for Circulating Tumor Cell Detection

Cancer is the leading cause of death in the world and early-stage diagnosis can help to facilitate the treatment process and to avoid fatality. Despite other nanomaterials graphene could offer a great platform for point-of-care diagnosis of cancer by immobilizing cancer detection biomarkers on graphene materials. The POCD could facilitate effectiveness in anticancer therapy. Hyeun et al. have fabricated functionalized GO for the detection of circulating tumor cells that originate from the primary tumor and float in the bloodstream. Detection of CTC is highly advantageous to identify the nature of the tumor and its type at the beginning of cancer. However, CTC availability is very low, and the biosensor is very specific to detect them. Here the GO has been functionalized with phospholipid-polyethylene-glyco-amine (PL-PEG-NH_2), N-γ-maleimidobutyryloxy succinimide ester (GMBS), and cancer cell-specific epithelial cell adhesion molecule (EpCAM) and achieved the detection limits of 3–5 cells/mL blood for pancreatic, breast and lung cancer in real time [94], as shown in Figure 21.21.

Xu et al. extensively discussed modified graphene materials and graphene as a signal amplifier or catalysts for prostate cancer diagnosis based on electrochemical, fluorescence and colorimetric sensors [95, 96]. Kovalska et al. fabricated a multi-layered graphene electrode biosensor for lung cancer point-of-care early-stage detection system based on an advanced e-nose approach. This demonstration is helpful for conceptual usage of concentration-dependent early-stage diagnosis of cancer [97].

21.5 Conclusions and Future Perspectives

In summary, we have well discussed the structure, synthesis and characterization of single, few-layered graphene, GO and RGO. The CVD method is one of the best of all to produce single-layered defect-free graphene which has been mostly used to fabricate ultra-sensitive electrical nanographene biosensors.

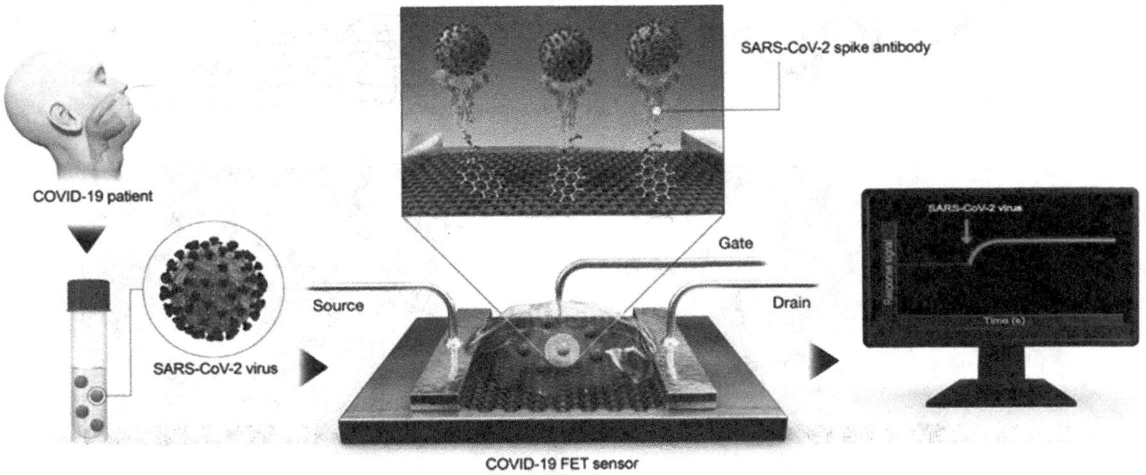

FIGURE 21.19 Schematic representation of graphene FET sensor for SARS-CoV-2 detection from COVID-19 patients and cultured virus. Here the graphene has been functionalized with SARS-CoV-2 spike antibody for virus detection. Image reprinted from [90].

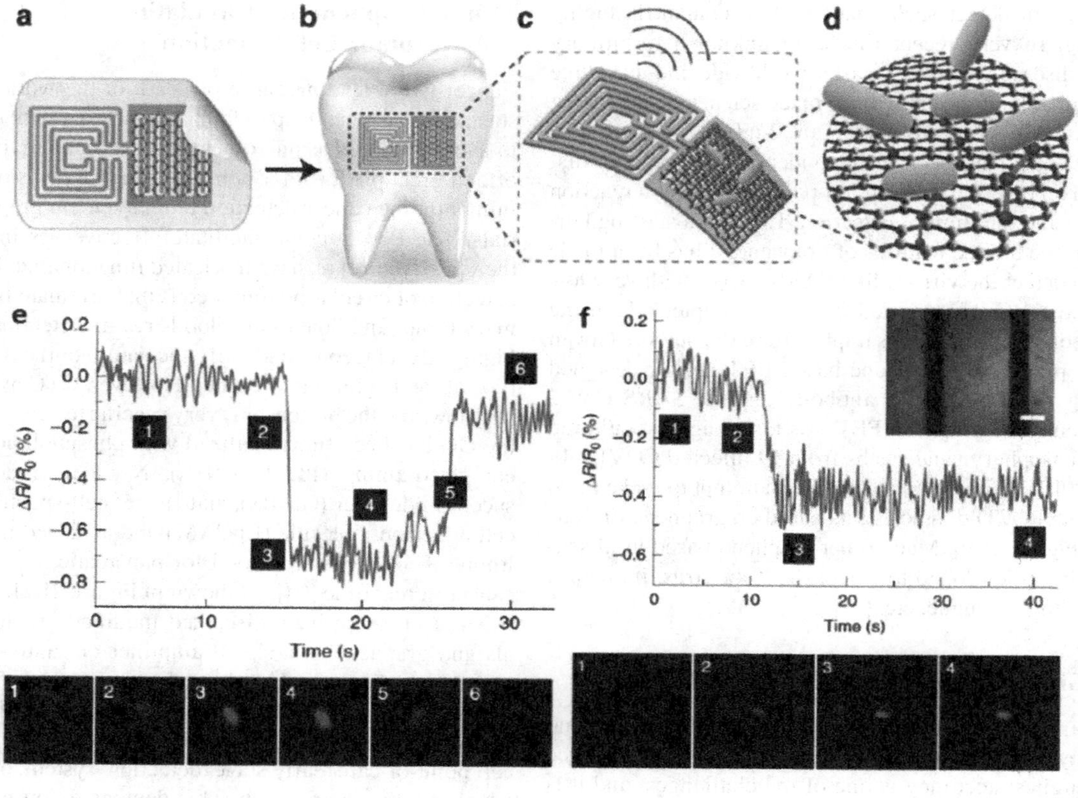

FIGURE 21.20 Graphene wireless nanosensor for single-bacterium detection with and without AMP. (a) Graphene-based wireless sensor has been printed on silk fibroin. (b) Biotransffered wireless sensor onto tooth. (c) Enlarged view to depict wireless readout. (d) Bacterial binding on AMP coated graphene nanotransducer. (e) Single *E. coli* binding and unbinding on graphene nanosensor and its electrical resistance (upper) and fluorescence (lower) recorded simultaneously versus time. (f) A single *E. coli* binding on an AMP-functionalized graphene nanosensor and its resistance (upper) and optical (lower) data. The inset is a fluorescent image of AMP-graphene with vertical electrodes. Image reprinted from [93].

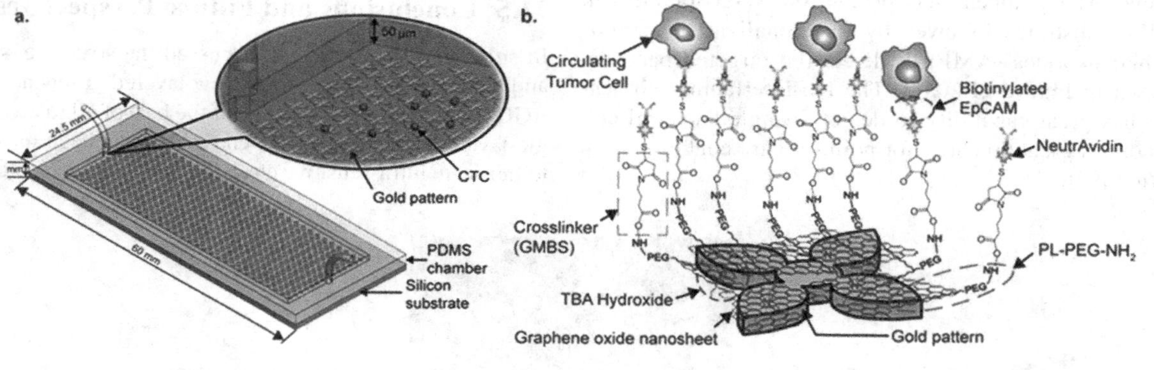

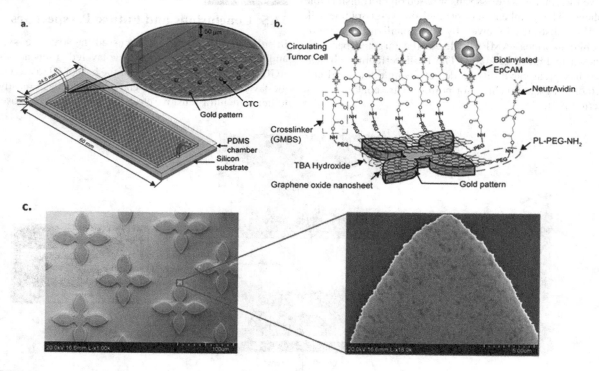

FIGURE 21.21 GO chip: functionalization and characterization for CTC detection. (a) GO chip diagram. (b) GO nanosheets adhere onto gold pattern (b) GO nanosheets adhere onto the gold patterns. The cell-specific EpCAM antibodies are linked to GO by PL-PEG-NH_2 followed by GMBS cross-linker and NeutrAvidin. (c) SEM image of gold patterns on the chip. The magnified inset shows the SEM image of gold patterns where GO nanosheets are adsorbed on it. Image reprinted from [94].

Solution-phase synthesis is highly desirable to produce GO and RGO for successful functionalization of biomarkers to identify the target molecules by colorimetric or electrical detections etc. The structural discussions are made to understand more about graphene to make bio or chemical modifications without sacrificing its inherent electrical and optical properties. The characterization of graphene is very important to confirm the formation of the desired product and its morphology on the nanoscale. The well-characterized graphene, GO and RGO, has been discussed and used to fabricate POCD tools based on electrical, FET and fluorescence sensor by functionalizing virus, bacteria and cancer cells markers. It is always reliable if the biosensors could offer reliable rapid and low detection limits in real-time. Such biosensors could prolong life by providing high-quality therapy for the particular disease on time. Research on the synthesis of high-quality graphene for biological applications is still highly desired to fabricate in economic and simple fabrication methods. The advancements of simple colorimetric and electrical graphene biosensors are yet to be developed further and commercialized by fulfilling high-quality clinical diagnosis demands. The wonderful features of graphene discussed here are ideal to develop any sort of biosensors to accomplish filling the clinical diagnosis research gaps.

REFERENCES

1. Fullerene Discoverers win Chemistry Nobel. (1996). Oct. 9, https://www.sciencemag.org/news/1996/10/fullerene-discoverers-win-chemistry-nobel.
2. Nobel Prize for Graphene. (2011). *Nature Materials,* Focus issue, *1*, 1–2
3. Https://www.theguardian.com/science/2010/oct/10/carbon-breakthroughs-two-nobel-prizes.
4. Https://www.natureindex.com/news-blog/journals-publish-most-nobel-prize-winning-research-papers-physics.
5. Geim, A. K., & Novoselov, K. S. (2007). The rise of graphene. *Nature Materials*, *6*, 183–191.
6. Gao, Y., Lihua, L., Wing, B. L., & Man, C. N. (2018). Structure of graphene and its disorders: a review. *Science and Technology of Advanced Materials*, *19*, 613–648.
7. Zheng, Z. S., Dustin, K. J., & James, M. T. (2011). Graphene chemistry: synthesis and manipulation. *Journal of Physical Chemistry Letters*, *2*, 2425–2432.
8. Santosh, K. T., Sumanta, S. N. W., & Andrzej, H. (2020). Graphene research and their outputs: status and prospect. *Journal of Science: Advanced Materials and Devices*, *5*, 10–29.
9. Xiluan, W., & Gaoquan, S. (2015). An introduction to the chemistry of graphene, *Physical Chemistry and Chemical Physics*, *17*, 28484—28504.
10. Enoki, T., Takai, K., & Kiguchi, M. (2012). Magnetic edge state of nanographene and unconventional nanographene-based host-guest systems. *Bulletin of the Chemical Society of Japan*, *85*, 249–264.
11. Takeshi, A., Atsuhiro, O., Shunichi, F., Hideki, K., & Yoshio, A. (2015). *Chemical Science of Π -Electron Systems*. ISBN 978-4-431-55356-4, Springer
12. Kuang, H., Gun-Do, L., Alex, W. R., Euijoon, Y., & Jamie, H. W. (2014). Hydrogen-free graphene edges. *Nature Communications*, *5*, 3040.
13. Neto, A. H. C., Guinea, F., Peres, N. M. R., Novoselov, K. S., & Geim, A. K. (2009). The electronic properties of graphene. *Reviews of Modern Physics*, *81*, 109–162.
14. Ahn, S., Sung, J. S., Kim, H. J., & Sung, Y. K. (2015). Emerging analysis on the preparation and application of graphene by bibliometry. *Journal of Material Science and Engineering*, *4*, 1–6.
15. Edward, P., Randviir, D. A. C. B., & Craig, E. B. (2014). A decade of graphene research: production, applications and outlook. *Materials Today*, *17*, 426–432.
16. Terrance, B. (2019). Graphene: the hype versus commercial reality. *Nature Nanotechnology*, *14*, 904–910.
17. Ruquan, Y., & James, M. T. (2019). Graphene at fifteen. *ACS Nano*, *13*, 10872–10878.
18. Reiss, T., Hjelt, K., & Ferrari, A. C. (2019). Graphene is on track to deliver on its promises. *Nature Nanotechnology*, *14*, 907–910.
19. Phaedon, A. (2010). Graphene: electronic and photonic properties and devices. *Nano Letters*, *10*, 4285–4294.
20. Taisuke, O., Aaron, B., Thomas, S., Karsten, H., & Eli, R. (2006). Controlling the electronic structure of bilayer graphene. *Science*, *313*, 951–954.
21. Brian, J., Schultz, R. V. D., Vincent, L., & Sarbajit, B. (2014). An electronic structure perspective of graphene interfaces. *Nanoscale*, *6*, 3444.
22. Yanxia, W., Shengxi, W., & Kyriakos, K. (2020). A review of graphene synthesis by indirect and direct deposition methods. *Materials Research Society*. doi:10.1557/jmr.2019.377.
23. Minghui, Y., Zhipan, Z., Yang, Z., & Liangti, Q. (2018). Graphene platforms for smart, energy generation and storage. *Joule*, *2*, 245–268.
24. McCreery, R. L. (2008). Advanced carbon electrode materials for molecular electrochemistry. *Chemical Reviews*, *108*, 2646–2687.
25. Syed, Z. H., Muhammad, I., Syed, B. H., Won, C. O., & Kefayat, U. (2020). A review on graphene based transition metal oxide composites and its application towards supercapacitor electrodes. *SN Applied Sciences*, *2*, 764.
26. Minghao, Y., & Xinliang, F. (2019). Thin-film electrode-based supercapacitors. *Joule*, *3*, 338–360.
27. Jun, C., Yizhou, W., Wenhui, L., & Yanwen, M. (2020). Nano-engineered materials for transparent supercapacitors. *Chemistry of Nano Materials*, 10, 38, 32536–32542.
28. Abid, H., Mohd, J., Ravi, P. S., Rajiv, S., & Shanay, R. (2021). Biosensors applications in medical field: a brief review. *Sensors International*, *2*, 100100.
29. Nan, L., Alex, C., Ting, L., & Lihua, J. (2017). Ultra-transparent and stretchable graphene electrodes. *Science Advances*, *3*, 1700159, 1–10.
30. Hongfei, L, Zijie, T., Zhuoxin, L., & Chunyi, Z. (2019). Evaluating flexibility and wearability of flexible energy storage devices. *Joule*, *2*, 613–619.
31. Maria, C., Florina, P., Lidia, M., Crina, S., & Stela, P. (2019). A brief overview on synthesis and applications of graphene and graphene-based nanomaterials. *Frontiers of Materials Science*. 13, 23–32.
32. Chul, C., Young, K. K., Dolly, S., Soo-Ryoon, R., Byung, H. H., & Dal-Hee, M. (2013). Biomedical applications of graphene and graphene oxide. *Accounts of Chemical Research*, *46*, 2211–2224.

33. XiaoYe, W., Akimitsu, N., & Klaus, M. (2017). Precision synthesis versus bulk-scale fabrication of graphenes. *Nature Reviews Chemistry*, *2*, 1–10.
34. Novoselov, K. S., Geim, A. K., Morozov, S. V., Jiang, D., Zhang, Y., Dubonos, S. V., & Grigorieva, I. V. (2004). Electric field effect in atomically thin carbon films. *Science*, *306*, 666.
35. Novoselov, K. S., Jiang, D., Schedin, F., Booth, T. J., Khotkevich, V. V., Morozov, S. V., & Geim, A. K. (2005). Two-dimensional atomic crystals. *Proceedings of the National Academy of Sciences, U. S. A.*, *102*, 10451.
36. Elena, S., Kwang, T. R., Sunmin, R., Janina, M., Philip, K., Louis, E., Brus, T. F. H., Mark, S. H., & George, W. F. (2007). High-resolution scanning tunnelling microscopy imaging of mesoscopic graphene sheets on an insulating surface. *Proceedings of the National Academy of Sciences*, *104*(22), 9209–9212.
37. Richard, V. N. (2012). Beyond sticky tape. *Nature*, *483*, S 33.
38. Robert, C., Sinclair, J. L. S., & Peter, V. C. (2019). Micromechanical exfoliation of graphene on the atomistic scale. *Physical Chemistry Chemical Physics*, *21*, 5716.
39. Yuan, H., Yu-Hao, P., Rong, Y., et al. (2020). Universal mechanical exfoliation of large-area 2D crystals. *Nature Communications*, *11*(2453), 1–9.
40. Hummers, W. S., & Offeman, R. E. (1958). Preparation of graphitic oxide. *Journal of American Chemical Society*, *80*, 1339.
41. Ayrat, M. D., & James, M. T. (2014). Mechanism of graphene oxide formation. *ACS Nano*, *8*, 3060–3068.
42. Hernandez, Y., Nicolosi, V., Lotya, M., Blighe, F. M., Sun, Z. Y., De, S., McGovern, I. T., Holland, B., Byrne, M., Gun'ko, Y. K., Boland, J. J., Niraj, P., Duesberg, G., Krishnamurthy, S., Goodhue, R., Hutchison, J., Scardaci, V., Ferrari, A. C., & Coleman, J. N. (2008). High-yield production of graphene by liquid-phase exfoliation of graphite. *Nature Nanotechnology*, *3*, 563–568.
43. Sungjin, P., & Rodney, S. R., (2009). Chemical methods for the production of graphene. *Nature Nanotechnology*, *4*, 217–224.
44. Wencheng, D., Xiaoqing, J., & Lihua, Z. (2013). From graphite to graphene: direct liquid-phase exfoliation of graphite to produce single- and few layered pristine graphene. *Journal of Materials Chemistry A*, 1, 10592.
45. Mark, L., Zˇeljko, S., & Stanko, T. (2015). Electronic and optical properties of reduced graphene oxide. *Journal of Materials Chemistry C*, *3*, 7632.
46. Rajesh, K. S., Rajesh, K., & Dinesh, P. S. (2016). Graphene oxide: strategies for synthesis, reduction and frontier applications. *RSC Advances*, *6*, 64993.
47. Raluca, T., Otto, T. B., Ioan, P., Cosmin, L., Simion, A., & Ioan, B. (2020). Reduced graphene oxide today. *Journal of Materials Chemistry C*, *8*, 1–28. doi:10.1039/x0xx00000x.
48. Yanyan, X., Huizhe, C., Yanqin, X., Biao, L., & Weihua, C. (2018). Liquid-phase exfoliation of graphene: an overview on exfoliation media, techniques, and challenges. *Nanomaterials*, *8*, 942.
49. Brisebois, P. P., & Siaj, M. (2019). Harvesting graphene oxide- years: 1859 to 2019 a review of its structure, synthesis, properties and exfoliation. *Journal of Materials Chemistry C*, *8*, 1517–1547.
50. Keith, R. P., Eswaraiah, V., Claudia, B., et al. (2014). Scalable production of large quantities of defect-free few-layer graphene by shear exfoliation in liquids. *Nature Materials*, *13*, 624–630.
51. Ji-Heng, D., Hong-Ran, Z., & Hai-Bin, Y. (2018). A water-based green approach to large-scale production of aqueous compatible graphene nanoplatelets. *Scientific Reports*, *8*, 5567.
52. Navneet, K., & Vimal, C. S. (2018). Simple synthesis of large graphene oxide sheets via electrochemical method coupled with oxidation process. *ACS Omega*, *3*, 10233–10242.
53. Songfeng, P., Qinwei, W., Kun, H., Hui-Ming, C., & Wencai, R. (2018). Green synthesis of graphene oxide by seconds timescale water electrolytic oxidation. *Nature Communications*, *9*, 145.
54. Juxiang, L., Yajing, H., Shi, W., & Guohua, C. (2018). Microwave-assisted rapid exfoliation of graphite into graphene by using ammonium bicarbonate as the intercalation agent. *ACS Sustainable Chemistry and Engineering*, *6*, 12261–12267.
55. Piers, T., Mark, H., Robert, D., & Carey, J. D. (2019). Controlled sonication as a route to in-situ graphene flake size control. *Scientific Reports*, *9*, 8710.
56. Ani, M. H., Kamarudin, M. A., Ramlan, A. H., Ismail, E., Sirat, M. S., Mohamed, M. A., & Azam, M. A. (2018). A critical review on the contributions of chemical and physical factors toward the nucleation and growth of large-area graphene. *Journal of Materials Science*, *53*, 7095–7111.
57. Andrew, J. S., Nils, E. W., Matthias, G. S., Michael, K. R., Thomas, W., Josef, R. W., Klaus, M., & Hermann, S. (2015). Chemical vapor deposition of high quality graphene films from carbon dioxide atmospheres. *ACS Nano*, *9*, 31–42.
58. Cheun, L. H. C., Wei-Wen, L., Siang-Piao, C., Abdul, R. M., Azizan, A., Cheng-Seong, K., Hidayaha, N. M. S., & Hashima, U. (2017). Review of the synthesis, transfer, characterization and growth mechanisms of single and multilayer graphene. *RSC Advances*, *7*, 15644.
59. Fujita, J-i., Ueki, R., Miyazawa, Y., & Ichihashi, T. (2009). Graphitization at interface between amorphous carbon and liquid gallium for fabricating large area graphene sheets. *Journal of Vacuum Science and Technology B*, *27*, 3063.
60. Saenger, K. L., Tsang, J. C., Bol, A. A., Chu, J. O., Grill, A., & Lavoie, C. (2010). In situ x-ray diffraction study of graphitic carbon formed during heating and cooling of amorphous-C/Ni bilayers. *Applied Physics Letters*, *96*, 153105.
61. Petr, M., Stanislav, C., Ladislav, L., & Ladislav, F. (2020). Graphene prepared by chemical vapour deposition process. *Graphene Technology*, *5*, 9–17.
62. Sukang, B., Hyeongkeun, K., Youngbin, L., et al. (2010). Roll-to-roll production of 30-inch graphene films for transparent electrodes. *Nature Nanotechnology*, *5*, 574.
63. Li, L., Jincan, Z., Haisheng, S., et al. (2019). Towards superclean graphene. *Nature Communications*, *10*, 1912.
64. Yingjun, L., Peng, L., Fang, W., Wenzhang, F., Zhen, X., Weiwei, G., & Chao, G. (2019). Rapid roll-to-roll production of graphene films using intensive joule heating. *Carbon*, *155*, 462–468.
65. Guowen, Y., Dongjing, L., Yong, W., Xianlei, H., Wang, C., Xuedong, X., Junyu, Z., Qian-Qian, Y., Hang, Z., Di, W., Jie, X., Shao-Chun, Li., Yi, Z., Jian, S., Xiaoxiang, X., & Libo, G. (2020). Proton-assisted growth of ultra-flat graphene films. *Nature*, *577*, 204–208.
66. Li, C., Li, D., Yang, J., Zeng, X., & Yuan, W. (2011). Preparation of single- and few-layer graphene sheets using Co deposition on SiC substrate. *Journal of Nanomaterials*, *2011*, 319624.

67. Narula, U., Tan, C. M., & Lai, C. S. (2017). Growth mechanism for low temperature PVD graphene synthesis on copper using amorphous carbon. *Scientific Reports*, *7*, 44112.
68. Yannick, Bl., Florent, B., Teddy, T., Anne-Sophie, L., Chirandjeevi, M., Christophe, D., & Florence, G. (2018). Review of graphene growth from a solid carbon source by pulsed laser deposition (PLD), *Frontiers in Chemistry*, *6*, 572.
69. Kumar, P. (2013). Laser flash synthesis of graphene and its inorganic analogues: an innovative breakthrough with immense promise. *RSC Advances*, *3*, 11987.
70. Wei, K., Hyun, K., Sang-Hoon, B., Jaewoo, S., Hyunseok, K., Lingping, O., Yuan, M., Kejia, W., Chansoo, K., & Jeehwan, K. (2019). Path towards graphene commercialization from lab to market. *Nature Nanotechnology*, *14*, 927–938.
71. Ganesh, G., Chun-Chao, C., & Yong-Chien, L. (2013). Facile synthesis of smart magnetic graphene for safe drinking water: heavy metal removal and disinfection control. *ACS Sustainable Chemistry and Engineering*, *1*, 462–472.
72. Al-Gaashania, R., Najjarc, A., Zakariaa, Y., Mansoura, S., & Atieh, M. A. (2019). XPS and structural studies of high quality graphene oxide and reduced graphene oxide prepared by different chemical oxidation methods. *Ceramics International*, *45*, 14439–14448.
73. Maxim, K. R., Sergei, A. R., Demid, A. K., et al. (2020). From graphene oxide towards aminated graphene: facile synthesis, its structure and electronic properties. *Scientific Reports*, *10*, 6902.
74. Tahriri, M., Del Monico, M., Moghanian, A., Yaraki, M. T., Torres, R., Yadegari, A., & Tayebi, L. (2019). Graphene and its derivatives: opportunities and challenges in dentistry. *Materials Science & Engineering C*, *102*, 171–185.
75. Ganesh, G., & Yong-Chien, L. (2012). Multi-functional graphene as an in vitro and in vivo imaging probe. *Biomaterials*, *33*, 2532–2545.
76. Behzad, D., Mir, K. R. A., Omid, R., Akram, T., & Amin, J. O. (2016). Synthesis and characterization of graphene and functionalized graphene via chemical and thermal treatment methods. *RSC Advances*, *6*, 3578.
77. Maocong, H., Zhenhua, Y., & Xianqin, W. (2017). Characterization techniques for graphene-based materials in catalysis. *AIMS Materials Science*, *4*(3), 755–788.
78. Ferrari, A. C., Meyer, J. C., Scardaci, V., Casiraghi, C., Lazzeri, M., Mauri, F., Piscanec, S., Jiang, D., Novoselov, K. S., Roth, S., & Geim, A. K. (2006). Raman spectrum of graphene and graphene layers. *Physical Review Letters*, *97*, 187401–187404.
79. Milan, B., Onejae, S., Sokratis, K., Eui-Hyeok, Y., & Stefan, S. (2010). Determination of edge purity in bilayergraphene using μ-Raman spectroscopy. *Applied Physics Letters*, *97*, 031908.
80. Jiang-Bin, W., Miao-Ling, L., Xin, C., He-Nan, L., & Ping-Heng, T. (2018). Raman spectroscopy of graphene-based materials and its applications in related devices, *Chemical Society Reviews*, *47*, 1822–1873.
81. Lukas, M., Erik, P., Tobias, F., et al. (2020). A swift technique to hydrophobize graphene and increase its mechanical stability and charge carrier density. *NPJ 2D Materials and Applications*, *11*, 1–7.
82. Blake, P., Hill, E. W., Neto A. H. C., et al. (2007). Making graphene visible. *Applied Physics Letters*, *91*, 6, Article ID 063124.
83. Panchal, V., Yang, Y., Cheng, G., Hu, J., Kruskopf, M., Liu, C. I., Rigosi, A. F., Melios, C., Walker, A. R. H., Newell, D. B., & Kazakova, O. (2018). Confocal laser scanning microscopy for rapid optical characterization of graphene. *Communication Physics*, *1*, 83.
84. Yi, S., & Aik, C. L. (2013). A facile method for the large-scale continuous synthesis of graphene sheets using a novel catalyst. *Scientific Reports*, *3*, 3037.
85. Pinshane, Y. H., Carlos, S. R. V., Arend, M. V. D. Z., William, S. W., Mark, P. L., Joshua, W. K., Shivank, G., Jonathan, S. A., Caleb, J. H., Ye, Z., Jiwoong, P., Paul, L. M., & David, A. M. (2011). Grains and grain boundaries in single-layer graphene atomic patchwork quilts. *Nature*, *469*, 389.
86. Jinfeng, C., Miao, D., & Guohua, C. (2012). Continuous mechanical exfoliation of graphene sheets via three-roll mill. *Journal of Materials Chemistry*, *22*, 19625.
87. Li, G., Jeffrey, R. G., & Nathan, P. G. (2010). Epitaxial graphene on Cu (111). *Nano Letters*, *10*, 3512–3516.
88. http://www.nanoscience.de/HTML/instrumentation/300mk-stm.html.
89. Ioannis, P., Ernestine, H., Patrik, G., Gabriele, S. K. S., Antonio, L., & Luigi, G. O. (2020). Graphene for biosensing applications in point-of-care testing. *Trends in Biotechnology*, *39*, 1065–1077.
90. Giwan, S., Geonhee, L., Mi, J. K., & Seung-Hwa, B. (2020). Rapid detection of COVID-19 causative virus (SARS-CoV-2) in human nasopharyngeal swab specimens using field-effect transistor-based biosensor. *ACS Nano*, *14*, 5135–5142.
91. Eleni, V., David, P., Kolleboyina, J., Martin, P., Ivo, F., Milan, K., Marián, H., Radek, Z., & Michal, O. (2020). Human virus detection with graphene-based materials. *Biosensors and Bioelectronics*, *116*, 112436.
92. Zhenglei, X., Minsi,P., Zhuliang, Z., Haotian, Z., Ruiyue, S., Xiaoxin, M., Lisheng, W., & Bihong, L. (2020). Graphene biosensors for bacterial and viral pathogens. *Biosensors and Bioelectronics*, *166*, 112471.
93. Manu, S. M., Hu, T., Jefferson, D. C., Amartya, S., David, L. K., Rajesh, R. N., Naveen, V., Fiorenzo, G. O., & Michael, C. M. (2012). Graphene-based wireless bacteria detection on tooth enamel. *Nature Communications*, *3*, 763.
94. Hyeun, J. Y., Tae, H. K., Zhuo, Z., Ebrahim, A., Trinh, M. P., Costanza, P., Jules, L., Nithya, R., Max, S. W., Daniel, F. H., Diane, M. S., & Sunitha, N. (2013). Sensitive capture of circulating tumour cells by functionalised graphene oxide nanosheets. *Nature Nanotechnology*, *28*(10), 735–741.
95. Li, X., Yanli, W., Santosh, P., Venkata, R. S. S. M., Ivan, M., Yan, L., Min, D., Shuzhen, R., Wen, L., & Gang, L. (2019). Graphenebased biosensors for the detection of prostate cancer protein biomarkers: a review. *BMC Chemistry*, *13*(112), 5–12.
96. Zhenglei, X., Minsi, P., Zhuliang, Z., Haotian, Z., Ruiyue, S., Xiaoxin, M., Lisheng, W., & Bihong, L. (2021). Graphene-assisted electrochemical sensor for detection of pancreatic cancer markers. *Frontiers in Chemistry*, *9*, 1–5.
97. Kovalska, E., Lesongeur, P., Hogan, B. T., & Baldycheva, A. (2019). Multi-layer graphene as a selective detector for future lung cancer biosensing platforms. *Nanoscale*, *11*, 2476.

22

3D-Printed Nanodevices of Pharmaceutical and Biomedical Relevance

Vaskuri G. S. Sainaga Jyothi

CONTENTS

22.1 Introduction

Nanotechnology is an emerging science in the field of pharmaceuticals sciences. It is widely applied for the delivery of drugs, diagnosis and biomedical devices [1]. The attention to nanotechnology is booming in the field of pharmaceutical and biomedical devices with the objective of targeted drug delivery, personalized medicine, controlled drug delivery with desired release profiles and point-of-care diagnostic devices [2]. However, the fabrication of pharmaceuticals and medical devices by conventional techniques and realizing all the intended functions is not possible. Hence a new technique called three-dimensional (3D) printing is adopted in the medical and pharmaceutical sciences to explore the unexplored paths.

3D printing is also called additive manufacturing and is a novel technique that prints the object in three dimensions by placing a layer upon a layer [3]. This technique enables the fabrication of the most complex structures which cannot be manufactured by the conventional 2D techniques. The execution of printing of these devices is enabled by designing the prototype *in silico* with the assistance of computer-aided design [3].

The first patent on 3D printing was filed in the year 1986, although research in this area has begun over the last decade [4]. This technology is being adopted in various fields and due to this, its research is being encouraged. The extensive research in 3D printing has opened new doors for pharmaceutical and biomedical applications. The technology is being deployed in the pharma industry for its decreased production expenditure and its speed and accuracy in printing, taking into consideration enormous advances in biosensors, drug delivery devices, implants and stents. With the advancement in this technology, point-of-care diagnosis and personalized medication are made possible [5–7].

The wide exploration of this technology is due to its advantages which involve fewer steps required to fabricate the structure, less labor needed for manufacturing, complicated designs that can be printed very easily and customization of the design that can be achieved as per the requirement [8]. Based on the application and nature of the material used, various technologies are being implemented in the 3D printing of pharmaceuticals.

22.2 Technologies Used for Fabrication

Numerous technologies have been explored for additive manufacturing of which the most widely utilized are the technologies for the execution of 3D printing, such as stereolithography (SLA), fused deposition modeling (FDM), selective laser sintering (SLS), pressure-assisted microsyringe extrusion (PAM), drop-on-powder (DOP) and digital light processing (DLP) (Table 22.1). These technologies enable the fabrication of various medical and pharmaceutical devices assisting in making personalized medicine possible.

DOI: 10.1201/9781003157823-22

TABLE 22.1
Various 3D Printing Technologies Used in the Pharmaceutical and Biomedical Fields

3D printing methods	Principle	Polymers used	Advantages	Limitations
Stereolithography (SLA)	Ultraviolet light photopolymerization	Polyethylene glycol diacrylate, polyethylene glycol dimethacrylate	Free of thermal energy	Only suitable for photocuring material
Fused deposition modeling (FDM)	Extrusion of filament	Polylactide-coglycoside, polylactic acid, polycaprolactone, polyvinyl alcohol	Solvent-free process, requires no post-processing, final products are mechanically strong	May not be appropriate for thermolabile drugs, filament preparation is prerequisite
Selective laser sintering (SLS)	Laser-induced sintering of powder	Polylactic acid, polymethylmethacrylate, polycaprolactone, polyvinyl alcohol	Single-step process and free of solvent	The laser light may degrade the drug
Pressure-assisted microsyringe extrusion (PAM)	Extrusion under high pressure	Poly-l-lactic acid, polylactic-co-glycolic acid, polycaprolactone	Processed at room temperature	Solvents are used, so drying step is needed
Drop-on-powder (DOP)	Fusing the powder by binder solution	Maltitol, maltodextrin or polyvinylpyrrolidone	High drug loading is possible, processed at room temperature	Solvents are used, drying of solvents is required
Digital light processing (DLP)	Light-induced photopolymerization as 2D layers	Poly(ε-caprolactone)	High-speed fabrication	Starting material should be photocurable

22.2.1 Stereolithography (SLA)

SLA is a laser-based technique where ultraviolet light is used for curing the photosensitive polymer (Figure 22.1). It was first patented in the year 1986 by Charles Hull [4]. The photosensitive liquid polymer (resin) is taken in a tank and subjected to a photopolymerization chain reaction with the aid of ultraviolet light building the entire structure. The polymer gets hardened and solidified by the focused energy from the ultraviolet light and the entire structure is built in a layer-by-layer fashion. All the layers were in a 2D fashion and were printed one upon the other, giving a 3D structure. The intact design of the product is designed with the assistance of computer-aided design. The first is a layer printed in the x- and y-axis and the platform moves down and the laser is focused on the next layer of the resin to form the next layer in the direction of the z-axis [9].

In general, the layers are printed from the bottom up where the layers are solidified from bottom to top. A build plate is placed below the liquid, the plate is lowered and the laser gets projected on the next layer of the resin and starts the polymerization process. However, a new approach is being used for the printing of the layers from the top downward, called the top-down approach. In this case, the build plate is placed in the resin from above and the light is focused from beneath the liquid through a non-adhering transparent plate. Even though the top-down approach leads to stress of the structure in separating the build plate from each fabricated layer, it leads to less use of polymer, fortification from the environment and accomplishment of a smoother surface [10]. After the complete fabrication process, the remnants of the resin need to be washed with solvents like alcohols. In addition, the engineered object needs to be photo-cured by ultraviolet to enhance its mechanical properties [11, 12].

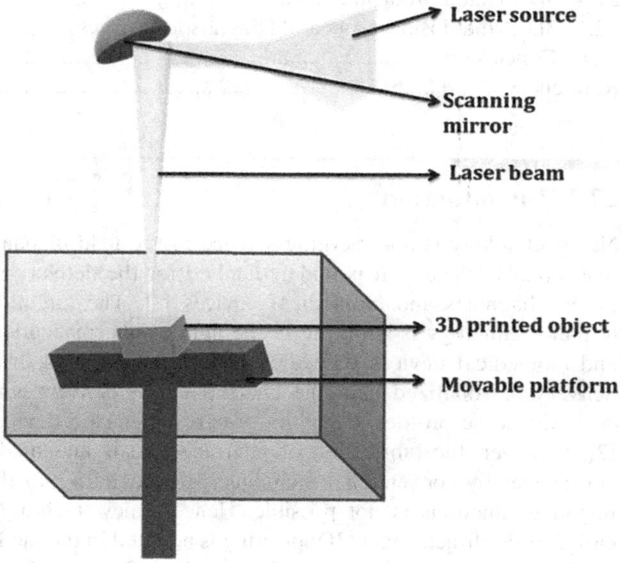

FIGURE 22.1 Schematic illustration of stereolithography printing process.

The materials used in the SLA technique should be photocurable polymers that enable cross-linking once the polymer is subjected to ultraviolet light and to be in liquid nature which solidifies by polymerization. The most widely used materials are low molecular weight polyacrylate or epoxy polymers. Polymers explored for pharmaceutical use include polypropylene fumarate-diethyl fumarate, polyethylene glycol diacrylate (PEGDA) [13], polyethylene glycol dimethacrylate [14] and poly-2-hydroxyethly methacrylate. However, the usage of these polymers in pharmaceuticals is limited by their nonexistence in the list of generally recognized as safe (GRAS) [15]. Therefore, the main challenge involved in this technique is the availability of the appropriate material for the fabrication process which is safe and non-toxic. Nevertheless, this technique is being employed for the development of orals systems [16], dermal patches [17] and biosensors. The main advantage of this technique is its good accuracy and good surface finish.

22.2.2 Fused Deposition Modeling (FDM)

FDM, also called fused filament fabrication (FFF), is an additive manufacturing technique where a 3D object is built by deposition of the melted polymer in a layer-by-layer manner (Figure 22.2). This is the most widely executed technique in 3D printing and was first patented and commercialized by Scott Crump in 1992 [18]. In this technique, thermoplastic polymer is fed into a nozzle which is heated up so the melted polymer in the semifluid state is extruded through a printing nozzle. The printer heads move and the extruded polymer is deposited as filament and gets solidified by cooling below its thermoplastic temperature. Then the subsequent layer of the polymer is deposited upon the first and thus the 3D structure is printed. Here, the build plate is present upon which the filament is formed and it moves down by printing layer upon layer [19].

It is the cheapest and most efficient technique in additive manufacturing where it facilitates the customization of the dose of the drug based on the patient requirement. However, in this technique, pre-processing of the polymer is to be carried out to incorporate the drug in the polymer. The polymer filaments can be soaked in a saturated solution of the drug in a methanol or ethanol solvent or by adopting other processing methods [20, 21]. Hot-melt extrusion (HME) is the widely employed process to incorporate the drug in the polymer and many studies have been conducted by coupling the HME with the FDM. The extruded filament from the HME is fed into the FDM to the heating unit and it is pushed into the nozzle for printing the material [19]. So, the critical parameters to be controlled while fabricating the process involve extrusion speed, temperature, nozzle diameter, print head moment speed and layer thickness [22]. The 3D objects printed by this technique exhibited good mechanical properties and are suppler than different designs.

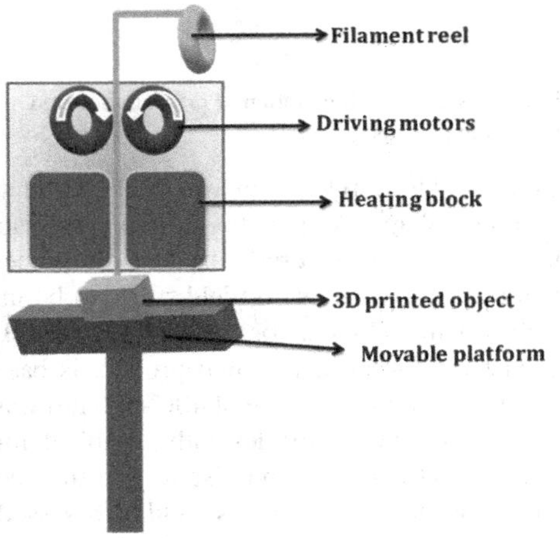

FIGURE 22.2 Schematic illustration of fused deposition modeling printing process.

Direct powder 3D printing which is a single-step FDM process is being studied to overcome the drawback of using two-step processes. In this methodology, the powder blend is fed into the cartridge and directly heated and extruded out from the nozzle enabling the printing of the object [23].

The materials used in this technique are thermoplastic which includes polylactide-coglycoside [24], polylactic acid [25], polycaprolactone [26], polyvinyl alcohol [27], acrylonitrile-butadiene-styrene, ethylene-vinyl acetate and other cellulose derivatives [22, 28, 29]. The main limitation of this technique is the execution of the process under a wide range of temperatures making it not suitable for the processing of thermolabile drugs.

22.2.3 Selective Laser Sintering (SLS)

SLS was first patented by Carl R. Deckard from the University of Texas. SLS is based on the use of a high-power laser to fuse the particles forming a layer subsequently forming a 3D structure (Figure 22.3). The laser used in his technique heats the powder to a temperature where the powder gets melted and the particles bind to each other. This fusion of particles forms a layer and the layer is positioned to project the laser on the new layer of powder. The fusion of particles occurs within the layer and also between the layers. The powder bed is adjusted and moves down every time while a new layer is deposited upon it [30]. This process is repeated which consequently forms a 3D object. The unused powder within the process is collected by sieving [31].

The key components in the process are the laser light, the powder bed and the spreading platform. The spreading platform forms a uniform layer of powder by spreading it and the layer is made uniform by a rollerblade. The laser light is passed through the layer of powder bed based on the program designed in the computer. It is a single-step and solvent-free process enabling it to be a green and more approachable technique [30]. The temperature usually maintained is between 40–50° C throughout the printing process which is below the melting point of the powder used, and after printing, the powder needs to be cooled slowly to evade stress. The entire process should be carried out in an inert atmosphere to avoid the oxidation of the raw materials [32].

The process parameters that influence the final object are laser focus, laser power and the speed of the laser which moves over the powder bed. The accuracy of the technique depends on the particle size of the powder bed and the laser system used [33]. The materials used in this technique are low melting powder with binding properties and thermostable polymer which can be heat processed. The polymers explored for the SLS are polyamides, polycarbonates, poly (ether-ether-ketone), polylactic acid, polymethylmethacrylate, polycaprolactone, polyvinyl alcohol and polyurethane [34].

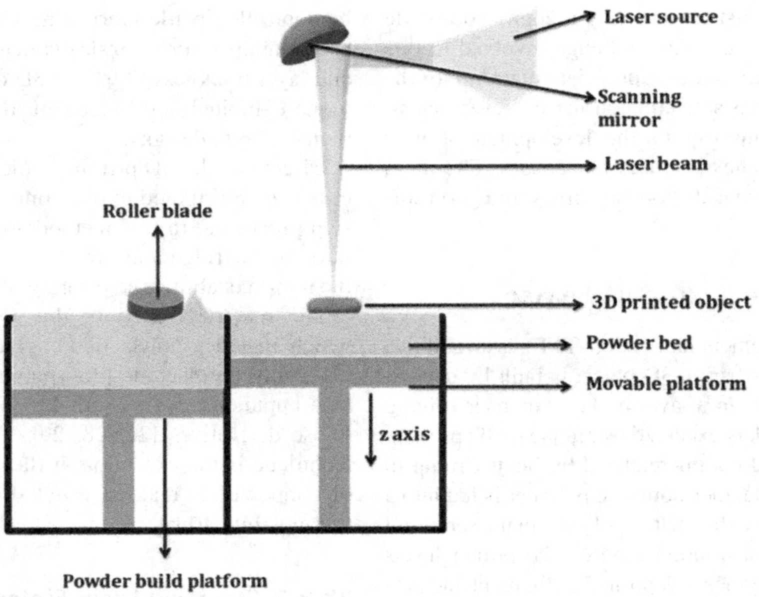

FIGURE 22.3 Schematic illustration of selective laser sintering printing process.

22.2.4 Pressure-Assisted Microsyringe Extrusion (PAM)

PAM is also known as semisolid extrusion technology where the semisolid material is extruded under high pressure and deposits as a layer with the assistance of the design developed in the modeling software [35]. It involves a syringe where the polymers are made into a paste with the appropriate solvent and fed into the syringe head (Figure 22.4). Based on the viscosity, material nature and design in the software, pressure is applied to the material to extrude out from the syringe. The extruded material is deposited on a platform in a layer-by-layer fashion making a 3D object. The extrusion is executed either as a mechanical, solenoid or pneumatic piston. The dimensions of the print head attached to the syringe fall in the range of 0.35 to 0.85 mm and a pressure of 0.4 to 3.8 bar is requisite for the extrusion process [36].

The main advantage associated with this technique is no involvement of a heating process which makes the technique favorable for the processing of thermolabile drugs and excipients [36]. However, the solvents utilized in the formation of paste of the raw materials need to be removed to ensure the object is free from the solvent. The main drawback is the presence of traces of solvents in the final pharmaceutical products. As the solvent is removed from the 3D-printed object, there might be chances of shrinkage of the product and deformation in the shape. The removal of solvents involves the drying step where the temperature maintained for the drying process may affect the drug and polymers or excipients' stability [37]. Hence analysis of product samples for their degradation is essential. Furthermore, if the layer deposited at the bottom does not have enough strength to bear the weight of the succeeding layers, the printed object might collapse.

As this technique involves the usage of raw materials under high pressure without any processing, the materials should have a few characteristic properties to aid the smooth running of the process. These involve extrudability, printability and ability to retain shape. As per the literature available on the PAM process, the raw materials used in the process should be shear thinning systems with yield stress to be maintained below 4,000 Pa and a loss factor that falls between 0.2 and 0.7 [38, 39]. The resolution of the final product is based on the size of the nozzle where a nozzle of 400–800 μm was used for pharmaceutical products. Nozzles with suitable diameters can be used based on the nature and viscosity of the material. As a result, the viscosity of the paste should be assessed to select the appropriate nozzle in order to evade clogging of the nozzles [40]. The materials used in this process are poly-l-lactic acid, poly lactic-co-glycolic acid, polycaprolactone and poly (3-hydroxybutyrate-co-3-hydroxyhexanoate) [41].

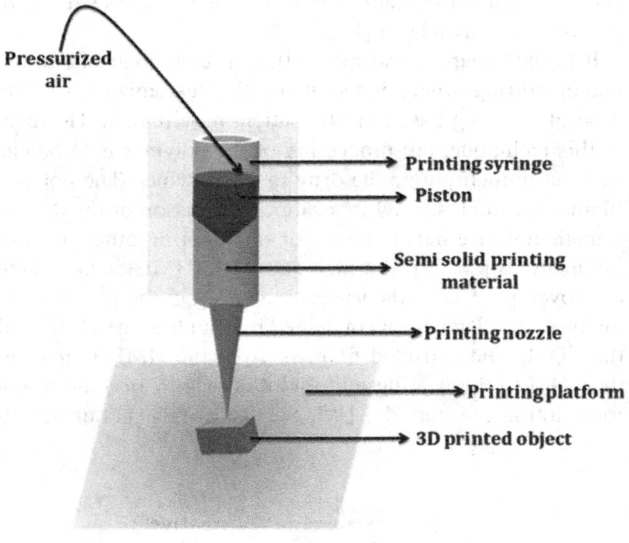

FIGURE 22.4 Schematic illustration of pressure-assisted microsyringe extrusion printing process.

22.2.5 Drop-on-Powder (DOP)

DOP, also called the drop-on-solid or binder jetting technique, uses the droplets of liquid to bind the powder particles within the layer of the powder bed. In this process, a thin layer of powder is evenly spread by a roller on a build platform and the particles of the powder are fused by spreading the binder solution through a print head (Figure 22.5). The movement of the print head over the powder bed is controlled by the aid of a computer. After printing the first layer, the build platform moves vertically downward and a new layer is printed over the previous layer. Thus the process is repeated and layers are printed one upon the other till the 3D object is printed [42]. After the completion of the printing process, the unused powder is collected and the solvent is removed by drying [43].

The binder solution used in the process may not necessarily aid in the binding of the particles but the solvent used in the binder solution may dissolve the powder particles and aid in solidification following evaporation of the solvent [44]. The drug and the excipients for the design of formulation can be included in different ways; the drug can be suspended or dissolved in the binder solution or can be incorporated in the powder bed or in both the binder solution and the powder bed [45].

Based on the mechanism involved in the ejection of binder solution from the print head, it is of two types, piezoelectric and thermal inkjet printer. In a thermal printer, thermal energy is used to eject the droplets from the nozzle. The print heads are implanted with resistors that are exposed to the binding solution and the induction of electric current results in the generation of heat. The resulting heat leads to the formation of a bubble and the expansion of the bubble effectuates the ejection of fluid from the nozzle forming a droplet. The main drawback associated with it is high temperature which may hamper the use of thermolabile drugs [46, 47]. The other mechanism involved is a piezoelectric element where its shape is being affected by the electric voltage. The change in shape leads to the ejection of fluid from the nozzle. Once the element retains its intact shape, the nozzle is filled with the fluid and can be reactivated for ejection. As this technique is free of high-temperature usage, it is suitable for thermolabile drugs [48, 49]. This technique is mainly utilized in pharmaceuticals in the fabrication of orodispersible films.

22.2.6 Digital Light Processing (DLP)

DLP also works with the same principle as that of SLA technology where a light source is used to photopolymerize the polymers to form the layers forming a 3D structure. However, the difference between the SLA and DLP is the way the light source is focused onto the photosensitive polymer. In SLA, a pointed light source is used for the photocuring of the resin whereas, in DLP, a 2D pattern of light is used projected on the layer of resin for photocuring. This way of projecting light aids in photopolymerizing the entire polymer layer in one go [50]. Hence DLP is much faster than the SLA in terms of printing 3D objects. Here, each 2D layer after being exposed to a light source is hardened in a single attempt instead of being exposed to several scans of laser light. The 2D layers are piled up one above the other to form a 3D structure [51, 52]. This technique is widely used to design biosensors [53].

22.3 3D-Printed Drug Delivery Devices

Fabrication of drug delivery devices requires new technological advances for the achievement of targeted drug delivery, desired release kinetics and personalized drug therapy with desired drug loading. The accomplishment of all these requirements by the conventional technologies cannot be realized as the provision for customization of each drug delivery device is not possible. Thus, 3D printing or additive manufacturing is an attractive alternative to overcome all the limitations associated

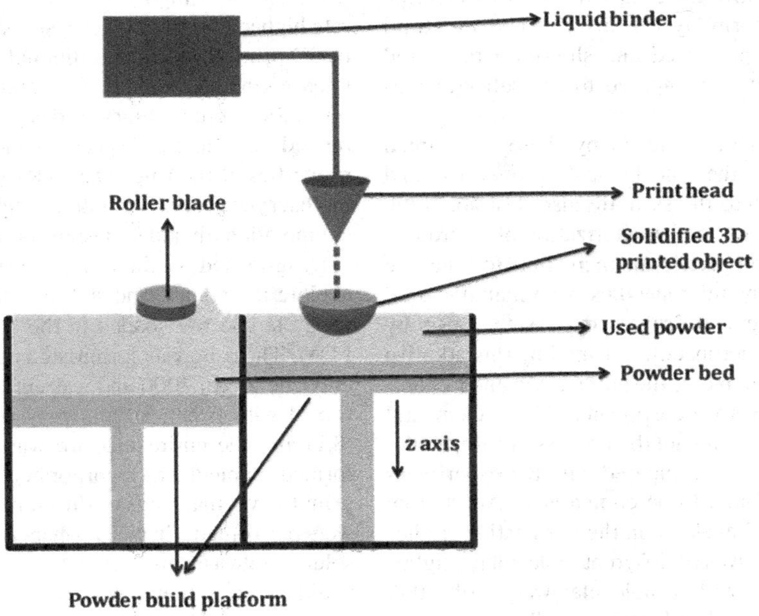

FIGURE 22.5 Schematic illustration of drop-on-powder printing process.

with traditional techniques. Fabrication of devices by conventional techniques for instance microneedles is linked with many practical concerns which can be surmounted by 3D printing. Control of drug loading and its release kinetics is essential for vaginal drug delivery which can be achieved by this 3D printing. The 3D-printing technique gives a stipulation for the customization of the design of the device as per the requirement with the aid of software connected with it. Hence it is being widely studied nowadays in the arena of drug delivery (Table 22.2).

3D printing is adopted for the manufacturing of microneedles. Microneedles are extensively studied in the transdermal delivery of drugs. However, the production of this microneedle involves a typical manufacturing process which results in scale-up difficulties. Uddin et al. fabricated a 3D-printed microneedle patch loaded with an anticancer drug for the transdermal delivery to the skin tumor. The study utilized the SLA technique for tailoring these microneedles. The needles were built by specifically photocuring the layers of resin controlled by the computer-aided design and it was coated with the drug cisplatin by the inkjet dispenser. The results showed good permeation of the drug with effective anticancer activity in the *in vivo* studies. This study highlighted the utility of 3D printing in designing microneedles for transdermal drug delivery [55].

In another study, microneedles were fabricated for the intradermal delivery of insulin. Insulin is injected subcutaneously, however, the intradermal delivery of insulin via microneedles showed relative bioavailability and pharmacological activity as compared with the subcutaneous injection [78]. Hence, 3D printing is adopted for the design of microneedles for the delivery of insulin. The needles were built by using biocompatible class 1 polymer via SLA technology. In this study, the needles were designed as spear and pyramid shapes, and then the needles were coated with the insulin with an inkjet printer. These polymer-printed microneedles showed good penetration ability with low force as compared to metallic microneedles with a uniform layer of insulin. The structural integrity of insulin was preserved and showed a rapid and longer duration of action as compared to the subcutaneous injection [54].

Delivery of drugs can be achieved by these 3D-printed microneedles, however, the mechanical properties and their ability to protrude into the skin are also desirable [79]. Xenikakis et al. performed the characterization of microneedles for their mechanical properties and their ablity to penetrate into the skin. In this study, microneedles were manufactured for the transdermal delivery of drugs using castable resin by SLA. The ability of the microneedles printed by this additive manufacturing technique to pierce the human skin and its complete mechanical properties were evaluated. Finite elemental analysis was performed to simulate the process of insertion of needles into the skin and was compared with the experimental outcome which was found to be comparable. Apart from the mechanical properties' evaluation, the permeation studies were carried out with two dyes of different molecular weights. FITC-dextran and calcein with a molecular weight of 4,000 and 622.54 Da were employed and the results showed the significant transport of dyes across the skin [58].

Delivery of drug and its simultaneous detection is the additional advantage that aids in quantifying the delivered amount of drug [80]. Yao et al. designed hydrogel microneedles performing multitask such as drug delivery and its simultaneous detection. The microneedles were fabricated by DLP using polyethylene glycol diacrylate photopolymer. Rhodamine B was used as a fluorescent dye and as a model molecule in this study. Various parameters were studied to evaluate their influence of which the period of exposure of the layer to the light source impacted the stiffness of the needles and the precision of the design [59].

The combination of microneedles with microfluidics results in the augmentation of drug delivery especially in the case of combination therapy [81]. Microfluidics enables the mixing of small volumes of fluids and its integration with the microneedles enables the effective co-delivery of drugs [82]. However, the manufacturing process varies widely for both techniques, so this hurdle can be overcome by 3D printing which enables the fabrication of two techniques in a single device [83]. An SLA 3D printer was used to fabricate this microfluidic-enabled microneedle. Biocompatible class IIa resin was used as a photopolymer which is a blend of methacrylic acid esters and a photoinitiator consisting of glycol methacrylate, methacrylic oligomer, phosphine oxide and pentamethyl-piperidyl sebacate. This device aids in the hydrodynamic mixing of two drug solutions by the microfluidics and microneedles enable the delivery of this mixed drug solution to the skin [56]. Thus microneedles fabrication is eased by the adoption of additive manufacturing technology and it is being explored to augment its versatility.

Intravaginal rings (IVRs) are the medical devices intended to be inserted in the vaginal cavity for the delivery of hormones or contraceptives and also bactericides and fungicides for the treatment of vaginal infections [84]. IVRs are conventionally tailored by injection molding and hot-melt extrusion [85]. However, these techniques suffer from the limitations of involvement of high temperature and incapability to incorporate higher drug loading. These roadblocks can be surmounted by 3D printing with the additional advantage of controlling the release kinetics. A study was conducted by Janusziewicz et al. on IVRs where a library of designs was developed and investigated the effect of aspects of the design on the mechanical properties of the ring. The IVRs were printed with urethane-methacrylate resin using digital light synthesis with the aid of computationally aided design software [62].

Drug-loaded vaginal rings were prepared in a study to explore the release and stability of the drug from the device. Progesterone was loaded in the ring and it was fabricated by FDM. The drug was formulated as solid dispersion with polyethylene glycol 4000 and was cut into pieces. The pieces were mixed with tween 80 and poly(lactic acid)/polycaprolactone (8:2) and the entire mixture was melted as a filament. The formed filament was incorporated into the FDM 3D printer to print the vaginal rings of different shapes. The release studies were performed where "O"-shaped rings exhibited a higher dissolution rate as compared to "M"- and "Y"-shaped rings which could be due to their spherical shape and high surface-area-to-volume ratio. The IVRs showed seven days sustained release of the drug following a diffusion-controlled pattern. Hence the

TABLE 22.2
3D-Printed Drug Delivery Devices

3D technology used	Drug	Device	Polymer	Category	Ref.
SLA	Insulin	Microneedle patch	Biocompatible class I resin	Intradermal insulin delivery	[54]
SLA	Cisplatin	Microneedle	Photopolymer resin	Transdermal delivery for skin tumors	[55]
SLA	–	Microfluidic-enabled microneedle	Class IIa biocompatible resin (mixture of methacrylic acid esters and photoinitiator comprised of [in % w/w])> 70% methacrylic oligomer, < 20% glycol methacrylate, < 5% pentamethyl-piperidyl sebacate (EG No. 255-437-1), and < 5% phosphine oxide (EG no. 278-355-8)	Transdermal drug delivery	[56]
SLA	Lidocaine	Polycaprolactone scaffold	Polycaprolactone	Drug delivery scaffold	[57]
SLA	FITC-Dextran and calcein as model dies	Microneedle	Castable resin	Transdermal drug delivery	[58]
DLP	Rhodamine B	Hydrogel microneedles	Polyethylene glycol diacrylate	Drug delivery and detection	[59]
DLP	–	Oral drug delivery	HTM 140 M V2 3D printing photopolymer	Reservoir devices for oral drug delivery	[60]
Continuous liquid interface production (CLIP)	Model proteins bovine serum albumin, ovalbumin and lysozyme	Microneedle patches	Resin composed of polyethylene glycol	Transdermal protein delivery	[61]
DLS	–	Intravaginal rings	Urethane-methacrylated resin	Intravaginal drug delivery	[62]
Two-photon polymerization (TPP)	Rhodamine B	Polymerized poly(ethylene glycol) dimethacrylate devices	Poly(ethylene glycol) dimethacrylate	Controlled drug delivery	[63]
Direct laser writing system	Doxorubicin	Magnetically powered, double-helical microswimmer	Methacrylamide chitosan, superparamagnetic iron oxide nanoparticles	Magnetically powered, double-helical microswimmer	[64]
DLP	Acetyl-hexapeptide 3	Microneedles	Polyethylene glycol diacrylate (PEGDA, Mw 700 Da) and vinyl pyrrolidone (VP)	Transdermal delivery of anti-wrinkle protein	[65]
Two-photon polymerization (TPP)	Cabotegravir sodium and ibuprofen sodium	Microneedle	PLA, poly (vinylpyrrolidone) (PVP), poly (vinyl alcohol) (PVA)	Transdermal delivery	[66]
SLA	Rhodamine B	Microneedle	Polylactic acid	Transdermal delivery	[67]
FDM and SLA	Salicylic acid	Patch/mask	Flex EcoPLA™ (FPLA) and polycaprolactone	Topical drug delivery	[17]
FDM	Fluorescein	Microneedle	Polylactic acid	Transdermal delivery	[68]
FDM	Progesterone	Vaginal rings	Poly(lactic acid) (PLA)/polycaprolactone	Vaginal drug delivery	[69]
FDM	Clotrimazole	Intravaginal ring	Thermoplastic polyurethane	Vaginal drug delivery	[70]
SLA	Lidocaine hydrochloride	Bladder device	Elastic resin	Intravesical drug delivery	[71]
PolyJet printer	Epirubicin	Microfluidic chip	Trihexyltetradecylphosphonium chloride	Electrotactic drug delivery	[72]
FDM	Sodium cromoglicate	Capsular devices	Polyvinyl alcohol	Oral delivery	[73]
SLA	–	Modular microreservoir	Dental SG resin	Reservoir-based drug delivery microsystems	[74]
Bioprinter	–	Self-healing ferrogel	Glycol chitosan, oxidized hyaluronate and superparamagnetic iron oxide nanoparticles	Magnetic field triggerable drug delivery system	[75]
FDM	5-fluorouracil	3D-printed reservoir of drug-loaded sponge	SYLGARD® 184 Silicone Elastomer prepolymer	Implantable and smart drug delivery system for the controlled release of therapeutic substances	[76]
Extrusion-based 3D printing	Doxorubicin hydrochloride	Hydrogel fiber scaffolds	Polydopamine/alginate	Local therapy	[77]

study concluded that personalized medication can be achieved by this 3D printing technology [69]. In another study on IVRs, clotrimazole-loaded rings were fabricated for the treatment of vaginal infections. Thermoplastic polyurethane was used as a polymer where the drug was incorporated into the polymer by hot-melt extrusion. The drug-infused polymer was printed as IVR by FDM 3D printing technology. The effectiveness of IVR as a fungicide and its release rate were studied which showed a drug release for seven days. These studies implied a prospective application of 3D printing for the fabrication of IVRs [70].

3D printing can also be adopted to design different drug delivery devices to achieve personalized medication and also controlled release of drugs. A study reported by Bozuyuk et al. designed microswimmers to control the release of drugs based on the trigger of light. The microswimmer was loaded with doxorubicin and it was magnetically powered by incorporating chitosan with a magnetic polymer nanocomposite [64]. In a study to design a personalized mask/patch, two techniques, FDM and SLA 3D printers, were adopted and their efficacy was compared. Salicylic acid was used in the mask for the acne treatment. A 3D scanner was used to take the total scan to design the mask as per the specifications of the person. In FDM, Flex EcoPLA™ (FPLA) and polycaprolactone polymers were used and blended with the drug salicylic in the hot-melt extrusion to form a filament and the formed filament was used in the FDM to print the mask. In the SLA technique, the drug was dissolved in the mixture of poly(ethylene glycol) diacrylate and poly(ethylene glycol) polymers and was fabricated with the aid of laser light. The diffusion studies revealed that the drug diffusion was faster with the mask prepared by SLA as compared to FDM. Hence this study concluded that SLA was more appropriate for this mask as compared to FDM [17].

Thus, the 3D printing technology was recognized as a special tool for the fabrication of various drug delivery devices. This technology augments the drug delivery to be more versatile and reliable and makes personalized medicine possible.

22.4 3D-Printed Medical Devices

3D printing technology was also adopted in the biomedical field where it eases the fabrication of biomedical devices. Biomedical devices like stents and implants involve careful tailoring as the structure involves complex geometries and entails vigilant customization as per the patient's requirement. Hence the wide application of 3D printing can be visualized in the vicinity of biomedical devices (Table 22.3).

A study conducted by Misra et al. fabricated a multi-drug-eluting stent by a 3D printer. This research work highlighted the need for customization of stents as per the patient requirement and studied the feasibility of its design by 3D printing technology. The stent design was prepared *in silico* and was printed by incorporating graphene nanoplatelets in the polycaprolactone. The sodium salt of phytic acid and niclosamide as anticoagulant and anti-restenosis agents were incorporated in the stent for effective percutaneous coronary intervention. The stent exhibited superior mechanical properties and *ex vivo* studies showed a plausible use for the patients [89].

A biodegradable stent was designed by FDM 3D printing technology which used polycaprolactone as a biodegradable polymer. Different parameters involved in the printing process were studied including the impact of flow rate, the temperature

TABLE 22.3

3D-Printed Biomedical Devices

S. no.	3D technology used	Drug	Device	Polymer	Ref.
1	FDM and SLA	Doxycycline, vancomycin and cefazolin	Polycaprolactone (PCL)-based femoral implants	Polyethylene glycol and polyethylene glycol diacrylate	[86]
2	SLA	Doxorubicin, ifosfamide, methotrexate and cisplatin	Biodegradable implants	Poly L-lactic acid	[87]
3	FDM	Gentamicin sulfate and methotrexate	Drug-laden 3D printed catheter	Polylactic acid	[88]
4	FDM	Niclosamide and sodium salt of phytic acid	Multidrug-eluting stent	Poly-l-lactic acid and poly-l-caprolactone	[89]
5	FDM	_	Biodegradable stents	Polylactide acid or polycaprolactone	[90]
6	FDM	Iodinated, gadolinium and barium contrast agents	Surgical hernia meshes	Polycaprolactone	[91]
7	FDM	5-fluorouracil	Drug-eluting stents	Polyurethane	[92]
8	FDM	Triamterene	Biodegradable drug-eluting tracheal stents	Polylactic acid	[93]
9	FDM	Diclofenac sodium	Implant	Polylactic acid, antibacterial polylactic acid, polyethylene terephthalateglycol and poly(methyl methacrylate)	[94]
10	FDM	Methylene blue, ibuprofen sodium and ibuprofen acid	Implantable drug delivery devices	Polylactic acid and polyvinyl acetate	[95]
11	3D rapid prototyping	Sirolimus	Drug-eluting, bioabsorbable vascular stent	Polycaprolactone	[96]
12	FDM	_	Biodegradable stent	Polycaprolactone	[97]
13	FDM	_	Biodegradable stent	polycaprolactone	[98]

of the nozzle and the speed of the printer on the precision of the stent. Although the results revealed that the flow rate and the temperature had an influence, the speed of the printer showed no impact [97, 98]. A drug-eluting bioabsorbable vascular stent was manufactured in a study where sirolimus was coated on the polymer. Polycaprolactone was used as a bioabsorbable polymer for its design. The release studies were performed where the drug showed sustained release kinetics. The stent was tested by implanting in a porcine femoral artery. Low fibrin score and reduced neointimal hyperplasia showed its successfulness as a drug-eluting stent for implanting in coronary diseases where the percutaneous coronary intervention is requisite [96].

Fouladian et al. synthesized the esophageal stent in the management of stenosis associated with esophageal cancer. Nevertheless, the tumor growth leads to restenosis after stent placement which can be overcome by inculcating the stent with an anticancer drug. Hence, the author incorporated 5-fluorouracil in the polyurethane polymer. This drug-eluting stent was fabricated by FDM; prior to its printing the drug was incorporated in the polymer and was processed to form a filament. The *in vitro* release studies of the drug showed a sustained release of the drug for 110 days and was stable to UV irradiation sterilization and accelerated stability conditions [92]. In another study, a biodegradable drug-eluting tracheal stent was 3D printed by FDM. This study highlighted the customization of the stent design as per the requirement. The stent was designed in such a way that the stent retains its high mechanical strength with low porosity. Based on the prerequisites, the design was optimized and programmed into machine command which is called slicing. Poly (lactic-co-glycolic acid) polymer was used for its fabrication and triamterene was incorporated in the stent as a model drug due to its low water solubility compared to the drugs widely used in the stents. This study showed the potential of the 3D printing technology in tailoring the stents as per the desired physicochemical properties and achieving the desired release of drugs [93].

The 3D printed implants are also extensively studied where the drug-loaded implants gain much attention as the technology gives the feasibility to incorporate the drug in the layers of the polymer of the implant. Antibiotic-laden implants were reported in joint replacement therapy and this technology aids in easing the drug loading within the implants [99]. A femoral implant infused with doxycycline, vancomycin and cefazolin was prepared by two different 3D-printing techniques namely SLA and FDM. The efficacies of these antibiotics were studied as the techniques involve exposure to high temperature and UV rays in the FDM and SLA respectively. Polycaprolactone was used in the FDM technique whereas polyethylene glycol and polyethylene glycol diacrylate were used in the SLA technique. The results found that the drugs were intact with no loss of their antibacterial efficacy [87].

A bioabsorbable catheter was fabricated using the FDM 3D printing technology. The catheter was impregnated with gentamicin sulfate or methotrexate. Polylactic acid bioplastic pellets were prepared and were coated with the drug. These pellets were processed as filaments and the filaments were printed in FDM. Both the drugs showed sustained release [89]. These studies showed the potential of 3D printing of personalized stents, implants and catheters.

22.5 3D-Printed Biosensors and Diagnostic Devices

Biosensors are widely used for the detection of an enzyme or protein related to a disease where it results in diagnosis. The 3D printing technology enabled the fabrication of these biosensors and turns the diagnosis into a point-of-care device (Table 22.4).

A microfluidic immunosensor was 3D printed to detect the biomarkers involved in prostate cancer. FDM was used in this study. RuBPY-silica nanoparticles were prepared and these nanoparticles were coated with poly(diallyldimethylammoniumchloride) and poly(acrylic acid). Three antigens, namely prostate-specific membrane antigen (PSMA), prostate-specific antigen (PSA) and platelet factor-4 (PF-4), are biomarkers in the detection of prostate cancer, hence their respective antibodies, namely anti-PSMA, anti-PSA and anti-PF-4asAb2, were

TABLE 22.4
3D-Printed Biosensors

S. no.	Purpose	Device	Ref.
1	Detection of enzyme activity	Electrochemical sensor	[100]
2	Detection of cancer	Supercapacitor-powered electro-chemiluminescent protein immunoarray	[101]
3	Detection of cancer	Biosensor with immunolabeling	[102]
4	Detection of malaria	Syringe test and well test	[103]
5	Detection of DNA damage	Stratospheric probe with carbon quantum dots linked fluorescence detection	[104]
6	Detection of cancer	Immunoarray for chemiluminescence detection	[105]
7	Detection of biomarkers for the preterm birth	Microfluidic devices for microchip electrophoresis	[106]
8	Detection of biomarkers for the preterm birth	Microfluidic devices with on-chip fluorescent labeling	[107]
9	Detection of cancer by immunosensor arrays	Immunosensor arrays	[108]
10	Detection of cancer	Modular 3D printed lab-on-a-chip	[109]
11	Diagnosis of anemia	Auto-mixing chip enabled smartphone diagnosis	[110]
12	Detection of influenza virus	Chip for electrochemical detection	[111]
13	Assessment of severe aortic stenosis	Flow circuit	[112]

used as detection antibodies. These antibodies were covalently tagged to a COOH group of poly(acrylic acid). These antibodies were coated on (RuBPY)-doped silica nanoparticles with the electro-chemiluminescent which generates light by the supercapacitor that is detected by the CCD camera. These antibodies were detected in the limit of 0.3–0.5 pgmL^{-1}. Thus, 3D printing aids in the generation of biosensors, however, a number of steps were involved in its synthesis [101]. In a similar study to detect cancer, microfluidics was incorporated by 3D printing the microfluidic device. The device consists of three parts which include a reagent reservoir, a mixing aid and a detection chamber for evaluating the chemiluminescence output with a CCD camera. PSA and platelet factor 4 were used as antibodies for the study. Detection of cancer biomarkers requires 30 min with minimal operation. This study showed the scope for further improving the device as a point-of-care diagnostic tool for cancer [105].

Detection of cancer can be aided by the specific proteins or enzymes expressed extensively as compared to the healthy tissues. These highly expressed molecules can be considered biomarkers. Metallothioneins is one of the widely studied molecular biomarkers where it was found comparable in malignant tumors. Based on the detection of metallothioneins, a biosensor was 3D printed by using PROFI3DMAKER. The antibodies of metallothioneins were produced and these antibodies were conjugated to CdTe quantum dots for the fluorescence detection of metallothioneins. The immobilization of metallothioneins in the sample was achieved by employing the magnetic nanoparticles. These magnetic nanoparticles and antibody-conjugated quantum dots were employed in the poly(dimethylsiloxane) biosensor chip which was fabricated by a 3D printer using acrylonitrile butadiene styrene as polymer [102]. This synthesized biosensor was tested with metallothioneins in cell lines of fibroblasts and spinocellular carcinoma where the biomarker was found in the levels of 37 nM and 90 nM, respectively. This shows the feasibility of 3D printing in the fabrication of biosensors.

3D printing technology was assimilated into the diagnostic devices for the detection of malaria. The basic principle associated with the diagnosis is the detection of *Plasmodium falciparum* lactate dehydrogenase (PfLDH) enzyme as a biomarker of malaria in the samples for analysis and utilizing the enzyme function for the generation of a visual color with the positive samples. Two diagnostic devices were fabricated with the 3D printing technology of which one method is a paper-based syringe test and the other is a magnetic bead-based well test. Both the tests were effective in detecting the PfLDH marker at nanogram concentration level with the sample volume as low as 20 μL. The syringe-based test showed superiority compared to the well test in terms of sensitivity. However, the syringe-based test requires more steps as compared to the well test which requires fewer steps. Hence the well test can be more adoptable on a clinical level as compared to the syringe test. The response produced with this test was the generation of color and the reagents responsible for the color generation were stable with higher temperatures. This technology aids in the rapid detection of malaria with many advantages relative to the well-established lateral flow immunochromatographic approaches [103].

The damage of DNA can be able to detect by implementing 3D technology. Carbon quantum dots were prepared and coated with polyethylene glycol as a detecting agent with the fluorescent probe. These synthesized carbon quantum dots were capable of detecting the damage to the DNA. The detector was tested with the DNA isolated from *Staphylococus aureus*. The DNA was incubated with the carbon quantum dot complex and was irradiated with UV radiation. With increase in exposure time, the fluorescence was increased. The UV irradiation resulted in damage of DNA which was reflected in the increased intensity of fluorescence. A stratospheric probe was fabricated by 3D printing technology using acrylonitrile-butadiene-styrene as polymer. The carbon quantum dots complex was evaluated in the stratospheric conditions by observing the fluorescence intensity at various conditions such as UV intensity, temperature, altitude, humidity and X-ray irradiation which showed increased intensity with intense of these conditions. This showed that the sensor was able to detect the damage to the DNA with the external conditions [104].

An electrochemical sensor with the ability to sense the enzyme activity in the close proximity of biological samples is required in the diagnostic tools for easing the process of diagnosis. A stud was reported which made an attempt to design the electrochemical sensor by 3D printing and was able to detect the enzyme activity in cancerous cell lines. The designed biocompatible chip consists of an electrochemical cell with an Ag/AgCl quasi-reference electrode and two Au electrodes. Polydimethylsiloxane was used as a polymer for casting the electrodes via an SLA 3D printer and the final sensor comprises of the electrode at the lower side toward the sample to be tested and the signal is processed to the opposite side of the chip. The sensor was tested for the detection of alkaline phosphatase enzyme which is secreted from cancer cells and is used as a biomarker for the detection of cancer cells from the normal cells. The results reflected the ability of the sensor to detect the enzyme activity which showed the practical feasibility in biosensing the cellular activity in the intact form of cells or tissues [100].

Additive manufacturing is also adopted in the detection of preterm birth biomarkers. A microchip electrophoresis device was fabricated using the SLA 3D printing technique which was able to detect the biomarkers involved in the preterm birth. Three biomarkers related to preterm birth which were proteins and peptides were selected for this study. The device was successful in detecting the biomarkers with a detection range of picomolar to nanomolar level [106]. In a similar study to detect the preterm birth biomarkers by the 3D printing technique, nine biomarkers molecules were selected for the fabrication of the device. In this study, a microfluidic device was designed that has the ability to retain peptides and proteins related to preterm birth and allowed the detection of molecules by fluorescence. SPE was implemented in the design of this device [107]. These two studies shed light on the utilization of 3D printing technology for the early and easy detection of preterm birth.

22.6 Conclusion

The 3D printing technology is widely adopted for biomedical and pharmaceutical applications. Various technologies are being utilized for implementing 3D printing for drug delivery and biomedical devices based on the suitability of the technique. Drug delivery devices that are difficult to fabricate, sustained release systems that are critical to controlling the release profile, biomedical devices that require precise customization and biosensors that require vigilant fabrication can be achieved by the 3D printing technology.

REFERENCES

1. Zahin N, Anwar R, Tewari D, Kabir MT, Sajid A, Mathew B, Uddin MS, Aleya L, Abdel-Daim MM. Nanoparticles and its biomedical applications in health and diseases: special focus on drug delivery. *Environmental Science and Pollution Research*. 2020;27: 19151–68.
2. Mehta M, Subramani K. Nanodiagnostics in microbiology and dentistry. *Emerging Nanotechnologies in Dentistry*: Elsevier; 2012:365–90.
3. Yeong WY, Chua CK, Leong KF, Chandrasekaran M, Lee MW. Indirect fabrication of collagen scaffold based on inkjet printing technique. *Rapid Prototyping Journal*. 2006;12(4):229–37.
4. Hull C. US 4575330 A. Apparatus for production of three-dimensional objects by stereolithography. 1986;11, United States.
5. Butscher A, Bohner M, Doebelin N, Hofmann S, Müller R. New depowdering-friendly designs for three-dimensional printing of calcium phosphate bone substitutes. *Acta Biomaterialia*. 2013;9(11):9149–58.
6. Saunders RE, Gough JE, Derby B. Delivery of human fibroblast cells by piezoelectric drop-on-demand inkjet printing. *Biomaterials*. 2008;29(2):193–203.
7. Xu T, Zhao W, Zhu J-M, Albanna MZ, Yoo JJ, Atala A. Complex heterogeneous tissue constructs containing multiple cell types prepared by inkjet printing technology. *Biomaterials*. 2013;34(1):130–9.
8. Han T, Kundu S, Nag A, Xu Y. 3D printed sensors for biomedical applications: a review. *Sensors*. 2019;19(7):1706.
9. Vithani K, Goyanes A, Jannin V, Basit AW, Gaisford S, Boyd BJ. An overview of 3D printing technologies for soft materials and potential opportunities for lipid-based drug delivery systems. *Pharmaceutical Research*. 2019;36(1):1–20.
10. Melchels FP, Feijen J, Grijpma DW. A review on stereolithography and its applications in biomedical engineering. *Biomaterials*. 2010;31(24):6121–30.
11. Lamichhane S, Bashyal S, Keum T, Noh G, Seo JE, Bastola R, et al. Complex formulations, simple techniques: Can 3D printing technology be the Midas touch in pharmaceutical industry? *Asian Journal of Pharmaceutical Sciences*. 2019;14(5):465–79.
12. Martinez PR, Basit AW, Gaisford S. The history, developments and opportunities of stereolithography. *3D Printing of Pharmaceuticals*: Springer; 2018:55–79.
13. Burke G, Devine DM, Major I. Effect of stereolithography 3D printing on the properties of PEGDMA hydrogels. *Polymers*. 2020;12(9):2015.
14. Arcaute K, Mann BK, Wicker RB. Stereolithography of three-dimensional bioactive poly (ethylene glycol) constructs with encapsulated cells. *Annals of Biomedical Engineering*. 2006;34(9):1429–41.
15. FDA. Generally recognized as safe (GRAS). Available from: https://www.fda.gov/food/ingredientspackaginglabeling/gras/.
16. Wang J, Goyanes A, Gaisford S, Basit AW. Stereolithographic (SLA) 3D printing of oral modified-release dosage forms. *International Journal of Pharmaceutics*. 2016;503(1–2): 207–12.
17. Goyanes A, Det-Amornrat U, Wang J, Basit AW, Gaisford S. 3D scanning and 3D printing as innovative technologies for fabricating personalized topical drug delivery systems. *Journal of Controlled Release*. 2016;234:41–8.
18. Crump SS. Apparatus and method for creating three-dimensional objects. *Google Patents*. 1992.
19. Ilyes K, Crişan A-G, Porfire A, Tomuţă I. Three-dimensional printing by fused deposition modeling (3dp-fdm) in pharmaceutics. *FARMACIA*. 2020;68(4):586–96.
20. Skowyra J, Pietrzak K, Alhnan MA. Fabrication of extended-release patient-tailored prednisolone tablets via fused deposition modelling (FDM) 3D printing. *European Journal of Pharmaceutical Sciences*. 2015;68:11–7.
21. Goyanes A, Buanz AB, Hatton GB, Gaisford S, Basit AW. 3D printing of modified-release aminosalicylate (4-ASA and 5-ASA) tablets. *European Journal of Pharmaceutics and Biopharmaceutics*. 2015;89:157–62.
22. Cui M, Pan H, Su Y, Fang D, Qiao S, Ding P, et al. Opportunities and challenges of three-dimensional printing technology in pharmaceutical formulation development. *Acta Pharmaceutica Sinica B*. 2021;11(8): 2488–2504.
23. Fanous M, Gold S, Muller S, Hirsch S, Ogorka J, Imanidis G. Simplification of fused deposition modeling 3D-printing paradigm: Feasibility of 1-step direct powder printing for immediate release dosage form production. *International Journal of Pharmaceutics*. 2020;578:119124.
24. Water JJ, Bohr A, Boetker J, Aho J, Sandler N, Nielsen HM, et al. Three-dimensional printing of drug-eluting implants: preparation of an antimicrobial polylactide feedstock material. *Journal of Pharmaceutical Sciences*. 2015;104(3):1099–107.
25. Melocchi A, Parietti F, Loreti G, Maroni A, Gazzaniga A, Zema L. 3D printing by fused deposition modeling (FDM) of a swellable/erodible capsular device for oral pulsatile release of drugs. *Journal of Drug Delivery Science and Technology*. 2015;30:360–7.
26. Khorasani M, Edinger M, Raijada D, Bøtker J, Aho J, Rantanen J. Near-infrared chemical imaging (NIR-CI) of 3D printed pharmaceuticals. *International Journal of Pharmaceutics*. 2016;515(1–2):324–30.
27. Goyanes A, Buanz AB, Basit AW, Gaisford S. Fused-filament 3D printing (3DP) for fabrication of tablets. *International Journal of Pharmaceutics*. 2014;476(1–2):88–92.
28. Melocchi A, Parietti F, Maroni A, Foppoli A, Gazzaniga A, Zema L. Hot-melt extruded filaments based on pharmaceutical grade polymers for 3D printing by fused deposition modeling. *International Journal of Pharmaceutics*. 2016;509(1–2):255–63.
29. Jacob S, Nair AB, Patel V, Shah J. 3D printing technologies: Recent development and emerging applications in various drug delivery systems. *AAPS PharmSciTech*. 2020;21(6):1–16.
30. Vaz VM, Kumar L. 3D printing as a promising tool in personalized medicine. *AAPS PharmSciTech*. 2021;22(1):1–20.
31. Charoo NA, Barakh Ali SF, Mohamed EM, Kuttolamadom MA, Ozkan T, Khan MA, et al. Selective laser sintering 3D printing–an overview of the technology and pharmaceutical applications. *Drug Development and Industrial Pharmacy*. 2020;46(6):869–77.

32. Jamróz W, Szafraniec J, Kurek M, Jachowicz R. 3D printing in pharmaceutical and medical applications: recent achievements and challenges. *Pharmaceutical Research*. 2018;35(9):1–22.
33. Olakanmi EO, Cochrane R, Dalgarno K. A review on selective laser sintering/melting (SLS/SLM) of aluminium alloy powders: Processing, microstructure, and properties. *Progress in Materials Science*. 2015;74:401–77.
34. Awad A, Fina F, Goyanes A, Gaisford S, Basit AW. 3D printing: Principles and pharmaceutical applications of selective laser sintering. *International Journal of Pharmaceutics*. 2020;586:119594.
35. El Aita I, Breitkreutz J, Quodbach J. On-demand manufacturing of immediate release levetiracetam tablets using pressure-assisted microsyringe printing. *European Journal of Pharmaceutics and Biopharmaceutics*. 2019;134:29–36.
36. Khaled SA, Burley JC, Alexander MR, Roberts CJ. Desktop 3D printing of controlled release pharmaceutical bilayer tablets. *International Journal of Pharmaceutics*. 2014;461(1–2):105–11.
37. Khaled SA, Burley JC, Alexander MR, Yang J, Roberts CJ. 3D printing of tablets containing multiple drugs with defined release profiles. *International Journal of Pharmaceutics*. 2015;494(2):643–50.
38. Khaled SA, Alexander MR, Irvine DJ, Wildman RD, Wallace MJ, Sharpe S, et al. Extrusion 3D printing of paracetamol tablets from a single formulation with tunable release profiles through control of tablet geometry. *AAPS PharmSciTech*. 2018;19(8):3403–13.
39. Cheng Y, Qin H, Acevedo NC, Jiang X, Shi X. 3D printing of extended-release tablets of theophylline using hydroxypropyl methylcellulose (HPMC) hydrogels. *International Journal of Pharmaceutics*. 2020;591:119983.
40. Lewis JA, Gratson GM. Direct writing in three dimensions. *Materials Today*. 2004;7(7–8):32–9.
41. Zhu M, Li K, Zhu Y, Zhang J, Ye X. 3D-printed hierarchical scaffold for localized isoniazid/rifampin drug delivery and osteoarticular tuberculosis therapy. *Acta Biomaterialia*. 2015;16:145–55.
42. Yuan S, Shen F, Chua CK, Zhou K. Polymeric composites for powder-based additive manufacturing: Materials and applications. *Progress in Polymer Science*. 2019;91:141–68.
43. Gaytan S, Cadena MA, Karim H, Delfin D, Lin Y, Espalin D, et al. Fabrication of barium titanate by binder jetting additive manufacturing technology. *Ceramics International*. 2015;41(5):6610–9.
44. Daly R, Harrington TS, Martin GD, Hutchings IM. Inkjet printing for pharmaceutics: a review of research and manufacturing. *International Journal of Pharmaceutics*. 2015;494(2):554–67.
45. Palo M, Holländer J, Suominen J, Yliruusi J, Sandler N. 3D printed drug delivery devices: perspectives and technical challenges. *Expert Review of Medical Devices*. 2017;14(9):685–96.
46. Azizi Machekposhti S, Mohaved S, Narayan RJ. Inkjet dispensing technologies: recent advances for novel drug discovery. *Expert Opinion on Drug Discovery*. 2019;14(2):101–13.
47. Acosta-Vélez GF, Wu B. 3D pharming: direct printing of personalized pharmaceutical tablets. *Polymer Sciences*. 2016;2(1):11.
48. Alomari M, Mohamed FH, Basit AW, Gaisford S. Personalised dosing: printing a dose of one's own medicine. *International Journal of Pharmaceutics*. 2015;494(2):568–77.
49. Vadodaria S, Mills T. Jetting-based 3D printing of edible materials. *Food Hydrocolloids*. 2020;106:105857.
50. Stansbury JW, Idacavage MJ. 3D printing with polymers: Challenges among expanding options and opportunities. *Dental Materials*. 2016;32(1):54–64.
51. Hornbeck LJ, editor. Digital light processing for high-brightness high-resolution applications. *Projection Displays*: International Society for Optics and Photonics; 1997:III.
52. Da F, Gai S. Flexible three-dimensional measurement technique based on a digital light processing projector. *Applied Optics*. 2008;47(3):377–85.
53. Comina G, Suska A, Filippini D. Autonomous chemical sensing interface for universal cell phone readout. *Angewandte Chemie*. 2015;127(30):8832–6.
54. Economidou SN, Pere CPP, Reid A, Uddin MJ, Windmill JF, Lamprou DA, et al. 3D printed microneedle patches using stereolithography (SLA) for intradermal insulin delivery. *Materials Science and Engineering: C*. 2019;102:743–55.
55. Uddin MJ, Scoutaris N, Economidou SN, Giraud C, Chowdhry BZ, Donnelly RF, et al. 3D printed microneedles for anticancer therapy of skin tumours. *Materials Science and Engineering: C*. 2020;107:110248.
56. Yeung C, Chen S, King B, Lin H, King K, Akhtar F, et al. A 3D-printed microfluidic-enabled hollow microneedle architecture for transdermal drug delivery. *Biomicrofluidics*. 2019;13(6):064125.
57. Asikainen S, van Bochove B, Seppälä JV. Drug-releasing biopolymeric structures manufactured via stereolithography. *Biomedical Physics & Engineering Express*. 2019;5(2):025008.
58. Xenikakis I, Tzimtzimis M, Tsongas K, Andreadis D, Demiri E, Tzetzis D, et al. Fabrication and finite element analysis of stereolithographic 3D printed microneedles for transdermal delivery of model dyes across human skin in vitro. *European Journal of Pharmaceutical Sciences*. 2019;137:104976.
59. Yao W, Li D, Zhao Y, Zhan Z, Jin G, Liang H, et al. 3D printed multi-functional hydrogel microneedles based on high-precision digital light processing. *Micromachines*. 2020;11(1):17.
60. Vaut L, Juszczyk JJ, Kamguyan K, Jensen KE, Tosello G, Boisen A. 3D printing of reservoir devices for Oral drug delivery: from concept to functionality through design improvement for enhanced Mucoadhesion. *ACS Biomaterials Science & Engineering*. 2020;6(4):2478–86.
61. Caudill CL, Perry JL, Tian S, Luft JC, DeSimone JM. Spatially controlled coating of continuous liquid interface production microneedles for transdermal protein delivery. *Journal of Controlled Release*. 2018;284:122–32.
62. Janusziewicz R, Mecham SJ, Olson KR, Benhabbour SR. Design and characterization of a novel series of geometrically complex intravaginal rings with digital light synthesis. *Advanced Materials Technologies*. 2020;5(8):2000261.
63. Do A-V, Worthington KS, Tucker BA, Salem AK. Controlled drug delivery from 3D printed two-photon polymerized poly (ethylene glycol) dimethacrylate devices. *International Journal of Pharmaceutics*. 2018;552(1–2):217–24.

64. Bozuyuk U, Yasa O, Yasa IC, Ceylan H, Kizilel S, Sitti M. Light-triggered drug release from 3D-printed magnetic chitosan microswimmers. *ACS Nano.* 2018;12(9): 9617–25.
65. Lim SH, Kathuria H, Amir MHB, Zhang X, Duong HT, Ho PC-L, et al. High resolution photopolymer for 3D printing of personalised microneedle for transdermal delivery of anti-wrinkle small peptide. *Journal of Controlled Release.* 2021;329:907–18.
66. Cordeiro AS, Tekko IA, Jomaa MH, Vora L, McAlister E, Volpe-Zanutto F, et al. Two-photon polymerisation 3D printing of microneedle array templates with versatile designs: Application in the development of polymeric drug delivery systems. *Pharmaceutical Research.* 2020;37(9): 1–15.
67. Krieger KJ, Bertollo N, Dangol M, Sheridan JT, Lowery MM, O'Cearbhaill ED. Simple and customizable method for fabrication of high-aspect ratio microneedle molds using low-cost 3D printing. *Microsystems & Nanoengineering.* 2019;5(1):1–14.
68. Luzuriaga MA, Berry DR, Reagan JC, Smaldone RA, Gassensmith JJ. Biodegradable 3D printed polymer microneedles for transdermal drug delivery. *Lab on a Chip.* 2018;18(8):1223–30.
69. Fu J, Yu X, Jin Y. 3D printing of vaginal rings with personalized shapes for controlled release of progesterone. *International Journal of Pharmaceutics.* 2018;539(1–2): 75–82.
70. Tiboni M, Campana R, Frangipani E, Casettari L. 3D printed clotrimazole intravaginal ring for the treatment of recurrent vaginal candidiasis. *International Journal of Pharmaceutics.* 2021;596:120290.
71. Xu X, Goyanes A, Trenfield SJ, Diaz-Gomez L, Alvarez-Lorenzo C, Gaisford S, et al. Stereolithography (SLA) 3D printing of a bladder device for intravesical drug delivery. *Materials Science and Engineering: C.* 2021;120: 111773.
72. Dalvand K, Ghiasvand A, Gupta V, Paull B. Chemotaxis-based smart drug delivery of epirubicin using a 3D printed microfluidic chip. *Journal of Chromatography B.* 2021;1162:122456.
73. Cotabarren I, Gallo L. 3D printing of PVA capsular devices for modified drug delivery: design and in vitro dissolution studies. *Drug Development and Industrial Pharmacy.* 2020;46(9):1416–26.
74. Forouzandeh F, Ahamed NN, Hsu M-C, Walton JP, Frisina RD, Borkholder DA. A 3D-printed modular microreservoir for drug delivery. *Micromachines.* 2020;11(7):648.
75. Ko ES, Kim C, Choi Y, Lee KY. 3D printing of self-healing ferrogel prepared from glycol chitosan, oxidized hyaluronate, and iron oxide nanoparticles. *Carbohydrate Polymers.* 2020;245:116496.
76. Shi K, Aviles-Espinosa R, Rendon-Morales E, Woodbine L, Maniruzzaman M, Nokhodchi A. Novel 3D printed device with integrated macroscale magnetic field triggerable anticancer drug delivery system. *Colloids and Surfaces B: Biointerfaces.* 2020;192:111068.
77. Wei X, Liu C, Wang Z, Luo Y. 3D printed core-shell hydrogel fiber scaffolds with NIR-triggered drug release for localized therapy of breast cancer. *International Journal of Pharmaceutics.* 2020;580:119219.
78. Yu W, Jiang G, Zhang Y, Liu D, Xu B, Zhou J. Polymer microneedles fabricated from alginate and hyaluronate for transdermal delivery of insulin. *Materials Science and Engineering: C.* 2017;80:187–96.
79. Nagarkar R, Singh M, Nguyen HX, Jonnalagadda S. A review of recent advances in microneedle technology for transdermal drug delivery. *Journal of Drug Delivery Science and Technology.* 2020:101923.
80. Dolatkhah M, Hashemzadeh N, Barar J, Adibkia K, Aghanejad A, Barzegar-Jalali M, et al. Graphene-based multifunctional nanosystems for simultaneous detection and treatment of breast cancer. *Colloids and Surfaces B: Biointerfaces.* 2020:111104.
81. Bittner B, Richter W, Schmidt J. Subcutaneous administration of biotherapeutics: an overview of current challenges and opportunities. *BioDrugs.* 2018;32(5):425–40.
82. Karnik R, Gu F, Basto P, Cannizzaro C, Dean L, Kyei-Manu W, et al. Microfluidic platform for controlled synthesis of polymeric nanoparticles. *Nano Letters.* 2008;8(9):2906–12.
83. Au AK, Lee W, Folch A. Mail-order microfluidics: evaluation of stereolithography for the production of microfluidic devices. *Lab on a Chip.* 2014;14(7):1294–301.
84. Thurman AR, Clark MR, Hurlburt JA, Doncel GF. Intravaginal rings as delivery systems for microbicides and multipurpose prevention technologies. *International Journal of Women's Health.* 2013;5:695.
85. Chen Y. *Development and evaluation of novel intravaginal rings fabricated via hot-melt extrusion-based technologies as innovative microbicides.* Doctoral dissertation, University of Manitoba, 2014.
86. Ranganathan SI, Kohama C, Mercurio T, Salvatore A, Benmassaoud MM, Kim TWB. Effect of temperature and ultraviolet light on the bacterial kill effectiveness of antibiotic-infused 3D printed implants. *Biomedical Microdevices.* 2020;22(3):1–14.
87. Wang Y, Sun L, Mei Z, Zhang F, He M, Fletcher C, et al. 3D printed biodegradable implants as an individualized drug delivery system for local chemotherapy of osteosarcoma. *Materials & Design.* 2020;186:108336.
88. Weisman JA, Ballard DH, Jammalamadaka U, Tappa K, Sumerel J, D'Agostino HB, et al. 3D printed antibiotic and chemotherapeutic eluting catheters for potential use in interventional radiology: in vitro proof of concept study. *Academic Radiology.* 2019;26(2):270–4.
89. Misra SK, Ostadhossein F, Babu R, Kus J, Tankasala D, Sutrisno A, et al. 3D-printed multidrug-eluting stent from graphene-nanoplatelet-doped biodegradable polymer composite. *Advanced Healthcare Materials.* 2017;6(11):1700008.
90. Guerra AJ, Cano P, Rabionet M, Puig T, Ciurana J. 3D-printed PCL/PLA composite stents: Towards a new solution to cardiovascular problems. *Materials.* 2018;11(9):1679.
91. Ballard DH, Jammalamadaka U, Tappa K, Weisman JA, Boyer CJ, Alexander JS, et al. 3D printing of surgical hernia meshes impregnated with contrast agents: in vitro proof of concept with imaging characteristics on computed tomography. *3D Printing in Medicine.* 2018;4(1):1–6.
92. Fouladian P, Kohlhagen J, Arafat M, Afinjuomo F, Workman N, Abuhelwa AY, et al. Three-dimensional printed 5-fluorouracil eluting polyurethane stents for the treatment of oesophageal cancers. *Biomaterials Science.* 2020;8(23):6625–36.

93. Feuerbach T, Kock S, Thommes M. Slicing parameter optimization for 3D printing of biodegradable drug-eluting tracheal stents. *Pharmaceutical Development and Technology*. 2020;25(6):650–8.
94. Arany P, Papp I, Zichar M, Csontos M, Elek J, Regdon G, et al. In vitro tests of FDM 3D-printed diclofenac sodium-containing implants. *Molecules*. 2020;25(24):5889.
95. Stewart SA, Domínguez-Robles J, McIlorum VJ, Mancuso E, Lamprou DA, Donnelly RF, et al. Development of a biodegradable subcutaneous implant for prolonged drug delivery using 3D printing. *Pharmaceutics*. 2020;12(2):105.
96. Park SA, Lee SJ, Lim KS, Bae IH, Lee JH, Kim WD, et al. In vivo evaluation and characterization of a bio-absorbable drug-coated stent fabricated using a 3D-printing system. *Materials Letters*. 2015;141:355–8.
97. Guerra A, Roca A, de Ciurana J. A novel 3D additive manufacturing machine to biodegradable stents. *Procedia Manufacturing*. 2017;13:718–23.
98. Guerra AJ, Ciurana J. 3D-printed bioabsordable polycaprolactone stent: The effect of process parameters on its physical features. *Materials & Design*. 2018;137:430–7.
99. Benmassaoud MM, Kohama C, Kim TWB, Kadlowec JA, Foltiny B, Mercurio T, et al. Efficacy of eluted antibiotics through 3D printed femoral implants. *Biomedical Microdevices*. 2019;21(3):1–10.
100. Ragones H, Schreiber D, Inberg A, Berkh O, Kósa G, Freeman A, et al. Disposable electrochemical sensor prepared using 3D printing for cell and tissue diagnostics. *Sensors and Actuators B: Chemical*. 2015;216:434–42.
101. Kadimisetty K, Mosa IM, Malla S, Satterwhite-Warden JE, Kuhns TM, Faria RC, et al. 3D-printed supercapacitor-powered electrochemiluminescent protein immunoarray. *Biosensors and Bioelectronics*. 2016;77:188–93.
102. Heger Z, Zitka J, Cernei N, Krizkova S, Sztalmachova M, Kopel P, et al. 3D-printed biosensor with poly (dimethylsiloxane) reservoir for magnetic separation and quantum dots-based immunolabeling of metallothionein. *Electrophoresis*. 2015;36(11–12):1256–64.
103. Dirkzwager RM, Liang S, Tanner JA. Development of aptamer-based point-of-care diagnostic devices for malaria using three-dimensional printing rapid prototyping. *ACS Sensors*. 2016;1(4):420–6.
104. Heger Z, Zitka J, Nejdl L, Moulick A, Milosavljevic V, Kopel P, et al. 3D printed stratospheric probe as a platform for determination of DNA damage based on carbon quantum dots/DNA complex fluorescence increase. *Monatshefte für Chemie-Chemical Monthly*. 2016;147(5):873–80.
105. Tang C, Vaze A, Rusling J. Automated 3D-printed unibody immunoarray for chemiluminescence detection of cancer biomarker proteins. *Lab on a Chip*. 2017;17(3):484–9.
106. Beauchamp MJ, Nielsen AV, Gong H, Nordin GP, Woolley AT. 3D printed microfluidic devices for microchip electrophoresis of preterm birth biomarkers. *Analytical Chemistry*. 2019;91(11):7418–25.
107. Bickham AV, Pang C, George BQ, Topham DJ, Nielsen JB, Nordin GP, et al. 3D printed microfluidic devices for solid-phase extraction and on-chip fluorescent labeling of preterm birth risk biomarkers. *Analytical Chemistry*. 2020;92(18):12322–9.
108. Sharafeldin M, Kadimisetty K, Bhalerao KS, Chen T, Rusling JF. 3D-printed Immunosensor arrays for cancer diagnostics. *Sensors*. 2020;20(16):4514.
109. Chiadò A, Palmara G, Chiappone A, Tanzanu C, Pirri CF, Roppolo I, et al. A modular 3D printed lab-on-a-chip for early cancer detection. *Lab on a Chip*. 2020;20(3):665–74.
110. Plevniak K, Campbell M, Myers T, Hodges A, He M. 3D printed auto-mixing chip enables rapid smartphone diagnosis of anemia. *Biomicrofluidics*. 2016;10(5):054113.
111. Krejcova L, Nejdl L, Rodrigo MAM, Zurek M, Matousek M, Hynek D, et al. 3D printed chip for electrochemical detection of influenza virus labeled with CdS quantum dots. *Biosensors and Bioelectronics*. 2014;54:421–7.
112. Harb SC, Xu B, Klatte R, Griffin BP, Rodriguez LL. Haemodynamic assessment of severe aortic stenosis using a three-dimensional (3D) printed model incorporating a flow circuit. *Heart, Lung and Circulation*. 2018;27(11):e105–e7.

23

Nanofluids: Basic Information on Preparation, Stability, and Applications

Rajyalakshmi Ch, Naresh Kumar Katari, K. Ganesh Kadiyala and G. Ramaswamy

CONTENTS

DOI: 10.1201/9781003157823-23

23.1 Introduction

23.1.1 Nanofluids

With the evolution of the microelectronics industry, high-heat flux devices have begun to be made, but the dissipation of heat is a significant issue for widespread usage in industry as well as in everyday life because heat accumulation causes electronic chips to slow down or even fail completely [1]. However, traditional coolants (water, oils, and ethylene glycols) have been shown to be ineffective due to their low thermal conductivity which shows poor heat dissipation.

As a result, J. C. Maxwell recommended in 1873 that extremely fine solid particles be added to base fluids to increase thermal conductivity and their heat dissipation capacity. Small solid particles have a higher thermal conductivity than the base fluid, resulting in an increase in the base fluid's overall heat dissipation capacity and thermal conductivity [2]. Although it is well known that micro- and millimeter-sized particles increase the thermal conductivity of base fluids, experiments have revealed that in addition to increased thermal conductivity, other issues such as abrasive wear of pipelines, blockage of channels, and sedimentation of particles limit their use in the micro industry. Later, nano-sized particles were produced to avoid these problems; with the growth of the nanoscale industry and nanotechnology, they acquired momentum for application in research [3]. Nanoparticles are minuscule powdered particles with a diameter of less than 100 nanometers. Masuda et al. employed ultrafine particles to improve liquid thermal conductivity in 1993 [4]. S. U. S. Choi proposed using nanoparticles in base fluids to boost thermal conductivity later in 1995 [5]. Al_2O_3, CuO, SiO_2, TiO_2, Cu, Ni, Al, and ZnO are examples of metallic and non-metallic particles [6].

Nanofluids are a special type of fluids prepared by dispersing nanometer-sized and high-conductivity materials (nanoparticles, nanofibers, nanotubes, nanowires, nanorods, nano-sheets, or droplets) in a base fluid, generally oil or water [6]. Nanofluids, in other terms, are nanoscale colloidal solutions that contain condensed nanomaterials. They are two-phase systems with a solid phase (one phase) and a liquid phase (second phase).

Nanofluids possess better thermophysical properties like viscosity, thermal diffusivity, thermal conductivity, and heat transfer coefficients when compared to base fluids, generally oil or water. It has been shown that they have a wide range of remarkable applications. Nanofluid suspensions have a higher heat transfer surface than colloidal solutions, giving them an advantage over colloidal solutions. Nanofluids are used in heat transfer intensification, electronics, transportation, heating buildings, industrial cooling, decreasing pollution, energy storage, nuclear cooling, solar absorption, magnetic sealing, biomedical use, nano-drug delivery, friction reduction, and other fields [7].

However, even with improved properties, researchers have a big difficulty in two-phase system stability [1]. We shall explore nanofluid synthesis methods, stability methodologies, and applications in this study. The stability of nanofluids is one of the most critical challenges and achieving the appropriate stability of nanofluids remains a major challenge. We will examine recent works in the methodologies for the synthesis of stable nanofluids and outline the stability mechanisms in this work. Nanofluids have got a lot of interest in recent years. A wide range of applications is the main driving force behind nanofluids research.

23.1.2 Preparation of Nanofluids

Preparation of the nanofluid is one of the significant steps that uses nanoparticles which helps to improve the thermal conductivity of the base fluid used. In producing the high-quality nanofluid, the distribution of the nanoparticles in a consistent way and appending the nanomaterials in the base fluid are difficult. Two general methods are being followed to prepare the nanofluids – the one-step method and two-step methods. Depending on the need for the nanofluids' composition, many methods like one-step and two-step methods have been developed to prepare the nanofluids. These one-step and two-step processes can be briefly elucidated below.

23.1.2.1 Two-Step Method

The two-step method is the utmost widely used standard method for the preparation of nanofluids. The required nanoparticles, nanofibers, nanotubes, and other nanomaterials that will be prepared in this method are first produced as dry powders by either chemical or physical processes. With the help of techniques like ultrasonic agitation, intensive magnetic force agitation, ball milling, homogenizing, and high-shear mixing, the nano-sized powder will be spread into a nanofluid in the second step.

This method is the most economical method to prepare nanofluids on a large scale since synthesis techniques of nanopowder have already been scrambled up to the level of industrial production. Owing to the huge surface activity and high surface area, the synthesized nanoparticles have the property to accumulate. Using surfactants is one of the essential techniques to enhance the stability of the nanoparticles in fluids. However, the surfactants' working at high temperatures is also a significant concern, particularly for high-temperature applications. Due to the difficulty in synthesizing the stable nanofluids by the two-step method (Figure 23.1), numerous advanced techniques are developed to produce nanofluids, together with the one-step process.

23.1.2.2 One-Step Method

Eastman et al. established a one-step method called the physical vapor condensation method for the preparation of Cu/ethylene glycol nanofluids to diminish the accumulation of nanoparticles [8]. In the one-step process, simultaneously manufacturing and dispersing particles into the fluid occurs. Here there is no involvement of the processes of drying, storage, transportation, and dispersion of nanoparticles. Therefore, the aggregation of nanoparticles can be minimized, and the stability of nanofluids will be increased [9]. Using the one-step processes (Figure 23.2), uniformly dispersed nanoparticles and the particles that are stably suspended in the base fluid can be prepared.

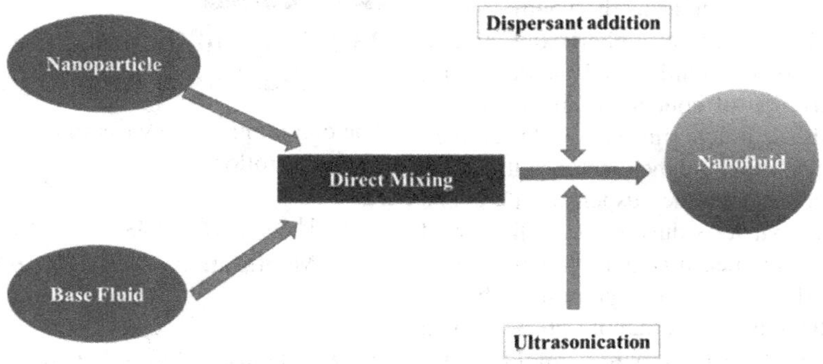

FIGURE 23.1 Method of preparation of nanofluids by a two-step method.

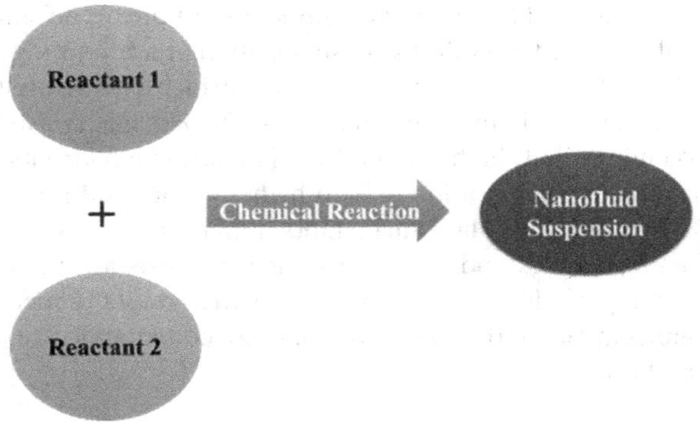

FIGURE 23.2 Method of preparation of nanofluids by a one-step method.

Another efficient method called vacuum-submerged arc nanoparticle synthesis system (SANSS) is used to prepare nanofluids using different dielectric liquids [10, 11]. The nanoparticles produced in this method look like needle, polygonal, square, and circular shapes. This method evades the unsought particle accumulation justly skillfully.

Zhu et al. published an innovative one-step chemical method to prepare copper nanofluids by reducing $CuSO_4$. $5H_2O$ by using $NaH_2PO_2.H_2O$ in the solvent of ethylene glycol under microwave irradiation [12]. Mineral oil-based nanofluids comprising silver nanoparticles with a narrow-size delivery can also be prepared by using this method [13]. The nanoparticles can be stabilized by Korantin, which is going to be coordinated to the surface of the silver particle via two oxygen atoms by making a dense layer surrounding the nanoparticles. The silver nanoparticles suspension thus formed by this method was stable for up to one month. By using this method, stable ethanol-related nanofluids comprising silver nanoparticles can be prepared through a microwave-assisted method [14].

A phase-transfer method has been established for the preparation of homogeneous and stable graphene oxide colloids. GONs (graphene oxide nano-sheets) were successfully transported from aqueous phase (water) to n-octane by alteration with oleylamine.

However, there are some disadvantages while using the one-step method to prepare nanofluids. The utmost significant one is that the reactants are left in the solution of nanofluids because of the unfinished reaction or less stabilization. Then, it is very tough to illuminate the nanoparticle effect due to lacking the removal of the formed impurity.

Wei et al. developed a method called continuous flow microfluidic micro-reactor to prepare copper nanofluids. This method is suitable for continuously synthesizing copper nanofluids, and the formed nanofluid microstructure and properties will be varied by altering a few parameters like flow rate, concentration of the reactants, and additive. Copper oxide nanofluids having a more solid volume portion (up to 10 vol%) can be prepared through a new precursor transformation technique by using ultrasonic and irradiation with microwave [15]. The precursor cupric hydroxide is completely changed to copper oxide (CuO) nanoparticles in the water by microwave irradiation. The growth aggregation of the nanoparticles can be prevented by using ammonium citrate results in the formation of the CuO aqueous nanofluid with a more excellent thermal conductivity than those prepared by using other dispersing methods.

There is another method to prepare nanofluids, i.e., the phase-transfer method. It is a facile way to get monodisperse decent metal colloids [16]. In a two-phase system (water cyclohexane), aq. formaldehyde is relocated to the cyclohexane phase with the help of dodecylamine. The intermediates formed during the process can reduce the silver or gold ions in an aqueous solution to prepare dodecyl amine-protected silver or gold nanoparticles in the cyclohexane solution at room temperature. This method is also applicable to prepare stable kerosene-based Fe_3O_4 nanofluids [17]. The current research studies show that nanofluids that are synthesized by the chemical solution method have higher conductivity enhancement and better stability than those prepared by the other techniques [18].

23.2 Stability Evaluation of Nanofluids

The methods involved in stability evaluation of the nanofluids are:

- Sedimentation and centrifugation method
- Zeta potential analysis
- Spectral absorbency analysis

23.2.1 Sedimentation and Centrifugation Methods

The synthesized nanofluids' stability can be found by using various methods. The simplest way among all is the sedimentation method [19, 20]. The sediment's weight or the volume of the nanoparticles of the nanofluid under the applied external field is assigned to the stability of the synthesized nanofluid. By special apparatus, the variation of concentration or the particle size of supernatant with sediment time can be determined [9]. Hu et al. used a sedimentation balance method. It is used to measure the stability of the graphite suspension [21]. The weight of the sediment nanoparticles during a specific period was calculated. Singh et al. applied a centrifugation method to detect the stability of silver nanofluids synthesized by the microwave synthesis in the solvent of ethanol with PVP as a stabilizing agent. It is found that the prepared nanofluids are stable for up to one month in a stationary state and over 10 h under the centrifugation (3,000 rpm) without sedimentation. Li prepared an aqueous polyaniline colloidal nanofluids, and the stability of the obtained colloids was evaluated by using the centrifugation method [22].

23.2.2 Zeta Potential Analysis

The zeta potential is the electric potential, and it displays the potential difference between the dispersion medium and stationary layer of nanofluid involved in the dispersed particle. The zeta potential value can be connected to the stability of colloidal distributions. Therefore, colloids with more zeta potential value are electrically stabilized, whereas colloids with lower zeta potential tend to flocculate or coagulate. The colloids with a zeta potential of 40–60 mV are supposed to be good stable, and the colloids with greater than 60 mV have outstanding stability. Kimet al. synthesized gold nanofluids with high stability even after one month [23]. The stability of the prepared gold nanofluids is owing to the high negative zeta potential of the prepared gold nanoparticles in the water. Zeta potential quantities were hired to study the absorption mechanisms of the surfactants on the surfaces of the multi-wall nanotube (MWNT) by using FTIR (Fourier transformation infrared spectra).

23.2.3 Spectral Absorbency Analysis

This is another well-organized way to assess the stability of the synthesized nanofluids. Overall, there is a linear connection between the intensity of the absorbency and the concentration of nanoparticles in the fluid. Huang et al. assessed the dispersal characteristics of Al2O3 (alumina) and the copper suspensions through the predictable sedimentation method by using absorbency analysis with a spectrophotometer after the recesses are deposited for 24 h [24]. The stability assessment of the FePt colloidal nanoparticle systems is evaluated through spectrophotometer analysis [25]. The deposit kinetics might also be calculated by probing the absorbency of a particle in the solution [26]. Hwang et al. calculated the stability of nanofluids with the help of a UV-vis spectrophotometer. It was whispered that the stability of nanofluids was strappingly affected by the features of the suspended particles and the particle morphology [27]. Besides, the addition of the surfactant can improve the stability of the colloidal suspensions.

23.3 Ways to Enhance the Stability of Nanofluids

The commonly used ways to improve the stability of nanofluids are as follows:

- Using surfactants in nanofluids
- Modification of the surface: Surfactant-free method

23.3.1 Using of Surfactants in Nanofluids

Surfactants that are used in nanofluids are called dispersants. The addition of dispersants to two-phase systems is an easy method and more economical to improve the stability of prepared nanofluids. Dispersants that are nothing but surfactants can noticeably mark the surface characteristics of the nanofluids in small quantities. Dispersants consist of two ends called the head and tail. The tail is a hydrophobic portion containing a long-chain hydrocarbon, and the head is a hydrophilic polar. Surfactants attach onto the surfaces of nanoparticles and enlarge the thermal resistance among the nanoparticles and the base fluid, which might reduce the enhancement of the effective thermal conductivity of the nanofluids.

23.4 Advantages of Nanofluids

Nanofluids have the following advantages, making them appropriate for a variety of applications [28].

i) The suspended nanoparticles improve thermal conductivity, resulting in increased heat transfer system efficiency.
ii) By altering the size, shape, substance, and volume fraction of nanoparticles, solar energy absorption will be optimized.
iii) Heat is transferred to a small region of fluid by heating within the fluid volume, allowing the peak temperature to be positioned away from surfaces that lose heat to the outside.
iv) Due to their small particle size, suspended nanoparticles increase the surface area and heat capacity of the fluid.
v) The fluid's mixing fluctuation and turbulence are intensified.
vi) Nanoparticle dispersion flattens the fluid's transverse temperature gradient.
vii) The properties of a fluid can be modified to suit different applications by altering the proportion of nanoparticles.
viii) Nanofluids increase the solar thermal application temperatures.

23.5 Applications of Nanofluids

The applications of nanofluids in heat transfer enhancement will be summarized as shown in Figure 23.3.

23.5.1 Heat Transfer Intensification

23.5.1.1 Electronic Applications

Dissipation of heat becomes more difficult due to poor design of electronic components and greater density of chips. Advanced and superior current electronic devices face thermal management challenges from the high level of heat generation and the reduction of available surface area for heat removal. So, as a result, for the smooth operation of modern electrical equipment, a reliable heat management system is essential [29].

Heat can be disposed of in two ways: by streamlining the cooling system's design and by utilizing liquids with high heat-dispersing limit. Nanofluids replace conventional coolants because they have a better heat-carrying capability.

Many studies have shown that nanofluids have a higher thermal conductivity than basic fluids. Convective heat transfer coefficients are predicted for nanofluids with better thermal conductivities than base fluids. Recent studies have shown that boosting the thermal conductivity of a coolant with nanofluids can increase the heat transfer coefficient. Jang and Choi [30] created a novel cooler that incorporates a microchannel heat sink and nanofluids. When compared with a gadget that utilized a pure form of water as the functioning medium, the cooling execution was improved. Thermal resistance and temperature difference between the warmed microchannel divider and the coolant were both brought down using these nanofluids.

The coming generation cooling gadgets for taking out ultra-high heat transition could be a consolidated microchannel heat sink with nanofluids. By substituting the base fluid, generally distilled water, with a nanofluid composed of distilled water and alumina nanoparticles at varied concentrations [31], Nguyen et al. had the option to concentrate on the heat transfer improvement of the fluid cooling system. The addition of nanoparticles to distilled water resulted in a significant increase in the cooling block's convective heat transfer coefficient, according to measured data. An increase in particle concentration has also resulted in a significant reduction in the junction temperature between the heated component and the cooling block. The performance of silicon microchannel heat sinks using nanofluids containing Cu nanoparticles was investigated [32]. When contrasted with a pure form of water as a coolant, it was found that nanofluids could further develop their potential performance. The rise was attributable to the coolant's increased thermal conductivity and the thermal dispersion effect of nanoparticles. Because the nanoparticle was small and the particle volume fraction was low, there was no additional pressure loss.

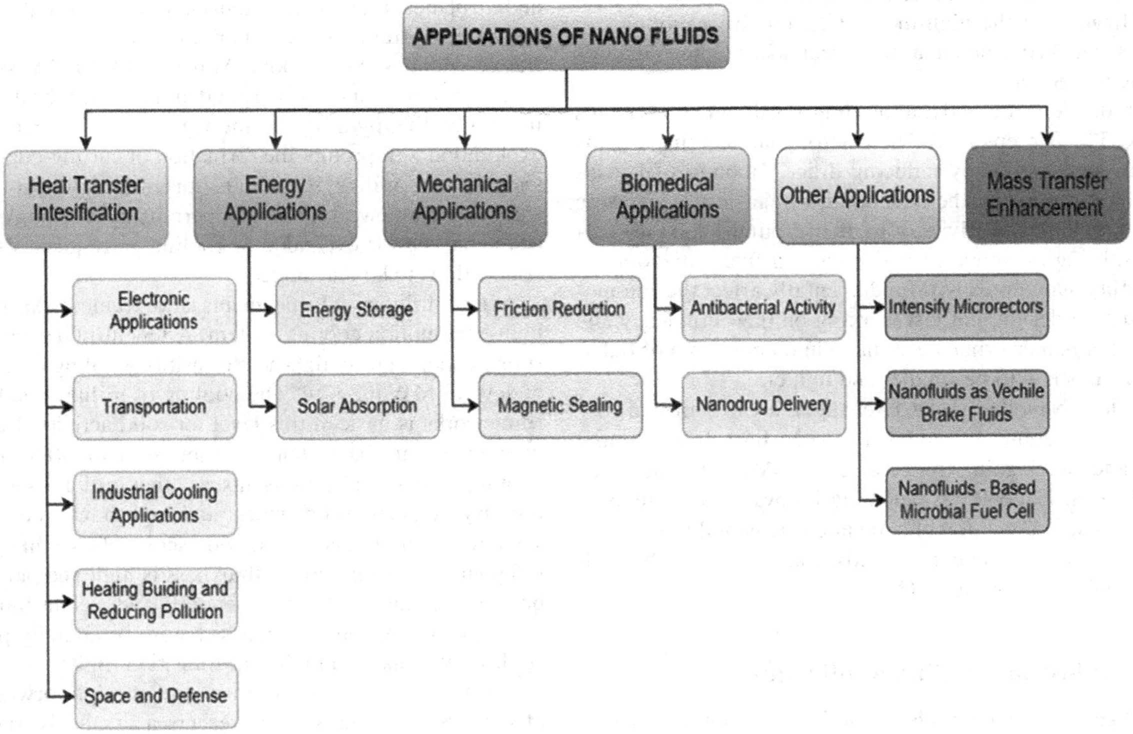

FIGURE 23.3 Summary of applications of nanofluids.

23.5.2 Transportation

The nanofluids actually have the capability to improve heavy-duty engine cooling systems and heat which is generated by the elements of automobiles. The nanofluids actually boost the heat-carrying rate of cooling jackets, lowering their weight and simplifying their design. This allows for more compact designs with lighter radiators for the same horsepower. It is advantageous in terms of strong performance and fuel economy. Because of their low-pressure operation, ethylene glycol-based nanofluids have got a lot of attention as an engine coolant with a 50–50 mixture of ethylene glycol and water, which is the most widely used automotive coolant [33]. Since the nanofluids have a high limit, they can be used to raise the run of the temperature of coolant working prior to dismissing additional hotness through the current coolant [34]. Kole et al. assessed the thermal conductivity and consistency of a motor coolant (alumina-nanofluid) produced using an ordinary motor coolant as the base liquid [35]. The produced nanofluid, which included simply 3.5% A12O3 nanoparticles by volume, had a more prominent thermal conductivity than the base liquid.

Tzeng et al. [36] applied nanofluids to the cooling of automatic transmissions. The experimental platform was a four-wheel-drive vehicle's transmission. CuO and Al_2O_3 nanoparticles were dispersed in engine transmission oil to make the nanofluids. CuO nanofluids provided the lowest transmission temperatures at both high and low rotating speeds, according to the findings. The use of nanofluid in the transmission has a significant advantage in terms of thermal performance. The use of nanofluids in transportation has been evaluated by Argonne National Laboratory researchers [37]. The utilization of the high-thermal conductive nanofluids in radiators can bring about a 10% decrease in the radiator's front-facing region.

The reduction in aerodynamic drag results in fuel savings of up to 5%. It prepares for new aerodynamic vehicle plans that limit emanations by reducing drag. Nanofluids likewise assisted with reducing the friction and wear, just as parasitic misfortunes and the activity of parts like pumps and the compressors, bringing about a 6% decrease in fuel utilization. In all actuality, nanofluids will fundamentally affect the automobile structure design, not just working on their efficiency and their economic performance. A nanofluid cooled motor radiator, for instance, will be smaller and lighter.

NanoHex (Nanofluid Heat Exchange), a C8.3 million FP7 project, has begun. It united 12 gatherings from around Europe and Israel, going from colleges to SMEs and enormous organizations. NanoHex is defeating innovative deterrents in the creation and execution of solid and safe nanofluids for further development of energy-effective, and biologically well-disposed labor and products [38].

23.5.3 Industrial Cooling Applications

The utilization of nanofluids in modern cooling will save a lot of energy and diminish pollutants. Substitution of cooling and warming water with nanofluids can possibly save one trillion Btu of energy [39] for US industry. The exhibition of polyalphaolefin nanofluids containing peeled graphite nanoparticle filaments in cooling was examined using a stream circle framework [33]. It was found that the particular fieriness of nanofluids was 50% higher than that of polyalphaolefin, and that it expanded with temperature. The thermal diffusivity of nanofluids was found to be four times more noteworthy. When nanofluids were utilized rather than polyalphaolefin, convective hotness move was improved by 10%.

23.5.4 Heating Buildings and Reducing Pollution

Building warming frameworks can profit from nanofluids. Kulkarni et al. explored how well they work with regards to warming structures in cool environments [40]. In bone-chilling environments, ethylene or propylene glycol joined with water in different sums is regularly utilized as a heat transfer liquid. The base liquid was picked as 60:40 ethylene glycol/water by its weight. The discoveries uncovered that utilizing nanofluids in heat exchangers decreased volumetric and mass stream rates, bringing about a decrease in general pumping power. Nanofluids need more modest warming frameworks that can deliver similar measures of nuclear power as greater warming frameworks while costing less. This decreases the fundamental gear cost, barring the expense of nanofluid. More modest warming units devour less energy, and a heat transfer move unit has less fluid and material waste to discard toward the finish of its life cycle, which decreases contamination.

23.5.5 Space and Defense

Since space, energy, and weight are restricted in space stations and airplanes, there is an extensive interest in high-effectiveness cooling frameworks that are lower in size. When we contrasted with base liquid alone, You et al. [39] and Vassalo et al. discovered that nanofluids expanded the critical heat transition in pool boiling by a significant degree [41]. Further nanofluid exploration will prompt the formation of cutting-edge cooling gadgets that utilize nanofluids for ultrahigh-heat-transition electronic frameworks, possibly permitting chip power in electronic parts to be expanded or cooling prerequisites for space applications to be rearranged.

A few military advancements and frameworks require a high-heat motion cooling system. A few military gadgets and frameworks require high-heat transition cooling to the degree of several MW/m2 [29]. The cooling of military gadgets and frameworks is basic at this level for solid activity. Nanofluids with enormous basic hotness motions can offer important cooling in such applications, just as other military systems like military vehicles, submarines, and high-power laser diodes. Therefore, nanofluids have a wide scope of uses in space and safeguarding, where power thickness is high and parts should be minimal and light. Space, energy, and weight imperatives on space stations and airplanes require profoundly proficient cooling systems with high heat-transition limits.

Because of the conservative and straightforward plans of hotness exchangers, this presents a dream for the lighter cooling systems, making space travel considerably more reasonable than it is currently. Nanofluids' high basic heat flux capacity limit makes them reasonable for military applications like submarines and high-power diode lasers. Subsequently, nanofluids can possibly be utilized in regions where parts should be lower in weight and force thickness is higher.

23.5.5.1 Nuclear Cooling Systems

The Massachusetts Institute of Technology is researching the usage of nanofluids in nuclear science. They are primarily concerned with these three issues:

i) Pressure water reactors for the main reactors.
ii) Coolant for core-cooling in an emergency.
iii) Coolant for molten core retention in a high-powered density of light water reactors after catastrophic accidents [42].

23.5.6 Energy Applications

Two outstanding qualities of nanofluids are used in energy applications: one is the greater thermal conductivities of nanofluids, which improve heat transmission, and the other is the absorption properties of nanofluids.

23.5.6.1 Energy Storage

The difference between energy sources and demands necessitated the construction of a storage mechanism. With the emphasis on efficient use and conservation of waste heat and solar energy in industry and buildings, thermal energy storage in the form of sensible and latent heat has already become a significant part of the management of energy. One of the most efficient methods of storing thermal energy is latent heat storage [43]. Wu et al. looked into the possibility of alumina-water nanofluids as a new phase change material for cooling system thermal energy storage.

The thermal response test revealed that adding alumina nanoparticles to water significantly lowered the supercooling degree, accelerated the initial freezing time, and decreased the total freezing time. The overall freezing period of alumina-water nanofluids may be lowered by 20% by adding 0.2 weight% alumina nanoparticles. By floating a small number of titanium dioxide nanoparticles in saturated barium chloride aqueous solution, Liu et al. discovered a new kind of phase change materials of nanofluids [44].

23.5.6.2 Solar Absorption

Solar energy is the best non-conventional energy source that is abundant. Solar collectors with direct absorption are a well-known technology. However, the efficiency is low due to the collection fluid's low absorption capacity. This innovation has as of late been joined with nanofluidic innovation. Otanicar et al. led studies on direct assimilation sun-based authorities utilizing nanofluids as the functioning liquid and tracked down a 5% expansion in sunlight-based warm gatherer productivity.

Solar energy is one of the most environmentally friendly types of renewable energy. The conventional direct absorption of sun-powered technology is a grounded innovation that has been proposed for an assortment of utilizations, including water warming; nonetheless, the effectiveness of these gatherers is restricted by the functioning liquid's retention properties, which are poor for average sun-oriented liquids. This technology has recently been integrated with emerging nanofluid and liquid-nanoparticle suspension technologies to generate a new class of nanofluid-based solar collectors.

Experimental results on solar collectors based on nanofluids generated from a variety of nanoparticles (nanotubes, graphite, and silver) were published by Otanicar et al. [45]. The use of nanofluids as the absorption media improved the efficiency of solar thermal collectors by up to 5%. They also compared the results of the experiments to a computational simulation of a solar collector using the direct absorption of nanofluids. The experimental and numerical results showed a quick increase in efficiency with increasing volume percent, followed by a plateau in efficiency as volume fraction increased.

The addition of nanoparticles boosted incoming radiation absorption by more than nine times over that of pure water, according to a theoretical assessment into the possibility of utilizing a non-concentrating direct absorption solar collector [46]. The efficiency of immersion of a solar collector using the nanofluid as the working fluid was actually found to be up to 10% greater than that of a flat plate collector under some similar operating conditions. The nanofluid-based sunlight-based gatherer had a somewhat longer recompense time than a conventional sun-powered authority given the current expense of nanoparticles, yet it saved a similar measure of cash toward the finish of its valuable life. The optical and heat properties of nanofluids comprised of watery suspensions of single-divider carbon nano horns were concentrated by Sani et al. The found varieties in optical qualities brought about by nanoparticles were promising, prompting a huge expansion in daylight retention. Both of these elements, just as the chance of compound functionalization of carbon nano horns, make this clever class of nanofluids especially engaging for working on the general proficiency of a solar-harvesting system.

23.5.7 Mechanical Applications

Why are nanofluids so effective at reducing friction? The fact that nanoparticles in nanofluids create a protective film on the worn surface with low hardness and elastic modulus can be attributed to the fact that few nanofluids have great lubricating qualities. A magnetic nanofluid is a special type of fluid. Magnetic liquid rotary seals use the magnetic characteristics of magnetic nanoparticles in liquid to function with no maintenance and exceptionally little leakage in several applications [29].

23.5.7.1 Friction Reduction

Advanced and superior lubricants can boost productivity by reducing energy consumption and increasing the reliability of many mechanical man-made systems. The goal of study is to reduce friction and wear. Nanoparticles have sparked a lot of attention in current years because of their greater load-carrying capacity, high extreme-pressure resistance, and friction-reducing quality. On a four-ball machine, Zhou et al. tested the tribological performance of Cu nanoparticles in oil. Cu nanoparticles demonstrated higher friction-reduction and anti-wear capabilities as an oil additive than zinc dithiophosphate, especially at high applied loads, according to the findings.

The nanoparticles could also improve the load-carrying capability of the base oil [47]. In this, the dispersion of solid particles was an important one. In the lubrication process of cast iron, water-based alumina and diamond nanofluids were applied. Using the nanofluid, the minimum quantity of lubrication benefited the grinding performance. Nanofluids showed the benefits of minimizing grinding forces, improving surface quality, and preventing burning of the workpiece. Surface-modified copper nanoparticles as a 50 cc oil addition were examined for their wear and friction properties. The characteristics of Cu nanoparticles improved as the oil temperature was increased.

It's possible that a thin Cu protective layer with reduced hardness and elastic modulus develops on the damaged surface, resulting in a good performance of Cu nanoparticles [48].

23.5.7.2 Magnetic Sealing

Magnetic fluid such as ferromagnetic fluid is the type of nanofluid that is unique. They're colloidal solutions of tiny magnetic particles like magnetite that are stable (Fe_3O_4). To meet the requirements of colloidal stability of magnetic nanofluids with nonpolar and polar carrier liquids, the properties of magnetic nanoparticles can be modified by altering their size and adjusting surface coatings [49].

Magnetic sealing in comparison to mechanical sealing provides a cost-effective method for environmental and risky gas sealing in a wide range of industrial equipment with long life, high-speed capability, low friction power loss, and excellent reliability [50]. When ferrofluid is injected into the gaps, it produces separate liquid rings that can support a pressure differential while leaking no liquid. Because the mechanical moving portions do not touch, the seals do not wear out as the shaft rotates. Magnetic fluids can be used to encapsulate liquids in a variety of applications due to their unique properties.

23.5.8 Biomedical Applications

Because certain types of nanoparticles have antibacterial or drug-delivery properties, nanofluids containing these nanoparticles will also have these characteristics.

23.5.8.1 Antibacterial Activity

Organic antibacterial compounds are often unstable, especially when exposed to extreme temperatures or pressures. As a result of their capacity to withstand extreme process conditions, inorganic materials like metals and metal oxides have got good attention in the last decades. ZnO nanofluids have antibacterial properties [51]. At high ZnO concentrations, electrochemical experiments reveal a direct connection between zinc oxide nanoparticles and the bacterial membranes. Jalal et al. used a green technique to make ZnO nanoparticles. By calculating the reduction ratio of bacteria treated with Zn-oxide, the antibacterial activity of ZnO nanoparticle suspensions against *E. coli* was assessed. Bacterial survival ratios fall as Zn-oxide nanofluid concentrations and time increase [52]. Further research has shown that zinc oxide nanoparticles have a wide spectrum of antibacterial properties on a variety of different microbes. The antibacterial activity of zinc oxide nanoparticles may be influenced by the size and presence of normal visible light [53]. The latest research found that Zn-oxide nanoparticles have outstanding anti-bacterial activities against *E. coli* O157: H7, a common foodborne pathogen, and that the inhibitory effects increased as Zn-oxide nanoparticle concentrations increased. The components of the cell membrane such as proteins and lipids were altered by Zn-oxide nanoparticles.

Zn-oxide nanoparticles have been shown to distort bacterial cell membranes, resulting in the loss of intracellular components and, eventually, cell death. They are thought to be an efficient antibacterial agent for agriculture and food safety [54].

23.5.8.2 Nano-Drug Delivery

Colloidal drug delivery technologies have been developed over the last few years to increase the efficiency and specificity of drug activity [55]. Nanoparticles, having a small size, remarkable surface, enhanced solubility, and multi-functional properties, open up a lot of new biomedical possibilities. Because of the unique features of nanoparticles, they are allowed to interact with complicated biological tasks in innovative modes [56]. Non-toxic carriers for medication and gene delivery are provided by gold nanoparticles. The gold core in these systems provides stability, while the monolayer enables for fine adjustment of surface features like hydrophobicity and charge.

The interaction of gold nanoparticles with thiols is another appealing aspect, since it allows for effective and selective intracellular release [57]. Nakano et al. suggested a drug-delivery method based on nanomagnetic fluid, which used a ferrofluid cluster made up of magnetic nanoparticles to target and concentrate medicines [58]. Magnetic nanoparticles have the capacity to load drugs and the biological features that can be conferred on them via an appropriate coating.

Carbon nanotubes have emerged as a novel and effective means of carrying and translocating medicinal compounds. CNT can be modified to deliver bioactive peptides, proteins, nucleic acids, and medicines to cells and organs. Because functionalized carbon nanotubes are non-immunogenic and have minimal toxicity, they have a lot of potential in nanobiotechnology and nano-medicine. Pastorin et al. used the 1,3-dipolar cycloaddition of azomethine yields to establish a unique approach for functionalizing CNTs with two distinct chemicals [59].

The addition of chemicals that target specific receptors on cancer cells will aid in the improvement of anti-cancer drug response. Pre-functionalized carbon nanotubes can adsorb commonly used aromatic compounds through easy mixing, generating assemblies on carbon nanotubes with PEG extending into the water to impart solubility and aromatic molecules densely capturing nanotube sidewalls, according to Liu et al. The research builds a new, simple-to-make SWNT-doxorubicin combination with extraordinarily drug loading capacity [60].

23.5.9 Mass Transfer Enhancement

The mass transfer enhancement of nanofluids has been examined in a number of studies. Kim et al. first looked at the influence of nanoparticles on bubble-type absorption in the ammonia-water system. The absorption performance is improved by 3.2 times when nanoparticles are added. They then estimated the effect of nanoparticles and surfactants on the absorption characteristics of ammonia-water [61]. The inclusion of surfactants and nanoparticles boosted absorption performance by 3.3 times, according to the findings.

Adding surfactants and nanoparticles to the ammonia bubble absorption process dramatically improved the absorption performance. The effects of thermo-diffusion and diffusion-thermo on convective instabilities in binary nanofluids were studied theoretically for absorption applications [62]. The heat gradient causes mass diffusion. Heat transfer is caused by a concentration gradient in diffusion-thermo [63]. Using carbon nanotubes and ammonia nanofluids as the working medium, Ma et al. investigated the mass transfer phenomenon of absorption. Carbon nanotubes and ammonia binary nanofluid absorption rates were greater than ammonia solution without nanotubes [64].

With the starting concentration of ammonia and the mass fraction of nanotubes, the effective absorption ratio of the carbon nanotubes and ammonia binary nanofluids rose. The inclusion of ferrofluids improved the mass transfer coefficient in gas/liquid mass transfer [65], and the magnitude of the augmentation relied on the amount of ferrofluid injected, according to Komati et al. For a volume proportion of the fluid of about 50%, the increase in mass transfer coefficient was 92.8%.

The types of nanoparticles and surfactants in the nanofluid, as well as the concentration of ammonia in the base fluid, were found to be the key parameters influencing the absorption process of ammonia on the influence of Al_2O_3 nanofluid. So yet, the mechanism that leads to increased mass transfer has remained a mystery. Existing research on mass transfer in nanofluids is insufficient. To elucidate certain key impacting aspects, a lot of experimental and modeling work needs to be done. So yet, the mechanism that leads to increased mass transfer has remained a mystery. Several experimental works should be carried out to elucidate some important factors.

23.5.10 Other Applications

23.5.10.1 Intensify Micro-Reactors

The discovery of more heat transfer enhancement in nanofluids could be also used in the domain of intensification in chemical reactors by combining reaction and heat transfer functions in multi-functional reactors. Fan et al. investigated an integrated heat exchanger operated by nanofluids, based on TiO2 material distributed in ethylene glycol. In the steady-state continuous testing, the net heat transfers coefficient raised by up to 35%. As a result, the temperature of selective reduction of an aromatic aldehyde by molecular hydrogen was improved, and the temperature of the reaction changed extremely quickly [66].

23.5.10.2 Nanofluids as Vehicular Brake F23luids

Now there is a strong demand for quality brake oil because the heat produced during the braking process is distributed throughout the hydraulic braking system [67]. Two nanofluids were synthesized based on copper oxide and aluminum oxide by using the arc-submerged nanoparticle synthetic method and the plasma charging arc system method [53] respectively. Both types of nanofluids have improved qualities over standard braking fluid, such as a greater boiling point, higher conductivity, and higher viscosity. The nanofluid brake oil will prevent the incidence of vapor lock and promote driving safety by having a greater boiling point, viscosity, and conductivity.

23.5.10.3 Nanofluid-Based Microbial Fuel Cells

Microbial fuel cells (MFC) generate electricity using energy-rich carbohydrates, proteins, and other natural products. The electrodes and electron mediator are responsible for MFC's remarkable performance. Sharma et al. developed a new microbial fuel cell (MFC) that used novel electron mediators and nanotube-based electrodes [68]. Nanofluids were created by distributing nanocrystalline platinum-attached carbon nanotubes in water to create new mediators. They also compared the working performance of the new *E. coli*-based fuel to *E. coli*-based microbial fuel cell with mediators of neutral red and methylene blue that had previously been reported. The performance of MFCs based on CNT-based nanofluids and CNT-based electrodes was compared to that of MFCs based on plain graphite electrodes. When compared to graphite electrodes, CNT-based electrodes showed a six-fold increase in power density.

23.5.10.4 Nanofluids with Unique Optical Properties

Different wavelengths of light are selected using optical filters. The optical filter based on ferrofluid has tunable features. An external magnetic field can also be used to appropriate wavelength. Philip et al. created a ferrofluid for wavelength selection in the UV, visible, and infrared areas [69]. Using appropriately adjusted ferrofluid emulsions, the required range of wavelengths, reflectivity, and bandwidth may be regulated.

Mishra et al. developed nanofluids using gold nanoparticles embedded in the polymer of polyvinyl pyrrolidone (PVP) in water [70]. They compared the changes in the apparent visible colors in the formation of Au-PVP nanofluids with different Au levels of 0.05, 0.10, 0.50, and 1.00 wt% Au. When gold concentration is changed from 0 to 1 wt%, the surface bands, which occur at 480–700 nm, change in relative position as well as intensity.

23.6 Limitations of Nanofluids

The development of nanofluid applications is hampered by several problems, one of which is the long-term stability of nanofluid in suspension. Nanofluids have the following disadvantages [28].

23.6.1 Lower Specific Heat

In order to exchange more heat, an ideal heat transfer fluid should have a greater specific heat value. Nanofluids have a lower specific heat than base fluids, according to previous research. It limits the use of nanofluid applications.

23.6.2 Increased Pressure Drops and Pumping Power

The efficiency of nanofluid application is determined by the pressure drop development and needed pumping power during coolant flow. Larger density and viscosity are known to result in pumping power and higher pressure drop. Several investigations have found that nanofluids have a significantly higher pressure drop than base fluids. Choi (2009) calculated a 40% improvement in pumping power as compared to pure water for a particular running rate in one of his experiments.

23.6.3 High Cost of Nanofluids

One-step or two-step procedures are used to make nanofluids. Both procedures necessitate expensive and modern equipment. This raises the cost of manufacturing nanofluids. This is one of the disadvantages of nanofluids.

23.6.4 Poor Long-Term Stability of Suspension

Because of the aggregation of nanoparticles due to very strong van der Waals interactions, the suspension is not homogeneous, and the long-term physical and chemical stability of nanofluids is an essential practical concern. For getting stable nanofluids, physical or chemical approaches like (i) addition of surfactant; (ii) surface modifications on suspended particles; and (iii) putting heavy force on suspended particle groups have been used. In comparison to fresh nanofluids, Lee and Choi discovered that Al2O3 nanofluids stored for 30 days show some settling. The setting of particles must be checked carefully since they may cause choked coolant pipes.

23.7 Conclusion

This chapter provides an overview of nanofluids, which consists of preparation procedures, stability estimation methodologies, and possible applications in heat transfer intensification, mass transfer enhancement, energy fields, mechanical fields, and biomedical fields, among others. The stability of nanofluids and the expense of their manufacture are important considerations when using them. As a result, they can be used as more efficient and compact heat transfer systems, while also keeping a cleaner and healthier atmosphere and providing novel applications. Although nanofluids have a lot of fascinating potential applications, there are several significant barriers to overcome before they can be commercialized.

In the future, the following critical topics should be given more attention. To begin, more experimental study is required to identify the primary parameters determining nanofluid performance. Because there has been a lack of consistency between experimental data from different groups thus far, it is critical to identify these elements in a systematic manner. The experimental data disparity could be explained by accurate and precise structural characterizations of the fluid suspensions.

Second, the increased viscosity caused by the usage of nanofluids has a significant drawback in terms of pumping power. Nanofluids with low viscosity and strong conductivity have a lot of potential applications. Modifying the interface characteristics of two phases to improve the compatibility of nanomaterials and base fluids could be the solution.

Third, the properties of the nanofluids are dependent on the shape of the adding components, hence, developing nanofluid production methods with a controllable microscopic structure would be a future research topic.

Fourth, stability of suspension is critical for scientific study and practical approaches. More attention should be given to the long-term stability of nanofluids.

Fifth, there has been little research on the thermal performance of nanofluids at high temperatures, which could expand the range of applications for nanofluids, such as high-temperature solar energy absorption and storage. High temperatures may hasten the breakdown of surfactants that are used as dispersants in nanofluids, resulting in several foams. These points must be considered and these aspects must be taken into consideration.

Hence, the property of the additive and the form have a significant impact on the properties of nanofluids. Nanofluids are incredibly complicated fluids that have a wide range of uses in research and engineering.

Despite the fact that numerous heat transmission processes have been explained, the use of nanofluid in actual applications is still critical due to sedimentation and clogging in the flow channel.

23.8 Future Scope

Nanofluids have shown considerable promise in a variety of fields. However, several obstacles remain in the way of its commercialization, as detailed here. To begin with, the lack of agreement among numerous academics on outcomes necessitated additional research and experimentation to determine the variables driving differences in results. Second, for research and practical purpose, the stability of nanofluid suspension is an important issue. Nanofluids should be made more stable using new approaches and procedures that do not compromise their thermophysical qualities.

Third, there is a paucity of research on nanofluids at higher temperatures, and dispersants degrade or even foam at higher temperatures, necessitating higher temperature research in order to improve nanofluid stability and surfactant behavior. Fourth, as the concentration rises, the viscosity rises, necessitating greater pumping force. Finally, the form and size of additives have a big impact on nanofluid characteristics. The degree of aggregation of nano-sized particles affects the thermal conductance of nanofluids. Nanoparticles are the future coolants with better thermal conductivity due to their applicability in numerous applications.

REFERENCES

1. Saini, A., Sharma, S., Gangacharyulu, D., & Kaur, H. (2016). Nanofluids: a review preparation, stability, properties and applications, *International Journal of Engineering Research & Technology*, 5(7):11–16.
2. Maxwell, J. C. (1873). *Maxwell_1873_Treatise_Preface*. Clarendon Press.
3. Mondragón, R., Segarra, C., Martínez-Cuenca, R., Juliá, J. E., & Jarque, J. C. (2013). Experimental characterization and modeling of thermophysical properties of nanofluids at high temperature conditions for heat transfer applications, *Powder Technology*, 249:516–529.
4. Masuda, H., Ebata, A., Teramae, K., & Hishinuma, N. (1993). Alteration of thermal conductivity and viscosity of liquid by dispersing ultra-fine particles. Dispersion of Al2O3, SiO2 and TiO2 ultra-fine particles, *Netsu Bussei*, 7(4):227–233.
5. Choi, S. U. S. (1995). Enhancing thermal conductivity of fluids with nanoparticles, in *Proceedings of the International Mechanical Engineering Congress and Exposition*, 66:99–105, San Francisco, CA (United States).
6. Salman, B. H., Mohammed, H. A., & Kherbeet, A. S. (2014). Numerical and experimental investigation of heat transfer enhancement in a microtube using nanofluids, *International Communications in Heat and Mass Transfer*, 59:88–100.
7. Yu, W., & Xie, H. (2011). A review on nanofluids: preparation, stability mechanisms, and applications, *Journal of Nanomaterials*, 2012, 1–17.
8. Eastman, J. A., Choi, S. U. S., Li, S., Yu, W., & Thompson, L. J. (2001). Anomalously increased effective thermal conductivities of ethylene glycol-based nanofluids containing copper nanoparticles, *Applied Physics Letters*, 78(6):718–720.
9. Li, Y., Zhou, J., Tung, S., Schneider, E., & Xi, S. (2009). A review on development of nanofluid preparation and characterization, *Powder Technology*, 196(2):89–101.
10. Lo, C. H., Tsung, T. T., & Chen, L. C. (2005). Shape-controlled synthesis of Cu-based nanofluid using submerged arc nanoparticle synthesis system (SANSS), *Journal of Crystal Growth*, 277(1–4):636–642.
11. Lo, C. H., Tsung, T. T., Chen, L. C., Su, C. H., & H. M. Lin (2005). Fabrication of copper oxide nanofluid using submerged arc nanoparticle synthesis system (SANSS), *Journal of Nanoparticle Research*, 7(2–3):313–320.
12. Zhu, H. T., Lin, Y. S., & Yin, Y. S. (2004). A novel one-step chemical method for preparation of copper nanofluids, *Journal of Colloid and Interface Science*, 277(1): 100–103.
13. Bönnemann, H., Botha, S. S., Bladergroen, B., & Linkov, V. M., (2005). Monodisperse copper- and silver-nanocolloids suitable for heat-conductive fluids, *Applied Organometallic Chemistry*, 19(6):768–773.
14. Singh, A. K., & Raykar, V. S. (2008). Microwave synthesis of silver nanofluids with polyvinylpyrrolidone (PVP) and their transport properties, *Colloid and Polymer Science*, 286(14–15):1667–1673.
15. Zhu, H. T., Zhang, C. Y., Tang, Y. M., & Wang, J. X. (2007). Novel synthesis and thermal conductivity of CuO nanofluid, *Journal of Physical Chemistry C*, 111(4):1646–1650.
16. Chen, Y., & Wang, X. (2008). Novel phase-transfer preparation of monodisperse silver and gold nanoparticles at room temperature, *Materials Letters*, 62(15):2215–2218.
17. Yu, W., Xie, H., Chen, L., & Li, Y. (2010). Enhancement of thermal conductivity of kerosene-based Fe3O4 nanofluids prepared via phase-transfer method, *Colloids and Surfaces A*, 355(1–3):109–113.
18. Wang, L., & Fan, J. (2010). Nanofluids research: key issues, *Nanoscale Research Letters*, 5(8):1241–1252.
19. Wei, X., & Wang, L. (2010). Synthesis and thermal conductivity of microfluidic copper nanofluids, *Particuology*, 8(3):262–271.
20. Li, X., Zhu, D., & Wang, X. (2007). Evaluation on dispersion behavior of the aqueous copper nano-suspensions, *Journal of Colloid and Interface Science*, 310(2): 456–463.
21. Zhu, H., Zhang, C., Tang, Y., Wang, J., Ren, B., & Yin, Y. (2007). Preparation and thermal conductivity of suspensions of graphite nanoparticles, *Carbon*, 45(1):226–228.
22. Li, D., & Kaner, R. B. (2005). Processable stabilizer-free polyaniline nanofiber aqueous colloids, *Chemical Communications*, 14(26):3286–3288.
23. Kim, H. J., Bang, I. C., & Onoe, J. (2009). Characteristic stability of bare Au-water nanofluids fabricated by pulsed laser ablation in liquids, *Optics and Lasers in Engineering*, 47(5):532–538.
24. Huang, J., Wang, X., Long, Q., Wen, X., Zhou, Y., & Li, L. (2009). Influence of pH on the stability characteristics of nanofluids, in *Proceedings of the Symposium on Photonics and Optoelectronics (SOPO '09)*, Wuhan, China.
25. Farahmandjou, M., Sebt, S. A., Parhizgar, S. S., Aberomand, P., & Akhavan, M. (2009). Stability investigation of colloidal FePt nanoparticle systems by spectrophotometer analysis, *Chinese Physics Letters*, 26(2):027501.
26. Zhu, D., Li, X., Wang, N., Wang, X., Gao, J., & Li, H. (2009). Dispersion behavior and thermal conductivity characteristics of Al_2O_3-H_2O nanofluids, *Current Applied Physics*, 9(1):131–139.
27. Hwang, Y., Lee, J. K., Lee C. H., Jung, Y. M., Cheonga, S. I., Leea, G., Ku, C., & Jang, S. P. (2007). Stability and thermal conductivity characteristics of nanofluids, *Thermochimica Acta*, 455(1–2):70–74.
28. Renuka Prasad, A., Singh, S., & Nagar, H. (2017). A review on nanofluids: properties and applications, *International Journal of Advance Research and Innovative Ideas in Education*, 3(3):3185–3209.
29. Wei, Y., & Xie, H. (2012). A review on nanofluids: preparation, stability mechanisms, and applications, *Journal of Nanomaterials*, 2012:1–17. doi:10.1155/2012/435873.
30. Jang, S. P., & Choi, S. U. S. (2006). Cooling performance of a microchannel heat sink with nanofluids, *Applied Thermal Engineering*, 26(17–18):2457–2463.
31. Bianco, A., Kostarelos, K., & Prato, M. (2005). Applications of carbon nanotubes in drug delivery, *Current Opinion in Chemical Biology*, 9(6):674–679.
32. Shokouhmand, H., Ghazvini, M., & Shabanian, J. (2008). Performance analysis of using nanofluids in microchannel heat sink in different flow regimes and its simulation using artificial neural network, in Proceedings of the World Congress on Engineering (WCE '08), III, London, UK.

33. Xie, H., & Chen, L. (2009). Adjustable thermal conductivity in carbon nanotube nanofluids, *Physics Letters Section A*, 373(21):1861–1864.
34. Yu, W., France, D. M., Choi, S. U. S., & Routbort, J. L. (2007). Review and assessment of nanofluid technology for transportation and other applications, *Techrep Marketing*, 78, ANL/ESD/07-9, Argonne National Laboratory.
35. Feng, X., Ma, H., S. Huang, Pan, W., Zhang, X., Tian, F., Gao, C., Cheng, Y., & Luo, J. (2006). Aqueous-organic phase-transfer of highly stable gold, silver, and platinum nanoparticles and new route for fabrication of gold nanofilms at the oil/water interface and on solid supports, *Journal of Physical Chemistry B*, 110(25):12311–12317.
36. Tzeng, S. C., Lin, C. W., & Huang, K. D. (2005). Heat transfer enhancement of nanofluids in rotary blade coupling of four-wheel-drive vehicles, *Acta Mechanica*, 179(1–2):11–23.
37. Singh, D., Toutbort, J., & Chen, G. (2006). *Heavy Vehicle Systems Optimization Merit Review and Peer Evaluation*, Annual Report, Argonne National Laboratory.
38. http://www.labnews.co.uk/feature archive.php/5449/5/keeping-it-cool (Accessed on 13 February 2022).
39. Routbort, J., et al. (2009). *Argonne National Lab, Michellin North America.* St. Gobain Corp.
40. Kulkarni, D. P., Das, D. K., & Vajjha, R. S. (2009). Application of nanofluids in heating buildings and reducing pollution, *Applied Energy*, 86(12):2566–2573.
41. Vassallo, P., Kumar, R., & D'Amico, S. (2004). Pool boiling heat transfer experiments in silicawater nanofluids, *International Journal of Heat and Mass Transfer*, 47(2):407–411.
42. Buongiorno, J., Hu, L. W., Apostolakis, G., Hannink, R., Lucas, T., & Chupin, A. (2009). A feasibility assessment of the use of nanofluids to enhance the in-vessel retention capability in light water reactors, *Nuclear Engineering and Design*, 239(5):941–948.
43. Demirbas, M. F. (2006). Thermal energy storage and phase change materials: an overview, *Energy Sources Part B*, 1(1):85–95.
44. Wu, S., Zhu, D., Zhang, X., & Huang, J. (2010). Preparation and melting/freezing characteristics of Cu/paraffin nanofluid as phase-change material (PCM), *Energy and Fuels*, 24(3):1894–1898.
45. Otanicar, T. P., Phelan, P. E., Prasher, R. S., Rosengarten, G., & Taylor, R. A. (2010). Nanofluid based direct absorption solar collector, *Journal of Renewable and Sustainable Energy*, 2(3):1–13.
46. Tyagi, H., Phelan, P., & Prasher, R. (2009). Predicted efficiency of a low-temperature Nanofluidbased direct absorption solar collector, *Journal of Solar Energy Engineering*, 131(4):0410041–0410047.
47. Zhou, J., Wu, Z., Zhang, Z., Liu, W., & Xue, Q. (2000). Tribological behavior and lubricating mechanism of Cu nanoparticles in oil, *Tribology Letters*, 8(4):213–218.
48. Yu, H. L., Xu, Y., Shi, P. J., Xu, B. S., Wang, X. L., & Liu, Q. (2008). Tribological properties and lubricating mechanisms of Cu nanoparticles in lubricant, *Transactions of Nonferrous Metals Society of China*, 18(3):636–641.
49. Vékás, L., Bica, D., & Avdeev, M. V. (2007). Magnetic nanoparticles and concentrated magnetic nanofluids: synthesis, properties and some applications, *China Particuology*, 5(1–2): 43–49.
50. Rosensweig, R. E. (1987). Magnetic fluids, *Annual Review of Fluid Mechanics*, 19:437–463.
51. Zhang, L., Jiang, Y., Ding, Y., Povey, M., & York, D. (2007). Investigation into the antibacterial behaviour of suspensions of ZnO nanoparticles (ZnO nanofluids), *Journal of Nanoparticle Research*, 9(3):479–489.
52. Jalal, R., Goharshadi, E. K., Abareshi, M., Moosavi, M., Yousefi, A., & Nancarrow, P. (2010). ZnO nanofluids: green synthesis, characterization, and antibacterial activity, *Materials Chemistry and Physics*, 121(1–2):198–201.
53. Jones, N., Ray, B., Ranjit, K. T., & Manna, A. C. (2008). Antibacterial activity of ZnO nanoparticle suspensions on a broad spectrum of microorganisms, *FEMS Microbiology Letters*, 279(1):71–76.
54. Liu, Y., He, L., Mustapha, M., Li, H., Hu, Z. Q., & Lin, M. (2009). Antibacterial activities of zinc oxide nanoparticles against Escherichia coli O157:H7, *Journal of Applied Microbiology*, 107(4):1193–1201.
55. Vonarbourg, A., Passirani, C., Saulnier, P., & Benoit, J. P. (2006). Parameters influencing the stealthiness of colloidal drug delivery systems, *Biomaterials*, 27(24):4356–4373.
56. Singh, R., & Lillard, J. W. (2009). Nanoparticle-based targeted drug delivery, *Experimental and Molecular Pathology*, 86(3):215–223.
57. Ghosh, P., Han, G., De, M., Kim, C. K, & Rotello, V. M. (2008). Gold nanoparticles in delivery applications, *Advanced Drug Delivery Reviews*, 1:1307.
58. Nakano, M., Matsuura, H., & Ju, D. (2008). Drug delivery system using nano-magnetic fluid, in Proceedings of the 3rd International Conference on Innovative Computing, Information and Control (ICICIC), Dalian, China.
59. Pastorin, G., Wu, W., & Wieckowski, S. (2006). Double functionalisation of carbon nanotubes for multimodal drug delivery, *Chemical Communications*, 11:1182–1184.
60. Liu, Z., Sun, X., Nakayama-Ratchford, N., & Dai, H. (2007). Supramolecular chemistry on watersoluble carbon nanotubes for drug loading and delivery, *ACS Nano*, 1(1):50–56.
61. Kim, J. K., Jung, J. Y., & Kang, Y. T. (2006). The effect of nanoparticles on the bubble absorption performance in a binary nanofluid, *International Journal of Refrigeration*, 29(1):22–29.
62. Kim, J. K., Jung, J. Y., & Kang, Y. T. (2007). Absorption performance enhancement by nano-particles and chemical surfactants in binary nanofluids, *International Journal of Refrigeration*, 30(1):50–57.
63. Kim, J., Kang, Y. T., & Choi, C. K. (2007). Soret and Dufour effects on convective instabilities in binary nanofluids for absorption application, *International Journal of Refrigeration*, 30(2):323–328.
64. Ma, X., Su, F., Chen, J., & Zhang, Y. (2007). Heat and mass transfer enhancement of the bubble absorption for a binary nanofluid, *Journal of Mechanical Science and Technology*, 21:1813.
65. Komati, S., & Suresh, A. K. (2008). CO_2 absorption into amine solutions: a novel strategy for intensification based on the addition of ferrofluids, *Journal of Chemical Technology and Biotechnology*, 83(8):1094–1100.
66. Fan, X., Chen, H., Ding, Y., Plucinski, P. K., & Lapkin, A. A. (2008). Potential of 'nanofluids' to further intensify microreactors, *Green Chemistry*, 10(6):670–677.

67. Nelson, I. C., Banerjee, D., & Ponnappan, R., (2009). Flow loop experiments using polyalphaolefin nanofluids, *Journal of Thermophysics and Heat Transfer*, 23(4): 752–761.
68. Wang, X. Q., &. Mujumdar, A. S. (2007). Heat transfer characteristics of nanofluids: a review, *International Journal of Thermal Sciences*, 46(1):1–19.
69. Philip, J., Jaykumar, T., Kalyanasundaram, P., & Raj, B. (2003). A tunable optical filter, *Measurement Science and Technology*, 14(8):1289–1294.
70. Mishra, A., Tripathy, P., Ram, S., & Fecht, H. J. (2009). Optical properties in nanofluids of gold nanoparticles in poly(vinylpyrrolidone), *Journal of Nanoscience and Nanotechnology*, 9(7):4342–4347.

24

Recent Trends in Nanomaterial-Based Electrochemical Biosensors for Biomedical Applications

Shikandar D. Bukkitgar, Nagaraj P. Shetti and Kakarla Raghava Reddy

CONTENTS

24.1 Introduction

Research related to the development of biosensors in recent years has seen rapid growth. A biosensor is an analytical device that converts any biological response to a measurable and processable signal. It is classified into four generations based on the degree of combination of its separate components, i.e. the attachment of the base transducer element with the bio-recognition or biomolecule as shown in Figure 24.1. In first-generation biosensors, the bio-receptors are physically trapped behind a membrane and subsequent immobilization of receptors on the transducers is obtained by a covalent bond on a suitable transducer interface. The electric response is due to the diffusion of product formed by the reaction to the surface of the transducer.

In the second generation, the individual components remain essentially distinct. To generate improved response a second-generation biosensor requires specific mediators between reaction and transducer. In third-generation biosensors, the bio-receptor molecule is an integral part of the base sensing element. Here, there is no involvement of product or mediator diffusion but the reaction itself develops the response. Finally, the fourth-generation biosensors are the latest development in the field of biosensing. Examples of such include the fabrication of micro, nano, and bio-nano electromechanical system. Oligonucleotide ligands known as aptamers are the molecular recognition element and behave like receptors.

Fabrication of the biosensors is majorly dependent on eight significant characteristics that have to be considered.

i) *Sensitivity* – Determines the response of the developed sensor to per unit change in concentration.
ii) *Selectivity* – Determines the accuracy of the sensor response to the only targeted analyte.
iii) *Linear range* – The range of concentration where the fabricated sensor shows an excellent response.
iv) *Response time* – Time required for the sensor to achieve 63% of its final response as the concentration of analyte takes place in step.
v) *Reproducibility* – The accuracy of the sensor output.
vi) *Detection limit* – The lowest concentration of the analyte where the response could be measured.

DOI: 10.1201/9781003157823-24

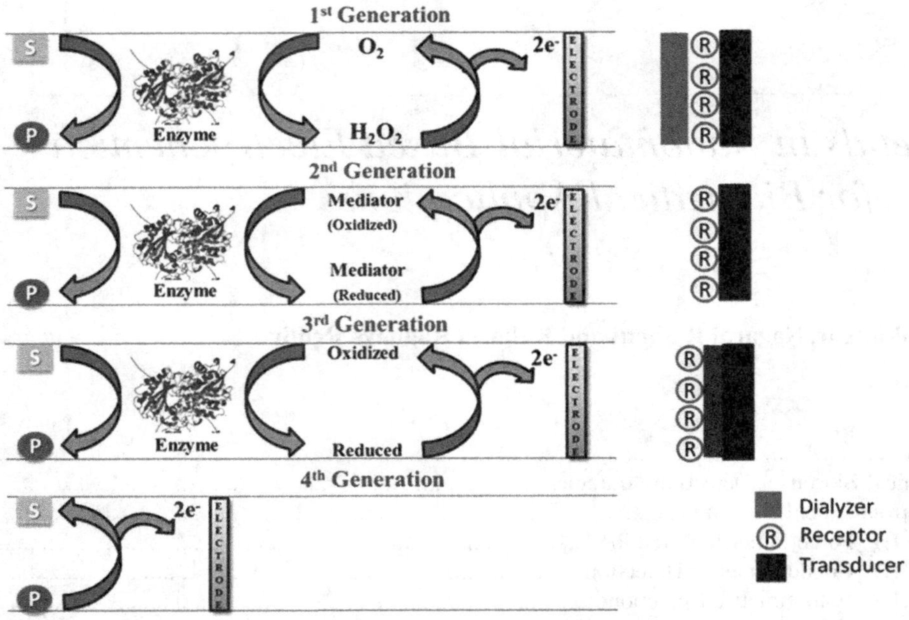

FIGURE 24.1 Types of biosensor generation.

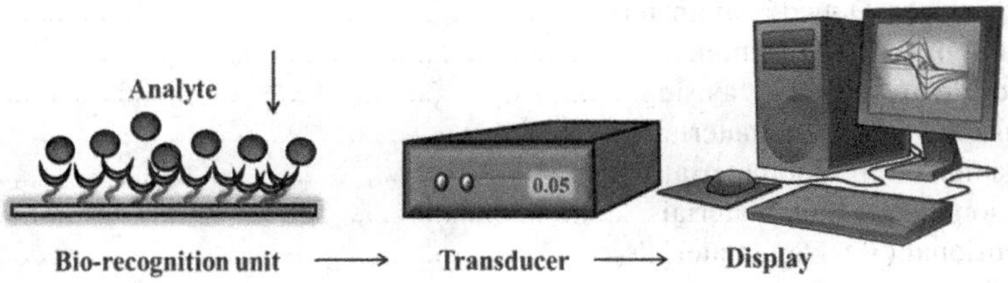

FIGURE 24.2 Components of a biosensor system.

vii) *Lifetime* – The time period of the sensor until which there is no significant decrease in the performance of the sensor.

viii) *Stability* – The change that happens in its baseline or sensitivity over a period of time.

In addition to the above characteristics, the major tasks for the development of a biosensor (Figure 24.2) once the analyte is identified are:

i) Recognition molecule or bio-receptor selection for that particular analyte.
ii) A method suitable for immobilization.
iii) Selecting or designing a suitable transducer to translate reaction to a measurable signal.
iv) Fabricating a complete device.

Typically, a biosensor device comprises a bio-receptor that is helpful in binding the analyte; an interface fabricated system where the specific biological event takes place giving rise to a detectable signal, and a transducer converting the signals produced to an electric signal which is then amplified by the detector and with reference sends the converted signals to a processing unit such as computer software to get a meaningful physical parameter that describes the process. The main purpose of using the biological recognition unit is to translate the biochemical domain that usually consists of the analyte concentration into physical or chemical output signal with a defined sensitivity. The recognition unit used in the sensor has significant importance in determining the selectivity toward the analyte that has to be measured. More or less selective to a specific analyte, there is a bio-recognition unit that is class-specific fabricated by using class enzymes.

The signals from the recognition unit in physical or chemical form are usually transferred to measurable signals, an electric signal by transducers. Biosensors may contain different types of transducer units based on the mechanism, as summarized in Table 24.1.

24.2 Electrochemical Biosensor Detection Strategies

24.2.1 Electrochemical Detection

Electrochemical transducers are the major component of electrochemical biosensor devices. With the assistance of the biochemical receptors, the electrochemical devices gather information regarding quantitative or semi-quantitative analysis. Different

TABLE 24.1
Types of Transducers

S. no.	Type	Measurement	Example
1	Electrochemical biotransducers	Produces an electrical signal that is proportional to the analyte concentration	Amperometry, potentiometric, photometric, conductometric piezoelectric, impedance
2	Optical biotransducers	Exploiting the interaction of the optical field with a bio-recognition element	Surface-enhanced Raman scattering biosensors, reflectometric interference spectroscopy biosensor, ellipsometric biosensors
3	FET-based electronic biotransducers	Changes in the surface potential induced by the binding of molecules	
4	Gravimetric/piezoelectric biotransducers	Response to a change in mass	
5	Pyroelectric biotransducers	Generate an electric current as a result of a temperature change	
6	Electrical capacitance as transducer	Change in the dielectric measurement constant of the medium	

analytical ways can be used to measure the electrical changes taking place during oxidation/reduction reactions of analytes. Current and potential are the measured properties during electrochemical analysis which follows the principle of change in the properties of solution due to consumption or production of electrons with reference to a stable reference electrode. Some electrochemical techniques focus on analyzing the changes on the electrode surface caused by molecular interactions and surface biofunctionalization. These do not rely on direct electron flow or redox reactions, but rather on measuring parameters such as capacitance, resistance, or impedance to analyze molecular interactions such as receptor-ligand, antigen-antibody, and others. Furthermore, these techniques have attracted the sensor industry due to their significant advantages, such as the simple conversion of biological interactions to simple electrical signals and the availability of a wide range of methods for measuring electrical properties, such as conductometry, voltammetry, amperometry, potentiometry, and impedance.

Electrochemical analysis is performed in an electrochemical cell containing the reference electrode, counter electrode, and working electrode. The measurements are highly influenced by the working electrode as different material used for the working electrode has different parameters such as potential window and capacity. The working electrode must be fabricated with materials that are chemically stable and conductive such as carbon, platinum gold, and more. In addition, the choice of electrode material has to be based on high reproducibility, cost-effective, non-toxic, and high S/N characteristics. During the reaction, there is electron flow between the working and counter electrode that closes the circuit in the cell. To avoid the kinetic limit, it should always be noted to have a much larger surface of the counter electrode than the working electrode.

Platinum wires are more efficiently used as counter electrodes. The working electrode reactions are balanced with constant potential produced in the cell by the reference electrode. Standard hydrogen electrode and silver wire coated with silver chloride is the most common reference electrode used in the electrochemical assay. There are miniaturized and various variations of a conventional electrochemical cell that use a three-electrode system. Further, microfluidic cells offer less interference and enhanced sensitivity with more advantageous properties like easiness in sampling. Along with these, lab-on-chip and screen-printed electrodes are of more interest considering point-of-care. These systems can be miniaturized to a few centimeters square platform with multi-laboratory functions using a very small amount of analyte. The screen-printed electrodes form a small measuring system that consists of three mini electrodes that are deposited or printed on polymer substrate allowing low cost and also mass production.

24.2.1.1 Potentiometric Detection

In potentiometric detections at zero current value, an electric potential difference is measured. The potential change occurring at the working electrode is measured in potentiometric detection with reference to other electrodes. Potentiometric detection of ionic species, especially inorganic anions in aqueous solution using ion-selective electrodes, has been the most attractive approach for a number of years and has also seen extensive development of ion-selective electrodes resulting in improved sensitivity and selectivity [1]. These potentiometric-based biosensors have the advantageous ability to detect several biological compounds that are significantly required for early diagnosis. Coupling with various other techniques allows the formation of sensitive tools for rapid and initial detection of critical biological samples.

There are various potentiometric biosensors but in the literature, we could find four main categories of potentiometric biosensors such as potentiometric biosensors amplified with molecularly imprinted polymer systems, photoelectrochemical potentiometric biosensors, light-addressable potentiometric biosensor, and wearable potentiometric biomarkers. The light-addressable potentiometric sensor is a method belonging to field-effect-based sensor devices consisting of semiconductor plates that are covered with an insulating layer and at the bottom surface there is an ohmic contact. The system consists of the light source and measurement electronics. The insulating layers that have been reported contain aluminum oxide, tantalum pentoxide, and Si_3N_4 with a thickness of 50–100 nm. In presence of a reference electrode, the analyte should be in contact with the insulating layer for the determination of the concentration [2, 3].

24.2.1.2 Conductometric Detection

Conductometric detections are based on the measurement of specific conductance of the analyte and are preferable due to their ability to detect both electroactive and electro-inactive species. They are more associated with capillary electrophoresis and the electrodes may be in contact with analyte solution or a thin layer can be used to insulate. In contact mode the electrode is in contact with the analyte solution directly leading to less response time, high sensitivity, however, they are also associated with contamination of samples and high risk of degradation of electrodes. The contact mode of conductometric detections was successfully applied for organic acids, circulating tumors, and alkali metal cations. On the other side, contactless conductometric detections do not suffer from electrode instability. A thin insulating layer is used to separate the electrode from the buffer or the sample solution. A thickness of the insulating layer should be observed to get good capacitive coupling.

Conducting polymers which are organic conjugated compounds that allow electron movement from one end to another due to their extended π-orbital system are also being used in conductometric-based detection. Reactions with redox-active agents allow the tuning of the conducting polymers to obtain unique electrical and optical properties. The ability to transport charge carriers along the polymer backbone is the key factor for their electric conductivity. The change in the conductivity of polymer due to the interaction with redox-active species is measured either in DC or AC conditions. They have many advantageous properties, such as compatibility with biomolecules, stability, ability to be coated on the surface of desired dimensions, easy processibility, and modification to bind to biomolecules together have allowed the development of conductometric-based biosensors for the detection of proteins, bacteria, and much more. In the detection of microorganisms, one can identify the change in the conductivity due to redox reactions at the polymer film interface due to the interaction of charged microorganisms with the polymer film.

Nanoparticles or nanofibers of conducting polymers are used they give additional surface area for the reactions and by increasing the sensitivity. Further, subsequent functionalization of these nanostructures can still increase the sensitivity. A biochemical phenomenon such as enzymatic reaction or binding process on the surface changes the electrical property of the material and hence conducting polymer does offer a unique transduction process and behaves like field-effect transistors. The use of conducting polymers is also cost-effective as fabricating can be done on an organic substrate like plastics, reducing the cost of materials as well as manufacturing processes. Hence, it can be easily integrated to develop a lab-on-chip, not only promising sensitivity but also low cost. However, compared to other electrochemical methods less sensitive response is a concern.

24.2.1.3 Voltammetric Detection

Voltammetric techniques work on the principle of potential application between a reference electrode and a working electrode. As a result of the oxidation/reduction reaction on the surface of the electrode, the resulting current between the working electrode and the counter electrode is measured. Voltammetric techniques are designed to have different approaches that may include amperometric detection based on measurement of current at specific time with constant application of potential or coulometry with measurement of current over a period of time. In addition, other approaches may include a variation of potential over time such as square wave voltammetry or differential pulse voltammetry. Cyclic voltammetric measurement can also be performed to investigate the redox behavior of the analyte. The cyclic voltammetric studies can reliably use to understand the reaction kinetics, concentration of analyte, diffusion properties, and the size of the active electrode surface. Voltammetric investigations are performed under no-current applied conditions and redox peak appearing at a specific potential for a specific analyte is characteristic and the intensity of the peak current can be directly related to the concentration of the analyte. These voltammetric investigations have been efficiently applied for the detection of pharmaceutical compounds that offer significant advantages in quality control and clinical applications [4–15].

In amperometric measurement, electric current versus time with constant application of the potential is the key principle of amperometric biosensors. It is based on the immobilization of the enzyme on the surface of the electrode and an electric current generated due to the oxidation or reduction reaction related to direct enzymatic reaction, and the signal produced is related to the concentration of the analyte. The relationship between the concentration of analyte and the current can be obtained by the Cottrell equation:

$$i = \frac{nFAc_0\sqrt{D}}{\sqrt{\pi t}}$$

where, i: current; n: number of electrons transferred; F: Faraday constant = 96,500 C mol^{-1}; A: area of planar electrode (cm^2); C_0: initial concentration of analyte; D: diffusion coefficient; t: time since potential applied.

24.2.1.4 Impedimetric Detection

Electrochemical impedance spectroscopy (ESI) depending on the AC potential frequency measures the impedance of a system. These techniques are very advantageous in characterizing the function and structure of the electrode surface, especially those modified with biological material. The presence or absence of the redox probe in the solution characterizes the classification of ESI as Faradaic and non-Faradaic. Non-Faradaic ESI is more prominently used in point-of-care devices as they require no reagents. For double-layer the capacitance and the resistance are changing during immobilization, hence label-free interactions and bio-recognition on the surface of the electrode can be detected. The results are expressed as Nyquist and Bode plots.

24.3 Types of Nanostructured Materials

Current biosensor technologies have provided excellent results, however, there are improvements needed in the areas of health care, biomedical application, clinical analysis, food processing, and environmental monitoring. The large gap in terms of detection of various chemical and biochemical substances directly in a very low level of concentration is yet to be filled. In addition, short-time analysis, reversibility, simple process, low cost, and determination of multi-analyte are the essential requirements for a biosensor in real situations [16]. The most alluring alternative to embedding the above properties is the application of recent progress in nanotechnology [17]. In recent years, it has been demonstrated that the use of nanomaterials has specifically improved the sensitivity and miniaturization of biosensor devices. Nano-systems with their submicron dimensions have also allowed rapid and simple analysis in vivo. Various characteristics of different nanomaterials and their use in biosensor devices are studied [18–23].

24.3.1 Metal and Metal Oxide-Based Nanomaterials

The morphological versatility of nanostructured metal and metal oxide has significantly allowed these materials to be some of the competitive materials in the field of biosensors. Further, the interfacial properties and the chemical stability of these materials, availability of numerous cost-effective methods for synthesis, and energy band alignment that is suitable for the immobilization of the enzymes have attracted considerable attention for researchers to adapt them into a biosensor system. These materials are also well-known for their catalytic activity that can drive many catalytic reactions. The use of these in electrochemical biosensing devices has allowed the development of non-enzymatic detection of metabolites, such as reactive oxygen species and sugar, enhancing the signal amplification that is catalytically driven.

Nanomaterial made of gold [24] and Fe_2O_4 [25] has been extensively used for their intrinsic enzyme-like activity for the fabrication of non-enzymatic electrochemical sensors. In addition, various nanostructures like nanowires, nanorods, and nanoparticles of NiO, CuO, Pd, and Pt have been successfully used as the fourth generation of electrochemical sensors for the detection of glucose [26–28]. Good stability, reproducibility, and reusability are the important characteristics when aiming to fabricate a glucose biosensor. An effective strategy to obtain this is to use conductive substrates for the growth of self-supported nanostructures. Zhao et al. recently used Cu foam as a substrate to grow vertically aligned CuO nanosheets electrochemically. Good conductivity, the transport of ions in a vertically aligned structure, and the controlled growth acquired through electrochemical synthesis were key features for the electrode to allow multi-test and long-term usage of the electrode [29].

Porous nanostructures have been extensively harvested for their great performance in the field of biosensing due to their advantageous properties, such as excellent electric conductivity, high surface area, and unique pore structures. Interlinked hierarchical porosity at various lengths exhibits excellent properties for ion and electron transport [30]. Metal-organic frameworks (MOFs) have been extensively studied for the detection of glucose. Ni-based MOFs were produced on Ni foam using the hydrothermal synthesis method (Figure 24.3A–B). The

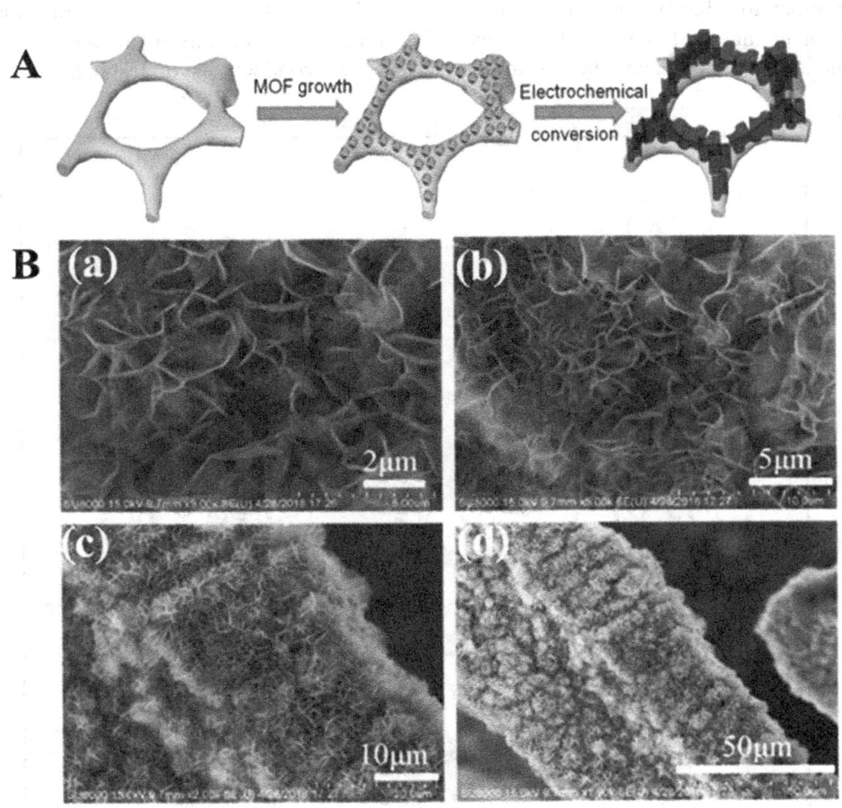

FIGURE 24.3 (A) Schematic diagram of preparing vertical aligned CuO nanosheets on Cu foam. (B) SEM images of vertical aligned CuO nanosheets prepared by an electrochemical conversion from Cu-MOF. Reproduced with permission from [29].

microscopic investigations revealed that the hierarchical porous formation enables better access through channels and availability of more active sites overall improving the performance of the sensor [31]. Hou et al. in their studies employed Ni nano-porous networks on laser scribed carbon paper producing disposable electrochemical sensors [32]. Similarly, electrochemical strips that could be disposed of were employed using Pt for the detection of glucose [33].

Nano-porous Pd has recently shown excellent catalytic properties. Commercializing nano-porous Pd has some issues that need to be resolved. Initially, the catalytic activity of the Pd is affected during catalyzation due to particle aggregation. Hence, the catalytical activity of nano-porous Pd has to be increased. Efforts have been made toward improving the catalytic performance of Pd nano-porous. Pd-coated nano-porous gold film was synthesized by Tavakkoli and showed improved stability toward the oxidation of glucose [34]. In addition, porous tubular Pd structures and modification of Pd particles on FCNTs have been also reported that have significantly improved the catalytical activity and high resistance toward the poisoning of the electrode due to chloride ions [35, 36]. Dealloying multicomponent metallic glasses to form a nano-porous Pd catalyst forming a three-dimensional, ligament-channel nano-porous structure, of size 11 nm with ligament size 7 nm, was reported by Yang et al., having improved catalytic properties toward glucose oxidation [37].

Detection of H_2O_2 has been extensively studied due to its significant importance in many biological activities such as cell growth, cellular signal transduction, and oxidative damage. Further, in environmental, pharmaceutical, and clinical research it's an essential mediator. It is not only a cause of different types of tissue and DNA damage, but also induces a defensive response during biotic and abiotic stresses at moderately elevated levels. Fe_3O_4, titanium oxide, and gold have dominated the research on the detection of H_2O_2. In recent literature we can find the use of CuO/g-C_3N_4 composite [38], copper bismuth oxide [39], $CuGa_2O_4$ [40], $MnFe_2O_4$ [41], nickel-cobalt double hydroxides micro-nano arrays [42], and cobalt nanorods supporting mesoporous gold nanoparticles [43] have been effectively used for the detection of H_2O_2.

Additionally, the metal and metal oxide nanoparticles conjugation with receptor molecules can be promoted directly due to their surface intrinsic properties. The pH of the surrounding environment mainly influences the charge on the surface of metal and metal oxide nanoparticles, giving them a characteristic isoelectric point (IEP). This further facilitates easy immobilization of the bio-receptors, circumventing the complications of further modifications [44–46]. Further, the metal and metal oxide nanoparticles have been utilized for the detection of pharmaceutical compounds and pesticides for monitoring clinical and agricultural processes [47–49]. They also have been applied for sensing applications in different areas [20, 22, 23, 50, 51].

24.3.2 Carbon and Nitrogen-Doped Nanomaterials

Fabrication of electrochemical sensors has been extensively evidenced in applications of carbon-based material for the development of biosensors with many advantageous properties (Figure 24.4). Major materials used in this field include carbon nanotubes (CNTs), graphene, and carbon nanofibers (CNFs) [4, 52–54]. Perfect arrangement in a honeycomb lattice, the sp^2 bonded carbon atoms are arranged in a planar sheet. Excellent electric conductivity in graphene and CNTs are obtained due

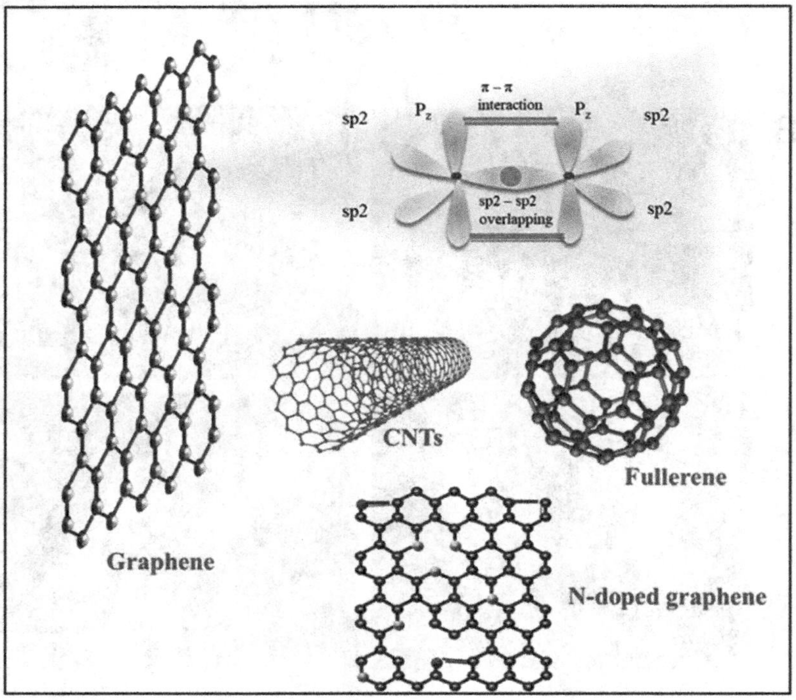

FIGURE 24.4 Carbon-based structures.

to this honeycomb lattice arrangement facilitating the delocalization of π electrons across the plane.

Many advantageous properties of graphene such as high specific area, easiness in modification, and high conductivity make it an ideal material for using it as a recognition component for a targeted molecule, or as a redox probe and as a nanocarrier in the electrochemical biosensing system that may be very powerful in enhancing the analytic performance. Further, chemical treatment of carbon nanomaterials can produce nanomaterials with oxygen groups such as graphene oxide, reduced graphene oxide, and CNTs which are acid treated. These are sometimes referred to as a defect and also have less conductivity as compared to that of pristine carbon due to the formation of sp_3-bonded carbon atoms interrupting the delocalization of the electrons. In addition, the result of lower conductivity is also due to the high oxygen-to-carbon ratio. Considering the field of electrochemical biosensing, these defects are considered to be beneficial. For instance, the covalent linking or electrostatic absorption of primary amine-containing molecules on the carboxylic group present on edges or basal planes of carbon nanomaterial gets easier [55–57]. This also promotes the fabrication of non-enzymatic electrochemical sensing devices as the deposition of various nanoparticles is anchored by oxygen-containing groups [58]. The negatively charged carbon nanomaterial suppresses the interference of other compounds by repelling the by-products [59, 60].

Special interesting properties advantageous for electrochemical sensing, such as functionality and reactivity, can be generated to carbon nanomaterials by doping with foreign atoms such as boron, nitrogen, and phosphorous. Doping carbon nanomaterial with nitrogen has been interesting and would be very useful due to the increase in conductivity. This is owed to the increased number of free-charge carriers. Nitrogen consists of a valence electron and is a similar size to carbon, so it forms strong valence bonds with carbon atoms [61]. In oxygen reduction reactions such as H_2O_2 oxidation, these nitrogen-doped carbons exhibit similar catalytic activity to Pt-decorated CNTs [62]. Weakening of O-O bonds and available sites for adsorption of oxygen is a key property due to the high positive charge density of the adjacent carbon atoms and the strong electronic affinity of nitrogen atoms. In another work, nitrogen doping into graphene has been shown to significantly contribute to the reduction of H_2O_2 [63]. The electronic properties of the carbon nanomaterial can be modulated by doping with nitrogen-producing metal-semiconductor transition [64]. Hence, these are also promising transducers in field-effect transistors (FET) biosensing systems [65, 66].

24.3.3 Conducting Polymer-Based Nanomaterials

Conducting polymers (Figure 24.5) are the organic polymers which consist of conjugated double bonds along the polymer chain and by inducing the delocalization of the electrons along the chain by doping electrical conduction can be obtained. Along with the advantageous properties, sensing manipulating the concentration of dopant concentration makes it easy for immobilization. For example, negatively charged nucleic acids can be easily electrostatically immobilized on positively charged conducting polymers [67]. In addition, the conducting polymers offer more advantageous functional groups such as –NH, –CN, and more specifically the –SH group facilitating the hybrids of conducting polymer and metal nanoparticles by self-assembly. Conducting polymers that contain sulfur groups show a high affinity toward noble metal nanoparticles and are very helpful in increasing the electrocatalytic activity and also serving as a platform for the immobilization of biomolecules. Conducting polymers with decorated nanoparticles can also be fabricated by spontaneous electrochemical reduction of metal ions [68–71].

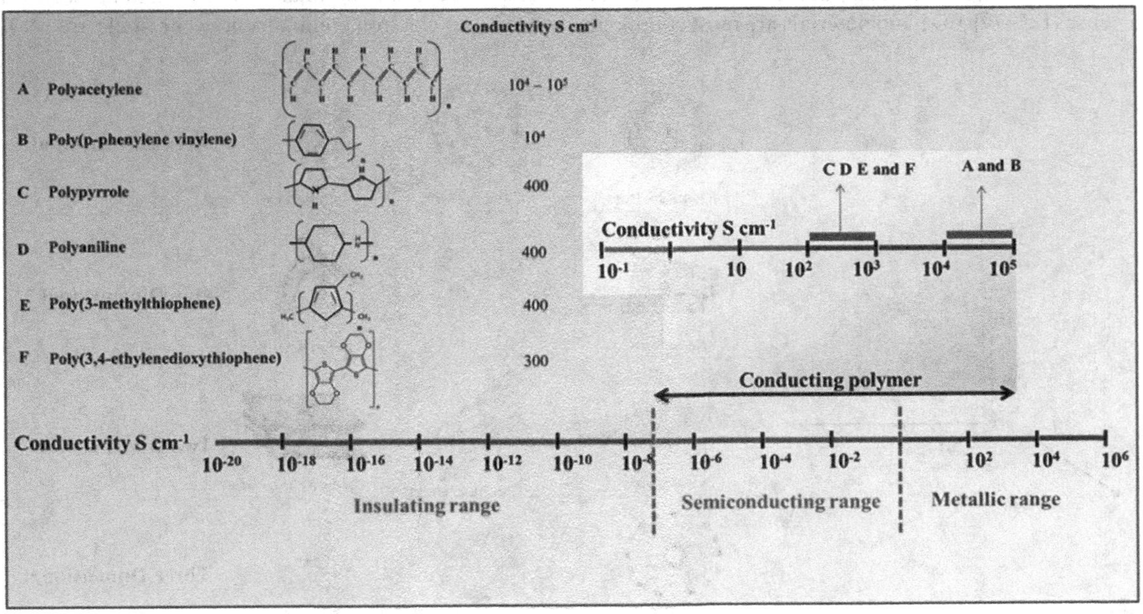

FIGURE 24.5 Conducting polymers in bio-sensing.

24.4 Nanostructure-Based Electrochemical Sensing

The properties of materials are defined by the arrangement of building blocks in three-dimensional structures as shown in Figure 24.6. The properties of nanomaterials differ greatly from those of bulk materials. For example, a metal is a conductor and cannot be capable of conduction when it has a single atom or small clusters of atoms. However, the single atoms and small clusters of atoms retain enhanced properties due to nano-structuring and quantum confinement. These nanostructure materials have been classified based on dimensions less than 100 nm [72, 73]. The nanostructures are mainly differentiated dimensionally as zero (0D), one (1D) two (2D), and three (3D) dimensional nanomaterials. Further, these are excellent candidates for designing electrochemical biosensors with high performance due to their unique properties. The analytical sensitivity of these materials can never be questioned due to their high surface-to-volume ratio and also the direct introduction of functionalities during modification or post-modification. The advancement in synthesis and fabrication methods has allowed the controlled design of these materials with different sizes, shapes, compositions, and arrangements [74]. Recently, electrochemical biosensing has witnessed a dramatic increase in the use of nanomaterial and nanostructure.

24.4.1 Zero-Dimensional (0D) Nanomaterials

Zero-dimensional nanomaterials are objects with all dimensions in the nano-scale range. 0D materials may be in triangular, cubic, and spherical forms and are the most basic and symmetric shape. Dimensions in these nanomaterials allow free movement of electrons due to confined electrons in 3D space and without electron delocalization. In recent years, several research groups have synthesized various structures such as hollow spheres, core-shell, quantum dots, flowers, onion, and nanolenses [75–79]. 0D nanomaterials are most commonly represented by nanoparticles having various compositions and are extensively used in electrochemical sensing as a catalyst, signal tracers, and labeling carriers. The most commonly used are noble metal nanoparticles due to their outstanding conductivity and the easiness of modification.

Various synthesis approaches have been reported in the literature suggesting the easiness in tuning the shape and size of the nanoparticles by altering the conditions of the reaction such as temperature, chemical composition ratio, and the precursor concentration [80, 81]. The sensitivity and reproducibility of an electrochemical sensor majorly depend on the high colloidal stability in biological buffer and size distribution of less than 5%. Hence a nanoparticle between 10–50 nm sizes has shown extensively successful results and has been widely used in electrochemical sensing [82, 83]. Semiconducting nanoparticles based on periodic table groups known as II–VI, III–V, or IV–VI semiconductor nanocrystals have also been used for the fabrication of electrochemical sensors. The most commonly used semiconducting nanoparticles for single analyte or simultaneous detection of analyte include PbS, ZnS, CdSe, CdS, etc. [84, 85]. Further, strong adhesion can be achieved by controlled nucleation and growth of metal nanoparticles by simple electrochemical reduction on electrochemical transducers.

Electrochemical biosensors have also seen extensive utilization of 0D nanomaterials such as magnetic nanoparticles [86]. Depending on the particle size, Fe_3O_4 exhibits superparamagnetic or ferromagnetic properties and the magnetic strength is higher than that of nanomaterials due to their dimensions in the micrometer range [87]. These nano-sized particles exhibit higher stability, higher surface-area-to-volume ratio, and show low susceptibility to particle clustering by applying a magnetic field and hence are favorable to a miniaturized system [88].

24.4.2 One-Dimensional (1D) Nanomaterials

1D nanomaterials are objects with one dimension outside the nano-scale and two dimensions within nano-scales. 1D nanostructures electron confinement occur in 2D are very sensitive

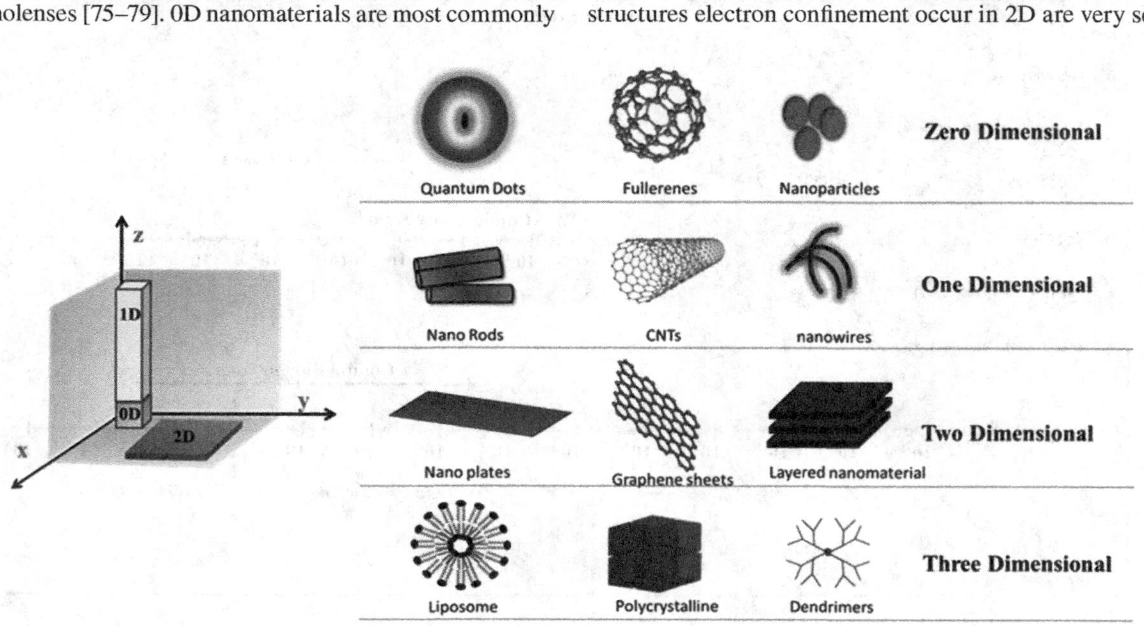

FIGURE 24.6 Classification of nanostructures.

to small variations at the surface due to the electron delocalization taking place along the long axis. Surface perturbations largely affect the conductance of these nanostructures and the unique electronic properties enable the development of label-free-based electrochemical sensors. Further, the nanostructures such as nanorods and nanotubes with their shorter macroscale length have shown applications as nanocarriers. 1D nanomaterials include structures such as nanotubes, nanorods, nanowires, and nanofibers.

Nanotubes such as CNTs have been widely applied in electrochemical investigations. The investigations have emphasized the role of oxygen-containing groups, sidewalls, and the CNT tips. It has been reported that the reaction mechanisms of the analyte necessarily facilitate the electroactive properties of CNTs rather than the nanotube tip and/or oxygen-containing surface groups [89]. Further, the orientation of the CNTs as reported by Liu et al. should also be considered to obtain better electrochemical performance. This has been demonstrated by the better electrochemical performance of vertically aligned short CNTs due to the high accessibility of the ends of the tubes to the sample when compared to CNTs which lie down on the electrode surface [90]. In addition, efficient mass transport through nonplanar radial diffusion has to be promoted by using the optimal density of CNT that avoids overlap of the diffusion layers from each CNT. This further facilitates exposure of CNT tips to the analyte and the sidewalls are embedded inside the insulator [91]. Nanorods, nanowires, and nanofibers have been also extensively used for the fabrication of biosensors. Nanorods can be synthesized with various techniques such as template-assisted electrochemical synthesis [92–99].

24.4.3 Two-Dimensional (2D) Nanomaterials

2D nanomaterials are objects in which two dimensions are not confined to the nano-scale. These structures have layered structures with weak van der Waals bonds between the layers and strong in-plane bonds. These nanostructures have a plate-like shape and are ideally single layered; however, nanosheets consist of a few layers that would be less than 10 layers. Recently, graphene, metal dichalcogenides, and hexagonal boron nitride have been extensively used in various applications including sensors. The conduction of electrons in 2D nanomaterials will be confined across the thickness but delocalized in the plane of the sheet.

24.4.4 Three-Dimensional (3D) Nanomaterials

These materials consist of objects in which no dimensions are confined to nano-scale; all three dimensions are above 100 nm. Since these materials consist of features of nanoscale or have nanocrystalline structure they are classified as nanomaterials. Considering the nanocrystalline structures, the bulk nanomaterials consist of nano-sized crystals arranged typically in different orientations. Whereas, the nano-scale features the dispersion of nanomaterials such as nanoparticles, nanowire bundles, nanotubes, and multi-nanolayers. These 3D structures have wide application in electrochemical biosensing and are used as scaffolds for cells, electrodes, and surfaces for bio-receptor immobilization.

Metallic nano-porous materials have been extensively used as electrode materials due to their high surface area. The fabricated nano-porous electrode possesses uniform pore structures, along with a large surface area that is electroactive and highly electric conducting. Further, the active surface when compared to the flat layer is 40 times higher. In addition, the nano-porous structure due to its entrapping ability allows higher adhesion stability. Nanofiber mats are another 3D material with high porosity. Conducting nanofibers are highly exploited as transducers due to their high electric conductivity and high porosity. The non-conducting nanofibers with the immense surface area are widely used as solid support and filtering membrane. Further, the nano-containers which are larger than 100 nm sizes have major ability to load functional species.

24.5 Transducer and Bio-Recognition Unit Integration

As per the definition of a biosensor by IUPAC, the transducers and the bio-recognition unit must be in intimate contact. However, literature studies have shown that a bio-recognition unit may be immobilized on other areas of the biosensor or may not be immobilized. Hence fabrication of biosensor just not deals with finding the suitable components but also there is a need to identify suitable immobilization strategies that allow us to develop bio-recognition elements with high stability and functionality providing specific reactions with the analyte. The most common methods/strategies involve covalent binding, adsorption/physisorption, use of thiol groups, embedment in gels/hydrogels, and membrane entrapment. In addition, antigen tags, DNA tags, histidine tags, and biotinylation biological functions can be used for immobilization. Furthermore, recent developments have shown the use of specific peptide sequences promoting binding to different sequences such as agarose, polystyrene, and cellulose.

24.6 Challenges and Application of Electrochemical Biosensors

Any analytical technique is judged to be advantageous over others if it can evade false-negative and false-positive results. The major concern in an electrochemical assay in real-world samples can be directly concerned with transducer interference with other components of samples. The major challenges to consider with this are nonspecific signal production due to other reactions and electrode fouling. Nonspecific signal production is usually due to the molecules in the sample that can undergo easy electrochemical reaction thereby producing the false-positive signal and also can misinterpret the actual analyte concentration in the sample by producing higher signals. Whereas, electrode fouling refers to nonspecific binding and adsorption of the molecules onto the surface of the electrode from the solution thereby decreasing the sensitivity of the electrode by blocking the electroactive surface and producing false-negative results.

The most common methods for avoiding the interface from the other molecules are mechanical polishing of the electrode surface before measurement or electrochemical cleaning of the electrode surface can avoid electrode fouling problems. In addition, the use of some protective films on the surface by electro-polymerization or the use of a membrane that pre-blocks the defined molecules can also allow solving electrode fouling problems. Further, the strategies such as removing undesired molecules from the sample by pre-treatment, use of blocking agents, coating, or use of a membrane on electrode surface that shows electrostatic repletion of interference.

There are many studies/research on avoiding interference problems and ensuring specific signal generation. Another problem related to electrochemical biosensors is the stability of long-term analysis. Retaining the electroactive surface of the electrode ensuring the functionality throughout storage is a challenge. In such cases, noble metal electrodes can outperform other electrodes. The surface of such electrodes can be easily generated by mechanically polishing the surface. Whereas with screen-printed and organic polymers the surface of the electrode has a single usage. However, the mass production of these electrodes has to be addressed. Some electrodes such as micro-fabricated electrodes under high potential and current flow suffer from detachment having a limited lifetime.

Electrochemical biosensors in recent years have seen extensive application in fields where on-site sensing or point-of-care is required. These sensors have advantages such as easiness in miniaturization, cost-effectiveness, and simplicity in operation. There are various fields such as health care, biomedical research, food analysis and safety, bioprocess industries, and environmental monitoring in which electrochemical biosensors have performed remarkably well. Commercialized electrochemical biosensors such as glucose sensors and biological oxygen demand sensors have proven to be successful due to their simplicity and low cost. Direct samples can be used in such commercialized products, thanks to their ability to allow the usage of sample matrixes with high optical density. The commercialization of electrochemical biosensors has now been concentrating on wearable and implantable sensors in the form of tattoos, contact lenses, eyeglasses, watches, etc.

24.7 Conclusion

Although there has been continuous research in the field of electrochemical biosensors, electrochemical sensing technology is facing problems related to applications for in-person health monitoring, autonomous sensors in the environmental application, and very few electrochemical-based sensors being commercialized. Research focusing on robust design, removing the need for calibration, and long-term usage is a must. In this chapter, we have focused on the fabrication of biosensors and the key components. The chapter discusses the initial basics of biosensors, key components, and use of nanostructured materials in biosensing, integration of biosensor units and challenges, and applications of electrochemical biosensors.

REFERENCES

1. Coppedè, N., Giannetto, M., Villani, M., Lucchini, V., Battista, E., Careri, M., & Zappettini, A. (2020). Ion selective textile organic electrochemical transistor for wearable sweat monitoring. *Organic Electronics*, *78*, 105579.
2. Adami, M., Alliata, D., Del Carlo, C., Martini, M., Piras, L., Sartore, M., & Nicolini, C. (1995). Characterization of silicon transducers with Si3N4 sensing surfaces by an AFM and a PAB system. *Sensors and Actuators B: Chemical*, *25*(1–3), 889–893.
3. Ismail, A. B. M., Harada, T., Yoshinobu, T., Iwasaki, H., Schöning, M. J., & Lüth, H. (2000). Investigation of pulsed laser-deposited Al_2O_3 as a high pH-sensitive layer for LAPS-based biosensing applications. *Sensors and Actuators B: Chemical*, *71*(3), 169–172.
4. Wang, Z., & Dai, Z. (2015). Carbon nanomaterial-based electrochemical biosensors: an overview. *Nanoscale*, *7*(15), 6420–6431.
5. Bukkitgar, S. D., Shetti, N. P., Malladi, R. S., Reddy, K. R., Kalanur, S. S., & Aminabhavi, T. M. (2020). Novel ruthenium doped TiO2/reduced graphene oxide hybrid as highly selective sensor for the determination of ambroxol. *Journal of Molecular Liquids*, *300*, 112368.
6. Shetti, N. P., Malode, S. J., Nayak, D. S., Bukkitgar, S. D., Bagihalli, G. B., Kulkarni, R. M., & Reddy, K. R. (2020). Novel nanoclay-based electrochemical sensor for highly efficient electrochemical sensing nimesulide. *Journal of Physics and Chemistry of Solids*, *137*, 109210.
7. Shetti, N. P., Malode, S. J., Bukkitgar, S. D., Bagihalli, G. B., Kulkarni, R. M., Pujari, S. B., & Reddy, K. R. (2019). Electro-oxidation and determination of nimesulide at nanosilica modified sensor. *Materials Science for Energy Technologies*, *2*(3), 396–400.
8. Bukkitgar, S. D., Shetti, N. P., Kulkarni, R. M., Reddy, K. R., Shukla, S. S., Saji, V. S., & Aminabhavi, T. M. (2019). Electro-catalytic behavior of Mg-doped ZnO nano-flakes for oxidation of anti-inflammatory drug. *Journal of The Electrochemical Society*, *166*(9), B3072.
9. Bukkitgar, S. D., Shetti, N. P., & Kulkarni, R. M. (2018). Construction of nanoparticles composite sensor for atorvastatin and its determination in pharmaceutical and urine samples. *Sensors and Actuators B: Chemical*, *255*, 1462–1470.
10. Bukkitgar, S. D., & Shetti, N. P. (2017). Fabrication of a TiO 2 and clay nanoparticle composite electrode as a sensor. *Analytical Methods*, *9*(30), 4387–4393.
11. Bukkitgar, S. D., & Shetti, N. P. (2016). Electrochemical oxidation of loop diuretic furosemide in aqueous acid medium and its analytical application. *Cogent Chemistry*, *2*(1), 1152784.
12. Bukkitgar, S. D., Shetti, N. P., Kulkarni, R. M., Halbhavi, S. B., Wasim, M., Mylar, M., ... & Chirmure, S. S. (2016). Electrochemical oxidation of nimesulide in aqueous acid solutions based on TiO2 nanostructure modified electrode as a sensor. *Journal of Electroanalytical Chemistry*, *778*, 103–109.
13. Bukkitgar, S. D., & Shetti, N. P. (2016). Electrochemical behavior of an anticancer drug 5-fluorouracil at methylene blue modified carbon paste electrode. *Materials Science and Engineering: C*, *65*, 262–268.
14. Bukkitgar, S. D., & Shetti, N. P. (2016). Electrochemical sensor for the determination of anticancer drug 5-fluorouracil at glucose modified electrode. *Chemistry Select*, *1*(4), 771–777.
15. Bukkitgar, S. D., Shetti, N. P., Kulkarni, R. M., & Nandibewoor, S. T. (2015). Electro-sensing base for mefenamic acid on a 5% barium-doped zinc oxide nanoparticle modified electrode and its analytical application. *RSC Advances*, *5*(127), 104891–104899.

16. Chokkareddy, R., Bhajanthri, N. K., & Redhi, G. G. (2017). An enzyme-induced novel biosensor for the sensitive electrochemical determination of isoniazid. *Biosensors*, *7*(2), 21.
17. Kilele, J. C., Chokkareddy, R., Rono, N., & Redhi, G. G. (2020). A novel electrochemical sensor for selective determination of theophylline in pharmaceutical formulations. *Journal of the Taiwan Institute of Chemical Engineers*, *111*, 228–238.
18. Shetti, N. P., Mishra, A., Bukkitgar, S. D., Basu, S., Narang, J., Raghava Reddy, K., & Aminabhavi, T. M. (2021). Conventional and nanotechnology-based sensing methods for SARS coronavirus (2019-nCoV). *ACS Applied Bio Materials*, *4*(2), 1178–1190.
19. Bukkitgar, S. D., Shetti, N. P., & Aminabhavi, T. M. (2021). Electrochemical investigations for COVID-19 detection-A comparison with other viral detection methods. *Chemical Engineering Journal*, *420*, 127575.
20. Bukkitgar, S. D., Kumar, S., Singh, S., Singh, V., Reddy, K. R., Sadhu, V., … & Naveen, S. (2020). Functional nanostructured metal oxides and its hybrid electrodes: recent advancements in electrochemical biosensing applications. *Microchemical Journal*, *159*, 105522.
21. Chokkareddy, R., Thondavada, N., Kabane, B., & Redhi, G. G. (2021). A novel ionic liquid based electrochemical sensor for detection of pyrazinamide. *Journal of the Iranian Chemical Society*, *18*(3), 621–629.
22. Shetti, N. P., Bukkitgar, S. D., Reddy, K. R., Reddy, C. V., & Aminabhavi, T. M. (2019). Nanostructured titanium oxide hybrids-based electrochemical biosensors for healthcare applications. *Colloids and Surfaces B: Biointerfaces*, *178*, 385–394.
23. Kumar, S., Bukkitgar, S. D., Singh, S., Singh, V., Reddy, K. R., Shetti, N. P., … & Naveen, S. (2019). Electrochemical sensors and biosensors based on graphene functionalized with metal oxide nanostructures for healthcare applications. *Chemistry Select*, *4*(18), 5322–5337.
24. Comotti, M., Della Pina, C., Matarrese, R., & Rossi, M. (2004). The catalytic activity of “naked” gold particles. *Angewandte Chemie International Edition*, *43*(43), 5812–5815.
25. Gao, L., Zhuang, J., Nie, L., Zhang, J., Zhang, Y., Gu, N., … & Yan, X. (2007). Intrinsic peroxidase-like activity of ferromagnetic nanoparticles. *Nature Nanotechnology*, *2*(9), 577–583.
26. Wang, G., He, X., Wang, L., Gu, A., Huang, Y., Fang, B., … & Zhang, X. (2013). Non-enzymatic electrochemical sensing of glucose. *Microchimica Acta*, *180*(3–4), 161–186.
27. Niu, X., Li, X., Pan, J., He, Y., Qiu, F., & Yan, Y. (2016). Recent advances in non-enzymatic electrochemical glucose sensors based on non-precious transition metal materials: opportunities and challenges. *RSC Advances*, *6*(88), 84893–84905.
28. Holade, Y., Tingry, S., Servat, K., Napporn, T. W., Cornu, D., & Kokoh, K. B. (2017). Nanostructured inorganic materials at work in electrochemical sensing and biofuel cells. *Catalysts*, *7*(1), 31.
29. Zhang, L., Liang, H., Ma, X., Ye, C., & Zhao, G. (2019). A vertically aligned CuO nanosheet film prepared by electrochemical conversion on Cu-based metal-organic framework for non-enzymatic glucose sensors. *Microchemical Journal*, *146*, 479–485.
30. Fang, B., Kim, J. H., Kim, M. S., & Yu, J. S. (2013). Hierarchical nanostructured carbons with meso–macroporosity: design, characterization, and applications. *Accounts of Chemical Research*, *46*(7), 1397–1406.
31. Zhang, L., Ding, Y., Li, R., Ye, C., Zhao, G., & Wang, Y. (2017). Ni-Based metal–organic framework derived Ni@ C nanosheets on a Ni foam substrate as a supersensitive non-enzymatic glucose sensor. *Journal of Materials Chemistry B*, *5*(28), 5549–5555.
32. Hou, L., Bi, S., Lan, B., Zhao, H., Zhu, L., Xu, Y., & Lu, Y. (2019). A novel and ultrasensitive nonenzymatic glucose sensor based on pulsed laser scribed carbon paper decorated with nanoporous nickel network. *Analytica Chimica Acta*, *1082*, 165–175.
33. Lee, S., Lee, J., Park, S., Boo, H., Kim, H. C., & Chung, T. D. (2018). Disposable non-enzymatic blood glucose sensing strip based on nanoporous platinum particles. *Applied Materials Today*, *10*, 24–29.
34. Tavakkoli, N., & Nasrollahi, S. (2013). Non-enzymatic glucose sensor based on palladium coated nanoporous gold film electrode. *Australian Journal of Chemistry*, *66*(9), 1097–1104.
35. Bai, H., Han, M., Du, Y., Bao, J., & Dai, Z. (2010). Facile synthesis of porous tubular palladium nanostructures and their application in a nonenzymatic glucose sensor. *Chemical Communications*, *46*(10), 1739–1741.
36. Chen, X. M., Cai, Z. M., Lin, Z. J., Jia, T. T., Liu, H. Z., Jiang, Y. Q., & Chen, X. (2009). A novel non-enzymatic ECL sensor for glucose using palladium nanoparticles supported on functional carbon nanotubes. *Biosensors and Bioelectronics*, *24*(12), 3475–3480.
37. Yang, C. L., Zhang, X. H., Lan, G., Chen, L. Y., Chen, M. W., Zeng, Y. Q., & Jiang, J. Q. (2014). Pd-based nanoporous metals for enzyme-free electrochemical glucose sensors. *Chinese Chemical Letters*, *25*(4), 496–500.
38. Atacan, K., & Özacar, M. (2021). Construction of a non-enzymatic electrochemical sensor based on CuO/g-C3N4 composite for selective detection of hydrogen peroxide. *Materials Chemistry and Physics*, *266*, 124527.
39. Sinha, G. N., Subramanyam, P., Sivaramakrishna, V., & Subrahmanyam, C. (2021). Electrodeposited copper bismuth oxide as a low-cost, non-enzymatic electrochemical sensor for sensitive detection of uric acid and hydrogen peroxide. *Inorganic Chemistry Communications*, *129*, 108627.
40. Yin, H., Shi, Y., Dong, Y., & Chu, X. (2021). Synthesis of spinel-type CuGa2O4 nanoparticles as a sensitive non-enzymatic electrochemical sensor for hydrogen peroxide and glucose detection. *Journal of Electroanalytical Chemistry*, *885*, 115100.
41. Zhao, X., Li, Z., Chen, C., Xie, L., Zhu, Z., Zhao, H., & Lan, M. (2021). MnFe2O4 nanoparticles-decorated graphene nanosheets used as an efficient peroxidase minic enable the electrochemical detection of hydrogen peroxide with a low detection limit. *Microchemical Journal*, *166*, 106240.
42. Zhao, J., Yang, H., Wu, W., Shui, Z., Dong, J., Wen, L., … & Huo, D. (2021). Flexible nickel–cobalt double hydroxides micro-nano arrays for cellular secreted hydrogen peroxide in-situ electrochemical detection. *Analytica Chimica Acta*, *1143*, 135–143.

43. Bach, L. G., Thi, M. L. N., Son, N. T., Bui, Q. B., Nhac-Vu, H. T., & Ai-Le, P. H. (2019). Mesoporous gold nanoparticles supported cobalt nanorods as a free-standing electrochemical sensor for sensitive hydrogen peroxide detection. *Journal of Electroanalytical Chemistry*, *848*, 113359.
44. Sastry, M., Rao, M., & Ganesh, K. N. (2002). Electrostatic assembly of nanoparticles and biomacromolecules. *Accounts of Chemical Research*, *35*(10), 847–855.
45. Jazayeri, M. H., Amani, H., Pourfatollah, A. A., Pazoki-Toroudi, H., & Sedighimoghaddam, B. (2016). Various methods of gold nanoparticles (GNPs) conjugation to antibodies. *Sensing and Bio-sensing Research*, *9*, 17–22.
46. Ansari, A. A., Alhoshan, M., Alsalhi, M. S., & Aldwayyan, A. S. (2010). Nanostructured metal oxides based enzymatic electrochemical biosensors. *Biosensors*, 302, 23–46.
47. Bukkitgar, S. D., Shetti, N. P., & Kulkarni, R. M. (2017). Electro-oxidation and determination of 2-thiouracil at TiO2 nanoparticles-modified gold electrode. *Surfaces and Interfaces*, *6*, 127–133.
48. Ilager, D., Seo, H., Kalanur, S. S., Shetti, N. P., & Aminabhavi, T. M. (2021). A novel sensor based on WO3·0.33 H_2O nanorods modified electrode for the detection and degradation of herbicide, carbendazim. *Journal of Environmental Management*, *279*, 111611.
49. Ilager, D., Seo, H., Shetti, N. P., & Kalanur, S. S. (2020). CTAB modified Fe-WO3 as an electrochemical detector of amitrole by catalytic oxidation. *Journal of Environmental Chemical Engineering*, *8*(6), 104580.
50. Kumar, D. K., Reddy, K. R., Sadhu, V., Shetti, N. P., Reddy, C. V., Chouhan, R. S., & Naveen, S. (2020). Metal oxide-based nanosensors for healthcare and environmental applications. *Nanomaterials in Diagnostic Tools and Devices*, *1*, 113–129.
51. Shetti, N. P., Bukkitgar, S. D., Reddy, K. R., Reddy, C. V., & Aminabhavi, T. M. (2019). ZnO-based nanostructured electrodes for electrochemical sensors and biosensors in biomedical applications. *Biosensors and Bioelectronics*, *141*, 111417.
52. Unwin, P. R., Guell, A. G., & Zhang, G. (2016). Nanoscale electrochemistry of sp2 carbon materials: from graphite and graphene to carbon nanotubes. *Accounts of Chemical Research*, *49*(9), 2041–2048.
53. Vasilescu, A., Hayat, A., Gáspár, S., & Marty, J. L. (2018). Advantages of carbon nanomaterials in electrochemical aptasensors for food analysis. *Electroanalysis*, *30*(1), 2–19.
54. Wang, Y., & Hu, S. (2016). Applications of carbon nanotubes and graphene for electrochemical sensing of environmental pollutants. *Journal of Nanoscience and Nanotechnology*, *16*(8), 7852–7872.
55. Raj, M. A., & John, S. A. (2013). Fabrication of electrochemically reduced graphene oxide films on glassy carbon electrode by self-assembly method and their electrocatalytic application. *Journal of Physical Chemistry C*, *117*(8), 4326–4335.
56. Mercante, L. A., Pavinatto, A., Iwaki, L. E., Scagion, V. P., Zucolotto, V., Oliveira Jr, O. N., ... & Correa, D. S. (2015). Electrospun polyamide 6/poly (allylamine hydrochloride) nanofibers functionalized with carbon nanotubes for electrochemical detection of dopamine. *ACS Applied Materials & Interfaces*, *7*(8), 4784–4790.
57. Zhang, D., Li, L., Ma, W., Chen, X., & Zhang, Y. (2017). Electrodeposited reduced graphene oxide incorporating polymerization of l-lysine on electrode surface and its application in simultaneous electrochemical determination of ascorbic acid, dopamine and uric acid. *Materials Science and Engineering: C*, *70*, 241–249.
58. Fang, H., Pan, Y., Shan, W., Guo, M., Nie, Z., Huang, Y., & Yao, S. (2014). Enhanced nonenzymatic sensing of hydrogen peroxide released from living cells based on Fe 3 O 4/self-reduced graphene nanocomposites. *Analytical Methods*, *6*(15), 6073–6081.
59. Tiwari, J. N., Vij, V., Kemp, K. C., & Kim, K. S. (2016). Engineered carbon-nanomaterial-based electrochemical sensors for biomolecules. *ACS Nano*, *10*(1), 46–80.
60. Gao, F., Cai, X., Wang, X., Gao, C., Liu, S., Gao, F., & Wang, Q. (2013). Highly sensitive and selective detection of dopamine in the presence of ascorbic acid at graphene oxide modified electrode. *Sensors and Actuators B: Chemical*, *186*, 380–387.
61. Pandikumar, A., How, G. T. S., See, T. P., Omar, F. S., Jayabal, S., Kamali, K. Z., ... & Huang, N. M. (2014). Graphene and its nanocomposite material based electrochemical sensor platform for dopamine. *RSC Advances*, *4*(108), 63296–63323.
62. Tang, Y., Allen, B. L., Kauffman, D. R., & Star, A. (2009). Electrocatalytic activity of nitrogen-doped carbon nanotube cups. *Journal of the American Chemical Society*, *131*(37), 13200–13201.
63. Wang, Y., Shao, Y., Matson, D. W., Li, J., & Lin, Y. (2010). Nitrogen-doped graphene and its application in electrochemical biosensing. *ACS Nano*, *4*(4), 1790–1798.
64. Wei, D., Liu, Y., Wang, Y., Zhang, H., Huang, L., & Yu, G. (2009). Synthesis of N-doped graphene by chemical vapor deposition and its electrical properties. *Nano Letters*, *9*(5), 1752–1758.
65. Xiao, K., Liu, Y., Hu, P. A., Yu, G., Sun, Y., & Zhu, D. (2005). n-Type field-effect transistors made of an individual nitrogen-doped multiwalled carbon nanotube. *Journal of the American Chemical Society*, *127*(24), 8614–8617.
66. Kwon, O. S., Park, S. J., Hong, J. Y., Han, A. R., Lee, J. S., Lee, J. S., ... & Jang, J. (2012). Flexible FET-type VEGF aptasensor based on nitrogen-doped graphene converted from conducting polymer. *ACS Nano*, *6*(2), 1486–1493.
67. Brandão, W. Q., Medina-Llamas, J. C., Alcaraz-Espinoza, J. J., Chávez-Guajardo, A. E., & De Melo, C. P. (2016). Polyaniline–polystyrene membrane for simple and efficient retrieval of double-stranded DNA from aqueous media. *RSC Advances*, *6*(106), 104566–104574.
68. Ferreira, V. C., Melato, A. I., Silva, A. F., & Abrantes, L. M. (2011). Attachment of noble metal nanoparticles to conducting polymers containing sulphur–preparation conditions for enhanced electrocatalytic activity. *Electrochimica Acta*, *56*(10), 3567–3574.
69. Ferreira, V. C., Melato, A. I., Silva, A. F., & Abrantes, L. M. (2011). Conducting polymers with attached platinum nanoparticles towards the development of DNA biosensors. *Electrochemistry Communications*, *13*(9), 993–996.
70. Janáky, C., & Visy, C. (2013). Conducting polymer-based hybrid assemblies for electrochemical sensing: a materials science perspective. *Analytical and Bioanalytical Chemistry*, *405*(11), 3489–3511.

71. Kim, W., Lee, J. S., Shin, D. H., & Jang, J. (2018). Platinum nanoparticles immobilized on polypyrrole nanofibers for non-enzyme oxalic acid sensor. *Journal of Materials Chemistry B*, *6*(8), 1272–1278.
72. Yurkov, G. Y., Gubin, S. P., & Ovchenkov, E. A. (2009). Magnetic nanocomposites based on the metal-containing (Fe, Co, Ni) nanoparticles inside the polyethylene matrix. *Magnetic Nanoparticles*, *4*, 1.
73. Tahir, M. B. (2018). Construction of MoS 2/CND-WO 3 ternary composite for photocatalytic hydrogen evolution. *Journal of Inorganic and Organometallic Polymers and Materials*, *28*(5), 2160–2168.
74. Zhang, S., Geryak, R., Geldmeier, J., Kim, S., & Tsukruk, V. V. (2017). Synthesis, assembly, and applications of hybrid nanostructures for biosensing. *Chemical Reviews*, *117*(20), 12942–13038.
75. Kim, Y. T., Han, J. H., Hong, B. H., & Kwon, Y. U. (2010). Electrochemical synthesis of CdSe quantum-dot arrays on a graphene basal plane using mesoporous silica thin-film templates. *Advanced Materials*, *22*(4), 515–518.
76. Zhang, G., & Wang, D. (2008). Fabrication of heterogeneous binary arrays of nanoparticles via colloidal lithography. *Journal of the American Chemical Society*, *130*(17), 5616–5617.
77. Wang, J., Lin, M., Yan, Y., Wang, Z., Ho, P. C., & Loh, K. P. (2009). CdSe/AsS core– shell quantum dots: preparation and two-photon fluorescence. *Journal of the American Chemical Society*, *131*(32), 11300–11301.
78. Gautam, U. K., Vivekchand, S. R. C., Govindaraj, A., Kulkarni, G. U., Selvi, N. R., & Rao, C. N. R. (2005). Generation of onions and nanotubes of GaS and GaSe through laser and thermally induced exfoliation. *Journal of the American Chemical Society*, *127*(11), 3658–3659.
79. Lee, J. Y., Hong, B. H., Kim, W. Y., Min, S. K., Kim, Y., Jouravlev, M. V., … & Kim, K. S. (2009). Near-field focusing and magnification through self-assembled nanoscale spherical lenses. *Nature*, *460*(7254), 498–501.
80. An, K., & Somorjai, G. A. (2012). Size and shape control of metal nanoparticles for reaction selectivity in catalysis. *ChemCatChem*, *4*(10), 1512–1524.
81. Lim, B., Jiang, M., Camargo, P. H., Cho, E. C., Tao, J., Lu, X., … & Xia, Y. (2009). Pd-Pt bimetallic nanodendrites with high activity for oxygen reduction. *Science*, *324*(5932), 1302–1305.
82. Bonanni, A., Pumera, M., & Miyahara, Y. (2011). Influence of gold nanoparticle size (2–50 nm) upon its electrochemical behavior: an electrochemical impedance spectroscopic and voltammetric study. *Physical Chemistry Chemical Physics*, *13*(11), 4980–4986.
83. de La Escosura-Muñiz, A., Parolo, C., Maran, F., & Mekoçi, A. (2011). Size-dependent direct electrochemical detection of gold nanoparticles: application in magnetoimmunoassays. *Nanoscale*, *3*(8), 3350–3356.
84. Kokkinos, C. T., Giokas, D. L., Economou, A. S., Petrou, P. S., & Kakabakos, S. E. (2018). Based microfluidic device with integrated sputtered electrodes for stripping voltammetric determination of DNA via quantum dot labeling. *Analytical Chemistry*, *90*(2), 1092–1097.
85. Kokkinos, C., Angelopoulou, M., Economou, A., Prodromidis, M., Florou, A., Haasnoot, W., … & Kakabakos, S. (2016). Lab-on-a-membrane foldable devices for duplex drop-volume electrochemical biosensing using quantum dot tags. *Analytical Chemistry*, *88*(13), 6897–6904.
86. Zhang, X., Ren, X., Cao, W., Li, Y., Du, B., & Wei, Q. (2014). Simultaneous electrochemical immunosensor based on water-soluble polythiophene derivative and functionalized magnetic material. *Analytica Chimica Acta*, *845*, 85–91.
87. Laurent, S., Forge, D., Port, M., Roch, A., Robic, C., Vander Elst, L., & Muller, R. N. (2008). Magnetic iron oxide nanoparticles: synthesis, stabilization, vectorization, physicochemical characterizations, and biological applications. *Chemical Reviews*, *108*(6), 2064–2110.
88. Jamshaid, T., Neto, E. T. T., Eissa, M. M., Zine, N., Kunita, M. H., El-Salhi, A. E., & Elaissari, A. (2016). Magnetic particles: from preparation to lab-on-a-chip, biosensors, microsystems and microfluidics applications. *TrAC Trends in Analytical Chemistry*, *79*, 344–362.
89. Gong, K., Chakrabarti, S., & Dai, L. (2008). Electrochemistry at carbon nanotube electrodes: is the nanotube tip more active than the sidewall?. *Angewandte Chemie*, *120*(29), 5526–5530.
90. Liu, J., Chou, A., Rahmat, W., Paddon-Row, M. N., & Gooding, J. J. (2005). Achieving direct electrical connection to glucose oxidase using aligned single walled carbon nanotube arrays. *Electroanalysis: An International Journal Devoted to Fundamental and Practical Aspects of Electroanalysis*, *17*(1), 38–46.
91. Li, J., Ng, H. T., Cassell, A., Fan, W., Chen, H., Ye, Q., … & Meyyappan, M. (2003). Carbon nanotube nanoelectrode array for ultrasensitive DNA detection. *Nano Letters*, *3*(5), 597–602.
92. Centi, S., Ratto, F., Tatini, F., Lai, S., & Pini, R. (2018). Ready-to-use protein G-conjugated gold nanorods for biosensing and biomedical applications. *Journal of Nanobiotechnology*, *16*(1), 1–11.
93. Parab, H. J., Chen, H. M., Lai, T. C., Huang, J. H., Chen, P. H., Liu, R. S., … & Hwu, Y. K. (2009). Biosensing, cytotoxicity, and cellular uptake studies of surface-modified gold nanorods. *Journal of Physical Chemistry C*, *113*(18), 7574–7578.
94. Castellana, E. T., Gamez, R. C., & Russell, D. H. (2011). Label-free biosensing with lipid-functionalized gold nanorods. *Journal of the American Chemical Society*, *133*(12), 4182–4185.
95. Kang, Z., Yan, X., Wang, Y., Zhao, Y., Bai, Z., Liu, Y., … & Zhang, Y. (2016). Self-powered photoelectrochemical biosensing platform based on Au NPs@ ZnO nanorods array. *Nano Research*, *9*(2), 344–352.
96. Arter, J. A., Taggart, D. K., McIntire, T. M., Penner, R. M., & Weiss, G. A. (2010). Virus-PEDOT nanowires for biosensing. *Nano Letters*, *10*(12), 4858–4862.
97. Wang, Y., Wang, T., Da, P., Xu, M., Wu, H., & Zheng, G. (2013). Silicon nanowires for biosensing, energy storage, and conversion. *Advanced Materials*, *25*(37), 5177–5195.
98. M. N, M. N., Hashim, U., Md Arshad, M. K., Ruslinda, A. R., Rahman, S. F. A., Fathil, M. F. M., & Ismail, M. H. (2016). Top-down nanofabrication and characterization of 20 nm silicon nanowires for biosensing applications. *PLoS One*, *11*(3), e0152318.
99. Nicolini, J. V., Ferraz, H. C., & de Resende, N. S. (2016). Immobilization of horseradish peroxidase on titanate nanowires for biosensing application. *Journal of Applied Electrochemistry*, *46*(1), 17–25.

25

Impact of Calcium Ions (Ca^{2+}) and Their Signaling in Alzheimer's and Other Neurological-Related Disorders

Neha Chauhan, Smita Jain, Kanika Verma, Swapnil Sharma, Raghuraj Chouhan and Veera Sadhu

CONTENTS

25.1 Introduction

Calcium diversifies the variety of physiological functions which also includes neural plasticity and apoptosis. Intracellular Ca^{2+} homeostasis is imperative for maintaining these functions; if this upkeep is dysregulated then it will proceed to Alzheimer's pathogenesis [1]. It has been well documented that an elevation of intracellular calcium results in characteristic lesions of Alzheimer's with the accumulation of amyloid-beta, tau hyper-phosphorylation and eventually approaching neural death and various issues [2]. The ER is a vital reserve for calcium, and any kind of disruption will mediate the signaling which is linked with Alzheimer's.

25.1.1 Possible Linkage between Calcium and AD

It can be acknowledged as a perpetual intracellular messenger. Calcium, being a divalent cation, has the potential of binding various channels, receptors and proteins. All these characteristics are of paramount significance with neurons, which further leads segments to the calcium cycle. It achieves the influx at intracellular buffering and plasma membrane level through calcium channels and provides efflux by the means of calcium plasma membrane transporters. The complete cycle includes various proteins and subcellular compartments. Specifically, two of the organelles play a pivotal role in executing calcium buffering, i.e., the mitochondria and endoplasmic reticulum (ER) [3]. NCX and ATPase calcium pumps are two major components in implying plasma membrane calcium efflux, where any type of disruption in this subtle equilibrium state might prove to be pernicious particularly for neuron and cell systems, which further augments to apoptosis/necrosis ultimately leading to neurodegeneration and neural disorders [4].

In 2008, Marambaud screened genes which are positioned in recognized AD linkage site and addressed a new channel

DOI: 10.1201/9781003157823-25

which is conducting calcium, with polymorphisms related with a higher probability for the blooming of sporadic AD, described as CALHM1, a glycoprotein with three transmembrane domains, present in every region of brain and neuronal lineage cells (Figure 25.1). Mainly, CALHM1 is confined in the ER but also existent in the PM, where it arbitrates a new Ca^{2+} influx into cytosol, which was unaltered with particular blockers of VGCC or SOCC but inhibited by non-specific blockers of cation channel like cobalt. CALHM1 is a multimeric complex having structural analogy with NMDAR in the ion selectivity region, providing gain to functional ion channels [5, 6].

Aβ production decreases as the influx of Ca^{2+} occur through CALHM1 followed by an increase in sAPPα (soluble amyloid precursor protein). The basic mechanism behind increase and decrease was not clear thus far but it is inferred that it includes Ca^{2+} dependent effects on enzyme α-secretase, cleaving APP83a.a from its carboxyl terminus and eventually halts the formation of Aβ. Conversely, after endogenous CALHM1siRNA knockdown increase in Aβ formation was observed combining with Ca^{2+} influxes. Contrarily, the above observation is conflicting with most of the studies.

With the assistance of extracellular spaces, Ca^{2+} enters inside cytosol by ROCs, ionotropic receptor-operated channels, VOCCs (voltage-operated Ca^{2+} channels) and also through SOCCs (store-operated calcium channels). ROCs are permeable to Ca^{2+} which involves NMDARs and some AMPARs (amino-3-hydroxy-5-methylisoxazole-4-propionate acid receptors). In intracellular stores like ER and mitochondria, IP3R and RyR help calcium enter the cytosol [7]. When the elevation of $[Ca^{2+}]_i$ begins, mitochondria provide protection against increasing $[Ca^{2+}]_i$ by rapidly sequestering Ca^{2+} buffers. Conclusively, gradual clearance of Ca^{2+} is performed by pumps of Ca^{2+} like Ca^{2+} ATPase (PMCA) [8]. Its activity is regulated by MICU1 and MICU2, injunction with the MCU make up the mitochondrial calcium uniporter complex and exchangers like Na+/Ca^{2+} exchanger (NCX) removes Ca^{2+} from the cytoplasm, it is having low affinity but raised capacity with PMCA for Ca^{2+} [9, 10]. Various CBPs (calcium-binding proteins) like calbindin have properties of buffers with that it also occurs during the removal of Ca^{2+} either into intracellular stores such as the ER via SERCA (sarco-endoplasmic reticulum calcium ATPase) or moves out across the PM via plasmalemmal calcium pumps and exchangers [11].

25.1.2 Calcium Homeostasis

There are a number of studies linking the connection between calcium homeostasis disruption and the augmentation of neurodegenerative disorders like AD. The pathogenesis linkage between calcium and AD has been reported by Khachaturian. Further, there are a lot more studies have been done on the same.

There are two main mechanisms for terminating calcium signals: (i) re-aggregation in intracellular calcium and (ii) upon

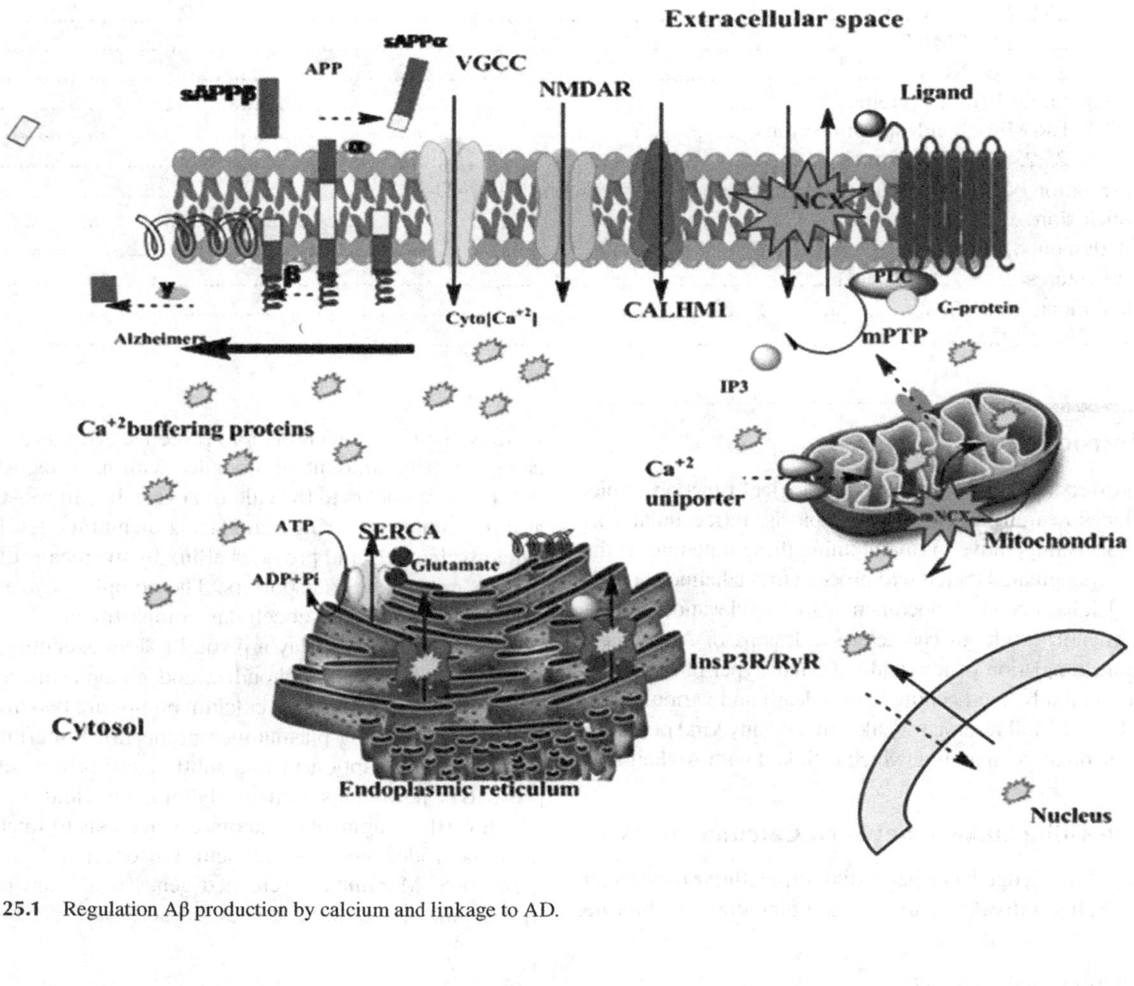

FIGURE 25.1 Regulation Aβ production by calcium and linkage to AD.

displacement from the cytosol into extracellular space [12]. Buffering capacity is also important in the propagation and production of cytosolic calcium signaling.

In resting conditions, the level of calcium inside the cytosol is maintained low, around 100 nM. After achieving stimuli the calcium level inside cytosol increases rapidly to many folds. Entry of extracellular calcium, where Ca^{2+} is about 1.8 mM. Majority of tissues except for muscle cells there is an activation of PLC which produces IP3 and DAG (diacylglycerol), out of which IP3 collaborate with calcium releasing channels those are present in ER, when IP3 binds its receptor Ca^{2+} is released into the cytosol, thereby activating various Ca^{2+}regulated intracellular signals. Subsequently, store calcium reduction induces the influx of calcium through the PM channel (capacitative calcium entry) [13]. An intracellular calcium signal is generated; it should be properly switched off. It is necessary specificity for controlling the function of cells with reduction of extra harmful stimulation (Figure 25.2).

Inside the cytosol, various calcium-binding proteins are available which are able to shape the signal. They are consisting of EF-hands domains which are conserved domains that have played a key role in calcium-binding. EF-hands are made up of two alpha helixes with 12 amino acid loops between these proteins; many have a function in buffering and chelating calcium ions, mainly paralbumin, calbindin and calretinin. CaM is a calcium sensor that recognizes the changes in structure. At the time when calcium binds to its EF-hands, their interaction occurs due to the exposure of hydrophobic domains with the modulation of effectors (different proteins). Therefore, this buffering capacity provides shape to cellular calcium signals in compartmentalization (spatial), also limiting calcium diffusing rate (temporal extension) [14].

25.1.3 Plasma Membrane

At the plasma membrane influx of the cellular Ca^{2+} is majorly established by ROCCs (receptor-operated calcium channel), SOCCs (store-operated calcium channels), VOCCs (voltage-operated calcium channel) and also in the rarest occurrence by sodium/calcium exchanger (NCX).

PM helps in maintaining ionic composition between two compartments allowing transmission of a signal from the extracellular environment to the cell as an ionic movement. Ion selectivity and their opening are the two bases of distinguishment between channels. The nervous system and cardiac system are excitable tissues, in which VOCCs play a significant role; their opening is primarily regulated by depolarization of the membrane. They are further classified into various subtypes with respect to their voltage and inhibitor sensitivity. P/Q in synapses regulates Ca^{2+} dependent releases of neurotransmitters and L-type in dendrites regulates calcium-dependent transcription (Figure 25.2A).

In addition to VOCCs, another set of channels, ROCCs (receptor-operated Ca^{2+} channels), are also present. These are usually interlinked through the binding of an extracellular ligand toward the equivalent peptide establishing a channel. Then, the same bond opens up the channel permitting Ca^{2+} entry into the cell. NMDAR and AMPAR are two main receptors that are activated through physiological ligand glutamate. ROCCs have prime importance in the neurons which depend upon their specific localization with the amount of calcium that they allow to enter, in the soma or at synapses. After that, it can easily generate a different or opposite signal as in LTD and LTP. These channels don't bind to extracellular ligands but can bind to other types of receptor-like GPCRs and RTKs.

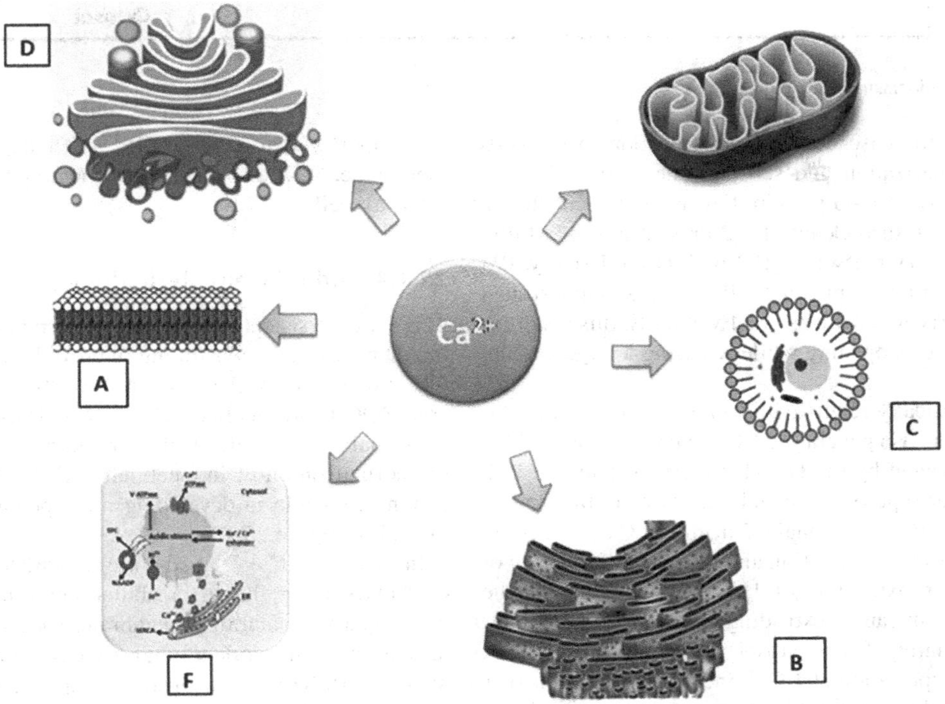

FIGURE 25.2 Ca^{2+}-handling organelles.

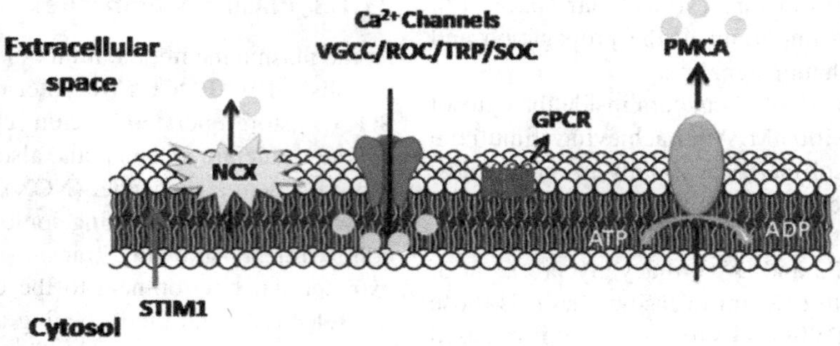

FIGURE 25.2A Plasma membrane.

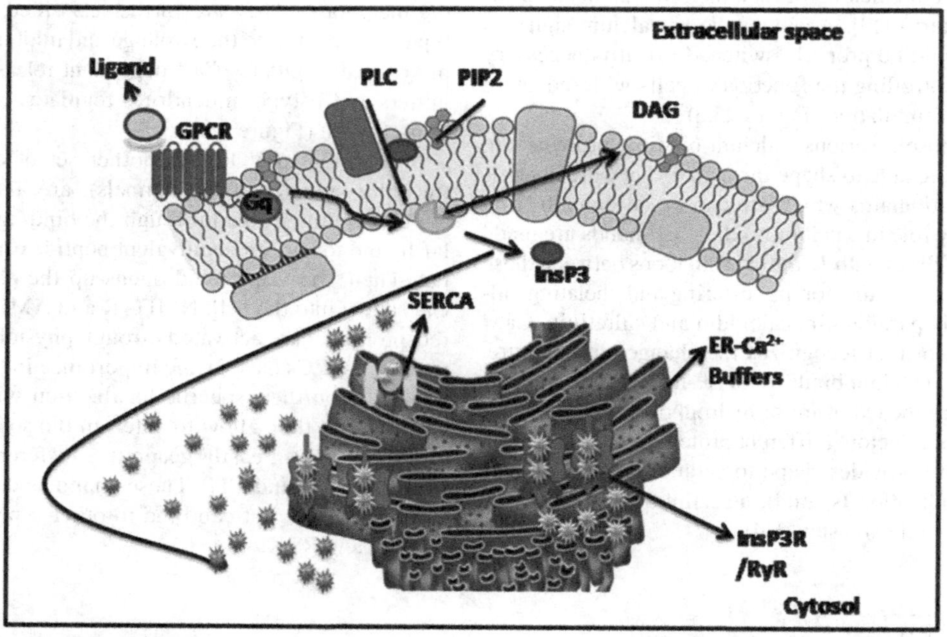

FIGURE 25.2B Endoplasmic reticulum.

RTKs initiate the cascade of phosphorylation in the cells during their dimerization and self-phosphorylation. GPCRs can activate the trimeric-G protein. Further, they activate the PLC-beta, which in turn cleaves PIP2 present in the PM giving rise to secondary messenger IP3 with DAG. Usually, IP3 is considered a prime agonist of IP3R which is demonstrated at membrane intracellular Ca^{2+}store. Eventually, this results in the inducement of its opening while releasing a large amount of Ca^{2+}.

SOCCs are basically regulated by a reduction in intracellular calcium stores, also generating CCE. The formation of this channel is augmented by the Orai-1 protein available in PM. STIM1 is another type of ER membrane protein, that senses ER (Ca^{2+}) via the EF-hand domain. During ER Ca^{2+} reduction, STIM1 augments oligomerization and allows the formation of clusters in closer proximity toward the PM. Contrarily to the PM, another protein causes extruding of Ca^{2+} to extracellular compartments mainly from cytosol. Systematically, PMCA relatively helps exportation of 1 Ca^{2+} including hydrolysis of 1 ATP molecule and also is vital in the maintenance of reduced cytosolic Ca^{2+}. While NCX displayed the ability to export Ca^{2+} and import Na^+ in ratios of 1:3. After depolarization of the membrane, NCX also displays the potential to work in reverse mode as well.

25.1.4 Endoplasmic Reticulum

There are persistent studies about aberrant calcium signaling in AD neurons, demonstrating improved intracellular calcium (Ca^{2+}) release from the endoplasmic reticulum (ER) and lowered SOC (store-operated Ca^{2+}entry). These alterations augment mainly as resultant of overloading Ca^{2+}. It implies that the normalization of intracellular Ca^{2+} homeostasis can play a significant role in developing more potent disease-altering therapies (Figure 25.2B).

In neurons, Ca^{2+} signals are modulated by organizing Ca^{2+}influx from the extracellular environment. It diffuses in cytoplasm containing membrane sheets which enclose a nucleus with a network of tubules and has three main compartments, SER, RER and nuclear envelope, having a role in folding, modification, transport and lipid synthesis. Representing the main $[Ca^{2+}]_i$ store, it exhibits a key role in cell calcium homeostasis with signaling. ATP-driven pumps for Ca^{2+} uptakes are known as SERCA.

Inside cytosol, [Ca^{2+}] is regulated in minimum levels in resting conditions at 50–300 nM. But increases massively in an amount from nM range to mM once after the activation is mediated by the receptor, electrical/synaptic. In the ER the lumen concentration of Ca^{2+} approaches 500 μM, then Ca^{2+} releases from two types of Ca^{2+} channels: IP3Rs and RyRs. In IP3Rs releasing of Ca^{2+} needs G-protein activation on the surface of the cell performed by PLC. Thereafter, phosphatidylinositol-4, 5-bisphosphate is cleaved through PLC as IP3 and DAG with release of mediated tyrosine-kinase-coupled receptors. In particular cells [Ca^{2+}] release from the ER is necessary which occurs in the form of fundamental events like blips (quarks) which define the activity of the individual IP3R/RyR, respectively. This will start either spontaneously or after minimum stimulation, puffs corresponding to Ca^{2+} wave oscillations. At higher stimulation puffs stimulate IP3R or RyR clusters opening and calcium wave as a propagating signal by the calcium released through calcium induction. Calcium signals are deciphered into the biological response by the spatio-temporal pattern, amplitude, frequency and kinetics of release and decay.

25.1.5 Nucleus

In AD, there are sequential cleavages of amyloid protein precursor by β, and γ-secretase produces amyloid β-peptide. It gives rise to oligomers inside the PM and generates pores that provide passage of Ca^{2+} to the cytoplasm. Phosphatidylserine (PtdS) binding facilitates the amyloid-β interaction with the PM. Mitochondrial impairment, i.e., reduction in ATP, might stimulate PtdS flipping from the inner portion of the PM to the surface of the cell. This results due to Ca^{2+} influx or release from the ER or mitochondria which results in activation of PLSCR1 (phospholipids scramblase). Interaction between amyloid-β with Fe^{2+} and Cu^{+} produces OH and H_2O_2 causes the peroxidation of membrane lipid which produces toxic aldehydes impairing the functioning of Na^{+} and Ca^{2+}pumps, resulting in membrane depolarization. After that both NMDAR and VDCC open and flux high toxic amounts Ca^{2+} into the cytoplasm. Amyloid-β will also affect mitochondria either directly or indirectly by increasing [Ca^{2+}] and oxidative stress which results in increased production of free radical generates ATP and Ca^{2+} overloads. Amyloidogenic processing produces AICD (APP intracellular cytoplasmic/c-terminal domain), the smaller cleavage product by γ-secretase which finally translocate to the nucleus. Thus, it alters the transcription of genes in a variety of ways, due to which it will lead to Ca^{2+} dysregulation (Figure 25.2C).

25.1.6 Golgi Apparatus

The Golgi apparatus is a type of organelle (Figure 25.2D). It is mainly divided into three types of different compartments, each having separate functionality: Golgi cis-, medial- and trans-. It exhibits a fundamental role as an intracellular Ca^{2+} store, which was previously displayed with the whole Golgi targeted aequorin Ca^{2+} probe [15]. Additionally, with respect to SERCA, GA showcases supplemental type ATP-dependent Ca^{2+} pumping, SPCA1 (secretary-pathways Ca^{2+} ATPase 1). This arrangement of twin pumps is considered heterogeneous within the GA, while the first one is applicable in the cis- and medial-GA, the other one is available in trans- and medial-GA [16, 17]. Following the similar approach, the IP3R has its presence in the cis- and, at relatively low degree, at the medial-Golgi level, but is absent in the trans-Golgi, which makes latter compartments relatively insensitive toward IP3 mediated stimulation. Contrarily, RyRs are divided thoroughly in the GA, subsequently, which makes the trans-Golgi a significant, dynamic Ca^{2+} store having the potential to release Ca^{2+} into the cytosol, especially tissues where expression of RyRs are present abundantly, such as the heart [16, 17].

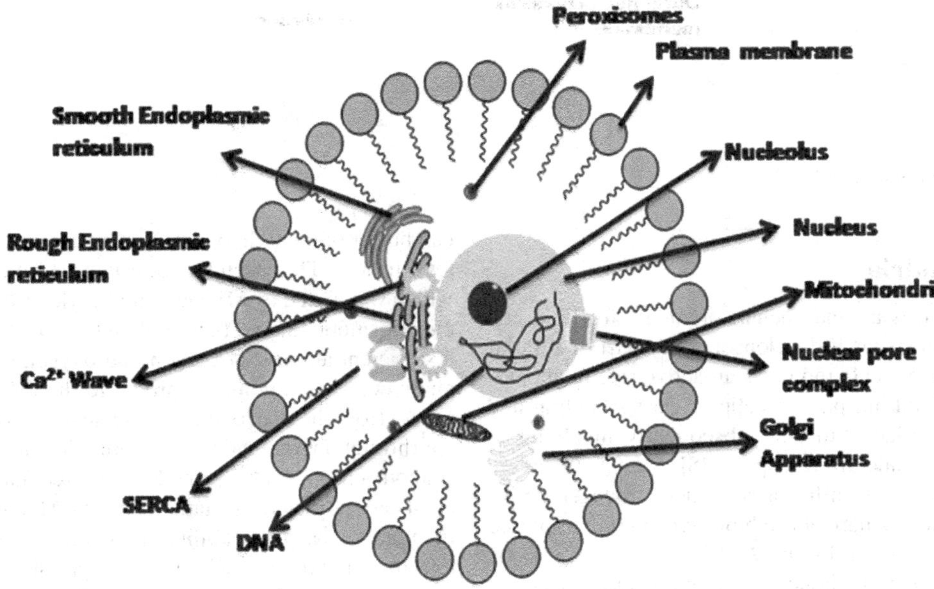

FIGURE 25.2C Nucleus.

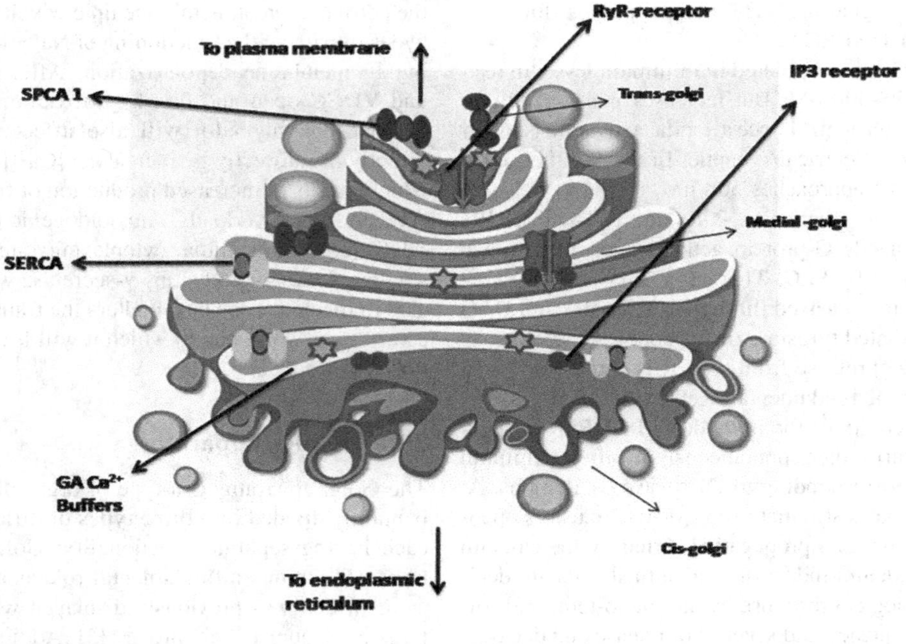

FIGURE 25.2D Golgi apparatus.

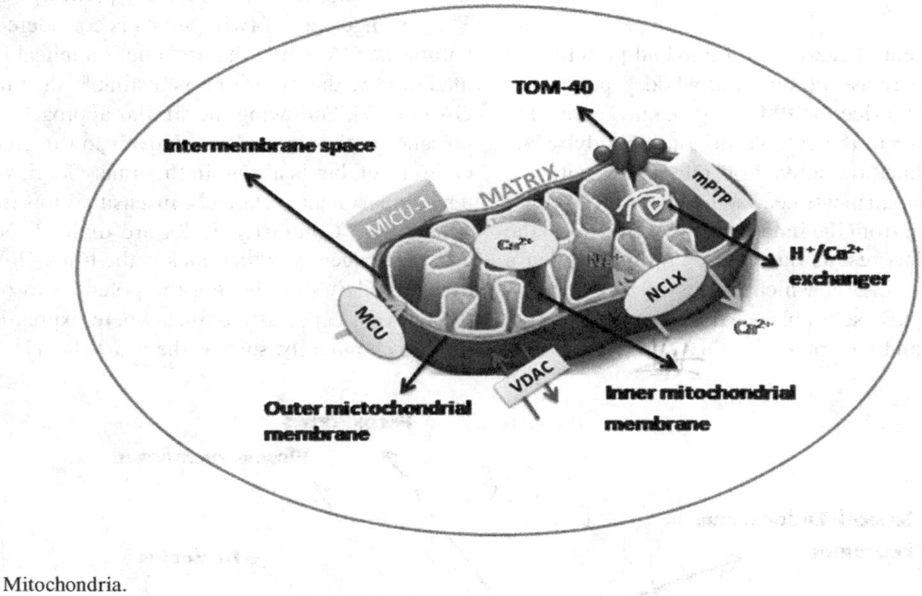

FIGURE 25.2E Mitochondria.

25.2 Mitochondria

The mitochondrion is a fundamental organelle for both cell survival and apoptosis, as it develops a major part of cellular energy in the form of ATP and plays an active role in apoptosis induction, also taking part in cellular Ca^{2+} signaling and acting as high localized buffers, subsequently involving in regulating cytosolic calcium transport [18]. Mitochondria are documented to have a significant role in neurodegenerative disorders; Alzheimer's patients exhibit substantial impaired mitochondrial functioning (Figure 25.2E).

In an isolated state, mitochondria can rapidly take up a large amount of Ca^{2+} from the extracellular medium, and then their can buffer them in their matrix shown in various seminal studies till now. This accumulation is based on thermodynamics which was stated by Mitchell and Moyle in 1967 on the basis of chemiosmotic theory. In mitochondria which are supplied with a sufficient amount of carbon and oxygen source, respiratory chain complexes pumping proton to the IMS (intermembrane space) from the matrix that gives rise to electrochemical gradient throughout the IMM (inner-mitochondrial membrane) that is about –180mV. This provides the force that drives the build-up of cations inside the matrix. The OMM (outer-mitochondrial membrane) has permeability up to minor solutes (>5 KDa) and ions (through the VDAC). Its overexpression favors Ca^{2+} deposition into mitochondria; it is not the limiting step for Ca^{2+} uptake for mitochondria. These high-capacity mechanisms are defined as the "mitochondrial Ca^{2+} uniporter", also known as MCU, primarily defined according to its dependency on an electrochemical gradient, ruthenium red sensitivity (displaying inhibitory actions), and activities when extramitochondrial $[Ca^{2+}]$ is in the mM range.

25.2.1 Vitality and Importance of Mitochondrial Ca^{2+} Uptake

The importance of $[Ca^{2+}]_{mito}$ uptake is versatile. There are supposed to be three main functions reported: those are ATP synthesis, buffering/shaping of cytosolic Ca^{2+} rises and cell death activation. In the synthesis of ATP, common consent enduring the stimulatory aspect of $[Ca^{2+}]_{mito}$ uptake on the activities of three main enzymes associated with Krebs cycle include isocitrate dihydrogen, α-ketoglutarate dehydrogenase and pyruvate dehydrogenase. Each enzyme increases its Vmax upon Ca^{2+} binding using separate mechanisms.

Moreover, Ca^{2+}exhibits modulation of different IMM located carriers for metabolites and ions. In recent times, a group of researchers stated that in regular conditions, the vital IP3-mediated Ca^{2+} releases from the ER, and this Ca^{2+} is distracted by mitochondria, an aspect which is essential for inhibiting pro-survival mitophagy, boosting effective mitochondrial respiration and retaining cellular bioenergetics [19]. There's already a validation that micromolar $[Ca^{2+}]$c inhibits mitochondrial movement over the binding of Ca^{2+} to the EF-hand domains from Miro1. This allows a high mitochondria Ca^{2+} uptake and higher ATP production at relatively high energy demanding sites, like those that are in close proximity to the synapses in neurons or ER [20, 21]. Over here, the mitochondria released ATP manages the activities of a variety of proteins, primarily IP3R and SERCA, thereby embodying the cytosolic Ca^{2+} signaling and providing a two-way link between Ca^{2+} release, i.e., from the ER and ATP production. Similarly, the latest developments showcased the process: upon ER stress, cells' ability to modulate the transfer of Ca^{2+} between mitochondria and ER for management of this peculiar circumstance citing an emergency.

Particularly, at initial aspects of ER stress, the networks of mitochondria and reticular are reorganized with regard to the perinuclear area, with a boost to their physical connections providing a relatively high $[Ca^{2+}]_{mito}$ uptake, better effective respiration and an elevation of ATP development. This mitochondrial metabolism stimulation implies a cell's effort to acclimate to ER stress conditioning, through an increase in respective bioenergetics, for supplying the increased energy demand. In a similar way, Sig-1R, defined as Sigma-1-receptor, is a typical MAM-residing ligand-operated receptor-chaperone, whose usually regular physiological conditions are linked to BiP at an inactive state. This linkage is Ca^{2+} dependent; since ER stress is linked to reduction in ER $[Ca^{2+}]$, Sig-1R detaches itself from BiP, further binds itself toward IP3R-3 and augments its degeneration, thereby encouraging the transfer of Ca^{2+} to mitochondria for metabolic activity and attenuating cell survivability [22].

In contrast, it is already a known fact that an overload of mitochondrial Ca^{2+} can result in cell apoptosis [3]. Redundant Ca^{2+} uptake disrupts the proton gradient between IMM on both sides, which results in bio-energetic impairment and apoptosis of cells due to necrosis. Although, $[Ca^{2+}]_{mito}$ also has the ability to activate cell apoptosis in finer ways, stimulating cells to apoptotic stimuli [23]. Especially, the physiological oscillations in mitochondria $[Ca^{2+}]$ do not induce mitochondria PTP opening per $[Ca^{2+}]_{mito}$ level elevates favoring the opening of PTP by binding itself toward cyclophilin-D, resulting in inducing pro-apoptotic factorial discharge in the cytosol (like cytochrome-c) and activating apoptotic-cascade. However, some observation has been made that VDAC1 selectively transmits apoptotic Ca^{2+} signaling to mitochondria, assembling a mediation via Grp75 with IP3Rs (especially IP3R-3) [18, 22]. This mediation is a not static one but tends to improve with apoptotic-stimuli. Moreover, Fis-1 (mitochondrial-fission-protein) has been showcased to transmit apoptotic signaling from mitochondria to the ER, through communication with Bap31 at the ER. Hence, it assists pro-apoptotic p20Bap31 cleavage, an incident correlated with Ca^{2+} discharge from ER and mitochondrial activation for apoptosis [24].

Finally, it's recorded that mitochondria are easily capable of actively buffering and are also shaped adequately for both bulk and local $[Ca^{2+}]_{cyto}$ rises. In the first case, the ER or plasma membrane close sites, where Ca^{2+} microdomains are built adjacent to the opening/mouth of Ca^{2+} channels. At the same time, the capability of mitochondria to actively detect Ca^{2+} microdomains in close proximity to ER is sporadically showcased, though, its respective actions close to the PM are highly contentious. Particularly, when mitochondria have the ability to buffer Ca^{2+} influx via VOCCs.

25.2.2 Different Proteins Differentially Involved in Mitochondrial $[Ca^{2+}]$ Uptake

There are varieties of proteins that are reported to participate in the process of mitochondrial Ca^{2+} uptake, though it is still very contradictive. For instance, via immune electron microscopy, RyR1 is discovered in IMM isolated from mitochondria of heart and are supposed to be working in opposite mode, and inter-mediating $[Ca^{2+}]$mito uptake [25] eliminating the likely contamination resulting from the SR, as cardiovascular cells showcase RyR2 instead of RyR1, but this interpretation further requires to be investigated. In another paper, it is also suggested that the down-regulation or overexpression of the UCP2/3 decreases/increases the deposition of Ca^{2+} in mitochondria [26]. It is also expressed that the modulation activity of uniporter is done by uncoupling proteins 2 and 3, and uptake of mitochondrial Ca^{2+} largely because of its release from ER subjected to cytosolic Ca^{2+}elevation [27]. The concise effect of UCP2/3 on mitochondrial Ca^{2+} uptake, is still not confirmable, particularly, the isolation of mitochondria UCP2/3 KO mice take up Ca^{2+} normally [28]. More importantly, it has been analyzed that the SERCA activity is inhibited by the UCP3 by reducing the capability of mitochondria to the production of ATP [29].

Earlier, it is supposed, the component exchanger Letm1 of PM H^+/K^+ as mitochondria H^+/Ca^{2+} electrogenic antiport with raised Ca^{2+} affinity, distinctive from the low-affinity MCU [30]. Contrarily, the other groups have an objecting view on this, exhibiting conclusive statements for Letm1 as an H^+/K^+

antiporter considering that MEFs and Letml KO phenotype can be protected via nigericin, a bonafide H^+/K^+ ionophore. Due to all the above-stated reasoning, the role of Letml role in mitochondria Ca^{2+} uptake remains a cause of contradiction and debatable. However, the latest RNAi screening process allows recognition of MCURl[31] re-establishment the part of Letml in mitochondrial Ca^{2+}, due to its silencing inducing a reduction in uptake level of the organelles. However, more research is necessitated.

25.2.2.1 MCU

Molecular identity with different accessory proteins playing the physiological role and mediating trafficking in neuronal ion channels, i.e., different cell organelles. It is widespread consent that the pivotal molecule which allows the swift deposition of Ca^{2+} throughout ion-impermeable inner channels is the electrophoretic pathway known as the MCU. Before its discovery, it was known that mitochondria were playing an essential role in the regulation of cellular [Ca^{2+}] homeostasis. In the 1980s it was reported that stimulation of receptors of PM which are coupled with PLC will produce IP3. Therefore, in the substantial overload of Ca^{2+}, high uptake of Ca^{2+} will occur inside the organelle and due to this, the fact of a mitochondrial role in Ca^{2+} homeostasis was denied.

In 2018, cryoelectron microscopy structure was reported of *Neurospora crassa* complete length MCU to a general resolution of 3.7 A. A tetrameric architecture was revealed by the above structure, with transmembrane and soluble domain employing distinctive symmetric arrangements inside the channels [32]. During mitosis, rapid mitochondrial Ca^{2+} transient is mediated by MCU. At early mitosis, there is a rapid fall in cellular ATP levels, and the mitochondrial Ca^{2+} transient boosts mitochondrial respiration to restore energy homeostasis. It is attained by mitosis-specific MCU phosphorylation and activation by the mitochondrial translocation of energy sensor AMP-activated protein kinase (AMPK) [33].

In 1972, soluble Ca^{2+} binding glycoprotein was extracted from the mitochondria of the liver of a rat [34]. After reconstitution in the lipid bilayer, this was capable of induction of Ca^{2+} current. However, this Ca^{2+} import was blocked by MCU inhibitor ruthenium red and by the antibodies directed against this protein [35]. Then in 1991 mitochondrial proteins (Mol. Wt. > 35 kDa) consisting of proteoliposomes were reported to take up Ca^{2+} in a ruthenium red-sensitive way [36], and a protein of 40 kDa generating Ca^{2+}conducting channels in membranes of lipid which was isolated from mitochondria of heart [37].

25.2.2.2 MICU1

MlCUl is the mitochondrial calcium uptake was firstly explained in 2010 [38]. It was also the very first element of the MCU complex, which are screened by MitoCarta. Its main function is to find out the mitochondrial Ca^{2+} entry proteins. It is a protein with a size of 54 kDa which is an associated domain with two highly conserved EF-hands with a single IMM-associated domain. The presence of the EF domain helps in the down-regulation of Ca^{2+} uptakes from the mitochondria, it acts as a positive regulator [13], and it is not going to alter membrane potential. MCU was physically linked with MICU1 [14, 15]. After knockdown stimulation of cell, there is no alteration in mitochondrial Ca^{2+} uptake but it results in elevation of basal mitochondrial Ca^{2+}content in resting condition. Therefore, when the cytosolic Ca^{2+} content is low activity of MCU is limited by M1CU1. It prevents the generation of ROS species and stress susceptibility due to the accumulation of mitochondrial Ca^{2+}.

M1CU1 has a TM domain that passes through the IMM; it faces the IMS, not the mitochondrial matrix [38]. Mitochondrial matrix Ca^{2+} is a sense by M1CU1; it also induces molecular changes in cytosolic Ca^{2+}. Activation of M1CU1 also occurs in an EF-hand domain-independent manner for stabilizing the closed state of the MCU complex; here MICU1 activates MCU in a cooperative manner at high cytosolic Ca^{2+} concentration [39]. Ca^{2+} binding to the MICU1-EF-hand domains represents the dependency of MCU on M1CU1. Further investigations will be required.

25.2.2.3 MICU2/3

M1CU2/3 (M1CU2 and M1CU3) are paralog of M1CU1 having ¼ sequence similarity with it. N-terminal targeting sequence for mitochondrial localization of both M1CU2 and M1CU3 are similar but still, M1CU3 is not included in MitoCarta due to its weak mitochondrial localization; due to this, till now M1CU2 has been characterized functionally [40]. MICU2 is basically present in visceral organs, while MICU3 is specifically present in skeletal muscle and neural tissues.

IMS consists of MICU1 and MICU2 which has two deeply conserved EF-hand domains. In the in vivo mouse model, liver silencing of MICU2 shows a reduction in mitochondrial Ca^{2+} uptake upon Ca^{2+} additions, without altering mitochondrial membrane potential. A similar effect is observed on M1CU1 silencing. In HeLa cells siMICU1 results in overexpression of M1CU2 which restores the phenotype (wt) rescuing the mitochondrial [Ca^{2+}] uptake with stabilization of M1CU1 level. Thereby further investigation is necessary to explain the clear function of M1CU2 whether it is complementary or redundant to M1CU1 and also to elucidate the role played by M1CU3 in CNS and skeletal muscle. Expression levels of these proteins are showing interdependency on each other.

25.2.2.4 MCUR1

MCUR1 exhibits a vital significance in the uptake of mitochondrial Ca^{2+}; it is a protein of 40 kDa which consists of two transmembrane domains at both NC termini facing IMS and a large helical coil domain span inside matrix (Figure 25.2F). It has been identified by RNAi screening 45 human IMM proteins [41]. In resting conditions or on stimulation MCUR1 down-regulation shows a reduction in mitochondrial Ca^{2+}. Consequently, AMP-kinase pro-survival autophagy and elevated ratio of AMP/ATP results in oxidative phosphorylation impairment [18]. MCU complex provides M1CU1, M1CU2, MCU, MCUb and EMRE by using quantitative MS but not MCUR1 thereby indicating that MCUR1 is not associated

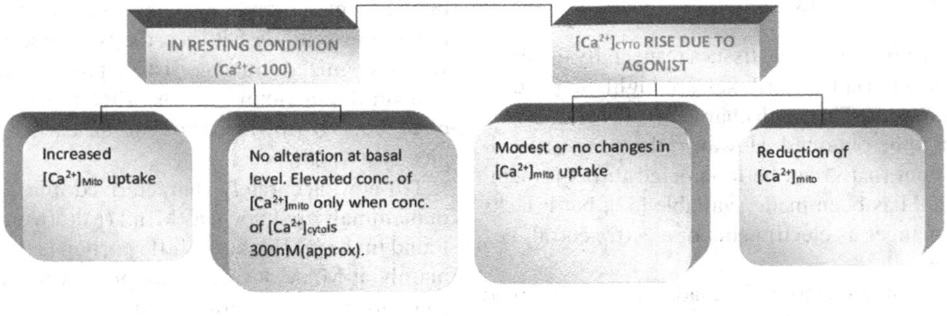

FIGURE 25.2F MCU1: loss in resting and with agonist.

with MCU complex. Therefore, further investigation is needed to understand its role clearly [42].

Overexpression of MCUR1 elevates the uptake of [Ca^{2+}]m in HeLa cells in the presence of histamine, but it is reduced by knocking down MCU. Conversely, overexpression of MCU increases the [Ca^{2+}]m but not the siMCUR1 cells. So, in MCUR1 knockdown cells, MCU levels rise [41]. Ultimately, it shows that for efficient [Ca^{2+}]m both MCUR1 and MCU are necessary and a close link exists between the two proteins.

25.2.2.5 EMRE

It is an essential MCU regulator, which is a metazoan-specific 10 kDa IMM protein. It has a trans-domain and is identified as a purified uniporter complex by using quantitative mass spectroscopy. The strong interdependence of EMRE on MCU is confirmed by the fact that reduction of MCU will reduce the EMRE with MCUb abundantly in a posttranslational way. In this, the MCU complex size reduces from 480 kDa to 300 kDa which can be detected by using blue native PAGE. This lack of MICU1 doesn't impair the MCU/EMRE interaction, which implies EMRE can behave like a vital bridge, linking MICU1/MICU2 Ca^{2+} sensing activities in the IMS to MCU oligomeric channel activities at the IMM. Due to lack of MCU, the stability EMRE is known to be weakened and the MCU complex comparably is decreased in size. Presently, EMRE has been studied in HeLa and HEK293T cells [43] and summarized in yeast [44]; more research is necessitated in different types of cell lines.

25.2.2.6 NCX

In relation to proteins playing a significant role in calcium homeostasis within the brain, a specific focus should be given to NCX. Notably, NCX is a transporter with the ability to carry sodium throughout the membrane primarily in exchange for calcium. NCX can be considered with two working modes: (i) calcium efflux/Na^{+} influx (forward type) or (ii) Ca^{2+} influx/Na+efflux (reverse type) depending on the gradient of electrochemical ion. The studies of the 1960s precisely elaborated the characteristics of mitochondrial capacity to accumulate an immense amount of Ca^{2+} and it was accomplished that this uptake must be stabilized via consequent Ca^{2+} release.

The implication with respect to the existence of an NCX was primarily established on the examination; in mitochondrial isolation, during accumulation of Na^{+} to an external solution, a Ca^{2+} release is activated [45–48]. Among the different testing of cations, solely Li^{+} was observed compelling in activating the same Ca^{2+} efflux and the quantity of Ca^{2+} release was recorded to be Na^{+} dependent. In contrast, ruthenium red inhibits this release, which is indicative of the mechanism that is mediated distinctly and is responsible for the uptake. Furthermore, the amplitude and direction of movement of ions were stringently reliant on the relative [$Ca^{2+]}$ and [Na^{+}] in the extramitochondrial and matrix environment (this means it can effectively work in reversal mode as well).

This NCX is recorded to be specifically effective in tissues that are in an excited state, like brain, heart and skeletal muscles; at the same time, in other non-excitable tissues, mainly the liver, the H^{+}/Ca^{2+} exchanger seems to be more prominent. In 1992, from the bovine heart, a mitochondria polypeptide of 110 kDa was purified and, when transferred in liposomes, catalyzed an NCLX [30]. In 2010, Sekler et al. showcased that the NCLX achieve the principle of evasive mitochondrial Na^{+}/Ca^{2+} antiport [48]. The group noted this mistargeted protein is not only capable of exchanging Na^{+}/Ca^{2+} but is also an effective mediator of Li^{+}/Ca^{2+} exchange, a rare characteristic of the not known NCLX of mitochondria, which is uncommon with the PM-NCX family [49, 50]. They further demonstrated that:

i) NCLX features as a monomer of 50 kDa, 70 kDa additionally and produces a 100 kDa dimer, which is equivalent to the 110 kDa polypeptide earlier isolated from mitochondria of the heart demonstrating NCLX activity [51].
ii) IMM contains endogenous NCLX.
iii) NCLX knockdown significantly decreases the Na^{+}-dependent Ca^{2+} discharge in the mitochondrial isolation.
iv) NCLX is susceptible to already established mitochondria NCX inhibitor CGP 37157.
v) NCLX catalytic mutant inhibits the Na^{+}/Ca^{2+} interchange in mitochondria secured from transfected cells. Hence, it acts as a dominant-negative which implies protein works as an oligomer.

The matter of stoichiometry of the NCLX mediated exchange is a debatable exercise, though significant proof suggests a 3 Na^{+}/1 Ca^{2+} ratio.

25.2.3 Other Efflux Proteins

With respect to other non-excitable tissues (especially the liver) many of the mitochondria Ca^{2+} effluxes are highly dependent on an H^+/Ca^{2+} antiporter. Thorough characteristics of the same antiporter are yet to be obtained. However, there is some evidence that points out that it can be modulated through membrane potential and has been made available [52], but it lacks clarity in this exchanger as electrogenic or electroneutral, i.e., 2 H^+ or 3 $H^+/1\ Ca^{2+}$.

Ultimately, a recommendation is made that Ca^{2+} could efflux from mitochondria via the larger size mPTP (permeability transition pore) [53]. Notably, PTP can be subjected to flickerings, both in intact cells and isolated mitochondria, thereby allowing the release of Ca^{2+} from mitochondria upon the concentration of $[Ca^{2+}]$ in the matrix being greater than the outside. The same process provides the Ca^{2+} homeostasis of organelle, playing a role as a fast, pro-survival Ca^{2+} release mechanism preventing irreversible PT and catastrophic. The molecular basis of PTP is still unclear. However, in 2013 it was reported that dimers of the F0F1 ATP synthase result in the formation of a channel with characteristics alike to that of the MMC (mitochondrial mega channel) [54].

25.2.4 ER-Mitochondria Connections

In the early 1960s, it was firstly reported that there is a close connection between the OMM and ER membrane [55, 56]. In 1990 Jean Vance introduced mitochondria-associated membranes (MAMs) as specific subdomains of ER membranes that mediate with OMM. Tighter linkages necessitated lipid exchange and synthesis between the two organelles [57]. ER portion linked with mitochondria can be isolated biochemically in the MAMs fraction, containing enzymes that showed involvement in lipid synthesis, mainly phosphatidylserine synthase [57, 58].

Clearly, ER/mitochondria binding is essential for the Ca^{2+} interaction between the two organelles. Although, Ca^{2+} low affinity of MCU, Ca^{2+} has been lapped up impromptu by mitochondria during release from the ER, all because of the close contact existence between the two organelles (i.e., endowed in IP3Rs and RyRs). In this stage, cytosolic microdomains of high $[Ca^{2+}]$, "Ca^{2+} hot spots", can be generated [59–61] and efficiently detected. FRET-based probes of Ca^{2+} finally proved the high $[Ca2+]_{cyto}$ microdomains closer to mitochondria are essential for organelle Ca^{2+} uptake particularly targeted to the OMM cytosolic side. Upon this approach, it exhibited that during stimulation of cell and release of Ca^{2+} from the ER, mitochondria are certainly being exposed to the $[Ca^{2+}]$ more elevated than with respect to that of the bulk cytosol (10–30 μM vs. 1–3 μM) [62, 63].

Mitochondrial membranes and ER interaction sites, where two membranes are proximally close but do not meld, thus maintaining organelle integrity, can be detected through fluorescence and electron microscopy in yeast and animal cells. With the stretches 100–300 nm between mitochondria and ER, this stretch is adequate to imply these two organelles bind themselves together with the help of proteins situated on the opposing membranes. The existence of trypsin-sensitive proteinaceous binding between the two membranes is at present well recognized, and the list of proteins with signaling and structural functioning at the ER/mitochondria boundary is continuously growing, showing an earlier unpredicted complexity [64, 65].

Till now, accurately characterized mitochondrial ER binds in mammals are known as Mfn2 (Mitofusin-2). This protein is found in the OMM, and a little portion is also found in the ER, mainly at MAM levels. Mfn2 present in ER has been anticipated to employ hetero- or homo-typic interactions [66]. Both Mfn1 homologous Mfn2 as well as Mfn2 are located in the OMM, consequently Affecting binding between two organelles that have been revealed to influence their Ca^{2+} shuttling [67, 68]. Mainly, Mfn2 KO MEFs show that an additional binding should exist. The PS1 and PS2 both have been found to be abundantly present in MAMs [69]. Although, it is earlier exhibited that solely PS2, and not the PS1, supports mitochondrial ER tethering and Ca^{2+} transfer, particularly, its FAD mutants are effective in this new development [70, 71]. There is also an observation of increased mitochondrial ER tethering in cells showcasing APP and FAD PS1 mutations also in fibroblasts from sporadic AD patients, though the same common effect mechanism is unclear [72].

Connecting sites can sustain distinct structural characteristics, ranging from discrete, punctuated regions to elaborate ones; here, ER membranes proactively encircle the mitochondrion. These contacts, though, are dynamic, but these emerge as comparatively stable structures, as these two organelles kept themselves bound to each other despite them continuously moving along the cytoskeleton. Outstanding functions characterized to date as intimate communication between mitochondria and ER are the lipid exchange and the Ca^{2+} transfer. Also, it is recommended that the mitochondria/ER axis is very significant for mitochondria biogenesis (particularly their fission) [61, 62], autophagosome formation [73], mitochondrial inheritance and mitochondrial quality control.

25.2.4.1 Peroxisomes

Peroxisomes are organelles covered with a single membrane having an elevated concentration of protein matrix which mainly contains a minimum of 50 enzymes included in various pathways (Figure 25.2G). From these enzymes, superior enzymes are those which are implicated in β-oxidation of fatty acid and hydrogen peroxide scavenging, and functioning of the animal cells are partially divided by mitochondria with ether phospholipids synthesis. Particularly, peroxisomes are capable of different variety of enzymes those are competent of disposition of oxygen peroxide, for example, peroxiredoxin V, glutathione peroxidase and catalase (for peroxisomal biochemistry comprehensive review). A number of organelles are distributed within the complete cytosol and these may show swift adaption of cellular requirement by raising their numbers or size; the elevation in number is augmented by lipids, with the help of activation of a ligand-dependent transcription factor PPAR and directed through a fission process the same as in mitochondria.

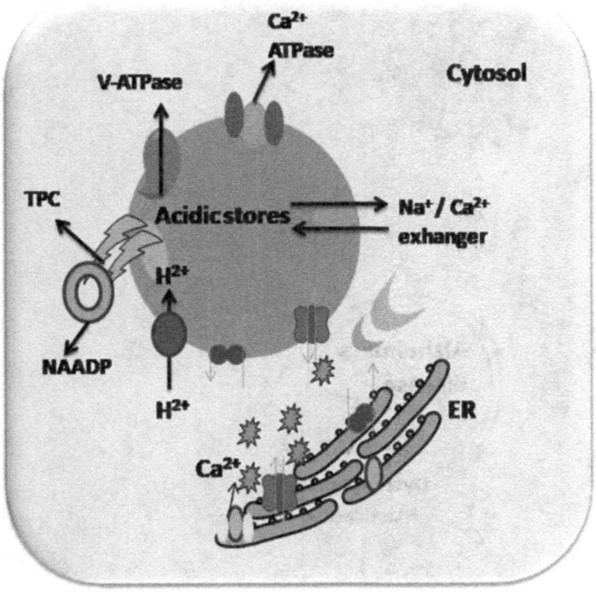

FIGURE 25.2G Acidic stores/peroxisomes.

Some evidence suggests that ER could be helpful in the production of peroxisomes [74]. Contrarily, researchers suggested this process might only provide a path to dividing peroxisomes including membrane constituents. Besides metabolic functioning, peroxisomes appeared as organelles that are decisive in cell fate determination, which also contributes to the regulation of cell differentiation and also morphogenesis and embryo development. Through particular binding proteins, the lipids synthesized through peroxisomes can be achieved by targeting them to the nucleus, so they activate lipid-dependent transcription factors (including PPARs) which in turn helps in the regulation of various genes expression involved in development and differentiation.

On the other hand, the peroxisomes are a very impermeable organelle using different carriers to transport several metabolites. With regards to peroxisomal Ca^{2+} handling and the issue relating to regulatory signal in the metabolic activity, the availability of data is very insufficient and contradictive, exhibiting organelles might be having differences in behavior caused by possible linkages to biogenesis and heterogeneity. The very process of characterization of peroxisomal Ca^{2+} displays vanadate sensitive Ca^{2+} ATPase into members. On the contrary, its studies were not executed in living cells and there is a possibility of contamination through different membranes. During the subsequent years, two groups produced a couple of different Ca^{2+} probes targeting primarily the peroxisome, thereby directly authorizing Ca^{2+} measurement of organelles in integral cells with the placement of two novel FRET-based Cameleon probes (along with different affinities for Ca^{2+}). In this, the very first group carried out the analysis of singular cells of Ca^{2+} dynamics inside the peroxisomal lumen. The other group produced peroxisome-targeting Ca^{2+} and accordingly analyzed the Ca^{2+}-peroxisome dynamic contrary to earlier. Contrarily, in mammals' cells, there is none of the modulating peroxisomal presently identified. So, peroxisomal signaling is yet to be analyzed [75].

25.3 Correlation between Calcium and Castigatory Dysregulation with AD

Mastication is considered a typical process modulated by the central nervous system which involves various aspects, i.e., gender, age, dental condition, and largely depends upon sensory feedback, but also the smell, taste, texture and food hardness to some extent [76, 77]. The neurodegeneration and brain atrophy in AD have a consequential effect on chewing, cortico-bulbar tract, smell and taste sensation which impairs sensory feedback and masticatory functioning [14]. Furthermore, bite force displays absolute relation with masticatory performance having the tendency to decrease with age because of atrophy of the jaw-closing nucleus, more evident in edentulous than dentate subjects. Skeletal muscle atrophy, strength decline and physical frailty might be present in AD; it is very much possible that the elderly with AD display relatively higher impaired bite force and basic factors for chewing function decline (Figure 25.3). Cognitive dysregulation is due to brain neurophysiological alteration, like neuronal, cell groups and synapses losses and diffused cortical atrophy [15]. Contrarily, intraneuronal alteration augments include extracellular deposition of β-amyloid protein and development of neuritic plaques taking place in the brain region involving learning, memory and emotional behavior, the region also correlated with chewing function [33, 34]. The mechanism of dementia still needs to be explored, but some risk factors have been recognized like age, female gender, heritage, presence of apoE4 allele, depression, trauma, alcoholism, atherosclerosis, etc. [35, 37]. Further epidemiological reports suggest a risk for developing dementia due to having fewer teeth [33]. Supporting it are the experiments in animal models displaying the positive relation between loss of natural teeth and decline in cognitive function [38, 39]. Additionally, Onozuka et al. exhibited a compelling correlation between declining chewing function, neuronal degeneration of the hippocampus and impairment of spatial memory in elderly mice [41]. This neuronal degeneration is a primary factor for age-induced cognitive impairment. Clinical reports established that count of teeth, median occlusal contact area and the average value of maximum bite force are lower in the elderly with cognitive impairment compared to the elderly with normal cognitive function. Subsequently, studies established a strong connection between a lesser number of teeth and an enhanced risk of augmenting dementia and Alzheimer's disease [35, 40, 44–46]. All these reports demonstrated that dental status might be a non-pharmacological approach to inhibit senile dementia and cognitive disorders, majorly Alzheimer's disease [41].

In 2016, Campos and his colleagues demonstrated chewing function in the elderly with AD and, correlating chewing function with cognitive status in their studies, they concluded that moderate AD is linked to impairment of chewing function [76]. The assembly of the crystals via interaction with matrix proteins and other ions occurs in the extracellular space and

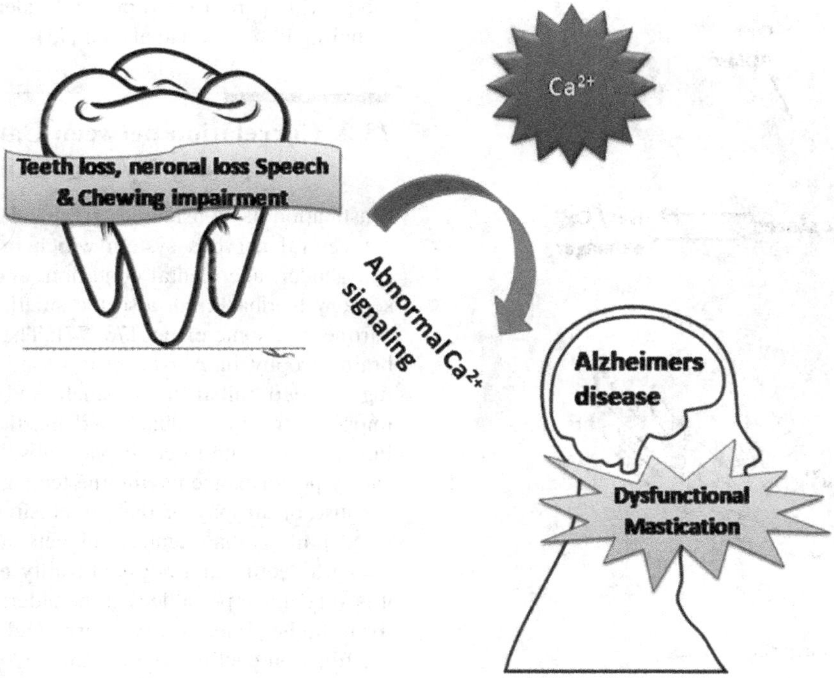

FIGURE 25.3 Ca^{2+} signaling: mastication and AD.

while this is beautifully controlled by the ameloblasts, once materials exit the cells there is limited direct control of the biochemical interactions that take place. Although ameloblasts are a clear director of the assembly and at the same time some achievements have been conquered to define their Ca^{2+} toolkit, it's now more evident that a panoply of intracellular Ca^{2+} signaling processes still needs to be elaborated further, which includes redox environment, mitochondrial Ca^{2+} signaling and the know-how of these processes' interaction for the formation of Hap crystals. The above data marks the significant importance of Ca^{2+} signaling molecules in enamel disease. The criticality of the same pathologies is remarkable. In the same it has beautifully elaborated two main linkages to disease to date: (i) Ca^{2+} uptake and Ca^{2+} extrusion are vital elements of healthy enamel; (ii) the intracellular Ca^{2+} machinery is seemingly less critical based on currently available data.

25.4 Conclusion

Alzheimer's disease is a multifactorial brain disorder that clears itself as a loss of memory. But a conclusive structure of this disease continues as an unsolved mystery. Significant evidence proposes that Ca^{2+}dyshomeostasis plays a crucial role in synaptic pathology in AD. Albeit, the extracellular deposition of amyloid plaques and intracellular neurofibrillary tangles are still supposed to be the root of AD, some of the latest research exhibited $[Ca^{2+}]_{cyto}$ dyshomeostasis as a principle source of AD. Not only is the dysregulation of Ca^{2+} recognized as secondary to the deposition of Aβ oligomers, but it is also recognized to pre-exist the AD pathologies, namely as tau hyperphosphorylation, generation of ROS, impaired synaptic plasticity and APP processing. Current therapeutic interventions should be developed to restore calcium levels to normal in AD. In conclusion, the Ca^{2+} dysregulation of $[Ca^{2+}]_{cyto}$ dyshomeostasis is previously associated with AD pathologies and novel therapeutic agents are required to modulate their level in successful eradication of AD.

Acknowledgments

The authors are sincerely thankful to Prof. Aditya Shastri, Vice-Chancellor of Banasthali Vidyapith, Rajasthan, for providing the necessary support for the completion of this study in a convenient manner.

Conflict of Interest

The authors declare that they have no conflict of interest.

Ethics Statement

The present chapter is an original piece of work and has not been considered elsewhere.

REFERENCES

1. Tsvetkov, P. O., Roman, A. Y., Baksheeva, V. E., Nazipova, A. A., Shevelyova, M. P., Vladimirov, V. I., … & Zernii, E. Y. (2018). Functional status of neuronal calcium sensor-1 is modulated by zinc binding. *Frontiers in Molecular Neuroscience*, *11*, 459.

2. Sotty, F., Danik, M., Manseau, F., Laplante, F., Quirion, R., & Williams, S. (2003). Distinct electrophysiological properties of glutamatergic, cholinergic and GABAergic rat septohippocampal neurons: novel implications for hippocampal rhythmicity. *The Journal of Physiology, 551*(3), 927–943.
3. Chokkareddy, R., Thondavada, N., Thakur, S., & Kanchi, S. (2019). Recent trends in sensors for health and agricultural applications. In *Advanced Biosensors for Health Care Applications* (pp. 341–355). Elsevier.
4. Wu, G., Xie, X., Lu, Z. H., & Ledeen, R. W. (2009). Sodium-calcium exchanger complexed with GM1 ganglioside in nuclear membrane transfers calcium from nucleoplasm to endoplasmic reticulum. *Proceedings of the National Academy of Sciences, 106*(26), 10829–10834.
5. Kang, H., Sun, L. D., Atkins, C. M., Soderling, T. R., Wilson, M. A., & Tonegawa, S. (2001). An important role of neural activity-dependent CaMKIV signaling in the consolidation of long-term memory. *Cell, 106*(6), 771–783.
6. Chakroborty, S., Kim, J., Schneider, C., West, A. R., & Stutzmann, G. E. (2015). Nitric oxide signaling is recruited as a compensatory mechanism for sustaining synaptic plasticity in Alzheimer's disease mice. *Journal of Neuroscience, 35*(17), 6893–6902.
7. Chan, S. L., Mayne, M., Holden, C. P., Geiger, J. D., & Mattson, M. P. (2000). Presenilin-1 mutations increase levels of ryanodine receptors and calcium release in PC12 cells and cortical neurons. *Journal of Biological Chemistry, 275*(24), 18195–18200.
8. Groten, C. J., Rebane, J. T., Hodgson, H. M., Chauhan, A. K., Blohm, G., & Magoski, N. S. (2016). Ca^{2+} removal by the plasma membrane Ca^{2+}-ATPase influences the contribution of mitochondria to activity-dependent Ca^{2+} dynamics in Aplysia neuroendocrine cells. *Journal of Neurophysiology, 115*(5), 2615–2634.
9. Patron, M., Checchetto, V., Raffaello, A., Teardo, E., Reane, D. V., Mantoan, M., ... & Rizzuto, R. (2014). MICU1 and MICU2 finely tune the mitochondrial Ca2+ uniporter by exerting opposite effects on MCU activity. *Molecular Cell, 53*(5), 726–737.
10. Kamer, K. J., & Mootha, V. K. (2014). MICU 1 and MICU 2 play nonredundant roles in the regulation of the mitochondrial calcium uniporter. *EMBO Reports, 15*(3), 299–307.
11. Li, S., Hao, B., Lu, Y., Yu, P., Lee, H. C., & Yue, J. (2012). Intracellular alkalinization induces cytosolic Ca2+ increases by inhibiting sarco/endoplasmic reticulum Ca2+-ATPase (SERCA). *PloS One, 7*(2), e31905.
12. Filadi, R., Zampese, E., Pozzan, T., Pizzo, P., & Fasolato, C. (2012). Endoplasmic reticulum-mitochondria connections, calcium cross-talk and cell fate: a closer inspection. In *Endoplasmic Reticulum Stress in Health and Disease* (pp. 75–106). Springer.
13. Wang, Y., Deng, X., Mancarella, S., Hendron, E., Eguchi, S., Soboloff, J., ... & Gill, D. L. (2010). The calcium store sensor, STIM1, reciprocally controls Orai and CaV1. 2 channels. *Science, 330*(6000), 105–109.
14. Bell, M., Bartol, T., Sejnowski, T., & Rangamani, P. (2019). Dendritic spine geometry and spine apparatus organization govern the spatiotemporal dynamics of calcium. *Journal of General Physiology, 151*(8), 1017–1034.
15. Pinton, P., Pozzan, T., & Rizzuto, R. (1998). The Golgi apparatus is an inositol 1, 4, 5-trisphosphate-sensitive Ca^{2+} store, with functional properties distinct from those of the endoplasmic reticulum. *The EMBO Journal, 17*(18), 5298–5308.
16. Lissandron, V., Podini, P., Pizzo, P., & Pozzan, T. (2010). Unique characteristics of Ca2+ homeostasis of the trans-Golgi compartment. *Proceedings of the National Academy of Sciences, 107*(20), 9198–9203.
17. Wong, A. K., Capitanio, P., Lissandron, V., Bortolozzi, M., Pozzan, T., & Pizzo, P. (2013). Heterogeneity of Ca2+ handling among and within Golgi compartments. *Journal of Molecular Cell Biology, 5*(4), 266–276.
18. Romero-Garcia, S., & Prado-Garcia, H. (2019). Mitochondrial calcium: transport and modulation of cellular processes in homeostasis and cancer. *International Journal of Oncology, 54*(4), 1155–1167.
19. Cárdenas, C., Miller, R. A., Smith, I., Bui, T., Molgó, J., Müller, M., ... & Foskett, J. K. (2010). Essential regulation of cell bioenergetics by constitutive InsP3 receptor Ca^{2+} transfer to mitochondria. *Cell, 142*(2), 270–283.
20. Cheng, K. T., Liu, X., Ong, H. L., Swaim, W., & Ambudkar, I. S. (2011). Local Ca2+ entry via Orai1 regulates plasma membrane recruitment of TRPC1 and controls cytosolic Ca2+ signals required for specific cell functions. *PLoS Biology, 9*(3), e1001025.
21. Yi, M., Weaver, D., & Hajnóczky, G. (2004). Control of mitochondrial motility and distribution by the calcium signal a homeostatic circuit. *Journal of Cell Biology, 167*(4), 661–672.
22. Hayashi, T., & Su, T. P. (2007). Sigma-1 receptor chaperones at the ER-mitochondrion interface regulate Ca^{2+} signaling and cell survival. *Cell, 131*(3), 596–610.
23. Chokkareddy, R., Thondavada, N., Kabane, B., & Redhi, G. G. (2021). A novel ionic liquid based electrochemical sensor for detection of pyrazinamide. *Journal of the Iranian Chemical Society, 18*(3), 621–629.
24. Iwasawa, R., Mahul-Mellier, A. L., Datler, C., Pazarentzos, E., & Grimm, S. (2011). Fis1 and Bap31 bridge the mitochondria–ER interface to establish a platform for apoptosis induction. *EMBO Journal, 30*(3), 556–568.
25. Beutner, G., Sharma, V. K., Giovannucci, D. R., Yule, D. I., & Sheu, S. S. (2001). Identification of a ryanodine receptor in rat heart mitochondria. *Journal of Biological Chemistry, 276*(24), 21482–21488.
26. Trenker, M., Malli, R., Fertschai, I., Levak-Frank, S., & Graier, W. F. (2007). Uncoupling proteins 2 and 3 are fundamental for mitochondrial Ca^{2+} uniport. *Nature Cell Biology, 9*(4), 445–452.
27. Westermann, B. (2010). Mitochondrial fusion and fission in cell life and death. *Nature Reviews Molecular Cell Biology, 11*(12), 872–884.
28. Brookes, P. S., Parker, N., Buckingham, J. A., Vidal-Puig, A., Halestrap, A. P., Gunter, T. E., ... & Brand, M. D. (2008). UCPs: unlikely calcium porters. *Nature Cell Biology, 10*(11), 1235–1237.
29. De Marchi, U., Castelbou, C., & Demaurex, N. (2011). Uncoupling protein 3 (UCP3) modulates the activity of Sarco/endoplasmic reticulum Ca^{2+}-ATPase (SERCA) by decreasing mitochondrial ATP production. *Journal of Biological Chemistry, 286*(37), 32533–32541.

30. Jiang, D., Zhao, L., & Clapham, D. E. (2009). Genome-wide RNAi screen identifies Letm1 as a mitochondrial Ca2+/H+ antiporter. *Science, 326*(5949), 144–147.
31. Chokkareddy, R., Thondavada, N., Kabane, B., & Redhi, G. G. (2018). *Current Advances in Biosynthesis of Silver Nanoparticles and their Applications* (pp. 165–198). Wiley-Scrivener Publishing LLC.
32. Yoo, J., Wu, M., Yin, Y., Herzik, M. A., Lander, G. C., & Lee, S. Y. (2018). Cryo-EM structure of a mitochondrial calcium uniporter. *Science, 361*(6401), 506–511.
33. Zhao, H., Li, T., Wang, K., Zhao, F., Chen, J., Xu, G., ... & Pan, X. (2019). AMPK-mediated activation of MCU stimulates mitochondrial Ca^{2+} entry to promote mitotic progression. *Nature Cell Biology, 21*(4), 476–486.
34. Sottocasa, G., Sandri, G., Panfili, E., de Bernard, B., Gazzotti, P., Vasington, F. D., & Carafoli, E. (1972). Isolation of a soluble Ca2+ binding glycoprotein from ox liver mitochondria. *Biochemical and Biophysical Research Communications, 47*(4), 808–813.
35. Panfili, E., Sandri, G., Sottocasa, G. L., Lunazzi, G., Liut, G., & Graziosi, G. (1976). Specific inhibition of mitochondrial Ca 2+ transport by antibodies directed to the Ca^{2+}-binding glycoprotein. *Nature, 264*(5582), 185–186.
36. Zazueta, C., Holguín, J. A., & Ramírez, J. (1991). Calcium transport sensitive to ruthenium red in cytochrome oxidase vesicles reconstituted with mitochondrial proteins. *Journal of Bioenergetics and Biomembranes, 23*(6), 889–902.
37. Saris, N. E. L., Sirota, T. V., Virtanen, I., Niva, K., Penttilä, T., Dolgachova, L. P., & Mironova, G. D. (1993). Inhibition of the mitochondrial calcium uniporter by antibodies against a 40-kDa glycorprotein T. *Journal of Bioenergetics and Biomembranes, 25*(3), 307–312.
38. Perocchi, F., Gohil, V. M., Girgis, H. S., Bao, X. R., McCombs, J. E., Palmer, A. E., & Mootha, V. K. (2010). MICU1 encodes a mitochondrial EF hand protein required for Ca 2+ uptake. *Nature, 467*(7313), 291–296.
39. Baughman, J. M., Perocchi, F., Girgis, H. S., Plovanich, M., Belcher-Timme, C. A., Sancak, Y., ... & Mootha, V. K. (2011). Integrative genomics identifies MCU as an essential component of the mitochondrial calcium uniporter. *Nature, 476*(7360), 341–345.
40. Mallilankaraman, K., Doonan, P., Cárdenas, C., Chandramoorthy, H. C., Müller, M., Miller, R., ... & Madesh, M. (2012). MICU1 is an essential gatekeeper for MCU-mediated mitochondrial Ca^{2+} uptake that regulates cell survival. *Cell, 151*(3), 630–644.
41. Kilele, J. C., Chokkareddy, R., & Redhi, G. G. (2021). Ultra-sensitive electrochemical sensor for fenitrothion pesticide residues in fruit samples using IL@ CoFe2O4NPs@ MWCNTs nanocomposite. *Microchemical Journal, 164*, 106012.
42. Plovanich, M., Bogorad, R. L., Sancak, Y., Kamer, K. J., Strittmatter, L., Li, A. A., ... & Mootha, V. K. (2013). MICU2, a paralog of MICU1, resides within the mitochondrial uniporter complex to regulate calcium handling. *PloS One, 8*(2), e55785.
43. Chokkareddy, R., & Redhi, G. G. (2020). A facile electrochemical sensor based on ionic liquid functionalized multiwalled carbon nanotubes for isoniazid detection. *Journal of Analytical Chemistry, 75*(12), 1638–1646.
44. Sancak, Y., Markhard, A. L., Kitami, T., Kovács-Bogdán, E., Kamer, K. J., Udeshi, N. D., ... & Mootha, V. K. (2013). EMRE is an essential component of the mitochondrial calcium uniporter complex. *Science, 342*(6164), 1379–1382.
45. Carafoli, E., Tiozzo, R., Lugli, G., Crovetti, F., & Kratzing, C. (1974). The release of calcium from heart mitochondria by sodium. *Journal of Molecular and Cellular Cardiology, 6*(4), 361–371.
46. Crompton, M., Künzi, M., & Carafoli, E. (1977). The calcium-induced and sodium-induced effluxes of calcium from heart mitochondria: evidence for a sodium-calcium carrier. *European Journal of Biochemistry, 79*(2), 549–558.
47. Chokkareddy, R., Kanchi, S., Thakur, S., & Hussein, F. H. (2021). Advanced applications of green materials in biosensor. In *Applications of Advanced Green Materials* (pp. 33–75). Woodhead Publishing.
48. Palty, R., Silverman, W. F., Hershfinkel, M., Caporale, T., Sensi, S. L., Parnis, J., ... & Sekler, I. (2010). NCLX is an essential component of mitochondrial Na+/Ca2+ exchange. *Proceedings of the National Academy of Sciences, 107*(1), 436–441.
49. Palty, R., Ohana, E., Hershfinkel, M., Volokita, M., Elgazar, V., Beharier, O., ... & Sekler, I. (2004). Lithium-calcium exchange is mediated by a distinct potassium-independent sodium-calcium exchanger. *Journal of Biological Chemistry, 279*(24), 25234–25240.
50. Drago, I., Pizzo, P., & Pozzan, T. (2011). After half a century mitochondrial calcium in-and efflux machineries reveal themselves. *The EMBO Journal, 30*(20), 4119–4125.
51. Li, W., Shariat-Madar, Z., Powers, M., Sun, X., Lane, R. D., & Garlid, K. D. (1992). Reconstitution, identification, purification, and immunological characterization of the 110-kDa Na+/Ca^{2+} antiporter from beef heart mitochondria. *Journal of Biological Chemistry, 267*(25), 17983–17989.
52. Bernardi, P., & Azzone, G. F. (1983). Regulation of Ca^{2+} efflux in rat liver mitochondria: role of membrane potential. *European Journal of Biochemistry, 134*(2), 377–383.
53. Bernardi, P. (1999). Mitochondrial transport of cations: channels, exchangers, and permeability transition. *Physiological Reviews, 79*(4), 1127–1155.
54. Giorgio, V., Von Stockum, S., Antoniel, M., Fabbro, A., Fogolari, F., Forte, M., ... & Bernardi, P. (2013). Dimers of mitochondrial ATP synthase form the permeability transition pore. *Proceedings of the National Academy of Sciences, 110*(15), 5887–5892.
55. Ruby, J. R., Dyer, R. F., & Skalko, R. G. (1969). Continuities between mitochondria and endoplasmic reticulum in the mammalian ovary. *Zeitschrift für Zellforschung und mikroskopische Anatomie, 97*(1), 30–37.
56. Morré, D. J., Merritt, W. D., & Lembi, C. A. (1971). Connections between mitochondria and endoplasmic reticulum in rat liver and onion stem. *Protoplasma, 73*(1), 43–49.
57. Chokkareddy, R., & Redhi, G. G. (2020). Ionic liquid and f-MWCNTs fabricated glassy carbon electrode for determination of amygdalin in apple seeds. *Electroanalysis, 32*(12), 3045–3053.
58. Stone, S. J., & Vance, J. E. (2000). Phosphatidylserine synthase-1 and-2 are localized to mitochondria-associated membranes. *Journal of Biological Chemistry, 275*(44), 34534–34540.

59. Rizzuto, R., Brini, M., Murgia, M., & Pozzan, T. (1993). Microdomains with high Ca2+ close to IP3-sensitive channels that are sensed by neighboring mitochondria. *Science*, *262*(5134), 744–747.
60. Rizzuto, R., Pinton, P., Carrington, W., Fay, F. S., Fogarty, K. E., Lifshitz, L. M., ... & Pozzan, T. (1998). Close contacts with the endoplasmic reticulum as determinants of mitochondrial Ca2+ responses. *Science*, *280*(5370), 1763–1766.
61. Csordás, G., Thomas, A. P., & Hajnóczky, G. (1999). Quasi-synaptic calcium signal transmission between endoplasmic reticulum and mitochondria. *The EMBO Journal*, *18*(1), 96–108.
62. Giacomello, M., Drago, I., Bortolozzi, M., Scorzeto, M., Gianelle, A., Pizzo, P., & Pozzan, T. (2010). Ca^{2+} hot spots on the mitochondrial surface are generated by Ca^{2+} mobilization from stores, but not by activation of store-operated Ca^{2+} channels. *Molecular Cell*, *38*(2), 280–290.
63. Csordás, G., Várnai, P., Golenár, T., Roy, S., Purkins, G., Schneider, T. G., ... & Hajnóczky, G. (2010). Imaging interorganelle contacts and local calcium dynamics at the ER-mitochondrial interface. *Molecular Cell*, *39*(1), 121–132.
64. Csordás, G., Renken, C., Várnai, P., Walter, L., Weaver, D., Buttle, K. F., ... & Hajnóczky, G. (2006). Structural and functional features and significance of the physical linkage between ER and mitochondria. *Journal of Cell Biology*, *174*(7), 915–921.
65. Filadi, R., Zampese, E., Pozzan, T., Pizzo, P., & Fasolato, C. (2012). Endoplasmic reticulum-mitochondria connections, calcium cross-talk and cell fate: a closer inspection. In *Endoplasmic Reticulum Stress in Health and Disease* (pp. 75–106). Springer.
66. Filadi, R., Pendin, D., & Pizzo, P. (2018). Mitofusin 2: from functions to disease. *Cell Death & Disease*, *9*(3), 1–13.
67. Hall, A. R., Burke, N., Dongworth, R. K., & Hausenloy, D. J. (2014). Mitochondrial fusion and fission proteins: novel therapeutic targets for combating cardiovascular disease. *British Journal of Pharmacology*, *171*(8), 1890–1906.
68. De Brito, O. M., & Scorrano, L. (2008). Mitofusin 2 tethers endoplasmic reticulum to mitochondria. *Nature*, *456*(7222), 605–610.
69. Area-Gomez, E., de Groof, A. J., Boldogh, I., Bird, T. D., Gibson, G. E., Koehler, C. M., ... & Schon, E. A. (2009). Presenilins are enriched in endoplasmic reticulum membranes associated with mitochondria. *American Journal of Pathology*, *175*(5), 1810–1816.
70. Zampese, E., Fasolato, C., Kipanyula, M. J., Bortolozzi, M., Pozzan, T., & Pizzo, P. (2011). Presenilin 2 modulates endoplasmic reticulum (ER)–mitochondria interactions and Ca^{2+} cross-talk. *Proceedings of the National Academy of Sciences*, *108*(7), 2777–2782.
71. Kipanyula, M. J., Contreras, L., Zampese, E., Lazzari, C., Wong, A. K., Pizzo, P., ... & Pozzan, T. (2012). Ca^{2+} dysregulation in neurons from transgenic mice expressing mutant presenilin 2. *Aging Cell*, *11*(5), 885–893.
72. Area-Gomez, E., del Carmen Lara Castillo, M., Tambini, M. D., Guardia-Laguarta, C., De Groof, A. J., Madra, M., ... & Schon, E. A. (2012). Upregulated function of mitochondria-associated ER membranes in Alzheimer disease. *The EMBO Journal*, *31*(21), 4106–4123.
73. Tyagi, S., Gupta, P., Saini, A. S., Kaushal, C., & Sharma, S. (2011). The peroxisome proliferator-activated receptor: a family of nuclear receptors role in various diseases. *Journal of Advanced Pharmaceutical Technology & Research*, *2*(4), 236.
74. Palty, R., Ohana, E., Hershfinkel, M., Volokita, M., Elgazar, V., Beharier, O., ... & Sekler, I. (2004). Lithium-calcium exchange is mediated by a distinct potassium-independent sodium-calcium exchanger. *Journal of Biological Chemistry*, *279*(24), 25234–25240.
75. Hamasaki, M., Furuta, N., Matsuda, A., Nezu, A., Yamamoto, A., Fujita, N., ... & Yoshimori, T. (2013). Autophagosomes form at ER–mitochondria contact sites. *Nature*, *495*(7441), 389–393.
76. Campos, C. H., Ribeiro, G. R., Costa, J. L. R., & Garcia, R. C. M. R. (2017). Correlation of cognitive and masticatory function in Alzheimer's disease. *Clinical Oral Investigations*, *21*(2), 573–578.
77. Liu, D., Deng, Y., Sha, L., Hashem, M. A., & Gai, S. (2017). Impact of oral processing on texture attributes and taste perception. *Journal of Food Science and Technology*, *54*(8), 2585–2593.

Index

Note: Locators in *italics* represent figures and **bold** indicate tables in the text.

9780367740245